Basic Concepts

ANALYTICAL

CHEMISTRY

Basic Concepts of
ANALYTICAL
CHEMISTRY

Third Edition

S.M. Khopkar

Professor Emeritus
Department of Chemistry
Indian Institute of Technology, Bombay
Mumbai-400076

PUBLISHING FOR ONE WORLD

NEW AGE INTERNATIONAL (P) LIMITED, PUBLISHERS

New Delhi • Bangalore • Chennai • Cochin • Guwahati • Hyderabad
Jalandhar • Kolkata • Lucknow • Mumbai • Ranchi
Visit us at **www.newagepublishers.com**

BRANCHES

- **Bangalore** 37/10, 8th Cross (Near Hanuman Temple), Azad Nagar, Chamarajpet, Bangalore-560 018
  Tel.: (080) 26756823, Telefax: 26756820, **E-mail: bangalore@newagepublishers.com**

- **Chennai** 26, Damodaran Street, T. Nagar, Chennai-600 017
  Tel.: (044) 24353401, Telefax: 24351463, **E-mail: chennai@newagepublishers.com**

- **Cochin** CC-39/1016, Carrier Station Road, Ernakulam South, Cochin-682 016
  Tel.: (0484) 2377004, Telefax: 4051303, **E-mail: cochin@newagepublishers.com**

- **Guwahati** Hemsen Complex, Mohd. Shah Road, Paltan Bazar, Near Starline Hotel, Guwahati-781 008
  Tel.: (0361) 2513881, Telefax: 2543669, **E-mail: guwahati@newagepublishers.com**

- **Hyderabad** 105, 1st Floor, Madhiray Kaveri Tower, 3-2-19, Azam Jahi Road, Near Kumar Theater
  Nimboliadda, Kachiguda, Hyderabad-500 027, Tel.: (040) 24652456, Telefax: 24652457
  E-mail: hyderabad@newagepublishers.com

- **Kolkata** RDB Chambers (Formerly Lotus Cinema) 106A, 1st Floor, S.N. Banerjee Road, Kolkata-700 014
  Tel.: (033) 22273773, Telefax: 22275247, **E-mail: kolkata@newagepublishers.com**

- **Lucknow** 16-A, Jopling Road, Lucknow-226 001
  Tel.: (0522) 2209578, 4045297, Telefax: 2204098, **E-mail:lucknow@newagepublishers.com**

- **Mumbai** 142C, Victor House, Ground Floor, N.M. Joshi Marg, Lower Parel, Mumbai-400 013
  Tel.: (022) 24927869, Telefax: 24915415, **E-mail: mumbai@newagepublishers.com**

- **New Delhi** 22, Golden House, Daryaganj, New Delhi-110 002
  Tel.: (011) 23262368, 23262370, Telefax: 43551305, **E-mail: sales@newagepublishers.com**

ISBN: 978 81-224-2092-0

$ 7.37

C-14-01-7367

Printed in India at Saras Graphics, Rai, Haryana.
Typeset at Pagitek Graphics, Delhi.

PUBLISHING GLOBALLY
NEW AGE INTERNATIONAL (P) LIMITED, PUBLISHERS
7/30 A, Daryaganj, New Delhi-110002
Visit us at **www.newagepublishers.com**

This edition is available for sale in India, Bangladesh, Bhutan, Maldives, Nepal, Pakistan and Sri Lanka only.

Dedicated to
my beloved brother Anna who left this
world so suddenly
&
To Aruna bhabhi who has all along
cheered me in my academic endeavour.

Preface to the Third Edition

Analytical chemistry is an unique experimental science wherein there are phenomenal developments each year. This results in progress in the formulation of better methods of analysis with more selectivity as well as sensitivity. In addition, this field of measurement science is flooded with the designing of newer sophisticated instruments with rapid mode of analysis. They provide improved methods of analysis with reliability and reproducibility.

Since the publication of second edition of this book almost a decade ago, a great revolution has taken place with the discovery of better methods with utmost precision and accuracy. To keep abreast with these modern developments this edition has been thoroughly revised and substantially enlarged with addition of recent methods or techniques of analysis and theories.

In all, ten new chapters have been added. They pertain to: Supercritical Fluid Chromatography and extraction; Capillary Electrophoresis and Chromatography; Electrophoresis in general; Raman Spectroscopy; Chemiluminscence, Atomic Fluorescence and Ionisation Spectroscopy; Refractometry and Interferometry; X-ray Spectroscopy covering XRD, XRF & PIX methods; Mosbaür Spectroscopy; Electron Microscopy including SEM, STM & ATM; and Chemical Sensors and the Biosensors. In view of new emerging theories of chromatography, the chapters on Principles of Chromatography as well as Ion Exchange Chromatography have been revised and enlarged.

Apart from additions of these new chapters, the already existing chapters are blended with additional information on the newer developments e.g. a new section on statistical methods of analysis or Rotaxanes in solvent extraction, or new hyphenated methods in the chromatography with special reference to gas chromatography have been added. So also application of IR spectroscopy in environmental pollution analysis or of NMR technique in MRI device in medicine are new additions. Various kinds of interferences mar the analysis by atomic absorption spectroscopy; hence detailed discussion on their origin and elimination is presented. The different modes of light scattering of radiation; or better electron spectroscopy methods like SIMS and UV photoelectron spectroscopy as well as recent modifications in voltammetric methods have been broughtout.

An endeavour is made to render this book as comprehensive treaties as possible but at no place it is done at the cost of explaining fundamental principles. At the same time, to retain it as concise monography, all minor details of instrumentation or technique have been avoided. The citation of literature pertains to only well known works which are easily available in a good library. At the end an exhaustive subject index has been added to cover every small topic of analytical chemistry.

I am quite confident the readers would enjoy reading and receive this contribution with the same zeal and enthusiasm as was done for the second edition. Further I hope the new version would kindle a great confidence in understanding basic concepts in the minds of specially the young students majoring in analytical chemistry.

I take this opportunity to thank Mr. Saumya Gupta, Managing Director and his staff at New Age International Pvt. Ltd., Publishers, for the excellent production of this work. I owe my thanks to my son Samir and daughter Dr. Supriya for time to time encouragement. Finally, but for the wonderful patience, forbearance and cheerfulness bestowed by my wife Dr. Sucheta, the mighty task of revision would not have been completed.

S.M. KHOPKAR
Professor Emeritus
Department of Chemistry
Indian Institute of Technology-Bombay
Powai, Mumbai - 400 076.

Preface to the First Edition

Analytical Chemistry is a discipline in its own right in chemistry. It cannot be absorbed into generally unrelated courses of study as it can lead to a lowering of the analytical ability of the research worker. Until recently, analytical chemistry was considered to be the branch of chemistry dealing merely with qualitative and quantitative analysis. Unfortunately, with the classical approach towards analysis, the subject became uninteresting and technique oriented, with less attention paid to fundamental aspects of analysis. However, with the advancement of science and great strides in instrumentation the scope has considerably expanded with more emphasis on understanding of basic principles. The present day stress is not only on qualitative or quantitative analysis but also on elucidation of structure. Today, analytical chemistry is a promising area with applications in physical, chemical, biological and environmental sciences. It is an interdisciplinary field in the real sense.

Existing books on analytical chemistry have either stressed too many details of classical gravimetric and volumetric analysis or have over emphasised the role of instrumental analysis. Most of such books have given meagre coverage to separation methods which are so vitally important for successful completion of quantitative determination. At the same time excessive weightage is given to laboratory practice with too little significance attached to basic postulates.

The present book is intended to serve as a textbook for postgraduate students majoring in analytical chemistry. The style and sequence of presentation in this book is based upon the author's teaching experience of 25 years in this subject at graduate as well as postgraduate level. The book is intended to serve as a reference manual for practicing analytical chemists, and research and development workers in laboratories whose main concern is to keep pace with developments in modern methods of chemical analysis. The present book is written by striking a balance between old and new methods of analysis. The classical volumetric and gravimetric methods are of course included but newer separation methods including solvent extraction as well as various novel chromatographic methods of separation are also adequately covered. Furthermore, in addition to normal optical methods of analysis mainly involving spectroscopy, the techniques used in elucidation of structure are also included. The electroanalytical methods of analysis, along with thermal and radiochemical techniques, are also presented.

A knowledge of basic concepts is absolutely essential for the use of modern instruments. Such knowledge can lead to an intelligent choice of methods of analysis and an understanding of probable limitations in terms of the sensitivity and selectivity of the methods. With this objective in mind, a great deal of stress is given to fundamental principles of each method of analysis with some indication of possible applications in chemistry. It is felt that once an analyst is conversant with basic postulates then he can utilise different techniques with great confidence. In future he can even suggest new modifications to the existing method, or invent new methods.

The literature in analytical chemistry is rapidly growing and it is almost impossible to cite exhaustive references for each technique. Therefore an attempt has been made to give at the end of each chapter references to well known books and authoritative monographs for further reading. Most of the chapters could have been written elaborately in the form of separate books, but to keep the size of the book within the bounds of a single volume, unnecessarily detailed descriptions of analytical procedures are avoided and only basic concepts are emphasised. Every chapter is provided with a large number of numerical problems some of which are solved and others are for providing practice and testing comprehension of basic principles. The answers to every problem are provided at the end of the book.

I take this opportunity to express my deep sense of gratitude to Prof. A.K. De, Director, Indian Institute of Technology, Bombay for all the encouragement he has given me during the writing of this book. Under his able guidance this Institute took great strides along the road to scholarly excellence. I am also thankful to the Coordinator(s) of the curriculum development programme of this Institute for providing me substantial financial assistance enabling me to prepare the manuscript of this book. I am thankful to Prof. T.S. West, Director, The Macaulay Institute of Soil Research, Aberdeen, United Kingdom, for agreeing to write the Foreword for this book in spite of his very busy schedule of work. I am indebted to a large number of my devoted research students especially Dr. S.N. Bhosale, Rajendra B. Heddur, Prakash Narayanan, B.S. Mohite and Dr. Raghunadha Rao and many others who have so assiduously helped me in the various phases of writing and typing of this book. I am extremely thankful to the staff of Wiley Eastern Ltd. Publishers for their excellent cooperation. I would like to express my appreciation to Mr. S.A. Talawdekar who has meticulously typed the entire book with utmost care. Finally I shall be failing in my duty if I do not thank my wife Sucheta and my children, Samir and Supriya, without whose encouragement, help, patience and sacrifice this book would not have seen the light of day.

S.M. Khopkar
Analytical Chemistry Laboratories
Department of Chemistry
Indian Institute of Technology
Mumbai 400 076.

1st March, 1984

Contents

Preface to the Third Edition   *vii*

Preface to the First Edition   *ix*

1. Introduction   1

   1.1 Importance of Analytical Chemistry 1
   1.2 Methods of Quantitative Analysis 2
   1.3 Criteria for Selection of Method for Analysis 3
   1.4 Chemical Analysis and Analytical Chemistry 3
   1.5 Quantitative Analysis with Scale of Operation 3
   1.6 Steps in Quantitative Analysis 3
   1.7 Methods of Analytical Determination 4
   1.8 Role of Instrumentation 5
   Literature 5

2. Reliability of Analytical Data and Statistical Analysis   6

   2.1 Errors in Chemical Analysis 6
   2.2 Classification of Errors 6
   2.3 Determining the Accuracy of Methods 7
   2.4 Improving the Accuracy of Analysis 7
   2.5 Statistical Analysis 8
   2.6 Criteria for Rejection of Results Q-test 10
   2.7 Presentation of Data 10
   2.8 Confidence Limit 10
   2.9 Q-test for Rejection of Result 11
   2.10 Standard 't' Test 11
   Problems 13
   Problems for Practice 14
   Literature 15

3. Sampling in Analysis   16

   3.1 Introduction 16
   3.2 Definitions of Terms 16

3.3   Theory of Sampling 16
3.4   Techniques of Sampling 17
3.5   Statistical Criteria of Ideal Sample 17
3.6   Stratified Sampling Versus Random Sampling 18
3.7   Minimisation of Variance in Stratified Sampling 18
3.8   Transmission and Storage of Samples 19
Problems 19
Problems for Practice 20
Literature 20

4. Gravimetric Analysis   21

4.1   Precipitation Methods 21
4.2   Purity of Precipitation: Coprecipitation 22
4.3   Optimum Conditions for Precipitation 23
4.4   Precipitation from Homogeneous Solution 23
4.5   Washing of the Precipitate 24
4.6   Ignition of the Precipitate 24
4.7   Role of Organic Precipitants in Gravimetric Analysis 25
4.8   Criteria for Choice of an Organic Reagent 26
4.9   Some Important Organic Precipitants 27
4.10 Electrogravimetry 28
Problems 28
Problems for Practice 29
Literature 30

5. Volumetric Analysis—Acid Base Titrations   31

5.1   Volumetric Analysis 31
5.2   Classification of Volumetric Methods 32
5.3   Acid-Base Titrations 32
5.4   Acid-Base Titration Curves 34
5.5   Acid-Base Indicators 37
5.6   Mixed Indicators 39
5.7   Fluorescent Indicators 40
Problems 40
Problems for Practice 41
Literature 41

6. Redox Titrations   42

6.1   Theory of Redox Titration Curves 42
6.2   Some Oxidising Agents as Titrants 45
6.3   Application of Iodine as Redox Reagent 47
6.4   Redox Indicators 49
6.5   Detection of End Point in Redox Titrations 50

Problems   51
Problems for Practice   52
Literature   52

7. Precipitation Titrations   53

7.1   Theory of Precipitation Titration Curves   53
7.2   Factors Influencing Solubility of the Precipitate   55
7.3   Titrations by Turbidity without an Indicator   56
7.4   Volhard's Method   56
7.5   Mohr's Method   56
7.6   Adsorption Indicators in Precipitation Titrations   57
7.7   Miscellaneous Indicators, for Titrations   59
Problems   59
Problems for Practice   60
Literature   60

8. Complexometric (EDTA) Titrations   61

8.1   EDTA and the Complexones   61
8.2   Complexometric Titration Curves   62
8.3   Metallochromic Indicators   68
8.4   Kinetics of Complexometric Titration   73
8.5   Selectivity in Complexometric Titrations   73
8.6   Typical EDTA Titrations   73
8.7   Advantages of Complexometric Titration   73
Problems   74
Problems for Practice   74
Literature   74

9. Solvent Extraction   76

9.1   Principles of Solvent Extraction   76
9.2   Classification of Extractions   77
9.3   Mechanism of Extraction   77
9.4   Extraction by Chelation   79
9.5   Extraction by Solvation   82
9.6   Extraction Equilibrium for Chelates   85
9.7   Extraction Equilibria for Solvation   88
9.8   Techniques of Extraction   89
9.9   Separation of Metals by Extraction   90
9.10   Extraction by Ion Pair Formation   90
9.11   Application of Solvent Extraction in Industry   93
9.12   Solid-Phase Extraction (SPE)   93
Problems   94
Literature   96

10. Supramolecules in Solvent Extraction   97

10.1 Origin of Macrocyclic and Supramolecular Compounds 97
10.2 Nomenclature of Compounds 97
10.3 Classification 98
10.4 Synthesis of Crown Ethers, Cryptands and Calixarenes 99
10.5 Factors Influencing Solvent Extraction 101
10.6 Solvent Extraction with Crown Ethers and Cryptands 103
10.7 Rotaxanes 105
10.8 Classification of Rotaxanes 105
10.9 Synthesis of Rotaxanes 105
10.10 Characterisations of Rotaxanes 108
10.11 Metal Complexes with Rotaxanes 109
10.12 Analytical Application 109
Literature 110

11. Principles of Chromatography   111

11.1 Non-Chromatographic Methods of Separation 111
11.2 Membrane Separation Methods 112
11.3 Chromatography as the Method of Separation 113
11.4 Historical Development 113
11.5 Classification of Chromatographic Methods 114
11.6 Fundamentals of Chromatography 117
11.7 Techniques in Chromatography 119
11.8 Dynamics of Chromatography 120
11.9 Van Deemters Equation 122
11.10 Resolution of Mixtures 123
11.11 Separations Characteristics 124
11.12 Special Features of Chromatographic Methods 125
Solved Problems 125
Problems for Practice 126
Literature 127

12. Ion Exchange Chromatography   128

12.1 Historical Development 128
12.2 Classification of Ion Exchange Resins 129
12.3 Mechanism of Ion Exchange 129
12.4 Synthesis of Ion Exchange Resins 131
12.5 Characteristics of Ion Exchange Resins 133
12.6 Ion Exchange Equilibria 136
12.7 Plate Theory for Ion Exchange Column 142
12.8 Techniques in Ion Exchange Methods 146
12.9 Analytical Applications 147
12.10 Application of Ion Exchange in Metal Separations 148
12.11 Cation Exchange Separations from Nonaqueous Media 150

12.12 Anion Exchange Separations of Metals 151
12.13 Liquid Ion Exchangers 151
12.14 Inorganic Exchangers 153
12.15 Chelating Resins 153
Solved Problems 154
Problems for Practice 155
Literature 156

13. Ion Chromatography

13.1   Ion Chromatography as a Separation Tool 157
13.2   Structure and Characteristics of Resins 158
13.3   Eluants Used in Separation 159
13.4   Suppressor Column in Chromatography 160
13.5   Detectors Used in Ion Chromatography 160
13.6   Instrumentation in Ion Chromatography 162
13.7   Analytical Applications of Ion Chromatography 162
13.8   Applications in Environmental Speciation 164
13.9   Application in Other Fields 164
13.10 Hyphenated Technique in Ion Chromatography 165
13.11 Ion Chromatography by Chelation 165
13.12 Ion Chromatography by Ion Pairing 165
13.13 Ion Chromatography by Ion Exclusion 165
13.14 Comparison of Ion Chromatography with Ion Exchange 166
Literature 166

14. Adsorption Chromatography

14.1   Principles of Adsorption Chromatography 167
14.2   Experimental Set-Up 168
14.3   Chemical Constitution and Chromatographic Behaviour 168
14.4   Uses of Adsorption Chromatography 169
14.5   Affinity Chromatography 169
14.6   Chiral Chromatography 170
Solved Problems 170
Problems for Practice 171
Literature 171

15. Partition Chromatography

15.1   Principles of Liquid-Liquid Partition Chromatography 172
15.2   Liquid-Liquid Partition Chromatography 174
15.3   Reversed Phase Partition Chromatography 175
15.4   Applications of Extraction Chromatography 175
15.5   Paper Chromatography 177
15.6   Techniques in Paper Chromatography 178
15.7   Thin Layer Chromatography (TLC) 178

15.8 Densitometric Analysis in Chromatography 180
15.9 Ion Pair Chromatography 180
Solved Problems 181
Problems for Practice 182
Literature 182

16. Gas Chromatography 183

16.1 Principles of Gas Chromatography 183
16.2 Plate Theory for Gas Chromatography 185
16.3 Instrumentation for Gas Chromatography 187
16.4 Working of Gas Chromatography 189
16.5 Applications of Gas Chromatography 189
16.6 Programmed Temperature Chromatography 190
16.7 Flow Programming Chromatography 190
16.8 Gas Solid Chromatography 190
16.9 Hyphenated Techniques in Chromatography 190
16.10 Advantages of Hyphenated Methods 191
16.11 Chemical Characterisation by Hyphenated Methods 191
16.12 Recent Development in Detectors in Gas Chromatography 192
16.13 Derivatisation in Chromatography 193
16.14 Head Space Gas Chromatography 194
Solved Problems 195
Literature 196

17. High Performance Liquid Chromatography 197

17.1 Principles of High Performance Liquid Chromatography 198
17.2 Components of High Performance Liquid Chromatography 198
17.3 Mobile Phase 199
17.4 Stationary Phase 199
17.5 Stationary Supports in HPLC 199
17.6 Detectors Used 200
17.7 Instrumentation for HPLC 201
17.8 Applications of HPLC Techniques 203
Problems 204
Literature 205

18. Supercritical Fluid Chromatography and Extraction 206

18.1 Characteristics of Supercritical Fluids 206
18.2 Mechanism of Working of SFC 207
18.3 Comparison with Other Chromatography Techniques 208
18.4 Selection of Supercritical Fluid for SFC 209
18.5 Applications of Supercritical Fluid Chromatography 209
18.6 Supercritical Fluid Extraction (SFE) 209

18.7   Selection of Supercritical Fluid for Extraction (SFE) 210
18.8   Applications of Supercritical Fluid Extraction 210
Literature 211

19. Exclusion Chromatography   **212**

19.1   Gel Permeation Chromatography 212
19.2   Applications of Gel Permeation Chromatography 213
19.3   Ion Exclusion 214
19.4   Mechanism of the Ion Exclusion Process 214
19.5   Merits and Demerits of the Ion Exclusion Technique 214
19.6   Applications of Ion Exclusion Techniques 215
19.7   Ion Retardation 215
19.8   Inorganic Molecular Sieves 215
Solved Problems 216
Literature 216

20. Electrochromatography   **217**

20.1   Principles of Electrochromatography 217
20.2   Electrochromatography Instrumentation 218
20.3   Curtain Electrochromatography 218
20.4   Applications of Electrochromatography 219
20.5   Reverse Osmosis 220
20.6   Electrodialysis 220
Solved Problems 221
Problems for Practice 222
Literature 222

21. Electrophoresis   **223**

21.1   Principles of Electrophoresis 223
21.2   Properties of Charged Molecule 224
21.3   Theory of Electric Double Layer 225
21.4   Classification of Electrophoresis Methods 226
21.5   Zone Electrophoresis 226
21.6   Isotachophoresis 227
21.7   Isoelectric Focussing 229
21.8   Immunoelectrophoresis 229
21.9   Techniques of Electrophoresis 229
21.10   Continuous Electrophoresis 230
21.11   Detectors used in Electrophoresis 231
21.12   Instrumentation 232
21.13   Applications in Inorganic Chemistry 232
20.14   Separation of Biological Products 233

21.15 Preparative Electrophoresis 236
Conclusion 236
Literature 236

22. Capillary Electrophoresis 237

22.1 Theory of Capillary Electrophoresis 237
22.2 Instrumentation 238
22.3 Sample Separation 239
22.4 Sample Detections 240
22.5 Applications of Capillary Electrophoresis 240
22.6 Capillary Electrochromatography 241
22.7 Miscellar Electrokinetic Capillary Chromatography (MECC) 241
Literature 242

23. Analytical Spectroscopy 243

23.1 Absorption Spectrophotometric Methods 243
23.2 Emission Spectroscopy Methods 243
23.3 Magnetic Resonance Spectroscopy 244
23.4 Electron Spectroscopy 244
23.5 Scattering Methods 244
23.6 Photoacoustic Spectroscopy and Related Methods 244
23.7 Origin of Spectra 245
23.8 Interaction of Radiation with Matter 246
23.9 Fundamental Laws of Spectroscopy 249
23.10 Validity of Beer's Lambert's Law 250
23.11 Deviation from Beer's Law 251
23.12 Relative Concentration Error and Ringbom's Plot 252
Solved Problems 253
Problems for Practice 253
Literature 254

24. Ultraviolet and Visible Spectrophotometry 255

24.1 Origin of Spectra and Electronic Transitions 255
24.2 Visual Colorimetry 257
24.3 Photoelectric Cells and Light Filters 259
24.4 Photoelectric Filter Photometry 262
24.5 Errors in Photoelectric Photometry 266
24.6 Spectrophotometry – Instrumentation 266
24.7 Functioning of Spectrophotometer 269
24.8 Simultaneous Spectrophotometry 270
24.9 Differential Spectrophotometry 272
24.10 Reflectance Spectrophotometry 273
24.11 Photometric Titrations 273

24.12 Composition of Coloured Complexes 276
24.13 Sandell's Sensitivity 277
24.14 Optical Fibres in Spectroscopy 278
24.15 Properties of Optical Fibres 278
24.16 Sensors with Optical Fibres 279
24.17 Errors in UV/Visible Spectrophotometry Measurements 279
24.18 Spectroelectrochemistry 280
24.19 Holographic Gratings 281
Solved Problems 281
Problems for Practice 283
Literature 284

25. Infrared Spectroscopy

25.1 Molecular Vibrations 285
25.2 Vibrational Frequency 287
25.3 Vibrational Modes 288
25.4 Special Features of IR Spectrometers 288
25.5 Comparison of IR and UV–Visible Spectrophotometers 288
25.6 Components of IR Spectrophotometers 288
25.7 Sample Handling Before Measurements 289
25.8 Infrared in Structure Elucidation 290
25.9 Spectral Interpretation of Typical Compounds 292
25.10 Quantitative Analysis by Infrared Spectroscopy 293
25.11 IR–Reflectance Spectrometry 294
25.12 Analysis of Air Pollutants by IR–Spectroscopy 295
25.13 Attenuated Total Reflectance 296
25.14 FT-IR Spectroscopy and its Advantages 297
Problems 297
Problems for Practice 298
Literature 298

26. Raman Spectroscopy

26.1 Raman Spectroscopy 300
26.2 Mechanism of Interaction 301
26.3 Stoke's and Antistoke's Lines 301
26.4 Depolarisation Ratio 303
26.5 Instrumentation 303
26.6 Type of Raman Radiations 303
26.7 Resonance Raman Scattering 304
26.8 Remote Raman Sensing 304
26.9 Analogy of Raman and Infrared Spectroscopy 305
26.10 Quantitative Analysis by Raman Spectroscopy 305
26.11 Applications of Raman Spectroscopy 306
Literature 307

27. Atomic Absorption Spectroscopy   308

27.1 Principles of Atomic Absorption Spectroscopy 308
27.2 Instrumentation for Atomic Absorption Spectrometer 311
27.3 Applications of Atomic Absorption Spectroscopy 314
27.4 Sensitivity and Detection Limits in Analysis 316
27.5 Spectral Interferences in AAS 316
27.6 Chemical Interference in AAS 317
27.7 Atomic Line Broadening 319
27.8 Background Correction 321
27.9 Merits and Demerits of AAS by Flame and Graphite Furnace AAS 322
Problems 323
Literature 324

28. Atomic Emission Spectroscopy   325

28.1 Principles of Emission Spectroscopy 325
28.2 Source of Excitations 326
28.3 Instrumentation for Atomic Emission Spectroscopy 328
28.4 Analysis by Emission Spectroscopy 328
28.5 Flame Emission Spectroscopy 328
28.6 Principles of Flame Photometry 329
28.7 Interferences in Flame Photometry 330
28.8 Evaluation Methods in Flame Photometry 331
Problems 332
Problems for Practice 332
Literature 333

29. Inductively Coupled Plasma–Atomic Emission Spectroscopy   334

29.1 Limitations of Flame Emission Spectroscopy 334
29.2 Principles of Plasma Spectroscopy 335
29.3 Process of Atomisation and Excitation 337
29.4 Plasma as an Excitation Source 337
29.5 Inductively Coupled Plasma Source 337
29.6 ICP-AES Instrumentation 339
29.7 Applications of Plasma Spectroscopy 340
29.8 Comparison of ICP–AES with AAS 340
29.9 Comparison of AFS, AAS and ICP-AES 341
Literature 342

30. Chemiluminescence, Atomic Fluorescence and Ionisation Spectroscopy   343

30.1 Luminescence 343
30.2 Chemiluminescence 344
30.3 Measurement of Chemiluminescence 344

30.4   Quantitative Analysis 344
30.5   Chemiluminescence Titrations 345
30.6   Chemiluminescence for Liquids 345
30.7   Electrochemiluminescence 346
30.8   Atomic Fluorescence 347
30.9   Principles of Atomic Fluorescence 348
30.10   Applications of Atomic Fluorescence 349
30.11   Instrumentation for Atomic Fluorescence Spectroscopy 350
30.12   Resonance Ionisation Spectroscopy 351
30.13   Laser Enhanced Ionisation Spectroscopy 352
Literature 353

31. Molecular Luminescence Spectroscopy   **354**

31.1 Principles of Fluorescence and Phosphorescence 354
31.2 Fluorimetry in Chemical Analysis 355
31.3 Luminescence Instrumentation 356
31.4 Applications of Fluorimetry 357
31.5 Fluorescence and Chemical Structure 358
31.6 Fluorescence Quenching and Inner Filter Effect 359
31.7 Phosphorescence Spectroscopy 361
31.8 Phosphorescence and Chemical Structure 361
31.9 Phosphorimetry in Quantitative Analysis 361
Problems 362
Literature 363

32. Nephelometry and Turbidimetry   **364**

32.1 Tyndall Scattering 364
32.2 Rayleigh Scattering 364
32.3 Raman Scattering 365
32.4 Turbidimetry 366
32.5 Nephelometry 366
Problems 367
Problems for Practice 368
Literature 368

33. Refractometry and Interferometry   **369**

33.1 Principles of Refractometry 369
33.2 Parameters Influencing Refraction 372
33.3 Significance of Critical Angle during Measurements 372
33.4 Refractometers 372
33.5 Qualitative and Quantitative Analysis 373
33.6 Analytical Applications 374
33.7 Interferometer Principles 375
33.8 Applications 376

33.9 Intrumentation 376
Literature 377

34. Polarimetry and Related Methods   **378**

34.1 Polarised Light 378
34.2 Applications of Polarimetry 379
34.3 Optical Rotatory Dispersion and Circular Dichromism 380
34.4 Instrumentation for ORD, CD 382
Problems 382
Problems for Practice 382
Literature 383

35. Nuclear Magnetic Resonance Spectroscopy   **384**

35.1 Principles of NMR Spectroscopy 385
35.2 NMR Spectra Measurement 388
35.3 NMR Instrumentation 388
35.4 Environmental Effects on NMR 390
35.5 Chemical Shift 391
35.6 Spin-Spin Splitting 392
35.7 Applications of NMR Spectroscopy 393
35.8 NMR of Other Elements Like C^{13}, F^{19} and P^{31} 394
35.9 Fourier Transform (FT) NMR 395
35.10 Two Dimensional FT–NMR 395
35.11 Magnetic Resonance Imaging (MRI) 396
Problems 397
Problems for Practice 398
Literature 399

36. Electron Spin Resonance Spectroscopy   **400**

36.1 Electron Spin Resonance Spectroscopy 400
36.2 Instrumentation for Electron Spin Resonance 402
36.3 Applications of ESR Spectroscopy 402
36.4 Spin Labelling ESR Spectroscopy 404
36.5 Multiple Resonance ENDOR and ELDOR Effect 404
36.6 Characterisation of Metal Complexes by ESR 405
36.7 Other Applications of ESR 405
Literature 405

37. X-Ray Spectroscopy   **406**

37.1 Basic Principles 406
37.2 X-ray Source 408
37.3 X-rays Absorption 409
37.4 Instrumentation 411

37.5   X-ray Absorption in Chemical Analysis 414
37.6   X-ray Emissions 414
37.7   X-ray Fluorescence Methods 414
37.8   Fluorescence Instruments 415
37.9   X-ray Fluorescence in Analysis 416
37.10   Particle Induced X-ray Emission (PIXE) 416
37.11   X-ray Diffraction Techniques 416
37.12   Analytical Applications 418
Literature 418

38. Mössbauer Spectroscopy   **419**

38.1   Historical Significance 419
38.2   Principles of Mössbauer Spectroscopy 420
38.3   Mössbauer Instrumentation 423
38.4   Interpretation of Spectra 424
38.5   Mössbauer Spectroscopy Applications 425
38.6   Analytical Science Applications 425
38.7   Chemical Sciences Applications 425
38.8   Physical Sciences Applications 426
38.9   Biological Sciences Applications 426
Literature 426

39. Electron Spectroscopy   **428**

39.1   Principles of Electron Spectroscopy 428
39.2   Auger Emission Spectroscopy 429
39.3   Instrumentation for Electron Spectroscopy 430
39.4   Applications of Auger Electron Spectroscopy 431
39.5   Electron Spectroscopy for Chemical Analysis (ESCA) 431
39.6   Chemical Shifts in ESCA 432
39.7   Analytical Applications of ESCA 432
39.8   Secondary Ion Mass Spectrometry 434
39.9   Electron Microprobe Analysis (EMA) 435
39.10   Ultraviolet Photoelectron Spectroscopy 436
39.11   Comparison of ESCA, UPS and Auger Spectroscopy 437
39.12   Comparison of Auger, ESCA and SIMS Techniques 438
Literature 439

40. Electron Microscopy   **440**

40.1   Classification of Electron Microscopy Methods 440
40.2   Scanning Electron Microscopy (SEM) 441
40.3   Working of SEM Instrument 441
40.4   Comparison of Scanning Probe Microscopy (SPM or STM) 442
40.5   Scanning Tunneling Microscopy (STM) 443

40.6 Basic Principles of STM 443
40.7 Atomic Force Microscopy (AFM) 443
40.8 Applications of Atomic Force Microscopy 444
40.9 Comparison of Electron Microscopy with Electron Spectroscopy 445
Literature 445

41. Photoacoustic Spectroscopy 446

41.1 Principles of Photoacoustic Spectroscopy 446
41.2 Instrumentation for Photoacoustic Spectrometry 447
41.3 Analytical Applications of Photoacoustic Spectrometry 449
Literature 449

42. Mass Spectrometry 450

42.1 Instrumentation for Mass Spectrometry 450
42.2 Resolution of Compounds 454
42.3 Mass Spectra and Molecular Structure 454
42.4 Fragmentation Patterns in Mass Spectra 455
42.5 Qualitative Analysis with Mass Spectrometry 456
42.6 Quantitative Analysis with Mass Spectroscopy 457
42.7 Applications of Mass Spectrometry 457
42.8 Laser Mass Spectrometry 458
42.9 Fast Atom Bombardment and Liquid Secondary Ion Mass
     Substituted Spectrometry 459
42.10 Electron Spray Ionisation Mass Spectrometry 460
42.11 Hyphenated Techniques in Conjunction with Mass Spectrometer 461
42.12 Tandem Mass Spectrometer 462
Problems 463
Problems for Practice 464
Literature 464

43. Thermoanalytical Methods 466

43.1 Thermogravimetric Analysis 466
43.2 Derivative Thermogravimetric Analysis (DTG) 468
43.3 Differential Thermal Analysis (DTA) 468
43.4 Thermometric or Enthalpimetric Titrations 470
43.5 Thermometric Titration—Applications 470
43.6 Enthalpimetric Titrations 471
43.7 Thermal Methods in Quantitative Analysis 472
43.8 Differential Scanning Calorimetry 473
43.9 Thermomechanical Analysis 474
Problems 475
Problems for Practice 475
Literature 476

44. Radioanalytical Techniques 477

44.1 Radiochemical Techniques 477
44.2 Source of Neutrons 477
44.3 Neutron Capture and γ-emission 478
44.4 Principles of Neutron Activation Analysis 478
44.5 Applications of Activation Analysis 480
44.6 Principles of Isotope Dilution Method 480
44.7 Applications of Isotope Dilution Technique 481
44.8 Substoichiometric Methods of Analysis 481
Problems 481
Problems for Practice 482
Literature 483

45. Electroanalytical Methods: Potentiometry 484

45.1 Electrochemical Cells 484
45.2 The Nernst Equation 486
45.3 Standard Electrode Potential 488
45.4 Effect of Current on Cell Potential 489
45.5 Classification of Electroanalytical Methods 489
45.6 Potentiometry for Quantitative Analysis 490
45.7 Various Electrodes in Potentiometry 491
45.8 Potentiometric Titrations 495
45.9 Potential Measuring Instruments 496
45.10 Ion Selective Electrodes 499
45.11 Applications of Ion Selective Electrodes 501
45.12 Advantages and Limitations of Ion Selective Electrodes 501
Problems 501
Problems for Practice 503
Literature 504

46. Polarography and Voltammetric Methods 505

46.1 Diffusion Limiting Current 505
46.2 Polarisable Dropping Mercury Electrode (DME) 506
46.3 Principles of Polarography 508
46.4 The Ilkovic Equation 508
46.5 Generation of Polarographic Waves 509
46.6 Concept of Half Wave Potential 510
46.7 Polarographic Maxima 512
46.8 Applications of Polarography 513
46.9 AC Polarography 515
46.10 Rapid Scan Polarography 515
46.11 Organic Polarograph 516
46.12 Pulse Polarography 517

46.13 Square Wave Polarography 518
46.14 Amperometric Titration 519
46.15 Biamperiometric Titration 521
46.16 Potentiostat 521
46.17 Hydrodynamic Voltammetry 522
46.18 Chronopotentiometry 523
Problems 525
Problems for Practice 526
Literature 527

47. **Conductometric Methods**   **528**

47.1 Principles of Analysis 528
47.2 Measurement of Conductance 530
47.3 Analytical Applications of Conductometry 531
47.4 Conductometric Titrations 531
47.5 High Frequency Titrations 532
47.6 Dielectric Constant 533
47.7 Measurement of Dielectric Constants 534
47.8 Direct Current Conductivity Analysis 535
47.9 Impedance 536
Problems 536
Problems for Practice 537
Literature 537

48. **Coulometry and Electro Deposition Methods**   **538**

48.1 Principles of Electrolysis 538
48.2 Electrolysis at Constant Potential 539
48.3 Electrolysis at Constant Current 542
48.4 Coulometric Methods of Analysis 542
48.5 Applications of Coulometry 543
48.6 Coulometric Titrations 544
48.7 Applications of Coulometric Titrations 545
48.8 Stripping Voltammetry 546
48.9 Conventional Stripping Analysis 547
Problems 548
Problems for Practice 549
Literature 549

49. **Chemical Sensors and Biosensors**   **550**

49.1 Classification of Sensors 551
49.2 Sensitivity and Limit of Detection 551
49.3 Signal and Noise 551
49.4 Definition of Sensors 552

49.5   Electrochemical Sensors   553
49.6   Gas Sensors   554
49.7   Voltammetric Sensors   555
49.8   Solid State Electrode Sensors   555
49.9   Optical Sensors   555
49.10   Thermal Sensors   557
49.11   Biosensors   557
49.12   Biocatalytic Biosensors   559
49.13   Mass Sensitive Sensors   559
49.14   Efficiency of Sensors   559
Literature   560

50. On-line Analysers–Automated Instrumentation Methods   **561**

50.1   Specifications of Methods of Analysis   561
50.2   Classification of Automated Methods   561
50.3   Automated Analysis   562
50.4   Selection of On-line Analysers   564
50.5   Applications of Automated Methods   564
50.6   Environmental Monitoring   566
50.7   Flow Injection Analysis (FIA)   568
50.8   Theoretical Considerations in FIA   569
50.9   Factors Affecting Peak Height   569
50.10   Applications of Flow Injection Analysis   570
Literature   570

Answers to the Problems   **571**

Appendix-1   580
Appendix-2   581
Appendix-3   583

Index   **587**

Chapter 1

Introduction

Analytical chemistry pertains to the determination of the chemical composition of matter. It was the main goal of analytical chemists. However, the identification of a substance, the elucidation of its structure and quantitative analysis of its composition are the aspects covered by modern analytical chemistry. The most difficult task for an analytical chemist is to explain what analytical chemistry is. It is an interdisciplinary branch of science wherein a large number of research workers have contributed to its development. For instance, most of the chromatographic methods were invented by biochemists, or biological scientists, while methods like nuclear magnetic resonance and mass spectrometry were discovered by physicists. A close look at the number of research papers published in journals indicate that 60% of such papers are published by persons who are not hardcore analytical chemists. Although a large number of research workers use these techniques of analysis in either inorganic, organic or bio-chemistry, they do not like to claim themselves as analytical chemists for various reasons. One of the principal reasons for such inhibition is the classical outlook of chemists who strictly followed the cookery book approach towards methods of analysis and this practice continued specially in pharmaceutical analysis using various specifications. The problem was further aggravated by extensive application of wet analysis methods involving volumetric or gravimetric methods. These classical methods of analysis have dominated the scene of analysis for the past few decades. Fortunately, with the discovery of modern methods of analysis, mainly involving instruments, these methods have been relegated. Nevertheless, these methods will not be phased out inspite of the greater advancement of newer methods of analysis for the simple reason that the new methods have their own limitations. They cannot be applied if the substance is present in very large concentration and further in order to standardise the newer methods it is absolutely essential to use classical gravimetric or volumetric methods of analysis. Another trend is to designate newer methods involving instruments as instrumental methods of analysis, which is also false as classical methods involving use of burette, pipette or weighing balance also involve instruments and can also be called as instrumental methods. In the present circumstances when speed, simplicity and sensitivity of analysis are of utmost importance, it is better to categorise such methods as modern methods of analysis instead of instrumental methods of analysis.

1.1 IMPORTANCE OF ANALYTICAL CHEMISTRY

No other branch of science finds so many extensive applications as analytical chemistry purely for two reasons: Firstly, it finds numerous applications in various disciplines of chemistry such as inorganic, organic, physical and biochemistry and secondly it finds wide applications in other fields of related sciences such as environmental science, agricultural science, biomedical and clinical chemistry, solid state research and electronics, oceanography, forensic science and space research. It would be worthwhile to consider an example in each area of research e.g. in environmental science the monitoring of SO_2, CO, and CO_2, can be done by fluorescence or infrared spectroscopy while analysis of dissolved oxygen or chlorine from water can be carried out by potentiometry or colorimetry. The analysis of pesticides or insecticides from crops by gas chromatography or high performance liquid chromatography, or ascertaining the ratio of potassium to sodium in fertilisers by atomic absorption or flame emission methods. There are instances of the use of analytical chemistry in agricultural sciences. The analysis of micronutrients such as iron, copper, zinc, molybdenum, boron and manganese by spectrophotometry is another example. In the field of biomedical research and clinical chemistry, one can cite several examples like spectral analysis of barbiturates, food poisons, presence of vanadium and arsenic in hair and nails. The spectra of cobalt in vitamin B12, iron in haemoglobin of blood after their isolation by electrophoresis or gel permeation. In the field of electronics the analysis of traces of elements such as germanium in semiconductors and transistors, determination of selenium and caesium in photocells is quite possible by newer methods like spectroscopy or neutron activation analysis. In the field of oceanography, earth sciences and planetary sciences, analytical chemistry is extensively used. The chemical analysis of sea water; or analysis of basaltic rocks for presence of manganese and aluminum or the rapid analysis of elements from lunar samples would not have been possible without spectroscopy. An array of examples can be cited in support of applications of analytical chemistry in various interdisciplinary areas. All these illustrations amply show that analytical science is truly interdisciplinary in nature.

1.2 METHODS OF QUANTITATIVE ANALYSIS

The two important steps in analysis are characterisation and determination of the constituents of a compound. The identification step is called qualitative analysis. While the second step of quantitative analysis is more complicated. Quantitative analysis can be classified depending upon the method of analysis, or it can be categorised according to the scale of analysis.

The second classification is discussed later in the chapter. The first stage can be subdivided as those involving classical methods like gravimetry or volumetry, or those involving sophisticated instruments. Unfortunately for getting reproducible results, one cannot always directly resort to these newer methods. Before that one must get a true representative sample for analysis and it should be free from interfering elements. The interfering elements are those which come in the way of quantitative analysis. However both these problems have been circumvented by the analytical chemist. Knowledge of sampling and excellent methods of separation such as solvent extraction, ion exchange and different methods of chromatography have significantly helped an analyst to mitigate these problems of interferences. No doubt there are a few methods which can be used for isolation and purification of sample e.g. atomic absorption spectroscopy. The next step of analysis may involve volumetric methods or gravimetry if the sample is present in milligram concentration. If the component to be analysed is present at very low concentration then one has to resort to optical methods or largely spectroscopic methods like UV-visible, IR spectroscopy or molecular luminescence or emission and absorption spectroscopy. The methods principally used for the elucidation of structure involve NMR and mass spectrometry. Thermoanalytical and radiochemical methods are equally useful for quantitative work. Another important set of methods include electroanalytical techniques such as potentiometry, voltammetry, polarography, conductometry and coulometry. The application of a particular method of analysis depends upon several parameters.

1.3 CRITERIA FOR SELECTION OF METHOD FOR ANALYSIS

From the foregoing description of methods, an analytical scientist is confronted with the problem of selection of the proper method from an array of methods for quantitative analysis. His choice will be influenced by several factors such as speed, convenience, accuracy, sensitivity, selectivity, availability of instruments, amount of sample, level of analysis. Of these the last consideration is of paramount importance. In addition to consideration of the concentration of the analyte, the background of the sample will have to be carefully examined e.g. for analysis of iron (III) from a sample of haematite or polluted water we cannot resort to the same method because we have to carefully watch interfering ions. In haematite we have manganese while in water we have calcium as interfering metal so we use colorimetry (thiocynate method) for the analysis of ore while 1-10-phenanthroline is used in spectrophotometry for analysis of iron in the water sample. Unfortunately one cannot set a hard and fast rule for the selection of such methods. The choice of method is a matter of judgement. Such judgement is difficult and can come only from one's personal experience. It is unhealthy to suggest only one selected method of analysis for a particular element. The knowledge of fundamental or basic concepts of analytical chemistry certainly provides and develops such a judgement. Thus, the analytical chemist should have an extensive knowledge and understanding of basic concepts underlying methods of analysis.

1.4 CHEMICAL ANALYSIS AND ANALYTICAL CHEMISTRY

Chemical analysis establishes the quantitative composition of the materials. The constituents to be detected or determined are elements, radicals, functional groups, compounds or phases. Analytical chemistry is concerned with much broader and more general aspects of analysis, while chemical analysis is confined to much narrower and more specific aspects of analysis. The determination of one constituent in the presence of several other similar materials is essential e.g. the careful control of conditions such as pH, complexation, change in oxidation state of metals etc. Several advances in analytical chemistry have been made possible due to the spectacular progress in separation science. Analysis is generally composed of quantitative analysis and qualitative analysis. The qualitative analysis is carried out before quantitative analysis. The constituents can be quickly detected by spectrographic methods or by spot tests with selective, specific and sensitive organic reagents.

1.5 QUANTITATIVE ANALYSIS WITH SCALE OF OPERATION

The amount of sample and range of relative amount of constituent determinable are important characteristics of quantitative analytical methods. The methods can be termed as macro, semimicro, or micro on the basis of the mass of the sample taken. A macro sample will be defined as one whose weight is greater than 0.100 gram and a semimicro as one between 0.100-0.010 gram but any sample less than 0.010 gram is termed as a microsample. In general a concentration between 0.01-0.001 gram is termed a microsample; while samples weighing less than 0.002 gram can be called submicro or ultramicro samples. Components which are between 100-1% of a sample are termed major constituents and those between 1-0.01% are minor constituents. Those below the range of 0.01% are termed as trace concentrations. A spectrophometric determination requires a macro sample but spectrographic determinations can be carried out with micro samples. Nowadays one can analyse fema (10^{-15}) or ato levels (10^{-18}) of analysis.

1.6 STEPS IN QUANTITATIVE ANALYSIS

The important steps involved in determination are: (1) Procurement of samples (2) Its conversion to measurable state (3) Measurement of desired constituent (4) Calculation and interpretation of numerical data. The most difficult problem is the isolation, before the measurement step. The important steps encountered are:

(a) *Sampling:* It should be representative of the mass of the material. This is possible provided the material is reasonably pure and is homogeneous in nature.

(b) *Conversion of desired constituent to measurable form:* This step involves method of separation. The selection of the separation technique for a specific situation will depend upon a number of factors. Such selection is generally decided on the basis of accuracy and the precision required.

(c) *Measurement of desired constituent:* Any physical or chemical property can be used as a means of qualitative identification and quantitative measurement or both. If the property is specific and selective for measurement, then separation and pretreatment of the sample can be minimised, e.g. analysis involving atomic absorption spectroscopy.

(d) *Calculation and interpretation of analytical data:* An analysis is not complete until the results have been expressed in such a manner that the person for whom the results are intended can understand their significance. In recent years greater attention has been paid to statistical techniques both in development and in assaying the value of the final analytical results. This has led to the establishment of a new branch of science termed as Chemometrics.

1.7 METHODS OF ANALYTICAL DETERMINATION

The chemical methods of determination are composed of volumetric and gravimetric methods of analysis, as they involve chemical reactions. The methods based on measurement of physical properties are called physicochemical methods. Physical methods of analysis are those methods which do not necessarily involve use of chemical reaction. Most of the physical methods of measurement are instrumental methods of analysis.

(a) *Chemical methods:* These methods comprise of gravimetric and volumetric methods of analysis. They are generally based upon stoichiometric equations of the type.

$$aC + bR \rightleftharpoons C_a R_b$$

In gravimetry, excess of the reagent (R) reacts with the constituent (C) to form a product ($C_a R_b$) which is solid and is weightable. The chemical reaction and separation should be quantitative and loss should be minimum i.e. with recovery of approximately 99.9% for the major components.

However in volumetry reagent (R) is added to constituent (C) until $C_a R_b$ is formed, the end point of the reaction is shown by an indicator. Many times the end point is reached before or after the equivalence point. Various reactions such as neutralisation, oxidation-reduction, precipitation and complex formation can be used in such titrations. The speed of analysis is of great importance. Both gravimetry and volumetric methods are useful for the determination of major components. For multianalysis or for samples many in number, titrimetry is most useful. On the whole, both methods are considered precise, accurate and practicable in routine analysis.

(b) *Physical methods:* The popularity that these physical methods or physicochemical methods enjoy arises from the new kind of determinations that they make possible due to their selectivity and from the increased speed and simplicity of analysis. They are used for the determination of minor or trace concentrations of elements or constituents in preference to the major constituent of the sample. It is generally believed that relative precision or accuracy is usually not so great as compared with chemical methods. However, this is not always true. As far as speed of analysis is concerned, physical and chemical methods cannot be compared as the former are very rapid and accurate. The majority of methods require the use of standards containing a known amount of the constituent which serves as the basis for comparison in the measurements as reference material.

1.8 ROLE OF INSTRUMENTATION

It is essential to distinguish between instrumentation for analytical techniques and operation of instruments. The former is of utmost importance to the analytical chemist. The chemist in general should understand the fundamental relation between chemical species and their characteristics. A good analytical chemist may not be an electronics expert but should know the scope and applicability of analysis and also the limitations of measurement in analysis. During analysis methods which demand the least skill from the analytical chemist are preferred. However, knowledge of the chemical reactions in a particular system is most essential. When the chemist resorts to the use of instruments, he should not end up as a black box operator. It is absolutely essential that he is able to interpret data obtained and arrive at the logical conclusion regarding the composition or structure of the analyte.

LITERATURE

1. I.M. Kolthoff, P.J. Elving, *"Treatise in Analytical Chemistry"*, Part I-III several volumes, Interscience (1959).
2. C.L. Wilson, D.W. Wilson, *"Comprehensive Analytical Chemistry"*, Elsevier (1958).
3. H.F. Walton, *"Principles and Methods of Chemical Analysis"*, 2nd Ed., Prentice Hall (1966).
4. R.A. Day, A.L. Underwood, *"Quantitative Analysis"*, 3rd Ed., Prentice Hall India (1977).
5. H.A. Latinen, W.E. Harris *"Chemical Analysis"*, McGraw Hill (1990).
6. J.G. Dick *"Analytical Chemistry"*, McGraw Hill (1973).

Chapter 2

Reliability of Analytical Data and Statistical Analysis

Analytical Science accumulates enormous amounts of data out of different measurements. This data has no value unless one examines how much of it is reliable and how much is reproducible. We would consider such reliability norms and measures in analysis.

2.1 ERRORS IN CHEMICAL ANALYSIS

The error which is equivalent to observed value and the true value in any chemical analysis is related by an equation $E = (O - T)$ ... where E = absolute error, O = observed value, T = true value. Usually the error of a measurement is in inverse proportion to accuracy of the measurement $i.e.$, the smaller the error the greater is the accuracy of an analysis. Errors are generally expressed relatively in terms of percentage or part per thousand

$$\text{percentage} \left(\frac{E}{T} \times 100 \right) = \% \text{ error}$$

$$\text{per thousand} \left(\frac{E}{T} \times 100 \right) = \text{ppt}$$

From the knowledge of the error, we can easily differentiate between the two terms generally used in analysis $i.e.$ precision and accuracy. The precision of the measurement of a quantity is meant the reproducibility or to the extent of agreement of the individual values among themselves. The differences between the individual values and the arithmetical mean is the measure of the precision of measurement. In other words, the smaller the difference between individual value and the arithmetic value, the greater is the precision. However, it should be remembered that precise value may not always be accurate value because a more or less constant error can affect them all. Accurate results are those where the individual value matches with the true value of analysis.

2.2 CLASSIFICATION OF ERRORS

Errors are classified into two groups, namely determinate errors and indeterminate errors. Determinate errors are those errors which have definite values, whereas indeterminate errors are those which have indefinite values and they follow no law and invariably have fluctuating values. Determinate errors are of three kinds depending upon their origin as methodic, operative and instrumental errors.

(a) *Methodic errors* are inherent in the method applied and are the most serious of analytical errors. Such errors have their sources in the chemical properties of the system. Various types of errors arise e.g. in precipitation of aluminium as aluminium hydroxide due to the presence of zirconium, titanium and iron in the solution which are also coprecipitated.

(b) *Operative errors* are those errors which must be blamed on the analyst, and on the method or the procedure. These are usually personal errors caused by a variety of reasons such as colour blindness, incorrect handling of instruments or careless weighing.

(c) *Instrumental errors* usually originate in the instrument itself. They arise from the effect of environmental factors on the instrument e.g. zero error in any reading instrument. Such errors can be minimised by calibration or use of proper blanks in case of spectrophotometric analysis.

(d) *Additive and proportional determinate* are other types of errors wherein the absolute value of an additive error is independent of the amount of constituent present in a determination, while the absolute value of a proportionate error depends upon the amount of impurity in a sample.

Indeterminate errors are those errors which either arise from variation in determinate errors or those which are erratic in occurrence. Those belonging to the first kind occur when the analyst cannot control conditions to exclude a determinate error or prevent it varying e.g. in gravimetric analysis of aluminium, aluminium oxide shows high weight due to the presence of water vapour and can be easily controlled whereas the analyst has no control over the erratic or random type of errors.

2.3 DETERMINING THE ACCURACY OF METHODS

It is possible to determine the accuracy of quantitative analytical methods by use of a sample containing a known amount of the constituent in a synthetic sample or by the use of a sample in which the content of the constituents has been added deliberately and determined by other methods for comparison. The first is called an absolute method while the second is called the comparative method. In the absolute method, the differences between the average of an adequate number of closely agreeing results and the amount of constituent present is measured. In the absence of diverse ions, this method is valid. In the comparative method, the sample is determined by two different independent methods and then the accuracy is verified, e.g. analysis of an element in question by volumetry, gravimetry and colorimetry. In most work, the allowable error is 1-10 ppt per component. However, in a few samples involving no separations the accuracy limit can be increased.

In actual practice, the accuracy of analysis is tested in various ways, namely by the replicate determination, summation, charge balance and by independent methods. In replicate determination two or more concurrent parallel determinations are made to gain confidence in the results, because a single determination is dangerous. A poor agreement in replicates indicates error. In the summation method, all results of individual constituents are summed upto see whether the sum is close to 100 because a good summation is a sufficient criterion of accuracy. The charge balance technique is applicable to solutions of electrolytes as well as solids composed of ions and charged components. Finally in an independent method the analyst checks the accuracy of the result by applying a known method of analysis.

2.4 IMPROVING THE ACCURACY OF ANALYSIS

There are several ways of enhancing accuracy of the analysis. One can use small blanks along with the sample during determination. The use of large blanks reduces the precision.

Another way is to give a correction e.g. in weighing of filter paper, or weight of precipitate unreacted or unignited. One can also use controls by use of standard addition, where a known amount of constituent is added to a sample which is to be analysed for the total amount of constituent present. Usually the difference between the analytical results of samples with and without added constituent gives the recovery of the amount of added component. The standard addition method clearly points out negative error but fails to pinpoint positive errors such as the interference of a foreign substance. This method is extensively used in instrumental analysis. In flame photometry an analysis of synthetic or standard sample provides a test of the accuracy of a procedure and helps to improve the accuracy of the values obtained. Finally if a sample cannot be made like the standard in chemical composition, then a standard is made synthetically similar to the sample. In determination of CaC_2O_4 by $KMnO_4$ it is advisable to standardise $KMnO_4$ in chemical composition against a solution of $Na_2C_2O_4$ in order to compensate for error.

An analytical chemist should strive hard to improve the existing methods by proper combination of methods of separation and determination to increase accuracy. Real advances can only be made by abandoning old methods and seeking newer ones. An analytical chemist should be an originator and innovator as well as a repair person.

2.5 STATISTICAL ANALYSIS

The evaluation and interpretation of analytical data is very important. Statistical methods are useful in analytical chemistry in several ways, as they measure the performance of an analytical procedure as there is need to have some measure of confidence in an analytical procedure. Such confidence depends upon the inherent performance, amount of information pertaining to performance and number of results that are used to get the average. Variation among averages does not necessarily suggest that the amount of element in the sample is changing. A large difference in averages may lead to doubt that the sample is contaminated. The main task of statistics is to recognise changes in composition or performance due to changes or differences in a set of readings. The various criteria used to evaluate analytical data are described below:

1. *Absolute and relative difference* in the values: Supposing we have two values 2.431 and 2.410 then the absolute difference between the two is 0.021 but the relative difference is expressed as

$$\frac{0.021 \times 1000}{2.4} = 8.7 \text{ ppt.}$$

2. *Mean value and deviation:* The mean can be expressed up to three significant figures. The difference between one value and the mean is the deviation (δ) of that value from the mean.

3. Average deviation of a single measurement is the mean of the derivatives of all the individual measurements *i.e.*

$$d = \frac{\sum \delta}{n}$$ if $\Sigma\delta$ is the sum of all deviations from the mean and n is the total number of measurements.

4. Average deviation of mean (D) is equal to the average deviation of a single measurement divided by the square root of the number of measurements made.

$$D = d/\sqrt{n}$$ if D = average derivation of mean, n = number of readings

d = average deviation of a single measurement.

5. Standard deviation is a precise and reliable measurement of deviation. Standard deviation of single measurement(s) is:

$$s = \sqrt{\frac{\sum \delta^2}{n-1}}$$ if $\sum \delta^2$ is sum of squares of all deviations from mean and n is total number of measurements.

6. Standard deviation of mean (s) is obtained from the standard deviation (s) by dividing it by the square root of the number of measurements.

$$S = \frac{s}{\sqrt{n}}$$ Usually $(2s)$ is taken as a reasonable limit within which the true value is likely to lie.

7. Median value: The median or median value of a series of readings is the reading of such magnitude that the number of readings having a greater numerical value than it, is equal to the number having a lower value. This value is half way in the series of ascending and descending order of readings.

8. Range: The range is the difference between the largest and smallest result. Such value of the range gives a quick estimate of standard deviation.

9. Relative standard deviation is also called the coefficient of variation (i.e. C.V.) and is expressed:

$$\frac{\text{Standard deviation}}{\text{average amount i.e., mean}} \times 100 = \frac{S \times 100}{m}$$

Example: A sample of hematite was analysed by eleven students. The values obtained for the percentage of iron were as follows: 22.62, 22.73, 22.75, 22.78, 22.79, 22.83, 22.84, 22.87, 22.87, 22.92 and 22.94.

(1) Is rejection of the first value (22.62) justifiable on the basis of huge error?
(2) What is the mean (m) value? What is the average deviation of the mean?
(3) What is the standard deviation (s) and standard deviation of mean?
(4) What is the median value and what is the range for measurement?
(5) What is the relative difference between the lowest and highest reported values?
(6) What is the relative standard deviation?

Answers:

(1) Rejection of the first value (22.62) is not justifiable as it involves no huge error because this value does not differ much from the mean value d = 0.193 < 4d (*i.e.* 0.088).

(2) The mean value is the average of eleven values: it is 22.813. The average deviation of the mean is

$$D = \frac{\sum \delta}{n} = \frac{0.242}{11} = 0.022$$

(3) The standard deviation is $(s) = \sqrt{\frac{\sum \delta^2}{(n-1)}} = 0.092$

(4) The median value is = 22.83 while the range is (22.94 − 22.62 = 0.32).

(5) The relative difference between the lowest and the highest reported value is 14 ppt.

(6) The relative standard deviation or coefficient of variation is $=\left(\dfrac{s}{m}\times100\right)=1.22\%$

2.6 CRITERIA FOR REJECTION OF RESULTS Q-TEST

The above example shows how statistical analysis is of great importance in interpretation of analytical data. It is to be noted that results appearing out of line may not necessarily be rejected e.g. for six replicate measurements, the gap between the extreme value and its nearest neighbour must exceed half the entire range of all six measurements before rejection of the result. If this is the case omit the doubtful value and determine in the usual way the mean and average deviation of the retained values. The differences (δ) $\geq 4d$ rejection is justified, if the deviation of the suspected value from the mean is at least four times the average deviation of the retained values. Some workers allow (δ) $\geq 2.5d$. The error of the rejected value is called huge error. In such examples d represents actual difference between the given values and the mean value in a particular case. However Q-test is most reliable basis of rejection.

2.7 PRESENTATION OF DATA

Finally rounding off the data to get rid of uncertainty in the last place is not always advisable for presenting data as every measurement can be used to contribute to the information and estimate. Sometimes rounding off conceals the original difference and throws away information on variation. Usually in published literature meaningless figures are trimmed but this hampers evaluation of the precision of a new method e.g. 10.37 and 10.52 has a difference of 0.15 units but on rounding off it appears 10.4 and 10.5 with a difference of 0.10 unit which is not a good representation of values.

We would finally consider the significance of confidence limit and the use of Q-test and student 't' test to evaluate the accuracy of the measurement. We use usual notations for the symbols used here in.

2.8 CONFIDENCE LIMIT

If 'n' is smaller than 'S' or s, it will not be reliable and will be uncertain. By calculating 90% confidence a limit for 'S' upper and lower limits of confidence limits can be obtained by dividing $(n-1)S^2$ by two numbers taken from Tables of *chi square* given at end of this book. Thus if $n = 7$, $(n-1) = 6$ then the upper confidence limit will be (if S = standard deviation, s = standard deviation of mean)

$$\text{Upper confidence limit} = 3.67S^2 = \frac{(n-1)S^2}{\text{chi square value at 95\% confidence limit}}$$

$$\text{Lower confidence limit} = 0.47S^2 = \frac{(n-1)S^2}{\text{chi square value for 90\% confidence limit}}$$

at 90% limit. For 'n' value chi square values of 95% or 90%, confidence limit are 3.94 and 18.1 resp. also available in *Appendix 3* at the end of book. Based on these confidence limits for 5% and 95% limit we can calculate upper and lower confidence limit.

Example: If the number of readings (n) are 11 and if standard deviation (s) is 0.308, calculate upper and lower confidence limits of n values.

$$\text{Upper C.L.} = \frac{(n-1)S^2}{\text{chi sq. value for 95\% CL}} = \frac{10(0.308)^2}{3.940} = 0.2407 \begin{pmatrix} \text{chi. sq. value 95\%} \\ CL = 3.94 \end{pmatrix}$$

$$\text{Lower C.L.} = \frac{(n-1)S^2}{\text{chi. sq. value 90\% limit}} = \frac{10(0.308)}{18.307} = 0.518 \begin{pmatrix} \text{chi. sq. value at 90\%} \\ CL = 18.307 \end{pmatrix}$$

2.9 Q-TEST FOR REJECTION OF RESULT

It can be found at 90% level by using equation for upper limit, Range, $Q_n = \left(\dfrac{M_2 - M_1}{M_n - M_1} \right) =$

lower limit $= \dfrac{M_n - M_{n-1}}{M_n - M_1}$. If Q_1, Q_n is larger than given values in Table in Appendix 4, it must be rejected. If readings are 6.593, 6.492, 7.098, 7.146 and 7.256 if $n = 5$ range is (7.256 − 6.942) = 0.668, 7.098, 7.148 and 7.256 if $n = 5$ range is 7.256 − 6.942 = 0.668, if $M_1 = 6.593$, $M_5 = 7.256$.

Now $Q_1 = \dfrac{M_2 - M_1}{M_5 - M_1} = \dfrac{(6.942 - 6.593)}{(7.256 - 6.593)} = 0.527 < 0.64,$

$$Q_2 = \frac{M_5 - M_4}{M_5 - M_1} = \frac{(7.256 - 7.146)}{(7.256 - 6.943)} = 0.17$$

If $M_2 = 6.942$, $M_1 = 6.593$, $(M_5 - M_1) = 0.668$ range. $M_5 = 7.256$, $M_4 = 7.146$.

For 90% confidence limit $Q_{90} = 0.64$ from Tables (Appendix 4). The lower value is less than 0.64 and therefore must be accepted.

2.10 STANDARD 't' TEST

First standard deviation (s) is calculated and a value of 't' is obtained from 't' Tables (Appendix 5). The limits are given by confidence level by an expression

$\left(mean \pm \dfrac{ts}{\sqrt{n}} \right)$. Suppose mean is 9.46 and standard deviation (s) is 0.17% and if $n = 6$ then $t = 2.02$ for 90% confidence limit from Table (Appendix 5) so we have $t = 2.02$.

$$\left(9.46 \pm \frac{2.02 \times 0.17}{\sqrt{6}} \right) = \text{mean } 9.46 \pm 0.14\%. \text{ This is acceptable value.}$$

The table for 't' 'Q', F and chi square values are given in Appendix 3-6.

Example: From following two sets of values, evaluate mean, standard deviation, confidence limit and rejection quotient 'Q'.

First Set (I)   =   107.861,   107.870,   107.881   and   107.892.
Second Set (II) =   107.797,   107.778,   107.779   and   107.780.
State which one is good set.

Answer: The mean value in (I) is 107. 876 while for (II) mean value is 107.778.

The standard deviation (S) $= \sqrt{\dfrac{\Sigma d^2}{n-1}} =$ for set I $= 0.12$ ppt for set II is 0.001 ppt

The deviation and average deviation of set I = 0.10 ppt and set II = 0.01 ppt
The confidence limit is calculated as shown below:

$$\text{Set (II)} = \frac{3.2 \times 0.013}{\sqrt{4}} = 0.021 \text{ for set (II)} = 0.002 \text{ for 95\% confidence limit}$$

$$\text{Set (I)} = \frac{5.8 \times 0.013}{\sqrt{4}} = 0.338 \text{ for set (I)} = 0.004 \text{ for 99\% confidence limit}$$

Now we calculate Q
$Q = a/b = \text{difference/average} = 0.011/0.031 = 0.35$ set (I)
$Q = a/b = 0.001/0.031 = 0.33$ for set (II) which is acceptable. So set (II) is good.

Example: Calculate 95% confidence limit for following readings for absorbance in spectrophotometric analysis of iron with 1-10 phenanthroline?

0.084, 0.089, 0.079 given 't' values for 3 readings = 4.30

Answer: Mean $= \dfrac{0.084 + 0.089 + 0.079}{3} = 0.0840$

Standard deviation (S) $= \sqrt{\dfrac{\Sigma \delta^2}{n-1}} = \dfrac{\sqrt{0 + 0.005 + 0.005}}{2} = 0.0050$

Now from Appendix the value for 95% confidence limit is 0.084.

if $n = 3$, $t = 4.13$ then C.L. $= \dfrac{t_{\text{value}} \times 0.0050}{\sqrt{n}} = \dfrac{ts}{\sqrt{n}} = \dfrac{4.3 \times 0.005}{\sqrt{3}} = 0.084 \pm 0.012$.

Example: For set of 10 results, if mean = 56.06 with standard deviation of 0.2% with use of t-values ($t = 1.833$ from Tables) calculate the confidence limit for 90% limit.

Answer: $\dfrac{\text{Mean} \pm \text{confidence t-value} \times \text{standard deviation}}{\sqrt{\text{No. of readings}}} = \dfrac{mts}{\sqrt{n}}$

By substituting above value in expression we have

$$= \frac{56.06 \pm 1.833 \times 0.21}{\sqrt{10}} = 5.6 \pm 0.12\%$$

Example: Which value should be rejected in the following set of results: 40.12, 40.15, 40.55, on basis of Q-test?

Answer: $Q = \dfrac{X_3 - X_2}{X_3 - X_1}$ if $X_1 = 40.12$, $X_2 = 40.15$, $X_3 = 40.55$

We have $Q = \dfrac{(40.55 - 40.15)}{(40.55 - 40.12)} = \dfrac{0.40}{0.43} = 0.93$ for $Q_{90}^3 = 0.94$

On referring Tables these values cannot be rejected if $Q = 0.94$ for 90% confidence limit.

Example: Calculate (a) average deviation, (b) Standard deviation (c) rejection of results (d) 90% confidence limit (e) t-distribution? The values are as follows: 7.146, 7.098, 6.942, 7.256 and 6.593? (n = 5)

Answer: Average deviation = 7.007 ± 0.92, standard deviation of mean = 7.007 ± 0.256. Mean = $0.192 \times 4 = 0.768$ limit of acceptance = 7.007 ± 0.768.

Both values are acceptable standard deviation: $0.256 \times 3 = 0.765$ limit of acceptance 7.007 ± 0.768 (See appendix).

The highest value = 7.256, lowest is = 6.593, range is 0.663.

So
$$Q_1 = \frac{M_2 - M_1}{M_5 - M_1} = \frac{6.942 - 6.593}{0.663} = 0.53$$

$$Q_2 = \frac{M_6 - M_4}{M_5 - M_1} = \frac{7.156 - 7.146}{0.663} = 0.17$$

Thus for 90% confidence limit $Q_{90} = 0.64$ so both values are good we had used confidence limit =

mean $\pm \dfrac{ts}{\sqrt{n}}$ (t is obtained from Tables in Appendix)

$$C.L. = 7.007 \pm \frac{2.13(0.256)}{\sqrt{5}} = 7.007 \pm 0.244.$$

There is 90% probability that distributor mean is within ± 0.244 of mean.

PROBLEMS

2.1 In the analysis of calcium the following values were obtained: 15.42, 15.56, 15.68, 15.51, 15.52, 15.53, 15.54, 15.53, 15.56. State which values are acceptable and which can be rejected.

Answer: The lowest value is 15.42. The range is (15.68-15.42) = 0.26. This is between the highest and lowest value.

$\dfrac{15.51 - 15.42}{0.26} = 0.35$ it is less than 0.41. Hence smallest value viz. 15.42 is accepted. Here 0.41 represents the average value of the difference in readings.

$\dfrac{15.68 - 15.56}{0.26} = 0.46 > 0.41$. Here the average value is 15.56 while the average difference is 0.41.

Since the difference in the highest value (15.68) is greater than the average difference viz. 0.41

the value is rejected. However, if we now take 15.56 as the highest value then the lowest value viz. 15.42 will also be rejected as it involves a huge error of 0.64. In this case the lowest value will be 15.51 which will have an average difference of 0.20 w.r.t. 15.56 and is therefore accepted.

2.2 Two students analysed a sample of gold and obtained the following results:

A-Set: 107.861, 107.870, 107.881, 107.892
B-Set: 107.777, 107.778, 107.779, 107.780.

Calculate the mean, standard deviation and rejection quotient (Q) in both set of values. *Answer:* The mean value for the first set (A) is 107.876 while the mean value for the second set (B) is 107.778. Then the deviation and average deviation of two sets are:

Set (A) 0.10 ppt and for Set (B) 0.01 ppt. By the usual criteria standard deviation for both sets is

$$S = \sqrt{\frac{\Sigma \delta^2}{n-1}}$$. We have for set (A) = 0.12 ppt while for set (B) it is 0.001 ppt.

Finally $Q = a/b$ = difference/range = 0.011/0.031 = 0.35 for set (A) and Q = 0.001/0.003 = 0.33 for set (B).

PROBLEMS FOR PRACTICE

2.3 The following absorbance values were obtained during spectrophotometric determination of zirconium as its alizarin complex: 0.2562, 0.2566, 0.2565, 0.2560, 0.2568 and 0.2566. Calculate the average derivation and standard deviation of the mean, of the results.

2.4 The percentage of silica in a sample of mineral was 3.685, 3.689, 3.681 and 3.692. Calculate the average deviation of the mean, standard deviation of the mean and the median value.

2.5 In gravimetric estimation of iron with cupferron following values were obtained: 0.6143, 0.6153, 0.6148, 0.6154, 0.6146, 0.6142. Calculate the mean average deviation of single measurement; and mean, standard deviation of the mean and difference between mean and median value.

2.6 The volumetric analysis of oxalic acid with potassium permanganate gave the following titre values: 22.62, 22.73, 22.75, 22.78, 22.79, 22.83, 22.84, 22.87, 22.92 and 22.94 ml. Can the first value be rejected as it involves huge error? Calculate the mean value, average deviation of mean, standard deviation of mean, median value, and relative error of the lowest reported value.

2.7 In quantitative analysis of nickel by complexometry titrations the following results were obtained.
Nickel Found (g): 19.2, 28.7, 37.4, 45.2, 54.2, 75.8, 96.8, 119.4.
Nickel Taken (g): 20.0, 30.0, 40.0, 50.0, 60.0, 80.0, 100.0, 120.0.
Calculate the standard deviation of a single determination and comment on the nature of systematic error in the determination of nickel.

2.8 In the gravimetric analysis of aluminium with 8-hydroxyquinoline the following data were obtained: 0.5000, 0.02851, 0.02856, 0.02873, 0.02852, 0.02831, 0.02856, 0.02851, 0.02853, 0.02855 and 0.02850. Calculate the mean value and standard deviation of the mean. State which values can be rejected because they involve huge error.

2.9 Calculate the mean, median for the following results obtained in analysis of zinc in a brass sample: 64.92, 65.05, 65.09, 65.11, 65.20, 65.22. Can any value be rejected?

LITERATURE

1. W.B. Guenther, *"Quantitative Chemistry — Measurement and Equilibrium"*, Addison Wesley (1962).

2. I.M. Kolthoff, E.B. Sandell, E.J. Meehan and S. Bruckenstein, *"Quantitative Chemical Analysis"*, 4th ed., Macmillan (1969).

3. K. Ecksehlanger, *"Errors, Measurements and Results in Chemical Analysis"*. Van Nostrand Reinhold (1970).

4. J.G. Dick, *"Analytical Chemistry"*, McGraw Hill Kogakusha (1973).

5. D.J. Pietryk and C.W. Franck, *"Analytical Chemistry; An Introduction,"* Academic Press (1974).

6. H.A. Laitinen and W.F. Harris, *"Chemical Analysis: An Advanced Textbook "*, McGraw Hill, 2nd Ed. (1982).

7. L.F. Hamilton, S.G. Simpson and D.W. Ellis, *"Calculations in Analytical Chemistry"*, 2nd Ed. McGraw Hill Co. (1978).

8. S.M. Khopkar, *"Analytical Chemistry Problems and Solutions"*, New Age International Publisher Ltd. (2001).

Chapter 3

Sampling in Analysis

3.1 INTRODUCTION

A complete knowledge of sampling processes and purpose of analysis eliminates probable error. The reliability of analytical data as discussed in the previous chapter also depends upon the kind of sample one is using in chemical analysis. The sample should be representative and should be homogeneous. The good analyst should be familiar with the importance of sampling, the statistical background of the sample, general methods of sampling, knowledge of proper transmission and storage of the sample, pitfalls leading to bias in sampling, applications of analysis and finally source for additional information about sampling.

3.2 DEFINITIONS OF TERMS

Some terms have a definite meaning in sampling techniques and they should be clearly understood e.g. A sample is a portion of material selected which possesses essential characteristics of the bulk of the original material. A sampling procedure consists of various steps. A sampling unit may be defined as a minimum sized package in the consignment of material taken from any bulk sample. While a gross sample is one which is prepared by mixing various portions or increments of samples. A sub sample is smallest portion of the main sample just like a gross sample. The analyte is the amount of sample taken for analysis.

3.3 THEORY OF SAMPLING

An ideal sample would be identical in all its respective properties of the bulk of the material from which it is removed. The factors to be noted are mainly the cost of the test and the value of the products, the permitted variation in the material, the accuracy of test method and the nature of the material used.

There are many obvious blunders in sampling which we may term as pitfalls in sampling. For instance, it is often dangerous to accept the material for analysis without some knowledge of its history. Random choice is one source of bias in sampling. The separation of particles on the basis of size can introduce a serious error. Operations that change the composition of the sample should be avoided. Adhesive tapes should not ordinarily be stuck on samples of minerals or ores because they give erroneous information about the presence of zinc oxide from the adhesive tape.

3.4 TECHNIQUES OF SAMPLING

Various techniques for sampling gases, liquids and solids are available. Thus gases are collected by flushing, displacement with a liquid and expansion into an evacuated container. The physical state of the contaminant will determine to some extent the method of sampling. In the sampling of pure liquids or homogeneous liquids obviously a simple procedure is used, as liquid is by and large homogeneous in character. However the sampling of solids introduces a variable that does not occur in the sampling of fluids, viz., particle size differences. Finely ground sample is usually preferable to heterogeneous solids. By and large hand sampling is usually not advisable. At times mechanical sampling machines are used extensively because they are more accurate and faster than hand sampling. In all the above cases thorough mixing is invariably essential. Fine material may be sampled by use of a sampling tube called 'thief' and on many occasions this thief serves as a quick measure for sampling of solids and liquids. In conclusion, we state that the bulk of material to be sampled should be subdivided into real or imaginary sampling units. By carrying out analysis of variance on stratified sampling it would be possible to decide whether it is necessary to continue stratification at several levels. Generally some stratification will be useful. For a particular material the size of sample should be maximum size of the particles. It is better to avoid excessive grinding of sample as sometimes coarse granules are one which differ in composition. Usually a gross sample should be collected and be reduced to appropriate size for analysis.

3.5 STATISTICAL CRITERIA OF IDEAL SAMPLE

The sampling scheme can be explicitly expressed as the sample mean which should provide an unbiased estimate of the population mean. The sample should also provide an unbiased estimate of population variance. This is possible provided every possible unit of pre-selected size has equal opportunity of being drawn. The sampling procedure should lead to estimates of the central value and dispersion that are as accurate as possible. This is possible by stratified procedure. Now the next important factor is the estimation of required sample size of particulate material. For solid particulate material the weight of sample is taken at random from bulk must be increased with increase in variation in composition, considering accuracy of analysis and nature of particle size. The main criterion is that the amount of sample should be drawn for which sampling error is less than the required limit. This is properly explained in problems provided at the end of the chapter. It will be seen from these problems that analytical error and sampling error can be easily identified.

Finally a few words about sampling units, although some discussion was already given. A random sampling procedure is one in which each portion of the whole is given an equal chance of appearing in the sample. In practice randomised selection is seldom used in sequential series, but at regular intervals samples are taken. A cyclic variation in quality would fall into phase and lead to a biased sample. This is called 'aliasing'. Uniform sampling procedure can give average composition more accurately than a random procedure. Further error can be minimised by increasing the frequency of the sampling. Generally each unit is analysed separately and the results are averaged. The (n) units are properly mixed and divided into n-parts before analysis; this averages the composition and changes the variance. Hence precise results are obtained by mixing groups of units and performing (n) analyses on mixed groups rather than by analysing the (n) units separately. Further the mixture is analysed by a method in which standard deviation. (σ) is proportional to the magnitude being determined. With a method of constant relative precision the analytical precision is independent of the amount analysed. Finally a single determination is carried out on the whole mixture by a method wherein the standard deviation (σ) is independent of the magnitude being determined, because a single determination on the combined sample will give more precise results than the average of 'n' separate units analysed.

3.6 STRATIFIED SAMPLING *VERSUS* RANDOM SAMPLING

When the material to be sampled is divided into sampling units there are two ways of random sampling. A number of samples can be chosen at random from the bulk of material or sampling can be stratified by first choosing from the amount the sampling units and then sampling within the units, again by random procedure. The stratified procedure gives results as precise as the simple random scheme and gives best results whenever the variance among sampling units is remarkable compared with the variance within such units. In order to reduce the sampling in the nonrandom sampling procedure, the population is divided by random procedure into a number of subdivisions and if the variance among sections is large compared with variance in sections a more accurate results will be obtained by accurately sampling each section. Such a stratified procedure includes a risk of bias. However such risk can be eliminated by making the procedure representative as far as possible. Now in statistical terms if the whole population to be sampled is homogeneous then simple random sampling is as good as stratified sampling, but if we have a different subpopulation which cannot be described by a single set of factors, then sampling the subpopulation by stratified procedure is accepted. If an estimate of the population mean is to be unbiased with minimum variance in the estimate then the number of samples taken for each stratum is taken in relation to the size of the stratum and also its standard deviation *i.e.*,

$$\frac{n_r}{n} = \frac{W_r(\sigma_0)_r}{\sum W_r(\sigma_0)_r}$$ from the 'r' stratum, 'n' is the total number of samples, 'W' is the weight of stratum

and $(\sigma_0)_r$ standard deviation with rth stratum. In calculation of population mean of the stratum must be the weighted average in relation to the size of strata. Thus

$$\bar{x} = \frac{\sum W_r \bar{x}_r}{\sum Wr}$$

If $\bar{x}_r$ is the mean of determined values in the 'r' stratum, $\bar{x}$ is estimate of population mean. If strata differ in size but not in variance, σ_r is the same for all.

$$\frac{n_r}{n} = \frac{W_r}{\sum W_r}$$

i.e., the number of samples within strata should be related to size of the strata. This is called representative sampling which gives an unbiased estimate of the population mean but leads to large variance in estimate, unless the variance is uniform in all the strata. The variance of the mean is independent of the variance among strata. Stratified procedure gives excellent results if the variance among the strata is appreciable compared with the variance within the strata.

3.7 MINIMISATION OF VARIANCE IN STRATIFIED SAMPLING

Consider a bulk of material to be analysed by taking (n_1) start, each of which provides (n_2) samples and (n_3) determinations are to be carried out on each sample. If strata are equal in size and in variance within strata, it is essential to look into minimisation of cost of determining an estimate of the population mean to a certain variance. If (C_1) is the relative cost of selecting the strata, is the cost within the sampling unit (C_3) and is the cost of performing determinations, the total cost (C) is:

$$C = n_1 C_1 + n_1 n_2 C_2 + n_1 n_2 n_3 C_3$$

Now if $S_1^2 S_2^2 S_3^2$ are quantities with σ_1^2, σ_2^2 and σ_3^2 as corresponding estimates then we have:

$$\sigma^2 = \frac{\sigma_1^2}{n_1} + \frac{\sigma_2^2}{n_1 n_2} + \frac{\sigma_3^2}{n_1 n_2 n_3}$$

Now it can be proved that to minimise the cost for (C) for fixed value of σ^2 the values of are given as

$$n_1 = \sqrt{\frac{\sigma_1^2 / C_1}{\sigma^2}} \left(\sqrt{\sigma_1^2}\, C_1 + \sqrt{\sigma_2^2}\, C_2 + \sigma_3^3 C_3 \right);$$

$$n_2 = \sqrt{\frac{\sigma_2^2 / C_1}{\sigma_1^2 C_1}}\, ; \quad n_3 = \sqrt{\frac{\sigma_3^2 / C_2}{\sigma_2^2 C_2}}$$

Here the optimum allocation of sampling after the first stage is independent of the desired overall variance σ^2. This is shown in problem 3.3 at the end of this chapter.

3.8 TRANSMISSION AND STORAGE OF SAMPLES

During transportation, the container, moisture, oxygen, carbon dioxide, light, heat and other environmental pollutants in the atmosphere may contaminate the nature of the sample. It is therefore necessary to consider both the nature of the material that is being sampled and the tests that are to be made on the samples, while transporting it. Utmost care should be taken while drawing a sample from a population with a finite number of units to be examined. Since the process of sampling involves various operations like crushing, grinding and subdivision, no contamination can be permitted during storage; in addition to each operation, storage and transportation also contributes to the overall variance. It is essential to recognise the problem of storage otherwise efforts may be expended on the sampling operation either increasing the cost or failing to achieve the desired level of accuracy and precision.

PROBLEMS

3.1 If the occurrence of defective units coming from an industrial unit is 2% and a random sample of 10,000 units is drawn, what is the expected number of faulty units in the sample? Calculate the standard deviation.

Answer: The expected number is $0.02 \times 10,000 = 200$

The standard deviation $= \sqrt{10,000 \times 0.2 \times 0.98} = 14$

Therefore relative standard deviation can be calculated as usual as described in Chapter 2.

3.2 A sample is composed of uniform grains of an alloy of density 5.2 and mixture of alloy of density 2.7. The percentage of metal in the alloy is 75 and 5 in the mixture. If the sample contains on average 60% of metal, how many grains must be taken so that the relative standard deviation is less than 1 ppt in average percentage due to sampling error?

Answer: The fraction (f) of alloy grain is calculated from the relation: $f/(1 - gf) = d_2\,(f - f_2)/d_1$ $(f_1 - f) - 1.903$ from which $f = 0.656$. The average density is given by equation $fd_1 + (1 - f)\,d_2 = 4.34$. The value $f_{av}\sigma_s$ of and n calculated from expansion

$$n = f(1-f)\,\frac{d_1\,d_2}{d_2}\,\frac{100(f_1 - f_2)}{\sigma_s\,f_{av}}$$

PROBLEMS FOR PRACTICE

3.3 If the standard deviation of a sampling unit is 0.07 and that of sampling within unit is 0.10 and of a single determination is 0.21 and if the relative price is in the ratio of 4:2:1, calculate the best sampling plan and minimum cost that will give standard deviation of the mean of 0.08.

3.4 What size of sample is necessary from a uniform solution of radioactive isotope of concentration of 10^{-8} M, if the standard deviation due to sampling should not exceed 0.1%?

3.5 A sample of mineral casserite (SnO_2) contains particles of density 4.5 and gauge of density 2.2 containing 60 and 15% of metal tin. If a sample of 0.20g containing about 40% tin is taken for analysis with catechol violet spectrophotometry at 560 nm and if particles have uniform diameter of 0.01 mm, what is the standard deviation due to sampling?

3.6 How large a sample of an equimolar mixture of oxygen and nitrogen measured at 273° absolute temperature and 10^{-6} mm pressure, must be taken if the sampling error is not to exceed 0.1%?

LITERATURE

1. D.J. Pictrzyk and C.W. Frank, *"Analytical Chemistry: An Introduction"*, Academic (1974).
2. D.A. Skoog and D.M. West, *"Fundamentals of Analytical Chemistry"*, 3rd Ed., Saunders Holt (1970).
3. K. Ecksehlanger, *"Errors, Measurements and Results in Chemical Analysis"*, Van Nostrand Reinhold (1970).
4. A.I. Vogel, *"The Textbook of Quantitative Inorganic Analysis"*, 4th Ed., Longman and Green (1980).
5. H.A. Laitnen and W.F. Harris, *"Chemical Analysis, An Advanced Text and References"*, McGraw Hill (1975).
6. F. Settle, *"Handbook of instrumental techniques in the analytical chemistry"*, Pearson Educational Publishers (2001).
7. Gy. Pierre, *"Sampling for analytical purposes"*, John Wiley & Sons (1998).

Chapter 4

Gravimetric Analysis

The gravimetric analysis, or quantitative estimation by weight, is the process of isolating and weighing an element or a definite compound of the element in as pure a form as possible. A large portion of the determinations in gravimetric analysis are concerned with the transformation of the element or radical into a stable, pure compound which can be readily converted into a form suitable for weighing. The weight of the element is readily calculated from the knowledge of the formula of the compound and atomic weights of the constituent elements. The separation of the element or of the compound containing the element may be effected in a number of ways such as those involving precipitation methods, volatilisation or evolution methods, electroanalytical methods or other miscellaneous methods. In practice the first two methods are most common. Gravimetric methods are time consuming but are reliable. The constituent may be examined for the presence of impurities and correction can be applied if necessary.

4.1 PRECIPITATION METHODS

Precipitation is carried out in such manner that promotes the separation by filtration e.g. silver is precipitated as AgCl dried at 130°C and weighed as chloride. Zinc is precipitated as Zn (NH$_4$) PO$_4$ 6H$_2$O is ignited and weighed as Zn$_2$P$_2$O$_7$ (pyrophosphate). The most important aspect to be borne in mind in such methods is that the precipitate must be insoluble and separable by filtration. Secondly, the physical nature of the precipitate should be such that it can be easily separated from the solution by filtration and can be washed free of soluble impurities, the particle size should be of good dimensions, and finally, the precipitate must be convertible into as pure a substance of definite chemical composition.

The solubility of a substance at a given temperature in a given solvent is defined as the amount of the substance that is dissolved by a known weight of that solvent when the substance is in an equilibrium with the solvent. It depends upon the particle size. If the substance dissolved is in excess, it results in a super saturated solution. This supersaturated solution is a solution which contains a greater concentration of solute that corresponds to the equilibrium solubility at that temperature under consideration (Fig. 4.1 region AB). As such supersaturation is an unstable state which may be brought to a state of stable equilibrium by the addition of a crystal of solute usually termed seeding (region BC). Generally the rate of precipitation (R) is or relative supersaturation is expressed as:

$R\alpha \left(\dfrac{A-B}{B} \right)$ Where A – total concentration mol dm^{-3}, B – equilibrium solubility mol dm^{-3}, and (A – B) is super saturation at the moment precipitation commences. Thus rate of precipitation

$$= K\left(\frac{A-B}{B}\right) \text{ where } K = \text{constant}$$

while
$$P = \left(\frac{W}{S} \times 100\right) \text{ if}$$

S = sample weight in grams; W = weight of residue
P = percentage of substance sought (Table 4.1)

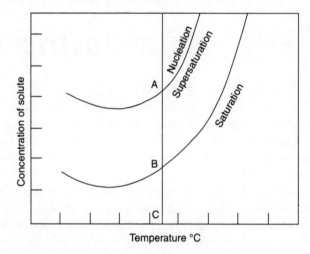

Fig. 4.1 Solubility behaviour

Table 4.1 Separation of Ba as insoluble sulphate

Reagent Concentration (N)	$\left(\dfrac{A-B}{B}\right)$	Nature of precipitate
7	175,000	Gelatinous
3	75,000	Gelatinous becomes turbid
1	25,000	Curdy and colloidal
0.05	1,300	Starry-shaped, feathery
0.005	125	Precipitate of crystals
0.001	25	2-3 hours for precipitation
0.0002	5	One month for precipitation

The precipitation is carried out in hot solution as solubility increases with rise in temperature. Precipitation is effected in dilute solutions and the reagent is added slowly with constant stirring, because initially formed particles serve as nuclei for precipitation. A suitable reagent is added to increase the solubility of the precipitate and this leads to formation of large primary particles.

4.2 PURITY OF PRECIPITATION: COPRECIPITATION

When a precipitate separates out from a solution it may not be perfectly pure. The contamination of the precipitate by substances which are normally soluble in mother liquor is termed coprecipitation. It is concerned with adsorption at the surface of the particles exposed to the solution and occlusion of foreign substances during the process of crystal growth from primary particles. Absorption is maximum in gelatinous precipitates and is least for microcrystalline precipitates e.g. In AgI an acetate of silver and $BaSO_4$ precipitate in alkali nitrates. Appreciable errors may also be introduced by post precipitation. This is the precipitation which occurs on the surface of the first precipitate after its formation. It occurs with sparingly soluble substances which form supersaturated solutions. They usually have an ion in common with the primary precipitate e.g. CaC_2O_4 precipitation in the presence of magnesium. The latter separates out as MgC_2O_4 gradually upon CaC_2O_4. The longer the time of contact the greater is the precipitation which generally occurs on the surface of the first precipitate.

However post precipitation and coprecipitation are not at all similar phenomena. For instance in postprecipitation contamination increases with time while in coprecipitation it decreases with time. Further contamination is enhanced on agitation of the solution only in postprecipitation but not in coprecipitation. The extent of contamination is very much larger in postprecipitation than in coprecipitation.

4.3 OPTIMUM CONDITIONS FOR PRECIPITATION

The general rules are as follows:

(a) Precipitation should be carried out in dilute solutions, as this minimises errors due to coprecipitation.

(b) The reagent should be mixed slowly with constant stirring, as it helps the growth of large regular crystals. To ensure complete precipitation, slight excess of the reagent should be added. The same order of mixing should be maintained in replicate determinations.

(c) Precipitation should be carried out in hot solutions provided the precipitate is stable at high temperature. This rule may not always be true in the case of many organic precipitants.

(d) Crystalline precipitates should ordinarily be digested for longer duration in a steam bath in order to avoid the problem of coprecipitation.

(e) The precipitate should be washed with dilute solutions as pure water tends to cause peptisation *i.e.*, hydrolysis.

(f) To avoid the problem of post precipitation or coprecipitation, it is frequently advisable to take recourse to reprecipitation.

4.4 PRECIPITATION FROM HOMOGENEOUS SOLUTION

In this method the reagent is generated slowly by homogeneous chemical reaction within the solution. Then it is allowed to react with the ion to form the precipitate. Such a precipitate is dense and filterable; coprecipitation is reduced to a minimum. A few examples of precipitation from homogeneous solution are:

(a) *Sulphates:* Dimethyl sulphate generates that sulphate radical. Also sulphamic acid on hydrolysis generates the sulphate ion

$$\left(CH_3\right)_2 SO_4 + 2H_2O \rightarrow 2CH_3OH + 2H^+ + SO_4^{2-}$$

$$NH_2 SO_3H + H_2O \rightarrow NH^+ + H^+ + SO_4^{2-}$$

(b) *Hydroxides:* pH is controlled slowly. NH_3 is generated from urea by the following reaction:

$$NH_2(CO) + H_2O \rightarrow 2NH_3 + CO_2 \uparrow \text{ at } 90\text{-}95 \text{ }^\circ C$$

Aluminium is precipitated by urea as $Al(OH)_3$ in succinic acid media or barium is precipitated as $BaCrO_4$ in ammonium acetate or nickel as glyoxime or aluminium as oxinate.

(c) *Oxalates:* Calcium is precipated as CaC_2O_4.

Thorium is also precipitated as $Th(C_2O_4)_2$ in presence of urea or ethyloxalate

$$(NH_2)_2 CO + 2HC_2O_4^- + H_2O \rightarrow 2C_2H_5OH + 2H^+ + C_2O_4^{2-};$$

$$(C_2H_5)_2 C_2O_4 + 2H_2O \rightarrow 2C_2H_5OH + 2H^+ + C_2O_4^{2-};$$

$$Th^{4+} + 2C_2O_4^{2-} \rightarrow Th(C_2O_4)_2$$

(d) *Phosphates:* Insoluble phosphates may be precipitated by generating the phosphate ion from trimethyl or triethyl phosphate by stepwise hydrolysis. Zirconium is precipitated as $Zr_3(PO_4)_4$ in $(CH_3)_3 PO_4$ in media containing sulphuric acid.

4.5 WASHING OF THE PRECIPITATE

The objective of washing the precipitate is to remove the contamination on the surface. The composition of the wash solution will depend upon the tendency of the precipitate to undergo peptisation. In such cases water is avoided. A solution of strong electrolyte is employed. This solution (wash) should have a common ion with the precipitate in order to reduce solubility errors and should be easily volatilised in the preparation for weighing. Hence ammonium salts are used to wash liquids. Hot solutions are preferred. Wash solutions are divided into three categories as follows:

(a) Solutions which prevent precipitation from becoming colloidal and passing through filter papers e.g. use of ammonium nitrate for washing precipitate of ferric hydroxide.

(b) Solutions which reduce the solubility of precipitate should be avoided.

(c) Solutions which prevent the hydrolysis of salts of weaker acids and weaker bases.

Washing with many small portions is much more effective than a single wash.

$$x_n = x_0 \left[\frac{u}{u+v} \right]^n = \text{concentration of impurities before washing}$$

x_n = concentration of impurities after washing

n = number of washings

u = volume of liquid remaining in precipitate, v = volume of solution used for each washing.

This equation is similar to the one used in solvent extraction by multiple extraction. This indicates that one should allow the liquid to drain off as far as possible in order to maintain (u) minimum and use a relatively small volume of liquid (v) ml and increase the number of washings (n).

4.6 IGNITION OF THE PRECIPITATE

The precipitate may contain adsorbed water, occluded water, sorbed water and essential water as water of hydration. Ignition temperatures are based upon knowledge of chemical properties of the substance. Heating should be continued till constant weight and should be uniform. The weight of filter paper ash should also be accounted for. The optimum temperature can be ascertained from the TGA curve.

4.7 ROLE OF ORGANIC PRECIPITANTS IN GRAVIMETRIC ANALYSIS

Those organic reagents which are used in gravimetric analysis are generally termed organic precipitants. The separation of one or more inorganic ions from mixtures may be made with the aid of organic reagents. These compounds have high molecular weight which in turn facilitates determination of a small amount of ion with a large amount of precipitate. The ideal organic precipitant should be specific in character. The precipitate so formed with organic reagents is weighed after drying or it is ignited and weighed as an oxide or the precipitate is dissolved in mineral acid and determined volumetrically. The optimum selectivity of reaction can be attained by controlling variables such as concentration of the reagents and pH of the solution and use of sequestering agents to eliminate interference from foreign ions. Most of the important organic reagents are chelating agents. When a polyfunctional ligand which can occupy more than two positions in the coordination sphere it combines with a central metal ion to form a coordination compound with a closed ring structure it is termed as chelate. A close membered ring incorporating a metal is thus called a chelate. A guide to qualitative predictions can be obtained from the study of formation constants of coordination compounds, the effect of the nature of the metal ion and nature of the ligand on the stability of complexes and precipitation equilibria involved in formation of uncharged chelates.

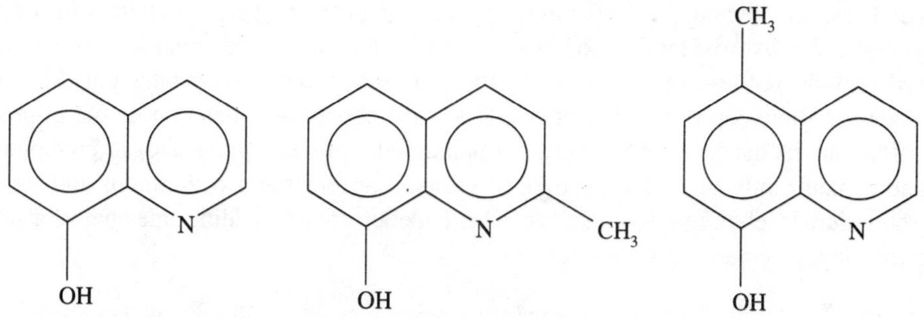

8-hydroxquinoline (I) can precipitate aluminum but not 2-methyl
8-hydroxquinoline (II) or 5-methyl-8-hydroxquinoline (III)

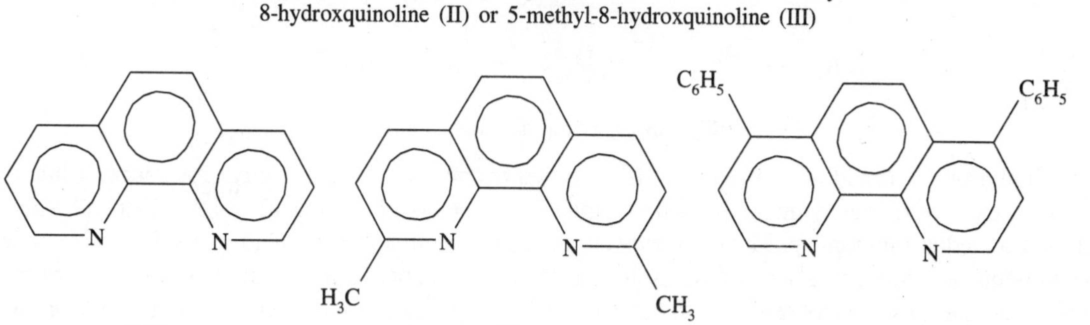

Fig. 4.2 Selectivity and substitution

Organic precipitants have a special place in inorganic analysis because the precipitants which they form are for the most part qualitatively different from the purely inorganic percipitate like $BaSO_4$. For instance, Ni (DMG)$_2$ where DMG is dimethylglyoxine is different. Organic compounds are classified as complexformers (mainly chelates), salt formers and lake formers. In order to form a chelate, the ligand must have a replaceable H-atom and a pair of unshared electrons for coordination. Organic reagents are preferred because they are selective. The substitution of the carbon atom can be varied widely. The word selective means the capacity of an organic reagent to combine with one or two metals to the exclusion of others under certain set conditions. Steric hindrance often causes the selectivity of the chelating agent.

Another example is the substituted derivatives of 1-10 phenanthroline (IV) or 2, 9 dimethyl 1-10 phenanthroline (V) or 4, 7 diphenyl-1-10 phenanthroline (VI). The reagents (IV) or (VI) are selective for iron (II) while reagent (V) is selective for copper on account of steric hindrance. The selectivity can be further improved by control of pH. Use of masking or sequestering agents also makes the reaction selective e.g. Al can be precipitated by oxine in presence of Cu and Zn if the latter elements are masked with KCN.

4.8 CRITERIA FOR CHOICE OF AN ORGANIC REAGENT

Various points must be considered while selecting an organic reagent for precipitation. It must be selective for a particular metal e.g., use of dimethylglyoxime or 1-nitro-2-napthol for precipitation for nickel or cobalt respectively; cupferron for iron; quinaldic acid for copper; mandelic acid for zirconium or N-phenyl N-benzoylhydroxylamine for niobium and tantalum. Since organic precipitates are not easily ionised they do not coprecipipitate impurities and other ionic precipitates e.g. Mg oxinate, or Mg (OX)$_2$ does not allow coprecptiation of Na, K as such coprecipitations is possible in precipitation of Mg (NH$_4$)PO$_4$ and its ignition to Mg$_2$P$_2$O$_7$. A small amount of metal gives a lot of precipitate e.g. Cu quinaldic acid contains only 14.94% Cu. These precipitates are light and voluminous and are adoptable at micro or semimicro levels. Organic reagents can be made selective by adding side chains or substituents in the aromatic ring, to create steric hindrance.

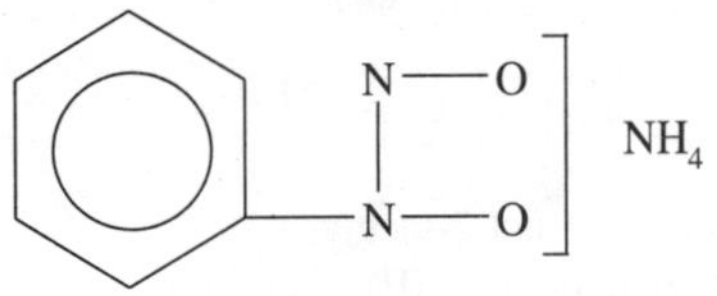

(VII) 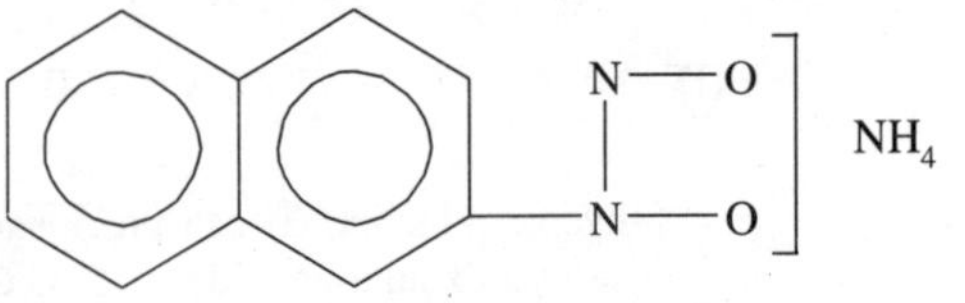   (VIII)

Fig. 4.3 Cupferron and neocupferron

Cupferron (VII) and neocupferron (VIII) are two examples. One is selective for copper while the other is selective for neodymium. Several precipitates can be dissolved in acid and the liberated reagent titrated by redox titration e.g. Metaloxinates on dissolution with a mineral acid like sulphuric acid can be subsequently titrated against KBrO$_3$ solution. Several reagents form coloured complexes which in turn favour their use in spot tests and in colorimetric analysis. Because of their covalent character most of the metal complexes with organic reagents are soluble in non-polar solvents. This technique is used in solvent extraction e.g. Fe (III) cupferron complex is soluble in ether which in turn facilitates its separation from other metals present in a large volume of the solution. Further, metal chelates are mostly anhydrous hence precipitates dry quickly. This can be accelerated by washing the precipitate with alcohol and not acetone as the latter dissolves the precipitated chelate compounds. They can be dried at 105-110°C due to their hydrophobic nature. The low solubility of reagent in water is a serious disadvantage. Although alcohols or acetic acid are used as solvents they do not give an idea about how much excess reagent should be added. Finally it is difficult to obtain organic reagents in pure form. Keto enol isomers cause errors in colorimetric analysis unless conditions are carefully standardised e.g. use of dithizone for zinc or cadmium.

4.9 SOME IMPORTANT ORGANIC PRECIPITANTS

Several organic reagents are used extensively in gravimetric analysis e.g.

1. Dimethylglyoxime (IX) (for Ni) Excess reagent should be avoided as it precipitates. Citrate and tartarate are used as the sequestering agents.

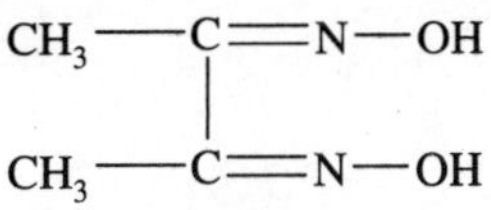

(IX) Dimethyl Glyoxime

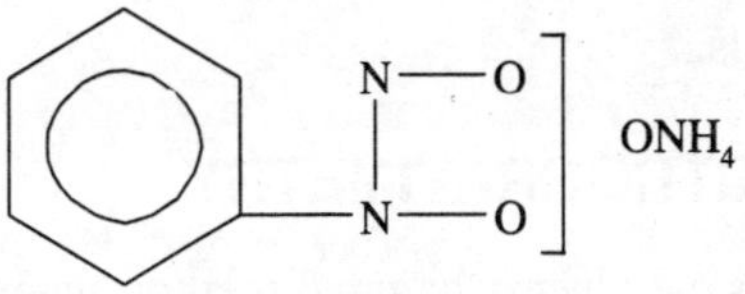

(X) Cupferron

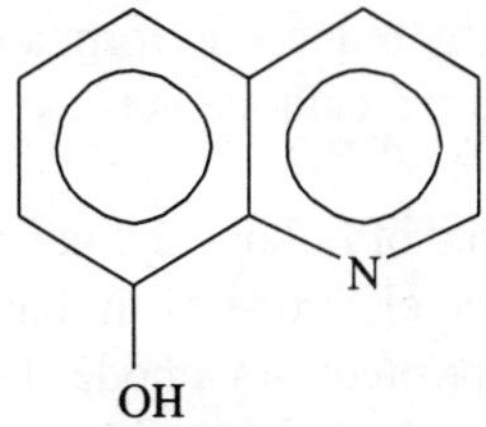

(XI) 8-Hydroxyquinoline

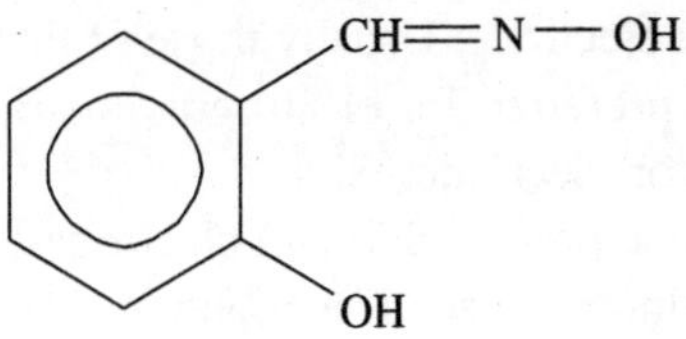

(XII) Salicyldioxime

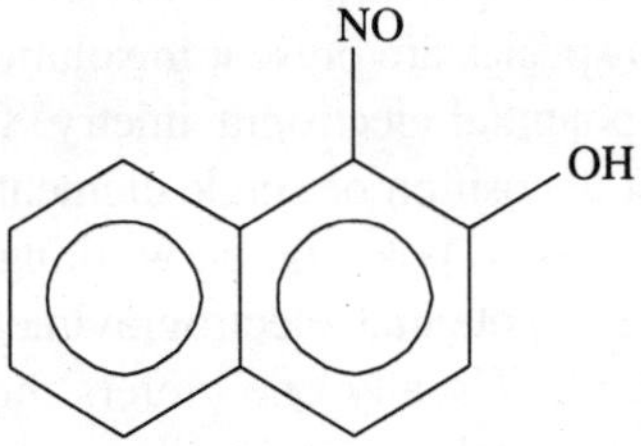

(XIII) 1-nitroso 2- naphol

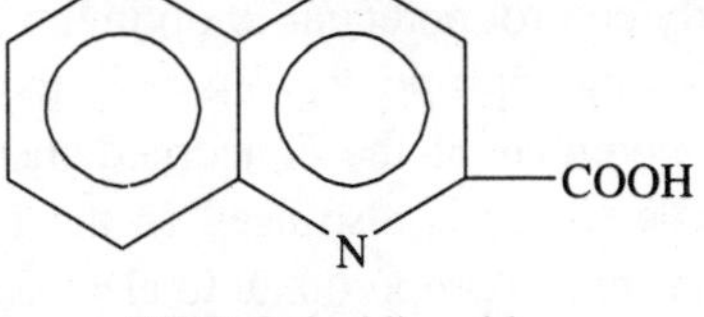

(XIV) Quinaldic acid

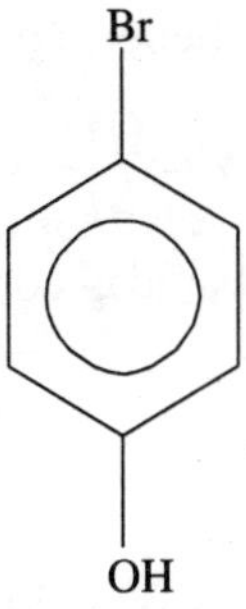

(XV) Mandelic acid

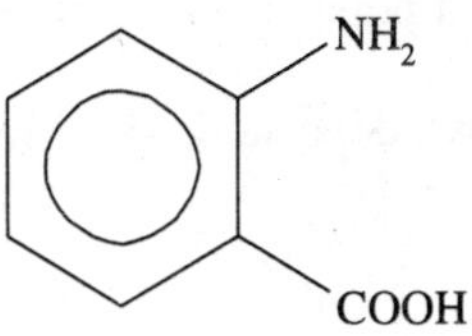

(XVI) Anthranilic acid

Fig. 4.4 Various organic reagents used in gravimetry.

2. Cupferron (X) (for Fe (III) & Cu (II)). It is useful in acid media, from cold solution, and its precipitate is ignited and then weighed as oxide.
3. 8-hydroquinoline (XI) (for Mg) the reagent is added in the cold and the precipitate is washed with warm water. The precipitate can be dissolved in acid and titrated.
4. Salicyldioxime (XII) (for Cu). It is weighed as such. Tartaric acid is used as the masking agent. The complex is soluble in alcohol but it is not stable for more than 73 days.
5. I-nitroso-2-naphthol (XIII) (for Co) used in acid media. This complex is ignited and weighed as Co_3O_4. The reagent is prepared in glacial acetic acid and distilled water.
6. Quinaldic acid (XIV) (for Cu). The method is made selective by use of complexing agents. In the complex only 15% copper is present.
7. Mandelic acid (XV) (for Zr). The precipitate is ignited and then weighed as oxide.
8. Anthralinic acid (XVI) (for Cu). Generally, the sodium salt is preferred. Useful for several metals.

4.10 ELECTROGRAVIMETRY

This is a classical electrochemical method wherein a substance is electrochemically generated to form coated layer on an electrode in an electrolytic cell. Such electrode is subsequently weighed and from the knowledge of original weight a difference in weight is established which pertains to material deposited on an electrode. Usually those of the metals which are easily reduced to form a metal coating on cathode are preferred for electrogravimetric investigations. Here the cathode acts as working electrode or indicator electrode.

Generally one prefers platinum electrodes. A cathode is composed of hollow cylindrical electrodes made from platinum gauze. The anode is also platinum hollow gauze electrode, with diameter slightly smaller than the cathode. By process of convection, material is deposited on cathode. In order to get reproducible results, the solution should be constantly stirred. One can control either a current or a potential. In case of species deposition with constant current electrogravimetry is preferred as it is easy task to adjacent the current. In situation where more than two species are present in solution which can be also deposited at cathode it is essential to use controlled potential electrogravimetry. This is useful as one can easily control potential to optimum level to permit deposition of single element at a time. It is done by imposing appropriate voltage in the circuit. The adjusted potential on working electrode is the deposition potential of the concerned metal. In controlled potential electrogravimetry saturated calomel electrode (SCE) is also used as the reference electrode. Usually one prefers three electrode assembly. The current is proportional to electroactive species. Time taken in controlled potential method is somewhat more in comparison to controlled current electrogravimetry. Metal interferences can be eliminated by controlling deposition potential at appropriate value, hence controlled potential electrogravimetry is most popular with analyst. One uses relationship as $W = \dfrac{QM}{nF}$ with symbols carry usual notation viz. W = wt. deposited, M = Mol. wt., then Q = current, F = Faraday constant, n = number of equiv. of formula wt.

PROBLEMS

4.1 In a sample of feldspar weighing 0.4150 gram, a mixture of chlorides of sodium and potassium weighed 0.0715 g. This on further treatment gave 0.1548 g. potassium hexachloroplatinate. Find out the percentage of Na_2O and K_2O in the sample and mineral.

Answer: Let x be weight of NaCl, $(0.0715-x)$ = weight of KCl. Since 2 KCl = K_2PtCl_6, so $(0.0715-x)$ gram of KCl = 0.1548 gm of K_2PtCl_6.

Now substituting values of molecular weights

$$(0.0715 - x) \times \frac{485.95}{2 \times 74.56} = 0.1548$$

Hence x = 0.0240 gm. The conversion factor of 2NaCl = Na_2O, 2KCl = K_2O.

$$\frac{0.0240 \times Na_2O}{NaCl \times 0.4150} \times 100 = 3.07\% \ Na_2O$$

$$\frac{(0.0715 - 0.24)K_2O}{2KCl \times 0.4150} \times 100 = 55.44\% \ K_2O$$

4.2 In a 0.5 gm sample of feldspar, the weight of (NaCl + KCl) = 0.1180 gm. Treatment with $AgNO_3$ gives 0.2451 gm of AgCl. Calculate the percentage of oxides of sodium and potassium in the ore.

Answer: Let a = wt. of KCl and b = wt. of NaCl

$$a + b = 0.1180 \text{ gm} \qquad \qquad ...(1)$$

No. of grams of AgCl recovered from (a) gram KCl = $a\left(\dfrac{AgCl}{KCl}\right) = 1.92\ a$

No. of grams of AgCl recovered from (b) gram of NaCl = $b\left(\dfrac{AgCl}{NaCl}\right) = 2.45\ b$

Hence $\qquad\qquad\qquad$ 1.922a + 2.452b = 0.2451 $\qquad\qquad\qquad$ ...(2)

Solving simultaneously Eqns. (1) and (2) we have

$$a = 0.0834 \text{ of KCl} \qquad b = 0.0346 \text{ gm of NaCl}$$

if mol wt. of KCl = 74.56

mol. wt. of NaCl = 58.5

$$\% \ K_2O = \left(\frac{K_2O}{2KCl}\right) = \frac{84.20}{2 \times 74.56} \times \frac{100}{0.5} = 10.5\% \ K_2O$$

$$\% \ K_2O = \left(\frac{Na_2O}{2NaCl}\right) = \frac{62.0}{2 \times 58.5} \times \frac{100}{0.5} = 3.67\% \ Na_2O$$

PROBLEMS FOR PRACTICE

4.3 A sample of calcium mineral gave the following analysis: CaO (45.18); MgO (8.10); FeO (4); SiO_2 (6.02); CO_2 (34.7) and H_2O (2.03) in percentage. On ignition no water was detected and CO_2 was 3.30%. If all iron is oxidised to the trivalent state, calculate the percentage of CaO and Fe_2O_3 in the burnt sample.

4.4 A platinum disc weighs 25.6142 gm in air against brass weights. What is its weight in vacuum if the platinum and brass are respectively 21.4 and 8.0 and the weight of one ml of air is 0.0012 gm?

4.5 What weight of ferrosoferric oxide (Fe_3O_4) will furnish 0.5430 grams of ferric oxide?

4.6 0.5 gram of iron ore (Fe_3O_4) is changed to ferric oxide of weight 0.411 gram. Calculate the percentage of Fe_3O_4 in the ore.

4.7 A sample of dolemite is ignited to give oxides of calcium and magnesium. Weight of mineral 1.0045 gram while weight of mixture of oxides is 0.5184 gram. What is the percentage of calcium and magnesium in the sample of dolomite?

4.8 A 340.0 mg sample containing KCl and NaCl gave 706.2 mg of silver chloride. Find out the percentages of each alkali metal.

LITERATURE

1. J.F. Flagg, *"Organic Reagents Used in Gravimetry and Volumetric Analysis"*, Interscience, New York (1948).
2. F. Feigel, *"Chemistry of Selective, Specific, Sensitive Reactions"*, Academic Press (1949).
3. F.J. Welcher, *"Organic Analytical Reagents"*, Van Nostrand (1959).
4. L. Erdey, *"Gravimetric Analysis"*, Pergamon Press (1963).
5. H.F. Walton, *"Principles and Methods of Chemical Analysis"*, 2nd Ed., Prentice Hall (India) (1966).
6. I.M. Kolthoff, E.B. Sandell and P.J. Elving, *"Treatise in Analytical Chemistry"*, Interscience, Part II all volumes (1959).
7. H.L. Gordan, M.L. Salulker and H.H. Willard, *"Precipitation from Homogeneous Solution"*, John Wiley (1959).
8. R.G.W. Holling, *"Oxine and its Derivants"*, Butterworth (1956).
9. A.I. Vogel, G.H. Jeffrey, G.H. Bassett, J.Mendham and R.C. Denney — *"Vogel's Textbook of Quantitative Chemical Analysis "*, — 6th Ed., Longmans, Wiley (1989).

Chapter 5

Volumetric Analysis— Acid Base Titrations

To measure a volume of liquid in volumetry is much faster than weighing a substance in gravimetric methods. Further, volumetric methods are faster if good standard indicators are available. These methods are as accurate as gravimetric methods of analysis. Volumetric analysis is also called titrimetry. A substance to be analysed is allowed to react with another substance of known concentration from a burette in the form of a solution. The concentration of the unknown *i.e.* analyte is then calculated. The criteria for selection of reagent are that the reagent must react rapidly, the reaction must be complete and no side reactions should be present and there should be a way of finding out excess reagent with the help of an indicator.

5.1 VOLUMETRIC ANALYSIS

The point at which just adequate reagent is added to react completely with a substance is called the equivalence point e.g. in $AgNO_3$ and NaCl titration the equivalent point is reached when one mole of $AgNO_3$ reacts with one mole of NaCl:

$$Ag^+ + Cl^- \rightleftharpoons AgCl$$

The concentrations of Ag^+ and Cl^- which remains unprecipitated must be equal at equivalence point and from a knowledge of the solubility product of AgCl, the concentrations of each is 1.2×10^{-5} molar at 25°C. The point at which the colour change of an indicator is apparent to the eye, it is called as end point. The equivalence point and end point are not the same. With Na_2CrO_4 as indicator in the above reaction, ordinarily red $AgCrO_4$ should show an end point when $[Ag] > 1.2 \times 10^{-5}$ M *i.e.* solubility product. However, in practice the end it is precipitate and a good amount of excess reagent used for such formation. Hence the actual difference between the end point and the equivalence point is called end point error. The end point error is a random error which is different for different systems. It is additive in nature and determinant in character. It can be evaluated, by use of potentiometric or conductometric methods, the end point error is reduced to zero.

All titrimetric methods depend upon standard solutions which contain an exactly known amount of the reagent in unit volume of the solution. The concentration is expressed in terms of normality (gm eq./litre). The standard solution is prepared by weighing the pure reagent. However, all reagents are not available in pure form e.g. $KMnO_4$ contains MnO_2 as an impurity which has to be filtered out first.

$KMnO_4$ is standardised against $Na_2C_2O_4$ standard which is available in highly pure form. Those substances which are available in pure form with definite chemical composition are called primary standards. All primary standards must be obtainable in pure form. They must react only under the condition of titrations and no side reaction should be present. They must not change weight on exposure to the atmosphere. Salt hydrates are not good as primary standards; for them the equivalent weight should be pretty high to reduce error due to weighing. Acids or bases should preferably be strong with high ionisation constants. The primary standards which are commonly used in volumetric titration are:

(a) *Acids:* C_6H_4 (COOK) (COOH): KHpthalate, C_6H_5COOH, constant boiling HCl; sulphuric acid SO_2 $(NH_2)OH$, K acid iodate *i.e.*, $KHIO_3$. Benzoic acid

(b) *Bases:* Na_2CO_3, MgO and $Na_2B_4O_7.10H_2O$, $K_2B_4O_7.4H_2O$; HgO

(c) *Oxidising agents:* $K_2Cr_2O_7$, $(NH_4)_2Ce(NO_3)_6$, $KBrO_3$, $KH(IO_3)_2$, I_2

(d) *Reducing agents:* $Na_2C_2O_4$, Fe $(N_2H_4N_2H_6)$ $(SO_4)_2$ $4H_2O$, Fe, $K_4Fe(CN)_6$.

(e) *Others:* NaCl, KCl.

5.2 CLASSIFICATION OF VOLUMETRIC METHODS

Volumetric methods have been broadly classified as:

(a) Acid-base titrations *i.e.*, those titrations involving strong or weak acids or bases.

(b) Redox titrations are those titrations which largely involve simultaneous oxidation-reduction reactions. The majority of titrations are covered by these two classes.

(c) Precipitation titrations are those titrations which involve formation of precipitate e.g. titration of Ag or Zn with $K_4Fe(CN)_6$ with a suitable indicator e.g. adsorption indicator.

(d) Complexometric titrations largely involving EDTA titrations. Such titrations are specific and also have applicability on account of existence of difference in pH of complexation. We would consider acid–base titrations.

5.3 ACID-BASE TITRATIONS

Acid–base titration gives a sharp enough end point with an indicator if the pH at the equivalence point is within 4-10 and also two acids or bases are present in equal concentrations, when the stronger one is titrated in the presence of the weaker, the end point is sharper if the ratio of acid dissociation constant is $>10^4$. During acid–base titration the pH of the solution changes in a typical manner, showing a sharp change in pH with volume of the titrant at equivalence point. The pH and sharpness of the curve vary for three different acids (Fig. 5.1). The analyst is usually interested in knowing the pH at the equivalence point, and the slope of titration curves to find the end point error. Such curves can be modified with use of nonaqueous solvents. Let us consider an acid–base reaction which is proceeding with a proton acceptor (in acid–base reactions protons are transferred from one molecule to another). In water the proton is usually solvated as H_3^+O. H_2O is added to base to lose (OH^-) or gain (H_3^+O). Acid base reactions are reversible. The reaction can be shown as:

$$HA + H_2O \rightarrow H_3^+O + A^- \text{ (base)}$$

$$HA + H_2O \rightarrow H^+B + OH^- \text{ (acid)}$$

Here $[A^-]$ is the conjugate base, (H^+B) is conjugate acid. Thus we say acid + base $\rightleftharpoons$ conjugate base + conjugate acid e.g.

$$HA + H_2O \rightarrow H_3^+O + A^- \text{ (base)}$$

$$HA + H_2O \rightarrow BH^+ + OH^- \text{ (acid)}$$

Hence
$$K_A = \frac{\left[H_3^+O\right]\left[A^-\right]}{[HA][H_2O]}$$

$$K_B = \frac{[HB]\left[OH^-\right]}{\left[B^-\right]}$$

If $K_W = \dfrac{\left[H^+\right]\left[OH^-\right]}{[H_2O]}$ is the ionic product of water, it is possible to give an expression for $[H^+]$ in terms of K_A, K_B and K_W for a combination of various types of strong and weak acids or bases. Most acid–base titrations are usually carried out at room temperature except titrations involving bases which contain CO_2 e.g. Na_2CO_3 which are carried out at ice cold temperature. Temperature affects acid–base titrations. The pH and colour change of an indicator depends indirectly upon temperature. This is due to change in acid–base equilibria with temperature, as K_A rises with temperature up to certain limit and falls with further increase in temperature. This is associated with a drop in the dielectric constant of water with rising temperature which makes it harder to separate the ionic charges. If the magnitude of the ionisation constant is small, it will be more temperature dependent. Further, at the equivalence point pH rises or falls much more steeply.

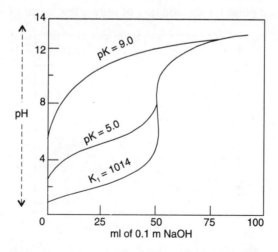

Fig. 5.1 Titration curve for weak acid and strong base

Table 5.1 Dissociation constants for acids and bases

Case	$[H^+]$	Approximation
General	$\sqrt{\dfrac{K_A K_W (a + K_B)}{K_b (a + K_A)}}$	$K_A \gg [H^+]$
Strong acid, strong base	$\sqrt{K_W}$	$K_H \gg [OH^-]$
Strong base, weak acid	$\sqrt{\dfrac{K_a K_W}{a}}$	$[HA] = [OH^-]$
Strong acid, weak base	$\sqrt{\dfrac{K_W K_a}{K_B}}$	$[B] = [H^+]$
Weak acid, weak base	$\sqrt{\dfrac{K_a K_W}{K_b}}$	$[B] = [HA]$
Weak dibasic acid	$\sqrt{K_1 K_2}$	$[H_2A] = [A]$

5.4 ACID-BASE TITRATION CURVES

It is useful to construct titration curves in considering the equilibrium aspects of acid–base reactions. In such curves, we observe gradual changes in pH both before and after the equivalence point and a large break near equivalence point for the addition of only a few drops of titrant.

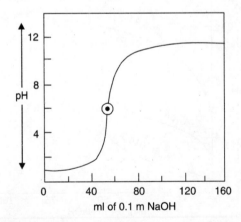

Fig. 5.2 Titration curve strong base and strong acid

We consider a few typical titration curves for (a) strong acid and strong base, and (b) strong base and weak acid in the following examples.

Example: 50 ml of 0.1 M HCl is titrated against 0.1 N NaOH. Calculate the pH at the start of titration and after addition of 10, 50, 60 ml of NaOH.

1. Initial pH: Since HCl is a strong acid it is completely dissociated, $\log(0.1)\left[H_3^+O\right] = pH$ *i.e.*, pH ~ 1.0.

2. For pH after the addition of 10 ml of base we use the expression

$$[R]_{V_T} = \frac{V_R M_R - V_T M_T}{(V_R + V_T)}$$

   if V_R, V_T are volume of reactant and titrant respectively.

   M_R, M_T are molarity of reactant and titrant respectively.

$$\left[H_3O^+\right] = \frac{50 \times 0.1 - 10 \times 0.1}{50 + 10} = 6.67 \times 10^{-2} M$$

$$pH = (2 - \log 6.67) = 1.18$$

3. pH at the equivalence point: This point is reached when 50 ml of NaOH is added, the solution is expected to be neutral $\left[H_3^+O\right] = [OH^-] = 1 \times 10^{-7}$, so pH = 7.0.

$$\left(as \ \sqrt{K_W} = 1 \times 10^{-7}\right)$$

4. pH after addition of 60 ml of base: At this point the solution is basic.

$$[V_R] = \frac{V_T M_T - V_R M_R}{(V_R + V_T)}$$

$$[R_T] = \frac{60 \times 0.1 - 50 \times 0.1}{50 + 60} = \frac{1}{110} = 9.1 \times 10^{-3} M = [OH^-]$$

$$pH = (3 - \log 9.1) = 2.04, \ pH = (14 - 2.04) = 11.96$$

Example: 50 ml of 0.1 M solution of weak acid HB ($K_a = 1 \times 10^{-5}$) is titrated with 0.1 M NaOH, calculate the pH at the start of the titration and after the addition of 10, 50, 60 ml of NaOH. Portray your result by drawing titration curves. Consider various dissociations:

$$HB + H_2O \rightleftharpoons H_3O^+ + B^-, \ K_a = 1 \times 10^{-5} \text{ (dissociation of weak base)}$$

$$B^- + H_2O \rightleftharpoons HB + OH^-, \ K_b = 1 \times 10^{-9} \text{ (dissociation of weak base)}$$

$$2H_2O \rightleftharpoons H_3O^+ + OH^-, \ K_W = 1 \times 10^{-14} \text{ (dissociation of water)}$$

$$HB + OH^- \rightleftharpoons B^- + H_2O, \ K_t = 1 \times 10^{+9} \text{ (reverse reaction of titration)}$$

So

$$K_a \times K_b = K_W \text{ and } K_t = 1/K_b$$

1. Initial pH of reaction [H₃O⁺] = [B⁻]

   [HB] = $0.1 - \left[H_3O^+\right] = 0.1$. We use the expression for K_a

$$K_a = \frac{[H_3O][B^-]}{[HB]} = 1 \times 10^{-5} \text{ if [HB]} = 0.1 \text{ and } K_a = 1 \times 10^{-5}$$

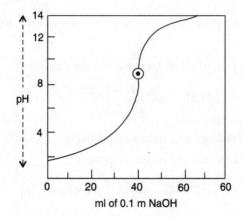

Fig. 5.3 Titration of strong base and weak acid

So, $\left[H_3^+O\right] = 1 \times 10^{-3}$; $pH = 1 - \log(1 - 10^{-3})$; $pH = 3.0$

2. pH after addition of 10 ml of base. If reaction proceeds well $K_t = 1 \times 10^9$

$$[R]_{V_R} = \frac{V_R M_R - V_T M_T}{(V_R + V_T)} = \frac{50 \times 0.1 - 10 \times 0.1}{50 + 10} = \frac{4}{60} \times [B] = \frac{1}{60}$$

$$K_a = 1 \times 10^{-5} = \frac{[H_3 O][B^-]}{[HB]} = \frac{[H_3 O^+][1/60]}{[4/60]}$$

since volume of solution cancels.

$$\left[H_3 O^+\right] = 4 \times 10^{-5} \quad \therefore pH = 5 - \log 4 = 4.40$$

Alternatively we can also use the relation $pH = pK_a - \log \dfrac{[HB]}{[B^-]}$

3. pH at the equivalence point: 50 ml of NaOH added: $[B^-] = 0.05$ M Since $[HB] = [OH^-]$. We use an expression for K_p:

$$K_p = 1 \times 10^{-9} = \frac{[HB][OH^-]}{[B^-]} = \frac{[OH^-]^2}{[B^-]} = \frac{[OH^-]^2}{0.05} \text{ as } [OH^-]^2 \ 5 \times 10^{-11}$$

$$2 \times pOH = (11 - \log 5) = 10.3 \text{ so } pOH = 5.15 \ pH = 8.85$$

4. pH after addition of 60 ml of base. We now have excess of NaOH

$$[V_T]_{V_R} = \frac{V_T M_T - V_R M_R}{V_T + V_R} = \frac{60 \times 0.1 - 50 \times 0.1}{110} = 9.1 \times 10^{-3} \text{ of } [OH]$$

$pOH = (3 - \log 9.1) = 2.4$; $pH = (14 - 2.04) = 11.96$
$pOH = 2.04$, so $pH = 11.96$

See Table 5.2 and Fig. 5.2 and 5.3 titration curves.

Table 5.2 Titration curve for strong acid and strong base

Vol. of NaOH added ml (V_R)	Total volume ml (V_T)	$[H_3O^-]$	pH	
0	50	0.1	1.0	
10	60	6.67×10^{-2}	1.18	
50	100	1×10^{-1}	7.0	Equivalence point
60	110	—	11.9	

Table 5.3 Titration curve for strong base and weak acid

Vol of NaOH added ml (V_R)	Total volume ml (V_T)	$[H_3O^-]$	pH
0	50	1×10^{-3}	3.0
10	60	4×10^{-5}	4.40
50	100	5×10^{-11} (OH)	8.85
60	110	9.1×10^{-3} (OH)	11.96

On examination of these titration curves (Fig. 5.2 and (5.3) we observe a gradual change in pH before and after the equivalence point. We have a large break at the equivalence point. The magnitude of change in pH at the equivalence point is less for strong acid and strong base (Fig. 5.2). For weaker acids (Fig. 5.3) there is small change in pH at the equivalence point. This is because the neutralisation reaction is not as complete at the equivalence point for weak acids as it is for strong acids. The equilibrium constant K_t for the reaction HB + OH $\rightleftharpoons$ B$^-$ + H$_2$O is smaller for the weak acid HB. The magnitude of this break relates to the sharpness of an indicator colour change and precision with which the titration end point can be determined.

The maximum slope occurs at equivalence point if the titration reaction is symmetrical. If the reaction is unequal then there will be a difference in equivalence point and influence point and the titration is likely to be erroneous. However, in the majority of cases such difference does not introduce significant error.

5.5 ACID-BASE INDICATORS

Acid–base indicators are substances which change colour or develop turbidity at a certain pH. They locate equivalence point and also measure pH. They are themselves acids or bases, are soluble, stable and show strong colour changes. They are organic compounds. A resonance of electron isomerism is responsible for colour change. Various indicators have different ionisation constants and therefore they show a change in colour at different pH intervals. Acid–base indicators can be broadly classified into three groups.

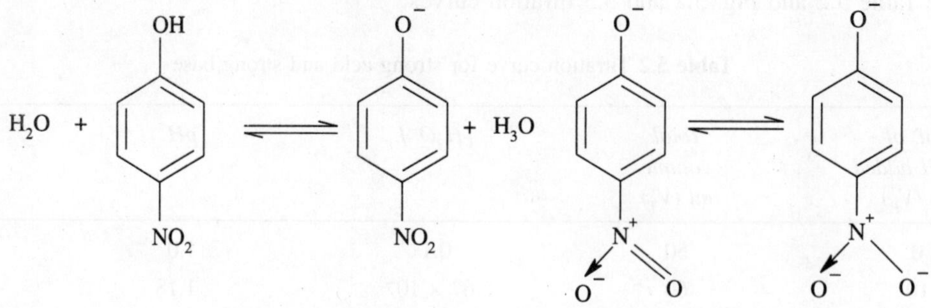

Fig. 5.4 Resonance electron isomerism for indicator

(a) The phthaleins and sulphophthaleins indicators
(b) Azo indicators
(c) Triphenylmethane indicators

The phthaleins are made by condensing phthalein anhydride with phenols e.g. Phenolphthalein changes at various stages of ionisation. At pH 8.0-9.8 it changes to red. Other members are o-cresolphthalein, thymolphthalein; a napthophthalein.

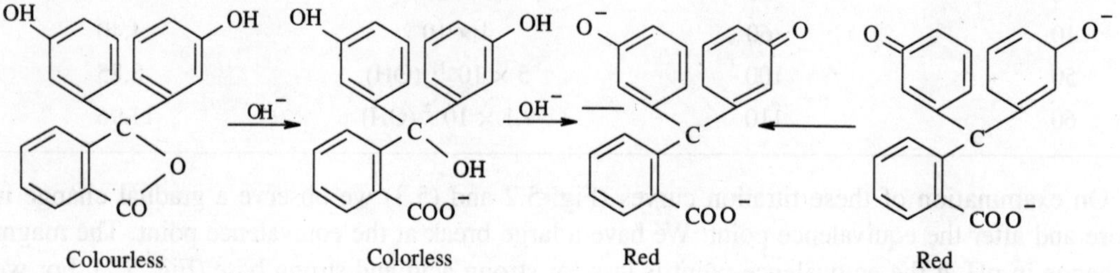

Colourless   Colorless   Red   Red

Fig. 5.5 Phenolphthalein – colour changes

All of them suffer from irreversible breakdown in strong alkalis. The sulphophthaleins show sharpness of colour change. They are made by condensing phenols with anhydride. They are more soluble in water. Thymol blue, m-cresol purple, chlorophenol red, bromophenol red, bromophenol blue and bromocresol red are other indicators in this series. The azo indicators are obtained by reacting aromatic amines with diazonium salt e.g. methyl yellow or p-dimethylaminoazobenzene.

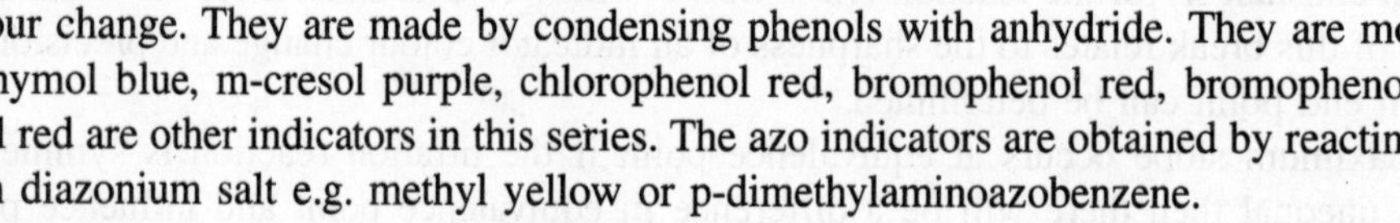

Red (acid)   Yellow (base)

Fig. 5.6 Methyl yellow – colour change

The effect of structure on ionisation is apparent. They show colour change in strongly acidic solution. Methyl orange is not soluble in water. Other indicators are methyl yellow, methyl red and TropeolinOO. The last group of indicators are triphenylmethane indicators e.g. malachite green, methyl violet, crystal violet. The azo indicators show increase in dissociation as temperature rises. Here a proton is separated from a tertiary ammonium ion leaving an uncharged residue

$$R-NH(CH_3)_2^+ \rightarrow R-N(CH_3)_2 + H^+$$

There is no separation of charges. In practice it has little influence in the removal of the proton. In p-nitrophenol, ionisation of the phenolic group causes separation of charges and the dissociation varies little with temperature. The phthalein and sulphophthalein derivatives show an appreciable change in ionisation with temperature due to their small ionisation constant.

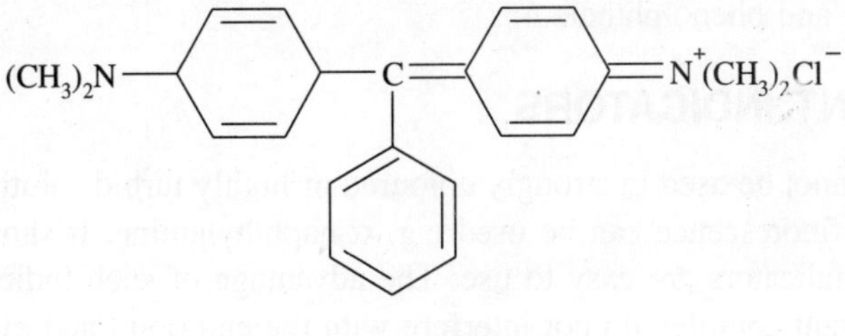

Fig. 5.7 Malachite green

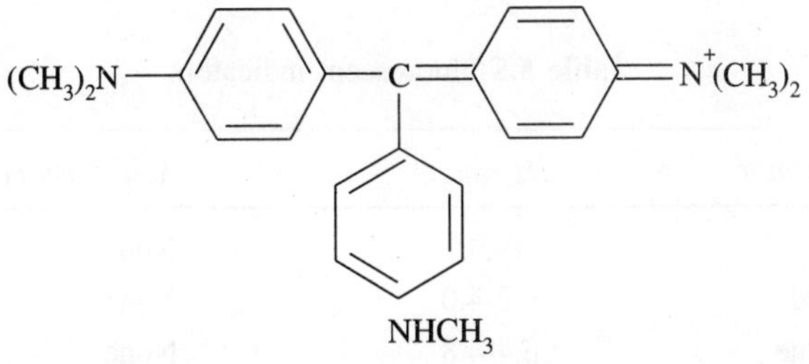

Fig. 5.8 Methyl violet

Table 5.4 Mixed indicators

S.No.	Indicator (I) + Indicator (II)	Ratio	Colour	pH	Solvent
1.	Methyl yellow + Methylene blue	1:1	B → G	3.25	Alcohol
2.	Methyl orange + Bromocresol green	1:5	O → BG	4.3	Water
3.	Methyl orange + Xylene cyanol FF	2:3	R → G	3.8	Alcohol
4.	Methyl red + Bromocresol green	2:3	R → G	5.1	Alcohol
5.	Phenol red + Bromothymol blue	1:1	Y → V	7.5	Water
6.	Thymol blue + Phenolphthalein	1:3	Y → V	9.0	Alcohol

B = blue, G = green, R = red, Y = yellow, V = violet, BG = blue green, O = orange.

5.6 MIXED INDICATORS

In the titration of H_3PO_4 with $NaHCO_3$, close control of pH is essential. For such a purpose a mixture of indicators to produce a colour change within a very narrow range of pH is used e.g. Bromocresol green (pK 4.9) methyl red (pK 5) give a red colour in acid and green in alkaline media with a sharp transition through grey at pH 5.1 as the two colours of indicators are complementary. Another example of a mixed indicator is modified methyl orange which is a mixture of methyl orange plus the dye xylene cyanol FF (blue) to give a purple red colour in acid and green in alkali and grey in the intermediate stage at pH 3.8. In addition, we have some indicators called universal indicators. Universal indicators are not used for titration but for pH measurements over a wide range with colours at pH 3.0 (red), pH 5 (orange), pH 6 (yellow), pH 8.5 (green), pH 9 (blue), pH 10 (violet). pH papers are coated with them to show such a change. The constituents of universal indicators are usually methyl orange, bromothymol blue, alizarin yellow G and phenolphthalein.

5.7 FLUORESCENT INDICATORS

Acid base indicators cannot be used in strongly coloured or highly turbid solutions. For this purpose an indicator which shows fluorescence can be used e.g. α-naphthylamine. It shows blue fluorescence in ultraviolet light. Such indicators are easy to use. The advantage of such indicators is that the intense colours of the metal titrant complex do not interfere with the end point and even colour blind persons can also see quenching or appearance of fluorescence. Important fluorescent indicators are listed in Table 5.5.

Table 5.5 Fluorescent indicators

S.No.	Name of Indicator	pH range	Acid form color	Base form color
1.	Rosin	0-30	None	Green
2.	Salicyclic acid	0.2-4.0	None	Blue
3.	α-napthylamine	3.4-4.8	None	Blue
4.	Dichlorofluoresein	4.0-6.6	None	Green
5.	β-naphthol	8.0-9.0	None	Blue
6.	Quinine	9.5-10.0	Blue violet	None
7.	Quinoline	6.2-7.2	Blue	None

PROBLEMS

5.1 How much of 0.6 N NaOH must be added to 750 ml 0.2 N NaOH in order to prepare a solution of 0.3N?

Answer: If x = no. of ml of 0.6 N NaOH to be added.

Net volume after dilution = $(750 + x)$ ml.

No. of gram mill. equivalent present = $[(750 \times 0.2) + (x \times 0.6)]$

Normality sought = $0.3 \, N = \dfrac{(750 \times 0.2) + (0.6\,x)}{(750 + x)}$ hence $x = 250$ ml.

5.2 If 0.5 gram of a mixture of K_2CO_3 and Li_2CO_3 requires 30 ml of 0.25 N acid solution for neutralisation, what is the percentage composition of the mixture?

Answer: Let x = gram of K_2CO_3, y = gram of Li_2CO_3

Then $x + y = 0.5$

No. of meq of K_2CO_3 present $= \dfrac{x}{K_2CO_3/2000} = \dfrac{x}{0.060}$

No. of meq of $Li_2CO_3 = \dfrac{y}{Li_2CO_3/2000} = \dfrac{y}{0.0375}$

No. of meq. of acid required $= 30 \times 0.25$

Hence $x/0.69 + y/0.0375 = 30 \times 0.25$

Solving the above simultaneous equations we have

$x = 0.247$ and $y = 0.253$. Hence percentage $\dfrac{100x}{0.5} = 49.4\%$ K_2CO_3 and $\dfrac{100y}{0.5} = 50.6\%$ $LiCO_3$.

5.3 A 1.2 gram sample of a mixture of $(Na_2CO_3 + NaHCO_3)$ is dissolved and titrated with 0.5 N HCl. With phenolphthalein the end point is at 15 ml while on after further addition of methyl orange a second end point is at 22 ml. Calculate the percentage composition of the mixture.

Answer: 15 + 15 = 30 ml acid is necessary to neutralise Na_2CO_3 completely. Total volume needed = 15 + 22 = 37 ml *i.e.*, (37 −30) = 7 ml acid is needed for neutralising $NaHCO_3$.

Therefore the composition (%) is $\dfrac{30 \times 0.5 \times 0.053}{1.2} \times 100 = 66.25\%$ Na_2CO_3

$\dfrac{7 \times 0.5 \times 0.042 \times 100}{1.2} = 24.50\%$ $NaHCO_3$

PROBLEMS FOR PRACTICE

5.4 How many ml of 0.1421 N lithium hydroxide is required to neutralise 13.72 ml of 0.06860 N phosphoric acid?

5.5 A sample of sulphuric acid containing H_2SO_4, SO_2 and SO_3 weighs 1.0 g and needs 23.4 ml of 1 N alkali for neutralisation. If another lot shows the presence of 1.3% SO_2, calculate the % of free SO_3, H_2SO_4 and combined SO_3 in the sample.

5.6 A mixture containing $(Li_2CO_3 + BaCO_3)$ weighs 1.0 g and needs 15 ml of 1 N hydrochloric acid for neutralisation. What is the % of $BaCO_3$ in the sample?

5.7 What volume of 6 N sulphuric acid should be mixed to prepare a litre of 3 N acid?

5.8 How many grams of potassium tetraoxalate must be dissolved in water and made up to 780 ml to make 0.03 N solution for use as an acid?

LITERATURE

1. C.L. Wilson and D.W. Wilson, "*Comprehensive Analytical Chemistry*", All Volumes, Elsevier (1958).
2. H.F. Walton, "*Principles and Methods of Chemical Analysis*", 2nd Ed., Prentice Hall (India) (1966).
3. R.L. Pecsok and L.D. Shields, "*Modern Methods of Chemical Analysis*", Wiley International (1968).
4. J.G. Dick, "*Analytical Chemistry*", McGraw Hill (1973).
5. H.A. Laitinen and W.E. Harris, "*Chemical Analysis: An Advanced Text and Reference,*" 2nd Ed., McGraw Hill (1995).
6. R.A. Day and A.L. Underwood, "*Quantitative Analysis*", 3rd Ed., Prentice Hall (India) (1977).

Chapter 6

Redox Titrations

The term oxidation refers to any chemical change in which there is an increase in oxidation number while the term reduction refers to any change resulting in decrease in oxidation number. Thus oxidation proceeds with loss of electrons while reduction is accompanied by gain of electrons. The oxidising agent is by definition the substance containing the atom which shows a decrease in oxidation number, while a reducing agent is the substance containing the atom which shows an increase in oxidation number. Oxidation and reduction must always occur together and must compensate each other. The term oxidising or reducing agent refers to the entire substance and not to just one atom. If a reagent acts as both oxidising and reducing agent we say it undergoes autooxidation or disproportionation.

Volumetric methods based upon electron transfer reactions are numerous and varied. The separation of oxidation-reduction into its components *i.e.*, half reactions is a suitable way of indicating the species that gain electrons and those that lose electrons, as oxidation-reduction reactions results from the direct transfer of electrons from donor to receiver. Many redox reactions could, in theory, be used in volumetric titrations provided that equilibrium is reached very rapidly following each addition of titrant and also that the indicator is capable of locating the stochiometric equivalence point with reasonable accuracy. Many standard redox titrations are carried out using colour change indicators. Two half reactions for any redox titrations system are in equilibrium at all points after the start of the titrations so that reduction potentials for the half cells are identical at all points. Thus the cell potential *i.e.* E cell, changes during the titration and furnishes a characteristics and relatively large ΔE cell (*i.e.* change) around the equivalence point for all titrations. The variation of E cell with volume of titrant shows that most redox titration systems are suitable and can help to ascertain the equivalence point of the titration. In redox titrations a table of standard electrode potential serves as guide to equilibrium conditions as on information on rate or mechanism of reaction is known. Many redox reactions are slow but a catalyst can often be found to accelerate them. Many redox titration procedures require heating and a pause for a minute after each addition at end point. The course of a titration reaction can be followed by plotting potential against titration volume.

6.1 THEORY OF REDOX TITRATION CURVES

In constructing titration curves in redox reaction titrations it is customary to plot a graph of e.m.f. of E cell (against SCE) versus volume of titrant. This is necessary because most redox indicators are sensitive but they are themselves oxidising or reducing agents and hence the change in potential of the system is considered during titration. Further, from the knowledge of standard electrode potential as well as the Nernst equation it is possible to calculate the value of such potential after each addition of titrant. Let us consider a typical titration curve construction by considering the potential of the oxidation reduction systems at the equivalence point (E_{eq}). The Nernst equation states

$$E = E_0 - \frac{RT}{nF}\log\frac{[\text{reductant}]}{[\text{oxidant}]}$$

For the simple reaction of

$$Fe^{2+} + Ce^{4+} \rightleftharpoons Fe^{3+} + Ce^{3+}$$

At equilibrium the electrode potentials for the two half reactions is the same $E_{Ce^{4+}} = E_{Fe^{3+}} = E_{\text{system}}$. This is the potential of the system. For the redox indicator the potential is the same as for the system $E_{\text{indicator}} = E_{Ce^{4+}} = E_{Fe^{3+}} = E_{\text{systems}}$. We have the half cell

SCE ∥ Ce^{4+} Ce^{3+} : Fe^{3+} Fe^{2+} ∥ Pt where SCE = saturated calomel electrode

We use $E_{Ce^{4+}}$ or $E_{Fe^{3+}}$ while calculating E_{system}.

We can apply the Nernst equation to the Fe (III)/Fe (II) couple or the Ce(IV)/Ce(III) couple: both would lead to the same answer. At equivalence point we calculate potential thus:

$$E_{eq} = E^0_{Ce^{4+}} - 0.0591\frac{[Ce^{3+}]}{[Ce^{4+}]} \text{ and also by}$$

$$E_{eq} = E^0_{Fe^{3+}} - 0.0591\log\frac{[Fe^{2+}]}{[Fe^{3+}]}. \text{ Adding both equations}$$

$$2E_{eq} = E^0_{Ce^{4+}} + E^0_{Fe^{3+}} - 0.0591\log\frac{[Ce^{3+}][Fe^{2+}]}{[Ce^{4+}][Fe^{3+}]}$$

Since at the equivalence point

$[Fe^{3+}] = [Ce^{3+}]$ and $[Ce^{4+}] = [Fe^{3+}]$ we reduce the equation to $E_{eq} = E^0_{Ce^{4+}} + E^0_{Fe^{3+}}$. We can calculate the equilibrium concentration of the reacting species from the equivalence point potential. Now using the above theory let us see how we can construct a titration curve for the above mentioned reaction.

Example: Depict a titration curve for the titration of 50 ml of 0.05 N Iron (III) with 0.10 N Ce (IV) if both solutions are made in 1M sulphuric acid.

$$\left[\text{Given } E^{Fe^{2+}}_{Fe^{3+}} = 0.68 \text{ V and } E^{Ce^{3+}}_{Ce^{4+}} = 1.44 \text{ V}\right]$$

(*a*) At the start of the titration we have the potential:

$$E = E^0_{Fe} - 0.059\log\frac{[Fe^{2+}]}{[Fe^{3+}]}$$

Since the solution contains no cerium ions it will have a very small concentration of Fe (III) due to air oxidation of Fe (II). Assuming 0.1% was in the ferric state,

$$\frac{[Fe^{2+}]}{[Fe^{3+}]} = 1000 : 1. \text{ So } E = (0.68 - 0.059\log 1000) = 0.50 \text{ V}.$$

(b) The addition of 50 ml of Ce (IV) solution to 50 ml of Fe (III). The concentration of Ce (III) is less than unreacted Ce (IV) or

$$[Ce^{3+}] = \frac{5 \times 0.1}{55} - [Ce^{4+}] = \frac{0.5}{55} \text{ and the concentration of Iron (III) is}$$

$$[Fe^{3+}] = \frac{5 \times 0.1}{55}[Ce^{4+}] = \text{ and the concentration of Iron (II) is}$$

$$\frac{0.5}{55} = [Fe^{2+}] = \frac{50 \times 0.5 - 5 \times 0.1}{55} + [Ce^{4+}] = \frac{2.0}{55.0}$$

Now potential

$$E = E^0_{Ce^{4+}} - 0.0591\log \frac{[Ce^{3+}]}{[Ce^{4+}]} \quad E^0_{Fe^{3+}} - 0.0591\log \frac{[Fe^{2+}]}{[Fe^{3+}]}$$

Since we know [Fe^{2+}] and [Fe^{3+}] we put them in the above equation

$$E = 0.68 - 0.0591\log \frac{[2/55]}{[0.5/55]} = +0.64 \text{ V}.$$

We could have used the Ce (IV)–Ce (III) system.

(c) The equivalent point potential can be calculated:

$$E_{eq} = \frac{E^0_{Ce^{4+}} + E^0_{Fe^{3+}}}{2} = \frac{0.68 + 1.44}{2} = +1.06 \text{ V}.$$

(d) Addition of 25.1 ml of Ce(IV) solution to 50 ml of 0.05 N Fe (III) solution. Solution contains an excess of Ce (IV) in addition to equivalent quantities of Fe (III) and Ce (III). Hence the concentration of Fe (II) is small.

$$[Fe^{3+}] = \frac{25 \times 0.1}{75.1} - [Fe^{2+}] = \frac{2.5}{75.10}$$

$$[Ce^{3+}] = \frac{25 \times 0.1}{75.1} - [Fe^{2+}] = \frac{2.5}{75.10}$$

$$[Ce^{4+}] = \frac{25.1 \times 0.1 - 50 \times 0.05}{75.10} + [Fe^{2+}] = \frac{0.01}{75.10}$$

We calculate potential from the Ce (IV)/Ce (III) couple

$$E = 1.44 - 0.0591\log \frac{[Ce^{3+}]}{[Ce^{4+}]} = 1.44 - 0.0591\log \left[\frac{2.5/75.10}{0.01/75.1}\right] = 1.30 \text{ V}$$

We can tabulate the results as shown in Table 6.1 (also see Fig. 6.1). Some important observations on this titration are as follows. The shape of the titration curve is the same as for other titration techniques. The equivalence point is marked by large change in the ordinate function. The titration curve is symmetrical about the equivalence point as the mole ratio of oxidant to reductant is equal to one.

Table 6.1 Titration curve for Fe (II) with Ce (IV)

	Vol. of 0.1 N Ce (IV) solution ml	E (volts) (vs SCE)
	0	—
	5	0.64
	10	0.66
	15	0.69
	20	0.69
	24.9	0.82
Eq. point	25.0	1.06
	25.1	1.30
	26.0	1.36
	30.0	1.40

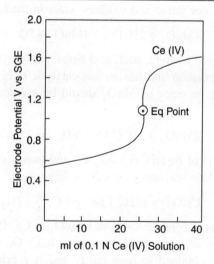

Fig. 6.1 Titration curve for Iron (II) with Cerium (IV)

6.2 SOME OXIDISING AGENTS AS TITRANTS

(a) *Ceric sulphate* is a very good oxidising agent used with 1-10 phenanthroline as an indicator. $Ce^{4+} \rightarrow Ce^{3+} + e^-$. Electrons from the 4f orbital are removed. The rate of reaction is affected by the nature of the solvent and by the process of complex formation. Ce (IV) during reaction exists as an anionic complex in media of H_2SO_4, HNO_3 and $HClO_4$. The formal potential of Ce (IV) – Ce (III) couple is 1.70 V in $HClO_4$, 1.60 V in HNO_3 and 1.42 V in H_2SO_4 solution. The stability is limited in HCl medium, with formal potential of 1.88 V. Sodium oxalate $(Na_2C_2O_4)$ is used as the primary standard for cerium (IV) in sulphuric acid. Ammonium hexanitrocerate *i.e.* $(NH_4)_2$ Ce $(NO_3)_6$ is used for the preparation of standard cerium (IV) solution. Ce (IV) solutions in $HClO_4$ are photosensitive due to the formation of the hydroxyl radical. Ce (IV) oxidations are used mainly in titrations of Fe (II), with ferroin as indicator in 0.5–8 M H_2SO_4 or $HClO_4$ or 0.5–3 M HCl medium, H_2O_2 can be titrated with Ce (IV) solution. Oxalates, arsenic (III) and ferrocyanide can be determined by direct titration. Some of them need catalysts e.g. $H_2C_2O_4$ against Ce (IV) in 0.5–2 N HNO_3 with $AgNO_3$ as catalyst to complete the reaction. Ce (IV) has the advantage over $KMnO_4$ as it permits the use of high concentrations of HCl as the reducing medium, because Ce (IV) forms a stable complex. Further the solution does not oxidise in dilute hydrochloric acid (0.5–3 M).

(b) *Potassium permanganate* is a powerful oxidising agent. No special indicator is required to note the end point. Its tendency to oxidise the chloride ion is the serious limitation. Solution have limited stability. It is employed is acid (0.1 N) media as $MnO_4^- + 8H^+ + 5e^- \rightleftharpoons Mn^{2+} + 4H_2O$, $E^0 = 1.51$ V. The oxidation reaction is rapid only for $H_2C_2O_4$ but it is slow at room temperature for others hence heating is required. The reaction with As (III) needs a catalyst. The permanganate end point is not permanent and slowly fades out due to the reaction:

$$2MnO_4^- + 3Mn^{2+} + 2H_2O \rightleftharpoons 5\ MnO_2 + 4H^+$$
$$\underset{\text{Coloured}}{\phantom{2MnO_4^- + 3Mn^{2+} + 2H_2O}} \qquad \underset{\text{Colourless}}{}$$

The aqueous solution is not stable and oxidises water in the following way

$$4MnO_4^- + 2H_2O \rightleftharpoons 4MnO_2 + 3O_2 + 4OH^-$$

Decomposition is catalysed by light, heat, acid bases, the Mn (II) ion, and MnO_4. The latter is obtained due to decomposition and has an autocatalytic effect. In preparation of a standard solution of $KMnO_4$, the presence of MnO_4 should be avoided. The solution is standardised against $Na_2C_2O_4$ in acid.

$$2MnO_4^- + 5H_2C_2O_4 + 6H^+ \rightleftharpoons 2Mn^{2+} + 10CO_2 + 8H_2O.$$

It is used for the analysis of Fe (II), $H_2C_2O_4$, Ca and many other compounds.

(c) *Potassium dichromate.* This reaction proceeds as follows:

$$Cr_2O_7^{2-} + 14H^+ + 6e^- \rightleftharpoons Cr^{3+} + 7H_2O, \ E^0 = 1.33 \text{ V}$$

It has limited application in comparison with $KMnO_4$ or Ce (IV) due to lesser oxidising characteristics and slowness of some of its reactions. $K_2Cr_2O_7$ is indefinitely stable and inert towards HCl. It can be obtained in high purity and is a primary standard. Diphenylamine sulphonic acid is used as internal indicator. It is mainly used for the analysis of Iron (III) with reactions

$$6Fe^{2+} + Cr_2O_7^{2-} + 14H^+ \rightleftharpoons 6Fe^{3+} + 2Cr^{3+} + 7H_2O$$

There are several other applications for this titrant.

(d) *Potassium bromate.* It is a strong oxidising agent. It reacts as

$$BrO_3^- + 6H^+ + 6e^- \rightleftharpoons Br^- + 3H_2O, \; E^0 = 1.44 \text{ V}$$

In acidic solution BrO_3^- reacts to give Br^- (aq) + $3H_2O$. $KBrO_3$ is a primary standard and is stable indefinitely. Methyl orange or methyl red are used as indicators but are not good but α-napthaflavone, and quinoline yellow are quite satisfactory. Potassium bromate finds extensive applications in organic chemistry e.g., titration with oxine. Most titrations involve back titrations with arsenous acid.

(e) *Potassium iodate:* has extensive uses in analytical chemistry.

$IO_3^- + 5I^- + 6H^+ \rightarrow 3I_2 + 3H_2O$ and the reaction in Andrew's titration is

$$IO_3^- + Cl^- + 6H^+ + 4e^- \rightarrow ICl + 3H_2O, \; E^0 = +1.20 \text{ V}$$

Andrew's titration as mentioned above is carried out from 6M hydrochloric acid in the presence of carbon tetrachloride. The end point is indicated by the disappearance of violet colouration. Vigorous shaking is necessary for getting an end point.

6.3 APPLICATION OF IODINE AS REDOX REAGENT

Such applications are extensive because its E^0 is intermediate hence it can act as an oxidising or reducing agent. If E^0 is independent of pH (pH < 8.0), the iodine-iodide couple is shown by the equation:

$$I_{2(S)} + 2e^- \rightarrow 2I^- \; E^0 = 0.535 \text{ V}$$

It is a weak oxidant and iodide is a relatively weak reductant. It is sparingly soluble in water by formation of triodide $[KI_3]$, $I_{2(S)} + 2e^- \rightleftharpoons 2I^- \; E^0 = 0.621 \text{ V}$. Hence it is the reaction at the beginning of $2I^-$ titration. Iodine can be purified by sublimation. It is dissolved in Kl solution and stored in a cool dark place. It is standardised against a solution of As_2O_3 as primary standard. The loss due to volatility and air oxidation of iodide promotes error in analysis. The standardisation can be carried out against $Na_2S_2O_3 \cdot 5H_2O$. The thiosulphate solution is again standardised against $K_2Cr_2O_7$ as the former is not a primary standard. The reaction is given below:

$$Cr_2O_7^{2-} + 14H^+ + 6I^- \rightarrow 3I_2 + 2Cr^{3+} + 7H_2O$$

The starch is used as an indicator. Iodide in concentrations $<10^{-5}$M can be easily detected by starch. The colour sensitivity depends upon the nature of the solvent used. The iodine-starch complex has limited water solubility and interferes, hence it should be added at the end point of the reaction. With formamide, microorganism attack on starch is the least. Let us consider a few selected reactions.

(a) *Iodine thiosulphate:* When a neutral or acidic solution of iodine in Kl is titrated with $Na_2S_2O_3$ we have $I_3^- + 2S_2O_3^{2-} \rightleftharpoons 3I^- + S_4O_6^{2-}$. During the reaction a colourless intermediate $S_2O_3I^-$ is formed

$$S_2O_3^{2-} + I_3^- \rightleftharpoons S_2O_3I^- + 2I^-$$

which proceeds to

$$2S_2O_3I^- + I^- \rightleftharpoons S_4O_6^{2-} + I_3^-$$

The colour thus reappears on addition of starch

The reaction proceeds well below pH = 5.0 while in alkaline solution hypoiodous acid (HOI) is formed.

(b) *Reaction with copper:* Excess KI reacts with Cu (II) to form CuI and release the equivalent amount of I_2 e.g.

$$2Cu^{2+} + 4I^- \rightarrow 2CuI + I_2 \; ; \; 2Cu^{2+} + 3I^- \rightarrow 2CuI + I_3$$

Here iodide serves as a reducing agent.

Further the reaction with copper is

$$Cu^{2+} + e^- \rightleftharpoons Cu^+ \; ; \; E^0 = 0.15 \text{ V}; \; I_2 + 2e^- = 2I^- \; E^0 = 0.54 \text{ V}$$

and $Cu^{2+} + I^- + e^- \rightleftharpoons CuI \quad E^0 = 0.86$ V

The best results are obtained in 4% KI. Optimum pH is 4.0, as in an alkaline media alkaline copper (II) hinders oxidation. Slow addition of $Na_2S_2O_3$ is useful as adsorbed iodine is only released slowly. The presence of chloride ions should be avoided as then iodide will not reduce Cu (II) quantitatively.

(c) *Dissolved oxygen: Winkler's method:* dissolved oxygen in water (DO) is determined by this method. It is based on the reaction between O_2 and suspended Mn(II) hydroxide in an alkaline media. On acidification oxidised manganese hydroxide is reduced to manganese (II) iodide. The iodine equivalent to the dissolved oxygen present is titrated with $Na_2S_2O_3$

$$2Mn(II) + 4H^- + O_2 \rightarrow 2MnO_2 + 2H_2O \; \text{ pH} = 9.0$$

$$MnO_2 + 4H^+ + 2I^- \rightarrow I_2 + Mn(II) + 2H_2O$$

The common sources of error are reducing agents like SO_3^{2-}, $S_2O_3^{2-}$. Fe(II) that, reacts with O_2 or oxidised manganese in alkaline media and oxidising agents that react with suspended Mn (II) hydroxide in alkaline solution, or iodine in acidic solution. Also dissolved oxygen in the reagent may not be properly reacted with manganese solution. Such problems can be overcome by use of a blank especially in environmental pollution analysis.

(d) *Water analysis by Karl Fischer method:* This involves titration of a sample in anhydrous methanol with a reagent made from iodine, SO_2 and pyridine in methanol. In titration the end point corresponds to the appearance of the first excess of iodine detected visually or by electrical means. The reaction path followed is

$C_5H_5N \cdot I_2 + C_5H_5N \, SO_2 + C_5H_5N + H_2O \rightarrow$

$2C_2H_5N \, H^+I^- + C_5H_5N \cdot SO_3$ ...(pyridine N-sulphonic acid)

$C_5H_5N \, SO_3 + CH_3 \, OH \rightarrow C_5H_5NHO \, SO_2OCH_3$ ...(pyridinium hydrogen sulphate)

$C_5H_5N \, SO_3 + H_2O \rightarrow C_5H_5NHO \, SO_2OH$ ...(pyridinium hydrogen sulphate)

Excess methanol must be used. The overall reaction is

$I_2 + SO_2 + H_2O + CH_3OH + 3py \rightarrow 2py \, H^+I^- + py \, HOSO_2 \, OCH_3$

The two reagent method is better where sample and pyridine methanol and SO_2 are titrated against iodine in methanol. The method is excellent for determination of moisture and H_2O from various materials.

6.4 REDOX INDICATORS

These are of two types: specific indicators and true redox indicators.

(a) Specific indicators are (like starch) ones which react with one of the component in titration. KSCN is another such specific indicator. These are, however, not very much preferred.

(b) True redox indicators respond to the potential of the system. The half reaction responsible for colour change can be shown as:

$$\text{In}_{ox} + ne^- \rightleftharpoons \text{In}_{red}; \; E = E_{\text{In}}^0 - \frac{0.0591}{n} \log \frac{[\text{In}_{red}]}{[\text{In}_{ox}]}$$

Usually a reactant concentration changes 100 fold.

i.e. $(\text{In}_{re})/(\text{In}_{ox})$ changes from 0.1 to 10 with the result that

$E = E^0 \pm \dfrac{0.591}{n}$. Thus an indicator will exhibit a detectable colour change when the titrant causes

a shift of about $(0.118/n)$ V or 0.059 V if $n = 2$. The potential at which a colour transition takes place depends upon the standard potential of a particular indicator system as shown in Table 6.2.

Table 6.2 Redox indicators

Sl. No.	Indicator	Colour change	Transition potential V
1	5 Nitro 1, 10-phen	B → Red violet	+ 1.25
2	1, 10 phen-Fe (II) complex	B → R	+ 1.11
3	Diphenyl amine	V → colorless	+ 0.76
4	Diphenyl amine sulphonic acid	R → V	+ 0.85
5	p-Ethoxychrysodine	Y → R	+ 0.76
6	Methylene blue	B → colorless	+ 0.53
7	Indigo tetrasulphonate	B → colorless	+ 0.36

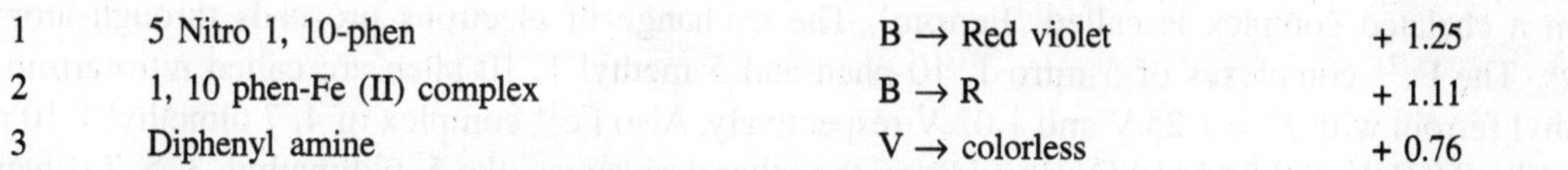

$E_0 = 0.76$ V

Colourless   Diphenylbenzidine (colourless)

$E_0 = 0.76$ V

Dimethylbenzidene (Violet)

Fig. 6.2 Diphenylamine colour change

These indicators function in the range up to +1.25 V. If one electron reaction is involved the difference in standard potential of reagent and analyst should be 0.4 V and for a two electron transfer reaction such difference should be 0.25 V. The end point can be located by potentiometric reactions.

Thus redox indicators are few because organic molecules suffer more radical change in such titration. Many organic oxidations involve loss of proton and electron. In such cases potential for half change depends on [H$^+$]. Indicators in which no hydrogen participate should have half change potential which is independent of pH. Most commonly used indicators are 1, 10-phenanthroline, triphenylemethane indicators and diphenylamine and chrysodine derivatives.

6.5 DETECTION OF END POINT IN REDOX TITRATIONS

Two types of indicators are used to detect the end point. They are either external indicators or internal indicators. In external indicators spot tests are used e.g. K$_3$Fe(CN)$_6$ is used in titration of iron with K$_2$Cr$_2$O$_7$. Also UO$_2$(NO$_3$)$_2$ is used as an indicator in the titration of zinc. External indicators are being replaced by more satisfactory internal redox indicators. The other kind of efficient indicator is the internal indicator. An ideal redox indicator should mark the sudden change in the oxidation potential in the neighbourhood of the equivalence point in redox reactions. The best indicators will be those which have an oxidation potential intermediate between that of the solution titrated and the titrant and which exhibits a sharp end point e.g. 1, 10-phenanthroline acts as

$$\left[(phen)_3 Fe^{3+} + e^-\right] \rightleftharpoons \left[(phen)_3 Fe^{2+}\right] \qquad E^0 = 1.06 \text{ V to } 1.11 \text{ V}$$
$$\underset{\text{Pale blue}}{} \qquad\qquad \underset{\text{Red}}{}$$

The complex salt obtained by mixing equivalent quantities of 1, 10-phenanthroline and FeSO$_4$ to form a chelated complex is called 'Ferroin'. The exchange of electrons proceeds through aromatic rings. The Fe^{2+} complexes of 5 nitro-1, 10-phen-and 5 methyl-1, 10-phen are called nitroferroin and methyl ferroin with E^0 = 1.25 V and 1.02 V respectively. Also Fe^{2+} complex of 4, 7 dimethyl 1-10 phen has E^0 = 0.921 V in 0.5 M H$_2$SO$_4$. In addition the other derivatives like 5, 6 dimethyl, 3, 5, 7 trimethyl; 3, 4, 6, 7, tetramethyl; 5 phenyl, 5-chloroferroin are extensively used. Next are triphenylmethane indicators which have varying color in neutral and alkaline solution e.g. Eroglaucine A (0.98 V) Eriogreen B (0.99 V) setogalucin O (0.99 V) Xylene cyanol FF all change from yellow to orange when oxidised. In such cases back titration is not possible as the colour change is not reversible. Diphenylamine in H$_2$SO$_4$ is the best available indicator. It reacts to form diphenzyl benzidine violet from diphenyl benzidine.

In HCl, H$_2$SO$_4$ E^0 = 0.68 V, but in H$_3$PO$_4$ E^0 = 0.60 V. Hence we add H$_3$PO$_4$ or F$^-$ to get a lower potential and good end point. It is relatively insoluble in H$_2$O hence it is dissolved in strong H$_2$SO$_4$. It can not be used in the presence of Na$_2$WO$_4$ as it gives a precipitate. Hg^{2+} lowers its rate of interaction with Fe^{3+} hence generally we remove it. The acid salt of diphenylamine sulfonic acid is preferred. It does not react with Hg. Chrysodine indicators are of the type shown below:

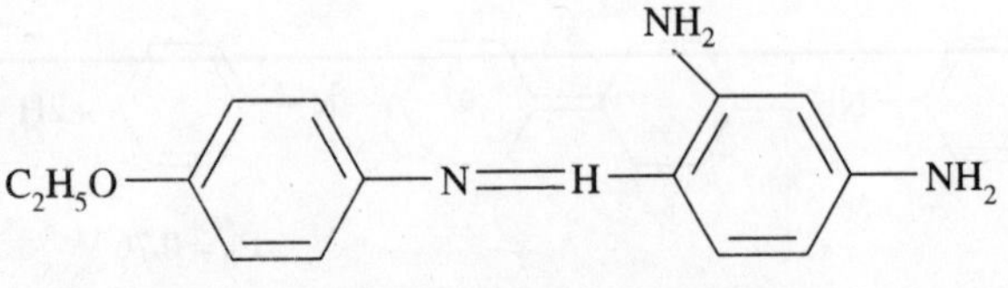

Fig. 6.3 Chrysoidine (p-ethoxychrysoidin)

It is an all purpose indicator. It is an absorption indicator. The change with $KBrO_3$ is as follows:

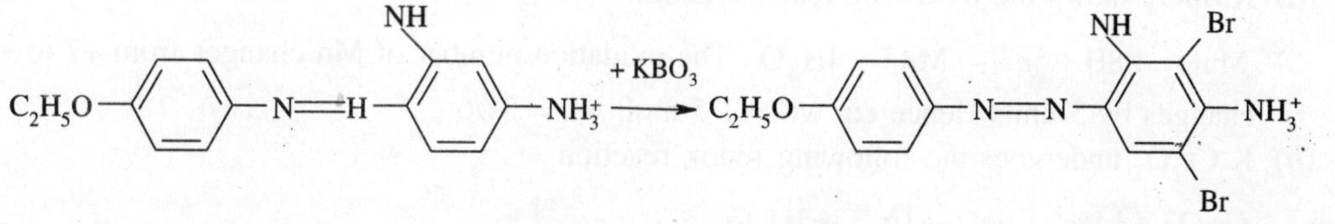

Fig. 6.4 Chrysodine—colour change

There are few indicators with low redox potential. They easily convert to the leuco form which is colourless e.g. Indigo, methylene blue.

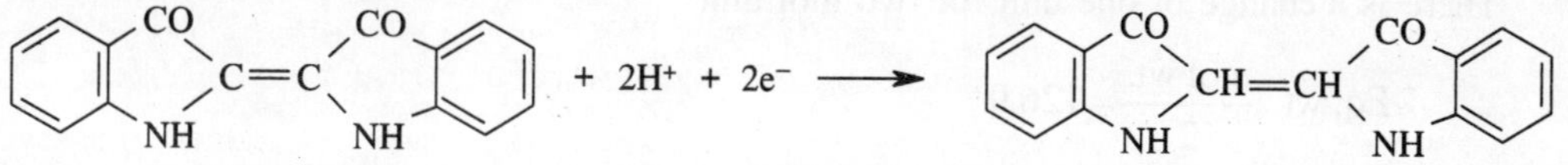

Fig. 6.5 Indigo—colour change

They are used in titanous chloride titration. Their potentials are −0.046 V and +0.110 V in 1M solution. One more indicator of the same type is 2, 6 dichlorophenol-indophenol. This is used to titrate ascorbic acid.

PROBLEMS

6.1 What is the equivalent weight of the following reducing agents?

(a) $FeSO_4 \cdot H_2O$
(b) $H_2C_2O_4 \cdot 2H_2O$
(c) $Na_2S_2O_3 \cdot 5H_2O$
(d) H_2O_2 (acid media)

Answer:

(a) $FeSO_4 \cdot H_2O$ shows the following redox reaction

$Fe^{2+} \rightarrow Fe^{3+} + e^-$. Since oxidation number changes by 1 unit, eq. wt. is equal to mol. wt. viz. 278.

(b) $H_2C_2O_4 \cdot 2H_2O$ undergoes the following redox reaction

$C_2O_4^{2-} \rightarrow 2CO_2 + 2e^-$. The oxidation number of C changes from + 3 to 4, the change for the oxalate radical is 2 hence eq. wt. is half mol. wt. viz. 63.3.

(c) $Na_2S_2O_3 \cdot 5H_2O$ undergoes transformation

$2S_2O_3^{2-} \rightarrow S_4O_6^{2-} + 2e^-$. oxidation number of S changes from +2 in ($Na_2S_2O_3$) to + 5/2 (in $Na_2S_4O_6$). The change is by 1/2 unit for 2S atoms. Hence mol. wt. = eq. wt. = 248.2

(d) For H_2O_2 transformation in acidic media is

$$5H_2O_2 + 2MnO_4^- + 6H^+ \rightarrow 5O_2 + 2Mn^{2+} + 8H_2O$$

Oxidation number change is from −1 to 0 *i.e.* 2 for a molecule hence equivalent weight is half molecular weight viz. 17.01.

6.2 Calculate the equivalent weight of the following oxidising agents.

(a) $KMnO_4$
(b) $K_2Cr_2O_7$
(c) I_2
(d) H_2O_2

Answer:

(a) $KMnO_4$ shows the following redox reaction

$MnO_4^- + 8H + 5e^- \rightarrow Mn^+ + 4H_2O$. The oxidation number of Mn changes from +7 to +2 *i.e.* changes by 5 units Hence eq. wt. = 1/5 mol. wt. = 31/6

(b) $K_2Cr_2O_7$ undergoes the following redox reaction

$Cr_2O_7^- + 14H^+ + 6e^- \rightarrow Cr^{3+} + 7H_2O$

Here the oxidation number of Cr changes from +6 to +3 *i.e.* by 3 units, for the dichromate ion the change is 6. $\therefore$ Eq. wt. 1/6 mol wt = 49.03

(c) I_2 is reduced to iodide as $I_2 + 2e^- \rightarrow 2I^-$

There is a change of one unit for two mol unit

$$\therefore \quad \text{Eq. wt.} \ \frac{\text{mol wt.}}{2} = 126.9$$

(d) $H_2O_2 + 2H^+ + 2e^- \rightarrow 2H_2O$ *i.e.* oxidation number of oxygen changes from −1 to −2. Per molecule change is two units. Eq. wt = 1/2 mol wt. = 17.01.

6.3 By adding excess of KI to $K_2 Cr_2O_7$ the liberated iodine was titrated with 48.8 ml of 0.1 N $Na_2S_2O_3$ to get an end point. Calculate the amount of $K_2Cr_2O_7$ present in solution.

Answer: $Cr_2O_7^- + 6I^- + 14H^+ \rightarrow 2Cr^{3+} + 3I_2 + 7H_2O$

$2S_2O_3^{2-} + I_2 \rightarrow S_4O_6^{2-} + 2I^-$ Eq. wt. $K_2Cr_2O_7$ = 49.03, Vol. of $Na_2S_2O_3$ × normality =

$\therefore$ Amt. of $K_2Cr_2O_7$ = 48.8 × 0.1 × 49.03 = 0.2393

PROBLEMS FOR PRACTICE

6.4 Calculate the amount of iron in a sample of hematite weighing 0.71 g if after reduction 48.08 ml of potassium permanganate is required to oxidise iron.

6.5 How many ml. of $K_2 Cr_2O_7$ solution containing 23 g of material per litre would react with 3.402 g of Mohr's salt ($FeSO_4 \cdot 7 H_2O$) in acid solution?

6.6 What is the purity of oxalic acid dehydrate if 0.2003 g/cm^3 needs 29.30 ml of $KMnO_4$ given if ml solution is equivalent to 6.02 X 10^{-3} g of Fe?

6.7 A sample of magnetite (Fe_3O_4) weighing 0.3g is titrated with mercurous solution. Calculate the volume of 0.1 M mercurous solution required if the sample is dissolved in HCl with CO_2 and slight excess of nitric acid?

LITERATURE

1. W.M. Latimer, *"Oxidation Potential"*, 2nd Ed., Prentice Hall (1952).

2. H.F. Walton, *"Principles and Methods of Chemical Analysis"*, 2nd Ed., Prentice Hall (India) (1966).

3. I.M. Kolthoff, E.B. Sandell, E.J. Meehan and S. Bruckenstein, *"Quantitative Chemical Analysis"*, 4th Ed., Macmillan. London (1969).

4. J.G. Dick, *"Analytical Chemistry"*, McGraw Hill (1973).

5. G.D., Christian S.E. O-Reilly, *"Instrumental Analysis"*, 2nd Ed., Ally & Bacon (1986).

Chapter 7

Precipitation Titrations

Precipitation titrations as a group are those titrations in which a titration reaction results in the formation of a precipitate or slightly soluble salt. The basic requirements are that the precipitation reaction attains fast equilibrium after each addition of titrant, interferences are absent and the indicator shows a clear end point. On the whole very few precipitation reactions are used in titrations. However, some of the oldest methods such as those involving Cl^-, Br^-, I^- by silver iodide, (also called argentometric methods) are very important. The main reason for lack of popularity of these methods is the lack of suitable methods indicators to detect the end point in precipitation reactions. Secondly, the composition of the precipitate is not always known. The factors influencing precipitation reactions have been already discussed elsewhere.

7.1 THEORY OF PRECIPITATION TITRATION CURVES

The basic theory is the same as given earlier. We illustrate it with a typical example.

Example: If 50 ml of 0.10 M NaCl is titrated with 0.1 M $AgNO_3$, calculate the chloride ion concentration during titrations at (a) start (b) after adding 10 ml (c) 49.9 ml (d) equivalence point and (e) 60 ml of 0.1 M $AgNO_3$ solution. (Given K_{sp} of AgCl = 1.56×10^{-10}). This is Mohr's method for analysis of chloride.

(a) At the beginning $[Cl^-]$ = 0.1 M; $-\log [Cl]$ = pCl = 1.00

(b) After adding 10 ml (V_R); in 50 ml (V_R) we have

$$[Cl^-] = \frac{V_R M_R - V_T M_T}{(V_R + V_T)} = \frac{[(50 \times 0.1) - (10 \times 0.1)]}{(50 + 10)} = 0.067 \quad pCl = 1.17$$

In the above equation V_R = volume of reactant, V_T = volume of titrant, M_R is molarity of reactant and M_T is molarity of titrant.

(c) Now after addition of 49.9 ml of $AgNO_3$

$$[Cl^-] = \frac{(50 \times 0.1) - (49.9 \times 0.1)}{(50 + 49.9)} = 1 \times 10^{-4} \quad \text{so pCl} = 4.0$$

(d) At the equivalence point, $\left[Ag^+\right] = [Cl^-] = [Cl^-][Cl^-] = [Cl^-]^2$

$[Cl^-]^2 = [1 \times 10^{-10}]$ So $[Cl^-]$ = 1×10^{-5}. Hence pCl = 5.0 since at equivalence point there is no excess of chloride nor silver ion and concentration is found from K_{sp} which represents the solubility product. Thus

$$K_{sp} = \left[Ag^+\right]\left[Cl^-\right] \text{ i.e., } \left[Ag^+\right] = \left[Cl^-\right] = \sqrt{K_{SP}} = \sqrt{1.56 \times 10^{-10}}$$

(e) After addition of 60 ml of $AgNO_3$ the concentration of Ag^+ ion

$$\left[Ag^+\right] = \frac{V_T M_T - V_R M_R}{\left(V_T + V_R\right)} = \frac{(60 \times 0.1) - (50 \times 0.1)}{(50 + 60)} = 9.1 \times 10^{-3}; \quad pAg = 2.04$$

Since $\quad$ pCl + pAg = 10, pCl + 2.04 = 10, or pCl = 7.96

We tabulate this data in Table 7.1 and from this we can plot the precipitation titration curve (Fig. 7.1). It should be noted that the end point and stoichiometric equivalence point should coincide *i.e.* the chloride ion should be removed within the limits of $\sqrt{K_{SP}}$ (AgCl) just at the end point. This example is discussed in details in a later section. It is interesting to note that the titration curve not only helps us to calculate the equivalence point but also to find out the concentration of cation or anion at any point during the titration.

Table 7.1 Titration curve

ml AgNO$_3$	[Cl$^-$]		pCl	[Ag$^+$]]	pAg	
0	0.1	1.0	–	–	–	
10	6.7 × 10^{-2}		1.17	2.5 × 10^{-9}	8.63	
20	4.3 × 10^{-2}		1.37	3.6 × 10^{-9}	8.44	
30	2.5 × 10^{-2}		1.60	6.2 × 10^{-9}	8.21	
40	1.1 × 10^{-2}		1.96	1.4 × 10^{-8}	7.85	
50	1 × 10^{-5}		5.0	1.5 × 10^{-5}	4.80	Eq pt
50.1	1 × 10^{-6}		6.0	9.9 × 10^{-3}	4.0	
60.0	1.1 × 10^{-8}		7.96	9.1 × 10^{-2}	2.04	

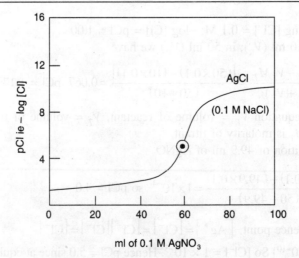

Fig. 7.1 Titration curve of NaCl against AgNO$_3$

7.2 FACTORS INFLUENCING SOLUBILITY OF THE PRECIPITATE

Precipitation is the most versatile method in gravimetric analysis. We mainly consider factors affecting solubility of precipitate. The important parameters are temperature, nature of solvent, interfering ions, pH, effect of hydrolysis, effect of complexes.

(a) *Temperature:* Solubility increases with temperature. Better precipitation occurs in hot solutions, while filtering one must take care not to filter the solution hot if precipitation is temperature dependent.

(b) *Nature of solvent:* Inorganic salts are more soluble in water. A decrease in solubility in the organic solvent can be used to separate two substances.

(c) *Common ion effect:* A precipitate dissolves less in a solution which contains one of the ions of the precipitate than in water, because of K_{sp} (*i.e.* solubility product); either cation or anion, whichever is added, has to decrease in concentration resulting in precipitation of additional salt e.g. In the precipitation of Fe $(OH)_3$ with NH_4OH from $FeCl_3$ solution if we add NH_4Cl as it will push the equilibrium so as to precipitate more iron with NH_4OH. This thus facilitates complete precipitation. In quantitative analysis a common ion is used in the wash solution during filtration.

(d) *Effect of diverse ion:* Many precipitates show an increased solubility when salts that contain no ions in common are present in the precipitate. This is also called the neutral salt or activity effect. The smaller the activity coefficients of two ions, the larger the product of molar concentrations of the ions.

(e) *Effect of pH:* The solubility of the salt of a weak acid depends upon the pH of the solution e.g. oxalates. Here $[H^+]$ ion combines with $\left(C_2O_4^{2-}\right)$ ion to form $H_2C_2O_4$ thereby enhancing the solubility of the salt. The separation of metal sulphides based upon the control of pH has been used in qualitative analysis e.g. metals forming less soluble sulphides (Group II) are precipitated by H_2S in 0.10 M HCl. Then pH is increased for precipitation of Group III metals.

(f) *Effect of hydrolysis:* If a salt of a weak acid is dissolved in water, it will result in a change in $[H^+]$, with the result that the cationic species of a salt undergo hydrolysis and this will also increase the solubility.

(g) *Effect of complexation:* The solubility of a sparingly soluble salt is a function of the concentration of the substance which forms complexes with the cation of the salt e.g. NH_3 influences AgCl precipitate. Many precipitates form soluble complexes with ions of the precipitating agent itself, where solubility first decreases (due to common ion effect) passes through minima and then increase due to complexation reaction.

A reaction which yields a precipitate can be made the basis for titration, provided if the reaction is fast and quantitative and the end point is clear. Many precipitation reactions are slow and lead to supersaturation. Unlike in gravimetry we cannot afford to wait till precipitation is complete. Further, the solubility product must be sufficiently low so that precipitation is quantitative within experimental error when only a drop of excess of reagent is added. The side reactions should be absent and there should be no coprecipitation. However the real restriction is the nonavailability of the right kind of indicators. All such reactions are generally classified on the basis of the type of indicators used to detect the end point.

7.3 TITRATIONS BY TURBIDITY WITHOUT AN INDICATOR

If NaCl is added to a solution containing silver, at first a suspension appears which subsequently coagulates. The speed at which coagulation takes place gives an indication of the nearness of the equivalence point. The addition of NaCl is continued till the end point is reached *i.e.* until there is no formation of AgCl precipitate in the supernatant liquid. However slight excess of NaCl is added at the end point e.g.

$$\left[Ag^+\right]=\left[Cl^-\right]=\sqrt{1.6\times10^{-10}}=1.2\times10^{-15}$$

The determination of Ag as AgCl can also be done by turbidmetric measurements by scattering of light.

7.4 VOLHARD'S METHOD

The titration of Ag with NH_4SCN with ferric alum as an indicator is an example of the class of titration involving the formation of a coloured substance in the solution. During titration AgSCN is formed while at the end point excess NH_4SCN reacts with Fe(III) to form deep red $[Fe\,SCN]^{2+}$. The amount of thiocyanate which will give a visible colour is very small. Thus the end point error is very small but the solution should be shaken vigorously at the end point as silver ions are absorbed on the precipitate and are then desorbed. In Volhard's method, one can easily determine chloride ions in acidic solution, otherwise in basic media Fe^{3+} will hydrolyse. An excess of $AgNO_3$ is added to chloride solution and part of it is unreacted. Ag solution is back titrated with ferric alum as indicator, but the method has a source of error. AgSCN is less soluble than AgCl ($K_{sp} = 1 \times 10^{-12}$)

$$AgCl + SCN^- \rightarrow AgSCN + Cl^-;\ Fe^{3+} + SCN^- \rightleftharpoons Fe\left(SCN\right)^{2+}\ \left(Red\right)$$

This will consume more NH_4SCN and chloride content will appear lower (~2% error). This error can be eliminated by filtering off AgCl precipitate before back titration or if a little nitrobenzene is added, it will adhere to AgCl and protect it from reaction with thiocyanate, however, nitrobenzene slows down the reaction. This can be avoided if $Fe(NO_3)_3$ and a small measured amount of NH_4SCN are added to the chloride solution at the start with HNO_3 and the mixture titrated with $AgNO_3$ till the red colour disappears.

7.5 MOHR'S METHOD

The titration of halides with $AgNO_3$ is carried out with Na_2CrO_4 as an indicator. This is an example of a titration which proceeds with the formation of a new coloured precipitate. At the end point excess Ag_2CrO_4 ions precipitate as brick red Ag_2CrO_4. The solution should be neutral or slightly alkaline, but not too basic so as to precipitate Ag as $Ag(OH)_2$. Further, with too much acid the end point will be delayed, due to lowering of CrO_4^{2-} concentration as per reaction

$$Ag^+ + Cl^- = \underset{\text{White}}{AgCl} \qquad K_{sp} = 1.8 \times 10^{-10}$$

$$2Ag^+ + CrO_4^{2-} = \underset{\text{Yellow}}{Ag_2}\underset{\text{Red}}{CrO_4} \qquad K_{sp} = 1.1 \times 10^{-12}$$

$H^+ + CrO_4^{2-} \rightarrow HCrO_4^-$. Under proper conditions Mohr's method is accurate and applicable at low Cl^- concentration. In titrations of this type the colour-indicating precipitate must be more soluble than the main precipitate formed during titration. Also it should not be so soluble as to need more reagent.

Hence organic reagents are used as indicators such as the one shown in Fig. 7.2. These are used in sulphate titration with $BaCl_2$ with end point by a precipitate red barium salt. This method is also used in water analysis. At pH 8.0 isopropylacohol is added at 25°C. The colour change is, however, not very sharp. Control of pH is essential to give adequate concentration of indicator anions without precipitating the unwanted substances. Lake forming organic reagents are also used as indicators in certain titrations e.g. titration of F^- ion with $Th (NO_3)_4$ with sodium alizarine sulphonate as an indicator. Thus ThF_4 precipitates and at the end point a red lake appears at a pH 2.9–3.4. $CH_2Cl\ COOH$ buffer is used if a small amount of F^- is present.

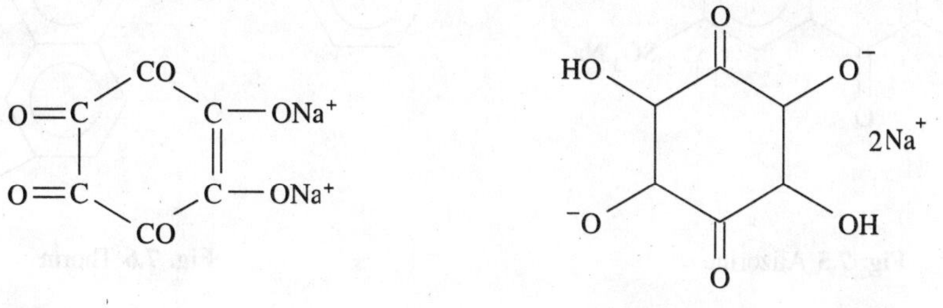

Na – rhodizonate   Na–salt of tetra hydroxyquinone

Fig. 7.2 Organic adsorption indicators

7.6 ADSORPTION INDICATORS IN PRECIPITATION TITRATIONS

If $AgNO_3$ is added to NaCl containing fluorescein at the end point colour changes from light yellow to rose pink. On standing, the precipitate appears coloured while the solution is colourless due to adsorption of indicator on the precipitate of AgCl. The colour of a substance is modified by adsorption on a surface. The reaction is represented with an anionic indicator, like fluoroscine:

If excess Cl: (AgCl) Cl^- + FL^- → no reaction (If FL^- = $C_{20}H_{11}O_5$ *i.e.* fluorocine)

If excess Cl: (AgCl)Ag^+ + FL^- → (AgCl) (AgFL) adsorption

In the case of a cationic indicator the reaction is:

If excess Cl: (AgCl) Cl^- + $(MV)^+$ → (AgCl) $(Cl^-MV)^+$ adsorption

If excess Cl: (AgCl) Ag^- + $(MV)^+$ → No action if MV = methyl violet

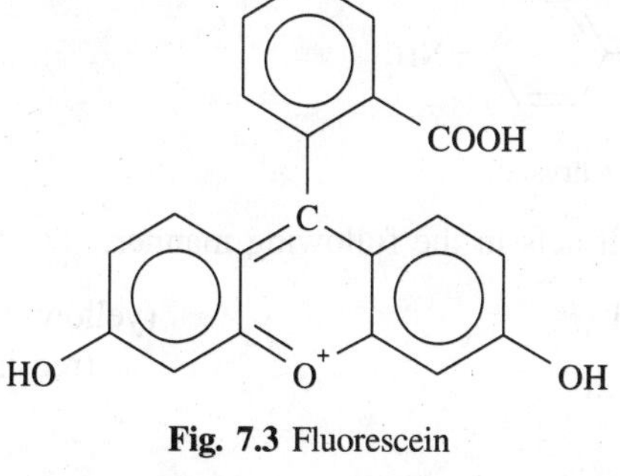

Fig. 7.3 Fluorescein 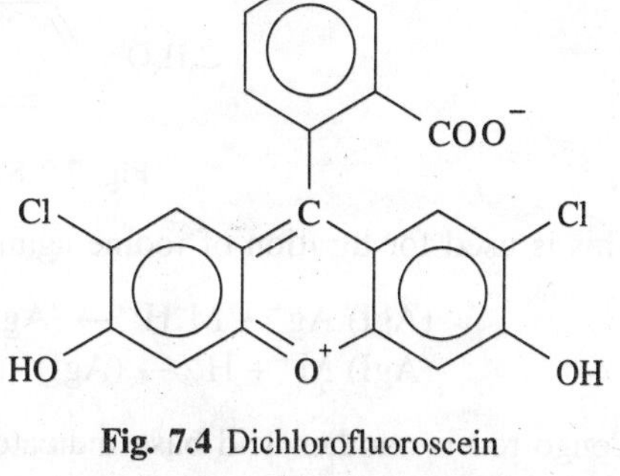   **Fig. 7.4** Dichlorofluoroscein

The adsorptivity of fluoroscein anions by substitution with bromoriodo groups is greatly enhanced as shown in Fig. 7.3 - 7.4.

This is because halogen atoms make these indicators more polarizable and therefore they are more strongly adsorbed. All adsorption indicators are ionic. In addition to these, adsorption indicators, alizarin and thorin are also used as indicators in titration of SO_4^{2-} with $BaCIO_4$ in organic solvent media such as acetone as shown in Fig. 7.5 - 7.6. Eosine and Erythrosine (Fig. 7.7 - 7.8) are also effective adsorption indicators.

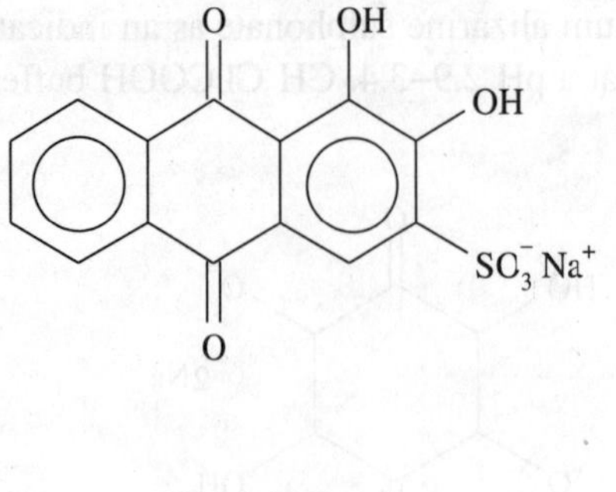

Fig. 7.5 Alizorine

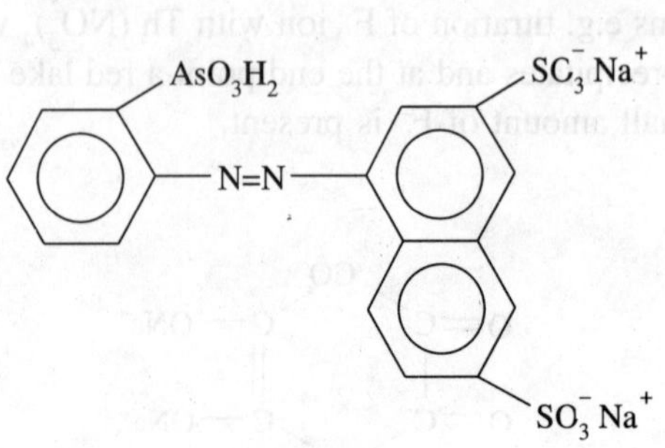

Fig. 7.6 Thorin

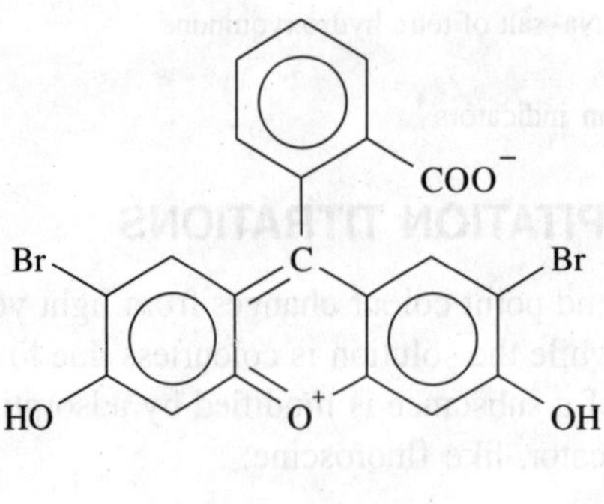

Fig. 7.7 Eosin

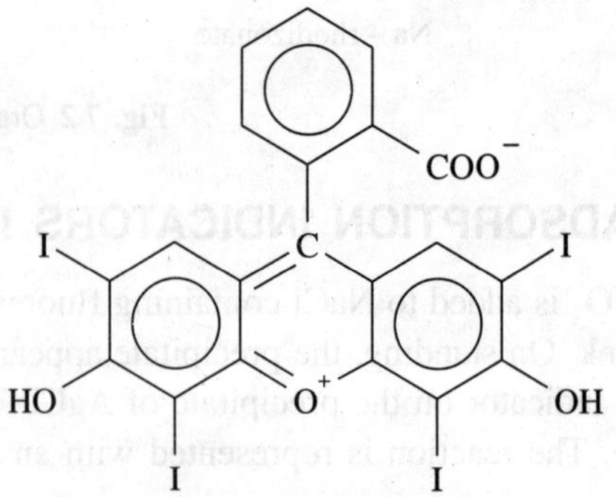

Fig. 7.8 Erythrosin

In addition to the above adsorption indicators, other indicators used in precipitation titration are chrysodine derivatives. They are acid base indicators and adsorption redox indicators (Fig. 7.9). They give reversible colour.

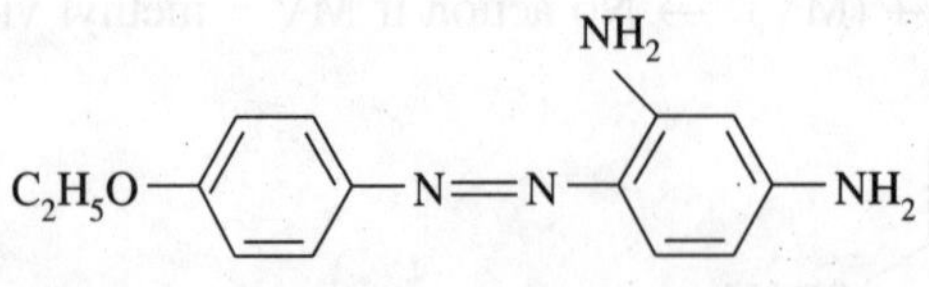

Fig. 7.9 Structure p-ethoxychrysodine

This is used for titration of iodine against Ag^+ ions. It acts in the following manner:

$$(AgI) \, Ag^+ + M^-H^+ \rightarrow (AgI) \, Ag^+ \, M^- + H \qquad \text{(yellow in base)}$$
$$(AgI) \, M^- + H^+ \rightarrow (AgI)^+ \, (M^-H^+) \qquad \text{(red in acid)}$$

Congo red is another acid base indicator.

The merits and demerits of adsorption indicators can be summarised as: They give very small end point errors, the colour change with adsorption indicators are very sharp. Adsorption, being a surface phenomenon, these indicators are best suited for precipitates having a large surface area, adsorption indicators lose their value if the precipitate coagulates in the presence of highly charged ions e.g. we cannot use then for Al (III) due to coagulation. Protective colloids can eliminate this problem, but their use has limitations for optimum control of pH, or concentration of precipitate, during adsorption and desorption procedures.

7.7 MISCELLANEOUS INDICATORS, FOR TITRATIONS

Zinc with $K_3 Fe(CN)_6$ reacts as follows:

$$2[Fe(CN)_6]^{3-} + 3Zn^{2+} + 2K^+ \rightarrow K_2Zn_3[Fe(CN)_6]_2$$

Diphenylamine is used as indicator with $K_3[Fe(CN)_6]$ as the titrant

$$[Fe(CN)_6]^{3-} + e^- \rightarrow [Fe(CN)_6]^{4-} \text{ if } [Fe(CN)_6]^{4-} \equiv [Fe(CN)_6]^{3-}$$

the indicator is colourless but if ferrocyanide concentration is high the indicator is oxidised to the blue form. At the end point and in the beginning of the titration the situation is different. Thus at the beginning, ferrocyanide concentration is low and redox potential is high, while at the end point a sudden drop is observed as ferrocyanide accumulates and so the indicator changes from blue to colourless. Cadmium, calcium can be titrated like zinc. Finally we mention the so called hydrolytic titration. In these titrations the end point is markedly shown by the hydrolysis of the excess of reagent which in turn causes a color change in the acid base indicator provided the reagent is a salt of a weak base or acid e.g. Ba against Na_2CO_3 to give $BaCO_3$; with excess of Na_2CO_3 it hydrolyses and increases the pH as noted by phenolphthalein as indicator, e.g. Al with NaOH in excess of chloride with bromophenol blue indicator.

PROBLEMS

7.1 A 1.5 gram feldspar sample is dissolved to give a 0.1801 gram mixture of chlorides of sodium and potassium. The salts are laxivated with water, mixed with 50 ml of 0.08333N silver nitrate to precipitate chloride as insoluble silver chlorides. The excess silver nitrate is titrated against 0.1N ammonium thiocyanate to give an end point at 16.47 ml. Calculate % of oxide of sodium and potassium in the mixture.

Answer: Let x = gram of KCl $(0.1801 - x)$ = wt of NaCl

$$\text{Total meq. of mixed halides} = \frac{x}{KCl/1000} = \frac{(0.1801 - x)}{NaCl/100}$$

$$= \frac{x}{0.0746} = \frac{(0.1801 - x)}{0.0585} = (50 \times 0.08333) - (16.47 \times 0.10) = 2.52$$

Therefore on solving we get $x = 0.152$ KCl, NaCl = $(0.1801 - 01.52)$

$$= 0.0281 \text{ g NaCl. Conversion factors for}$$

2KCl = K_2O is 0.528, 2 NaCl = Na_2O is 0.53

$$\% \text{ amount of NaCl} = \frac{0.0281 \times 0.530 \times 100}{1.5} = 9.88\% \text{ Na}_2\text{O}$$

$$\% \text{ amount of KCl} = \frac{0.152 \times 0.528 \times 100}{1.5} = 5.3\% \text{ K}_2\text{O}$$

7.2 A sample of $BaCl_2$ weighs 0.500 gram. 50 ml of 0.21 N $AgNO_3$ is added to precipitate AgCl. The excess silver nitrate is titrated with 0.28 N potassium thiocyanate to give 25.5 ml at the end point. Find the percentage of $BaCl_2$ in the sample.

Answer: meq. of $AgNO_3$ added = 50 × 0.21 = 10.50

meq. of KSCN required = 25.5 × 0.28 = 7.14

Total meq. needed = (10.5–7.14) = 3.36 ml

$$\text{Amount of BaCl}_2 = \frac{3.36 \times \frac{BaCl_2}{2000} \times 100}{0.50} = \frac{3.36 \times 0.104 \times 100}{0.50} = 69.88\% \text{ BaCl}_2$$

PROBLEMS FOR PRACTICE

7.3 A sample of a silver bangle weighing 0.5 g and consisting of 90% silver is analysed by Volhard's method. What is the minimum normality of thiocyanate solution which may or may not require more than 50 ml in the titration?

7.4 A sample of feldspar contains 7.58% Na_2O and 9.93% K_2O. What is the strength of $AgNO_3$ if it takes 22.71 ml to precipitate all chlorides as AgCl? (Given the weight of sample is 0.15 gram).

7.5 What weight should be taken for titration of chloride so that the volume of 0.1 M $AgNO_3$ will be equal to the percentage of chloride ion and the percentage of NaCl?

7.6 Calculate the error per thousand in Mohr's titration of 4.0 meq of Cl^- with 0.1 M $AgNO_3$ to a final volume of 100 ml in presence of 0.10×10^{-3} M of K_2CrO_4 at end point.

7.7 A sample weighing 0.340 gram containing a mixture of chlorides of sodium and potassium needed 48.2 ml of 0.10 M $AgNO_3$ in Mohr's titration. Calculate the percentage composition of NaCl and KCl in the mixture.

LITERATURE

1. H.F. Walton, *"Principles and Methods of Chemical Analysis"*, 2nd Ed., Prentice Hall (India) (1966).
2. R.L. Pesok and L.D. Shields, *"Modern Methods of Chemical Analysis"*, Wiley International (1968).
3. J.G. Dick, *"Analytical Chemistry"*, McGraw-Hill, Kogakusha (1973).
4. H.A. Laitinen and W.E. Harris, *"Chemical Analysis: An Advanced Text and References"*, McGraw-Hill (1975).
5. D.A. Skoog and J.J. Leoray, *"Principles of Instrumental Analysis"*, Harcourt Brace (1994).
6. G.D. Christian, *"Analytical Chemistry"*, 5th Ed., John Wiley (1994).
7. A.I. Vogel, G.H. Jeffrey, J. Bassett, J. Mendham and R.C. Denney, *"Vogel's Test Book of Quantitative Chemical Analysis"*, John Wiley (1989).
8. S.M. Khopkar, *"Analytical Chemistry – Problems & Solution"*, New Age International (P) Publishers (2002).

Chapter 8

Complexometric (EDTA) Titrations

eactions involving complex ions or undissociated neutral molecules in solution fall under the class of complexometric titrations. The basic requirement is that such a complex should be extremely soluble. The complexes of metals with EDTA (disodium salt) fall into this category. In addition to EDTA titration, titrations involving mercuric nitrate and silver cyanide are also known as complexometric titrations.

8.1 EDTA AND THE COMPLEXONES

EDTA also called by several other names such as Versene, Complexone III, Sequesterene, Nullapon, Trilon B, Idranat III. It has the structure as shown in Fig. 8.1.

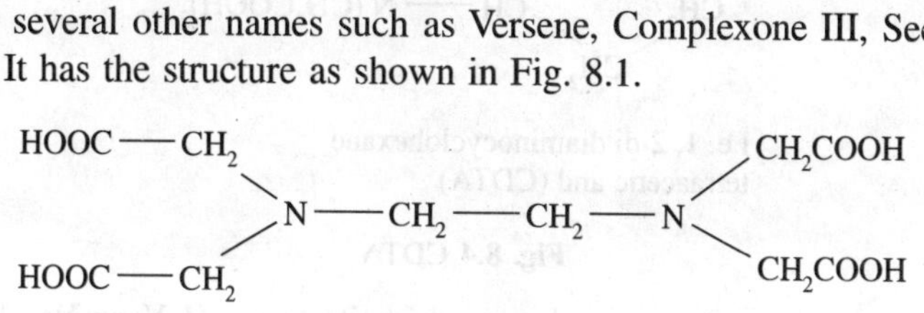

Fig. 8.1 EDTA

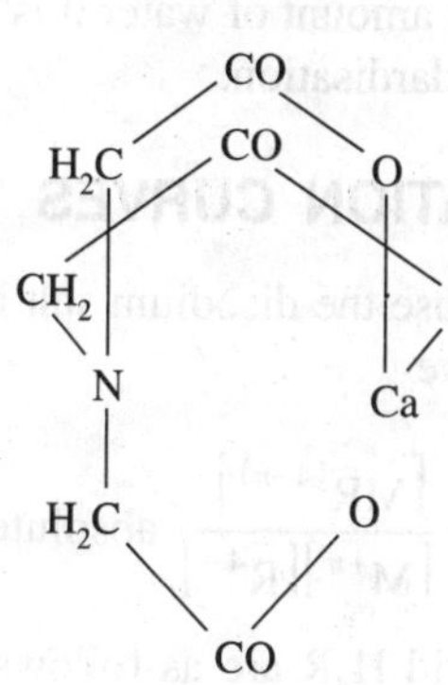

Fig. 8.2 Ca (NTA) complex

It has both donor oxygen atoms and donor nitrogen atoms and it can form six (or five) membered chelate rings simultaneously. Another agent is nitriliotriacetic acid $N(CH_2COOH)_3$. Its calcium complex is shown (Fig. 8.2). Schwarzenbach exploited a number of amine carboxylic acids as titrating and masking agents.

Flaschka reviewed its use as a masking agent. Several metals from complexes at differing pH. The complexing action is proportional to the activity of the free anion Y^{4-} (if acid is H_4Y with ionisation constants $pK_1 = 2.0$, $pK_2 = 2.64$, $pK_3 = 6.16$, $pK_4 = 10.26$) showing that activity of Y^{4-} varies from pH 1.0 to 10.0 and universally changes as the square of $[H^+]$ in pH 3.0–6.0. The complexes of metals with higher charge are more stable. Only Be^{2+}; UO_2^{2+} do not form complexes with EDTA. In addition other complexing agents are used in titrimetry (Fig. 8.3 and Fig. 8.4) are EGTA and CDTA.

$$CH_2 - O - CH_2 - CH_2 - N(CH_2COOH)_2$$
$$|$$
$$CH_2 - O - Ch_2 - CH_2 - N(CH_2COOH)_2$$

i.e. 1, 2-di (2-aminoethoxylethane)
N N N' N' (tetraacetic acid (EGTA)

Fig. 8.3 EGTA

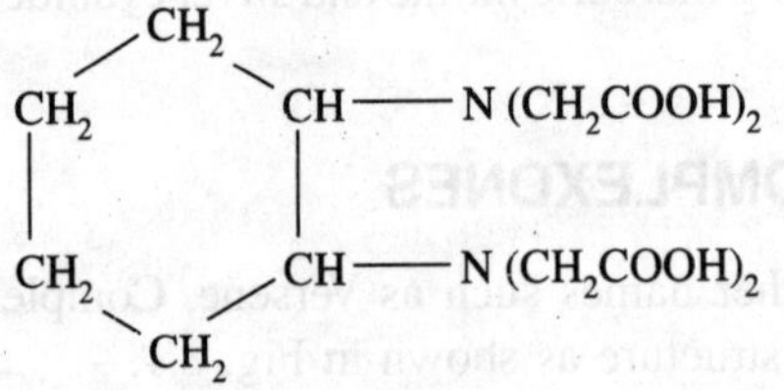

i.e. 1, 2-di diaminocyclohexane
tetraacetic and (CDTA)

Fig. 8.4 CDTA

Similarly triethylene tramine (trien) is also used in titrations. H_4Y or $Na_2 H_2Y$ are its formulae. EDTA dissolves readily in water. It is available in high degree of purity. The dihydrate can be used as a primary standard but due to an uncertain amount of water it is better to standardise it. Cadmium metal is a good primary standard for such standardisation.

8.2 COMPLEXOMETRIC TITRATION CURVES

EDTA is hexadentate, but in titration we use the disodium salt hence it behaves as quadridentate, H_4R. During the complexation reaction we have

$$M^{n+} + R^{4-} \rightleftharpoons [MR^{(4-n)}]^- \text{ so } K_{abs} = \frac{[MR^{(4-n)}]^-}{[M^{+n}][R^{4-}]} \text{ absolutely stability constant}$$

The four dissociation constants of acid H_4R are as follows:

Case (i) $H_4R + H_2O \rightleftharpoons H_3O^+ + H_3R^-$ $K_1 = 1.02 \times 10^{-2}$

Case (*ii*) $H_3R^- + H_2O \rightleftharpoons H_3O^+ + H_2R^{2-}$ $K_2 = 2.1 \times 10^{-3}$

Case (*iii*) $H_2R^{2-} + H_2O \rightleftharpoons H_3O^+ + HR^{3-}$ $K_3 = 6.9 \times 10^{-7}$

Case (*iv*) $HR^{3-} + H_2O \rightleftharpoons H_3O^+ + R^{4-}$ $K_4 = 5.5 \times 10^{-11}$

In actual practice beyond pH ~ 10.0, H_4R only exists as R^{4-}; while at lower pH, HR^{3-} exists. H_3O^+ competes with metal ions for EDTA. As the pH goes down the equilibrium is shifted from case (*iii*) to case (*i*). The fraction of EDTA in R^{4-} can be obtained from the expression.

$$C^2 = [\text{Analytical con}] = [R^{-4}] + [HR^{3-}][H_2R^{2-}] + [H_3R^-]$$

Substituting values of concentrations in terms of formation constants.

$$\alpha_4 = \frac{[R^{-4}]}{C_a} = \frac{K_1K_2K_3K_4}{[H_3O^+] + [H_3O^+]K_1 + [H_3O^+]K_1K_2 + [H_3O^+]K_1K_2K_3 + K_1K_2K_3K_4}$$

Here C_a represents total concentration of uncomplexed EDTA.

Table 8.1 Values of α_4 for EDTA at different pH

pH	α_4	pH	α_4
2.0	3.7×10^{-14}	7.0	4.8×10^{-4}
2.5	1.4×10^{-12}	8.0	5.4×10^{-3}
3.0	2.5×10^{-11}	9.0	5.2×10^{-2}
4.0	3.6×10^{-9}	10.0	3.5×10^{-1}
5.0	3.5×10^{-7}	11.0	8.5×10^{-1}
6.0	3.7×10^{-5}	12.0	9.8×10^{-1}

If the fraction of EDTA in the R^{4-} form is given a symbol (α_4) $[R^{-4}] = \alpha_4 C_a$ ...etc. may be calculated at any desired pH for a metal complex (i.e. chelon) whose dissociation constant is known. At very high pH, $[H_3O^+]$ is negligible. Such constants are usually available in the literature (Table 8.1). The substitution of $\alpha_4 C_a$ absolute stability constant (K_{ab}) gives:

$$K_{ab} = \frac{[MR^{(4-n)-}]}{[M^{n+}][\alpha_4 C_a]}$$

where K_{eff} is called the effective or conditional stability constant. K_{ab}, K_{eff} also varies with pH due to its dependence on α. Thus $K_{eff} = (\alpha_4 \times K_{ab})$ is used in usual calculations. With this K_{eff} at constant pH it is possible to construct a titration curve as shown in the following example.

Example. 50 ml of solution of 0.01 M calcium chloride buffered at pH 10.0 is titrated with 0.01 M EDTA solution. Calculate the values of pM *i.e.* pC at various stages of titration and plot the titration curve. (Given K_{ab} for CaR^{2-} is 5.0×10^{-10}, α_4 at pH 10 is 0.35, so $K_{eff} = 1.8 \times 10^{10} = \alpha_4 \times K_{ab}$).

The various points for titration curves are calculated thus:

(*a*) At the start of titration

$$[Ca^{2+}] = 0.01; \ pCa = 2.0$$

(b) After addition of 10 ml of 0.01 M EDTA using equation

$$[M]_R = \frac{V_R M_R - V_T M_T}{V_R + V_T}$$

if V_R, V_T are volume of reactant and titrant and M_R, M_T are molarity of reactant and titrant

$$[Ca^{2+}] = \frac{(50 \times 0.01) - (40 \times 0.01)}{(50 + 40)} = 0.0067 \text{ M} ; \quad pCa = 2.17$$

Similar calculations can be made at various intervals before the equivalence point.

(c) At equivalence point the concentration of calcium will be

$$[M]_R = \frac{V_R M_R - V_T M_T}{(V_R + V_T)} = \text{Since } [Ca^{+2}] = [EDTA^{4-}] = Ca$$

or $\quad [CaR^{-2}] = 0.005 - [Ca^{2+}] = 0.05$

or $\quad [CaR^{2-}] = \dfrac{0.5}{100} = 5 \times 10^{-3} \text{ M}$; since $K_{eff} = 1.8 \times 10^{10} = \dfrac{K_{ab}}{[Ca^{2+}]}$

or $\quad [CaR^{2-}] = 1.8 \times 10^{10} [Ca^{2+}] \text{ Ca}$

So $\quad [Ca^{2+}] = 5.2 \times 10^{-7}$ or $pCa = 6.28$

(d) Finally after addition of 60 ml of excess of EDTA we have

$$[Ca^{2+}] = [R]_T = \frac{V_T M_T - V_R T_R}{(V_T + V_R)} = \frac{60 \times 0.01 - 50 \times 0.01}{110} = 4.55 \times 10^{-3} \text{M}$$

$$(Ca\ EDTA) = \frac{10 \times 0.01}{110} = 9.1 \times 10^{-4} \text{M. Hence } [CaR^{2-}] = \frac{0.5}{110} = 4.55 \times 10^{-3} \text{M}$$

$$K_{eff} = 1.8 \times 10^{10} = \frac{4.55 \times 10^{-3}}{[Ca^{2+}]\, 9.1 \times 10^{-4}} = \frac{[Ca\, R^{2-}]}{[Ca^{2+}][Ca\, EDTA]}$$

$$[Ca^{2+}] = 2.8 \times 10^{-10} \text{ so } pCa = 9.55$$

Table 8.2 represents a titration curve table while Fig. 8.5 represents a typical titration curve for calcium against EDTA at pH 10.0. The equations used in such calculations are:

$$[R]_{V_T} = \frac{V_R - M_R - r/t\, V_T M_T}{(V_R + V_T)} \qquad ...(i)$$

where t or T is titrant and R or r is reactant

$$[T]_{V_T} = \frac{V_T M_T - t/r\, V_R M_R}{(V_R + V_T)} \qquad ...(ii)$$

where V_T or V_R are volumes of titrant and reactants resp.

$$\frac{MR^{n-4}}{[M^{n+}][R^{-4}]} = K_{abs} \qquad ...(iii)$$

where M_T and M_R are molarities of titrants and reactants

$$C_{EDTA} = [R^{4-}]\left[1 + \frac{[H_3O^+]}{K_4} + \frac{[H_3O^+]^2}{K_3K_4} + \frac{[H_3O^+]^3}{K_2K_3K_4} + \frac{[H_3O^+]^4}{K_1K_2K_3K_4}\right] \qquad ...(iv)$$

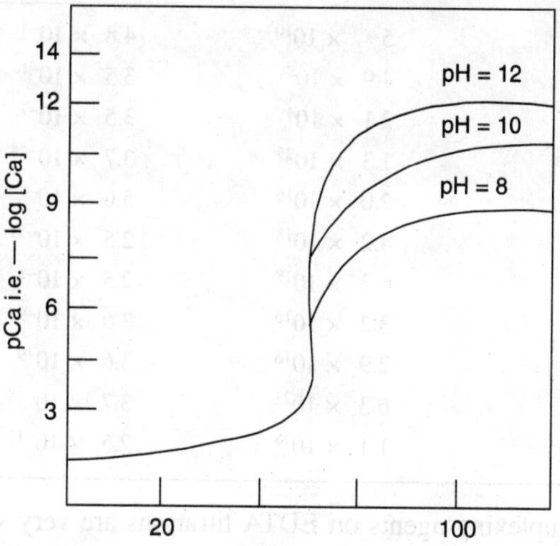

Fig. 8.5 Titration curve for calcium

This titration curve is of familiar shape with a sharp break in the value for pCa at the equivalence point. Also in Fig. 8.5 curves are shown at pH 12 and pH 8.0. For them K_{eff} is 4.9×10^{10} and 2.6×10^8 respectively. The curves are the same only up to equivalence point. A large break is obtained at pH 12 because of the large value of K_{eff}.

Table 8.2 Titration curve for Ca (II) by EDTA

S.No.	ml of EDTA	[Ca(II)] (M)	pCa	Equation No. used
1	0	0.010	2.0	–
2	10	0.0067	2.17	i
3	20	0.0043	2.37	i
4	30	0.0025	2.60	i
5	40	0.001	2.96	i
6	49.9	1.0×10^{-9}	5.0	i
7	50	9×10^{-7}	6.28	iii and iv
8	50.1	2.8×10^{-8}	7.55	ii and iii
9	60	2.8×10^{-10}	9.55	ii and iii

The optimum conditions for complexometric titration of common metals are listed in Table 8.3. One common observation is that the minimum solution pH is based on a conditional or effective formation constant of approximately not less than 108 and with the assumption that no competitive complexation conditions exist in the reaction vessel.

Table 8.3 Optimum conditions for complexometric titrations of selected elements

Metal ion	Minimum pH	K_{aba}	$\propto_4$	K_{eff}
Ca^{2+}	7.3	5×10^{10}	4.8×10^{-4}	1.04×10^8
Mg^{2+}	10.0	4.9×10^8	3.5×10^{-1}	1.71×10^8
Fe^{2+}	5.1	2.1×10^4	3.5×10^{-7}	0.735×10^8
Fe^{3+}	1.5	1.3×10^{25}	3.7×10^{-14}	4.810×10^8
Co^{2+}	4.1	2.0×10^{16}	3.6×10^{-9}	0.72×10^8
Ni^{2+}	3.2	4.2×10^{18}	2.5×10^{-11}	1.05×10^8
Cu^{2+}	3.2	6.3×10^{18}	2.5×10^{-1}	1.57×10^8
Zn^{2+}	4.1	3.2×10^{16}	3.6×10^{-9}	0.0115×10^8
Cd^{2+}	4.0	2.9×10^{16}	3.6×10^{-9}	0.104×10^8
Hg^{2+}	2.2	6.3×10^{21}	3.7×10^{-14}	2.33×10^8
Pb^{2+}	3.3	1.1×10^{18}	2.5×10^{-11}	0.27×10^8

The effect of other complexing agents on EDTA titrations are very significant, on account of the tendency of ions to precipitate as hydroxide or oxide at the pH required for titration. Sometimes use of a masking reagent is essential to keep the ion in solution, e.g. Zn^{2+} is titrated at pH 10 in the presence of a high concentration of ammonium ion. Ammonia not only buffers the solution at the requisite pH but also it prevents hydrolysis. The reaction with EDTA is

$$\left[Zn(NH_3)_4 \right]^{2+} + HR^{3-} \rightarrow ZnR^{2-} + 3NH_3 + NH_4^+$$

Thus completeness of reaction and the end point not only depend upon pH but also on concentration of ammonia present. The influence of auxiliary complexing agent is considered and a quantity (β_4) is defined as the fraction of metal in uncomplexed form

$$\beta_4 = \frac{[M^{n+}]}{C_M} \text{ for zinc it is } \beta_4 = [Zn^{2+}]/C_{Zn} \text{ or } [Zn^{+2}] = \beta_4 C_{Zn}$$

Thus C_M is the sum of concentrations of species containing the metal ion excluding that combined with EDTA. Thus

$$Zn^{2+} + R^{4-} = ZnR^{2-}; \quad K_{ab} = \frac{\left[Zn R^{2-} \right]}{[Zn^{2+}][R^{-4}]} = \frac{\left[Zn R^{2-} \right]}{\beta_4 C_{Zn} \alpha_4 X_a} \text{ if } Zn = \beta_4 C_{Zn}; \ R^{4-} = \alpha_4 X_a$$

$$\therefore \quad K_{eff} = K_{ab} \alpha_4 \beta_4 = \frac{\left[Zn R^{2-} \right]}{C_{Zn} C_a} \text{ with usual notations.}$$

It is worthwhile to consider how the complexometric titration curve is calculated in such an instance from the following example.

Example: Calculate pZn of solutions prepared by mixing 40, 50, 60 ml of 0.001 M EDTA with 50 ml of 0.0001 M Zn^{2+} solution and hence draw a titration curve. (Assume Zn^{2+} and EDTA solutions are 0.1 M in NH_3 and 0.17 M NH_4Cl is added to get pH 9.0). The formation constants are:

$$K_1 = 190, K_2 = 210, K_3 = 250 \text{ and } K_4 = 110$$

(a) Let us first consider calculation of condition or effective stability constant form β_4.

$$\beta_4 = \frac{1}{1 + K_1(NH_3) + K_1 K_2 (NH_3)^2 + K_1 K_2 K_3 (NH_3)^3 + K_1 K_2 K_3 K_4 (NH_3)^4}$$

If we take $\beta_4 = \dfrac{1}{1 + (19 + 420) + (1.04 \times 10^4) + (1.14 \times 10^3)} = 8 \times 10^{-6}$

If $K_{ab} = 3.2 \times 10^{16}$ and α_4 at pH 9.0. 5.2×10^{-2} (Tables 8.3 and 8.1)

$$K_{eff} = 5.2 \times 10^{-2} \times 8 \times 10^{-16} \times 3.2 \times 10^{16} = K_{ab} \alpha_4 \beta_4 = 1.33 \times 10^{10}$$

(b) Now we calculate pZn after addition 40 ml EDTA using eq. (i) given earlier we have

$$C_{Zn} = \frac{V_R M_R - V_T M_T}{V_R + V_T} = \frac{50 \times 0.001 - 40 \times 0.001}{(50 + 40)} = 1.11 \times 10^{-4} M$$

$$\beta = \frac{[Zn^{2+}]}{C_{Zn}} \text{ or } [Zn^{2+}] = \beta \times C_{Zn} = (8.0 \times 10^{-6})(1.11 \times 10^{-4}) = 8.9 \times 10^{-10}$$

If $Zn = 8.9 \times 10^{-10}$. So $pZn = 9.05$.

(c) Calculation of pZn after 50 ml of EDTA addition.
This is the equivalence point $[ZnR^{2-}] = 5 \times 10^{-4} M$
If $C_M = C_a[Zn R^{2-}] = 5 \times 10^{-4} - C_M = 5 \times 10^{-4} M$
Once again using equation $\beta = [Zn^{2+}]/C_{zn}$
$[Zn^2] \ ] \ \beta \times C_{zn} = (8.0 \times 10^{-6}) (1.94 \times 10^{-7}) = 1.55 \times 10^{-12} M$
$C_{zn} = 3.76 \times 10^{-10}$, $[Zn^{2+}] = 3.01 \times 10^{-15}$. Hence $pZn = 1.81$

(d) Finally calculation of pZn after addition of 60 ml of EDTA. Here we have excess of EDTA using the equation (ii) given earlier.

$$C_{EDTA} = \frac{V_T M_T - V_R M_R}{V_T + V_R} = \frac{60 \times 0.001 - 50 \times 0.001}{60 + 50} = 9.1 \times 10^{-5} M$$

Hence $[ZnR^{2-}] = \dfrac{50 \times 0.001}{110} = 4.55 \times 10^{-4}$ M again using equation

$[Zn^{2+}] = \beta \times C_{Zn}$ and $K_{eff} = [ZnR^{2-}]/C_{Zn} C_T [Zn^{2+}] = (3.76 \times 10^{-10})(8 \times 10^{-6})$ as $C_{Zn} = 3.76 \times 10^{-10}$

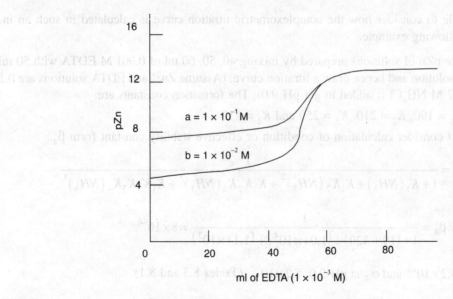

Fig. 8.6 Titration curve for Zn vs EDTA

Fig. 8.6 shows two curves for EDTA at pH 9.0. The equilibrium concentration of ammonia was 0.1 M (*a*) (Fig. 8.6 (*a*) and 0.001 M (*b*) (same Fig. 8.6 (*b*)). Thus ammonia has the effect of decreasing the change in pZn in the equivalence point region. Even the concentration of the auxiliary complexing agent is kept minimum, the magnitude of $_\beta C_{zn}$ is not affected.

8.3 METALLOCHROMIC INDICATORS

Most complexometric titrations are performed with indicators which are themselves chelating agents and whose metal complexes have a different colour from the reagents themselves. They are called metallochromic indicators. Most commonly used indicators are shown in Table 8.4 and structures (I) to (XXI). Most indicators are IDA-iminodiacetate or SP *i.e.* sulphophthalem type or purines or azocompounds. The usefulness of an indicator depends upon stability. The indicator should be half complexed and half free at pM value of inflection in the curve *i.e.* log K_M^1 = pH at equivalence where in K_M^1 is apparent stability constant of metal indicator complex defined as

$$K_M^1 = [MZn]/[M][\text{Total uncomplexed Zn}]$$

Table 8.4 Metallochromic indicators

Structure No.	Name	Structure No.	Name	Structure No.	Name
I	Eriochrome Black	VIII	Thorin	XV	Pthalein complex
II	Erichrome Blue	IX	Calgmite	XVI	Tiron
III	Pyrecatechol violet	X	SPADNS	XVII	Calcein
IV	Acid chrome Blue	XI	Sulphon Black	XVIII	Zincon
V	Xylenol Orange	XII	Varamine Blue	XIX	Bromopyrogallol Red
VI	PAN	XIII	Patton Reeder	XX	Erichrome Red
VII	Murexide	XIV	Pyrocatechol Red	XXI	Thymolpthelin

The structure of important metallochromic indicators are shown in the in Figs. 8.7 to 8.17.

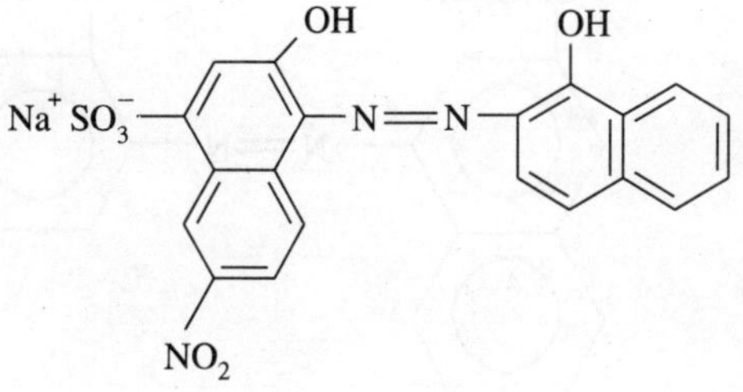

(I) Eriochrome black T (pH 8.0 – 10.5)

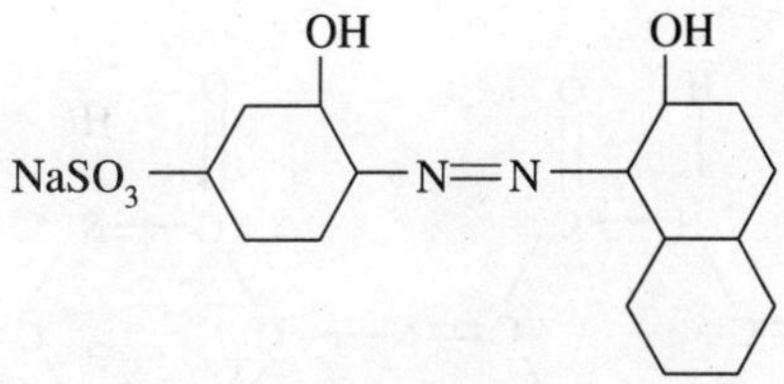

(II) Erichrome Blue black

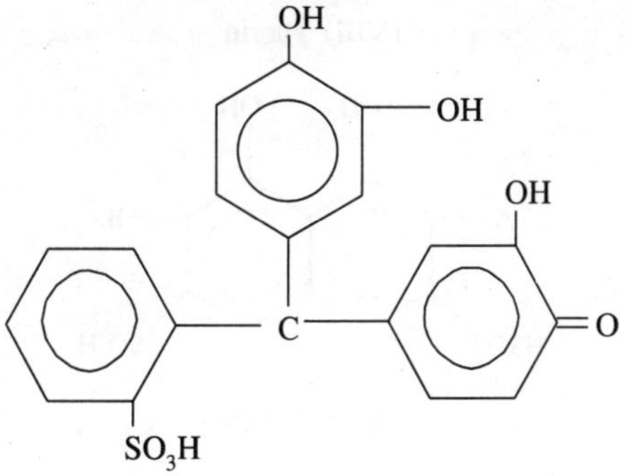

(III) Pyrocatechol violet (pH 1.0 – 5.0)

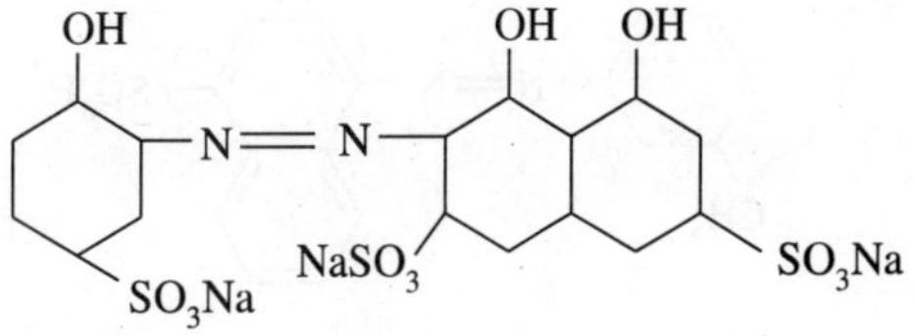

(IV) Acid Chrome black

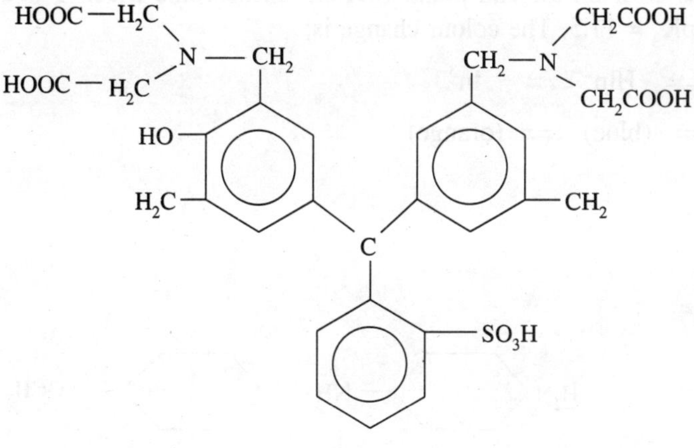

(V) Xylenol orange (pH 1.0 – 5.0)

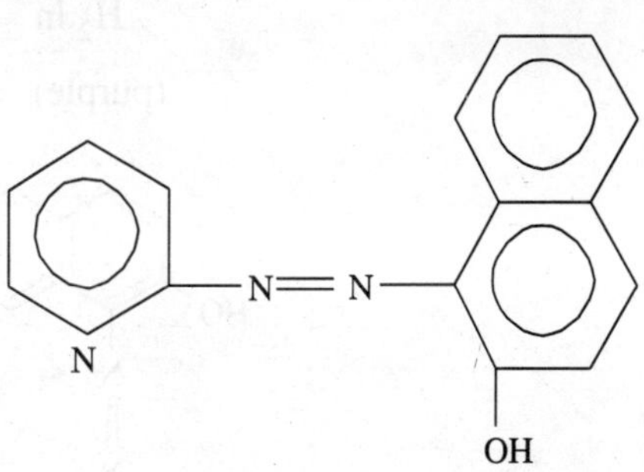

(VI) PAN 1.2 (pyridylazo) napthol

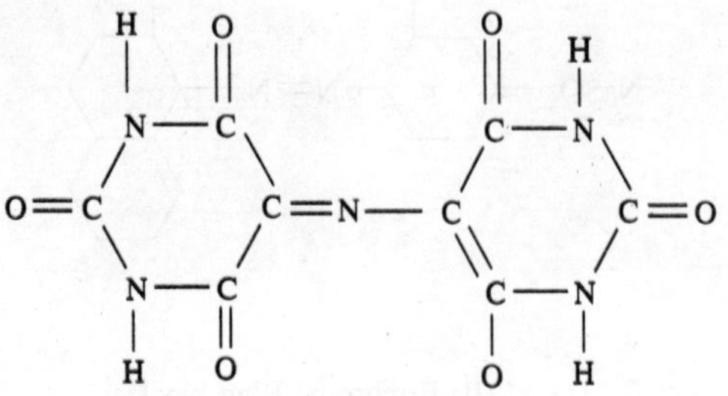

(VII) Murexide (pH 6.0 – 13.0)

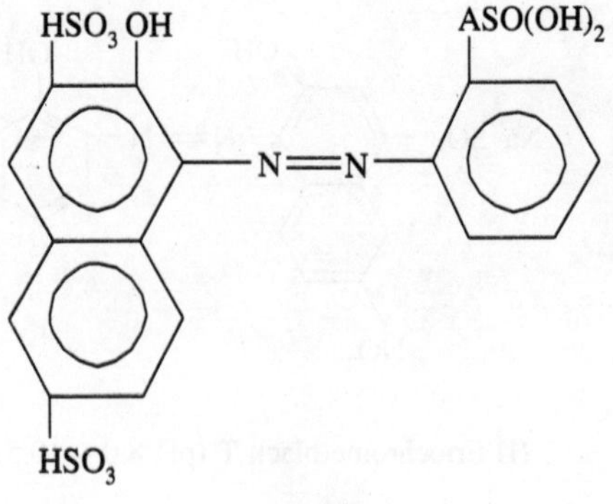

(VIII) Thorin

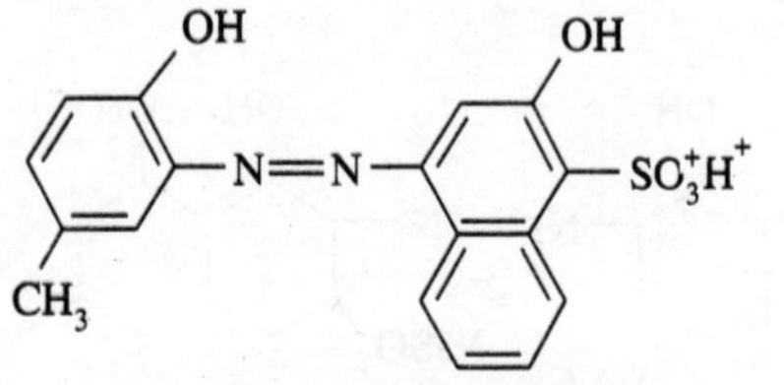

(IX) Calmagite (pH 9.0 – 11.5)

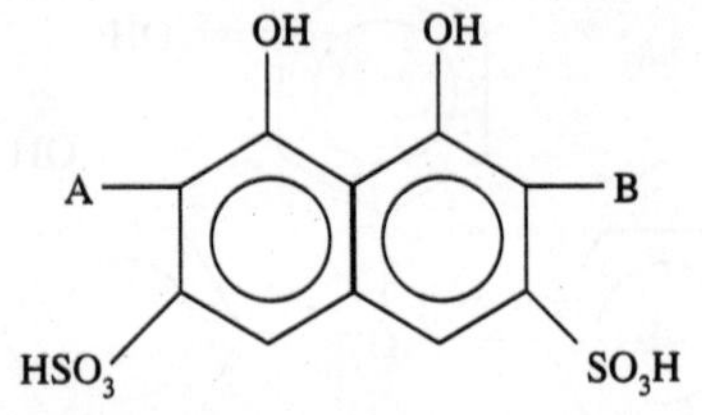

(X) SPADNS

Since all these indicators are weak acids the value of K'_{Zn} depends upon the acidic ionization constants of the reagent and on the pH. If log K_{zn} equals pM at the equivalence point and if the amount of indicator is small, the curve of colour change *versus* equivalence of the titrant will be symmetrical. A large amount of indicator will cause as error on the end point. EBT *i.e.* Eriochrome Black T is a tribasic acid its pK_1 is large, $pK_2 = 6.9$, $pK_3 = 11.5$. The colour change is:

$$H_2 \, ln^- \;\rightleftharpoons\; Hln^{2-} \;\rightleftharpoons\; ln^{3-}$$

$$(\text{purple}) \;\rightleftharpoons\; (\text{blue}) \;\rightleftharpoons\; (\text{orange})$$

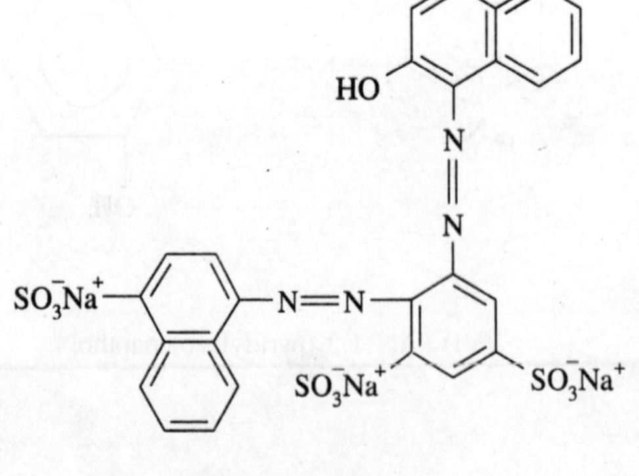

(XI) Sulphon black

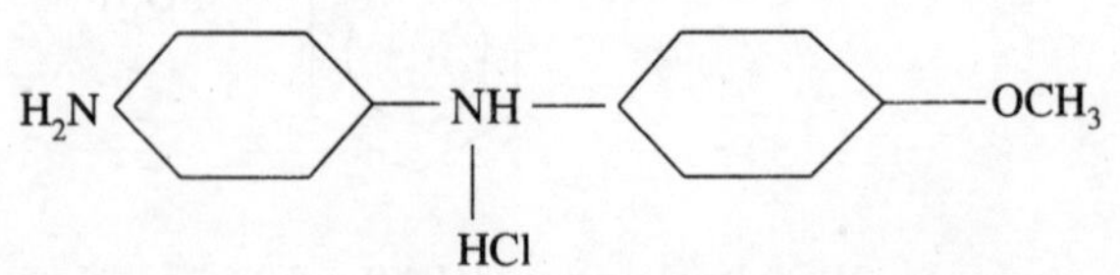

(XII) Varamine blue

(XIII) Patton and Reeders' Dye

(XIV) Pyrocatehol Red

(XV) Phthelein Complex

(XVI) Tiron

(XVII) Solochrome or Calcein

(XVIII) Zincon

The metal complex is purple red but the indicator is inefficient at pH < 8.0. Above pH 6.0 xylenol orange is useless as an indicator. Murexide has a broad pH region, where $pK_1 = 0$, $pK_2 = 9.2$ and $pK_3 = 10.5$. Calcein and a calcein blue are fluorescent indicators effective in neutral pH. Sometimes a too strong or too weak complex is formed with EBT in direct titration. A strong complex is blocked and prevented from functioning as indicator e.g. Cu, Co, Ni form stable metal complexes with EBT and we use KCN to mask these metals. Such a reaction happens in water analysis when the sample is contaminated with copper. The best indicator for water analysis is no doubt Patton and Reeder's Indicator. On the contrary if the metal indicator complex is weak, then excess of EDTA is added and back titrated it with standard solution of zinc or magnesium.

(XIX) Bromophyrogallol red

(XX) Eriochrome red B

(XXI) Thymolphthalexone

Table 8.6 Metal titration with indicator number

Metal	Ind. Number	Metal	Ind. Number	Metal	Ind. Number	Metal	Ind. Number
Al	III, XVIII	Ch.	XVIII	Mg	XIX, I, II, III, XV	Th	III, X, XIII, V
Ba	I, XV	CO	XIX, VII, XVIII	Mn	XIX, VII, I, XVIII	T_1	III
Bi	XIX, III, V	Cu	VI, VII, XII	Hg	V	V	I
Cd	XIX, I, XVIII	Fe	II, XVI, XII, XVIII	Ag	V	Zn	I, VI, III, V, VIII
Ca	I, VII, XIII	Pb	XIX, I, II, III, V, XVIII	Ni	XIX, VIII, XVIII	Zr	X

Substitution titration involves the addition of Mg(EDTA)$_2$ complex to a calcium salt when we get Ca(EDTA)$_2$ and free Mg^{2+} (which in turn forms a coloured complex with EBT) which is titrated with titrant EDTA. This gives a sharp end point. Mercury can be similarly titrated by using Magnesium or zinc complexes of Magnesium or zinc EDTA. Details are provided in Table 8.6.

8.4 KINETICS OF COMPLEXOMETRIC TITRATION

Some EDTA complexation reactions are very slow e.g. Cr^{3+}, Co^{3+}, Al^{3+}, Zr^{4+} and sometimes Fe^{3+}, Bi^{3+} are also slow. Such titrations are performed at 40–60°. In slow reactions the elements are determined by back titration e.g. Cr(III) is titrated with excess of EDTA at pH 1.0–4.0 at 40–50°C by back titrating with EDTA with Zn or Mg salt. EDTA forms a good complex with Cr(III) obtained immediately from Cr(IV). Fe(III) rapidly forms EDTA complex at pH 3.0 rather than at pH 1.0. For aluminium complexes there is hydrolysis at pH > 4.0 but at pH < 3.0 the complex is extremely stable. However in such a case, the order of addition of the reagents is equally important and must be controlled.

8.5 SELECTIVITY IN COMPLEXOMETRIC TITRATIONS

Since a large number of metals can be titrated with EDTA, the question of selectivity is of paramount importance. At times preliminary separation is desirable e.g. anion exchange separations. At times solvent extraction is resorted to. Complexometric titrations can be made more selective by control of pH, use of auxiliary complexes or sequestering agents, appropriate choice of titrant and control of differential reaction rates. Stable complexes are formed usually at low pH e.g. Fe (pH 2.0) Al^{3+}, Zr^{4+}, Bi^{3+} are titrated at low pH to avoid hydrolysis. Zn, Cd and Pb are titrated at pH 5.0. In titration of calcium with interference from Zn, Cd, these are masked with KCN. One can titrate Ca, Mg at pH 10.0 by addition of glycollic nitrile which releases or demasks complexes of Zn, Cd which in turn can be titrated against EDTA. BAL or 2, 3 dimercaptopropanol can be fruitfully used as a masking agent for Zn, Bi, Pb, Hg. Thiourea thiglycollic acid, thiosemicarbazide can be used for masking elements which form insoluble sulphides. EDTA can be used to titrate Ca in the presence of Mg with zircon or murexide as indicators. Cd, Zn can be titrated from mixtures with NH_3–NH_4Cl buffer, as Cd $(NH_3)_2$ is less stable than the Zn $(NH_3)_2$ complex hence we titrate Cd only.

8.6 TYPICAL EDTA TITRATIONS

A common example of a typical EDTA titration is the determination of calcium and magnesium in hard water. The total (Ca+Mg) is found by titrating at pH 10 with Eriochrome black T as indicator. Then magnesium is precipitated as $Mg(OH)_2$ at pH 12.0 and then calcium titrated with EDTA with murexide as an indicator. Free copper should be masked with H_2S. EBT is ground to fineness with NaCl in solid form is usually preferred. The end point in Ca $(EDTA)_2$ titrations is not very satisfactory. Hydroxynapthol blue is a good indicator. It should not coprecipitate with Mg for which EDTA is recommended. However Patton-Reeder indicator is best for the determination of calcium in hard water.

Another typical example is the titration of a mixture of Mg, Cu, Zn without separation involving masking-demasking reactions during titration with EDTA. First total metal is titrated at pH 10 with EBT indicator. Then KCN is added to mask Zn and Cu and again titrated with EDTA to determine Mg from the solution. After the end point, formaldehyde is added to dissociated the Zn $(CN)_4$ complex and liberate free zinc. The titration is then continued to give the amount of Zn is solution. The amount of Cu is found by difference from the first titration.

8.7 ADVANTAGES OF COMPLEXOMETRIC TITRATION

There are several advantages of complexometric titrations in comparison with other kinds of titrations.

Since EDTA is stable, soluble and has definite composition it is preferable to other titrants. The selectivity in such titrations can be easily accomplished by judicious choice of pH e.g. At pH 11, Mg, Cr, Ca, Ba can be titrated while at pH 4.0–7.0 Mn, Fe, Co, Ni, Zn, Cd, Al, Pb, Cu, Ti and V can be titrated and finally at pH 1.0–4.0 metals like Hg, Bi, Co, Fe, Cr, Ga, In, Sc, Ti, V and Th can be titrated. Further EDTA (disodium salt) or Na_2H_2Y is itself a primary standard and does not need further standardisation. Complexes are readily soluble in water. An equivalence point is readily reached in such titrations and finally it is possible to apply such complexometric titrations for the determination of several metals at a semimicro scale of operation.

PROBLEMS

8.1 Ascertain the minimum pH needed for the titration of Mg^{2+}, Zn^{2+} and Fe^{3+} assuming one part in 10^4 remains uncomplexed by Chelon when 0.05 ml excess 0.02 M EDTA is added after equivalence.

Answer: For all three kinds $K_f = [MR]/(M)(R)^{-4} = 10^{-2}/10^{-6}(R)^{4-}$

i.e. $[R^{4-}] = 10/K_f = \alpha_0$ [EDTA excess]

$$= \alpha_0 \frac{(0.05)(0.02 \text{ M})}{100 \text{ ml}} = 10^{-5} \alpha_0, \text{ since } \alpha_0 = 10^9/K_f \text{ if } \alpha_0 = \alpha_4$$

For Mg^{2+}, $\alpha_0 = 10^9/10^{8.7} = 10^{0.3} \times 2.0$

Two drops will do so if, $\alpha_0 = 1$ pH ~ 12.0. Then for

$$Zn^{2+}, \alpha_0 = 10^9/10^{16.5} = 10^{-7.5} \alpha_0 \text{ at pH 4.5}$$

Fe^{3+}, $\alpha_0 = 10^9/10^{25} = 10^{-16}$ *i.e.* complexing is quantitative well below pH 2.0.

PROBLEMS FOR PRACTICE

8.2 Calculate points on a titration curve of 25 ml of 0.02 M Zn (II) by 0.02 M EDTA at pH 10.0 and 0.10 M free ammonia.

8.3 A sample of an alloy containing iron and nickel was analysed complexometrically. If it has 30% nickel and 0.02 M EDTA is used, what size of samples should be taken to achieve ±2 ppt precision in titrations? (Hint: Fe^{3+} is analysed at pH 1.0 while Ni^{2+} at pH 5.5).

8.4 A 50 ml sample of well water is titrated against 0.0102 M EDTA and the end point is reached at 29.85 ml. What is the hardness of water in ppm of $CaCO_3$?

8.5 A 0.5 g sample of bronze containing zinc is dissolved in 100 ml of acid. An aliquot of 25 ml needed 10.25 ml of 0.01 M EDTA. What is the percentage of zinc in the bronze sample?

8.6 A sample containing 500 mg of $CaCO_3$ and 300 mg of $MgCl_2$ was dissolved in acid and was made to 2000 ml. It was titrated with EDTA. What is the hardness of water?

8.7 A standard solution of EDTA is prepared. One ml of this solution complexes with Mg in 10 ml solution containing 300 mg $MgCl_2$ per litre. If 100 ml of well water needs 8.60 ml of standard EDTA, calculate the hardness of water.

LITERATURE

1. F.J. Welcher, *"The Analytical Uses of Ethylenediaminotetraacetic Acid"*, Van Nostrand (1958).

2. A. Ringbom, *"Complexation in Analytical Chemistry"*, Interscience, New York (1963).

3. G. Schwarzenback and H. Flaschka, *"Complexometric Titrations"*, Metheuen and Co. Ltd., London (1969).

4. I.M. Kolthoff, E.B. Sandell, E.J. Mehan and S. Bruckenstein, *"Quantitative Chemical Analysis"*, 4th Ed., Macmillan, London (1969).

5. J.A. Dean, *"Chemical Separation Methods"*, Van Nostrand Reinhold, London (1970).

6. A.I. Vogel, *"The Textbook of Quantitative Inorganic Analysis"*, Longman and Green (1980).

7. H. Flaschka, *"EDTA Titrations"*, Pergamon Press (1959).

8. G.D. Christian, *"Analytical Chemistry"*, John Wiley (1994).

2. A. Ringbom, "Complexation in Analytical Chemistry," Interscience, New York (1963).
3. G. Schwarzenbach and H. Flaschka, "Complexometric Titrations," Methuen and Co. Ltd., London (1969).
4. I.M. Kolthoff, E.B. Sandell, E. Meehan and S. Bruckenstein, "Quantitative Chemical Analysis," 4th Edn., Macmillan, London (1969).

Chapter 9

Solvent Extraction

Amongst the various methods of separation, solvent extraction, or liquid-liquid extraction, is considered to be the most versatile and popular method of separation. The main reason for its usefulness is that separations can be carried out on a macrolevel as well as on a microlevel. One does not need any sophisticated apparatus or instrument except a separatory funnel. It is based on the principle that a solute can distribute itself in a certain ratio between two immiscible solvents such as benzene, carbon tetrachloride, or chloroform. In limiting cases, the solute can be more or less transferred into the organic phase. The technique can be used for the purpose of preparation, purification, enrichment, separation and analysis on all scales. It has come to the forefront in analytical chemistry four decades ago, as it is elegant, simple, rapid and is applicable at tracer and macrogram concentrations of metal ions.

9.1 PRINCIPLES OF SOLVENT EXTRACTION

As per the Gibb's phase rule: $P + V = C + 2$
where P = number of phases, C = number of components, V = degree of freedom

In solvent extraction we have two phases namely the aqueous and the organic phase, the component is ($C = 1$) solute, in solvent and water phase and at constant temperature and pressure $V = 1$, thus we have $2 + 1 = 1 + 2$ *i.e.*, $P + V = C + 2$.

According to the Nernst distribution law,

If $[X_1]$ is the concentration of solute in phase 1 and if $[X_2]$ is the concentration of solute in phase 2 at equilibrium X_1, X_2 then $K_D = \dfrac{[X_2]}{[X_1]}$

where K_D is called the partition coefficient. This partition or distribution coefficient is independent of the total solute concentration in either of the phases. In the above expression for K_D we have not considered the activity coefficient of the species in the organic as well as in the aqueous phase.

We therefore use the term distribution ratio (D) to account for the total concentration of species in two phases. In the circumstance we have the distribution ratio (D):

$$D = \frac{\text{Total concentration of species in the organic phase}}{\text{Total concentration of species in the aqueous phase}}$$

Now assuming there is no association, dissociation or polymerisation in both the phases then, under idealised conditions, K_D would be equal to D. In practical work, instead of using the term K_D or D one prefers to use the term percentage extraction (E). This is related to distribution ratio (D) by the expression

$$D = \frac{\left(\dfrac{V_W}{V_0}\right)E}{(100-E)}$$

where V_W is volume of aqueous phase and V_0 is volume of organic phase.

9.2 CLASSIFICATION OF EXTRACTIONS

There are several ways in which various extraction systems can be classified. The classical way was classification on the basis of the nature of the extracted species formed such as chelate extraction or ion association system. However, present classification has a more scientific basis *viz.* the process of extraction. For instance if the extraction of a metal ion proceeds by formation of chelate or close ring structure we call it chelate extraction, e.g. the extraction of Uranium (VI) with 8-hydroxyquiniline in chloroform; or the extraction of Iron (III) with cupferron with carbon tetrachloride as the solvent.

The next class of extraction can be termed as extraction by solvation because the extracted species gets solvated in the organic phase. The classical example of this class is the extraction of Iron (III) from hydrochloric acid with diethylether or the extraction of Uranium (VI) from nitric acid media with tributylphosphate.

The third category of extraction is the process involving ion pair formation. Such extraction proceeds with the formation of neutral uncharged species which in turn gets extracted into the organic phase. The best example of this class is the extraction of scandium with trioctylamine or uranium with trioctylamine. In both cases an ion pair is formed between Sc or U in mineral acid along with high molecular weight amine. All extractions with crown ethers and cryptands also fall under this category.

The last category of extraction can be termed as synergic extraction. As the name implies, there is an enhancement in extraction on account of the use of two extractants e.g. extraction of uranium with tributylophosphate (TBP) as well as 2-thenoyltrifluoroacetone (TTA). Although either TBP or TTA are individually capable of extracting uranium, if a mixture of these two extractants is used we get enhanced extraction. Such extraction is therefore termed as synergistic extraction.

9.3 MECHANISM OF EXTRACTION

Solvent extraction process proceeds in three stages:

1. The formation of uncharged complex, which may include any class of extraction as described above.
2. The distribution of the extractable complex.
3. Interaction, if any, of the complex in the organic phase.

1. *Formation of uncharged complex:* In order to get a clear insight into the technique let us consider each step in details. The formation of uncharged complex is the most important step in extraction. It is obvious that if the complex is charged it will not be extracted hence it is essential to have the extracted complex without any charge. Such uncharged complexes can be formed by the process of chelation (*i.e.*, neutral chelates), solvation or ion-pair formation. Thus if M is a metal ion with valency 'n' and if R^- is an anion of the extracting ligand (HR) then, through coordination we have

$$M^n + nR^- \rightleftharpoons MR_n \text{ (neutral complex)}$$

Now in solvation phenomena or in extraction involving ion-pair formation, the complex formed may be anionic or cationic, which in turn gets associated with cation or anion respectively to give an uncharged complex which in turn gets extracted into the organic phase

$$M^{n+} + bB \rightleftharpoons MB_b^{n+}$$

$$MB_b^{n+} + nX^- \rightleftharpoons \left(MB_b^{n+}, nX^-\right)^0 \text{ (cationic complex)}$$

where B = neutral ligand, X = anion, M = metal with valency 'n'

$$Mn + (n+a)X^- \rightleftharpoons MX_{n+a}^{a+}$$

$$MX_{a+n}^{a-} + aY^+ \rightleftharpoons \left(aY^+, MX_{n+a}^{a-}\right)^0 \text{ (anionic complex extracted)}$$

where X = anionic ligand, Y = cationic with valency 'a' or a proton.

The extraction of copper with 1, 10-phenantroline in chloroform belongs to first category of cationic complex.

$$Cu^{2+} + \text{o-phen} \rightleftharpoons \left[Cu(phen)^{++}\right]$$

$$Cu(phen)^{++} + 2ClO_4^- \rightleftharpoons \left[Cu(phen)^{++}; 2ClO_4^-\right]_0$$

while extraction of iron (III) with ether is anionic complex

$$Fe^{3+} + 4Cl^- \rightleftharpoons FeCl_4^-$$

$$FeCl_4^- + H^+ \rightleftharpoons \left\{H^+, FeCl_4^-\right\}_0$$

both cation and anion are linked to the ether molecule as:

$$\left\{H(ether)^+, FeCl_4^- (ether)_4\right\}_0$$

At this stage it is necessary to examine the nature of metal complexes and various factors governing their formation. Let us first consider coordination complexes. The formation of a complex by a metal ion depends upon its tendency to fill up vacant or unoccupied atomic orbitals and accomplish stable electronic configuration. The bonds formed may be covalent as in $[Fe(CN)_6]^{4-}$ or electrostatic as in $[Ca(H_2O)_6]^{2+}$. During the process of polarisation, ion deformation is favoured by highly charged metal cations, by large ligands and by metal ions with non inert gas atom type electronic configuration. A charged complex is nullified by another charged ion only to facilitate extraction.

The stability of such a coordinated complex depends upon the acidity of the metal ion, the basicity of the ligand, stereochemical considerations and special features related to the configuration of the resultant complex. Thus, if the metal has high cationic charge the greater is acidity and hence valency e.g. Fe (III) is rapidly extracted in comparison with Fe (II) by diethyl either or ethyl acetate from hydrochloric acid. The Born equation states

$$F = \frac{Z^2}{2r}\left(1 - \frac{1}{\epsilon}\right)$$

where ϵ = dielectric constant.

      r = ionic radii, Z = ionic charge

      F = Boltzmann constant

Thus the stability of metal complex increases with ionic potential (i.e., $Z^2/2r$). Vacant atomic d-orbitals, generally favour coordination in transition elements. Complexes or more electronegative elements tend to be more stable. We can give a scale of selectivity of various ligands forming complexes such as:

$$CN^- > SCN^- > F^- > OH^- > Cl^- \ Br^- > I^- \qquad \text{(For anions)}$$

$$NH_3 > RNH_2 > R_2 NH > R_3N \qquad \text{(For neutral ligands)}$$

The complex formed may or may not be chelate, it may be simple coordination complex with a heteropolyacid. One point in common is that all kinds of these complexes are extractable.

9.4 EXTRACTION BY CHELATION

The most important class of complex is the chelate. Chelating ligands play an important role in extraction of metals because they comprise of an impressive array of extracting and sequestering agents. Metal chelates represent a type of coordination compound in which a metal ion combines with a polyfunctional base capable of occupying two or more positions in the coordination sphere of the metal ion to form a cyclic compound. The functional group of the base must be so situated in the molecule that it permits the formation of a stable ring. If $B_1 - (X)y - B_2$ is a chelate where B are donor atoms such as O, S, N and X is a halide anion, and M is a metal ion we have the structure given in Fig. 9.1.

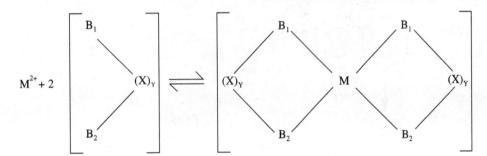

Fig. 9.1 Formation of metal chelate

The general characteristics of the chelate will depend upon the basic nature of the chelating ligand, the acidic nature of metal ion, and factors inherent in the metal chelate itself. The chelates are classified according to the type of basic groups and charge present. If both basic groups are uncharged it results in a neutral chelate, while multiple negative charges on the chelating agent results in negatively charged chelates. Further, depending upon the number of donor atoms in the ligand it is called bidentate, tridentate, hexadentate if the number of donor atoms is 2, 3 and 6 respectively. Copper tartarate is tetracoordinated, while Ca $(EDTA)_2$ complex is a hexacoordinated complex. Thus denticity refers to the ligand and not to the complex.

All types of chelating agents find useful applications in metal extraction procedures. Neutral chelates are easily extracted in organic solvents while certain cationic chelates like Cu (2-9 dimethyl) 1-10 phenanthroline, pair with the perchlorate anion to form uncharged extractable species. Anionic chelates are generally used as sequestering agents, to mask interference of cations during extraction.

There are several factors which must be considered which influence chelate formation. For instance, basic strength of the functional groups, the electronegativity of the bonding atom, and size and number of the chelate ring formed determines the effectiveness of the chelating agent. The basicity of the chelate is measured by the pK_a value of the chelating ligand. This value also decides the stability of the chelated complex, e.g., in gravimetry the substituted oxine derivatives precipitate at a lower pH as the basic strength of the reagent decreases, however, concentration of reagent anion is also important. Acetylacetone (HAA, pK_a = 9.7) should form more stable chelates than 2-Thenoyl-trifluroacetone (HTTA, pK_a = 6.2, Fig. 9.2).

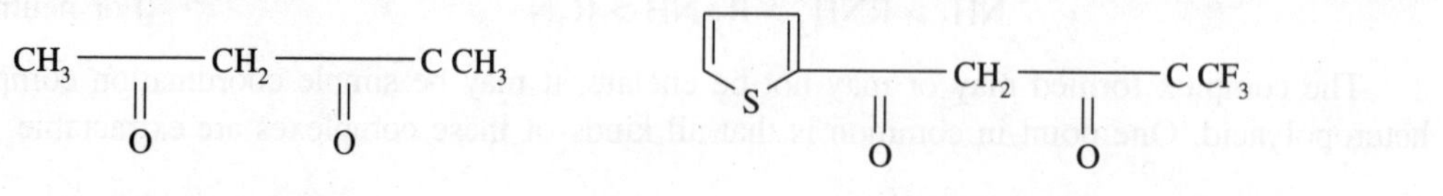

CH$_3$ ———— CH$_2$ ———— C CH$_3$ [thiophene] ———— CH$_2$ ———— C CF$_3$
        ‖   ‖   ‖   ‖
        O   O   O   O

(HAA) (pKa - 9.7) (I)   (HTTA) (pKa - 6.2) (II)
Acetylacetane   2-Thenoyltrifluoroacetone

Fig. 9.2 β-diketones

The reagent anion concentration of TTA (II) is greater and inspite of the low value of pK it favours extensive chelation at low pH. The electronegativity of the basic groups in the reagent is also of great importance. For transition elements, atoms of lower electronegativity tend to form stronger bonds. N and S are better than oxygen e.g., dithizone with sulphur as the donor atom forms a better chelate than diphenylcarbazone with oxygen as donor atom.

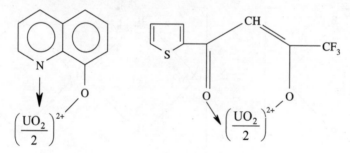

Fig. 9.3 Uranium chelated complex with oxine and TTA

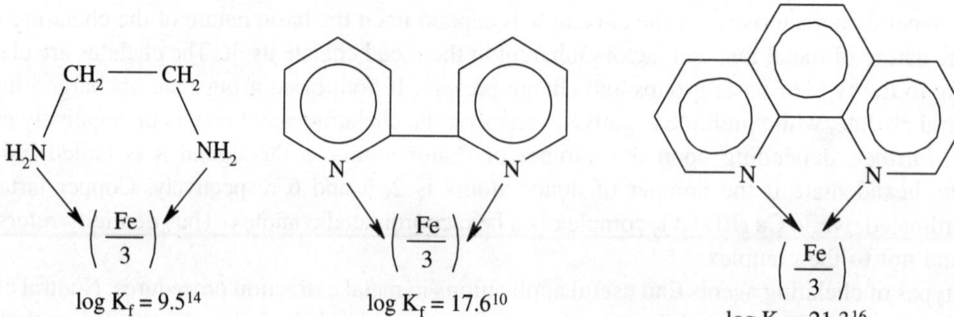

CH$_2$ ——— CH$_2$

H$_2$N   NH$_2$

$\left(\dfrac{Fe}{3}\right)$

log K$_f$ = 9.5[14]   log K$_f$ = 17.6[10]   log K$_f$ = 21.3[16]

Fig. 9.4 Heterocyclic compounds

Apart from basicity, the size of the chelated complex is also important e.g., 5 or 6 membered ring chelates are most stable as they have minimum steric hindrance and no strain (Fig. 9.3). Chelate stability rises with the number of rings (Fig. 9.4). They are formed as an increasing number of water molecules are displaced from the coordination sphere of metal by the molecule of the polyfunctional ligand. Resonance effects and sterochemical effects are also of prime importance e.g., β-diketone like HTTA has better stability. The chelate stability in the complexes shown in Fig. 9.4 increases in order I > II > III.

A substituent in a sterically sensitive position alters stability (Fig. 9.5). However, at times, selectivity of the chelating ligand is increased at the cost of stability by effective substitution, in a sterically sensitive position. Two examples are cited in Fig. 9.5 and 9.6. In the first, 8-hydroxquinoline on substitution of halogen in 5, 7 position or methyl in 2-position becomes selective for particular metals. While in Fig. 9.6, the 1-10 phenanthroline which can easily complex with copper and iron cannot complex with Fe if in position 2-9; a CH_3 group is introduced. Similarly by substitution in the 4, 7 position by diphenyl group only iron (III) gets complexed. This steric hindrance indirectly enhances selectivity.

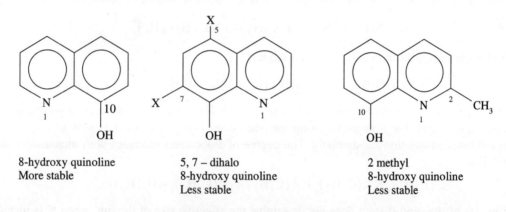

Fig. 9.5 Oxine and its derivatives

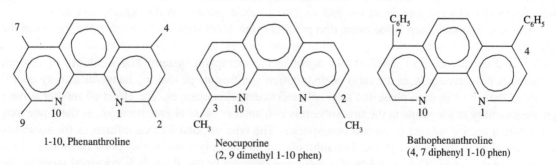

Fig. 9.6 Phenanthroline and its derivatives.

In extraction involving ion pair formation of type proceeds as:

$$A^+ + B^- \;\left(A^+, B^-\right)^0 \; i.e., \; K = \frac{\left[A^+, B^-\right]}{\left[A^+\right]\left[B^-\right]}$$

9.5 EXTRACTION BY SOLVATION

The value of the ion pair formation constant (K) is related to the dielectric constant $(\in)$, and temperature (T) by the expression

$$K = \frac{4\pi Ne^2}{1000 \in K_T} Q(b), \text{ where } b = \frac{e^2}{a \in K_T}$$

where N = Avogadro's number, e = charge, K = Boltzmann constant, T = absolute temperature. $Q(b)$ = calculable functions, a = empirical parameter dependent upon the distance between charge centre of the paired ions when in contact. Since the organic solvents used in extraction have a dielectric constant $(\in)$ less than 40, ion association will be extensive in such solvents. The decrease in the dielectric constant of the aqueous phase by addition of salt favours ion pair formation. In solvents with a high dielectric constant, ion association increases with temperature. While in solvents with a low dielectric constant ion pair formation decreases with temperature. Apart from the dielectric constant, consideration of structural influences other than size also plays important role in determining the extent of ion pair formation. Thus ion association decreases with the increase in size of anion as:

$$\left[Na^+ \, NH_2^- \right] < \left[Na^+ \, NH^- \, C_6H_5 \right] < \left[Na^+, N(C_6H_5)_2^- \right]$$

It also changes with the nature of donor atom:

$$\left[Na^+ OC_6H_5^- \right] < \left[Na^+ NHC_6H_5^- \right] < \left[Na^+ SC_6H_5^- \right]$$

The order of donor atoms for association is O < N < S. The substitution of negative groups in anions decreases association, but a negative group introduced on one of the alkyl groups of a quarternary amine enhances association substantially. The degree of association increases with increasing concentration. Thus

$$(C_4H_9)_4 \, N < (C_4H_9)_4 \, P < (C_4H_9) As < (C_4H_9)_3 < (C_4H_9)I$$

The size of the central atom does not determine the effective size of the ion, when K is increased. Finally, ions with long hydrocarbon chains form aggregates *i.e.* micellar ions, which in turn interact with oppositely charged ions in an ion pair to give a species soluble in the organic solvent.

The next important step in the extraction process is the distribution of the extractable species in the organic phase. Such distribution depends upon various factors, such as basicity of ligand, sterochemical factors and the presence of a salting out agent in the extraction system. The solubility of a metal complex is not necessarily in the ratio of distribution coefficient, as, due to change in activity coefficients of the solute in each phase the total solute concentration changes. The effect of another solvent on the solubility of the solute in the first solvent is also another cause of enhancement, as the solute may react with a second solvent to form a new species. The type of bond formed influences the solubility of the complex in the organic phase. The solubility of an electrolyte in a highly polar medium is aided by electrostatic forces. The solubility of a substance in water or the alcohols is governed more by the ability of the substance in water or the alcohols to form a hydrogen bond than by its polarity as measured by the dipole moment. The solubility of the solute in an organic solvent parallels the electron densities in the nucleus of these aromatic compounds. Metal salts do not dissolve as they are strong electrolytes. The solubility characteristics of a chelate or ion associated (*i.e.*, solvated or ion paired) complex is of importance while considering the mechanism of extraction. Extracted metal chelates are invariably neutral, and are least soluble in the aqueous phase. For them, the theory of "like dissolves like" is largely applicable. The use of solubility parameters (δ) is a useful guide for selectivity. The solubility parameters (δ) represent the square root of heat of vaporisation of liquid per ml which is a measure of cohesive energy density. The solubility increases with increasing similarity of the solubility parameter values, since the heat of mixing of solutes and also solvents depends upon the difference in their δ-values as per the Hilderbrand theory of regular solutions. An increase in solubility of chelate in an organic solvent results from substitution of a hydrocarbon group in the chelate (see Fig. 9.7 and Fig. 9.8).

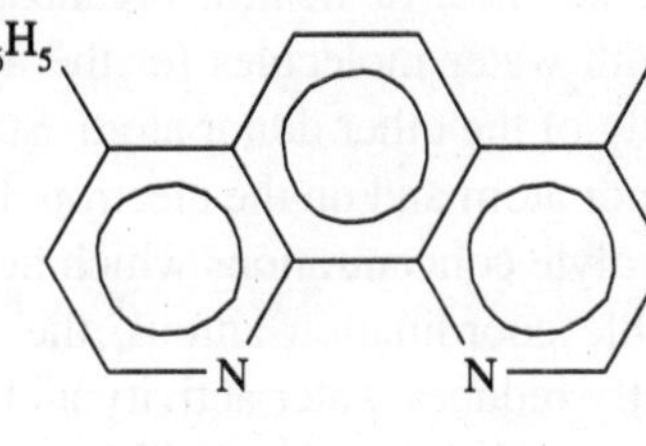

Bathophenanthroline

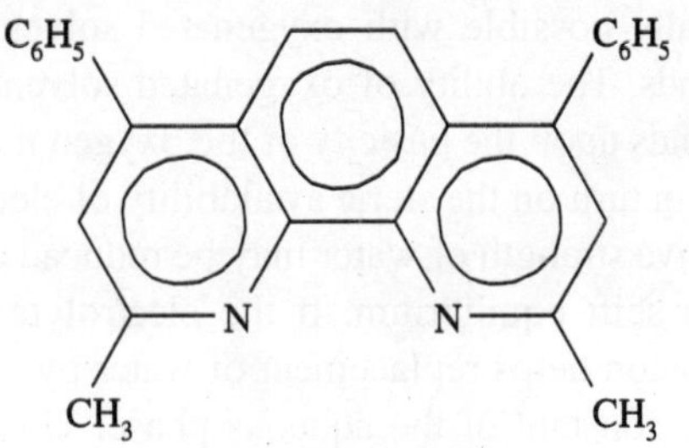

Bathocuporoine

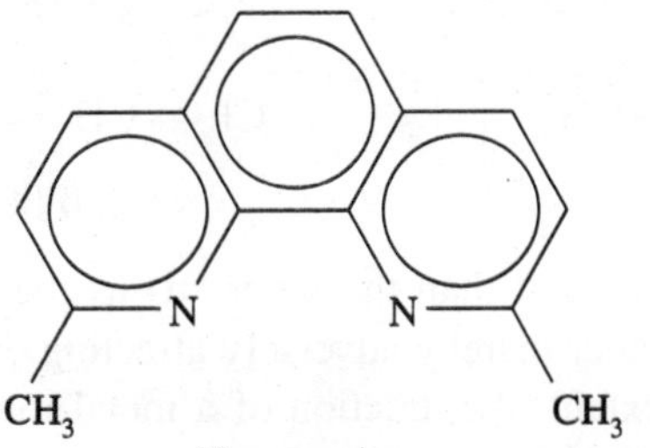

Neocuporine

Fig. 9.7 Bathophenanthroline and Bathocuporine

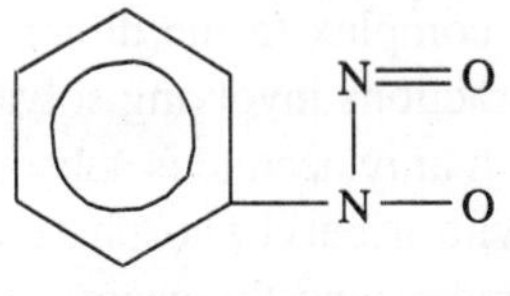

Cupferron (I)

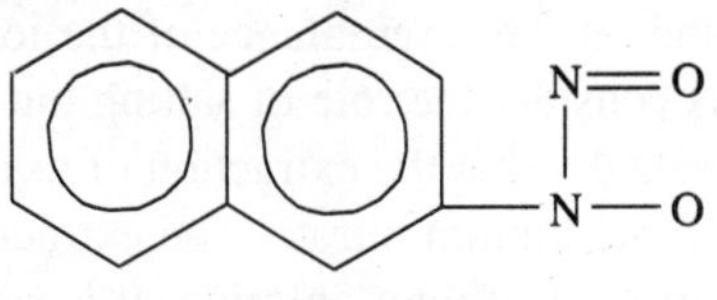

Neocupferron (II)

Fig. 9.8 Cupferron (I) and Neocupferron (II)

The latter is more soluble in organic solvent due to structural resemblance. In many divalent metal ions with coordination number six, water occupies two coordination positions thereby lowering the solubility of the chelate in an organic solvent. However the displacement of water molecules in the coordination sphere of the metal ion helps in enhancing its solubility e.g. Cu, Mg extraction in oxine chloroform solution in the presence of butylcellusolve or tributylamine. These oxygen containing solvents replace water molecules from the coordination sphere.

The consideration of solubility characteristics of solvated species or ion paired complexes is also of great importance. The species formed by either mechanism can be considered as polar molecules. In such cases, solubility depends upon the structural resemblance to the solvent. If a large organic group is present in the complex, more solubility will be observed e.g. the complex of Cu^+ with neocuproine (*i.e.*, 2, 9, dimethyl 1-10 phenanthroline) in the presence of chlorate is soluble in chloroform.

In cases of extraction by solvation, the solvent itself participates in extraction of the complex. The basic character of the oxygen atom enables the incorporation of the solvent molecule in the coordination sphere of the metal ion to form a solvated complex which is extractable e.g. Co^{+2} can be extracted in octanol (R-OH) in the presence of the perchlorate ion as $\left[Co(R\text{-}OH)_6^{2+}, 2ClO_4^- \right]^O$. Such solvation is generally possible with oxygenated solvents like alcohols, ketones, or neutral organosphorphorus compounds. The ability of oxygenated solvents to compete with water molecules for the acidic metal ion depends upon the basicity of the oxygen molecule or basicity of the other donor atom. Such basicity depends in turn on the steric availability of electrons on the donor atom and on the electron density. The competitive strength of water may be reduced using high electrolyte concentrations which help by mass action to shift equilibrium. If the electrolyte possesses suitable coordination anions, the high anion concentration helps replacement of water by the anion. It greatly reduces water activity and lowers the dielectric constant of the aqueous phase. This will be further considered under the role of salting out agents. The observed decrease in basic strength of a series of oxonium solvents follows the order, alcohols < ethers < ketones. The decrease in electron density at the oxygen atom adversely affects solubility e.g.

$$CH_3 - CH_2 - \ddot{O} - CH_2 - CH_3 \qquad\qquad Cl - CH_2 - CH_2 - \ddot{O} - CH_2 - CH_2 - Cl$$

Diethyl ether (I) $\qquad\qquad\qquad\qquad$ **β β′ dichlorodiethyl ether (II)**

The former (I) is better extractant than the latter (II) as the electron density of the oxygen atom decreases in substituted diethylether thereby adversely affecting extractability. Using an inert diluent in solvent extraction reduces the extent of extraction of a metal ion in proportion to the decrease in the concentration of an extractant in the diluent.

Solubility of metal salts in oxygen containing solvents depends upon the specific interaction of the solvent oxygen and the solute. The extent to which dilution of the oxygenated solvent may be carried out without affecting good extraction will depend on the strength of the bonding of the metal ion and solvent oxygen and on the resemblance of the ion association complex to the diluent.

Finally, let us consider the role of salting out agents in extractions involving solvation. It has been observed experimentally that the extraction of uranyl nitrate with amylacetate is substantially enhanced by the addition of ammonium nitrate. The extraction of gold with tributyl phosphate in HCl is greatly enhanced by addition of lithium chloride. It is interesting to understand the exact role of NH_4NO_3 or LiCl in extraction of uranium or gold. At the outset we term them as salting out agents. These are electrolytes whose addition greatly enhances the extractability of the complex. The main functions of a good salting out agent are:

(*a*) Providing a higher concentration of the complex anion which by mass action increases concentration of the complex and improves extraction,

(*b*) Binding of water molecules by the ions of the salting out agent so it becomes unavailable as a free solvent.

(*c*) A decrease in the dielectric constant of the aqueous phase with increasing salt concentration, which will favour the formation of ion association complexes.

A less important point in consideration of the mechanism of extraction is the interaction in the organic phase. This interaction owes its importance to an effect on the concentration of the complex and extent of extraction. In extraction by solvation, polymerisation can sometimes occur. At higher concentrations polymerisation is fast e.g. $[Cu(SCN)]^+$ polymerises in TBA. This polymerisation tends to reduce the activity of the extractable species in the organic phase but the overall equilibrium is shifted in favour of higher distribution ratios. Dissociation of the ion association complex may also occur in dilute solutions e.g. polar solvents which would result in increased extraction. For covalent metal chelates such problems never arise. Sometimes extraction of a buffer compound affects the overall extractibility of the metal ion with chelating ligand but such cases are rare.

9.6 EXTRACTION EQUILIBRIUM FOR CHELATES

Consideration of an extraction equilibrium gives quantitative information regarding experimental parameters which influence the selectivity of extraction. To make the point clear we shall consider two extraction systems one involving chelation and another involving solvation.

Let us consider extraction by chelation. The reaction occurs when an aqueous phase containing a metal ion comes in contact with an organic phase containing a chelating ligand. The chelating ligand distributes between two phase. If M is a metal ion with the valency 'n' and HR is a chelating ligand then:

$$HR \rightleftharpoons HR_{org} ; K_{D_R} = \frac{[HR]_0}{[HR]} \qquad ...(9.1)$$

The chelating ligand dissociates as

$$HR \rightleftharpoons H^+ + R^- ; K_a = \frac{[H^+][R^-]}{[HR]} \qquad ...(9.2)$$

The chelating anions combines with metal ion M to form the chelate.

$$M^{n+} + n R^- \rightleftharpoons MR_n : K_f = \frac{[MR_n]}{[M^+][R^-]_n} \qquad ...(9.3)$$

This chelates distribution in two phase, hence

$$MR_n \rightleftharpoons MR_{n_{org}} \quad K_{D_X} = \frac{[MR_n]_0}{[MR_n]} \qquad ...(9.4)$$

However the distribution ratio (D) can be evaluated if the metal chelate MR_n in the organic phase and M^{n+} in aqueous phase.

$$D = \frac{|M_0|}{|M|} = \frac{[MR_n]_0}{[M^{n+}]} \qquad ...(9.5)$$

Combining equations (9.1) to (9.5) gives

$$D = \frac{[MR_n]_0}{[M^{n+}]} = \frac{K_f K_a^n K_{D_X}}{K_{D_R}^n} \times \frac{[HR]_0^n}{[H^+]^n} = K^* \left[\frac{[HR]_0}{[H^+]} \right]^n \qquad ...(9.6)$$

In the above expression, K_f is the formation constant of the metal complex, K_a is the acid dissociation constant, K_{D_x} is the distribution coefficient of the metal complex, K_{D_R} is the distribution coefficient of the chelating ligand (HR).

The above equation was experimentally verified by Kolthoff and Sandell. Furman, Mason and Pekola also verified it for dithizone and cupferron complexes of metals such as lead and copper respectively.

We can easily furnish an interpretation of the above mathematical expression in actual practice. The equation shows the importance of chelate stability (K_f) and solubility of chelate in the organic phase $\left(K_{D_x}\right)$. Since K_a^n is in the numerator, an acidic reagent with high K_a, is relatively soluble. Further, the extraction depends upon concentration of (HR), i.e., the chelating ligand. Further $\left(H^+\right)^{n-}$ represents the pH of solution. Increase in reagent concentration increases (D) as much as would a rise of one unit in pH. An increase in the pH range of good extraction is achieved by using higher reagent concentration. This permits extraction at a lower pH e.g. the substitution of CF_3 in acetylacetone to give hexafluoroacetylacetone increases its acidity. Since the metal concentration $[M^{n+}]$ does not figure in the equation, the overall extraction does not depend upon initial metal concentration i.e. both trace and macro accounts of metal can be extracted. The nature of solvent has not been a critical factor in achieving success of an extraction. The selectivity of extraction is evaluated in terms of the separation factor (α) which is expressed as:

$$\alpha = \frac{D_1}{D_2} = \frac{K_{f1}K_{D_{X1}}}{K_{f2}K_{D_{X2}}} \qquad \qquad ...(9.7)$$

The case of separation of two metal ions with the same extractant depends upon the difference in stability of chelates (K_f) and on the relative solubilities $\left(K_{D_x}\right)$ of chelates in the organic phase. Steric hindrance by effective substitution of ligand and use of sequestering agents can enhance selectivity by indirectly increasing the magnitude of the separation factor (α). There will be no possibility of separation if $\alpha = 1$ as $D_1 = D_2$ and is maximum if α is large.

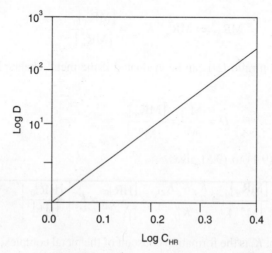

Fig. 9.9 Nature of excited species

The selectivity of extraction is also evaluated in terms of the value of pH 1/2 (pH Half) (Fig. 9.10). Supposing we plot a graph of log D against log C_{HR} we get a linear relationship from the slope of which 'n' can be calculated (Fig. 9.9).

$$D = \frac{K_F K_a^n K_{D_X}}{K_{D_R}^n} \frac{[HR]_0^n}{[H]^n}, \text{ if } K^* = \frac{K_f K_a^n K_{D_X}}{K_{D_R}^n}$$

$$D = K^* \frac{[HR]_0^n}{[H^+]^n} \qquad \qquad \qquad ...(9.8)$$

Now if we keep the reagent concentration $[HR]$ constant then $D = K^* [H^+]^{-n}$.

Taking the log of both sides $\log D = \log K^* - npH$

When $V_w = V_0$ we have

$$D = \left(\frac{E}{100 - E}\right) = K^* [H^+]^{-n}$$

$$\log D = \log E - \log (100 - E) = \log K^* - npH$$

When $E = 50$ the pH for 50% extraction is

$$pH_{1/2} = \frac{1}{n} \log K^* \qquad \qquad \text{If } [HR] \text{ varies} \qquad \qquad ...(9.9)$$

$$pH_{1/2} = \frac{1}{n} \log K^* + \log (HR) \qquad \qquad \qquad ...(9.10)$$

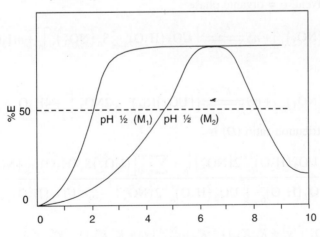

Fig. 9.10 Significance of pH 1/2

This shows pH at half extraction is constant and for the same concentration of a given chelating agent (HR) the magnitude of pH is dependent only on the valence of the ion (n) and stability constant (K^*). For two metals having the same valence $(n = n)$, $pH_{1/2}$ will vary linearly with chelate stability (K_f).

This of course depends upon the assumption that the partition coefficient (K_{d_X}) of the metal chelated does not vary much or varies in proportion to change in formation constant (K_f). Thus $\text{pH}_{1/2}$ is used as the measure of the relative stability constants of chelates in aqueous solutions. The more stable the chelate, the lower the pH for 50% extraction.

9.7 EXTRACTION EQUILIBRIA FOR SOLVATION

Let us consider extraction equilibria for extraction by solvation in various steps.

(a) Formation of complex

$$\left[UO_2 \left(H_2O_6 \right)_6 \right]^{2+} + 2\left(NO_3^- \right) \xrightleftharpoons{K_f} \left\{ UO_2 \left(H_2O \right)_6^{2+}, 2\left(NO_3^- \right) \right\}$$

If water activity is very low

$$\left[UO_2 \left(H_2O \right)_6 \right]^{2+} + 2\left(NO_3^- \right) \xrightleftharpoons{K_f'} \left\{ UO_2 \left(H_2O \right)_4, \left(NO_3^- \right)_2 \right\} + 2H_2O$$

(b) Distribution of complex

$$\left[UO_2 \left(H_2O \right)_6^{2+}, 2\left(NO_3^- \right) \right] \xrightleftharpoons{K_D} \left\{ UO_2 \left(H_2O \right)_6^{2+}, 2\left(NO_3^- \right) \right\}_0$$

or

$$UO_2 \left(H_2O \right)_4 \left(NO_3 \right)_2 \xrightleftharpoons{K_D'} \left[UO_2 \left(H_2O \right)_4 \left(NO_3 \right)_2 \right]_0$$

(c) Reactions in the organic phase

$$\left\{ \left[2NO_3^- \, UO_2 \left(H_2O \right)_6 \right]^{2+} \right\}_0 + nS \xrightleftharpoons{K_S} \left[UO_2 S_n \left(H_2O \right)_{4-n} \left(NO_3 \right)_2 \right] + \left[(n+2)H_2O \right]$$

If S = basic solvent, o = organic phase

$$\left[UO_2 \left(H_2O \right)_4 \left(NO_3 \right)_2 \right] + nS \xrightleftharpoons{K_S'} \left[UO_2 \left(H_2O \right)_{4-n} S_n \left(NO_3^- \right)_2 \right]_0 + nH_2O$$

If S is strong base

$$\left[UO_2 \left(H_2O \right)_4 \left(NO_3 \right)_2 \right] + 6S \xrightleftharpoons{K_S'} \left[\left(UO_2 \right) \left(S_6 \right)^{2+}, 2NO_3^- \right]^2 + 4H_2O$$

(d) The overall distribution ratio (D) is

$$D = \frac{[U]_0}{[U]} = \frac{\left[UO_2 \left(H_2O \right)_6^{2+} \, 2\left(NO_3^- \right) \right]_0 + \sum_{n=0}^{n=4} \left[\left(UO_2 \right) S_n \left(H_2O \right)_{4-n} \left(NO_3 \right)_2 \right]}{\left[UO_2 \left(H_2O \right)_6^{2+} \right] \left[UO_2 \left(H_2O \right)_6^{2+} \, 2\left(NO_3^- \right) \right] + ... + UO_2 \left(H_2O \right)_4 \left(NO_3 \right)_2}$$

i.e., $\quad D = \left[NO_3^- \right]^2 \gamma^3 \pm K_f K_D \left(1 + K_S \, la_W^{N+2} \right) \gamma_0 + K_f' K_D' \left(1 + K_S' \, la_w^n \right) \gamma_0'$

$$D = K^* \left[NO_3^- \right]^2 \gamma \pm \gamma^3$$

The above expression leads to the conclusion that extraction depends upon concentration of the complexing and auxiliary complexing anion.

9.8 TECHNIQUES OF EXTRACTION

The three basic techniques used in liquid-liquid extraction are batch extraction, continuous extraction and counter current extraction. Of these, batch extraction is the simplest, consisting of extracting the solute from one immiscible layer by shaking two layers until an equilibrium is attained, after which layers are allowed to settle and separate. This method is mostly used for analytical separations. Completeness of extraction depends upon the number of extractions carried out. The best results are obtained with a relatively large number of extractions with a small volume of solvent. This can be proved mathematically in the following way.

If 'V' ml of solution (phase-1) containing 'W' gram of solute is extracted with 'S' ml of another solvent (phase-2) immiscible with the first and W_1 is the weight of solute remaining unextracted in phase-I, then

$$\therefore \quad \text{concentration in phase-1} = \frac{W_1}{V} \text{ gm/ml} = C_1 \qquad \qquad ...(9.11)$$

$$\therefore \quad \text{concentration in phase-2} = \left(\frac{W - W_1}{S}\right) \text{ gm/ml} = C_2 \qquad ...(9.12)$$

Distribution ratio $(D) = C_2/C_1$

$$D = \frac{C_2}{C_1} = \frac{(W - W_1)/S}{W_1/V} \text{ i.e.,} \qquad \qquad ...(9.13)$$

$$W_1 = W\left[\frac{V}{(DS + V)}\right] \qquad \qquad ...(9.14)$$

Another extraction of phase-1 is made with another partition of the solvent, say 's' ml; W_2 will be the unextracted material in phase-1, then

$$W_2 = W_1\left[\frac{V}{DS + V}\right] \qquad \qquad ...(9.15)$$

$$W_2 = W_1\left[\frac{V}{DS + V}\right]^2 = W\left[\frac{V}{DS + V}\right]\left[\frac{V}{DS + V}\right]; \ W_n = W\left[\frac{V}{DS + V}\right]^n \qquad ...(9.16)$$

This shows for complete extraction 's' should be as small as possible and 'n' should be as large as feasible for a given amount of solvent. Thus the best results are obtained with a relatively large number of extractions (n) with small volume of solvent (s). Batch extractions may also be used with advantage when the distribution ratio is large. The usual apparatus for batch extraction is a separatory funnel.

Continuous extraction is used when the distribution ratio is relatively small, so that a large number of batch extractions would normally be necessary to effect quantitative separation. High efficiency in continuous extraction depends on the viscosity of the phases and other factors affecting the rate of attaining equilibrium, the values of D, the relative volumes of two phases and any other factors. Efficiency of extraction can be improved by employing a large area of contact. The half extraction volume is the volume of extracting solvent necessary to decrease the amount of extracted constituent to one half the former value. This volume may be used to obtain the distribution factor (K) from the relationship.

$$K = \frac{0.693 \ W}{V}$$ where K = concentration of solute in the extracting solvent divided by the concentration

of original solution. The half extraction volume in ml is 'V' and the volume of the original solution is 'W'.

Continuous counter current extraction involves a process whereby the two liquid phases are allowed to flow counter to each other. It is used for separating materials for isolation or purification purposes. This is specifically applicable to the fractionation of organic compounds and less applicable to inorganic compounds.

In general the choice of method to be employed will depend upon the value of the distribution ratio of the solute of interest, and the separation factor of interfering materials. Usually the last process is used for fractionation purposes.

9.9 SEPARATION OF METALS BY EXTRACTION

Solvent extraction finds extensive applications in analytical chemistry not merely for separation but also for quantitative analysis. The latter is specially true if one employs a chelating ligand as an extractant which in turn gives a coloured complex in the organic phase which can be spectrophotometrically measured e.g. the extractive photometric determination of uranium, iron, cerium or copper with 2-thenoyltrifluoroacetone. One can give numerous examples to demonstrate applications of extraction by chelation in analytical chemistry. The principal ligands used can be broadly classified as β-diketones, oxine and its substituted derivatives, oximes and dioximes; cupferron and analogous compounds, napthols and azonaphthols; dithizone and dithiocarbamates; dithol and xanthates and supra molecules. Amongst these acetylacetone, benzoylacetone, dibenzoylmethane, furoyl, thenoyl and thiothenoyltrifluoracetone have been extensively used for extractive separation and direct colorimetric determination of elements. 8-hydroxyquinoline, its dihalo derivatives, quinalidine and mercaptoquinoline find wide utility. Amongst oximes dimethylglyoxime, furildioxime, α-benzildioxime and cuporine are mainly used for extractive separation. Also, cupferron, N-phenyl N-benzoylhydroxyxylamine (BPHA) have been used for quantitative separations, 1-nitroso 2-naphthol, 1-(2-pyridylazo) naphthol (PAN), 4-(2-pyridylazo) resorcinol (PAR) find extensive use in extractive spectrophotometry of many metals. The sulphur containing ligands like dithizone, cupral, dithol and xanthates are being extensively utilised in the study of metal pollutants in water. Crown ethers cryptands and calixarenes are good extractants.

The next important class of extractant is extraction by solvation. There are several kinds of solvating solvents like ether, esters and neutral organophosphorus compounds which are principally used for the purpose of separations e.g. diethyl ether, ethyl acetate, methyl isobutyketone (MIBK), 4-methyl 3-pentane 2-one (i.e.) mesityloxide. These are all called oxygenated solvents. In addition, extractants like organosphosphorus acids, esters and oxides, tributylphosphate (TBP), tributylphosphine oxide (TBPO) and triotylphosphine oxide (TOPO) are extensively used for the quantitative extraction of transition elements and inner transition elements. The methods of separation are clear cut and can be used at microgram concentrations of metal ions. Many of the extractants (e.g. TBP) have even been used for separation on pilot plant scale. In the purex process, uranium is extracted on an industrial scale by using TBP as an extractant. The latest trend is to use these solvating solvents for reversed phase extraction chromatography as the stationary phase.

9.10 EXTRACTION BY ION PAIR FORMATION

The third large group of extractants is extraction by ion pair formation. Most of the high molecular weight amines or so called liquid ion exchangers belong to this group. The mechanism of extraction can be described as follows:

$$R_3N_{org} + H^+ + A^- \rightleftharpoons R_3NH^+A^-_{org} \text{ (extraction)}$$

$$R_3NH^+ A^-_{org} + B^- \rightleftharpoons R_3NH^+B + A^-_{org} \text{ (anion extraction)}$$

In such a system the best separations are possible with a good diluent. Emulsions can be eliminated with high molecular weight alcohols. The control of temperature and activity is most important in accomplishing quantitative separations. The structure and branching of the amine and the nature of the diluent affects the process of extraction. It has been found that tertiary amines are the best extractants for extraction of anionic complexes of metals with mineral acids, while the primary and secondary amines are good for the extraction of anionic complexes with organic carboxylic acids. The quarternary ammonium salts are more efficient than secondary amines. The principal amines used for such extractions are triocylamine (TOA), trisooctylamine (TIOA), Amberlite LA-1, LA-2 Primene JMT, Aliquat 336S, Arquad-6C, etc. A typical example of metal extraction is that of iron in chloride media. We have:

$$Fe^{3+}_{eq} + 4Cl^- \rightleftharpoons FeCl^-_4 \text{ (Complex formation)}$$

$$R_3N_{org} + H^+ + Cl^- \rightleftharpoons \left[R_3NH^+_{org}Cl^-_{org} \right] \text{ (Salt formation)}$$

$$\left[R_3NH^+Cl^-_{org} \right] + \left[FeCl^-_4 \right] \rightleftharpoons \left[R_3NH^+FeCl^-_{4\,org} \right] + Cl^- \text{ (Extraction)}$$

Thus it is possible to extract chloro, sulphato complexes of metal ions. One of the redeeming features of all these liquid anion exchangers is that they can be reused after regeneration like solid anion exchangers. Further, it is possible to have selectivity of separation by controlling various parameters such as pH for complexation, the variation in concentration of an extractant or the complexing ligand or by selective stripping. It is possible to achieve the most difficult separations in the organic acids like malonic, ascorbic, tartaric and citric acids. The use of a liquid cation exchanger can also be considered as an example of extraction by ion pair formation. Extractants like HDEHP *i.e.* bis (2-ethylhexyl) phosphoric acid and DNS, *i.e.*, dinonyl napthalene sulphonate, are two examples. Since the mechanism of extraction is similar to cation exchange, they are called liquid cation exchangers. The mechanism of extraction can be briefly depicted as:

$$M^m + m(HX)_{2\,org} \rightleftharpoons M(HX_2)_{m\,org} + mH^+$$

$$m(HX_2)_{m\,org} + CaCl_2 \rightleftharpoons M(CaX_2)_{org} + HCl$$

In such systems extraction varies with basicity and steric availability of the cation exchanger. The extractions with crown ether are also an example of ion pair formation.

Finally, let us consider synergic solvent extraction. Such an extraction involves the use of two extractants. The extractants may be a chelating ligand and solvating solvent or may comprise both chelating agents or two solvating solvents. One can use, e.g. dialkyl phosphoric acid in conjunction with neutral organophosphorus esters. The optimum conditions for such extraction are that (HX) the chelating ligand should neutralise the metal charge by chelating; the solvent (S) should coordinate less strongly than the chelating agent; the maximum coordination number of the metal and the geometry of the molecule should also be favourable. In such extractions, chelating extractants like HTTA, IPT and solvating extractants like S = TBP, TBPO, MIBK are used. The synergism in such a combination increases with the basicity of the solvent in the order TOPO < TBP < TPP etc. Chelating ligands like

TTA show the largest tendency to form an addict with neutral ligands. The mechanism of such an extraction can be explained as:

$$M_{aq}^{n+} + nHR_{org} \rightleftharpoons MR_{n\,org} + nH^+$$

$$M R_{n\,org} + nB_{org} \rightleftharpoons M R_n B_{n\,org} \,(Adduct)$$

$$\beta_n = \frac{[MR_n B_n]_{org}}{[MR_n]_{org}[MR_n B]_{org}^n}$$ is called the "stabilization factor" or adduct formation constant (β_n).

Another example of synergic extraction is the use of TBP + HDEHP where the effectiveness of extraction depends upon basicity of $\rightarrow$ P = O group and it increases in the order

$$(RO)_3 PO < R(RO)_2 PO < R_2 (RO)PO < R_3PO$$

Phosphate < phosphonate < phosphinate < phosphine oxide

Before concluding this chapter let us consider what is meant by antisynergism. This is the reverse of the above phenomenon. It was noted in TBP-TTA or DBP-TBP systems that extractions decreased due to the addition of excess solvent(s) as the latter reduces the concentration of free chelating agent by increasing interaction between HX and S through hydrogen bonding.

Table 9.1 briefly illustrates four extraction systems with optimum conditions of extraction.

Table 9.1 Typical solvent extraction systems — applications

Sr. No.	Class	Extractant	Metal	pH/(H⁺)	Separated From	Solvent
1.	Chelate extraction	Thenoyltrifluoro acetone (TTA)	U (VI)	3.4-8.0	Th Ce	Benzene
2.	Chelate	Dimethyl-glyoxime	Ni (II)	7.3-8.4	Co	Chloroform
3.	Chelate	Cupferron	Fe (III)	(1 + 9M) H_2SO_4	Cu	Chloroform
4.	Chelate	Dithizone	Zn (II)	4.0-11.0	Cd Cu	Chloroform
5.	Solvation	TBP	Ga	6M HCl	Fe Interfere	Kerosene
6.	Solvation	TOPO	Ti (IV)	8 M HCl		Xylene
7.	Solvation	HDEHP	Ag (I)	0.1 M HCl		Toluene
8.	Ion pair formation	MDOA	Cu (II)	7 M HCl		Chloroform
9.	Ion pair	TOA	Pt (IV)	2M HCl	Rb, Ru	Xylene
10.	Ion pair	Aliquot 336	Be	pH 5.0	Oxalate media	Xylene
11.	Ion pair	Amberlite LA-I	Zr (IV)	pH 3.0	Ti Th Zr	Xylene
12.	Synergic extraction	TTA + TBP	U (VI)	pH 4.5	Several metals	Benzene
13.	Crown cyptand system	Cyptand 222	Pb (II)	pH 4.0	Eosin counter anion	Dichloro ethane
14.	Crown ether	DC18C6	Rb	pH 6.0	Picrate counteranion	Dichloro-methane

In conclusion one can say that no other method of separation is as versatile as solvent extraction. Therefore it finds extensive application in analytical chemistry for quantitative separation and determination and also in inorganic chemistry at the preparative level.

9.11 APPLICATION OF SOLVENT EXTRACTION IN INDUSTRY

In the preceeding sections we have considered a brief review of applications of solvent extraction. Here we propose to consider specific illustrations with optimum conditions of extraction. They are listed in Table 9.1. The kind of extractant, metal extractions, pH of extraction, solvent used and interferences are best summarised in this table.

Several systems involving macrocyclic and macrobicyclic polyethers and supramolecular compounds are discussed in the next chapter.

9.12 SOLID-PHASE EXTRACTION (SPE)

Solvent extraction techniques are applicable provided one of the liquids is immiscible with the other. Further, any system causing emulsion formation is a hindrance during quantitative analysis. Such problems can be partially mitigated by use of solid phase extraction. Hydrophobic organic functional groups are chemically bonded to a solid surface like silica. These groups react with hydrophobic organic compounds through van der Waal's forces and extract metals from an aqueous phase in contact with a solid. Usually the powder phase is put in a syringe and it is forced out by a plunger. Thus compounds are extracted and separated from the sample. They are subsequently eluted with solvents like methanol and analysed. Various extractants can be used for chemical bonding. On bonding stationary supports become liphophobic. On washing with water soluble methanol, such a phase permits the aqueous phase to diffuse into the bonded phase. Polymer based supports which are stable are also used. Solid phase extractants are made in the form of filter disks of PTFE of (*i.e.*, polytetrafluoroethylene). however, the system has not yet become popular. Reversed phase partition chromatography, also called extraction chromatography, differs from solid phase extraction in many ways. In the former the stationary support is made hydrophobic by exposing it to vapours of diethy-dicholorosilane followed by coating with a suitable extractant like HDEHP, TBP or TOPO. It is packed into a column and operated just like column chromatography. Elution or stripping is usually carried out with mineral acids such as hydrochloric or nitric acids. The support consists largely of silica but other inert supports like PTFE are also used. The principle merit of such separation is the use of a small volume of eluant, say, 20-40 ml and multielement separations. As many as four to six metals can easily be separated. The process is rapid as the time required for separation and analysis is just 20-30 minutes. This is followed by analysis. The details are spelled out elsewhere. With respect to speed this technique is comparable to ion chromatography.

Uses: This technique reduces the interferences for lower sample (final) volume, increases sensitivity and provide compatible sample for instrumental analysis. SPE also removes sample particulates. SPE works also on the principle of differential migration, where mobile phases elute the analyt. So the main purpose of using SPE is to (*i*) remove interferences using cartridge (*ii*) analyt concentration is conditioned, (*iii*) provides phase exchange from emulsions to SPE cartridge (*iv*) promotes solid phase derivatisation and facilitates (*v*) simple storage and transport.

In practice SPE is used to isolate proteins from fluids, enrichment of polynuclear aromatics, soil gas analysis and organics in water are analysed. SPE uses small plastic disposable cartridge (e.g. barrel of syringe) packed with sorbent (<1 gm). The packing is contained in barrel by frits. SPE gives reduced thermal plates due to short length. Cost and non-efficiency is of great value. SPE disk are also used if volume is large with disk of 0.4-9.6 cms diameter coated with sorbent material is best. Faster flow ensures rapid extraction.

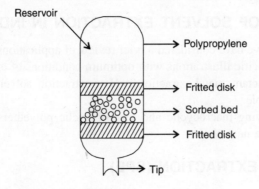

Fig. 9.11 Solid phase extractor cartridge

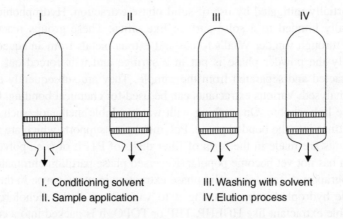

I. Conditioning solvent   III. Washing with solvent
II. Sample application   IV. Elution process

Fig. 9.12 Steps of operation of SPE

A solid phase microextraction (SPME) uses fused silica fibre coated with a stationary phase. Fiber is immersed in analyt. The organics diffuse partition into polymeric coating as function of D (disi. ratio). Convection current accelerates process of partition. Convection process is the most important. The stages of operation are: conditioning of solvent followed by sample application with wash with solvent and finally the process of elution (See Fig. 9.12). SPE not only isolates the analyst but it also concentrates the analyst into smaller volume. Syringe is used to push or pull the sample to cartridge as shown in Fig. 9.11. The solid phase contains bonded silica or special material. Florisil or alumina are also used. Retention data determines loading and elution quantum.

PROBLEMS

9.1 In solvent extraction of uranium with 8-hydrocxyquinoline in chloroform the volumes of the aqueous and organic phase were 25 ml when the percentage extraction was 99.8%. Calculate the distribution ratio.

Answer: $V_0 = 25$ ml, $V_W = 25$ ml, $E = 99.8$

$$D = \frac{\left(\dfrac{V_W}{V_0}\right)E}{(100-E)} = \frac{\left(\dfrac{25}{25}\right)99.8}{100-99.8} = \frac{99.8}{0.2} = 499$$

9.2 In the extraction of cerium (IV) with 2-thenoyltrifluoroacetone in benzene the distribution ratio was 999.0. If the volume of the organic phase was 10 ml and that of the aqueous phase was 25 ml, what was the percentage extraction?

Answer: Since $D = 999$, $V_0 = 10$ ml, $V_W = 25$ ml. Assume $E = x$ is percentage of extraction

$$D = \frac{(V_W/V_0)}{(100-E)} = 999 = \frac{\dfrac{25}{10}x}{(100-x)} = \frac{2.5x}{(100-x)}$$

Solving for x we have $x = 99.75\%$.

9.3 A certain extraction system has a distribution ratio of 10. If 300 mg of solute is dissolved in 100 ml of solvent, A, find the amount of solute extracted by one extraction with 100 ml of solvent B and by two extractions with 50 ml of the solvent B which is immiscible with solvent A, if the solute has a mol. wt. of 71.0.

Answer: Moles of solute present $\dfrac{0.30}{71} = 4.22 \times 10^{-3}$

If x = moles extracted then $\left(4.22 \times 10^{-3} - x\right)$ is unextracted

$$D = 10 = \frac{\dfrac{x/100}{(4.22 \times 10^{-3} - x)}}{100} = 3.84 \times 10^{-3} \text{ moles of solute}$$

$$\% \text{ of solute extracted} = \frac{3.84 \times 10^{-3}}{4.22 \times 10^{-3}} \times 100 = 91.0\%$$

Now if x = moles extracted in first 50 ml solvent B
$\quad\quad y$ = moles of extracted in second 50 ml solvent B

$$D = 10 = \frac{\dfrac{x/50}{(4.22 \times 10^{-3}) - x}}{100} \quad \because i.e, \; x = 3.52 \times 10^{-3} \text{ solute } i.e. \; 0.70 \times 10^{-3} \text{ remains in solvent A}$$

$$D = 10 \; \frac{\dfrac{y/50}{(0.70 \times 10^{-3}) - y}}{100} \quad \text{as } y = 0.58 \times 10^{-3} \text{moles}$$

Total moles extracted 4.10×10^{-3}

$$E = \frac{4.10 \times 10^{-3} \times 100}{4.22 \times 10^{-3}} = 97.2\% \text{ extracted.}$$

9.4 Calculate the distribution ratio (D) when iron (III) is extracted from hydrochloric acid with tributyl phosphate if V_0 = 10 ml, V_W = 25 ml, E = 99.8.

9.5 What will be the percentage extraction of vanadium with mesityl oxide if volume of organic phase is 10 ml, the volume of aqueous phase is 10 ml and distribution ratio is 1247?

9.6 If after one extraction 90% of zinc is extracted with trioctylamine in the xylene phase and only 5% of cadmium is in the xylene phase, calculate the separation factor $\left(\alpha_{Cd}^{Zn}\right)$ assuming the volume ratio was unity and volume ratio was 1/2.

9.7 When nitric acid solution of uranylnitrate is equilibrated with twice its volume of ethylacetate saturated with HNO_3, 99% uranylnitrate is extracted. Calculate the distribution ratio of uranylnitrate between nitric acid and nitric acid saturated ethylacetate.

9.8 What is the minimum value of (D) which would permit extraction of 99.9% of iodine from 50 ml of aqueous phase with five successive 50 ml of portions of carbon tetrachloride?

9.9 If D = 100 for extraction of cerium (IV) from HCl by MIBK, calculate the percentage extraction by 1, 2, 3 extractions of 100 ml of Ce (IV)-HCl solution by 30 ml of MIBK.

9.10 A 100 ml volume being 0.5 M with respect to $FeCl_3$ and 5 M with respect to HCl is shaken with 10 ml of ethylacetate when the distribution ratio is 17.6. Determine the degree of extraction and decrease in the concentration in acid phase after one extraction.

LITERATURE

1. A.K. De, *"Separation of Heavy Metals"*, Pergamon Press, London (1961).
2. G.H. Morrison and H. Freiser, *"Solvent Extraction in Analytical Chemistry"*, John Wiley, New York (1958).
3. J. Starry, *"Solvent Extraction of Metal Chelates"*, Pergamon Press Lts. (1964).
4. Y. Marcus and A.S. Kertes, *"Ion Exchange and Solvent Extraction of Metal Complexes,"* Wiley Interscience (1969).
5. J.Korkish, *"Modern Methods for the Separation of Rare Metal Ions"*, Pergamon Press (1969).
6. A.K. De, S.M. Khopkar and R.A. Chalmers, *"Solvent Extraction of Metals"*, Van Nostrand Reinhold (1970).
7. Yu. V. Zolotov, *"Extraction of Metal Compounds"*, Ann Arbor Publication (1972).
8. T. Braun and G. Ghersini, *"Extraction Chromatography"*, Elsevier Publishing Co., (1975).
9. M.S. Cresser, *"Solvent Extraction in Flame Spectroscopy Analysis"*, Butterworth (1978).
10. T. Sekine and Y. Hasegawa, *"Solvent Extraction Chemistry"*, Marcel Dekker and Co. (1980).
11. S.M. Khopkar, *"Analytical chemistry of macrocyclic and Supramolecular compounds,"* 2nd Ed. Springer Verlag (2005).
12. S.M. Khopkar, *"Solvent extraction Separations with Liquid ion exchangers"*, New Age International Pub. Ltd. (2007).

Chapter 10

Supramolecules in Solvent Extraction

In the preceding chapter we had considered the theory and applications of solvent extraction for the separation of elements. One distinct anomaly was noted in such procedures. There were innumerable methods for solvent extraction of d-block (transition) elements or for p-block (main group) elements. Unfortunately scant attention had been paid to the problem of solvent extraction of s-block (alkali and alkaline earth) elements. This has been primarily due to a dearth of suitable extractants. Fortunately, this problem has been simplified with discovery of supramolecular compounds. The task of extracting main group elements and specially toxic metals in the atmosphere was eased with their discovery. There were exciting inventions in the field of supramolecular compounds which largely included calix (n) arenes and rotaxanes. This chapter presents an account of applications of these compounds in solvent extraction separation of elements.

10.1 ORIGIN OF MACROCYCLIC AND SUPRAMOLECULAR COMPOUNDS

In 1967 Predersen discovered the first synthesis of DBI8C6. He was studying the reaction between dichlroethylether and catechol and discovered a new product along with bisphenol which reacted readily with sodium or potassium. Subsequently, in 1968, Lehn discovered macrobicyclic compounds called the cryptands. Cryptand 1, 1, 1, and Cryptand 2, 1, 1 were first synthesised while Guitche developed the synthesis of calixarenes or calixresorcinol by base catalysed or acid-catalysed reactions between p-tertiary butylphenol and formaldehyde or interaction of resorcinol with acetaldehyde to yield calix (6) arene or calix (6) resorcinarene resp. Rotaxane is recent discovery.

10.2 NOMENCLATURE OF COMPOUNDS

All these compounds were cyclic in nature and it was difficult to name them using I.U.P.A.C. nomenclature.

In structure (1) we have a cyclic structure with 6-oxygen atoms and 18-carbon atoms. It is, therefore, designated as 18C6; if it has a substituent like DB on the ring then it is called DB18C6 or DC18C6. If we were to write its IUPAC nomenclature it would be quite cumbersome. Since they mainly contain oxygen carbon linkages they are called macrocyclic polyethers. In structure (II) we have a cyclic structure with a bridge between two nitrogen atoms via oxygen atoms. It is like a crown ether with the replacement of two oxygen atoms by two nitrogen atoms. Since they are bicyclic in nature with carbon-oxygen and carbon-nitrogen links they are termed macrobicyclic polyethers. We have a bridge between three pairs of two oxygen atoms all along hence we call it cryptand 222. If there was one bridge we would have designated it as cryptand (III). Finally in structure (IV) we have a supramolecular (IV) 2, 2, 1 compound containing 4-phenyl rings with 4 hydroxyl groups. The phenylring has tertiary butyl as p-substituent. We therefore call this compound p-tertiary butyl calix (4) arene. If there were 6 or 8 phenyl rings we would designate them calix (6) arene or calix (8) arene. The number of phenyl rings are denoted in brackets. For *n*-rings we write calix (*n*) arenes.

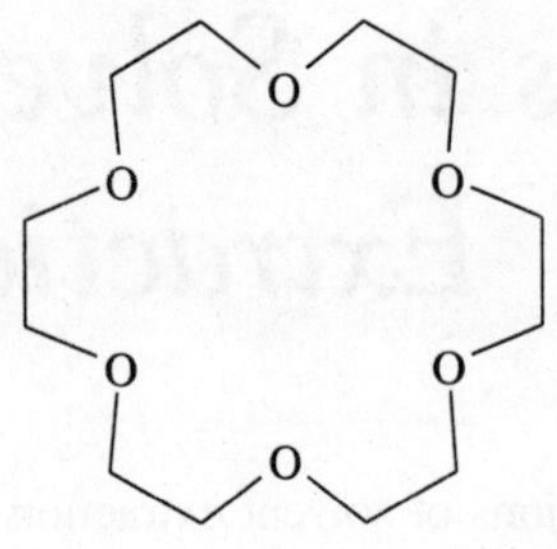

18 C6
(1)

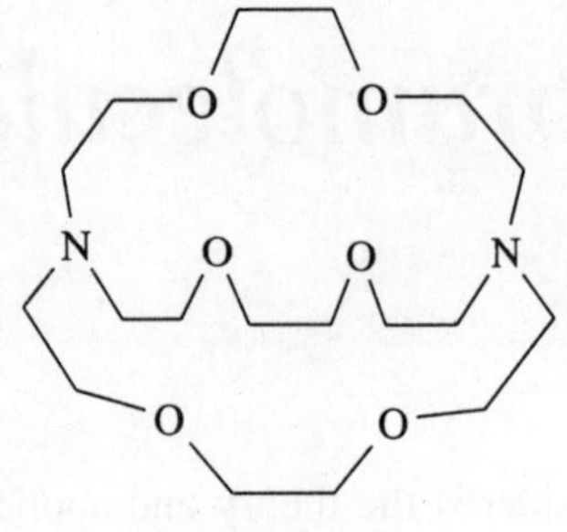

Cryptand 2,2,2
(II)

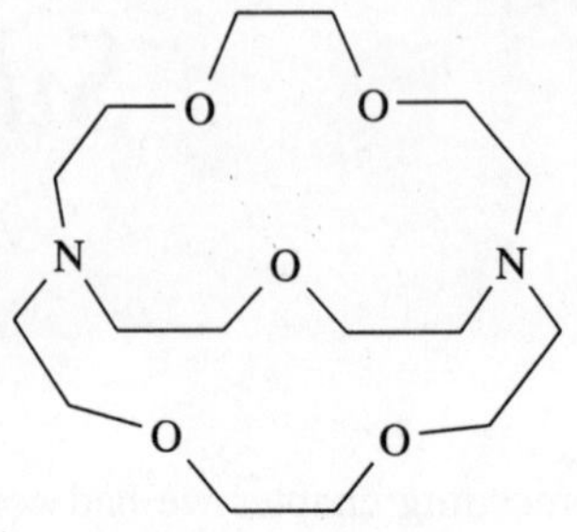

Cryptand 2,2,1
(III)

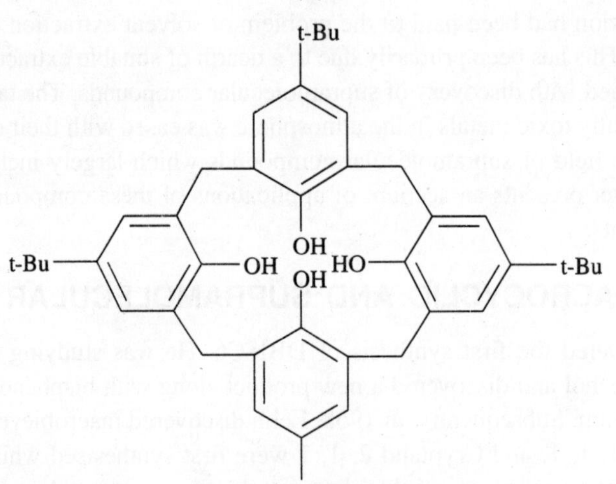

Calix (4) arene (p-t-butyl)
(IV)

Fig. 10.1 Crown, cryptands calix (*n*) arene

10.3 CLASSIFICATION

They are grouped into macrocyclic, macrobicyclic and supramolecular compounds on the basis of their cyclic structure. Most of the crown ethers which are cyclic in nature are termed coronands. We have a group of crown ethers, known as podands, with an open chain: they are useless for solvent extraction. Compounds which are spherical in nature with a bridged atom along with a donor atom like nitrogen are termed as cryptands. Those compounds which are not only three dimensional cyclic structures but also have large molecular weights are called supramolecular compounds. Apart from these principal groups of compounds we have compounds containing sulphur as a donor atom called thia crown and nitrogen and sulphur macro cyclic polyethers called aza crown ethers one with armed chains called lariat crown ethers.

10.4 SYNTHESIS OF CROWN ETHERS, CRYPTANDS AND CALIXARENES

Pedersen, Lehn and Cram were awarded the Nobel prize in 1989 for their discovery of macrocyclic and macrobicyclic polyethers. The discovery in fact was accidental. An attempt is made to describe the synthesis of these three kinds of compounds. The synthesis of DB18C6 envisaged by Pederson is as follows:

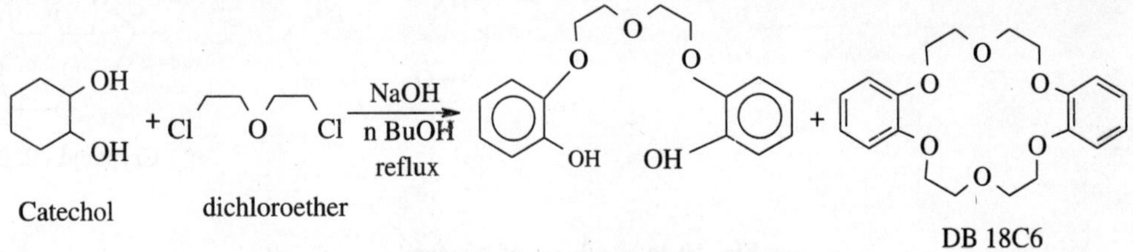

Fig. 10.2 Synthesis of DB 18 crown 6

Thus, catechol when reacted with diethyldichloroether in the presence of butanol and sodium hydroxide yielded, in addition to bisphenol, a new product called DB18C6. It has the characteristic property of reacting readily with alkali metals.

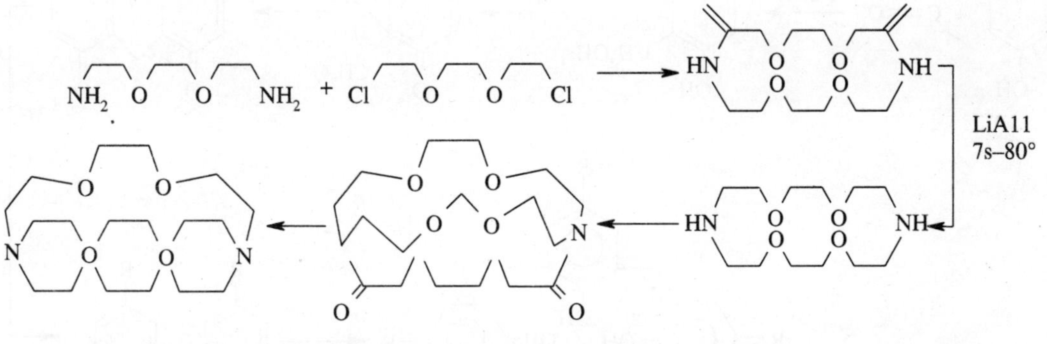

Fig. 10.3 Synthesis of cryptand 2, 2, 2

The above reaction shows a synthesis of cryptand which involves several stages before the final (Fig. 10.3) or Fig. 10.3(*a*) product is obtained. The synthesis of supramolecular compounds like calixarenes is interesting. (Fig. 10.4) They are obtained by base catalysed (*i.e.* calix (n) arene) or acid catalysed reactions *i.e.* calix (n) resorcinarene. In both cases, the starting material is p-tertiary butyl

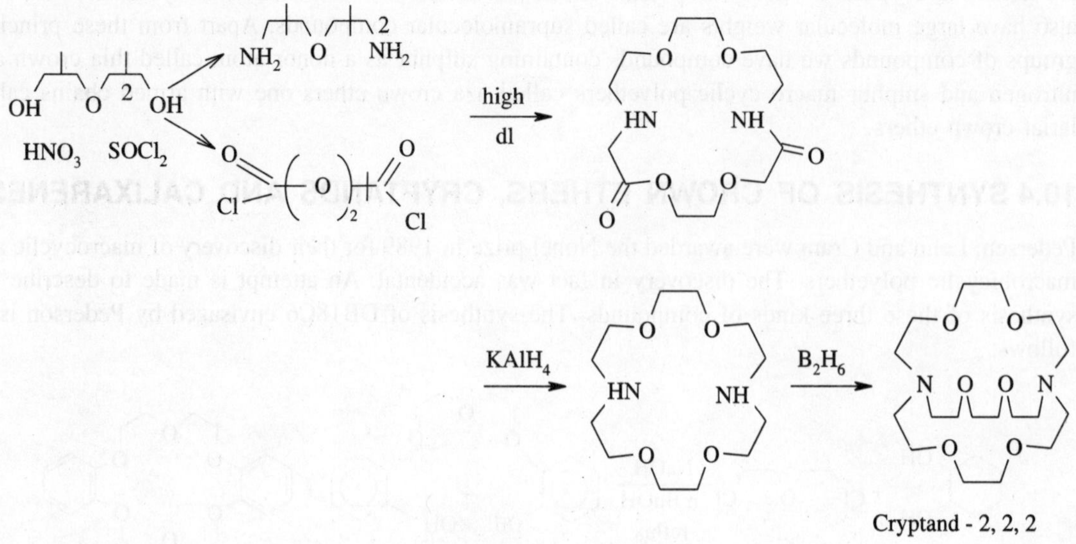

Fig. 10.3a Synthesis of cryptand

phenol but in one case it is condensed with formaldehyde or paraformaldehyde, while in the other case we have the interaction of resorcinol with aceteldehyde. The various steps involved in such synthesis of calix (n) arene or calix (n) resorcinorene are summarised in Fig. 10.4.

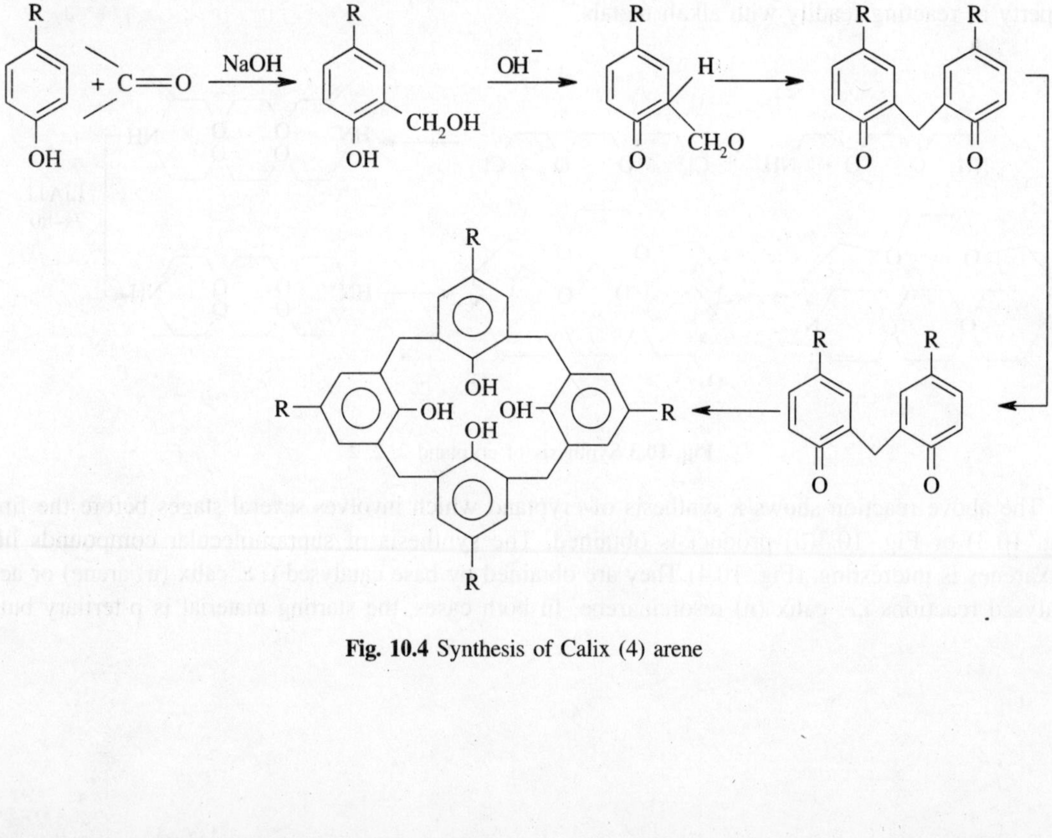

Fig. 10.4 Synthesis of Calix (4) arene

Calix(n) resorcinarene is obtained as per reaction in Fig. 10.5.

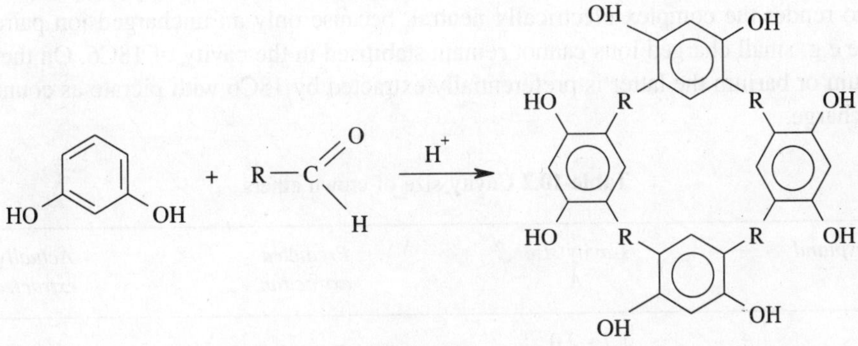

Fig. 10.5 Synthesis of calix (4) resorcinarene

The most important part is cyclisation. All these compounds have been characterised by determination of melting point, solubility considerations, C, H, N, analysis and spectral characterisation by UV, Visible, IR, NMR and mass spectrometric methods. An extensive study has been made of their complexing characteristics and the stability of metal complexes.

10.5 FACTORS INFLUENCING SOLVENT EXTRACTION

Various parameters such as size and charge of cation, cavity of crown ether and cryptands, nature of crown ether, type of counter anion, kind of diluent and temperature, greatly influence the extraction process. Such factors are valid whether it is extraction of metals by crown ether or cryptand or calix (n) arenes or calix (n) resorcinarene.

(a) *Size and charge of cation:* The size of cation is determined by the cation radius while the cation charge is decided by the number of protons and electrons.

Table 10.1 Size of cations

Cation	Ionic Radii Å	Ionic diameter Å
Li^+	0.60	1.48
Na^+	0.95	1.99
K^+	1.33	2.66
UO_2^{2+}	0.83	1.66
Cs^+	1.69	3.44
Ce^{2+}	1.05	2.10
Sr^{2+}	1.11	2.32
Ba^{2+}	1.35	2.72

Size is of utmost importance. If the size of the cation matches with that of the cavity of the crown ether, the metal gets embedded in cavity and easily gets extracted in the presence of the counter anion. The cavity offers great electrostatic stability to the complex. The size of a few cations is listed in Table 10.1 while cavity size of selected crown ethers are listed in Table 10.2.

Since a cavity size of 15C5 viz. 1.7 – 2 Å does not match the ionic size of caesium viz. 3.44 Å so it does not get extracted by it. Depending upon the charge on the metal-crown complex, a counter ion is attached to render the complex electrically neutral, because only an uncharged ion paired complex is extractable e.g. small charged ions cannot remain stabilised in the cavity of 18C6. On the other hand with potassium or barium the latter is preferentially extracted by 18C6 with picrate as counter ion as it has a large charge.

Table 10.2 Cavity size of crown ethers

Crown/cryptand	Cavity size Å	Predicted extraction	Actually extracted
15C5	1.7 - 2.0	–	–
18C6	2.6 – 3.2	K, Na, Sr, K	
DB18C6	2.6 – 3.2	Ba, Pb, Na	Na
DB24C8	4.0	Cs, Rb	Cs, Ba
Cryptand 222	2.8	Pb, Tl	Na
Cryptand 221	2.2	Ni, Cu	–
Cryptand 211	1.6	–	

(b) *Cavity and nature of crown ether and cryptands:* In Table 10.2 we have noted the cavity of crown ethers or the cryptands. It is only synchronisation or matching of hole or cavity of crown ether of metal with atomic size which promotes complex formation. As regards the nature of crown/cryptand a significant role is played by the donor atom. Oxygen in crown, sulphur in thia crown and nitrogen in cryptands are donor atoms which influence complexation. A large size donor like sulphur favours complexation with metal, as it increases cation-dipole interaction. Substituents on the benzene ring like DC18C6, DB24C8 influence their solubility in organic solvents. The substituted compounds do not offer better extraction.

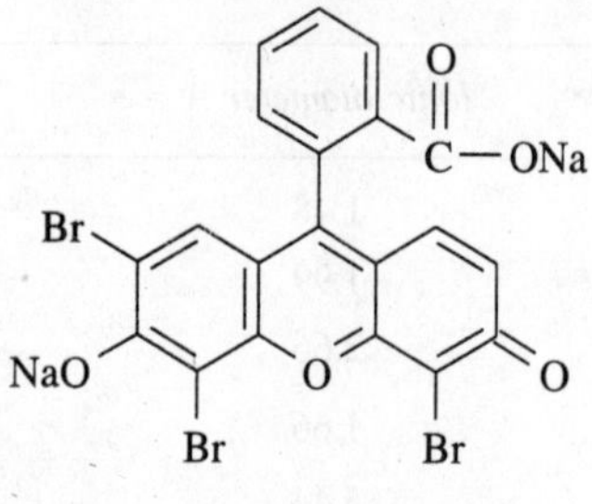

Eosine (I) 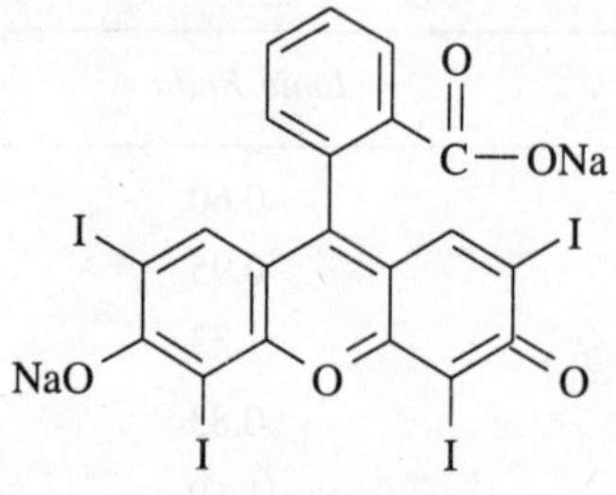   Eryothrosine (II) 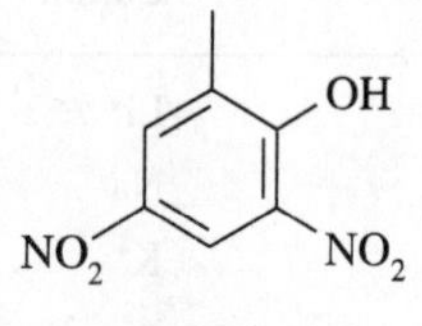   Picric acid (IV)

Fig. 10.6 Counter anions

(c) *Nature of counter anion:* All extractions, especially with crown ethers and cryptands, are governed by the nature of the counter anion. In normal extractions, like copper with 1–10 phenonthroline, we use perchlorate as a counter ion. In extraction of iron with ether we use chloride as a counter ion but in the case of extraction of metals with crown ethers we need a large size counter anion. This is because the crown metal complex or cryptand metal complex is so large that it needs far neutralisation of charges by large anion. The commonly used counter ions are picric acid, metanil yellow, rose bengal, eosine, fluoresceine, erythrosine, bromothymol blue and tropeoline. However, the most commonly used counterions, especially for extractions with cryptands, are eosine (I), erythrosine (II), while extractions of alkali metals and alkaline earths give best results with picric acid (III) as the counter anion, as shown in Fig. 10.6.

Planar counter anions are preferable for extraction. Long chain compounds like HDEHP are good for extractions with cryptands. Selectivity is substantially changed with change of counter-anion. The stereochemistry of the counter anion is the most important aspect. Organophilicity is also important. A high molar volume of counter anion favours extraction. A bulky counter anion with anionic metal complex can also serve as a counter anion e.g. SbI_4^- acts as a counter anion for extraction with K18C6 cation. Their concentration is also important e.g. 1×10^{-3} M erythrosine effective for extraction of thallium with cryptand 222.

(*d*) *Diluent used:* In usual solvent extraction procedures solvents with a low dielectric constant *i.e.* nonpolar solvents are usually preferred as they promote association of cation and anionic species to form an extractable neutral complex. On the other hand in the extraction of metals with calixarene or crown ether, solvents with a high dielectric constant offer better results. This is because of better solubility of crown ether and the presence of strong interaction between oxygen and hydrogen atoms of the diluent. Protic solvents offer better interaction between oxygen and hydrogen atoms of the diluent as protic solvents solvate anions. Extractability is influenced by the Gibbs free energy of hydration: if it is low it permits ion pair formation of the complex due to transfer of anions from water and solvent. Finally, since most extractions are carried out at room temperature, dehydration causes difference in entropy with no change in the enthalpy of the system.

(*e*) *Extraction of complexes by crown, cryptands and calix (n) arenes:* A large molar volume of complex favours better extraction.

$$D = K_{ex}[A^-]^n[R]_{org}$$

This shows more extraction is possible with greater concentration of crown ether or cryptand (**R**) and is also proportional to concentration of counter anion (A^-). The magnitude of K_{ex} will increase if the concentration of complex in the organic phase viz. $[M\,R\,A_n]_{org}$ is also increased.

10.6 SOLVENT EXTRACTION WITH CROWN ETHERS AND CRYPTANDS

It is worthwhile comparing solvent extraction of metals with these two versatile extractants.

Crown ethers like 18C6, DB18C6, DC18C6 or DB24C8 have been extensively used for solvent extraction of not only alkali and alkaline earths but also transition metals like zinc, uranium, molybdenum and main group metals. The counter ions used are picric acid for alkali and alkaline earths while many transition metals can be extracted from mineral acid media. Anionic complexes such as iodo or halo or thiocyanate complexes of main group elements have been used for solvent extraction in conjunction with cationic crown ethers embedded with potassium ion. Most extractions have been carried out with solvents like dichloroethane or nitrobenzene. The stereochemistry is simpler as complexes are planar.

Cryptands can extract metals like thallium, lead, nickel, copper, uranium, cobalt, etc. with nonpolar solvents like xylene and toluene. The counter anion used in most cases is eosine, erythrosine or rose-bengal. These complexes are three dimensional structures with better selectivity. A significant feature being their application for separation of metal pollutants from the aquatic environment.

Finally calix (n) arene is a special class of compound. Their striking feature is that they hardly need counter anions, as their derivatives like acetato, nitro, sulphato or keto are directly used for the purpose of complexation. Metals like, palladium, cobalt have been extracted by a hexacetato derivative of calix (6) arene from alkaline media and the metal could easily be stripped with acid. Table 10.3 summarises some significant extractions.

Table 10.3 Solvent extraction with crown ethers and analogues compounds

Sr. No.	Extractant (M)	Metal	pH	Counter anion	Diluent	Stripping agent
1.	18C6 (0.01)	Rb	3-7	Picric acid	Dichloro ethane	2M HNO$_3$
2.	DB24C8 (0.02M)	Cs	2.10	Picrate	DEDC	3M HC10$_4$
3.	18C6 (0.01M)	Ca	3.0-9.0	Picric	DMDC	1M.CH$_3$COOH
4.	DB18C8 (0.01M)	Ba	2-9	Picric	CHCl$_3$	1M HNO$_3$
5.	18C6 (0.02M)	Ga	6MHCl	Chloride	DMDC	0.1 M HCl
6.	Cryptand 222 (0.01M)	Tl	4.0-10.0	Erythrosine	CHCl$_3$	0.1 M H$_2$SO$_4$
7.	Cryptand 222B (0.01M)	Pb	4.5-10.5	Eosine	Toluene	1M HCl
8.	DC18C6 (0.04M)	Sc	2.4-4.5	Picric	DMDC	0.1M HNO$_3$
9.	DC18C6 (0.02M)	Nb	6-8 MHCl	Chloride	DEDC	0.1M H$_2$SO$_4$
10.	Calix (6) arene (0.001M)	Co	7.0-8.0	–	DMDC	1M HNO$_3$

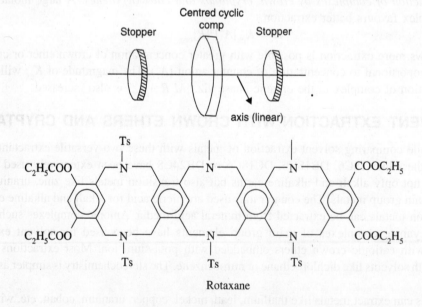

Fig. 10.7 Geometry of Rotaxane

10.7 ROTAXANES

The discovery of Rotaxanes is of recent origin (1995). Lindey synthesised it for the first time by extending template effect. In such compounds a cyclic molecule (called as wheel) is threaded with bulky groups or stoppers on an axle to lock the wheel in a place. A cyclic molecule is capped at both the ends of axle by stopper or caps which are bulky in nature. The Rotaxanes were compounds with new ribbons and thread containing hydrogen bond, dipole-dipole interaction and stacks of π-electrons. In 1979 Bartsch obtained Rotaxane by oligomerisation of ethylene oxide with dipotassium tetraethylene glycol with 30-50 member crown ether as central cyclic compound with termination of triphenyl anions membrane as the linear component acting as the axle.

These are molecular structures embodying non covalent interaction of component which have ordered assemblies. They were characterised. They substantially differ from those of parent compounds in their mixtures. They have noticeable solubility in solvents. Their spectra by IR or NMR does not vary from the mixtures of the constituents. They have very high molecular weights.

10.8 CLASSIFICATION OF ROTAXANES

These are compounds containing cyclic molecules (as wheel) pierced in centre by linear components (as axle) with no covalents bond between these two entities. We have three kinds of rotaxanes as are simple rotaxane, semirotaxane and pseudo-rotaxane as shown in (a), (b) and (c) in Fig. 10.8.

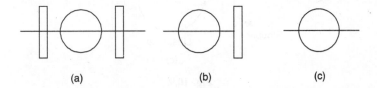

(a)   (b)   (c)

Fig. 10.8 (a) Rotaxane, (b) Semirotaxane, (c) Pseudorotaxane.

The compounds (b, c) are inclusion compounds capable of undergoing dissociation. Rotaxanes are also classified as homo and hetero rotaxanes. In former the cyclic linear structure components are same while in heteroform they are different. The so called 'Polyrotaxanes' are made out of macrocycles (wheel) threaded by linear polymer (axle) molecule without any chemical bonding between cyclic and linear components. In clipping process (b) (Fig. 10.9) there is direct reaction between dumbell shaped component and linear part and latter is clipped around dumbell shaped component. In threading mechanism (c) is most common. In the synthesis the interaction between cyclic and linear component proceeds by covalent bonding of two stoppers to prevent further threading and slipping. Thus for self assembly of rotaxane the slipping macrocycle over stopper with the covalent bond is essential. In nomenclature of rotaxanes first total number of groups are mentioned then linear component, followed by cyclic part and finally stopper is mentioned in structure.

10.9 SYNTHESIS OF ROTAXANES

For the process of self assembly of rotaxane the slipping of the macrocycle over stopper (with covalent bond) is essential. There is a reaction between molecular thread compounds like three and five π-electron rich aromatic component and also π-electron deficient tetra cationic cyclophane to give pseudorotaxanes (Fig. 10.10 to 10.12). A template directed synthesis provides better yield. In the

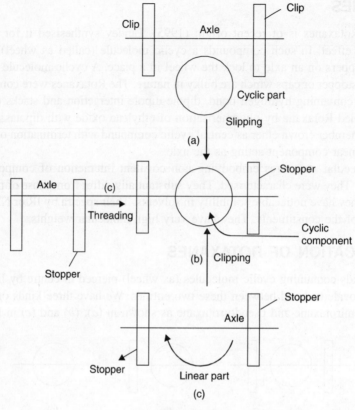

Fig. 10.9

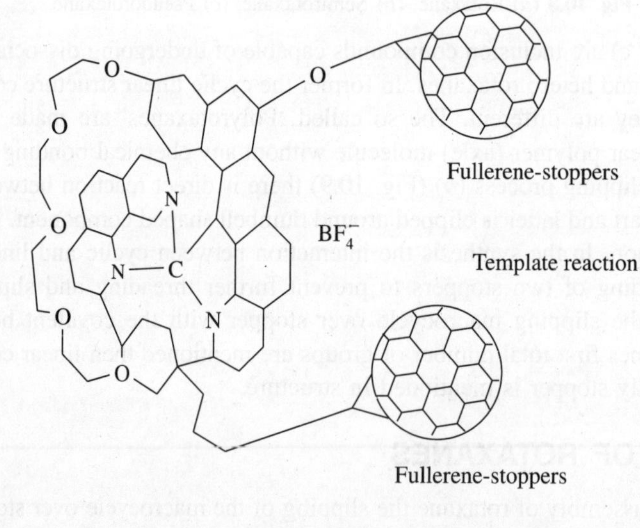

Fullerene-stoppers

BF$_4^-$

Template reaction

Fullerene-stoppers

Fig. 10.10 Rotaxane

slipping process cyclic component (wheel) and dumb shaped linear part (axle) are prepared separately and then associated with each another at latter stage. The ring size is critical in the synthesis of polyrotaxanes. For better results cavity size of macrocycles such as crown ethers *i.e.* 30 C10 should be bigger than that of the linear component to facilitate threading. In order to ensure polymerisation cyclic component as rule must possess 22 members in the ring to facilitate threading. There should be no chemical reaction between cyclic and linear part to arrest process of polymerisation. There should be donor centre on macrocycle. The details of synthesis are described in Table 10.3. The information must cover a kind of central component (wheel) threaded on linear part (axle) and include stoppers used to prevent slipping of bulky central molecule from the axle. The characteristics and synthesis are coupled together.

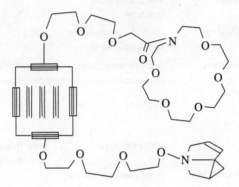

Fig. 10.11 L27 Rotaxane

Table 10.3 Synthesis of Rotaxanes

Type	Cyclic component (Wheel)	Linear component (Axle)	Stoppers for capping	Rotaxane formed
A. Rotaxane:				
	1-10 phenanthroline α, β-cyclodextrin cyclophane	Esters, polymers diphenyl phenanthroline π-electron deficient pyridium.	Porphorine dendritic part	[2] Rotaxane [3] Rotaxane [4] Rotaxane
	Hydroquinone DB24C8 34C10	polyethylene glycol. hexasiloxanes Dialkylamino hexa fluoro phosphate	Tetrabenzoyl trisisopropylene Glucose	
	π-Electron rich hydroquinone.	polyethylene glycol.	Fullerenes	Polyrotaxanes.
B-Polyrotaxane:				
	α, β cyclodextrin Crown ethers like 60C20, 36C12, 32C10, 30C10, 42C14.	Ethylene glycol Polyurethane Triglyco hexamethylenedisocyanate	Sebacoyl chloride 2, 4 dinitrofluoro benzene.	

(Contd.)

Type	Cyclic component (Wheel)	Linear component (Axle)	Stoppers for capping	Rotaxane formed
	α-cyclo dextrin	Polyethylene bromide	Large stoppers	
	Cyclobisparaquat p-phenylene	Polyethylene glycol	Adamantoyl	
C-Substituted Rotaxanes:				
	Cyclophane	1-4 Substituted benzene	–	Polyamide rotaxanes
	β-cyclodextrin	methyl crylolyamine deaniconic acid	30C11	
	α-cyclodextrin 3OC10	1-2 dibromodecane decaediol-sebacoyl chloride/difunctional group.	phenolable β-12	
d-Metal Complexes (Sodium, cobalt, palledium, copper, zinc)				
	Na	α-cyclodextrin	poly (phenylene benzimedzole) 15C5	–
Co	α, β cyclodextrin	α, ω diamino alkane	–	
Pd	DB24C8	1, 2 bis (4, 4' dipyridinum)	Tetrabutyl benzyl	H-bonding π-π stacking
Cu	Cycloposphorene	–	Fullerene	–
Zn	Phenonthroline	5, 5' (2, 2 bithrophenysl)	Metalled porphyrin	–

10.10 CHARACTERISATIONS OF ROTAXANES

UV, visible spectroscopy was used for their characterisation. NMR technique was used to characterise chained compounds. Voltammetry was mainly used to study their metal complexes. In chromatography, gel permeation technique was used while crystal structure was studied by single crystal X-ray diffraction method. The polyrotaxanes were characterised by inclusion of large bulky group as central cyclic wheel (e.g. 30C10, 24C8, 60C20, 32C10, 36C12 and 60C10). Crown ethers as the central component with linear axle being polymer glycol, propylene chain, polymethylacrate. The stoppers consisted of tetraphenyl methane, 1-phenyl alanine, 2-4 dinitrofluorobenzene. They were obtained by threading. Polyrotaxane have very large molecular weight ranges from 6000-19,000 with 20-50 degree of polymerisation (Table 10.4) A first polyrotaxane was prepared by reaction of β-cyclodextrin and inclusion complex of diammes with diacid chloride with chain like structure. They could swell and exhibit molecular symmetry with macrocyclic polyether as central component. The π-electron rich hydroquinone and π-electron deficient tetraatomic cyclophane were used for synthesis of rotaxanes. The transition metals induced template effect. Several substituted Rotaxane were synthesised as conjugated, amide based, comblike structure. The pseuderdodecane were also obtained which were called as polyamide rotaxanes. The crown ethers, α, β cyclodextrin were used as the central cyclic component as wheel tetra cationic cyclophane were also used as the wheel. Linear component was usually 1-4 substituted benzene, amide based linkage and π-electron rich compounds and carboxylic acids were also used. The conjugated rotaxane exhibited the hydrophobic properties. Polyester rotaxane were also synthesised in similar manner.

Table 10.4 Characteristics of Polypsenol Rotaxane

Structure	Mol Wt.	Yield	Polymerisation degree	Structure
a (chemical structure)	11300-19100	83%	50	a
b (chemical structure)	18600-25400	89%	13	b
c (chemical structure)	7100-10300	47%	65	c
d (chemical structure)	7500-15600	72%	24	d
e (chemical structure)	6600-13,200	20%	20	e

10.11 METAL COMPLEXES WITH ROTAXANES

It was observed that none of the inner transition (f block) metals formed any complex with rotaxane. Similarly with the exception of selenium (IV, VI) none of the p-block or main group elements formed any complex with Rotaxane. However transition elements like iron (III), cobalt (III), ruthenium (III). palladium (II) copper (II), silver (Z), zinc (II) and mercury (II) formed stable complexes with it. Within alkali and alkaline earths only sodium formed stable complexes.

The templates effect was exploited to bring together cyclic and linear component. α-cyclodextrin as wheel with α-ω diaminoalkane bis chlorobisethylene diamine as linear part readily complex with cobalt. Good amount of work has been carried out on Group VIII metal complexes with Rotaxane.

10.12 ANALYTICAL APPLICATION

Polyrotaxanes behave like polymers and hence used as blending and phase adhesion promotion. They have definite m.p. good solubility in solvents and water. They could be used therefore in chromatographic column. They are used as the insulated wires and molecular straws. In analytical chemistry they were used mainly in chromatographic and solvent extraction separations.

The chiral stationary phases were useful for analysis of drugs. They were used for isolation of enantiomeric compounds.

However only significant contribution in solvent extraction consisted of use of (2) rotaxane for the quantitative extraction of Selenium in the presence of surface active reagent like (CTNB) *i.e.*, cetyl trimethyl ammonium bromide from 3M hydrochloric acid with chloroform as the diluent. From the organic phase, selenium (IV) was determined spectrophotometrically at 284 nm against a reagent blank. It can also be determined from the organic phase with ICP-AES. The method permitted separation of selenium (IV) and tellurium (IV) with thioglycolic acid as selective stripping agents for tellurium followed by extraction of selenium (IV) with rotaxane. Only aluminium, cobalt, chromium, tantalum and manganese showed very strong interference in any concentration. We expect in future fascinating applications of Rotaxane in analytical chemistry in quantitative analysis.

LITERATURE

1. S.M. Khopkar, *"Analytical Chemistry of Macro cyclic and Supramolecular Compounds"*, 2nd Ed. Narosa–Springer Verlag (2005).
2. M. Hiraka, *"Crown Ether and Analogous Compounds"*, Elsevier (1992).
3. C.A. Melson, *"Coordination Chemistry of Macrocyclic Compounds"*, Plenum Press (1979).
4. S.M. Khopkar, M.N. Gandhi, *"Crown ethers and Cryptands is Solvent Extraction"*, (Unpublished Monograph).
5. M. Hiraoka, *"Crown Compounds, Their Characterisation and Applications"*, Elsevier, Amsterdam (1982).
6. G.W. Gokel, S.H. Korzeniowskir, *"Macrocyclic Polyethers Synthesis"*, Springer Verlag, (1982).
7. R.M. Izatta, G.D. Christenson, *"Synthetic Multidentate Macrocyclic Compounds"*, Academic Press, London, (1978).
8. S.M. Khopkar, *"Solvent extraction Separation of Elements with Liquid Ion Exchangers"*, New Age International Publishers Ltd. (2007).
9. Y.K. Agarwal et al., *"Topics in Inorganic Chemistry"*, Vol. 21 p. 43 (2001).

Chapter 11

Principles of Chromatography

Analytical science in recent years has progressed spectacularly with the discovery of newer separation methods. The subject has grown into a separate entity called "Separation Science". Several journals have been devoted exclusively to this field. It is an important and integral component of analytical science because few instrumental methods can be directly used for quantitative analysis on account of the presence of interfering substances or ions. These interfering constituents must be separated before analysis. Hence, the progress in development of newer methods of separation has rendered quantitative methods of analysis more reliable. These separations are primarily based upon the creation of a second phase with a different concentration of desired constituent or fully use differences within a single phase. Such separation depends upon the existence of a difference in physical or chemical properties of the two constituents to be separated. The methods include exploitation of differences in properties such as volatility, solubility, partition ratio, exchange equilibrium, surface activity, molecular geometry and kinetic energy. Various chromatographic methods based upon such differences are discussed in the following chapters. However, some of the non-chromatographic methods including solvent extraction are of equal importance. So also membrane separations, distillation, zone refining, inclusion compounds, foam separation, thermal diffusion. Most of them are not useful for the separation of constituents at microgram or tracer concentration, but are useful for bulk separation. Since they are worth considering, they have been included mainly to understand the principles of their operation.

11.1 NON-CHROMATOGRAPHIC METHODS OF SEPARATION

Distillation is the simplest and the most classical method of separation used for the separation of two volatile liquids. With the discovery of gas chromatography, this method has been neglected but the technique of fractional distillation is often very useful. Vacuum fusion involves melting the sample in a suitable container by means of high-frequency heating. The gases are collected and then analysed. Zone refining is another method of purification of materials (e.g., semiconductors like germanium). It provides ultrapure samples. It works on the principle of exploiting the difference in solubility of an impurity in the liquid and the main solid component. On cooling, impurities remain in the melt while the major component freezes out in the pure zone. Here one uses a difference between the temperature and velocity of flow of solid and liquid freezing interface to separate them. Zone levelling is a process wherein one obtains a uniform concentration over a great length of the column of material. It is useful for preparing reference samples. The use of inclusion compounds is not a common method. It involves two molecular species which provide space in their structure in which other molecules can be easily entrapped. The foam separation technique is based upon stable foam to separate components in solution. It is based on the tendency of surface active solutes to collect at the gas–liquid interface. This is similar to the process of froth flotation. The technique of thermal diffusion uses the principle of application of a temperature gradient to a homogeneous system creating a concentration gradient; sometimes a partial unmixing occurs. The process of thermal diffusion offers a separation method wherein the liquid phase depends largely on sharp differences in diffusivity between molecules. This has been used successfully for the separation of isotopes of gases like He^3 and He^4 or $C^{13}H_4$ and $C^{12}H_4$.

11.2 MEMBRANE SEPARATION METHODS

These are promising methods in biosciences. We would take only the bird's eye view of these methods which include dialysis, electrodialysis, electrodecantation, gas permeation through solids. Some of these methods are discussed in a later chapter on Electrochromatography. The membrane is a barrier which separates two fluids. The membrane is semipermeable in nature and favours selective transport of some components permitting separations on a molecular scale. Transfer through the membrane is by diffusion. Such transfer may be initiated by dialysis *i.e.*, a concentration gradient between the phases and separation by membrane or by ultrafiltration involving hydrostatic pressure gradient or electrodialysis where an electric field is applied across the membrane. In dialysis, the separation is based upon the exploitation of difference in relative rates of diffusion of two species through the membrane. A selective mass transfer is attained. No convective exchange between solutions is favoured. Such methods have been used in the rayon industry to regenerate sodium hydroxide for its separation from steep liquors. Reverse osmosis or ultrafiltration is a process in which a solution is forced under pressure through a membrane with accompanying separation of components. The driving force is the energy due to pressure difference. The technique involves movement of solvent. This is a technique of flow of liquid against a concentration gradient. Similarly electrodialysis is a process in which electrolytes are transferred through a membrane by application of electrical energy. The electrolyte is driven from dilute to concentrated solution. Both techniques are discussed in detail in subsequent chapters. Electrodecantation or electrogravitation separation is based upon a stratification effect that may occur when colloidal dispersions are subjected to an electrical field between membranes permeable to the electrical current and impermeable to the colloid. The difference in density causes a formation of film or more dense liquid to pass on the concentration side. It is used for natural rubber latex concentration. Finally, gas permeation through solid facilitates separation. This involves the phenomenon of condensation of penetrant dissolving on a surface layer, which in turn migrates under the influence of the concentration gradient via the voids of the film and evaporates from the opposing surface down stream. Unfortunately, none of these techniques have relevance for separation at tracer concentrations of the elements.

The most promising method of separation is obviously chromatography. With the discovery of rapid methods of separation involving gas chromatography, high performance liquid chromatography or ion chromatography or capillary electrophoresis, there is now a preference for rapid instrumental chromatographic methods. An exception to this generalisation is solvent extraction which has remained in use in spite of the advancement in instrumentation. This is because the methods are simple, selective, rapid and inexpensive. Further, with the rapid development of new extractants the utility of these methods has grown specially for the separation of constituents at microgram concentrations. There is a great similarity between solvent extraction and ion exchange methods.

Analytical chemistry is primarily the chemistry of the identification and separation of substances— the resolution of the substances into their components and their subsequent determination. Separation is always a must when we consider the purification and isolation of substances. Progress in recent years has shown that chromatographic methods are the best choice for separations.

11.3 CHROMATOGRAPHY AS THE METHOD OF SEPARATION

In the earlier sections, we have learnt about different nonchromatographic separation methods. Thus amongst classical methods we have to consider simple techniques like filtration, distillation, froth flotation, ringoven or the use of molecular seives. Amongst modern methods, we have the solvent extraction, dialysis, reverse osmosis and numerous chromatographic methods of separations including ion exchange technique.

The separation of two or more components was possible by fractional distillation, fractional crystallisation as it was used for the separation of lanthanide series of elements or even fractional precipitation which was used for lanthanide separations by exploiting difference is basicity. However none of them provided rapid method for the isolation. The separations usually were confined to two or at the best three components. During separation, thermal dissociation or association was noticed. The worst among them was the time required for carrying out such lengthy separation.

Although great strides were made in advancement in the applications of instrumental methods of quantitative analysis for determination, several of these methods demanded the species to be in pure form with no interferences at tracer level. In order to eliminate these interferences, a fast method was needed for purification. There was no means to test the homogeneity of a substance. Limited methods were available to establish identity of substance, no method was available for concentration of a product at the infinite dilution or for the separation of mixtures or isolation of compounds.

Tswett developed for the first time a columnar method of the separation of pigments into different color bands. Owing to these colors the methods were designated as 'Chromatography' meaning color bands display. However subsequent discoveries led to the separation of even colorless components into several segments but the name chromatography still continued to be used in chemical separation. The history of their discovery is quite interesting which would be further discussed.

11.4 HISTORICAL DEVELOPMENT

The credit for the discovery of the chromatography technique goes to Russian botanist Tswett. In 1903, he was trying to isolate plant products like chlorophyll by process of filtration through porous media like calcium carbonate when he discovered that several color bands were visible in the column containing porous media. Each color fragment was isolated and analysed and to his delight he found that each of them contained different chemicals with different characteristics. The color bands were designated as the chromatographs. He was awarded Nobel Prize for his discovery in 1938. Similarly, in 1859, in America, another chemist Day was isolating petroleum products and he observed various bands when the sample was passed through the column. In fact in ancient times the testing of dyes was conventionally carried out on the filter paper. It was then observed that different dyes migrated at different speeds on filter paper by process of capillary movement. Although the period 1903-1931 witnesssed brisk activity in this field, real impetus was given in 1946 when Tesilus discovered what is known as reversed phase extraction chromatography. Around the same time Martin & Synge (1941) discovered the technique of chromatography. For their invention, all three of them were awarded Nobel prizes. The ion exchange chromatography was discovered by agricultural chemist Thompson and Way. They were busy experimenting with partition chromatography, when the phenomena of gas partition was overlooked. However with announcement of the discovery of gas chromatography, brisk activity resulted in replacement of classical methods of separation like fractional distillation or fractional precipitation. In fact nowadays two coveted prestigious international awards are instituted in America called S. Dal Nogare Award in gas chromatography and M.S. Tswett Awards in the field of chromatography for identifying excellence in the field of chromatography.

Subsequent to these discoveries there was no looking back as enormous progress was made in this field specially for separations of chemically similar species at microgram concentrations within a short interval of time.

11.5 CLASSIFICATION OF CHROMATOGRAPHIC METHODS

The chromatographic methods have good speed, resolving power and ability to handle small amounts of material. The operations are simple with equipment supported by practical theory. As regard sample size, some forms of chromatography are able to achieve routine separations. In particular, the analysis below the nanogram level (*i.e.*, 10^{-9} gms) is easily accomplished.

Every chromatographic system consists of moving or mobile phase in intimate contact with fixed or stationary phase. The latter is composed of the stationary or all non-moving portion of chromato-graphic column or bed. A sample component undergoes an equilibrium distribution between these two phases. This equilibrium in turn decides the velocity with which each component migrates on column. The band-broadening and dispersion of each component in the direction of migration also occurs. Thus differential migration and band-broadening decide the extent of separation of the sample. One of the ways of classifying different chromatographic methods is based on phases used or mechanism control-ling separation or kind of technique used and or sample development. Such classification is presented in Table 11.1.

Table 11.1 Classification of chromatographic methods

S.No.	Mobile phase	Stationary phase	Mechanism	Technique	Sample Development
1.	Gas	Liquid (GLC)	Sorption	Column	Elution
2.	Gas	Solid (GSC)	Adsorption	Open bed	Frontal
3.	Liquid	Liquid (LLC)	Partition	Column	Displacement
4.	Liquid	Solid (LSC)	Exclusion	Batch	Displacement
5.	Supercritical fluid	Liquid (SFLC)	Gas-liquid	Column	Elution
6.	Supercritical fluid	Solid (SFSC)	Gas-Solid	Column	Elution

Table 11.2 Alternative way of classification on basis of mechanism

S.No.	Kind	Adsorption	Partition	Exclusion	Ion Exchange
1.	Support	Solution-solid	Solution-liquid	Solution-solid	Resin material
2.	Phases	Gas-Solid	Gas-Liquid	Gas-Solid	Inorganic substances
3.	Technique	Adsorption Chromatography	Gas Chromatography	Exclusion Chromatography	Ion Exchange Chromatography
4.	Mode	Gas-Solid Chromatography	Liquid-liquid partition	–	Inorganic Exchangers

In Table 11.1 classification is based on mobile phase which covers gas or liquid to form gas chromatography or liquid chromatography, with solid or liquid as stationary phase. We have gas-liquid, liquid-liquid or liquid-solid chromatography. The later addition in supercritical fluid (SFC) chromatography. In gas chromatography, stationary phase is nonvolite as high boiling liquid is coated on porous support, while inert gas is used in mobile phase. The separation occurs due to difference in vapour pressure of components. In liquid-liquid partition chromatography stationary phase is liquid coated on porous support (Table 11.3). While mobile phase is second liquid which is immiscible with first liquid or stationary phase. The separation occurs due to difference in equilibrium distribution of sample components in two phases. In liquid-solid chromatography the separations originates due to difference in adsorption from the stationary phase. GSC *i.e.*, gas-solid chromatography uses solid stationary phase (e.g. granulated activated charcoal GAC) and gas mobile phase. The use of supercritical fluid as mobile phase leads to newer methods. This new technique has just come into existence very recently.

The classification based on a technique is one which involves use of either a column (Table 11.2) or open bed. The column may be packed column or capillary column also called as open tubular columns. The paper, thin layer techniques are examples of open bed chromatography. Flow of phase is due to capillary wetting. Open bed chromatography is restricted to use of liquid as mobile phase. The technique is simple, flexible and best for characterization. Column chromatography has better capacity ease of operation and excellent efficiency. It is used in preparation and quantitative work. Other ways of classification is based upon mode of injection of sample. Elution, frontal development, displacement development are its examples. These terms would be discussed in more details in the section on Technique.

Table 11.3 Modern classification

Sr. No.	Class	Example	Abbreviation
1.	Adsorption chromatography	Columnar method	LC
		Gas-solid chromatography	GSC
2.	Partition chromatography	Liquid-liquid partition	LLPC
		Paper	PC
		Thin layer	TLC
		Reversed phase partition *i.e.*, Extraction chromatography	RPPC
3.	Ion exchange chromatography	Cation exchange	CEC
		Anion exchanger	AEC
		Inorganic exchanger	IE
		Liquid exchangers	LIE
		Ion chromatography	LAE/LCE
4.	Exclusion chromatography	Gel-permeation	GP
		Ion exclusion	IExd.
		Molecular sieve	MS
5.	Electrochromatography	Zone electrophoresis	ZE
		Boundary layer method	BLE
		Curtain chromatography	CC
		Capillary electrophoresis	CZE

However, the most logical way of classification is based upon the mechanism of retention of solute or analyte in the stationary phase. It forms the backbone of the chromatographic separations (Table 11.3). The process involved are either, the sorption by adsorption, partition and exclusion e.g. GLC, LLPC involve partition, while adsorption deals with sample fixation on stationary support involving some forces like London forces, dipole induced dipole interaction or the molecular interaction. The distribution ratio is most important factor in the consideration of partition chromatography. While exclusion mechanism is based upon segregation of species by taking the advantage of the difference in geometry of the molecules or species as seen in Gel permeation chromatography. The stationary phase is porous media. Small species permeate faster in comparison to large size particles e.g. proteins or polymers. More than one retention mechanism is possible in few separations. So we have many modes of classification but one based upon the mechanism of retention is usually considered best. It is briefly summarised in Table 11.3.

It is interesting to note that HPLC *i.e.* high performance liquid chromatography or IC-ion chromatography are both highly instrumental techniques, do not figure in the above classification. The reason is simple; they are techniques and not a class of chromatography. For instance, we can use HPLC for any four kinds of techniques described in Table 11.3.

It is worthwhile to familiarise with these six important techniques. The details of each technique will be considered in individual chapters.

(*a*) *Adsorption Chromatography:* This method is based upon exploitation of the difference in adsorbility of solute to the stationary support which is usually packed in a column e.g. various fatty acids can be easily separated by adsorption chromatography. The separation of inert gases on activated charcoal is an example of gas-solid adsorption chromatography.

(*b*) *Partition Chromatography:* This, by and large, is the largest class of chromatography. All methods like LLPC, PC, TLC, GLC and RPPC fall under this category. We exploit the difference in the partition coefficient or distribution ratio of individual species in the mobile and stationary phase. Two phases are immiscible with one another. Those components having a large value for partition coefficient travel down the column slower within the mobile phase. However, in the case of GLC they are returned strongly on the column by a high boiling liquid phase and in RPPC they are retained on the column because of high partition coefficient in the hydrophobic phase on the stationary support. It is thus possible to effect their clear cut separation. In PC and TLC we have plates instead of columns.

(*c*) *Ion Exchange Chromatography:* This method is based upon differences in the exchange potential between various ions for an ion exchange resin packed in a column. For instance, in cation exchange, usually cations with high valency are strongly retained while monovalent ions are easily removed. In the case of anion exchange methods, anions with a high charge are strongly retained in relation to ones having a low charge. Inorganic exchangers or liquid ion exchangers are a special class of separation.

(*d*) *Exclusion Chromatography:* This case of separation is based fundamentally upon exploitation of the difference in size or molecular geometry of the components. In gel permeation, small constituents are retained in inter shell space or pores while large size components emerge first. While in ion exclusion the phenomenon is slightly different. A molecular sieve is the simplest example of separation based upon the size difference of solutes.

(*e*) *Electrochromatography:* Most of the methods under this category were classified as electrophoretic methods; several such methods were not carried out in columns, plates, or capillary tubes. Whenever columns, paper plates or capillary tubes were used they were termed electrochromatographic methods. In such separations, we usually exploit the difference in mobility of different ions when an external DC potential is applied. Ions with better mobility like Na^+ move faster under the influence of direct electric current, while those with low mobility move sluggishly thereby facilitating good separation. Only in recent times has capillary zone eletrophoresis come into the limelight. This technique, interestingly, permits separation of metals at low concentration viz. 10^{-21} grams.

Thus chromatography methods have a few things in common. There is a stationary support in the form of a column packed with inert material, then a phase moving down the column, called the mobile phase and the phase which usually adheres to the stationary support called the stationary phase. Separation is feasible on account of a differential migration front, developed by exploiting differences in adsorbility, partition coefficient, exchange potential, molecular size or ionic mobility. Further, most terms used while describing chromatographic methods are common to all the methods.

11.6 FUNDAMENTALS OF CHROMATOGRAPHY

In order to understand basic theory of chromatography, let us consider chromatographic behaviour of the solute. Such behaviour can be explained in terms of retardation factor or in terms of retention volume. The latter term explains time period for which the solute resides in mobile phase. The different substance have different values of retardation factor or retention volume. These values can be varied by changing experimental conditions. Now we try to find out, what partition coefficient is.

(a) *Partition coefficient:* It is $K_d = C_S/C_M$ if C_S = concentration in stationary phase, C_M = concentration in mobile phase. When partition isotherm is linear, K_d does not depend upon concentration.

(b) *Retardation factor:* Now let us consider retardation factor (R). It is ratio of the displacement velocity of solute to that of the velocity of standard substance. R represents fraction of time spent by species in mobile phase *i.e.* $R = \dfrac{tM}{(tM + tS)}$ if t_M, t_S, represents time of retention in stationary and mobile phase. We sometimes use R_f value, $R_f = \dfrac{\text{distance travelled by solute front}}{\text{distance travelled by mobile phase (solvent)}}$. It is used in paper and TLC methods.

Now if R is equilibrium fraction of solute in the mobile phase, then $(1 - R)$ is fraction of solute present in stationary phase. So $\left(\dfrac{R}{1-R}\right)$ is same as fraction in mobile and stationary phase. So

$$\frac{R}{(1-R)} = \frac{C_M}{C_S} \cdot \frac{V_M}{V_S}$$ where V_M, volume of mobile phase, V_S, volume of stationary phase. C_M, C_S are

concentration in mobile and stationary phases. So $R = \left[\dfrac{V_M}{V_M + K_d V_S}\right] = \left[\dfrac{1}{1 + \left(K_d \dfrac{V_S}{V_M}\right)}\right]$. This equation

is similar to one used in solvent extraction. Let us consider what is meant by retention volume.

(*c*) *Retention Volume:* At appearance of peak, the half solute is eluted in retention volume (V_{max}) while other half remains on the column in mobile phase (V_M) plus volume of stationary phase (V_S) $V_R C_M = V_M C_M + V_S C_S$. The same on rearranging and putting partition coeffficient we have

$V_R = (V_M + K_d \cdot V_S)$. This "retention volume" is related to retention time as

$V_R = t_R F_C$, if F_C is rate of flow of the mobile phase. It is related to R as

$V_R = \dfrac{L \cdot F_C}{R_v}$, if L = length of column, v = linear flow rate or average velocity.

$V_R = L F_C \left(\dfrac{V_M + K_d V_S}{v V_M} \right)$.

(*d*) *Column capacity:* Now let us consider what column capacity is. It is given as $k = \dfrac{C_S V_S}{C_M V_M}$.

If $\beta = V_M / V_S$. So $k = \dfrac{(V_R - V_M)}{V_M} = K_d / \beta$ is relation to retention volume.

(*e*) *Temperature effect:* Since temperature influences K_d, we must keep it constant, otherwise we

have $K_d = \dfrac{e^{\Delta S°/RT}}{e^{-\Delta 1 t°/RT}} = a e^{-\Delta H°/RT}$ if $\Delta S°$, $\Delta H°$ are entropy and enthalpy respectively.

R = universal gas constant. Usually the entropy ($\Delta S°$) decreases with an increase in enthalpy ($\Delta H°$). Resolution is affected by temperature. The value of partition coefficient and concentration dependence influences solute behaviour on the column. A plot of stationary phase to mobile phase concentration provide partition isotherm. These are shown in Fig. 11.4 representing (*a*) linear isotherm, (*b*) classical isotherm and (*c*) anticlassical isotherm, but linear isotherm is considered to be best for chromatography.

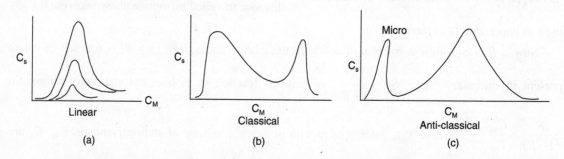

Fig. 11.1 Elution patterns

The first is symmetrical bell shaped Gaussian error function curve where $C_S \propto C_M$. In (*b*) we have classical or Langmur curve which is observed in absorption chromatography while in (*c*) we have micropeak and then broad peak. Such curve is observed when solute has low solubility in stationary phase (Fig. 11.1).

11.7 TECHNIQUES IN CHROMATOGRAPHY

One can develop a chromatogram by any of the three modes viz. frontal analysis, elution development and displacement development.

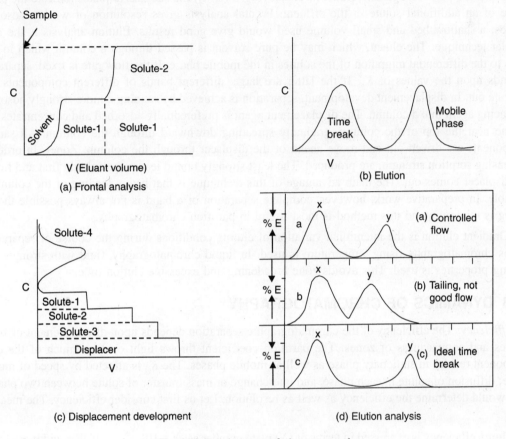

Fig. 11.2 Development of chromatograph

Frontal analysis (*a*) (Fig. 11.2) consists of passing solution continuously through a column on adsorbent when strongly adsorbed species occupy active centre of adsorbent while weakly adsorbed species discharges in travelling front. So first pure solvent comes out then weakly bound species and finally strongly bound species comes out from column. Here a shallow bed and small sample volume is needed. However, most common process is use of Elution analysis. (*b*) Due to differential migration of solute in mobile phase, which in turn related to partition coefficient (K_d) exhibits behaviour of symmetrical bell shaped curve. Those species strongly bound come latter while weakly bound species emerge out first. This occurs due to what is called as "zone spreading". Finally, in the displacement development (*c*) Convex adsorption isotherm are obtained. At constant speed, for displaces agent, front of displacer or moves down the column. This permits heavy loading of the column in preparative work. Here complete band separation does not take place, recovery in pure form is lower and stationary phase gets saturated with displacer. So a more stronger displacer replaces solute bound on column and itself occupies this place on the column. In flow chromatography, the effluent is collected in different fractions and each component is detected in the effluent stream. In frontal analysis, the sample solution is passed continuously through the absorbent. The active centres of the adsorbent are occupied by more strongly adsorbed components and the least strongly adsorbed component accumulates in the travelling front. In the first stage pure solvent and then least strongly adsorbed solute will emerge from the column. If we plot %E against volume of eluant (Fig. 11.2(d)) the successive plates indicate the emergence of an additional solute in the effluent. Frontal analysis gives resolution of weakly adsorbed solutes; a shallow bed and small volume used would give good results. Elution analysis is the most popular technique. The eluant, which may be pure solvent is passed through a column which in turn leads to the differential migration of the solutes in the mobile phase. If the flow rate is fixed, separation depends upon the values of K_d. If the latter are large, different bands of different components will separate out. In displacement development, separation is achieved by running a more strongly adsorbed displacing agent into a column. The displacement agent is preferentially adsorbed and concentrates into a zone near the top of the column, gradually spreading downward through the column. All sample components are driven out and move ahead of the displacer through the column. Zones, in order of decreasing sorption strength, are produced. The least strongly bound leaves the column first and finally the displacer comes out. The main advantage of this technique is that heavy loading of the column is possible. In preparative work, however, complete separation of a band is not always possible though tailing is minimum and this method is most useful in partition chromatography.

Gradient elution is the intentional variation of eluting conditions during the course of separation. In gas chromatography rising temperature is used. In liquid chromatography, fluid with stronger displacing properties is used. This avoids zone broadening and excessive elution time.

11.8 DYNAMICS OF CHROMATOGRAPHY

(a) *Efficiency:* The efficiency of the chromatographic separation depends upon disengagement of zones centres and compactness of zones. The partition coefficient throws light on how much of the each component resides in stationary phase as well as mobile phases. The K_d is affected by speed of mobile phase; diffusion of solute in each phase, and interchange or mass transfer of solute between two phases. This would determine the efficiency as well as resolution. Let us first consider efficiency. The measure

of column efficiency is expressed in terms of (N) plate number as $N = 16\left(\dfrac{V_R}{W}\right)^2$. If W = width of elution

peak in volume units (Fig. 11.3).

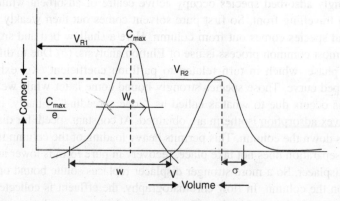

Fig. 11.3 Dynamics of technique

An alternative way of expression is as $N = 8\left(\dfrac{V_R}{W_e}\right)^2$, where W_e is

width of C_{max}/e if C_{max} is peak elution concentration (or $0.368\, C_{max}$). The width is expressed in terms of the standard deviation as $W = 4\sigma$ if σ = standard deviation. Another mode of measuring efficiency of the column is plate height (H) or height equivalent theoretical plate

(HETP), $H = \dfrac{L}{N}$ if L = length of column. Irrespective of any tech-

nique used it is presumed that a chromatographic column is made up of small (N) number of plates with height of each plate as (H) (Fig. 11.4). Now the efficiency is related to N, larger value of number of plates indicates better efficiency of separation. Further if H is small zone spreading including diffusion is uniform thereby improving the column efficiency. This concept was carried from field of distillation practice is related to N, larger value of number of plates indicates better efficiency to chromatography. A column is divided into number

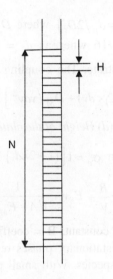

Fig. 11.4 Column plates

of plates with complete equilibrium between two phase for solute. N and H are the useful index of column efficiency. The distribution of solute during separation occurs in thin plates or zones. The rate of sorption and desorption in each zone lags behind change of concentration of solute in mobile phase.

(b) *Zone spreading:* The phenomenon of zone spreading is of utmost importance. In terms of standard deviation (σ), plate is defined as $H = \sigma^2/L$ if L = length of column and σ = the variance. Zone spreading is due to molecular diffusion. Inequality in velocity of flow leads also to zone spreading, but diffusion is more important. The difference in fast and slow stream paths arises by simple diffusion eddy diffusion and local non equilibrium during mass transfer in phases. When solute migrates from high to low concentration region, it leads to variation in concentration. The concentration profile after a finite time is Gaussian and variance is given as $\sigma_d^2 = 2rD_M t = 2r\,D_M L/r$ wherein D_M is diffusion coefficient of solute in mobile phase, r = obstructive factor (~0.6) and t = time spent ($t = L/r$) if L = distance zone has migrated, V = linear flow rate of mobile phase. The longitudinal diffusion also affects value of H. The variance is $2\gamma D_M t/(1+k)$ or $2\gamma D_M t(1-R)/R$.

(c) *Eddy diffusion:* The term eddy diffusion deals with diffusion process. Part of mobile phase takes longer time for diffusion due to tortuous and constricted path in the particular medium and hence lags behind the average while other portion resorts to shorter and smooth route and moves faster than average. Thus mobile phase is more rapid in some regions. The longitudinal diffusion also affects value of H. The variance due to Eddy diffusion is $\sigma_V^2 = 2\lambda d_p L$ wherein L-migration distance, d_p is particle size of substrate and λ = packing factor. Finally non-equilibrium effects arises during mass transfer between two phases. Mobile phase near initial edge brings small concentration of solute and desorption occurs. Forward flow of mobile phase partially obstructs equilibrium. At leading age reverse situation prevails. The mobile phase brings another end where solute concentration is much larger than the expected equilibrium concentration and sorption proceeds, accelerating equilibrium process. The spreading is: $\sigma_N^2 = 2\omega R(1-R)_V L t_s$ where t_s = is average residence time of solute in stationary phase

$\left(t_s = d_p^2 / 2D_s\right)$ where D_s = diffusion coefficient in stationary phase, ω = numerical coefficient (ω is 1/16 whereas d_p = diameter of substrate) and dv = thickness of liquid film containing substrate. The coupling between nonequilibrium in mobile phase and eddy diffusion is $\sigma v^2 = 1/\left[(l/2 \times dv) + D_M / vdv^2\right]$.

(d) *Height of the plate:* So plate height $H = 2rD_M / V + R(1-R)d_1^2 v / D_s + 1/\left[(1/2 \lambda d_i) + D_M / d_1^2 v\right]$

when $\sigma_c^2 = 1\left[(1 + 2\lambda d_i) + D_M / vd_V^2\right]$. Combining these two equations we have finally the

$$H = \frac{B}{V} + E_S v + \frac{1}{1/A + E_{M/v}} \quad \text{wherein:}$$

A = constant, B = coefficient for diffusion term, E_M and E_S mass transfer for coefficient in mobile and stationary phases respectively. The effective plate height depends on the retardation factor of the species. With small partition ratio the zone movements are faster and plate height is large. The reduced velocity $\bar{V}$ is given as $\bar{V} = d_r v / D_M$. This is very large in liquid chromatography. Hence

$H = 2rD_M / V + 2\lambda d_v + R(1-R)r d_v^2 / D_S$. For the low flow expression will be $H = 2rD_{M/v} + R(1-R)Vd_v^2 / D_S + d_v^2 r / D_M$ when $h = H/d_p$ if h = reduced plate height.

(e) *Plate height Efficiency:* This equation for plate height facilitates us to select operating conditions favouring reduction of plate height and enhancing efficiency of chromatographic separation. Here

$$V_{optimum} = \sqrt{D_M / \left[R(1-R)d_v^2 / D_S\right]}.$$ This is compromise velocity. There is limit to reduce plate height upto certain limit.

11.9 VAN DEEMTERS EQUATION

So far we had considered what is meant by longitudinal diffusion, mass transfer into the stationary phase and mobile phase contribution to the plate theory. A special case arises specially in case of gas chromatography where Van Deemters equation has direct implications.

(a) *Equation:* According to this equation plate height (H) is given as $H = A + \dfrac{B}{\gamma} + C \cdot \gamma$ where A, B, C are constants, while γ = rate of flow of liquid/gas in the column. Alternatively it is defined as

$H = H_L + H_S + H_M + H_{SM}$ wherein H = sum of all terms namely

H_L = longitudinal diffusion, H_S = mass transfer in the stationary phase,

H_M = effects of mobile phase and H_{SM} = mass transfer across mobile phase. (See Fig. 11.5 & 11.6)

where $H_L = \dfrac{2\lambda D_M}{V}$, $H_S = \dfrac{qK'd_V^2}{D_S(1+k)^2}$; $H_M = \dfrac{1}{1/A\,dx + D_M / \Omega r d_2^2}$ and $H_{SM} = \dfrac{K'}{30(H+K')^2} \cdot \dfrac{d_v^2 v}{PM}$

(b) *Significance of height:* The significance of the most of the terms is explained in earlier section. Experimentally it is possible to evaluate a relationship between H and v.

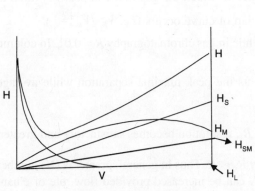

Fig. 11.5 Plate height

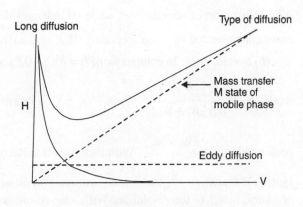

Fig. 11.6 Type of diffusion

In above Van Deemter's equation the three constants (viz, A, B, C), represent the plate height contribution from eddy diffusion, longitudinal diffusion and the sum of stationary and mobile phase mass transfer. Fig. 11.5 shows form of misequation. Longitudinal diffusion is not working at reduced velocity (v).

(c) *Optimum conditions:* We need minimum (H) and maximum (N) for efficient separation. We can express H_{min} as $H_{min} = A + 2\sqrt{BC}$ and $v_{opt} = \sqrt{B/C}$ since at $v_{opt} - B/v = cv$. Flow velocity if kept large then v_{opt} gives better results.

11.10 RESOLUTION OF MIXTURES

The purpose of chromatography is to carry out both the qualitative as well as quantitative analysis. Such separation depends upon disengagement of zone centres and the compactness of bands. The concept of the zone spreading in terms of plate theory is not sufficient. The plate height is a simple measure of separation power of column in the presence of a certain level of the disengagement. A specific term separation factor or separation coefficient does characterise the separation. Temperature also influences separability of species. A consideration of degree of separation or resolution for two adjacent bands is most important. The bandwidth and separation of band centres influences the nature of separation. A difficult separation need increase of distance between two bands and simultaneous decrease in band width (*i.e.*, W_e).

(a) *Resolution:* We have resolution, (R) as $R_S = \dfrac{2(t_2 - t_1)}{W_1 + W_2}$,

where W_1, W_2 are bandwidth of elution curve while t_1, t_2 are the values of bands 1 and 2. Since W_0 is equal to four standard deviation σ_1 for given band $R_S = \dfrac{t_2 - t_1}{4\sigma_2}$. High value of R gives better separations. This expression is also written $R_S = \Delta z/4\sigma$ where $\Delta z = (t_2 - t_1)$ and σ is standard deviation of system. Thus resolution is a measure of degree of separation of zones. It is usually defined from graph shown in Fig. 11.7.

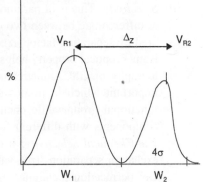

Fig. 11.7 Resolution of peaks

ΔZ is separation of zone centres while σ- indicates degree to which the gap between peaks is filled to cross contaminated by zone spreading. If $R = 1$ a overlap of curve occurs (*i.e.*, $V_{R_1}/V_{R_2} = 1$).

(b) *Special case:* In column work $R = (0.2 - 0.5)$, while in gas chromatography $R = 0.01$. In column

work, $R_S = \dfrac{\left(V_{R_2} - V_{R_1}\right)}{0.5\left(W_1 + W_2\right)}$, where $(V_{R_2} - V_{R_1})$ represents the peak maxima separation while average

peak width is $0.5\dfrac{(W_1 + W_2)}{2}$. With increase on value of R_S, resolution becomes satisfactory and greater.

Further if $K = K_d\,(V_S/V_M)$ here $K =$ partition ratio usually $R_S \propto K$. Thus in general, greater is (N) number of plates, better in the resolution. With long columns N can be increased provided flow rate of eluant is also faster (e.g. HPLC). The large number of plates is a distinct advantage for the purpose of efficient separations.

11.11 SEPARATIONS CHARACTERISTICS

So far we have considered the various factors influencing resolution. The most important parameters influencing separations are: (*a*) adoptability (*b*) load capacity (*c*) fraction capacity (*d*) selectivity and (*e*) speed (*f*) convenience. Each factor is very briefly considered.

(a) *Adoptability:* This relates to ability of particular method to permit the separation depending on physicochemical property of the material to be separated. The property such as volatility, nonvolatility in macromolecular compounds decide e.g. first can be separated by gas chromatography while macromolecular substances can be separated by exclusion chromatography, while nonvolatile matter can be resolved by liquid chromatography or HPLC.

(b) *Load Capacity:* This represents maximum quantity of mixture which can be separated chromatography is rarely used on preparative scale like process of distillation. This decides ultimately the selection of proper method of separation.

(c) *Fraction Capacity:* It is the maximum number of components that can be separated in single separation. Peak capacity is same as fraction capacity. This represents maximum number of separable bands. It can vary from 10 to 100. This number depends upon efficiency of separation. For conventional methods like distillation, extraction, dialysis, zone melting, the fraction capacity is restricted to only two.

(d) *Selectivity:* This is of paramount importance during separation. It is a property which helps us to differentiate between two components. It is related to mechanism of separation. It is measured by separation factor (α). The molecular properties like weight, fundamental group, shape, isomer, optical property helps us to distinguish between two components e.g. in liquid chromatography or GSC, functional group difference facilitates separation. Polarity for instance is important which facilitates separation of complex mixtures into group—which have same functional group while chemical functional group selectivity is most useful for separating components with different functional groups.

(e) *The Chemical Selectivity:* It includes acid base equilibria, complexation, precipitation e.g. ion exchange separations (pK value difference). Molecular geometry facilitates the separation in Gel permeation technique. Isomer selectivity in fact is a special class of functional group selectivity e.g. substituted pyridines separated by gas-solid method. The biochemical selectivity is based on highly specific reaction between biological molecules e.g. biopolymer. Optical selectivity is related to optical activity property of substances to isolate optical antipodes. Optical isomers can thus be separated by GLC. Finally other kind of selectivity has less implications e.g. o and p-hydrogen separated by taking advantage of magnetic properties of ortho and para isomers.

(f) *Speed:* is most vital factor in deciding separation method. For the routine analysis, speed of separation is most important unless otherwise prohibited by cost, fast methods are always preferred during separation.

(g) *Overall feasibility:* This term is related to several factors like fraction capacity, speed, selectivity factor taken together with other variables. In conclusion we can say that GLC considers molecular weight volatility is of importance. Liquid-liquid chromatography considers voltality, molecular structure and weight and functional group present while in ion exchange molecular structure, chemical group attached, nonvoltility are important. In exclusion method size, weight, shape and nonvoltility decides the method. In electrophoresis, thermal stability, molecular size, molecular weight are important.

11.12 SPECIAL FEATURES OF CHROMATOGRAPHIC METHODS

Chromatography as compared to other separation methods has advantage of separating multicomponent mixtures. The methods are very simple and fast in relation to other separation techniques. Much of the sophisticated methods of determinations prelude need of prior separations at microgram concentration, where chromatography fits in well. The history of its discovery is very fascinating. Large number of scientists have contributed to its discovery mostly by accident and rarely by planned research work. There are several ways of classifications based on difference in the mobile phase, stationary phase, technique used, sample development mode. However, the one based on mechanism of sorption involving partition, exclusion, adsorption is most accepted way of classification. The principles of chromatography involves consideration of differential migration of component on column/plate by exploiting intrinsic difference in mechanism of sorption. Various techniques like frontal analysis, elution or displacement development are efficient and are used to remove component from column. The dynamics of chromatography involves consideration of plate theory giving stress on number (N) and thickness (HETP) of plate in terms of efficiency. With special reference to gas chromatography the Van Deemter Equation provides means to ascertain thickness plate (H) in terms of eddy diffusion, mass transfer of mobile and stationary phase. The efficiency of separation is dwelt upon resolution. Several factors are considered while ascertaining resolution. Finally in any situation, consideration of separation of characteristics from the viewpoint of adoptibility, load and fraction capacity are of utmost importance specially the selectivity of method. Such selectivity depends upon difference in kind of the functional group present on component, variation in molecular weight or size, presence of isomerism, or optical activity or biological selectivity. No matter whatsoever chromatographic method is used, consideration of various factors discussed in this chapter, are valid for any chromatographic separation method.

SOLVED PROBLEMS

11.1 The substance X and Y had retention times of 16.4 and 17.63 minutes on a 30 cm column. Unretained species had 1.30 minutes retention time. The peak widths (W_e) were 1.11, 1.21 mm. resp. Calculate column resolution and average number of plates and plate height and time needed to elute substance (Y).

Answer: For column resolution, we use expression $R_S = \dfrac{2\Delta_Z}{W_x + W_y}$.

where W_x and W_y are width of peaks and Δ_z is the difference in time of their arrival at the detector

$$R_S = \frac{2(17.63 - 16.40)}{(1.11 + 1.21)} = 1.06$$

In order to obtain N we have the equation with usual notation

$$N = 16\left(\frac{t_R}{W}\right)^2 = 16\left(\frac{14.40}{1.11}\right)^2 = 3493, \text{ if } t_R = 16.4, W = 1.11$$

$$N = 16\left(\frac{17.63}{1.21}\right)^2 = 3397, \text{ if } t_R = 17.36, W = 1.21$$

The height of plate is calculated as $N_{\text{ave}} = \dfrac{3493 + 3397}{2} = 3445$

$$H = L/N = 30.0/3445 = 8.708 \times 10^{-3} \, \text{cm}$$

Finally from the equation

$$(t_R)_r = \frac{16 R_S^2 H}{u}\left(\frac{\infty}{\infty - 1}\right)\frac{\left(1 + K_y'\right)}{\left(K_y'\right)^2}, \text{ if } (t_R)_y \text{ is time required to elute } y \text{ with velocity } u'.$$

$$(t_R)_1 / (t_R)_2 = (R_S)_1^2 / (R_S)_2^2 = 17.63 / (t_R)_2 = (1.06)^2 / (1.5)^2 = 35.3 \text{ minutes.}$$

11.2 Calculate the minimum plate height (H) from the data $A = 0.01$ cm, $B = 0.30$ cm^2/S, $C = 0.155$ and $u = 0.287$ ml/mt.

Answer: The Van Deemter equation states

$$H = A + \frac{B}{u} + C \times u$$

wherein A = eddy diffusion, B = longitudinal diffusion, C = non-equilibrium mass transfer and u = flow rate. On substituting the values

$$H = 0.01 + \frac{0.03}{0.287} + (0.0150 \times 0.0287)$$

$$H = 0.01 + 1.045 + 4.305 = 5.36 \text{ cm is plate height.}$$

PROBLEMS FOR PRACTICE

11.3 If in a given system $L = 22.6$, $U = 0.287$ ml/mt, $V_m = 1.26$ ml, $V_s = 0.148$ ml and retention times (t_R) for compounds *A, B, C, D* respectively are 4.2, 6.4, 14.4 and 15.4 with width of peak base

(W) are nil, 0.45, 1.07 and 1.16. Calculate the number of plates from each peak, and plate height for the column.

11.4 Calculate the average number of plates and average plate height of the column if retention times (t_R) for compounds A, B, C, D are 2.5, 10.7, 11.6 and 14.0 minutes with width of peak base (W) of nil, 1.3, 1.4 and 1.8.

11.5 From the data in problem 11.3 calculate for A, B, C and D capacity factor, and partition coefficient. Also calculate for B and C the resolution, selectivity factor (α), length of column to give $R_S = 1.5$. Finally for compounds C and D calculate the resolution, and length of column if $R_S = 1.5$, and the time required to separate C and D when the resolution is 1.5.

LITERATURE

1. J.M. Miller, *"Separation Methods in Chemical Analysis"*, John Wiley & Sons (1975).

2. J.C. Giddings, *"Dynamics of Chromatography Part I. Principles & Theory"*, Marcel Dekker Inc. New York (1965).

3. B.L. Krager, L.R. Snyder and C. Harvath, *"Introduction to Separation Science"*, John Wiley & Co. (1973).

4. E. Heftman, *"Chromatography 2nd Edition"*, Van Nostrand Reinhold & Co. (1967).

5. Z. Devyl, *"Separation Methods"*, Elsevier Publishing House (1984).

6. E.W. Berg, *"Physical and Chemical Methods of Separations"*, McGraw Hill & Co. (1963)

7. R.Stock, C.B.F. Rice, *"Chromatographic Methods"*, Chapman & Hall Sci. (1974)

8. E. Lederer & M. Lederer, *"Chromatography–Review of Principles and Applications"*, Elsevier Publishing Co. (1957).

9. J.A. Dean, *"Chemical Separation Methods"*, Van Nostrand Reinhold & Co. (1969).

Chapter 12

Ion Exchange Chromatography

Amongst the various chromatographic methods, the most frequently used methods are ion exchange chromatography as well as gas chromatography. The former has very wide applications in Inorganic Chemistry while the latter is very useful in the separation of organic compounds. The cleancut separation of chemically similar elements like rare earths, or zirconium and hafnium or niobium tantalum was possible by ion exchange chromatography. The softening or deionisation of water was facilitated by the ion exchange process. The task of removal of total salt concentration and removal of interfering ions, separation of ions from nonelectrolytes or isolation of trace concentration of precious metals was possible by ion exchange methods. The principal advantage of ion exchange method over other methods of separation is that this procedure is simple, rapid and inexpensive. For instance, the resin can be reutilised after regeneration. These resins are easily available in any particle size with varying exchange sites. Most of them are insoluble in water and therefore can be packed into the column, which permit reasonable flow rate of the solution provided appropriate particle size of resin is selected.

12.1 HISTORICAL DEVELOPMENT

The discovery of ion exchange materials dates back to 1835 when an agricultural chemist Thompson and Way reported observations that when the solution of ammonium sulphate or chloride was passed on soil, the ammonia was adsorbed by soil while equivalent amount of calcium or magnesium was released from the soil. This may be ascribed to base exchange properties of soil. They discovered soil contained $Na_2O\ Al_2O_3\ SiO_2\ xH_2O$ or Zeolite, where the replaceable ion was sodium. Some attempts were made to prepare potable or sweet water from sea water, where the chemists poured solution in wooden container, which on standing for some time gave sweet water. This was perhaps due to the presence of cellulose material in wood acting itself as exchanger. Some efforts to separate ammonia from urine containing other amino acids was also successful when the mixture was passed upon Zeolite or permitite in column.

However, real breakthrough was attained after preliminary work in 1935 by Gans as well as Adams and Holmes who prepared synthetic materials having exchange properties like zeolite or permutite. The synthetic materials were copolymers of phenol and formaldehyde. Unfortunately their stability was limited. In 1944, a French chemist De'Alilo prepared a copolymer of styrene and divinyl benzene where in one could vary % of DVB from 1-16 maximum. This in turn influenced not only its exchange capacity but also porosity of the resin.

This discovery gave rise to surge of activity in the utilisation of ion exchange materials for various purposes. With discovery of synthetic materials, one could achieve not only separation of cations but also separation of anions by ion exchange methods. In fact in biochemistry separation of amino acids, peptides, nucleotides could be best accomplished by ion exchange chromatography in preference to the technique of electrophoresis.

12.2 CLASSIFICATION OF ION EXCHANGE RESINS

The ion exchange resins have been broadly classified as cation exchange resins — *i.e.* those of the resins which are capable of exchanging cations and anion exchange resins, which are ones capable of exchanging anions. These resins have been further classified depending upon the structure and presence of functional group as strongly basic or strongly acidic resins respectively. However we shall consider this subclassification after studying synthesis of resins. Both cation as well as anion exchange resins are organic in nature. The inorganic exchangers like permutite and zeolites are one which are naturally occurring and these were the first materials discovered. The heteropoly acids of the metals like zirconium (IV), tungsten (VI), tin (II) and antimony (V) account for a class of the synthetic inorganic exchangers.

The liquid ion exchangers, as the name implies are in liquid state and the liquid cation exchangers like DNS, HDEHP (*i.e.*, dinonyl sulphonic acid or bis 2 (ethyl hexyl) phosphoric acid respectively) can exchange cations. The liquid anion exchangers like Amberlite LA-2, Primene JMT are composed of high molecular weight amines and are capable of exchanging not only simple anions but also anionic complexes of several transition elements. They would be considered in later section of this chapter. The chelating resins like Amberlite A-1 which consist of polyfunctional group which can occupy more than two positions in the coordination sphere of cation and which are capable of complexing with metals to form tetra or hexa-cordinated complexes. Inspite of so much of advancement in coordination chemistry these chelating resins have not become very popular with the analytical chemist. The commercially available cation exchangers are: Dowex 50W-X8, Amberlite 1RC-120 while the anion exchange resins are Dowex 1-X8, Dowex 21K, Amberlite-1RA 400. The typical liquid cation exchangers are DNA or HDEHP while typical liquid anion exchangers are Amberlite LA-1, Primene-JMT. Aliqat 336S. The chelating resin is like Dowex A. Finally inorganic ion exchangers are somewhat like zeolite (Na_2O, $Al_2O_3\ SiO_2\ xH_2O$) and permutite.

12.3 MECHANISM OF ION EXCHANGE

The various explanations which have been put forward to explain the exchange mechanism can be grouped into three theories:

(*a*) Crystal lattice exchange

(*b*) Double layer

(*c*) Donnan membrane

(*a*) According to crystal lattice theory, Pauling and Bragg drew an analogy between ion exchange resin and ionic solids. In ionic solids, the constituents of the crystal lattice are present as ions instead of molecules. A crystal of KCl contains $K^+\ Cl^-$ ions, each ion being surrounded by a fixed number of ions. When placed in a medium of high dielectric constant such as water, the net attractive forces binding the ion to the crystal are reduced to such an extent that exchange of this ion for another ion in solution becomes quite feasible. Such exchange depends upon the nature of forces binding the ion to the crystal, concentration of exchanging ions, size of the two ions, accessibility of lattice site and solubility effects. Now if $NaNO_3$ is added to KCl solution, K^+ exchanges with the Na^+ ion and Cl^- with the NO_3^- ion. This exchange is similar to the exchange of crystal lattice ions and ions of an electrolyte solution. The mechanism of ion exchange in resins, although noncrystalline, is quite similar to the exchange of a crystal lattice ions. The resin is a high molecular weight insoluble polymer *i.e.*, electrolyte. The functional groups - HSO_3, - COOH, - OH are responsible for exchanging ions. The capacity of the sulphonic type resin can be stated by knowing the sulphur content of the resin. The exchange of ions with these resins occurs throughout the entire gel structure of the resin and is not merely restricted to surface effects. In anion exchange resins, exchange is due to covalent absorption of acid. They are polyamines, the amine content of the resin is the measure of the total exchange capacity.

(*b*) In double layer theory, the electrokinetic properties of colloids have been extended to ion exchange systems to interpret the various factors involved. The Helmholtz double layer consists of two rigid electrical layers at the surface of a colloid, similar to the plates of a condenser. According to Gouy and Stern, the double layer actually consists of an inner fixed layer which is diffuse and mobile outer layer of charges. The charged layers originate from absorbed ions which may be quite different from ions present in the inner layer and influence the electrokinetic properties of the colloidal system. The concentration of ions in the outer diffuse layer is related to the concentration and pH of the external solution. On addition of a foreign ion to the external solution, new ions will enter the outer layer replacing some of the ions previously held in this layer. A new equilibrium is set up and the overall exchange is stoichiometric. Exchange at the crystal lattice centre and in a double layer is only somewhat similar. The relation between the theoretical total exchange capacity and pH or concentration is not the same for the two systems. In the former, it is assumed that there are a few exchange sites which must be satisfied independent of pH and concentration change, but in the latter exchange, this does not apply as the capacity of different double layers depends both on pH and concentration (Fig. 12.1).

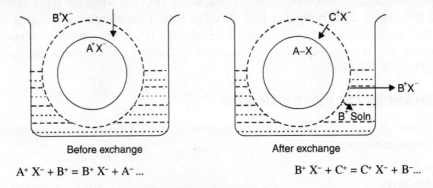

Before exchange

$A^+ X^- + B^+ = B^+ X^- + A^-$ ...

After exchange

$B^+ X^- + C^+ = C^+ X^- + B^-$ ...

Fig. 12.1 Double layer theory

(*c*) Finally the Donnan membrane theory deals with unequal distribution of ions on two sides of a membrane. One side contains an electrolyte whose ions cannot penetrate through the membrane. In ion exchange equilibria, the interface between solid and liquid phases may be considered as a membrane. The nondiffusible ion is the colloidal micelle to which the exchangeable ions are attached e.g. NaCl in contact with cation exchange resin (H^+ form). Equilibrium is set up as in a heterogeneous reaction between the solution phase and the resin phase wherein

$H_2^+R^-$ denotes the resin and the subscripts R and S indicate the resin and solution phases respectively. Hence we have

$$[Na^+]_R [Cl^-]_R \rightleftharpoons [Na^+]_S [Cl^-]_S$$

$$[H^+]_R [Cl^-]_R \rightleftharpoons [H^+]_S [Cl^-]_S$$

To maintain electroneutrality

$$[Na^+]_S = [Cl^-]_S \text{ and } [Na^+]_R = [Cl^-]_R$$

Hence $\qquad [Na^+]_S [Cl^-]_S = [Cl^-]_S [Cl^-]_S = [Cl^-]_S^2$

Now if $\qquad [Na^+]_R > [Cl^-]_R$ for cation exchanger

and $\qquad [Cl^-]_S^2 = [Na^+]_R [Cl^-]_R$ and then

$$[Cl^-]_S > [Cl^-]_R \text{ because it is a cation exchanger.}$$

Thus activity of NaCl is greater on the side that is free of nondiffusible ions in the solution phase. Now if a second cation K$^+$ is added to the system the following conditions must hold.

$$[Na^+]_R [Cl^-]_R \rightleftharpoons [Na^+]_S [Cl^-]_S$$

and $\qquad [K^+]_R [Cl^-]_R = [K^+]_S [Cl^-]_S$ dividing

$$\frac{[Na^+]_R}{[K^+]_R} = \frac{[Na^+]_S}{[K^+]_S}$$

Comparison of Theories

This equation means that an exchange of ions must take place till activity ratios are equal in both the phases. In conclusion, we can say that all the theories are essentially similar as the law of electroneutrality governs the exchange of ions. They differ only in the position and origin of the exchange sites. Thus the exchange site is an ionic grouping which gives rise to an electrostatic bond with an ion of opposite charge. The strength of bond decides the readiness of an exchange. There are several factors upon which the strength of such bond is dependent.

12.4 SYNTHESIS OF ION EXCHANGE RESINS

Adams and Holms carried out the condensation of phenol with formaldehyde to form a compound which in turn was subsequently sulphonated.

In Fig. 12.2 the condensation product of phenolsulphonic acid and formaldehyde acts as resin. However real impetus to use of cation and anion exchange resin was given by De'Alielo who synthesised a copolymer of styrene and divinyl benzene as shown in Fig. 12.3.

This was starting material for synthesis of both cation and anion exchangers. This product on sulphonation with chloro sulphonic acid provides cation exchange resin as shown in Fig. 12.4.

From above copolymer of DVB and styrene by process of chloro methylation as well as amination individually for such synthesis resorted as follows (12.5) to procure anion exchange resin as:

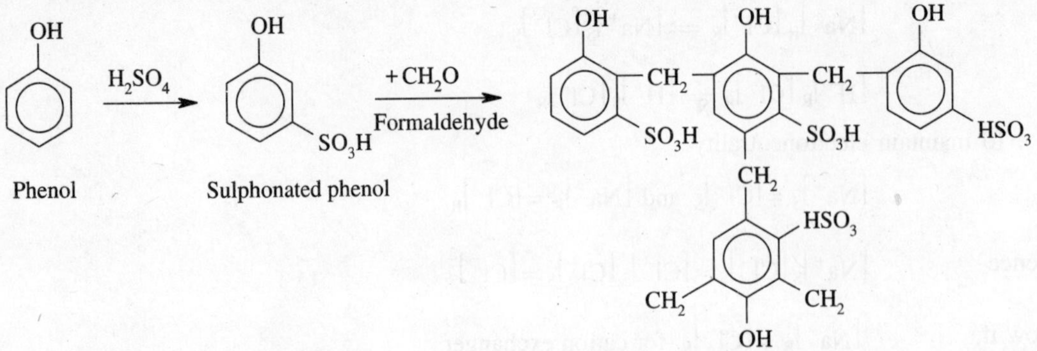

Fig. 12.2 Copolymer phenol and formaldehyde

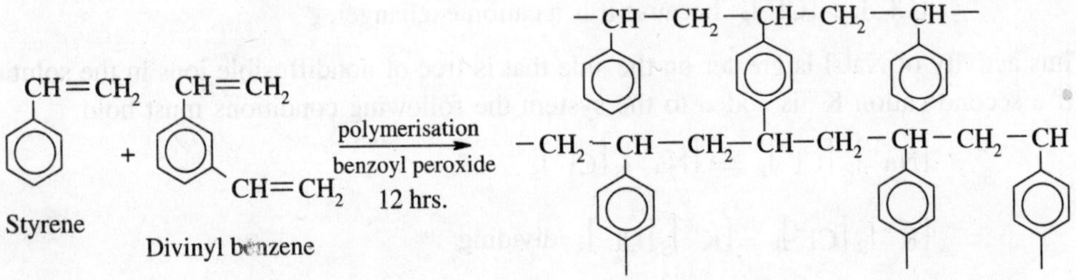

Fig. 12.3 Copolymer of styrene and DVB

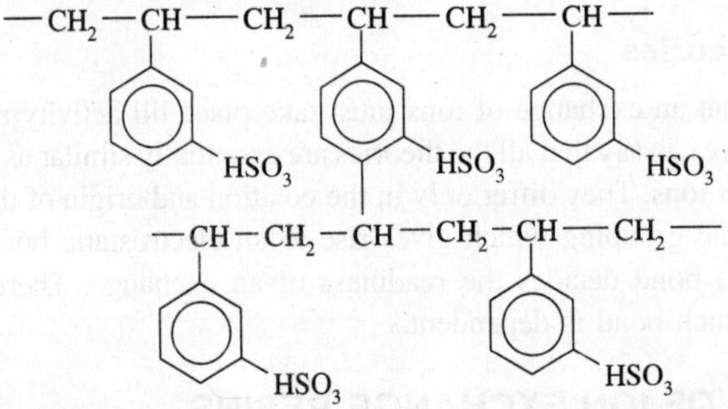

Fig. 12.4 Structure of cation exchange resin

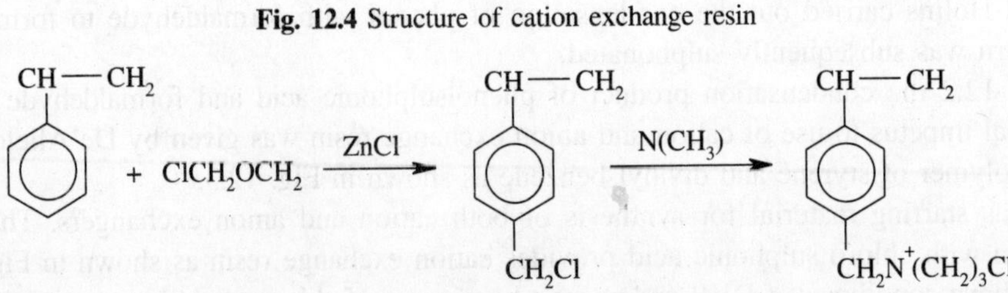

Fig. 12.5 Structure of typical anion exchange resin

A cation exchange resin (Fig. 12.3) contains $-HSO_3$ of strong acid. Such resin is called as strongly acidic cation exchanger. If in place of $-SO_3H$ if we have COOH as in Fig. 12.6, it is called weakly acidic cation exchange.

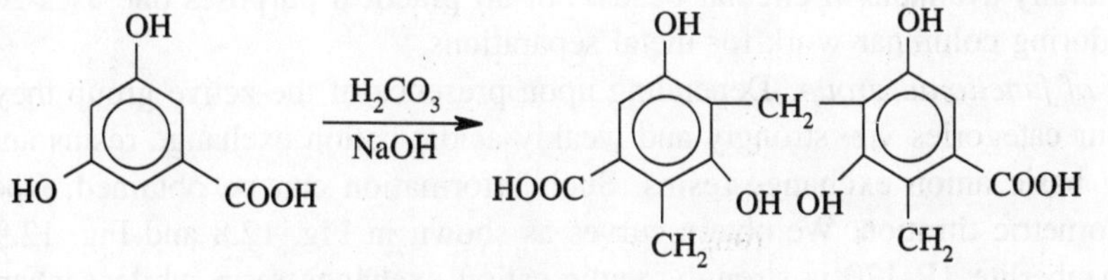

Fig. 12.6 Weakly acidic cation exchange resin

The structure shown in Fig. 12.4 represents strongly basic anion exchange resin as during amination we use quarternary amine or ammonia. The weakly basic group is synthesised from m-phenylene diamine and formaldehyde as:

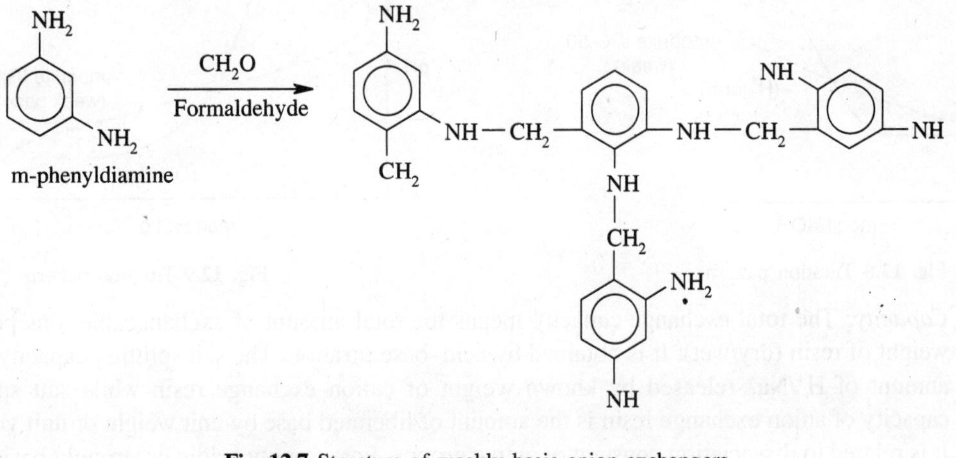

Fig. 12.7 Structure of weakly basic anion exchangers

However, the most commonly used cation exchange resins are always made from copolymer of styrene and divinyl benzene (DVB) containing strong acidic group. The structure and the presence of characteristic functional group present in ring imparts different properties to the different ion exchange resin.

12.5 CHARACTERISTICS OF ION EXCHANGE RESINS

We would consider both cation and anion exchange resin from the view point of their physical and chemical properties. Such properties include: (*i*) color, density, (*ii*) particle size, (*iii*) presence of functional group, (*iv*) exchange capacity, (*v*) selectivity, (*vi*) degree of cross linking (*vii*) porosity, (*viii*) swelling, (*ix*) rate of exchange, (*x*) chemical resistance, (*xi*) effect of temperature, (*xii*) effect of media.

(*a*) *Color:* The color of resin is generally brown for strong resins, but few are colorless e.g. Dowex 50W-X8. For anion exchange resin color varies from pale yellow to brown. The density of dry cation exchange resin is 1.4 gm/ml while that of anion exchange resin is 1.2 g/ml but for wet resin density always increases from 1.1-1.3 g/ml. On an average they contain water 40-60%.

(*b*) *Particle size:* This is most critical as it influences the plate height (HETP) as well as total number of plates (N) in the column. With smaller grain size exchange is always better though it is slow. Mesh size varies from 16-400 (US standard) or (1.168-0.038 mm). Polystyrene resins are generally available in circular beads. For all practical purposes one uses 20-50 mesh size resins during columnar work for metal separations.

(*c*) *Nature of functional group:* Depending upon presence of the active group they are classified into four categories viz. strongly and weakly acidic cation exchange resins and strongly and weakly basic anion exchange resins. Such information can be obtained experimentally by potentiometric titration. We obtain curves as shown in Fig. 12.8 and Fig. 12.9. Here in Fig. 12.8, Amberlite 1R-120 is strongly acidic cation exchange resin while Amberlite 1RC-50 is weakly acidic cation exchange resin. Similarly in Fig. 12.9 Amberlite 1RA-400 is strongly basic while Amberlite 1R4B is weakly basic ion exchange resin.

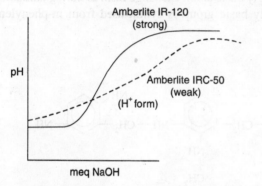

Fig. 12.8 Titration pattern

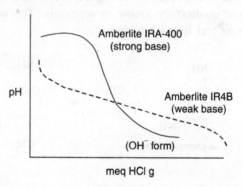

Fig. 12.9 Titration pattern

(*d*) *Capacity:* The total exchange capacity means the total amount of exchangeable ions per unit weight of resin (dry/wet). It is obtained by acid–base titration. The salt splitting capacity is the amount of H^+/Na^+ released by known weight of cation exchange resin while salt splitting capacity of anion exchange resin is the amount of liberated base by unit weight or unit volume. It is related to dissociation constant of active groups. For strongly acidic or strongly basic resin the salt splitting capacity and total capacities are identical. The washing effect must be also considered in relation to capacity. The breakthrough capacity usually depends on the pH, grain size, column size, flow rate in column.

(*e*) *Selectivity:* The different ions are sorbed at different strength e.g. multivalent ions are strongly held in comparison to mono or divalent ions by cation exchange resin. The order of sorption for cation resin is:

$$Li^+ < H^+ < Na^+ < NH_4^+ < K^+ < Rb^+ < Cs^+ < Ag^+ \qquad \text{(For monovalent cations)}$$

$$Mn^{2+} < Mg^{2+} < Zn^{2+} < Co^{2+} < Cu^{2+} < Cd^{2+} < Ni^{2+} < Ca^{2+} < Sr^{2+} < Pb^{2+} < Ba^{2+}$$
$$\text{(divalent cations)}$$

$$Al^{3+} < Sc^{3+} < Y^{3+} < Eu^{3+} < Sn^{3+} < Nd^{3+} < Pr^{3+} < Ce^{3+} < La^{3+} \qquad \text{(trivalent cations)}$$

The sorption strength of resin increases with decreasing diameter of hydrated cation. It also depends upon kind of active group present and phenomena of complexation. The weakly basic resins are good for multivalent anions. The order for sorption of anions on resin is:

$$OH^- < F^- < CH_3COO^- < HCO_3^- < Cl^- < NO_2^- < HSO_3^- < CN^- < Br^-$$
$$< NO_2^- < ClO_2^- < HSO_4^- < I^-$$

(e.g. Dowex-1 resin). For Dowex 2 resin order is slightly altered for OH^- ion. Here the valency, diameter of hydrated ion, strength of acid and structure of ion decides the selectivity scale. For weakly basic anion exchange resin the order is: $HCO_3^- < CH_3COOH < F^- < Cl^-$ $< SCN^- < Br^- < I^- < NO_3^- < OH^-$.

(*f*) *Cross linking and swelling*: One can vary crosslinking from 1-16%. When the resin is immersed in solution, it swells and gets saturated with that solution. The swelling decreases with increasing crosslinkages and decreasing size of hydrated species (Table 12.1) is shown.

Table 12.1 Swelling of cation exchange resin

% DVB	Swelling (mg of water per mill equi. of resin)				
Ionic radii (°A) →	9.0 (H⁺)	6.0 (Li⁺)	4.2 (Na⁺)	3.0 (K⁺)	2.5 (Cs⁺)
4	417	357	303	294	233
8	219	196	172	167	144
16	128	119	99	95	86

Table 12.2 Swelling of anion exchange resin

% DVB	Swelling (mg of water per mill equi. of resin)				
Ionic radii (°A)	3.5 (F⁻)	3.0 (Cl⁻)	3.0 (Br⁻)	3.0 (I⁻)	Resin capacity
4	345	264	181	85	3.47
8	185	139	100	67	3.10
16	128	114	87	65	1.91

The resin with low (%DVB) can swell more readily than with high %DVB (Table 12.2). The mechanical resistance is good and it has better selectivity. Low pore size give low swelling and low rate of exchange. On an average good resin contains 8% DVB. Otherwise this is mentioned as Dowex 50W-X12 (meaning 12% DVB). The swelling is measured by weighing, the saturated resin with solution followed by driving out solution out of resin on centrifuge in vacuum and weighing resulting dry resin. The difference between the two figures gives swelling capacity of the resin.

(*g*) *Porosity:* Resins as a rule are porous. Porocity determines its selectivity. Pore size is obtained by vapour pressure measurement. For resins, the pore size varies from sample to sample hence mean value is accepted for actual pore size.

(*h*) *Resistance:* Let us consider chemical resistance of resin. Most of the styrene based resin are safe in strong acid or strong base; but condensation type of resin (e.g. phenol formaldehyde) are affected by alkalies. So also anion resin containing –COOH group. The deciding strength is >1 M for base. The storage of anion exchange resin must be done in the chloride and not in hydroxyl form as such resins release OH⁻ ion in water. Few resins are affected by strong oxidising agents and also are unaffected by reducing agents. The strong oxidising agents which attack resin are $KMnO_4$, $K_2Cr_2O_7$, H_2O_2, hot nc. HNO_3. The rapid swelling spoils these resin. Those resin beads with large size (20-60°A) split readily in small pieces.

(i) *Rate of Exchange:* On immersing resin in solution, an equilibrium is established between resin and the hydrophilic solution. Such rate of exchange depends on size of exchanging ion and grain size of the resin as the process of the diffusion is in continuous operation during exchange reaction.

(j) *Effect of temperature:* At high temperature, the resin tends to pulverise with loss in swelling and splitting of active groups. Polystyrene base resin are stable even at high temperature. Anion exchange resins are more sensitive to high temperature. The selectivity for ions decreases with rise in temperature, but diffusion is accelerated for ions with temperature. On thermal heating, one can observe loss of crosslinking and also loss of the ionogenic groups. Thus it loses % DVB at 50°C. For OH⁻ anion exchange resin loss of ionogenic group is significant (~15% loss). They dissociate in following manner

(a) $CH_2N\,Me_3OH \rightarrow CH_2OH + NMe_3\,(60\%)$

(b) $CH_2N\,Me_3OH \rightarrow CH_2NMe_3 + MeOH$ for anion exchange resin. Weak base anion exchange resin are more stable at high temperature.

(k) *Resins in nonaqueous solvents:* Solvents like methanol, acetone, acetic acid, benzene, pyridine all have definite effect on swelling capacity of resin. It depends upon polarity and dielectric constant of the solvent. The exchange capacity lowers as few adsorb better in these solvents. Water behaviour is altogether different with the mixtures of solvents. Selectivity can be improved in non aqueous solvents.

(l) *Exchange characteristics:* The electroneutrality is followed and accordingly equivalent amount of cation/anion are exchanged. It is a stochiometric exchange. The exchange reactions are invariably reversible. The conversion to suitable form (H⁺ with hydrochloric acid for cation) or (OH⁻ with ammonia and not sodium hydroxide for anion) is rapid. After conversion resin should be thoroughly washed with deionised water before further use for exchange reactions-Chelating resins like Dowex A-1 are useless as their conversion reactions are extremely slow.

12.6 ION EXCHANGE EQUILIBRIA

Let us consider the following reaction:

$$A + B_R = A_R + B \qquad \qquad ...(12.1)$$

assuming A^+ and B^+ are monovalent ions and suffix 'R' denotes molar concentration of species B or A respectively in the resin phase. By the law of mass action, we have selectivity coefficient K_B^A for ions with same charge as

$$K_B^A = \frac{[A]_R[B]}{[A][B]_R} \qquad \qquad ...(12.2)$$

In an event exchange occurs between two ions of different valency as 'a' and 'b' we have exchange as

$$A^{b+} + B_R^{a+} \rightleftharpoons A_R^{b+} + B^{a+} \qquad \qquad ...(12.3)$$

So by application of the law of mass action we have

$$K_B^A = \frac{[A]_R^b [B]^a}{[A]^b [B]_R^a} \qquad \qquad ...(12.4)$$

where square brackets indicate molar concentration.

Such selectivity coefficient (K_B^A) can be evaluated from experimental data for strongly acidic or strongly basic resin. The number of the equivalent of exchangeable ions in the resin is equal to total exchange capacity. For ions of same valency selectivity coefficient is independent of units used to measure concentration in both the phases. Now $K_B^A > 1.0$ if 'A' has more affinity of resin but if $K_B^A < 1.0$ indicates ion 'B' is held strongly by resin and needs strong external solution as an eluant to accomplish exchange reaction while $K_B^A = 1.0$ indicate no selectivity for both ions in column. In exchange of ions of different valency, K_B^A depends upon the molar concentration as shown in equation no. 12.4. The electrolyte penetration in resin changes its swelling. Van der Waal's adsorption and variations in the degree of dissociation in the resin phase or in the external solution is of considerable significance. For other systems such influence is totally negligible or absent.

Let us now consider cation exchange equilibria (of strongly acidic resin) for exchange of ions with same and different valency. As before we have reaction as shown below:

$$[A]_R^{z+} + [C]_R^z = [A^{z+}] + [C^-]^z \qquad \qquad ...(12.5)$$

$$[B^+]_R + [C^-]_R^z = [B^{+z}] + [C^-]^z \qquad \qquad ...(12.6)$$

Where $[C^-]$ represents activity of anion present while suffix 'R' represents resin phase. Now by dividing first equation with second one we have.

$$\frac{[A^{z+}]_R [B^{+z}]}{[B^{z+}][A^{z+}]_R} = \frac{rB_R}{rA_R} \cdot \frac{r_A}{r_B} \qquad \qquad ...(12.7)$$

$$K_B^A = \frac{rB_R}{rA_R} \cdot \frac{r_A}{r_B} \qquad \qquad ...(12.8)$$

This equation contains activity coefficient (γ) of ion in resin phase as well as in solution phase.

So we determine selectivity coefficient by considering ratio of activity coefficient of ions in resin and solution phase respectively. The dilution of external solution has no effect on equilibrium. The relative amount of ion in resin phase is marginally changed. K_B^A changes if A_R or B_R changes. So for the selectivity coefficient, chemical structure of resin usually influences the selectivity coefficient viz. K_B^A .

Let us now consider equilibria with reference to exchange of cations in different valency state where selectivity coefficient not only depends upon the activity coefficients in two phases but also depends upon the total ion concentration of ions in resin phase. An equation for exchange of bivalent

(A^{2+}) ion with monovalent (B^+) is given as $K_B^A = \eta_R \dfrac{r_{B_R}^2}{rA_R} \cdot \dfrac{r_A}{r_B}$ \qquad ...(12.9)

(a) where n_A is total ion concentration in the resin phase. In this expression, concentration is in terms of equivalent concentration.

Bonner Affinity Series

(a) Let us consider what is meant by affinity series. The selectivity scale for univalent ions is

$$Li^+ > H^+ > Na^+ > NH_4^+ > K^+ > Rb^+ > Cs^+ > Ag^+ > T^+ \text{(monovalent ions)}$$ on resin with 8% DVB. The corresponding scale for divalent ions on Dowex. 50W-X8, resin is:

$$UO_2^{2+} > Mg^{2+} > Zn^{2+} > Co^{+2} > Cu^{2+} > Cd^{2+} > Ni^{2+} > Ca^{2+} > Sr^{2+} > Pb^{+2} > Ba^{2+} \text{ (divalent ions)}$$

Thus selectivity coefficient (K_B^A) decides the position of ion in the affinity series. The change in concentration affects K_B^A of external solution, even it reverses order in the scale. The uptake of ions decreases with increase in concentration. The aromatic resins extracts readily complex ions from external solution.

As regards equilibria for cation exchange from weakly acidic solutions (e.g. carboxylate, phosphate etc.) they have higher affinity for protons. They exhibit high affinity for s-block alkali and alkaline earth elements.

(b) *Selectivity coefficient, Separation factor and Partition ratio in ion exchange:* This selectivity also depends upon the pH of the external solution. It is worthwhile to know the difference between selectivity coefficient (K_B^A) and separation factor (α_B^A) in terms of partition coefficient or ratio (D). The ratio of the concentration of an ion (B) in exchanger to its concentration in solution, is called 'partition ratio' or partition coefficient (K_D)

So $K_D = \dfrac{[B]_r}{[B]_s}$. It is independent of ionic concentration. Now if we consider total solution concentration in resin and solution phase, we define partition ratio (D) as

$$D = \frac{\text{Total conc. in resin phase}}{\text{Total conc. in soln. phase}} \qquad \text{...(12.10)}$$

The partition ratio for ion A is D_A, for ion B is D_B. The ratio of $\alpha_B^A = \dfrac{D_A}{D_B}$ is called "separation factor".

If $\alpha > 1$ or $\alpha < 1$ separation is feasible but if $\alpha = 1.0$, no separation of two ions 'A' and 'B' is feasible. The thermodynamic stability constant is defined for such reaction as:

$$A + B_R \rightleftharpoons A_R + B \qquad \text{...(12.11)}$$

So, $$K_B^A = \frac{[A]_R[B]}{[A][B]_R} \qquad \text{...(12.12)}$$

but we account for γ-activity coefficient. So thermodynamic stability constant is

$$K_B^A = \frac{[A]_R[B]}{[A][B]_R} = \frac{X_B}{X_A} \cdot \frac{\gamma A_R}{\gamma B_R} \cdot \frac{[A]_R[B]}{[A][B]_R} \qquad \text{...(12.13)}$$

if X_B X_A is activity coefficient in solution and γA_{r_1} or γB_r is active coefficient in resin phase for both ions. It is assumed both ions have same charge. These constants can be easily evaluated by simple experimental procedure. The enthalpy changes in ion exchange do not exceed 2-3 kcal/gm e.g. the temperature has insignificant effect on selectivity (ΔH) but entropy term $T\Delta S$ is always greater than enthalpy. They originate from changes in hydration of ions in the solution. The selectivity coefficient obtained by activity coefficient of ions in the solution and resin phase is called "Apparent equilibrium constant" which can be experimentally determined. It depends upon concentration of ion and type of resin used. The selectivity coefficient can be interpreted in terms of following parameters such as with highly cross-linked structure selectivity is high. The difference in volumes of two hydrated species influences selectivity when resin absorbs one of the ions to form ion pair, there is decrease in activity of ion in resin. Temperature decreases selectivity. With high capacity resin, the absorption of ions is small. Same is the story with highly crossed-linked resins.

(c) *Equilibrium at trace concentration:* We consider an equation as follows:

$$[B]^b + [A]_R^a \rightleftharpoons [B]_R^a + [A]^b \qquad \ldots(12.14)$$

'B' is present in trace concentration then we have K_B^A as

$$K_A^B = \frac{[B]_R^a [A]^b}{[B]^b [A]_R^a} \qquad \ldots(12.15)$$

if a, b are absolute values of charges of ions A and B. We use D_B as dist. coefficient of ion B in resin phase. So we have

$$D_B = K_a^{B1/a} \left(\frac{[A]_R}{[A]} \right) b/a \qquad \ldots(12.16)$$

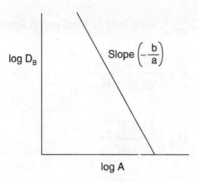

Fig. 12.10

provided if we assume $a \neq b$ and A is present in large excess. Since activity coefficient ratio $\left(\dfrac{\gamma_B}{\gamma_A} \cdot \dfrac{b}{a} \right)$ is constant.

So $$D_B = \text{constant} \times \frac{1}{[A]b/a} \qquad \ldots(12.17)$$

So this ratio is independent of concentration as the sorption isotherm is linear. A plot of log D_B against logA will be a straight line with slope $(-b/a)$. If we consider activity coefficient, then we have magnitude of D_B as

$$D_B = \text{constant} \times \frac{1}{[A]^{b/a}} \cdot \frac{\gamma_B}{\gamma_A^{b/a}} \qquad \text{...(12.18)}$$

(d) *Cation exchange equilibrium in the presence of complexing agents:* The ion exchange (both cation and anion) behaviour in the presence of complexing ligand is more complicated. It is used to increase separability (e.g. separation of lanthanide with ammonium citrate as complexing ligand) of ions.

$$D_M = \frac{K_{N/4}^M \cdot K_{\text{complex}}}{K^3} \left\{ \frac{(NH_4^+)(H^+)}{(NH_4^+)(H_3\text{cit})} \right\}^3 \qquad \text{...(12.19)}$$

As ammonium citrate and citric acid are used as eluants buffered at pH 6.0.

K_{comp} is instability constant of metal (M) as $\log D_M \propto \log[NH_4]$.

The separation factor for two adjacent metals M_1 and M_2 in the series is given as

$$\alpha_{M_2}^{M_1} = D_{M_1} / D_{M_2} = \frac{K_1}{K_2} \frac{(K_{\text{complex}})_1}{(K_{\text{complex}})_2} \qquad \text{...(12.20)}$$

Now for sorption of anionic complexes (e.g. chloride) the fraction of suitable metal (F) is given as

$$F = \frac{\left[MCl_{z+a}^{a-} \right]}{-mM} \qquad \text{...(12.21)}$$

If M forms complex as $\left[MCl_{z+a}^{a-} \right]$, then mM is total metal ion concentration. So we have

$$K = \frac{\left[Cl_{z+a}^a \right]_R [Cl^-]^a}{\left[MCl_{z+a}^a \right][Cl^-]_R^a} \qquad \text{...(12.22)}$$

so

$$D_M = \frac{\left[MCl_{z+a}^a \right]}{mM} \qquad \text{...(12.23)}$$

or

$$D_M = k \cdot F \cdot \left\{ \frac{[Cl^-]_R^a}{[Cl^-]} \right\}^a \qquad \text{...(12.24)}$$

(e) *Anion exchange equilibria in complex salt solutions:* Although most significant contribution were made in separation of cations by cation exchange chromatography and anion separations by anion exchange methods, the real challenging task was focused on the separation of cations by the anion exchange chromatography. Many transition metals form anionic negatively charged complexes with mineral acids as chloro, sulphato or nitrito complex (e.g. HCl, H_2SO_4 or HNO_3 respectively). Such complexes in turn can be sorbed on strongly basic anion exchange resin like Dowex 1-X8, or, Dowex 21K. Such complexes can be subsequently eluted successively by taking the advantage of their stability on column. Apart from complexing reactions with mineral acids good amount of contribution is made with organic dicarboxylic acid like oxalic, malonic, succinic acids and other organic acids as citric and tartaric acids. These acids are capable of forming anionic complexes with transition metals at varying pH. Therefore one can exploit the difference in their stability to accomplish selective sorption or selective elution with specific eluants on the column at specific pH of solution.

Let us consider equilibrium of metal chloride complex like $\left[MCl_{z+a}^{a-} \right]$ in reaction

$$\left(MCl_{z+a}^{a-} \right) + \left(Cl^{-a} \right)_R \rightleftharpoons \left(MCl_{z+a}^{a-} \right)_R + \left(Cl^{-a} \right) \qquad ...(12.25)$$

By application of law of mass action

$$K = \frac{\left[MCl_{z+a}^{a-} \right]_R \left[Cl^- \right]^a}{\left[MCl_{z+a}^{a-} \right] \left[Cl^- \right]_R^a} \text{ so the distribution coefficient is written as} \qquad ...(12.26)$$

$$D_M = \frac{\left[MCl_{z+a}^{a-} \right]_R}{mM} \text{ if mM is total metal concentration.} \qquad ...(12.27)$$

Assuming M^{a+} ion metal with valency 'a' and Cl^- is anion in resin.

If mM total metal ion if $\left(MCl_{z+a}^{a-} \right)$ is anionic chloro complex of metal M^+ with valency 'm'.

So,
$$D_M = kF \left\{ \frac{[Cl^-]_R}{[Cl^-]} \right\}^a \qquad ...(12.28)$$

If F-fraction of suitable metal complex, K = constant.
It is worthwhile comparing D of metals in different acid media.

Table 12.3 Ion exchange distribution in nitric acid

Cation	HNO_3 (Nitric acid) (M)			
	0.5	1.0	2.0	4.0
Fe (III)	362	74	14	3.1
Ga (III)	445	94	20	5.8
Th (IV)	10^4	1180	120	25
U (VI)	60	24	11	6.6
Zr (IV)	10^4	6500	650	31

When we examine data in Table 12.3 and 12.4 we can device separations e.g. separation of Th and Zr on anion exchange resin from 0.1N H_2SO_4 separation of Ga and U (VI) by cation exchange resin with 0.5 M HNO_3 as eluant.

Table 12.4 Ion exchange distribution in sulphuric acid

	H_2SO_4 (Sulphuric acid)			
Cation	0.1	0.5	1.0	2.0
Th (IV)	35	480	230	50
U (VI)	520	90	27	9.3
Zr (IV)	1350	210	47	11
Cr (VI)	12000	4400	2100	800
Mo (VI)	530	480	230	50

12.7 PLATE THEORY FOR ION EXCHANGE COLUMN

A less rigorous approach was made by Martin and Synge who developed a theory for chromatographic columns similar to the plate theory for solvent extraction and distillation. We have considered only the final equation in plate theory as it is covered in earlier chapter from first principles in separation methods. For calculating the separation, a resin column is considered to be made up of a large number of plates each successively equilibrating with the portions of the solution. A column is supposed to consist of a large number of theoretical plates. Within a plate the concentration of solute is considered to be uniform, both in the resin and in the interstitial void. If a small fraction of the column is loaded with the solutes to be separated, a constant fraction of the solute is partitioned between the aqueous phase and the resin phases as the fluid moves downward, from upper to lower plates (see Fig. 12.11).

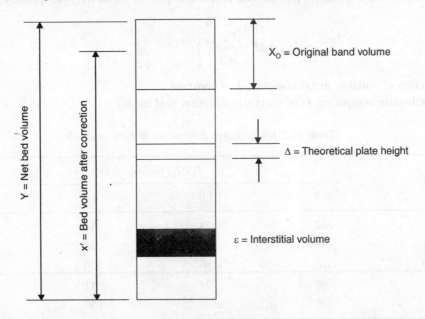

Fig. 12.11 Plate theory

They are large on account of the formation of complex ions. A survey of the relative affinity of different ions with equal valencies gives a scale as per Bonner affinity series (of section 12.6).

$$Li^+ > H^+ > Na^+ > NH_4^+ > K^+ > Rb^+ > Cs^+ > Ag^+ > Tl^+ \quad \text{(Monovalent)}$$

$$UO_2^{++} > Mg^{2+} > Zn^{2+} > Co^{2+} > Cu^{2+} > Cd^{2+} > Ni^{2+} > Ca^{+2} > Sr^{2+} > Pb^{2+} > Ba^{2+} \quad \text{(Divalent)}$$

A selectivity sequence for organic cations has been given as:

$$NH_4 < (Me)_4\,N < (Et)_4\,N < (Me)_3\,(N\ amyl) < Phen\ (Benz)\,Me_2N$$

Where Me = methyl, Et = ethyl, amyl = amyl, Phen = phenyl and Benz = benzoyl. For ions within the same group of the periodic table those ions with a larger radii are held more strongly than ions with smaller radii.

The uptake of certain ions decreases with the increase in their concentration. In the case of equilibrium at tracer concentration we have:

$$bA + aB_R \rightleftharpoons aB + bA_R \qquad \qquad ...(12.29)$$

$$\therefore \qquad K_B^A = \frac{[B]_R^a\,[A]^b}{[A]_R^b\,[B]^b} \qquad \qquad ...(12.30)$$

where a and b are the absolute values of charges of ions A and B.

$$D_B = K_D^{1/a}\left[\frac{[A]_R}{[A]}\right]^{b/a} \qquad \qquad ...(12.31)$$

where we consider the overall distribution of the total amount of B of solute, per gram of dry resin, divided by the total amount of the same constituent in ml of external solution. In the above expression, it is presumed $K_D = [B]_R/[B]$. Since A is present in large excess, the amount of A in the resin phase $[A]_R$ is constant and K_D is independent of trace concentration if $(\gamma B/\gamma A)^{b/a}$ is constant.

$$D_B = K_D^{b/a}\left[\frac{[A]_R}{[A]}\right]^{b/a} = \text{constant}\,\frac{1}{[A]^{b/a}}\frac{\gamma_B}{\gamma_A^{b/a}} \qquad ...(12.32)$$

This shows that the distribution coefficient for the trace ion (B) is independent of its concentration for a given concentration of gross component *i.e.* the uptake of trace ion is directly proportional to its concentration in the external solution.

In actual practice, we use the term elution constant (E) which is related to distribution coefficient by

$$K_D = \frac{\text{Concentration of ion in the resin phase}}{\text{Concentration of ion in the solution}} \qquad ...(12.33)$$

$$E = \frac{d \cdot A}{V_{max}} \qquad \qquad ...(12.33a)$$

$$K_D = \frac{M_R/\text{mass of resin}}{M_S/\text{volume of solution}} = \frac{M_R}{M_S}\cdot\frac{\text{Volume of solution}}{\text{Weight of resin}} \qquad ...(12.34)$$

where M_R and M_S represents the concentration of the metal ion M in resin.

According to the plate theory, elution curves are bell shaped and maximum peaks move down the column at a constant velocity which is inversely proportional to (f/c). After the passage of more than 25 plates, the shape of the elution curve can be calculated according to the following equation.

$$C = C_{max} \exp\left[\frac{-N(\bar{V} - V)^2}{2\bar{V} V}\right] \qquad \qquad ...(12.35)$$

where C = concentration of solute; N = column parameter or the number of theoretical plates; N_0 = initial number of plates, $N = (N - 1/2 N_0)$: average value, V = eluant volume; $\bar{V}$ = peak elution volume.

The peak elution volume, $\bar{V}$, is the volume of an eluant required to elute the maximum of the elution band. The expression for the peak elution volume is given by:

$$\frac{\bar{V}}{X'} = f/c \qquad \qquad ...(12.36)$$

where $X' = (X - 1/2 X_0)$, X = total column volume, X_0 = volume of original band and X' = corrected column volume, X_0 = is too small. Thus the elute volume required to reach the maximum solute concentration in the elute is directly proportional to the corrected column volume. If V is known for one column, peak elution volume can be easily calculated for columns of other dimensions. The weight distribution coefficient (D_W) can be obtained from batch equilibrium studies. Hence

$$\frac{f}{c} = \frac{\bar{V}}{X'} = \rho_R = \epsilon \qquad \qquad ...(12.37)$$

where ϵ = void fraction and ρ_R = density of resin.

Thus we can calculate D_W from column experiments. The volume distribution coefficient (D_V) is defined as the amount of sorbed solute per cm^3 of exchanger bed divided by the amount of solution. The relationship between D_V and D_W is

$$D_V = D_W \cdot \rho_R, \text{ if } \rho_R = \text{ bed density of resin} \qquad \qquad ...(12.38)$$

Hence
$$\frac{\bar{V}}{X} = Dv + \epsilon \qquad \qquad ...(12.39)$$

If the molarity ratio K^* or the sorption coefficient K_e are known, they can be used to calculate $\bar{V}$ to give the expression

$$\frac{\bar{V}}{X} = K^* \frac{W}{\rho} \rho_R + \epsilon = K_e \frac{\Sigma_{\rho R}}{\rho} + \epsilon \qquad \qquad ...(12.40)$$

where ρ is density of external solution, W is the swelling of resin calculated as grams of solvent per gram of resin and Σ is the exchange capacity expressed as milliequivalent per gram of dry resin.

It is always possible to calculate the number of theoretical plates from the column experiment

$$N' = 8\left(\frac{\bar{V}}{\beta}\right)^2 \qquad \qquad ...(12.41)$$

where β is the width of the elution curve.

The number of plates can be obtained from the maximum elute concentration using the equations.

$$C_{max} = \frac{m}{V}(N'/2\pi)^{1/2} \qquad \qquad ...(12.42)$$

$$N' = 2\pi \left(\frac{C_{max}\bar{V}}{m} \right)^2 \qquad \qquad ...(12.43)$$

Thus the height of theoretical plate Δ is of course equal to the ratio between the height of the resin bed and the total number of theoretical plates (HTP). A decrease in HTP gives a narrow peak which leads to improved separation.

As regards calculation of separation conditions, it is possible to know the peak elution volume $\bar{V}$, for a given solute and the number of theoretical plates (N') in a given column and this data can be used to calculate the elution curve for all types of columns provided particle size, eluant concentration, temperature and flow rates are the same. On knowing D_w, its peak elution volume $\bar{V}$ can be found. The plate theory can be used for predicting column dimensions suitable for the separation of different solutes from a single elution curve for one of the solutes, if D_w for the other solute is known. The purity of the solutes is calculated from the overlap of the elution bands for the two solutes to be separated (see Fig. 12.12).

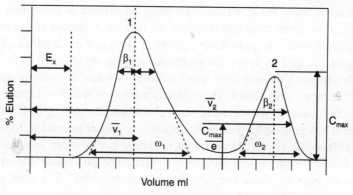

Fig. 12.12 Elution curves for solute (1) and solute (2)

The important conclusions are: the height of a peak decreases inversely as the square root of the column length and the width of the elution curve (β) (Fig. 12.12) increases directly as the square root of the column length.

The proper conditions for separation can be predicted by plate theory though these are approximate.

The height of the theoretical plate decreases with a decrease in the particle size. The HETP is equal to the diameter of the resin particles. Several difficult separations have been achieved, even though the HETP has been ten times the diameter of the particle because of a channeling effect. A value which is three times the diameter is to be considered as extremely good. Plate theory can be adopted even for nonequilibrium conditions but the HETP is higher in such cases. This is due to a nonequilibrium between solution and resin particles due to slow particle diffusion in the liquid.

$$\Delta = \Delta_{min} + \Delta_{pd} + \Delta_{fd} \qquad \qquad ...(12.44)$$

if Δ_{min} HTEP is in equilibrium conditions, Δ_{pd} particle diffusion, and Δ_{fd} film distribution.

Glueckauf gave the following formulae for the above terms:

$$\Delta_{min} = 1.64\,r \quad \text{(Height at equilibrium)} \qquad \qquad ...(12.45)$$

$$\Delta_{pd} = \frac{Dv}{(Dv + \in)^2} \frac{0.142\,r^2\,\overline{F}}{Ds} \quad \text{(particle diffusion)} \qquad ...(12.46)$$

$$\Delta_{fd} = \left(\frac{Dv}{Dv + \in}\right)^2 \frac{0.266\,r^2\,\overline{F}}{D_L\left(1 + 70\,r\overline{F}\right)} \quad \text{(Film diffusion)} \qquad ...(12.47)$$

where D_R and D_S are diffusion constants in resin and solution phase resp. $\overline{F}$ the linear flow velocity of the solution in the column; r is the particle radius; c is the concentration of solute, and f is the amount of solute in column equilibrium. In the above discussion, it is assumed that the sorption isotherms are linear. The theory also predicts elution curves with a monosymmetric shapes.

12.8 TECHNIQUES IN ION EXCHANGE METHODS

We have already considered the important characteristics of ion exchange resins. Let us now consider various techniques employed in ion exchange experiments. The first technique involves 'sorption', when solution is passed on resin. We avoid calling it either adsorption or absorption as both or none of them exists in operation. In the technique of sorption, we take small aliquat of solution containing ions or constituents of mixture which is passed on column at flow rate of 2 ml/mt. All ions capable of exchanging on resin are retained by it and the molecular species are released in the outgoing solution called 'Effluent'. After washing column with small quantity of water, we start process of 'desorption' or what is called as "Elution" with eluant. The eluant is a solution which is capable of removing ions or species from resin in column. If we are using same eluant with increasing concentration like say 1M and 4M eluant we term this process as "Gradient elution". An eluant is specific for one ion but not for second, we term it as 'specific eluant'. If an eluant is eluting two or more ions under different set of conditions, we call this as the process of 'selective elution'. The process of converting resin into the original form after work for subsequent use is called as "regeneration". The resin is restored in same form.

In mode of operation, we have two categories as the "batch" method and the "columnar" technique. In first, we do not use column but ordinary beaker wherein resin is kept and analyte is added, to react and after certain interval, solution is decanted and species which have exchanged by the resin is removed by suitable eluting agent (s). In column operation, we pack the column (usually of size of 1.4 × 15 cms) with wet and washed resin in the column, then washed with water and charged it with ions separated by the process of sorption followed by technique of elution to remove the sorbed ions. In ion exchange chromatography several ions are removed successively from the column by process of selective elution.

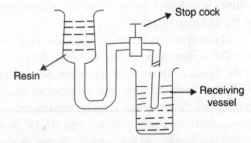

Fig. 12.13 Ion exchanger column

Such selectivity largely depends upon concentration of eluant, pH or molarity of eluant, temperature of elution, flow rate of solvent through column. After desorption or elution of particular ion from column it is quantitively analysed preferably by instrumental methods like spectrophotometry, conductometry, polarography or suitable radiochemical method if present in microgram concentration. The batch method is rarely used for quantitative separation and analysis. If one resorts to columnar methods, care should be taken to saturate the ion exchange column with appropriate ions, removal of excess of solution by washing, following sorption and elution process properly; washing column with water and regenerate to its original for reuse in second cycle. The problem of void volume must be carefully controlled. The void volume is the space between the beads and dead space from bottom of bed to tip of delivery tube. When large number of samples are to be collected over a period of time, it is better to use 'fraction collector' which operates automatically based upon volume control or drop control or time resolved method to collect equal volume in each fraction.

Since ion exchange is reversible process, an equilibrium state is reached and needs more amount of resin if solution is highly concentrated. The rate of exchange is controlled by film and particle diffusion process. The reasonably slow process ensures complete sorption followed by elution process. Finally the regeneration is reverse process, complete regeneration proceeds with large volume of regenerating solution. The total number of exchanging groups in the column is expressed as 'milli equivalents' and is called as total exchange capacity but breakthrough capacity is lower than total exchange capacity and is dependent on kind of resin, its affinities and concentration, particle size, flow rate and several other factors influencing elution.

12.9 ANALYTICAL APPLICATIONS

Ion exchange resins are mainly used in two [A] Non chromatography work as well as in for the conventional Ion exchange chromatography for the separations of components.

[A] *Non chromatographic application:* This area primarily covers the use of ion exchange resin to establish total salt concentration or removal of the interfering ions of opposite charge or the isolation of trace constituents.

(i) *Total salt concentration:* Ion exchange is primarily used for the concentration into small volume from bulk of the solution present. For instance, a column containing cation exchange resin in H^+ form is taken. The solution containing sodium sulphate and copper sulphate (both M/40) are taken. Then 25 ml of each solution is taken and mixed, diluted to known volume and passed on column at rate of 5-10 ml/mt. The acid present in combined effluent is titrated with N/10 sodium hydroxide solution and concentration of sulphate calculated.

(ii) *Removal of interfering ions:* Supposing a mixture containing cations and few anions is passed on cation exchanger column, in one step all anions which interfere and which have negative charge pass through resin leaving behind cations on the column.

(iii) *Isolation of trace metals:* One of the beautiful part of the entire ion exchange separation is, it permits separation not only at macro but microgram levels. This method mainly facilitates concentration of species in small volume of the solution. The similar technique can be resorted for determination of anions also.

(iv) *Deionisation of Water:* A significant application of both the anion and cation exchange method is the deionisation of water. Water has temporary hardness due to the presence of $Ca(HCO_3)_2$ or $Mg(HCO_3)_2$ which can easily be removed by boiling, or addition of lime water. However, permanent hardness which is attributed to the presence of sulphates and chlorides of calcium and magnesium cannot easily be eliminated. The commonly used method for softening of such water is the lime soda ash process wherein $Ca(OH)_2$ removes temporary hardness while Na_2CO_3

removes insoluble salts of calcium and magnesium as carbonates. Unfortunately, this is not feasible on a large scale.

In ion exchange methods one uses a cation exchange resin in the hydrogen form H_2R, while the anion exchange resin is used in the OH form e.g. RNH_3–OH. The overall transformation occurs as follows:

$$CaCl_2 + H_2R \rightarrow CaR + 2HCl$$
$$MgCl_2 + H_2R \rightarrow MgR + 2HCl$$
$$CaSO_4 + H_2R \rightarrow CaR + H_2SO_4 \qquad \text{(Removal of cations)}$$
$$MgSO_4 + H_2R \rightarrow MgR + H_2SO_4$$

Then we pass it on an anion exchange column to undergo the reaction

$$HCl + RNH_3 - OH \rightarrow RNH_3 - Cl + H - OH \qquad \text{(Removal of anions)}$$
$$H_2SO_4 + RNH_3 - OH \rightarrow RNH_3 - SO_4 + H - OH$$

The greatest advantage of the technique is to reutilise the resin after regeneration

$$CaR + 2HCl \rightarrow H_2R + CaCl_2 \qquad \text{(Regeneration of cation exchanger)}$$
$$Mg-R + 2HCl \rightarrow H_2R + MgCl_2$$

Similarly an anion exchanger is also regenerated as

$$RNH_3-Cl + NH_4OH \rightarrow RNH_3-OH + NH_4Cl \qquad \text{(Regenerates anion exchanger)}$$
$$RNH_3-SO_4 + NH_4OH \rightarrow RNH_3-OH + (NH_4)_2 SO_4$$

The principal merit of deionisation is that it is rapid and is not time consuming like distillation. The water so obtained after treatment is equivalent to conductivity water. The most significant advantage is that potable water contains traces of iron and manganese, such ionic impurities can be removed only by this method.

12.10 APPLICATION OF ION EXCHANGE IN METAL SEPARATIONS

Ion exchange methods have several applications. They were first used for deionisation or softening of water, then for the determination of total salt concentration and are also used for the removal of interfering ions of opposite charges from the solution. Anion exchange techniques are useful for the separation of anions.

However, the most significant use of ion exchange methods is in metal separations. Both cation exchange as well as anion exchange methods have been used to remarkable advantage. Thus cation exchange chromatographic methods have been used for the separation of magnesium, calcium, strontium and barium on Dowex 50 with 1.5 M ammonium lactate buffered at pH 7. The outstanding separation was the separation of lanthanides using ammonium citrate-citric acid, buffered at pH ~ 5.5 on Dowex 50. The elements were eluted in the reverse order of their atomic number. This was possible due to the effect of lanthanide contraction. From atomic number 57-71 the atomic size decreased but the size of hydrated cationic series on the contrary increased. The net result was that when a complexing agent like ammonium citrate was passed, hydrated cations having the largest size *viz.* Lu (71) readily formed complexes and came out first from the column. On the same basis, elements with similar chemical properties such as zirconium, hafnium or niobium, and tantalum have been separated by cation exchange methods.

The outstanding revolution in analytical chemistry was however been made by anion exchange methods. Not only anions but cations also have been separated by anion exchange methods by converting them into suitable negatively charged complexes. The simplest separation of anions is of chloride, bromide and iodide. They are separated on a Dowex 21 K by making use of sodium nitrate (0.5 M) as the eluting agent. By a stepwise process, it is possible to desorb them sequentially from the column. The separations of polythionate anions like $S_2O_6^{2-}$, $S_3O_6^{2-}$, $S_4O_6^{2-}$, $S_5O_6^{2-}$ have also been carried out using 1, 3, 6 and 9M hydrochloric acid respectively as the eluants.

Anion exchange separation of metals is carried out in various acid media due to their capacity to form anionic complexes, e.g. chloride, nitrate, sulphate media.

Ion Exchange Chromatographic Separations

The real crux lies in the fact that it offers the excellent method for separation of several ions by exploiting the difference in affinity of ions toward resin e.g. separation of alkali metals is depicted in Fig. 12.14. The technique of gradient elution was used. Ammonium chloride in increasing concentration from 0.05-0.75 M was used as eluant. Finally caesium was eluted with 4-5M hydrochloric acid.

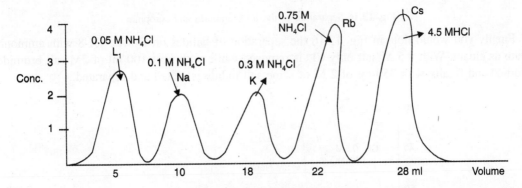

Fig. 12.14 Separation of alkali metals

In Fig. 12.15 depicts separation of alkaline earths on Dowex 50 with 1.5 M ammonium lactate (pH-7) in different volumes of the eluant.

Here one uses 1.5M ammonium lactate buffered at pH 7.0. This is an example of complexation of alkaline earths with lactate at pH 7.0 when in 20 ml/Mg, 100 ml Sr and 225 ml. Ba and 500 ml Ra is eluted. As volume of eluant increases strongly bound cations are easily eluted from the column.

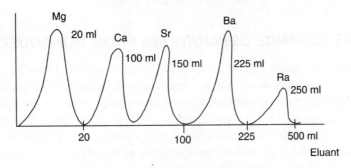

Fig. 12.15 Separation of alkaline earth

In Fig. 12.16 we consider separation of thorium, uranium and europium by anion exchange chromatography in sulphate media.

It was observed that all these trace metals formed anionic sulphato complex with 0.1-1.5 M sulphuric acid. When passed on anion exchange column containing Dowex 1-X8 all anionic complexes were sorbed. Since sulphato complex of europium was weak, first it is eluted, then thorium which formed moderate complex while strong uranium complex was eluted last. Fig. 12.16 is an instance of anion exchange separation of anionic sulphato complexes.

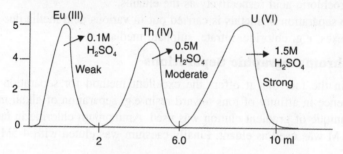

Fig. 12.16 Separation of thorium, uranium and europium

Finally you would note in Fig. 12.16 the separation of halides on Dowex 1X-8 with ammonium nitrate as eluant. With 0.5 M salt only Cl⁻ is removed while with next 100 ml of 2M salt bromide is desorbed and finally with 350 ml of 2 M of same salt iodide is eluted and separated.

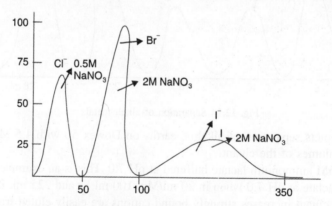

Fig. 12.17 Separation of halides

12.11 CATION EXCHANGE SEPARATIONS FROM NONAQUEOUS MEDIA

The specific advantages in such a system are that, the higher separation factor for two ions reveals that they are amenable to easy separation, and secondly the total volume of eluant required is relatively small thereby facilitating rapid separations. Further, the volume required for maximum recovery of metal *i.e.*, peak elution volume (V_{max}) is small which in turn increases the efficiency of separation. The ratio in which two ions are separated is much larger and separations are feasible at lower metal concentration. Many nonaqueous solvents like methanol, ethanol, propanol, acetone, dioxane, tetrahydrofurone are used in conjunction with mineral acids as eluants. Such solvents influence water solvent distribution, swelling, solvation, solubility and dissociation of the complex. The choice is further decided by the nature and structure of the nonaqueous solvent, its dielectric constant and solubility in water. The separation of manganese from iron, cobalt, nickel, copper and gallium would not have been possible in the absence of mixed solvents. Thus manganese is eluted with 2M hydrochloric acid containing 80% acetone while uranium or similar ions are separated with 0.5 M hydrochloric acid containing 90% acetone. (Table 12.5).

Table 12.5 Metal separations by ion exchange methods in mixed solvents

Sr. No.	Resin	Metal	Eluant for first ion	Eluant for other ions	Separated from
1.	Dowex 50WX12	Co (25 mg)	1 M NaCl	Cu (O.5 M H_2SO_4)	Separation from several metals
2.	Dowex 50WX8	Tl (20 mg)	1 M CH_3 COONH$_4$ (300 ml)	Pb (1 M amm. acetate (200 ml)	Separation in varying ratio
3.	Dowex 50WX8	Sc (5 mg)	1M H_2SO_4	Y (4M HCl)	La, Ce, Sm sepn.
4.	Dowex 50WX8	Mn (5 mg)	2M HCl+80% acetone	U (0.5 M HCl + 90% acetone)	Fe, Co, Cu, Ga.
5.	Dowex 21K	In (15 mg)	2M HCl	Zn with water	Anionic complex pH 2.2
6.	Dowex 21K	Th (10 mg)	0.25 M H_2SO_4,	Zr (1M H_2SO_4)	Ti Zr, Hf
7.	Dowex 21K	Ga (7 mg)	0.25 M HNO_3	Tl (No complex)	Sequential separation of Ga, In, Tl, Al.
				Al (0.25M NH_4NO_3 In (1M HNO_3)	Malonic complex pH 4.3

12.12 ANION EXCHANGE SEPARATIONS OF METALS

It is possible to carry out separations of metals on an anion exchange resin by converting them into negatively charged anionic complexes. Such anionic complexes can be formed with mineral acids like hydrochloric, nitric and sulphuric acid and also with organic acids like malonic, citric and tartaric acids. A typical example of separation is of Ga, In, Tl and Al from malonic acid media. At pH 4.3, 0.1 M of malonic acid forms complexes with all three metals except Tl. So, when a mixture of all these ions in 0.1M malonic acid buffered at pH 4.3 is passed on a Dowex-21K an anion exchange column, thallium is not sorbed as it remains cationic while all others forming anionic malonate complexes are sorbed. Then aluminium is eluted with 200 ml of 0.25 M sodium nitrate followed by elution of gallium with 100 ml of 0.25 M nitric acid and finally indium with 110 ml of 1M nitric acid (Table 12.5, Fig. 12.18). Thus it is possible to attain sequential separations on an anion exchange column of chemically similar methods.

12.13 LIQUID ION EXCHANGERS

These are organic exchangers similar to solid resin but are present in liquid form. Like solid analogue, we have liquid anion exchangers and liquid cation exchangers. The long chain amines constitute liquid anion exchangers. Since their behaviour is similar to solid ion exchange resin, they are termed as liquid

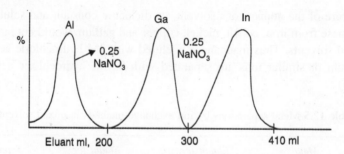

Fig. 12.18 Separation of aluminium, gallium indium, thallium.

ion exchangers. The high molecular weight amines (HMWA) constitute class of liquid anion exchangers depending upon the type of amine we have primary, secondary, tertiary and quarternary amines. They have 18 to 27 carbon atoms. The aliphatic amines are also used e.g. Amberlite LA-1 and LA-2. The first is dodecenyl (trialkyl methyl) amine while latter is lauryl (trialkyl methyl) amine. The liquid anion exchanger form salt as

$$RNH_3\text{-}X + \left[MCl_4^- \right] \rightleftharpoons \left[RNH_3\text{-}MCl_4 \right]_R + X^-$$
anionic complex

$$...(i)$$

While liquid anion exchangers are acids insoluble in water. Total numbers of carbon atoms are 10 to 28. They all involve transfer of species of ions across liquid-liquid interface. The system is similar to solvent extraction by ion pair formation. They do have small solubility in water, large specific exchange capacity, large solubility in organic solvents, high selectivity, stability and small surface activity. They are more useful than solid analogues. They have faster exchange rate, possess no interstitial water, have no problem with suspended matter, provide facility for true counter current process, have greater permselectivity. Both these resins have several analytical applications e.g. uranium was separated from iron, aluminium, calcium on Aliquat 336S as in sulphate complex. The reaction is as follows:

$$\left[UO_2(SO_4)_2 \right]^{2-} + RNH_3\text{-}X \rightleftharpoons RNH_3 - UO_2(SO_4)_2 + X^- \text{ (salt of extractant is used)}$$

While liquid cation exchangers are similar to solid cation exchange resins as H_2R. They all have the replaceable group as H^+. Since these resins are present in the liquid form, they are called liquid cation exchangers. The simplest example is HDEHP (i.e., bis 2 (ethyl hexyl) phosphoric acid and DNN (i.e., dinnonyl napthalene sulphonic acid). They act like solid cation exchangers during separation of ions:

$$(H\text{-}DEHP)_0 + CaCl_2 \rightleftharpoons (Ca\text{-}DEHP)_0 + 2HCl$$

This represent exchange of calcium on cation exchange resin in the liquid form. One has to use inert diluents like xylene, toluene or benzene for the purpose of separation. Unlike solid cation exchange resins they can be converted back to H^+ form as $(Ca\text{-}DEHP)_0 + HCl \rightleftharpoons CaCl_2 + (H\text{-}DEHP)_0$ (regeneration). In above equation subscription o-represents, organic phase. Many useful separations of rare earths were accomplished by these liquid cation exchangers.

The metal can be recovered from organic phase with Na_2CO_3 as

$$RNH_3\text{-}UO_2(SO_4)_2 + Na_2CO_3 \rightleftharpoons UO_2SO_4 + Na_2SO_4 + RNH_3\text{-}CO_3.$$

The resins are also used in liquid membrane and ion selective electrode.

12.14 INORGANIC EXCHANGERS

The first ion exchange material discovered by Thomas and Way was zeolite ($Na_2OAl_2O_3$. 24 H_2O). Although organic resins were good they could not withstand high temperature and were degraded on the exposure to radioactive radiation. This led to discovery of synthetic inorganic ion exchangers, which had property of not swelling or shrinking in concentrated solutions. However, they showed slow reaction.

12.14.1 Hydrous oxides

Hydrous oxides of chromium, zirconium, tin and thorium behave like anion exchangers. They are like polymers crosslinked by oxygen bonds between the metal atoms. OH^- group is hydrolysed in acid media and they behave like anion exchanger. At low pH, they show better exchange capacity e.g. ZrO_2 behaves as cation and anion exchanger in pH region of pH = 5-8. Dehydration needs high temperature, but exchange form is totally lost. Silica gel has ion exchange property wherein exchange is complete in 5 mts. Hydrated Al_2O_3 has anion exchange property and can exchange anions easily.

12.14.2 Salts of multivalent metals

Hydrous phosphate oxide behaves as cation exchanger e.g. $xZrO_2.yP_2O_5.z2H_2O$. The property depends upon method of preparation of zirconium phosphate. Large (p/2r) ratio favours large capacity. High temperature improves selectivity coefficient. On drying salt increases exchange capacity. The separation of two ions by these materials is accelerated by ratio of high selectivity coefficient. Apart from Zr, one can use salts of Ti, Sn, Sb (V). The antimonate, arsenae, molybdate, tungstate can replace phosphate to give new exchangers.

12.14.3 Salts of heteropoly acids

$(NH_4)_3PO_412MoO_3\ 3H_2O$ is ammonium molybdato phosphate is best known salt of heteropoly acids. It behaves like cation exchanger; such property depends upon ability of other cations to replace ammonium ion. They are insoluble and P can be replaced by As, Si and W whereby their properties are not changed easily. The alkali metals are readily absorbed by them. H^+ ion play minor role in exchange reaction. They are used in column also. Sodium was separated from potassium by this exchanger. Thallous tungsten phosphate was used for the separation of calcium from fission products.

12.15 CHELATING RESINS

They are also called as the specific ion exchangers. Anions of weak acids tend to associate with H^+ ions forming weakly dissociated complex. A compound dipicryl (Fig. 12.19) was synthesised as a specific resin for potassium. Many compounds forming chelates are incorporated into resin by polycondensation with phenols and aldehydes. Anthralinic acid (Fig. 12.20) is specific for zinc. The chelating resins can be also introduced into styrene type resin e.g. iminodiacetic acid groups are used.

All are cation exchange resins. The mobility of counter ions in resin is considerably reduced. Selectivity is inversely proportional to rate. Similar to chelating cation exchanger one can synthesise chelating anion exchanger. They included

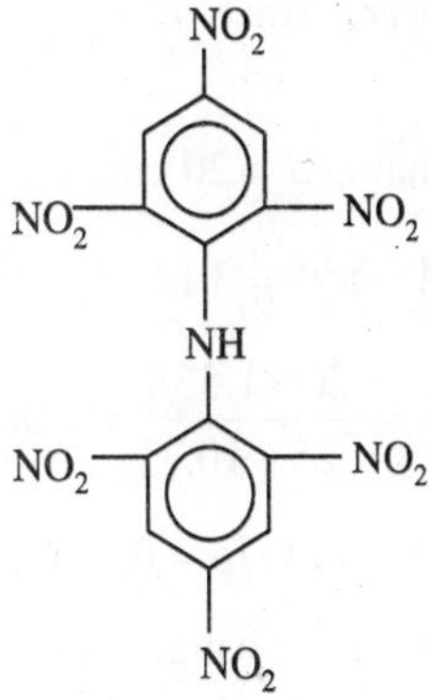

Fig. 12.19 D-picrylamine (I)

strong base quarternary phosphonium groups and tertiary sulphonium group. Aliphatic polyamines are not weakly basic and can be condensed with aldehydes e.g.

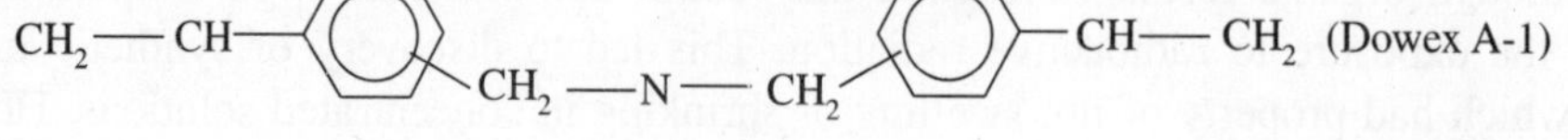

$$CH_2 — CH \text{—} \bigcirc \text{—} CH_2 — N — CH_2 \text{—} \bigcirc \text{—} CH — CH_2 \text{ (Dowex A-1)}$$

Fig. 12.20 Anthralinic acid.

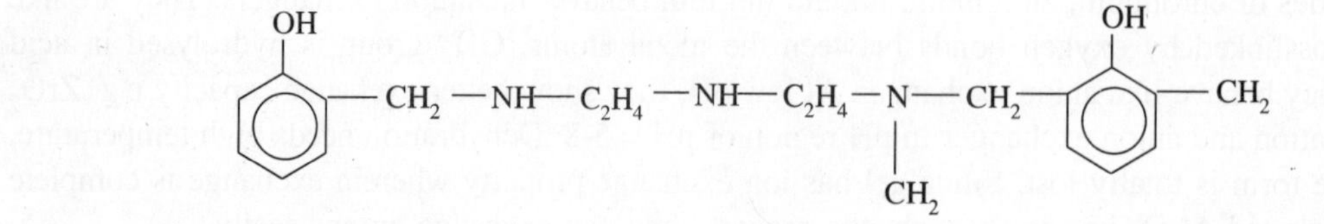

$$\underset{OH}{\bigcirc} — CH_2 — NH — C_2H_4 — NH — C_2H_4 — N — CH_2 \text{—} \underset{OH}{\bigcirc} — CH_2$$
$$| \\ CH_2$$

Fig. 12.21 Polyfunctional resin.

Recently resins with quaternary phosphonium groups and tertiary sulphonating group were synthesised. Like solid resin, addition polymer resins are chloromethylated and then treated with NH_3. As regards applications of dipicrylamine resin, it was used for selective sorption of potassium. The phenol condensation product of pyrogallol was used to separate bismuth and lead. A phenol condensed resin containing dimethyl group was used for the separation of nickel. Resorcinol condensation produces with 8-hydroxyquinoline was good for separation of copper and cobalt. The polystyrene condensed iminodiacetic acid resin was used for separation of alkaline earths, copper and mercury. Ion exchange resins containing both acidic and basic groups are capable of binding many ions at varying pH of solution. They are called amphoteric resins.

SOLVED PROBLEMS

12.1 In cation exchange studies of copper on a column of 1.4 × 20 cm filled with Dowex 50W-X8, the peak elution volume was 25 ml. If the free column volume was 10.3 ml and the density of resin 0.80 gm/ml, calculate the elution constant and weight and volume distribution coefficient.

Answer: Since $E = \dfrac{d \cdot A}{(V_{max} - v)} = \dfrac{1}{D_V} = \dfrac{1}{\rho D_W}$

We have $d = 20$ cm, $V_{max} = 25$ ml, $v = 10.3$ ml, $A = \pi\gamma^2 = 4$, $\rho = 0.80$ density of resin.

$A = \pi\gamma^2 = 3.14 \times 0.70 \times 0.70 = 1.538$, so we have

$E = \dfrac{20 \times 1.538}{(25 - 10.3)} = 2.089$, $D_V = E^{-1} = 0.478$ if $\rho = 0.80$

$D_W = D_V / \rho = 0.597$.

12.2 In anion exchange chromatographic studies of lead in malonate media with 1.4 × 24 cm column packed with Dowex 21K, the elution constant was found to be 0.60. Calculate the peak elution volume in such elution.

Answer: $d = 24, A = \pi\gamma^2 = 3.14 \times (0.70)^2 = 1.54$

Hence on substituting these values we have,

$$E = \frac{dA}{V_{max}} = 0.60 = \frac{24 \times 1.54}{V_{max}}$$

$$= 0.60 \times V_{max} = 24 \times 1.54$$

or $V_{max} = \dfrac{24 \times 1.54}{0.60} = 61.6$ ml peak elution volume.

12.3 Calculate the density of resin Dowex 1-X8 in a particular elution, when the peak elution volume is 75 ml, column height is 18 cms., column diameter is 0.80 cms and the weight distribution coefficient D_W is 2.6.

Answer: $V_{max} = 75$ ml, $d = 18$ cm, $\gamma = 0.80$

$$E = \frac{18 \times 2}{75} = 0.48, \text{ now } D_V = \frac{1}{F} = 2.08$$

$$D_V = \rho D_W \text{ or } \rho = \frac{D_V}{D_W} = \frac{2.08}{2.60} = 0.80 \text{ gm/ml.}$$

PROBLEMS FOR PRACTICE

12.4 A standard solution of sulphuric acid 0.05 M (100) ml is synthesised by passing Na_2SO_4 through Amberlite IRC-120 cation exchange resin (H^+ form). The eluted H_2SO_4 is used as an acid, 25 ml of Na_2SO_4 solution is passed through the column, with subsequent passing of 50 ml of water and the total volume is diluted to 100 ml. How many grams of Na_2SO_4 should be dissolved in volume of water to produce a standard solution?

12.5 Two anion exchange columns packed with De Acidite FF' are taken. Through the first column, 100 ml of standard 0.1 mg/litre of fluoride ion is passed. Through the other column one litre of an unknown solution of fluoride is passed and is then eluted with the zirconium fluoride complex and the latter is analysed spectrophotometrically as alizarine complex at 530 nm. If the unknown solution has an absorbance of 0.560, with the standard solution giving a reading of 0.480, calculate the concentration of the unknown solution in ppm.

12.6 To 100 ml of 0.1N HCl was added 5 gm of Zeokarb-225 in its sodium form; when equilibrium was reached, $[H^+]$ decreased to 0.015N. Determine the exchange capacity of resin.

12.7 500 ml of 0.05N solution of calcium was passed through a column containing 5 grams of resin. In the effluent of 50 ml the amount of calcium detected was 0.003, 0.015, 0.025, 0.050 and 0.050. Determine the exchange capacity of the exchanger for calcium ions.

12.8 Two grams of Agrion-20 (Ca-form) was added to 25 ml 1N solution of NaOH. When equilibrium was reached the solution was titrated with 0.8 N HCl. Calculate exchange capacity of resin.

LITERATURE

1. R. Kunin, *"Ion Exchange Resins"*, John Wiley & Co. (1958).
2. R. Kunin, *"Elements of Ion Exchange"*, Reinhold & Co. (1960).
3. G.H. Osborn, *"Synthetic Ion Exchangers"*, Chapman and Hall Co. (1961).
4. O. Samuelson, *"Ion Exchange Separations in Analytical Chemistry"*, John Wiley & Co. (1963).
5. J. Inczedy, *"Analytical Applications of Ion Exchangers"*, Pergamon Press Ltd. (1966).
6. F. Helfferich, *"Ion Exchange"*, McGraw Hill & Co. (1962).
7. W. Rieman, H.F. Walton, *"Ion Exchange in Analytical Chemistry"*, Pergamon Press Ltd. (1970).
8. J.E. Salmon and D.K. Hale, *"Ion Exchange–A Laboratory Manual"*, Butterworth Publications (1959).
9. H.F. Walton, W. Reiman, *"Ion Exchange in Analytical Chemistry"*, Pergamon Press (1970).
10. C. Calman, T.R.E. Kressman, *"Ion Exchangers in Organic and Biochemistry"*, Butterworth Publications (1957).
11. C.B. Amphlet, *"Inorganic Ion Exchange"*, Elsevier Publications (1964).
12. J.A. Kitchner, *"Ion Exchange Resins"*, John Wiley NY (1961).
13. F.C. Nachod, J. Schubert, *"Ion Exchange Technology"*, Academic Press (1956).
14. Y. Marcus, A.S. Kertes, *"Ion Exchange and Solvent Extraction of Metal Complexes"*, Wiley Interscience (1969).
15. S.M. Khopkar, *"Solvent Extraction Separations with Liquid Ion Exchangers"*, New Age International Publishers (P) Ltd. (2007).

Chapter 13

Ion Chromatography

The techniques of gas chromatography and high performance liquid chromatography have become very popular with chemists and analysts for several reasons. One of the principal reasons is the speed of analysis. One can easily perform separations in a few minutes. The speed or rapidity of separation and subsequent analysis may be attributed to efficient detectors which are available in both systems. In comparison, ion exchange chromatography, inspite of being a selective and versatile method of separation, lagged behind only because following separation, analysis of the eluate was cumbersome and time consuming. In many instances, one had to resort to dry ashing before analysis could be carried out by spectrophotometry or polarographic techniques. Those using radiotracers had some edge in rapid analysis. However, with the discovery of efficient detectors, ion exchange chromatography has become a very popular tool with the analytical chemist.

13.1 ION CHROMATOGRAPHY AS A SEPARATION TOOL

This is an abridged version of cation or anion exchange chromatography. Since both cations or anions are used, it is broadly termed as ion chromatography. Further in addition to the phenomenon of the exchange of ions, other mechanisms such as ion pair chromatography (MPIC) and ion exclusion phenomena (HPICE) are also used. Hence it is given the broad nomenclature of ion chromatography by deleting the word exchange.

The technique can be broadly classified into three groups. In real time used ion exchange chromatography, there is an exchange between a mobile phase and a stationary phase containing the exchangeable group where the stationary phase is resin; in ion pair chromatography there is adsorption on a stationary support of a neutral resin of the component which is followed by elution with a complex forming agent; in ion exclusion chromatography the separation of ions is feasible by exploiting the difference in geometry of molecules with resin as the stationary phase. The principal advantages of ion chromatography are its selectivity, speed of analysis, sensitivity of detection and facility of simultaneous analysis. One can summarise several merits of this technique. Separation is possible in the minimum amount of time; one can handle nano (10^{-9}) concentrations, very small volumes of an aliquat are required and the technique facilitates multi element separation, with the possibility of simultaneous analysis with greater selectivity of separation. The eluants used are very dilute or weak. One can even use a mixture of eluants. It is possible to use a variety of resins with a variety of detectors in the system for subsequent analysis.

13.2 STRUCTURE AND CHARACTERISTICS OF RESINS

Unlike conventional ion exchange resins used in analysis, in ion chromatography specially developed anion and cation exchange resins are used. Structurally they are very similar to resins used in normal columnar methods but frequently differ in cross linkage, particle size and exchangeable group. Typical anion exchange resins are listed in Table 13.1. The structure of an anion resin (Fig. 13.1) as well as a cation exchange resin (Fig. 13.2) is also depicted in the corresponding diagrams.

Table 13.1 Characteristics of an anion exchange resin

Column Resin	Particle size (um)	% Cross linking	Nature of particle	Hydrophobicity
HPIC-AS1	25	5	medium	Hydrophobic
AS2	25	5	medium	Hydrophillic
AS3	25	2	small	Hydrophillic
AS4	15	3.5	small	Hydrophillic
AS5	15	1.0	small	Hydrophobic
AS6	10	5	large	Hydrophillic
AS7	10	5	large	Hydrophillic

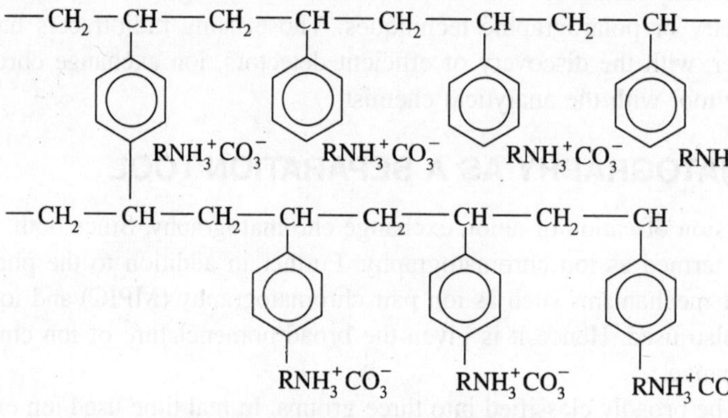

Fig. 13.1 Structure of anion exchange resin

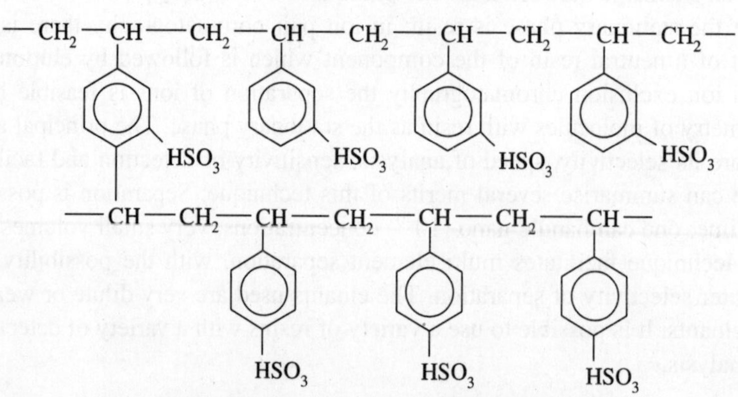

Fig. 13.2 Structure of cation exchange resin

13.3 ELUANTS USED IN SEPARATION

Irrespective of whether it is anion exchange separation or cation exchange separation, the eluants used are not same as in columnar methods. Since it is necessary to detect a signal, anions which give a strong signal are preferred. Typical eluants used in anion exchange chromatography are listed in Table 13.2 while eluants used in cation exchange separations are listed in Table 13.3.

Table 13.2 Eluants for anion separations

Eluant	Eluting ion	Suppressor used	Nature of the eluant
$Na_2B_4O_7$	$B_4O_7^-$	H_3BO_3	too weak
NaOH	OH^-	H_2O	weak
$NaHCO_3$	HCO_3^-	H_2CO_3	weak
Na_2HCO_3	HCO_3^-/HCO_3^{2-}	H_2CO_3	moderate
Na_2CO_3	CO_3^{-2}	H_2CO_3	strong
$NaNO_3$	NO_3^-	–	quite strong
NaI	I^-	AgI	very strong

Table 13.3 Eluants for cation separation

Column	Eluant	Separated ions in order of elution			
HPIC–CS$_2$	0.01 M$H_2C_2O_4$ +0.0075 M Citric acid (pH 4.2 with LiOH)	Fe(II) Co(II)	Cu(II) Pb(II)	Ni	Zn
HPIC–CS$_2$	0.04M Tartaric acid +0.012 M Citric acid (pH 4.3 with LiOH)	Ni Fe(II)	Zn Cd	Co (III) Mn(II)	Pb (II)
HPIC–CS$_2$	0.2M $(NH_4)_2SO_4$ +0.01M H_2SO_4	Al	Fe(III)		
HPIC–CS$_2$	0.01M $(NH_4)_2 SO_4$ +0.05M H_2SO_4	UO_2^{2+}	Zn		

However, in the separation of cations with dicarboxylic acid, the pH of elution as well as the pK value of the acid is of prime importance while determining the kind of eluant. Such eluants to be used are listed in Table 13.4.

Table 13.4 Eluant sequence and pK of organic mono and dicarboxylic and analogous acids

Acid	pK_1	pK_2
Oxalic	1.27	4.29
Malonic	2.88	5.68
Succinic	4.21	5.64
Tartaric	3.04	4.37
Mandellic	3.36	–
Hippuric	3.64	–
Benzoic	4.20	–

13.4 SUPPRESSOR COLUMN IN CHROMATOGRAPHY

This is the most important component of a chromatography column. In usual separations, the eluant, after emerging from the column with the ion in consideration, is directly taken for analysis. Such analysis involves volumetric, spectrophotometric, polarographic methods which by and large are time consuming. The effluent from the column is once again passed on the suppressor column to lead to elution of ions (anion/cation) as the case may be. These eluants are capable of giving a good signal in electrical conductivity measurement. Supposing one is using $NaHCO_3$ or $NaCO_3$ as the eluant on the anion exchange, when it is passed through the suppressor column it gives out strong anions to give sharp signals in the detector. The overall mechanism of suppression can be summarised as:

$$\text{Resin } SO_3^- + NaHCO_3 \longrightarrow \text{Resin } SO_3Na + CO_2 + H_2O \text{ (supression)}$$

$$\text{Resin } SO_3^+ + Na_2CO_3 \longrightarrow \text{Resin } SO_3Na + CO_2 \text{ (supression)}$$

$$\text{Resin } SO_3^-H^+ + NaCl \longrightarrow \text{Resin } O_3^+Na^+ + HCl \text{ (strong anion released)}$$

$$\text{Resin } SO_3^+H^+ + NaBr \longrightarrow \text{Resin } SO_3^+Na^+ + HBr \text{ (strong anion released)}$$

The net reaction can be summarised as:

$$RNH_3\text{-}X + H_2CO_3 \longrightarrow RNH_3\text{-}CO_3 + HX \text{ (strong signal)}$$

A similar reaction can be depicted for a cation exchange column

$$H_2R + NaCl \longrightarrow Na\text{-}R + HCl \text{ (high conductivity)}$$

$$H_2R + NaBr \longrightarrow Na\text{-}R + HBr \text{ (high conductivity)}$$

The suppression reaction is shown pictorially for an anion membrane suppressor in Fig. 13.3.

13.5 DETECTORS USED IN ION CHROMATOGRAPHY

One can use different kinds of detectors including conductivity detectors, photometric detectors, fluorimetric and amperiometric detectors as shown in Table 13.5.

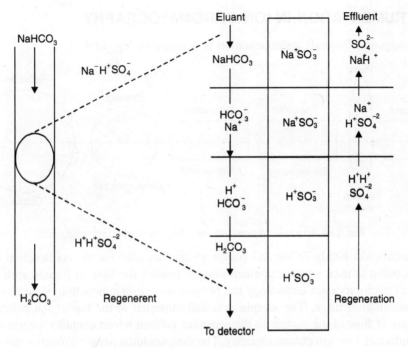

Fig. 13.3 Mechanism for supression columnar reaction

Table 13.5 Different kinds of detectors

Detector	Measured	Applications
Conductivity	Electrical conductivity	Separations of cations and anions with eluants with pK < 7.0
Photometric	Absorbance in uv/visible	Chromophore formed with PAR, Tiron etc.
Fluorimetric	Fluoroscene intensity	Amino acids, polyamines reaction with o-phenythaldehyde
Amperiometry	Diffusion current	Anion and cations with eluant of pK < 7.0

The working of the conductivity detector is simple. Anions like Cl^- and Br^- give strong conductivity *i.e.* HCl/HBr have high electrical conductivity in comparison with NaCl/NaBr when they emerge from the suppressor column and are instantaneously measured. The intensity of conductivity is obviously related to the number of ion separated. In a photometric detector the sample after elution is passed through a premixed column to interact with a chromogenic ligand like 4 (2 pyridyl azo) resorcinol *i.e.* (PAR) to form a coloured complex with the cation which shows strong absorbance at 520 nm. An amperiometric detector is used if ions undergo oxidation/reduction reactions to generate a diffusion current. Finally in a fluorescence detector the intensity of delayed emission which is generally used for analysis of an organic compound is measured e.g. phenol aminoacid. Refractive index and AAS detectors are less frequently utilised. A plot of the intensity of physical property *versus* time is obtained and the area under the curve is measured.

13.6 INSTRUMENTATION IN ION CHROMATOGRAPHY

A schematic diagram of overall instrumentation is depicted in Fig. 13.4.

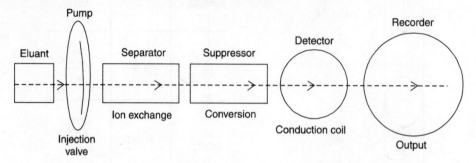

Fig. 13.4 Schematic diagram of ion chromatography instrumentation

Inert materials like Kel-F, Teflon and polypropylene are used for the construction of columns. A double piston pump is used which can electronically control the flow of liquids with accuracy. The advancement of microprocessor technology has boosted the use of conductivity detection of ions in the nanogram concentration range. The sample is usually injected at the top of the column. The pump controls the rate of flow of the sample to the separator column which contains the ion exchange resin specially manufactured for ion chromatography. The sample eluted passes through a suppressor column to release the ion/solution with high conductivity. The signal is fed to a recorder. From the elution profiles the concentration of ions is measured.

13.7 ANALYTICAL APPLICATIONS OF ION CHROMATOGRAPHY

The significant separation of anions as well as cations is possible with the use of ion chromatography. It has also facilitated the study of environmental speciation. No other method provides such an excellent means for analysis of anions. No doubt, the ion selective electrode provides a means of analysis of anions by ion analysers like Orion 801. But ion chromatography facilitates the separation of anions (as many as seven) in a period of just 20 minutes in a volume of 40 ml of solution. Some interesting separations of anions from tap water and rain water are described in Fig. 13.5 and 13.6 respectively.

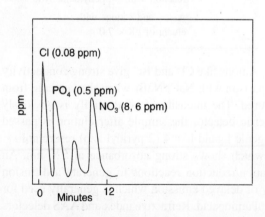

Fig. 13.5 Separation of chloride, phosphate and nitrate in tap water

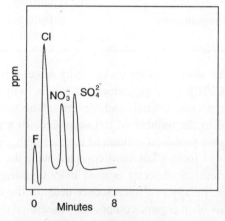

Fig. 13.6 Separation of fluoride, chloride, nitrate and sulphate in rain water

The technique is also used for analysis of cations. For instance separation of cations from drinking water (Fig. 13.7) and from a sample of soil is shown in Fig. (13.8).

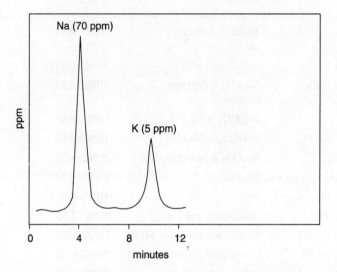

Fig. 13.7 Separation of alkali metals in drinking water

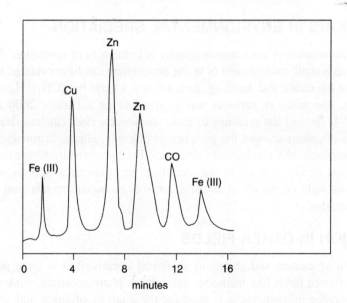

Fig. 13.8 Separation of multicomponent mixture

From these figures one notices that a small volume (50 μl) is required for work. The overall separation is feasible in 8-16 minutes. The flow rate is 2ml/min i.e. 32-40 ml of weak (~0.0075M) eluant is needed and ions handled are at ppm concentrations. Table 13.6 summarises applications of ion chromatography in environmental analysis.

Table 13.6 Environmental applications of ion chromatography

Sr. No.	Separation	Eluant	Column	Detector
1.	Cl NO$_3$ SO$_4$	NaHCO$_3$ + Na$_2$CO$_3$ pH 4.2	HPIC–AS4	Conductivity
2.	Fe Cu Zn Ni	H$_2$C$_2$O$_4$ + LiOH	HPIC–CS2	Photometric
3.	F Cl PO$_4$ SO$_4$ NO$_3$	Na$_2$CO$_3$ + Ethylene diamine	HPIC–AS3	Conductivity
4.	I, CN	NaHCO$_3$ + Na$_2$CO$_3$	HPIC–AS5	Conductivity
5.	SeO$_2$ HPO$_4$, HA$_5$O$_4$	NaHCO$_3$ + Na$_2$CO$_3$	HPIC–AS4	Conductivity
6.	F Cl SO$_4$ NO$_2$	Na$_2$CO$_3$ + NaH$_2$BO$_3$	HPIC–AS3	Amperiometric
7.	F Cl SO$_4$ NO$_3$	Na$_2$CO$_3$	HPIC–AS3	Conductivity
8.	Li Mg Sr Ca Ba	HCl	HPIC–CSI	Photometric
9.	Fe Cu Zn Co Pb	oxalic-citric pH 2.0	HPIC–CS2	Photometric
10.	Ni Co Zn Pb Cd Mg	Tartaric + citric acids pH 4.3	HPIC–CS2	Photometric
11.	Pt Cu Cd Co Zn Ni	Oxalic acid pH 4.8	HPIC–CS2	Amperiometric
12.	Al from alkali metals	(NH$_4$)$_2$ SO$_4$ + H$_2$SO$_4$	HPIC–CS2	Photometry

13.8 APPLICATIONS IN ENVIRONMENTAL SPECIATION

The most significant application of ion chromatography is in the area of speciation. Toxic metal pollutants when present in an aquatic environment or in the atmosphere can be considered hazardous depending upon the nature of the cation and anion eg. mercury is not toxic but (CH$_3$)$_2$ Hg is. This is due to a difference in species. The study of aerosols was carried out on a Dionex 2000 instrument with a conductivity detector. It showed the presence of iron, aluminium, zinc, calcium, lead and nickel. The simultaneous analysis of anions showed the presence of chloride, nitrate, nitrite and sulphate by judicious work with photometric and conductivity detectors. Further correlation revealed the presence of ZnCl$_2$, NiCl$_2$, AlCl$_3$ and Fe$_2$O$_3$. This facilitated identification of chemical species in aerosol samples. This in turn has thrown light on the effect of these aerosols on human metabolism and physiological activity in living organisms.

13.9 APPLICATION IN OTHER FIELDS

Apart from separation of cations and anions in analytical chemistry, it is quite possible to use ion chromatography in different fields like medicine, agriculture, pharmaceutical, mining and petrochemical industries. For instance in medicine it is used for the analysis of sugar and amino acids. In the chemical industry for the analysis of impurities, in agricultural products for analysis of micronutrient in soils, in the petrochemical industry for the determination of sulphur compounds and in mining for analysis of inorganic anions and heavy metals.

Thus, this technique has very wide and interesting applications. It provides a sensitive method of analysis, especially in environmental monitoring. As many as 8-10 ions can be separated in 6-8 minutes. No sample preparation is required. It can be fully computerised with a microprocessor attachment to furnish chromatograms.

13.10 HYPHENATED TECHNIQUE IN ION CHROMATOGRAPHY

The combination of two instrumental technique leads to what is called as the "hyphenated methods". This is quite common with gas chromatography which is usually coupled with mass spectrometry (GC-MS) or inductively coupled plasma atomic emission spectroscopy (GC-ICP). It is discussed in detail in the latter chapter on mass spectrometry. We consider at this stage only those methods which are common hyphenated methods with the ion chromatography. The mass spectrometry or ICP-AES is coupled with IC to get better selectivity. We thus have hyphenated IC-ICP ie. ion chromatographic-inductively coupled plasma method or IC-MS *i.e.* ion chromatography coupled with mass spectrometry. In the former IC-ICP has advantages over simple use of ICP for complicated samples, specially at low concentrations (5-10 µg/L) or when interfering metals are present in the analyt. In IC usually chelation is preferred for reaction to proceed *i.e.*, IC-ICP is used for isolation and determination of lanthanide elements at microgram level or transition (d-block) metals from marine environment.

IC-MS is another technique very frequently used by chemists. This provides quantitative analysis and provides a tool for identification of the analyte. The volatile eluant or an eluant which can be suppressed before entering mass spectrometer is used to isolate the compounds. The oligosaccharides or sulphonic acids are best analysed by ICP-MS methods. The commercial instruments for the purpose are also available; but one must be very conversant with both IC and ICP or MS methods of analysis as the individual techniques.

13.11 ION CHROMATOGRAPHY BY CHELATION

Ion exchange resins containing the chelating groups are not routinely used. They are best for the usual complexation of transition and inner transition elements but the reaction is slow. However anionic metal complexes with dicarboxylic acid like malonic or succinic acids are quite selective. The chelating resins are used for metal concentration. In such separations however s-block metals show strong interference. Ion chromatography has selectivity with interference-free determination of elements.

13.12 ION CHROMATOGRAPHY BY ION PAIRING

It is also known as ion-interaction or dynamic ion exchange chromatography. A neutral hydrophobic stationary phase and mobile phase with hydrophobic ion (ion pairing agent) with charge opposite to charge of analyte is used. Ion pairing agent is associated with stationary phase and behaves as ion exchange functional group. For anions, quaternary ammonium ions are used in reversed phase HPLC with macroporous styrene DVB polymers as the support.

The technique is flexible and selective. Such selectivity is altered by changing ion pairing agent. The capacity can be increased by increasing concentration of the ion-pairing agent. Ion pairing technique is compatible with organic solvent modifiers e.g., methanol, acetonitrile in mobile phase. Organic modifiers can change selectivity. One demerit of ion pairing is that it impairs working of conductivity or the amperiometric detectors. The combination of technique of ion pairing chromatography with suppressed conductivity detection is termed as mobile phase ion chromatography (MPIC). Large hydrophobic ions is ideal for achieving good separations.

13.13 ION CHROMATOGRAPHY BY ION EXCLUSION

This phenomena is complementary to ion exchange. The stationary phase is resin, with high exchange capacity and same charge as of analyte. The weakly ionised species are easily first separated; white eluting strongly sorbed ionised species. The process depends on establishment of electrical neutrality between dilute mobile phase and stationary phase with high ion exchange sites. The high concentration of sites rules high concentration of counterions in stationary phase in order to maintain electrical neutrality. Diffusion process forces lead to equalising concentration in mobile. The stationary phases latter charged with same analyte ions. The difference in exchange potential facilitates neutral molecules to enter into stationary phase, while analyte is repelled or excluded. The weakly ionised species exhibit in between behaviour and are isolated from one another based on the extent of ionisation. Organic acids are best isolated with cation exchanger in hydronium ion form. Weak base microporous anion exchanger can separate weak bases.

13.14 COMPARISON OF ION CHROMATOGRAPHY WITH ION EXCHANGE

This is a process wherein charged analyte solute flows and competes with mobile phase ion of like charge for sites having opposite charge on stationary phase. Such sites are termed as functional groups. The functional groups are paired with eluant ions by maintaining electrical neutrality in the stationary phase. On sorption new ions compete with eluant at functional group sites. Analyte ions competing successfully are retained by resin little longer in comparison to other ions. The process involves rapid movement of analyte in two phases viz. mobile and stationary phase. More time is spent by ions in stationary phase due to immobilation. The ions travel in mobile phase. The key to separation is differential affinities for functional group for various ions. The difference in affinity leads to selectivity effect. It is decided by many factors viz. kind of functional group, nature of the stationary phase, characteristics of eluants and its concentration, and presence of non ionic species, kind of solvent, or ion pairing agent used and the temperature. This is possible by designing appropriate ion exchange column suiting the various parameters. The functional group type controls exchange capacity. It is independent of selectivity and can be raised or lowered without influencing selectivity of the resin.

Thus ion chromatography is dependable technique for separation of anions. In combination with inductively coupled plasma atomic emission spectroscopy, it takes care of the isolation and determination of cations in one step.

LITERATURE

1. S.M. Khopkar and M.N. Gandhi, *"Indian Journal of Environment Protection"*, **14**, 256 (1994).
2. J. Weiss, *"Handbook of Ion Chromatography"*, Dionex Publication (1985).
3. A. Tarter, *"Advanced Ion Chromatography"*, Wiley Interscience (1989).
4. S.M. Khopkar, *"Environmental Pollution Analysis"*, Wiley Eastern Ltd., New Delhi (1994).
5. J. Janta, *"Principles of Chemical Sensors"*, Plenum Press Ltd., New York (1990).
6. S. Cherian, M.N. Gandhi, S.M. Khopkar, *"Indian Journal of Environmental Protection"*, **12**, 3, 24 (1992).
7. W.F. Rich, F. Smith, L. McNeill, G.T. Sidebottom, *"Ion Chromatographic Analysis of Environmental Pollutants"*, Vol. 2, Ann Arbor Science, Michigan (1979).
8. E. Sawicki, J.D. Mulik, E. Wittgenstein, *"Ion Chromatographic Analysis of Environmental Pollutants"*, Ann Arbor Science (1978).
9. S.M. Khopkar, *"Environmental Pollution, Monitoring and Control"*, New Age International Publisher (P) Ltd. (2004).

Chapter 14

Adsorption Chromatography

This technique has been used on a large scale for preparative work. A large number of organic compounds have been separated by it. In comparison, it is not much used for separation of inorganic substances. Tswett used finely powdered adsorbent for chromatographic separation of pigments into different zones which were visually seen in the column. The different portions were separated into fragments and were eluted individually. Subsequently liquid or flowing chromatograms were developed. This was thus a truly color chromatography technique.

The forces acting between a solute and stationary support are of three kinds viz. the weak van der Waal's interaction, or dipole induced dipole interaction or London's forces. Due to difference in such forces for individual components a migration front is developed leading to separation. This is explained in separation of inert gases from the atmosphere on activated charcoal.

14.1 PRINCIPLES OF ADSORPTION CHROMATOGRAPHY

Adsorption chromatography is based on the retention of solute by surface adsorption. This technique is useful in the separation of nonpolar substances and constituents of low volatility. The ionic organic compounds are not separated in this way. In liquid-solid chromatography a solid substrate serves as the stationary phase. Separation depends on the equilibrium established at the interface between the grains of the stationary phase and mobile liquid phase and on the relative solubility of the solute in the mobile phase. Competition between solute and solvent molecules for adsorption sites establishes a dynamic process in which solute and solvent molecules are continuously coming in contact with the surface, residing there momentarily and then leaving to re-enter the mobile phase. While being desorbed, solutes are forced to migrate by forward flow of the mobile phase, only those molecules with a greater affinity for the adsorbent will be selectively retarded. The intra-molecular forces arises because, by its nature, the surface introduces a discontinuity into the system. There is a surface energy effect at any interface. Relatively weak London forces exist between all surfaces and any adsorbed molecule, or between nonpolar surfaces and polar adsorbed molecules. They induce dipoles in large nonpolar molecules and increase the existing dipole moments of polar compounds. Between strong electron donors and acceptors there exist charge transfer forces which culminate in the special case of hydrogen bonding. The strong forces arise between covalent and ionic bonded atoms. Adsorption is a physical phenomenon.

In adsorption chromatography, the partition coefficient (K_d) is equal to the concentration of solute in the adsorbed phase divided by its concentration in the solution phase. Adsorption isotherms gives the amount of solute adsorbed per unit weight of adsorbent (x/w) from a solution of concentration c at equilibrium, thus $K_d = \left(\dfrac{x/w}{c}\right)$. Adsorption isotherm curves are bell shaped and never linear. Adsorption linear capacity is defined as the maximum column loading possible without loss in linearity. It is high for adsorbents of high surface area and solids of uniform size. High temperature and strong eluants favour linearity of the isotherm. One factor in the separability of a pair of compounds is their difference in retention volume. A major theoretical problem is the dependence of solute retention volume on solute molecular structure and experimental conditions. The minimum plate height occurs at a low velocity of adsorption chromatography as a natural consequence of a small diffusion coefficient in the liquid phase. The degree of adsorption is expressed as the specific retention volume, as the volume of liquid passing through the column per gram of adsorbent.

14.2 EXPERIMENTAL SET-UP

A long column with a stop cock provision at the bottom is used to control the flow of the liquid. The height of the column should be 4–10 times the diameter. At least, half the column should be filled with adsorbent. Chromatobars are obtained by moulding the mixture of adsorbent e.g. H_2SiO_3 with plaster of paris to get self supporting columns where zones can be visually located by streaking with a reagent. Such chromatobars are used for the separation of terpenes. In working with volatile solvents, it is better to apply pressure to the top of the column instead of a vacuum to the receiver. This not only prevents evaporation of solvent but also prevents uneven flow rate of an eluant.

As regards adsorbents, several inorganic as well as organic adsorbents are used e.g. alumina, bauxite, magnesia, magnesium silicate, calcium hydroxide, silica gel and Fuller's earth. Amongst organic adsorbents, charcoal, sucrose and activated carbon are commonly utilised. Equilibrium is attained as the adsorbed layer consists of a monolayer covering the entire adsorbent surface volume (V_a) given by

$$V_a = \left(3.5 \times 10^{-8}\,\text{cm} \times \text{surface area in } \text{cm}^2/\text{g} - 0.01\%\ (H_2O)\right)$$

The adsorbent should be uniform in size. Impurities cause irreversible adsorption or tailing on a part of the substance to be separated. Better separations are obtained by pretreating the adsorbent with a substance which itself is strongly adsorbed. The step-graded adsorption column consists of segments filled with adsorbent with different degrees of saturation. This arrangement allows separation of components of widely differing adsorbility with a high total yield in one single chromatogram and efficient separations are obtained.

The choice of the mobile phase is governed by competition between solute and eluant molecules for a place on the adsorbent surface. Solute-eluant parameters reflect free energy for adsorption-desorption equilibria. When one uses esters they should be free from acids or alcohols. Chloroform should be free from acid. The general rule is that the eluting power of solvents being proportional to the dielectric constant is not always true. Portraying elution curves is of great significance.

14.3 CHEMICAL CONSTITUTION AND CHROMATOGRAPHIC BEHAVIOUR

The relationship between the chemical constitution of a substance and its behaviour on a chromatographic column depends upon whether or not adsorption, partition, or ion exchange are operating. The majority of organic separations undergo true adsorption. According to Tswett, the more double bonds and hydroxyl groups a molecule contains the more strongly it is adsorbed e.g. aldehyde is adsorbed less as compared with the corresponding alcohol as the latter contains -OH grouping which is conducive to adsorption. In the same way esters adsorb more strongly than hydrocarbons. The adsorption scale is: acid and bases > alcohols > aldehydes, ketones > halogen substituted esters > unsaturated hydrocarbons > saturated hydrocarbons. Hydrogen bonding decreases adsorbility, e.g. 2-hydoxyanthriquinone is more strongly adsorbed than tetrahydroxy anthriquinone because the latter has H-bonding. Hydrogen bonding has an influence on the eluting power of solvents. Conjugated double bonds have significant effects on the nature of adsorption. A solvent which contains a high proportion of groups for which the adsorbent has a strong affinity also increases adsorbility. In the case of strongly polar substances, the presence of one or more double bonds does not increase adsorption because the form of the surface of adsorption can explain the difference in behaviour of the substances which are being adsorbed. Thus the adsorption of an organic substance is a function of its chemical structure, the nature of adsorbent and of the solvent.

Other factors affecting relative positions of zones on column during adsorption are solvent effect, presence of impurities, pH, change in concentration, change in adsorbent, temperature, etc. At a high concentration a substance adsorbs more rapidly than at a low concentration, because increase in concentration affects the rate of movement of zone and changes the position of the zone of adsorption on column.

14.4 USES OF ADSORPTION CHROMATOGRAPHY

Adsorption chromatography finds extensive application in the separation of organic compounds. One must make a judicious selection of solvent, adsorbent concentration and temperature, e.g. in the separation of derivatives of alcohol aldehydes, ketones and acids, all these factors must be controlled. Several examples can be cited from separations of organic compound.

14.5 AFFINITY CHROMATOGRAPHY

This is also called biospecific adsorption chromatography or bioaffinity chromatography. The formation of specific dissociable complexes of biological macromolecules serves as the basis of their isolation and purification by this technique.

The name biospecific adsorption is preferred because a series of adsorbents like synthetic inhibitors bound by hydrophobic hydrocarbon chains adsorb macromolecules. The basic principle consists of exploitation of the property of biologically active substance to form stable, specific and reversible complexes. If one of the components is immobilised, a specific sorbent is formed for the second component of the complex. Thus affinity chromatography is based on the exceptional ability of biologically active substances to bind specifically and reversibly to other substances termed ligands or affinity ligands. If an insoluble affinant is made by covalent coupling to a solid support, and a solution of biologically active products is passed through a column containing this affinant, components having no affinity to the support would pass through the column unretarded, while products showing affinity for the insoluble affinity ligand are sorbed on the column. However, they can be subsequently released from the columns with attached affinant, dissolving the afliant or changing the solvent concentration. Such dissociation is possible by altering the pH, temperature, and ionic concentration.

The isolation and analysis of insoluble starch substances have been carried out by affinity chromatography. The isolation of antibodies on cellulose columns bound with a covalently attached antigen has also been carried out. Enzymes like tyrosinase have also been isolated. Nonspecific adsorption has been noted with hydrophobic or ionogenic groups. Apart from isolation of biologically active substances, it has been used for quantitative determination of antibodies by means of solid carriers with bonded antigens. Affinity chromatography is ideal for the study of interactions in biological processes. This has also been used for the study of the mechanism of enzymatic processes and elucidation of molecular structure. The use of affinity chromatography for the determination of inhibitor constants of enzymes has a great future. Affinity chromatography opens a new avenue for the study of interactions of biologically active substances and also in the future useful for the elucidation of the effect of microenvironment on the formation of these complexes. The applications of affinity chromatography is becoming varied as these methods make use of specific interactions of biologically active substances. It is used for low molecular weight enantiomeric pairs. It is also used to study the possibility of substituting natural peptide chains of enzymes with modified synthetic peptides. The problems of mechanism of enzymatic activity can be studied by this technique. It is also used for the elimination of undesirable substances from the blood of living organisms. Thus there are innumerable applications in biosciences and biotechnology. Affinity chromatography is used for isolation and purification of biological macromolecules like proteins and nucleic acid using the principal of biospecific adsorption. Specific ligands like proteins, dyes, nucleic acid are bound to the stationary support and is used as stationary phase. The ligands are chosen on strength of their binding affinity for analyte. On sorption on column, the substrate is attached to a ligand while residual solution passes through. Finally bound substrate is eluted from stationary phase with suitable eluant.

Agarose, polysaccharide, polyamides are used as the stationary supports which are linked to specific ligand via space arm which is a short alkyl chain used to eliminate any steric interferences. The eluant has double role to play in this technique viz. to permit firm binding of ligand and substrate and the breaking of the ligand substrate interaction. The difference between starting and ending elution is related to pH changes or the change in ionic concentration.

14.6 CHIRAL CHROMATOGRAPHY

The chiral compounds are best separated by this method. One can use chiral mobile phase additive and chiral stationary phases. Enantiomeric resolution is due to preferential complexation between chiral resolving agent and one of the isomers. The latter must possess chiral characteristics itself. They are solid supports like gel. Four chiral agents available are chiral stationary phase attractive interaction as H-bonding, π-π or dipole interactions leading to the complexation of solute chiral stationary phase. D-napthalene and cellulose acetate act as chiral stationary phase. Inclusion complex by fitting in cavity (e.g. crown ether) are also used. α, β cyclodextrin is commonly used. The chiral ligand exchange involves metal complexes (e.g. Cu) with both solute and selective ligand which is bound to stationary phase. The proteins bound to a modified silica support is also used. The complexation between proteins and chymotrypsin results in hydrophobic and polar interactions. Phenomenal activity in this area is expected in future.

SOLVED PROBLEMS

14.1 Calculate the absorbent activity (ϕ) of the fully activated alumina whose surface area was measured by nitrogen adsorption which was found to be 155 m^2/g, when $K_d = 68$ with napthalene-pentane as the standard eluant system. (Deactivated alumina has 3% water and $K_d = 3.3$).

Answer: According to Snyder, the surface energy function or adsorbent activity (ϕ and solute eluant parameter (f) are related as: $\log K_d = \log V_a + \phi f$. Now in the example $V_a = (3.5 \times 10^{-8}$ cm) $(155 \times 10^4$ cm^2/g) $= 0.0542$ cm^3/g because V_a the adsorbent surface volume is given by $V_a =$

$(3.5 \times 10^{-8} \text{ cm})$ (surface area in cm^2/g) for calcined adsorbent and $V_a = (3.8 \times 10^{-8})$ (surface area) $- 0.001$ (%H_2O) when some quantity of water is added. Now with $\phi = 1.0$ for a calcined adsorbent, the eluant solute parameter (f) is evaluated from the expression $\log K_d = \log V_a + \phi f$, i.e. $\log 68 = \log 0.0542 + \phi f$, i.e., $f = 3.10$. Now for a deactivated alumina of 3% water.

$$V_a = (3.5 \times 10^{-8})(155 \times 10^4) - 0.001(3) = 0.0242 \text{ cm}^3/\text{g}.$$

The experimental value of $K_d = 3.3$ for deactivated alumina $\log 3.3 = \log 0.0242 + 3.10$. On simplification $\phi = 0.69$.

PROBLEMS FOR PRACTICE

14.2 Calculate the partition coefficient (K_d) for the system of a particular solute-eluant water deactivated silica $(V_a = 0.342 \text{ ml/g})$ from the data given.

Adsorbed phase (g/g) $\times 10^8$:	549	153	38.0	15.3	3.82
Non-adsorbed phase (g/ml) $\times 10^6$:	161	52.5	13.4	5.36	1.32

14.3 What are the K_a values for the adsorbents X and Y (a) X 4.6% H_2O-SiO_2 (b) X 15% H_2O-SiO_2? The experimental data value for K_d with pentane as eluant from X 6.9%, H_2O-SiO_2 is 45.6 if $V_a = 0.23$ and $\phi = 0.69$.

14.4 With activated silica gel of surface area 1.55 m^2/g, $K_d = 68$ with napthalene solute and pentane as an eluant $f = 3.10$. If k_d values found were 30, 17, 7.1 and 3.3 with 0.5, 1.0, 2.0 and 3.0% water, calculate the adsorbent surface volume and adsorbent activity of alumina.

LITERATURE

1. E. Lederer and M. Lederer, "*Chromatography, Theory and Practice*", 2nd ed., Elsevier Publishing Co. (1957).
2. E. Heftman, "*Chromatography*", Reinhold (1969).
3. J.A. Dean, "*Chemical Separation Methods*", Van Nostrand Reinhold (1970).
4. L.R. Snyder and J.J. Kirkland, "*Introduction to Modern Liquid Chromatography*", Wiley (1979).
5. L. Turkova, "*Affinity Chromatography*", Elsevier Publishing House, Amsterdam (1980).
6. G.D. Christian, "*Analytical Chemistry*", 5th ed. John Wiley, (1994).
7. B.L. Karger, L.R. Snyder, C. Horvath, "*Introduction to Separation Science*", John Wiley (1973).

Chapter 15

Partition Chromatography

Chromatography is an analytical method for the purification and separation of organic and inorganic compounds. It is useful for fractionation of complex mixtures and separation of closely related compounds. In 1941 Martin and Synge developed partition chromatography while Gordon discovered paper chromatography. More recently the most extensively used gas and liquid chromatographic methods have been based on phenomena of partition of solutes.

15.1 PRINCIPLES OF LIQUID-LIQUID PARTITION CHROMATOGRAPHY

In liquid-liquid partition chromatography, a separation is effected by distribution of the sample between a stationary liquid phase and a mobile liquid phase with limited miscibility. When a solution of a substance is shaken with an immiscible solvent, the solute will distribute itself between the two phases and when equilibrium is reached the partition coefficient (K_d) is expressed as

$$K_d = \frac{\text{conc. of solute in solvent A}}{\text{conc. of solute in solvent B}}$$

Martin and Synge developed a fractional liquid-liquid partition method by packing columns with silicagel, placing the solution of a mixture on a column and developing with a water immiscible solvent like butanol. The liquid held in the column is called the stationary phase and the mobile phase is an eluant travelling down the column. The separations are based upon the exploitation of difference in partition coefficients with respect to two solutes are thus termed partition chromatography.

The general theory of partition chromatography is similar to fractional distillation. The equation to correlate the rate of movement of a zone with the partition coefficient is given by

$$R_f = \frac{A_M}{A_M + K_d A_S}$$

where A_M = area of mobile phase and A_S = area of stationary phase and R_f is the retardation factor. Thus

$$R_f = \frac{\text{distance travelled by the solute front}}{\text{distance travelled by solvent front}}$$

The above expression can easily be derived by simple approximations.

According to Cremer and Muller, the molecules of a given solute are continually moving from the stationary phase to the mobile phase and back, depending on K_d i.e., the partition coefficient which remains constant in each phase. However, all the molecules do not have the same energy. Some will spend more time than average in the mobile phase and some less thus producing a band with the characteristic concentration curve similar to the distribution curve. Consider a paper strip (Fig. 15.1). Let x be distance travelled by the solute band and let $(x + y)$ be the distance travelled by the solvent. Now since $x \propto$ solubility of solute in the mobile phase, $y \propto$ solubility of solute in the stationary phase

$$\frac{x}{y} \propto \frac{1}{K_d} \left(\text{where } K_d = \text{partition coefficient} \right)$$

and $\dfrac{x}{y} = \dfrac{K}{K_d}$ if K is constant

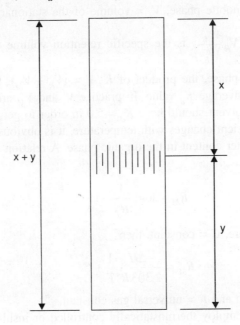

Fig. 15.1 Chromatographic fronts

$$R_f = \frac{x}{x + y} \quad i.e., \quad \frac{1}{R_f} = \left(\frac{x + y}{x} \right) = \left(1 + \frac{y}{x} \right)$$

Since $A_M / A_S = \text{constant} = K$

$$\frac{1}{R_f} = \left[1 + K_d \frac{1}{K} \right] = \left[1 + K_d \frac{A_S}{A_M} \right] = \left(\frac{A_M + K_d A_S}{A_M} \right)$$

$$\therefore \qquad R_f = \frac{A_M}{A_M + K_d A_S}$$

The volume necessary to elute the peak of a band of a solute may be obtained e.g. x = length of the column with peak band, y = section of the column used to collect effluent, $K_d \dfrac{1}{K} = \dfrac{y}{x}$ but $y \propto$ volume of effluent at peak and $x \propto$ volume of feed on column.

$$\therefore \qquad \frac{\text{peak effluent volume}}{\text{column volume}} = \frac{y}{x} = K \times K_d$$

This is also called the peak effluent volume.

The efficiency of separation depends upon disengagement of zone centres and maintenance of narrow and compact zones. A low flow rate is favourable for disengagement of such zones. Increase in height also improves separation. The relation of the column retention volume V_R to partition coefficient K_d is given by an equation.

$$V_R = V_M + K_d V_S$$

where V_M = volume of the mobile phase, V_s = volume of the stationary phase. If $(V_R - V_M)$ is the adjusted retention volume, $\left(V_R \dfrac{V_M}{W_L} \right)$ is the specific retention volume and W_L = weight of solvent comprising of the stationary phase, the product of $K_d V_s = (V_R - V_M)$. For convenience, we assume sample components have an average K_d value. In practice V_s and V_M are fixed within narrow limits. Thus V_M / V_s = 5/10 *i.e.*, the solvent should have K_d = 5.0 in order to get better separations.

Since the partition coefficient changes with temperature, it is obvious that R_f must also vary similarly due to change in the water content in the organic phase. A relation between R_M and temperature variation is obtained.

$$R_M = \log \left(\frac{1}{R_f} - 1 \right)$$

If T = absolute temperature, C = constant, then

$$R_M = \frac{-\Delta H}{2.303\,R} \frac{1}{T} + C$$

where $-\Delta H$ = heat of partition and R = universal gas constant.

Many workers therefore employ thermostatically controlled or insulated chromatographic chambers, to control temperature as temperature affects equilibrium.

Further, the movement of solvents during development depends upon the viscosity of the solvent and slows down with time *i.e.*,

$h^2 = (Dt - b)$ where h = height in mm

D = constant for paper, t = time in seconds

D is called the diffusion coefficient.

When a solvent is allowed to run over a strip of filter paper in paper chromatography, the paper dehydrates the solvent to some extent (~20% of the water is absorbed). In a saturated atmosphere this dehydration is made up by some water vapour condensing on the paper and the composition of the solvent remains uniform. Hence the atmosphere must be kept saturated with solvent. In a nonvolatile solvent it will precede the main portion of the solvent flowing over the paper and thus two liquid fronts will be developed. This phenomenon of mixing is called demixion.

15.2 LIQUID-LIQUID PARTITION CHROMATOGRAPHY

This consists of a column of a finely divided solid support on which the solvent is fixed and made immobile. A second phase, *i.e.*, mobile phase, which is immiscible with a stationary liquid substrate flows over the latter. The components of the sample mixture participate in a partition between the stationary and mobile phase when they migrate down the column. Those components which partition more readily into the stationary phase are retarded in the passage through the column with respect to solute which partition more into the mobile phase.

The support consists of solid porous material with a mesh size of 100-300 mesh holding about 50% weight of stationary liquid. Kieselguhr, diatomaceous earth, cellulose powder and silicagel have proved specially suitable e.g. silicagel absorbs about 70% by weight of water without becoming wet in the real sense. Cellulose powder is also useful due to its fibrous nature.

As regards partitioning liquids, the selection of solvent is confined to a hydrophilic solvent on a stationary support and a hydrophobic solvent serving as a mobile phase. Solvents can be classified on the basis of H-bonding. Solvents which are either donors or acceptors of electron pairs have the ability to form intermolecular hydrogen bonding. The solvent pair selected should have low mutual solubility. Water can be bound as the hydrophilic solvent. Substances with medium porosity can then be separated. Buffers can be added to the water phase for adjusting pH for complexation. A hydrophilic organic solvent can also be used in place of water are formamide > methanol > acetic acid > ethanol > 1-propanol > acetone > n-propanol > t-butanol > phenol > n-butanol > amyl alcohol > ethylacetate > ether > butylacetate > chloroform > benzene > toluene > cyclohexane > petroleum ether > petroleum > paraffin oil in order of decreasing ability of the compound to form H-bonds.

15.3 REVERSED PHASE PARTITION CHROMATOGRAPHY

This is also called extraction chromatography. As the name implies, the phases used in liquid-liquid partition chromatography are reversed. This means that instead of a hydrophilic stationary phase, we use the hydrophobic phase as the stationary phase on a suitable stationary support. Similarly instead of using the hydrophobic phase as the mobile phase, we make use of a hydrophilic phase as the mobile phase. There is a special technique employed to coat the stationary phase with a hydrophobic — water repellent material. The stationary support consists of silica gel, alumina, KelF, Teflon or Kieselghur. They are rendered hydrophobic by exposure to vapours of diethyl dichlorosilane. The stationary support is subsequently treated with a suitable hydrophobic solvent or extractant, preferably on a rotary vacuum evaporator, in order to get a uniform coating. The drying process involves simply immersing the stationary support which has already been made hydrophobic into an extractant. Several solvating solvents like TBP, TBPO, TOPO, HDEHP, MIBK, ethylacetate, mesityloxide, known from solvent extraction, can be used as the hydrophobic phases. Similarly, those solvents which are used in the solvent extraction process for ion pair formation such as high molecular weight amines like TOA, TIOA, MDOA, TBA, Amberlite LA-1, LA-2, Aliquat 336S, Arquad 6C can also be used as the hydrophobic phase. Recently there has been a growing trend to use liquid cation exchangers like HDEHP, DNS as the hydrophobic phases. The principal advantage of reversed phase extraction chromatography is the use of previous knowledge or history of an element from the point of the extraction for development of new separations. With knowledge of batch extraction, it is fully known in what acidity and what concentration of an extractant a particular metal is extractable. Further, from extraction coefficient data, it is quite feasible to develop clearcut separation by fully exploiting the acidity at which two metals are extracted. Also it is possible to devise separation by fully utilising the difference in molarity of acids with which metals are backwashed or stripped from the column into the effluent stream.

15.4 APPLICATIONS OF EXTRACTION CHROMATOGRAPHY

Extraction chromatography is used extensively in metal separations at microgram concentrations. This technique combines the advantages of solvent extraction and column chromatography to effect selective separations. With this technique of RPPC, it has been possible to accomplish separations of difficulty, separable metal ions e.g. with TBP as the stationary phase, excellent separations have been devised for iron (III), chromium (VI), germanium (IV), molybdenum (VI), vanadium (V), gold (III) and mercury (II) from multicomponent mixtures. It has also been possible to use Amberlite LA-1 as the stationary phase for the separation of scandium and indium by extracting their anionic malonate complexes. Similarly with TOPO as the stationary phase, it is quite possible to separate bismuth as well as tin from mixtures and principally from low fusible alloys such as Wood's metal solder, type metal, white metal and various bronzes.

In addition to the above, RPPC has been used with a silica gel or Kieselguhr coated column with TOA/TBP for the separation of As from Ge, Sc from Ca, Nb from Mo; with TBP/TOA as the stationary hydrophobic phase, it has been possible to separate Co from Ni and Fe from Mo. One can safely draw an analogy between extraction chromatography on the one hand and anion exchange chromatography on the other hand.

Like the use of extraction chromatography in inorganic separations, it has been extensively used in organic separations e.g. a system of silica gel with H_2O is useful for separation of hydrophilic substances and substances of medium polarity. Methanol, polyethylene glycol and formamide have been used a stationary phase and a hydrophilic solvent as the mobile phase for the separation of lipids. Of course, in this technique RPPC is useful for the separation of epoxy and hydroxy esters and polyunsaturated esters. It can be used to improve the technique of solvent extraction. Thus if batch extraction fails in solvent extraction, RPPC works successfully because it gets the advantages of solvent extraction, inspite of having a small extraction coefficient and also in column chromatography where because of a large number of theoretical plates in a particular column, it is possible to have intimate contact of the hydrophilic mobile phase and hydrophobic stationary phase to attain an equilibrium. Further, the time required for the total operation is relatively small, hence fast selective separations are possible. A few typical examples of separations of metals by extractions of metal by extraction chromatography are listed in Table 15.1 and described in Reference No. 10.

Table 15.1 Metal separations by extraction chromatography

Sl. No.	Extractant (Stationary Phase)	Metal Septd. (main)	Stripping agent for metal ion	Septd. from	Stripping agents for other ion	Remarks
1.	TOPO	Sn	2M HNO_3	Al Sn (IV) Cr (VI)	2M HCl 2M HNO_3 H_2O	Sequential separation
2.	TOPO	Ge	2M HCl	Sb (III) Ga (III) TI (III)	5M HCl 1M HCl 0.5 M H_2SO_4	Separations are at tracer concentration
3.	Amberlite LA-1	Th	0.25M H_2SO_4	Ce (III) Zr (IV)	H_2O 2M HCl	Separation at (pH - 5.0)
4.	Aliquat 336	Be	0.1M H_2SO_4	Zr (IV) Hf	6M HCl 2M $HClO_4$	pH = 5.5

(Contd.)

Sl. No.	Extractant (Stationary Phase)	Metal Septd. (main)	Stripping agent for metal ion	Septd. from	Stripping agents for other ion	Remarks
5.	TBP	Hg	0.1 M sodium citrate pH 5.0	Cu Fe (II) Cr (VI)	H_2O 1M HCl 0.1 M HCl	Several separations
6.	Aliquat 336	TI	1M $HClO_4$	Al Ga/In	water 0.5M HCl	Anionic citrate complex pH 3.5-6.0
7.	HDEHP	Pb	0.1M HCl	Ge Sn	0.01 M HCl 5M HCl	Volume = 60 ml. Extd. from 0.1M HCl
8.	HDEHP	Bi	1M HCl	Cd/Cu	0.01M HCl	Extracted from 0.1M HCl
				Pb Sn (IV)	0.1 M HCl 5M HCl	Septd. from low fissile alloys

15.5 PAPER CHROMATOGRAPHY

In 1944 Consden, Gordon and Martin introduced this technique employing a sheet of filter paper as the stationary support and a mobile liquid phase which percolates within the porous structure of the paper. A small amount *i.e.*, 1 µl of sample is placed and irrigated by a solvent system. It is useful for qualitative and quantitative analysis of samples though the recovery is not necessarily in a purified form. A long development time and less well defined zones are limitations of this method.

The packing of fibre paper constitutes a porous medium for the retention of the stationary phase while the space between fibres provides a passage for the mobile phase. The varieties of paper commercially available are Whatman 1, 2, 31 and 3 MM. Acetylacetate papers, Kieselguhr papers, silicon treated papers and ion exchange papers are also used. Papers of pure cellulose or modified cellulose papers and glass fibre papers are also available. Hydrophobic substances can be separated on the last two kinds of papers. Acetylated or silicon treated papers can be used for hydrophobic substances while for corrosive reagents, glass fibre paper can be used. While selecting paper one must consider the degree and clarity of separation, diffusiveness of spots, formation of a brown front, extent of tails and 'comets' formed and rate of movement of solvent especially in development. Sometimes the R_f value may differ from paper to paper. The impurities on filter paper are Ca^{2+}, Mg^{2+}, Fe^{3+} and Cu^{2+} ions. Papers should preferably be water repellant.

As regards the solvent system used in partition chromatography, these are of three types: (a) aqueous stationary phase, (b) hydrophilic organic solvent as stationary phase and (c) hydrophobic organic solvent as the stationary phase as is used in RPPC. The aqueous stationary phase involves the use of polar or ionic solvents. The stationary phase is obtained by exposing the paper to an atmosphere of water in a closed chamber. The paper can be dipped in a buffer solution and dried before use. When the stationary phase is hydrophilic organic solvents, either of the two methods may be used for conditioning. In the case of a volatile solvent, the method of exposure is used, whereas for a less volatile solvent the conditioning is carried out by dipping the paper in solvent. Formide is a suitable hydrophilic solvent, the mobile phases may consist of cellosolve, carbitol, glycerol and benzoyl alcohol. In the third category, specially treated papers are used as in RPPC work. The papers are rendered hydrophobic *i.e.*, water repellant by treatment with dimethyldichlorosilane vapour or acetylation of the cellulose. It is also possible to impregnate the paper with a hydrophobic solvent dissolved in a volatile diluent and allowed to evaporate in air. Phenol, butanol, benzyl alcohol, propanol, furfuryl alcohol, acetone, pyridine and THF are the most commonly used solvents for LLPC work with paper. The R_f value of a molecule in the unionised form is greater than that in the ionised molecule the increase in acidity increases the R_f value due to formation of coordination complexes between metal and mineral acids. Comet formation is a phenomenon whereby instead of a round spot, large tails are obtained on the chromatographic paper. Addition of acid in solvent inhibits comet formation.

As regards the mechanism, separations are possible because of continuous partitioning of the substances between the aqueous phase and organic mobile phase running down the paper. Solute migration is initiated from a small compact spot or narrow streak. Differential migration of solute molecules begins when the initial zone is covered by the solvent front by the capillary nature of the driving force as wetting liquid flows into empty capillary or pore spaces. Surface tension is the driving force for capillary movements. Thus liquid-liquid partition is predominant mechanism of paper chromatography.

15.6 TECHNIQUES IN PAPER CHROMATOGRAPHY

The process of the removal of mineral acids from paper is called desalting. Solutions are placed on paper by means of a micropippete at a distance of 2-3 cm from one end, in the form of a horizontal line or streak. After drying it is put for development (Fig. 15.2) in a chamber which is presaturated with water or solvent as the case may be. Such presaturation can be done 24 hrs. before development. There are three modes of development. In descending development, (b) liquid is allowed to run down the paper under gravity employing a glass trough to hold the solvent, while in ascending development (a) the solvent runs in an upward direction by capillary action. The R_f value must be same by both ascending and descending techniques. The third is called radial development or circular paper chromatography (c) where circular paper is used. The following conditions must be maintained to obtain reproducible R_f values. The temperature should be controlled within ±0.5°C. The time of running should be constant. The paper should be equilibrated with the atmosphere in the chamber for at least 24 hours prior to the irrigation. The same batch of paper must be used. R_f should not differ by more than ±0.02°.

An atomiser is generally employed for spraying the reagent when fronts are made visible. An atomiser with small jets is more useful. Papers can be directly exposed to developing gases. In the case of carbohydrates, we use R_G values instead of R_f values. After development of the spot or front the metal can be determined colorimetrically or by reflectance spectroscopy. It is possible that dissolved material form paper can be transferred and directly determined. In addition to separation and quantitative analysis, paper chromatography is quite useful for identification, e.g. If we plot a graph of R_M against the number of cations in a homologous series, it is possible to identify an unknown member of the homologous series.

15.7 THIN LAYER CHROMATOGRAPHY (TLC)

In 1938, TLC was developed by Ismailoff and Schraiber. An adsorbent is coated on a glass plate which serves as a stationary support on which the mobile phase percolates and develops the chromatogram. It is open column chromatography. This method is simple, rapid in separation and sensitive. The speed of separation is fast and it is easy to recover the separated compounds from plate.

Silica gel is used as the coating material but cellulose powder, Kieselguhr or diatomaceous earth can also be used. For hydrophilic stationary phases, plaster of paris is used as the binder; starch, colloidion plastic dispersion, and hydrated silica can be also used. For spreading the binder and solid a commercial applicator is used in order to get a uniform layer. Ready to use thin layer plates are also available as precoated glass plates, chromatotubes, etc. The amount of moisture in the plates should be controlled in order to get reproducible results.

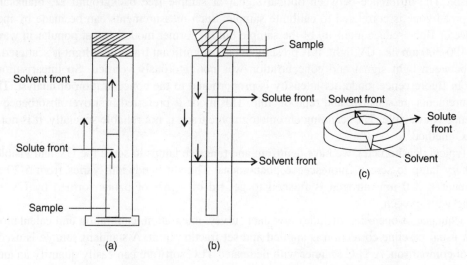

Fig. 15.2 Ascending (a), descending (b) and circular paper chromatography (c).

The choice of a solvent system and the composition of the thin layer is determined by the principle of chromatography to be used. The sample to be separated is applied with a microsyringe near one end of the chromoplate (1 μ ml *i.e.*, 0.01-10 μ g of substance). The solvent should be nonpolar and volatile. The layers can be divided into different columns by scrubbing the adsorbent. Secondly, chromatoplates are developed by the ascending technique at room temperature to a height of 15-18 cm. The time required for full development is 20-40 minutes. All development techniques used for paper chromatography can be applied for TLC. Superior resolution is obtained in TLC due to extremely small diffusion rates in the adsorbent layer. The development is done on plates which have been impregnated with paraffin, silicon oil. Polyethylene glycol is another impregnant. Solvents used are acetic acid or acetonitrile. A long development time is sometimes necessary.

Coloured substances can be visualised directly against the colour of the stationary phase. Suitable reagents can be used for spraying. Chromic acid is used for charring organic compounds. Radiochemical methods can be also used for the purpose of development. Reagents can be used for streaking or brushing along one edge of the plate to locate band positions. After development the chromatogram can be removed by scraping.

Photodensitometry of plates is used for quantitative analysis. Such analysis can be carried out with UV, visible, IR or fluorescence spectrophotometry or by colorimetric reaction with a suitable chromogenic agent. The details are spelled in next section viz. 15.8.

Finally, as regards applications of TLC, it is useful for a large number of sample separations. Compounds which are not volatile or too labile for gas liquid chromatography can be analysed by TLC. It can be used to check impurities in solvents. Forensic chemists use TLC for various separations. Useful separations of plasticisers, antioxidants, separation of ink and dye formulations have been accomplished by TLC. It has extensive applications in inorganic separations. A new technique, high pressure thin layer chromatography, HPTLC, is emerging in pharmaceutical analysis.

15.8 DENSITOMETRIC ANALYSIS IN CHROMATOGRAPHY

This is mainly used in planar chromatographic techniques like TLC. The separation lines are scanned with light beam of fixed wavelength and width in the presence of diffusively reflected light with photosensor. The difference between optical signal of sample free background *i.e.* blank and from sample spread zone is correlated to calibrate sample. Such measurements can be made by measuring absorbance or fluorescence intensity of the sample but the former mode is most popular in wavelength range of 190-300 nm *i.e.*, UV light. It is presumed that insignificant fraction of light is scattered. A rigid relation between light signal and concentration was not rigorously verified. So linearisation is not possible. In fluorescence mode, its intensity is proportional to the concentration of analyst. Therefore the measurements are sensitive and reproducible. This mode is preferred first over absorbence mode if fluorescence is available. The scanning chromatographic track is not reliable specially. It is not reliable in HPTLC method.

In a typical densitometer, we have deutrium and tungsten lamps to serve the UV and visible region and mercury lamp to screen fluorescence phenomena. The slit bandwidth varies from 5-20 nm. For measurement < 220 nm nitrogen is flushed to get rid of signal of ozone formed by UV lamp by interacting with oxygen.

The sequence of operation includes raw data survey, integration, calibration and calculation of the results. A usual baseline correction is applied and set fraction limit. A standard sample is used for the purpose of comparison. A TLC scanner with dedicated TLC software can easily quantify an analyte on the plate. The quantification via image processing is thus possible. The real strength of TLC densitometer depends upon the use of selective UV radiation till 190 nm with good monochromatocity. The quantitative analysis method by TLC densitometer via image processing will progress in future. The analysed results provide simple means of isolating compounds in a mixture in semiquantitative form with no special sample preparation. One can however, get reproducible results. The hyphenated mode as HPLC-TLC, FTIR-TLC, TLC-Raman, TLC-MS are also very useful.

15.9 ION PAIR CHROMATOGRAPHY

The separation of samples that dissociate in aqueous solution can be optimised by adjusting the pH of the stationary or mobile phase to suppress the dissociation and eluting the solute as sharper zones, e.g. Acetic acid is added to the solvent in TLC or paper chromatography to get sharper peaks. Similarly, bases such as ammonia or organic bases are added to get sharper peaks. One can add counter ions to the stationary or mobile phase to promote salt formation between phases and samples. The "ion pairs" formed alter the retention behaviour of ionic compounds. Thus addition is useful to carry out effective separation.

Ion pair chromatography is thus based on the formation of an ion pair between sample and either mobile or a stationary phase. It has extensive applications in HPLC for separation of ionic compounds. Silica gel is used as the support for an aqueous stationary phase which contains counter ions and buffers. An immiscible solvent is used as an eluant. In paired ion chromatography (PIC) similar to reversed phase methods, a nonpolar phase is used as the stationary phase. Thus for the separation of acids, organic bases like tetrabutyl ammonium phosphate are added to the eluant like methanol; for bases, organic acids like 1-heptane sulphonic acid are used. In soap chromatography, organic counter ions with long carbon chains ($>C_{10}$) are used. During separation the degree of dissociation of the sample and counter ion is controlled by varying the pH of the stationary or mobile phase. Consider an example. If we take a carboxylic acid (R-COOH) it dissociates as

$$RCOOH^- \rightleftharpoons RCOO^- + H^+$$

By adding acid, the equilibrium can be shifted to the left, *i.e.*, dissociation is suppressed which results in a sharp peak. Strong acid dissociation however cannot be suppressed but addition of counter ions leads to formation of ion pairs with different partition coefficients. The dissociation equilibrium of an added counter ion is

$$R_4NCl \rightleftharpoons R_4N^+ + Cl^- \text{ leads to ion pair formation as}$$

$$R_4N^+ + RCOO^- \rightleftharpoons \{RCOO^-, NR_4^+\} \text{ ion pair}$$

The pH is so adjusted to effect complete dissociation of sample and counter ion (viz. RCOOH and R_4N X if X = Cl). A strong acid like $HClO_4$ and strong base like R_4NCl are used. Selectivity can be enhanced by varying the temperature. In practice ion pair chromatography using reversed phases is simpler as the counter ions can be added as an eluant.

This technique has been used for the separation of biogenic amines and their metabolic products, caboxylic and ascorbic acids and dye intermediates. The most fascinating development is the separation of the lanthanide series of elements with HDEHP *i.e.* bis (2ethylhexyl) phosphoric acid in dodecane coated on stationary silanised silica. Subsequently HNO_3 or HCl are used as the eluant for stepwise separation.

SOLVED PROBLEMS

15.1 In paper chromatographic separation of silver, lead and mercury the solvent front was 18 cms while fronts due these elements were respectively 16, 12 and 6 cms. Calculate R_f value of all metals.

Answer: We should calculate R_f value from relationship

$$R_f = \frac{\text{distance travelled by solute}}{\text{distance travelled by solvent}}$$

$$R_f Ag = 16/18 = 0.88. \quad R_f Pb = 12/18 = 0.66$$

$$R_f Hg = 6/18 = 0.33$$

15.2 In a particular thin layer chromatographic separation, the R_f value of an unknown compound was 0.809. The fronts due to compounds A, B, C were 24, 28, 30 cms and the solvent front was 34 cms. Identify the unknown compound.

Answer: First calculate the R_f value for all compounds

e.g., $$R_{f_A} = \frac{24}{34} = 0.70; \ R_{f_B} = \frac{28}{34} = 0.82, \ R_{f_C} = \frac{30}{34} = 0.88$$

It appears from the R_f values that the compound B with $R_f = 0.82$ is similar to the unknown compound with $R_f = 0.81$.

PROBLEMS FOR PRACTICE

15.3 In partition chromatographic analysis of an organic compound, the solvent front was 18 cms while the fronts due to compounds X, Y, Z and W were respectively 16.6, 14.3, 10.2 and 5.7 cms. If the R_f value of the unknown compound was 0.79, match this compound with X, Y, Z and W.

15.4 A mixture of uranium, magnesium and aluminium is separated by paper chromatography and analysed spectrophotometrically. With standards of 100 ppm the absorbance for uranium with oxine was 0.720, for magnesium with magneson, 0.550 and for aluminium with Alizarine it was 0.800. With unknown mixtures absorbance due to uranium, magnesium and aluminium was respectively 0.150, 0.840 and 0.750. Calculate the percentage composition of the mixture.

15.5 In the analysis of dicarboxylic acid by paper chromatography the following data was obtained:

Acid conc. µg:	5	10	15	20	25
Wt. of paper mg:	78	123	154	172	194

If 100 mg of the sample was dissolved in 10 ml distilled water and three 0.02 ml samples were chromatographically analysed. The spots has masses of 105, 98 and 109 mg. Calculate the percentage of dicarboxylic acid in the sample.

LITERATURE

1. O.C. Smith, "*Inorganic Chromatography*", Van Nostrand (1953).

2. J.M. Bobbitt, "*Thin Layer Chromatography*", Reinhold, New York (1963).

3. E.W. Berg, "*Physical and Chemical Methods of Separation*", McGraw-Hill (1963).

4. E.Stahl, "*Thin Layer Chromatography: A Laboratory Handbook*", Academic Press New York (1964).

5. E. Heftman, "*Chromatography*", Reinhold and Co. (1967).

6. J.A. Dean, "*Chemical Separation Methods*", Van Nostrand Reinhold, London (1970).

7. J.J. Kirkland, "*Modern Practice of Liquid Chromatography*", Wiley Interscience (1971).

8. G. Zweig and J.R. Whitaker, "*Paper Chromatography and Electrophoresis*", (1964).

9. S.M. Khopkar, "*Solvent extraction separation of elements with Liquid ion exchangers*", New Age International Publishers (P) Ltd. (2007).

10. S.M. Khopkar et al, Analyst **108**, 194 (1981); Anal. letter **17**, 1753 (1981), J. Radio anal. chem. **84**, 33 (1984); Analyst **110**, 1295 (1985); Chemia Analyte **30**, 55 (1985), Analyst **113**, 1782 (1988); Analy. Letter **23**, 635 (1990), Bult Bismuth Inst. **57**, 5 (1989).

Chapter 16

Gas Chromatography

In gas chromatography, the mobile phase is a gas and stationary phase is high boiling liquid where solutes are separated as vapours. Separation is feasible by partitioning of the sample between a mobile gas phase and a thin layer of nonvolatile, high boiling liquid held on a solid support. In contrast, gas solid chromatography employs a solid sorbent. The idea of fractionating gases by passage over a solid or immobilised liquid with a selective action on the individual components was first suggested in 1941, but the method became popular only after 1955. Immobilised liquid as a stationary phase has a much wider utility than solid. These techniques were referred to as gas liquid chromatography, vapor phase chromatography or simply as gas chromatography.

Gas chromatography was simplified when gas liquid chromatography was invented with good instrumentation fifty years ago. It became important in a short span of time. A large number of scientific papers appeared every year in this field. Gas chromatography was at that time used for separations of components which were considered difficult and time consuming. Progress was so spectacular that with a little refinement, it could be used for the separation of non-volatile compounds. Spectacular advances in the field of instrumentation made this technique very popular. In order to understand the utility of this technique fundamental principles are first considered.

16.1 PRINCIPLES OF GAS CHROMATOGRAPHY

In chromatography, we use the term R or R_f, the retardation factor. R_f = distance travelled by zone/distance travelled by mobile phase and R = rate movement of zone/rate movement of mobile phase. With reference to partition chromatography, we have an expression

$$R = \left[\frac{V_m}{V_m + KV_S} \right] \qquad ...(16.1)$$

where V_m, V_s are the volumes of the mobile and stationary phase and K = partition coefficient or distribution ratio. Here β = phase ratio = V_m/V_s, $K = (a_s/a_m)$ if 'a' refers to activities, or $K = C_s/C_m$ if C represents concentration

$$K = \frac{\text{wt. of solute per gram in stationary phase}}{\text{wt. per ml of mobile phase}} \qquad ...(16.2)$$

if W_s represents the weight of the stationary phase in column. This is true if K is expressed as ml/g. Thus the retardation experienced by a component will be decided by its K-values.

In the gas chromatogram in Fig. 16.1, we have a Gaussian error function curve which is symmetrical. If a substance has no affinity for the stationary phase it will have $K = 0$ and therefore it will not be retained by the column. t_o or OA (Fig. 16.1) is the time of unabsorbed component swept through the column. This is termed 'gas hold up time' or gas hold up or dead volume if y-axis is volume and not time. Now the uncorrected retention volume, V_a, is the volume of the mobile phase which is required to elute the band *i.e.*, we express it as $V_R = t_R F_m F_m$ if F_m is the volumetric flow of the mobile phase. $V_o = t_o F_m$ for the unreacted component. Since $V_R = (V_R - V_o) = K V_S$ if V_R is adjusted retention volume and V_S is the volume of the stationary phase.

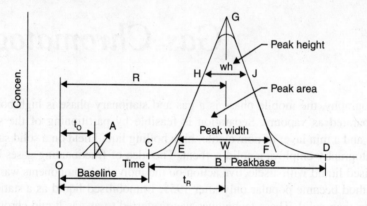

Fig. 16.1 Gaussian error function curve

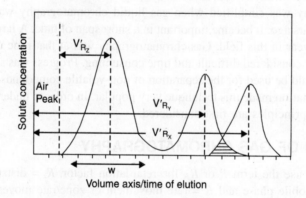

Fig. 16.2 Theory of gas chromatography

There is a pressure difference due to the difference in the rate of flow of gas. The average pressure of gas in the column can be found by considering the compressibility correction (j). We have

$$\bar{U} = j U_o \qquad \qquad ...(16.3)$$

where $\bar{U}$ and U_o are the average ($\bar{U} = L/t_o$) and outlet line linear velocities of gas. The compressibility correction is evaluated as:

$$j = \frac{3}{2} \frac{\left(P_j/P_o\right)^2 - 1}{\left(P_j/P_o\right)^3 - 1} \qquad \qquad ...(16.4)$$

where P_j, P_o are the inlet and outlet pressures in the column.

The net retention volume V_N is adjusted retention volume V_R^1 so that

$$V_N = jV_R^1 \qquad \qquad ...(16.5)$$

It is proportional to the weight of the liquid phases in the column. If W_S represents the weight of the stationary phase packed in the column, the specific retention volume V_R will be

$$V_R^o = \frac{V_N}{W_S} \times \frac{273.16}{T} \qquad \qquad ...(16.6)$$

where T is the absolute temperature V_R^o is the volume of the mobile phase. The partition ratio K will be

$$K' = \left(\frac{t_R - t_o}{t_o} \right) = \left(\frac{V_R - V_o}{V_o} \right) = \frac{KV_S}{V_o} \qquad \qquad ...(16.7)$$

Where t_R, t_o are the retention times of the component and unretained component thus:

$$t_R = t_o(1 + K') = L(1 + K')/V_1 \qquad \qquad ...(16.8)$$

$$V_R = V_o(1 + K') \qquad \qquad ...(16.9)$$

where L = length of the column and V = velocity of the mobile phase or separation factor. Therefore relative retention (α) or separation factor is obtained (Fig. 16.2) as

$$\alpha = \left(t_{R_1} - t_o\right)/\left(t_{R_2} - t_o\right) = \left(V_{R_1} - V_o\right)/\left(V_{R_2} - V_o\right) \qquad \qquad ...(16.10)$$

$$\alpha = V_{R_1}^1 / V_{R_2}^1 = V_{R_1}^o / V_{R_2}^o = K_1'/K_2' = K_1/K_2 \qquad \qquad ...(16.11)$$

This separation factor (α) can easily be calculated experimentally in GLC. An unsymmetrical chromatogram is useless for analytical separations. It can overlap on adjacent peaks or t_R may change with sample size, or the area under curve becomes uncertain. t_R should be independent of the amount of sample injected. The forces involved in chromatography, in general are, van der Waal's, London, dispersion forces, inductive forces, H-bonding, charge transfer or covalent bonding. The first three are present in GLC.

16.2 PLATE THEORY FOR GAS CHROMATOGRAPHY

The number of theoretical plates (N) can easily be calculated in GLC. The expression is used to evaluate the efficiency of the column. The greater the value of N, the better is the efficiency of separation. We therefore have the number of theoretical plates as given by:

$$N = 16\left(\frac{V_R}{W_V} \right)^2 = 16\left(\frac{t_R}{W_1} \right) \qquad \qquad ...(16.12)$$

Here W_V and W_1 refer to the width of the elution band. We can also express it as

$$N = t_R^2 / \sigma^2 \qquad \qquad ...(16.13)$$

if σ = standard deviation of the peak. It can also be given as

$$N = 5.54\, t_R^2\, (W_n)^2 \qquad \qquad ...(16.14)$$

If W_p = peak width half-way up to the peak

$$N = 2\pi (h_{tR}/A)^2 \text{ when } h = \text{height and } A = \text{area of the peak} \qquad ...(16.15)$$

For practical purposes, only equations (16.12) and (16.14) are used.

The HETP $i.e.$, height equivalent of theoretical plate is a measure of the length of the column needed for the equilibrium process to proceed. It is given by

$$H = HETP = L/N \qquad \qquad ...(16.16)$$

If L = length, N = number of plates which should be large.

The resolution R_S can be calculated from the following equations

$$R_S = 2(tR_2 - tR_1)/(W_1 + W_2) = 1.18(t_{R2} - t_{R1})/\left[W_h(1) + W_h(2)\right] \qquad ...(16.17)$$

W_h – refers to width of half height. It is measured across the peak half way between the base line and peak maximum. Finally, for two closely spaced peaks, if $W_1 = W_2$ we have the resolution:

$$R_S = 0.25 \frac{(a-1)}{a} \frac{K_2'}{(1+K_2')} N_2^{1/2} \qquad \qquad ...(16.18)$$

where the subscript 2 refers to the second component. If $\alpha = 1$ then $R = 0$.

In summary, it must be noted that the K value ($i.e.$ distribution ratio) of two components to be separated must differ widely. The success or failure of separation depends not only on the difference between the 't_R' values of two components but also on the extent to which peaks get broadened from the initial narrow band of injected sample. Narrower wide peaks can demonstrate good separation provided there is a substantial difference in the magnitude of 't_R' for two components.

Table 16.1 Specification for gas chromatography columns

Details	Open tabular column	Analytical column	Packed GC column
Material of construction	Plastic Glass	Stainless steel Aluminium	Copper co-alloys
Outer diameter	1/16″	1/4″	1/16″
Internal diameter	0.01″	0.15″	0.047″
Column length	3600″	600″	600″
Rate of flow of gas	25 cm/sec	7 cm/sec	10 cm/sec
Sample volume	0/01 µl	20 µl	1 µl
Nos. of plates (N)	3×10^5	2.5×10^3	5×10^4
Nos. of plate per inch (average)	25	4	4

The elution time of individual components can be altered by adjusting the column temperature. A lower operating temperature leads to increased retention. Among members of homologous series, each solute has the same specific retention volume as the column temperature corresponding to that boiling point.

Retention behaviour is expressed as relative retention both for sample components and the reference substance as they are analysed under identical conditions. Relative relations are independent of column length, carrier flow rate, compressibility factor and ratio of liquid substrate to solid support but not independent of temperature.

In open tubular columns, retention occurs as the walls wet with a partition liquid. In separation the effective height of a plate depends upon K and the plate is different for different species. An asymmetrical peak is ideal. Asymmetry is in two forms: distorted forward edge called 'leading' and elongated rear edge called 'tailing'. A leading curve is caused by too low a column temperature and also by compounds with a wide boiling range. The use of a polar liquid phase will deactivate the support and minimise the cause of tailings. Instantaneous vaporization of the sample also prevents the tailing effect. Finally the efficiency of the separation of a particular pair of components is defined as resolution. A chromatograph with narrow peaks will offer the best resolution.

16.3 INSTRUMENTATION FOR GAS CHROMATOGRAPHY

A gas chromatograph consists of the following important components: (*i*) pressure regulator, (*ii*) sample injection port, (*iii*) stationary support in column, (*iv*) stationary phase, (*v*) detectors, (*vi*) signal recorder. (Fig. 16.3)

1. *Pressure regulator:* Pressure is adjusted within the limits of 1-4 atmospheres while the flow meter measures 1-1000 litres per minute of gas. Flow valves are adjustable by means of a needle valve mounted on the base. The trap containing a molecular sieve is interposed to filter out contaminating impurities. The carrier gas may be He, N_2, H_2, Ar but He is preferred for thermal conductivity detectors due to its high thermal conductivity.

2. *Sample injection port:* Samples are injected by a macrosyringe through a self sealing silicon rubber septum into heated metal. The metal block is heated by electrical heaters. Sample size varies from 0.5 – 10 µl.

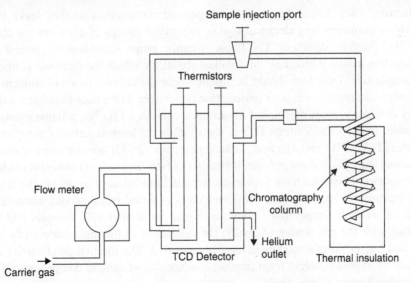

Fig. 16.3 Typical gas chromatograph

3. *Chromatographic column:* These are made of tubing coiled into an open spiral. Copper is the material of construction for high temperature operations. Stainless steel is used. Column diameter varies from 1/16 to 3/16″. Short columns (3-4 cm in length) are also used but separation in such cases is not possible. The usual size is 2 meters. (Table 16.1)

4. *Stationary support:* Structure and surface characteristics are very important. The first contribute towards efficiency while the latter govern the degree to which it enters into separation. The surface holds to immobile liquid phase as the thin film over its surface. The commonly used supports are diatomaceous earth and kieselguhr or chromsorb, porapak, carbowax, dexil apiezon etc.

5. *Stationary phase:* The versatility of gas liquid chromatography lies in the availability of an almost infinite variety of liquid partition materials. Limitations are their volatility, thermal stability and ability to wet the support. Liquid phases can be divided into nonpolar, intermediate polarity, polar carbowaxes and hydrogen bonding compounds like glycol. The maximum temperature of the liquid phase is determined by its volatility as excess volatility shortens column life. Loading of the column is usually expressed as percent by weight. 15% means a 100 gm column has 15 gram stationary phase. Glass beads and polymers have low specific surface area. Open tubular columns and support coated open tubular columns are two more forms of column (Table 16.2).

Table 16.2 Various stationary phases used in GC

Type	Kind of phase employed
GLC	Squalene, silicon oil—nonpolar polyethylene, glycol, p-oxyelipropionibile polar coated on chromsorb, glass teflon beads
GSC (usual)	Silica gel, alumina, charcoal, molecular sieve, inorganic salts, mineral, porous polymers
GSC (Reverse phase)	Silica alumina coated with organic/inorganic compounds, complexes

6. *Detectors:* They detect the arrival of separated components as they leave the column and provide a corresponding electrical signal. Two major groups of detectors are constant temperature and pressure detectors. The linear dynamic range represents the largest signal which a system can handle divided by the smallest signal for which the response is linear with respect to sample size. Detectors should be situated near the column to avoid condensation of liquid, as sample decomposition prior to detection is possible. For a packed column a thermal conductivity detector is most useful, but for an open column a FID *i.e.*, a flame ionisation detector is suitable. In a packed column FID is useful if the effluent is suitably attenuated by a stream splitter. TCD, FID and electron capture detectors (ECD) are the most commonly available detectors but TCD is most popular. It consists of four heat sensing elements made of thermistors or resistance wires which are kept under tension to avoid sagging during the heating operation. The thermistors are electronic semiconductors of fused metal oxides whose electrical resistance varies with temperature. They are useful in smaller cell volumes and have no direct contact with the gas stream. Actually, the difference in thermal conductivity between carrier gas and carrier sample mixture is usually measured. The thermal conductivity of a carrier gas should be widely different from thermal conductivity of sample components; H_2, and He meet these stipulations (Table 16.3).

Table 16.3 Thermal conductivity of carrier gases for thermal conductivity detector

Carrier gas	Thermal conductivity cal/cm sec/deg	Carrier gas	Thermal conductivity cal/cm sec/deg
Ar	3.9	CH_4	7.2
He	33.6	C_2H_6	4.3
Ne	10.9	$CH_2 = CH_2$	4.1
N_2	5.6	$CH \equiv CH$	4.4
H_2	39.6	NH_3	5.1
O_2	5.4	SO_2	1.9
N_2O	3.5	NO	5.5

7. *Signal recorder:* In the area of readout, the choice of recorder determines the ultimate accuracy of the chromatogram. Full scale pass response should be 1 second. Recorder sensitivity is 10 mV and ranges from 1-10 mV. Sometimes amplification of the signal is essential. In direct channel operation two electrometers are built into the signal unit.

16.4 WORKING OF GAS CHROMATOGRAPHY

A sample is injected into a heating block by syringe where the compound is vaporised and carried as vapor along with the carrier gas into the column inlet. Solutes are adsorbed at the head of the column by the stationary phase and the solute travels at its own rate through the column in proportion to its value of K_d. Solutes are eluted in increasing order of their partition coefficient K_d and then enter the detector. Here they register a series of signals resulting from concentration changes and different rate of elution. On the recorder, they appear as a plot of time against composition of the carrier gas stream.

There are several advantages of GLC: we can use longer columns to achieve high separation efficiency; good speed results from low viscosity of gases and vapors and fast partition equilibrium between thin films of gas and liquid; the sensitivity is higher; the gas phase reacts less with the stationary phase of the solutes than most liquid phases do. The technique is however limited to volatile materials.

16.5 APPLICATIONS OF GAS CHROMATOGRAPHY

In reaction gas chromatography, the substance is passed through a reaction zone within the closed system between the introduction point and the detector. The reaction occurs ahead of the injection port. The reaction takes place instantaneously and products have normal times e.g. 8-10 seconds for reaction.

In the subtractive process the subtraction of a specific group of compounds is possible by including a reagent in the chromatographic train that selectively retains these components. For comparison, dual column recording instruments are useful. The compound may be converted to another form with a markedly different retention time e.g. H_2O on passing CaC_2 forms $CH \equiv CH$ acetylene.

In pyrolysis of nonvolatile materials the substance is cracked in the carrier gas and volatile degradation products are carried directly into the gas chromatograph. It is useful for identification and structural analysis of polymers. In elemental analysis and class reactions pyrolysis is carried out to the extreme where organic products break down to CO_2 and H_2O. When a class of compounds is not readily chromatographed because of its thermal instability or lack of volatility, its conversion into derivatives that are more readily chromatographed is often possible e.g. fatty acids are converted in to methyl esters by BF_3 methanol reaction. Similarly the acetylation of fatty alcohol, sterols and hydroxy compounds can also be accomplished with acetic anhydride and pyridine.

16.6 PROGRAMMED TEMPERATURE CHROMATOGRAPHY

The separation of constituents in samples composed of compounds with a wide range of boiling points can be improved and accelerated by raising the temperature of the central column at a uniform rate during the analysis. Temperature has a great effect upon chromatographic behaviour. Liquids with low boiling points emerge first while those with high boiling point will not be retained by the column on increasing temperature and come out later. This happens due to a change in K_d with change in temperature. Thus, compounds with an extremely wide boiling range are separated in less time and peaks are sharp and uniform. Peak heights can be used for quantitative analysis e.g. separation of butanol, pentanol, hexanol, octanol and decanol. A sub ambient programme extends the technique to extremely volatile compounds.

16.7 FLOW PROGRAMMING CHROMATOGRAPHY

As an alternative to temperature programming, the carrier gas flow rate is progressively increased during an analysis thus sweeping components through the column more rapidly. Peak height is directly related to retention time. The height of a late emerging peak would be raised by increasing the gas flow as analysis proceeds. It is useful for the separation of thermally unstable and volatile liquid phases. A drawback to the flow programme is that higher gas flow rates mean lower column efficiency, an increase in plate height for a given separation and a somewhat poor resolution. This is more prevalent in packed columns than in open tubular columns.

16.8 GAS SOLID CHROMATOGRAPHY

The separation of permanent gases and certain light hydrocarbons is achieved by GSC and temperature programming. A silica gel column is used. CO_2 is separated from C_2H_2, H_2, O_2, N_2 and CO on a 4 ft column.

The output of the detector record must be linear for concentration and carrier gas flow must be constant so that the time abscissa may be converted to volume of carrier gas. Thus we can use the peak area for measurement by triangulation, automatic integrated device and planning. The internal standard and normalisation methods are the most common methods and are useful for quantitative evaluation.

The important merits and demerits of GSC over GLC are as follows: In GSC physical adsorption is rapid and it provides specific relative retentions. Columns can be operated in GSC at high temperature, while in GLC the boiling point of the stationary phase determines the temperature of operation. Non-volatility is an advantage in GSC and glass capillary columns are freely used for large samples with no band spreading. The only limitation of GSC is that for elution, a small size is required to obtain symmetrical peaks with good reproductibility. The number of adsorbents are limited. The phenomenon of chemisorption or chemical interactions prevents the use of high temperature for operation. The adsorbent often gets fouled during operation.

16.9 HYPHENATED TECHNIQUES IN CHROMATOGRAPHY

The chromatography is interfaced with another method like mass spectrometry (MS) or plasma spectroscopy. We restrict our discussion to the interfacing of GC with MS. The ultimate aim is to operate both methods without degrading the performance of either of the said techniques. Since chromatography is the ideal separation method while mass spectrometry is the ideal analysis technique, when both techniques are used and are compatible with each another, we get excellent results. GS-MS has several functions. The gas at the exit is attenuated, excess carrier gas is flushed out. A pressure drop is attained in the instrument by a valve or capillary. A needle valve gives favourable results. The degree of enrichment in GC-MS depends upon the kind of column used. Low flow capillary columns are useful. Flow rates of 1-25 ml/min in a column or 0.1-2 ml/min in a mass spectrometer are not very suitable. Intermediate values say 1-2 ml/min are preferred. An enrichment device is used in separators. Such enrichment is around five to six fold. A jet orfice system is used to remove carrier gas by effusion. One can use a membrane separator which takes advantage of the great difference in permeability of organic molecules and carrier gas at a membrane surface. The carrier gas is not adsorbed while organic molecules are adsorbed and the gas permeates through the membrane. In hyphenated GC-MS technique, magnetic sector and quadrupole mass spectrometers are usually used. A low resolution of mass spectra is thus required. We would consider hyphenated HPLC in the next chapter. Such techniques as HPLC-MS are also known. LC-MS is attractive as it handles nonvolatile and thermally unstable compounds. Enrichment is to the extent of 10^5. The flow rate is 0.4 ml/min and may reach upto 2 ml/min. Considerable progress has been made in the field of hyphenated techniques.

16.10 ADVANTAGES OF HYPHENATED METHODS

We have already noted the use of more than one technique or instrument. It facilitates more than one analysis which can be carried out on same sample at same time. It furnishes reliable methods of analysing complex samples. These methods are useful in the qualitative, quantitative and structural analysis. The use of twin methods is quite useful for the analysis of the real samples. The separation methods and spectrometric techniques are coupled together to procure reliable information. The electroanalytical techniques are also coupled with the separation methods such as GC, LC and HPLC. A sample must be subjected to separation of the components to facilitate subsequent analysis. Two instrumental methods if interfaced successively provides a good and reliable information. The multiple techniques can be used in two different ways: In first method, the typical samples are taken and each is analysed by usual technique. It is a multidisciplinary approach e.g., UV, visible IR, NMR, MS are fruitfully used for quantitative analysis. In second method, a single sample is taken and analysed by more than one analytical method e.g., GC-IR can be coupled together. The hyphenated techniques may consist of GC-MS or GS-IR which can operate both methods in one step.

Such technique must be optimised. The quantum of sample will decide the method of analysis. However, time factor is also of importance. A method which saves time is always preferred. If linking with computer is possible, it will be a still better method. The overall advantages can be summarised as: (1) The combined approaches provides more information than individual method (2) Multidisciplinary analysis is essential in analytical work (3) hypenated techniques provides time savings as two techniques can be simultaneously operated (4) The sample must be really representive (5) A time resolved or temperature resolved spectra or chromatogram is obtained (6) If reaction promotes change by way of phase separation or an evaporation such changes must be recorded by recording analytical data. Let us consider few examples of characterization by typical hyphenated techniques.

16.11 CHEMICAL CHARACTERISATION BY HYPHENATED METHODS

The UV, visible IR, PMR, NMR and MS techniques are used as complementary tools. Each spectra provides valuable information in its own way. A good quality spectra must be first obtained. One can derive information from spectras about molecular formula, functional groups present, geometrical isomers or stereochemistry present and their connectivity with each another method or technique. The computerised library database on spectras provides valuable information. The conditions under which spectra has been recorded must be carefully noted. Instrumental parameter must be also recorded. If source of sample is known valuable help is rendered for its identification. The elemental analysis is starting data for analysis and characterization. The existence of unsaturation informs the presence of double bond. From such double bond equivalent as DBE = $(2n_4 + 2 + (n_2 - n_1)/2$ if n_4 = nos. of tetravalent ions. n_3 is nos. of trivalent ions, n_1 = nos. of monovalent ions like hydrogen or halogens present e.g. Benzene has $DBE = \dfrac{(14-6)}{2} = 4$ *i.e.,* three double bonds and ring is present. The IR spectra provides knowledge on functional groups present in unsaturated aromatics, or aliphatic groups. A Raman spectroscopy is useful for group identification specially when there is a weak absorbances. While UV spectrum provides leads for absence or presence of aromatics, conjugated or ketonic structures. mass spectrum provides valuable information specially on m/e ratio to arrive at the molecular formula. Both NMR and PMR gives essential information about type of protons and carbon present and their connection with adjacent atoms. The data at the end of the analysis must be self consistent. The multidisciplinary approach is always more reliable; but one must be very careful while dealing with aromatic system other than benzene e.g. Natural products containing heterocyclic aromatic rings with nitrogen, oxygen or sulphur. The spectrometry is most powerful tool but many times fails in characterisation of mixtures, although it is the most reliable technique for pure samples with one component.

16.12 RECENT DEVELOPMENT IN DETECTORS IN GAS CHROMATOGRAPHY

In the earlier discussion, we have seen three important detectors viz. ECD, FID and TCD which are most commonly used in analysis. The detector is device that senses presence of analyte from the carrier gas and convert it to the electrical signal. The sensitivity in GLC is defined as the change in concentration level or change in the response with change in detected quantity. As seen earlier in TCD, temperature of filament varies with type of analytical gas, leading to the increase in resistance. TCD is not very sensitive in comparison with other detectors. In FID, organic compounds are ignited in flame to form ions which are collected and then converted to current. It has detection limit at picogram concentration (1×10^{-11}) *i.e.* ppb. In ECD, electromagnetic species travel through detector and in process gains low energy electrons thereby diminishing the current. It is the best detector for pesticide analysis.

Nitrogen-Phosphorous detector (NPD) generates current. The rise in the flame is enriched by vapourised alkali metals. While in flame photometric detector (FPD), the combustion of sulphur and phosphorus in flame produces chemiluminisence species which are monitored at particular frequency. In the electrolytic conductivity detector (ELCD) halogen sulphur or nitrogen compounds mix with reaction gas, which on mixing with liquid result in conducting solution which is then measured. In photoionisation detector, (PID) molecules are ionised by excitation with photons in UV lamp. The charged particles produce the measurable current. In mass selective detector, molecules are bombarded by electrons to produce fragments which transmit through mass filter and are segregated as per *m/e* ratio. In hyphenated GC-MS technique such detector is very useful. In IR detector (IRD) molecules absorb IR radiation at particular frequency typical of bonds in molecule. It is best for analysis of isomers. Finally in atomic emission detector (AED) molecules are energised by plasma source and are isolated into excited atoms. The returning electron in stable state emits light which is specific to particular element.

Several GC models are fitted with double detectors to use them at same time. The characteristics of all detectors is provided in Table 16.4.

Table 16.4 Various detectors in GLC

S.No.	Type of Detector	Limit of analysis	Compounds analysed	Remarks
1.	FID	10-100 pg	Hydrocarbons	Good
2.	TCD	5-10 ng	Organic Compounds	Good
3.	ECD	0.05-1 pg	Chlorinated Pesticides	Limited use
4.	NPD	0.1-10 pg	Organophosphorus compounds	Specialised
5.	FPD	10-100 pg	Thio compounds	Specific
6.	FPD	1-10 pg	Phosphorus compounds	Specific
7.	PID	2 pgc/sec	UV ionised substances	Satisfactory
8.	ELCD	2.5 pg	Organo halogen compounds	Good
9.	FTIR	1000 pg	Organic derivatives	Universal
10.	Mass Selective	10 pg-10 ng	Any species	Common
11.	AED	0.1-20 pg	Any elemental form	Versatile

The carrier gas H_2/air flow at rate of 20-60 m/ml or 70-100 m/ml respectively.

16.13 DERIVATISATION IN CHROMATOGRAPHY

The compounds which are not thermally stable or have no low detection level have to be derivatised before running into GLC or HPLC. Such process involves chemical reaction between the analyte and a reagent which changes chemical and physical characteristics of analyte. The idea is to change molecular structure or polarity of analyte sample, to stablise the analyte which improves its detectivity and alter matrix for good separation. Such reaction must be fast enough for analysis with no side reaction or residue. In GC, thermal stability of analyte is raised. The samples with high molecular weight or those with polar groups are usually derivatised. Polar groups are thereby replaced by nonpolar groups. Peaks are refined, detectivity is enhanced. In GC, the process is performed by precolumn derivatisation process.

While on HPLC derivatisation is used to improve sensitivity in analysis. The derivatisation agent should have two organic groups which control volatility and detectibility. Such reagent should have no side reaction, be nontoxic and can be capable of being automated.

There are several derivatisation reactions. E.g. for gas chromatography the silylation reagents are used, as they are reactive, they do not form byproducts and are capable of derivatisation of most of the functional groups. Trimethysilyl compounds are stable with better hydrolytic stability. Those compounds containing electron capture groups (eg. Cl^-, Br^-, I^-) are preferable to raise detectibility. For HPLC derivatisation chromophoric groups are used which have high value for molar absorptivity. One can also use fluorinating agent to produce fluorescence. Variety of derivatisation reagents can be used. For example in GC alkyl silanisation of alcohol, mercaptans, carboxylic acid and amines can be derivatised. By haloalkylacyl process, amides, carbonyl, mercaptans can be derivatised. The esterification of carboxylic acid is best process of derivatisation. The alkylates of alcohols formation is possible. The ketones form oxime derivatives and finally penta-fluorophenyl derivatives of alcohols, amines, phenols, ketones can be also obtained. In HPLC derivatisation the flourinating reagents are used for phenylisocyanate (alcohol) dinitrobenzoyl chloride (aldehydes), Dansyt chloride (amines), napthyl isocyanate (phenols). All detection are carried out in UV region.

In post column detection OPA (orthophthalaldehyde) is used for amines or amino acids in solution containing thiol. While carbamates, pesticides can be also derivatised. It is two step process after separation of carbamates; first they are hydrolysed at high temperature and methylamine so produced is derivatised by OPA to gel compound with absorption maxima at 465 nm. Finally let us consider merits and demerits of precolumn and post column derivatisation. In precolumn, there is no time limit for reaction, many reactants are there but automation is complicated, derivatisation reaction at times is not quantitative or leads to formation of byproducts and excess reagent generally interfere in analysis. In post column derivatisation, automation is easy, nonquantitative reaction can be used and there is no interference in chromatographic reaction. The only limitation is reaction must be rapid, lead to bond broadening and have fewer chemical reaction at final reaction stage.

16.14 HEAD SPACE GAS CHROMATOGRAPHY

It is also termed as headspace analysis. It is primarily used for analysis of volatile compounds in solid or liquid solution which contain both mixture of volatile as well as non-volatile components. The difference lies in the technique of sample injection. Otherwise the methodology is analogous to conventional gas-liquid chromatography. The gaseous sample in the headspace above non-gaseous component is injected into gas chromatographic column. At the low pressure, components do not exist as gas above liquid sample, therefore they are not injected into the column. The sample in column follows three stages for liquid solution. A known measured volume of liquid sample is placed in a vial. A cap of container (septum) part is put in constant temperature bath at 80-90°C. On attaining thermal equilibrium, volatile component partitions between gaseous phase and liquid phase. The concentrations of gaseous phase is directly related to its concentration in the liquid phase. A partition coefficient of each can be written for volatile part. It is the concentration of component in gaseous phase divided by the corresponding concentration in the liquid phase which gives partition coefficient. On attainment of the thermal equilibria, the vial is pressurised by mobile phase gas by introducing metal tube through septum in the cap of sample vial and into gaseous head space within the vial. The pressure of gas in sample vial and time of pressurisation is monitored. Later, the tube is withdrawn and second tube is introduced to inlet column via septum in headspace vial. Within the vial, the pressure promotes part of gaseous phase to be injected, the period of which is controlled. The second tube is then removed and chromatogram recorded as usual.

If the sample is solid multispace extraction (MHE) is followed wherein volatile component of sample is fed to column. Repetitive extractions of components (volatile) from the sample in head space chromatography is resorted to. As quantum of volatile component decreases with each extraction, work is continued of extraction till peak becomes insignificant. If the extraction is continued until such time, volatile component is finished. The sum of chromatographic peak areas is ascertained for quantitative analysis to know the total concentration.

Therefore we conclude that head space analysis involves chromatographing vapors obtained from a sample by heating it in a partially filled vial sealed with septum cap. After equilibrium under rigid conditions the proportion of volatile component in headspace above the sample are representative of those in the bulk sample. The head space vapors which are under slightly more pressure are sampled by altered and automated injection system or gas syringe and injected onto column (Fig. 16.4).

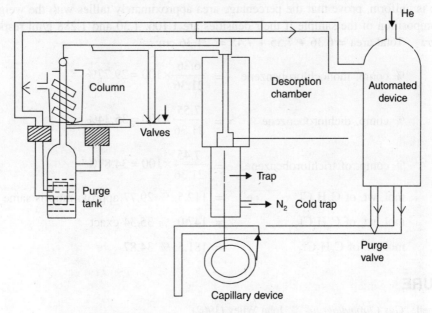

Fig. 16.4 Head space gas chromatograph

The process is useful for syringe and injected onto the column as shown in Fig. 16.4. The headspace technique is most useful for analysis of mixtures of volatile and non-volatile components. The residual monomers in polymers, alcohols or solvents in blood sample are used specially to detect and punish drunken drivers; also flavour and perfumes in manufacturing units. This technique simplifies the chromatograms and protects the chromatographic column from the fouling or contamination by non-volatile samples. Several organic compounds like hexane, acetone, dichloromethane, benzene, ethanol, toluene, MIBK were characterised in presence of nonvolatile components by this technique.

SOLVED PROBLEMS

16.1 In gas chromatographic separation of benzene, toluene and xylene the areas under the peak were noted to be 31.0, 14.5 and 53.2 cm. respectively. Calculate the percentage composition of the sample.

Answer: The total area = 31.0 + 14.5 + 53.2 = 98.7

$$\text{Composition of benzene} = \frac{31.0}{98.7} \times 100 = 31.40\%$$

$$\text{Composition of toluene} = \frac{14.5}{98.7} \times 100 = 14.69\%$$

$$\text{Composition of xylene} = \frac{53.2}{98.7} \times 100 = 53.9\%$$

16.2 When equal volumes of monochlorobenzene, dichlorobenzene and trichlorobenzene were chromatographed, the areas under each peak were 6.36, 7.55 and 7.45 cm². If the detector response is uniform, prove that the percentage area approximately tallies with the weight percentage composition of the sample if their densities are 1.106, 1.30 and 1.288 g/ml respectively.

Answer: Total area = 6.36 + 7.55 + 7.45 = 21.36 cm².

$$\text{\% comp. monochlorobenzene} = \frac{6.36}{21.36} \times 100 = 29.77\%$$

$$\text{\% comp. dichlorobenzene} = \frac{7.55}{21.36} \times 100 = 35.34\%$$

$$\text{\% comp. of trichlorobenzene} = \frac{7.45}{21.36} \times 100 = 34.87\%$$

mol. wt. of C_6H_5Cl = 112.5, % 29.77 approx. ratio is same if not

mol. wt. of $C_6H_4Cl_2$ = 147.0, % 35.34 exact

mol wt. of $C_6H_3Cl_3$ = 181.5, % 34.87

LITERATURE

1. J.H. Purnell, "*Gas Chromatography*", John Wiley (1962).

2. D.W. Grant, "*Gas Liquid Chromatography*", Van Nostrand Reinhold (1971).

3. S. Dal Nogare, R.S. Juvet, "*Gas Liquid Chromatography Theory and Practice*", Inter Science (1962).

4. W. Jennings, "*Gas Chromatography with Glass Capillary Column*", 2nd Ed. Academic Press (1980).

5. A Zlatkis, C.F. Poole (Ed)., "*Electron Capture Theory and Practice in Chromatography*", Elsevier, N.Y. (1981).

6. W. Jennings, "*Analytical Gas Chromatography*", Academic Press, New York (1987).

7. A.B. Littlewood, "*Gas Chromatography Principles, Techniques and Applications*", Academic Press (2nd Ed), (1972).

8. C.J. Cowper, A.J. Deruse, "*The Analysis by Chromatography*", Pergamon Press (1983).

Chapter 17

High Performance Liquid Chromatography

As discussed in an earlier chapter, gas-liquid chromatography has remained a widely accepted tool of separation because of the advancement in instrumentation. Liquid chromatography was initially less popular due to the slow method of separation as well as long time of analysis and was largely used for the separation of thermally unstable and nonvolatile compounds. Several devices were used in liquid chromatography to expedite separation, including high pressure head on column, coarse packing material as stationary support. However, with the discovery of high pressure pumps as well as knowledge of effluent analysis with refractive index or UV detectors, the technique of high pressure liquid chromatography, later termed high performance liquid chromatography (HPLC), attained greater significance in separation science. In fact, this is not a special class of chromatography. It is a rapid technique or methodology applicable to all kinds of chromatographic methods such as adsorption, partition, ion exchange and exclusion chromatography. It promoted the discovery of ion chromatography and supercritical fluid chromatography. An endeavour is made in this chapter to explain this new technique of HPLC which has become so versatile in analytical chemistry.

Chromatographic methods such as adsorption, partition and ion exchange are examples of liquid column chromatography. These liquid chromatographic methods employ a glass tube with varying diameter and sufficiently long columns. Particles of varying dimensions are employed as the stationary support. The head of liquid in the column is just sufficient to develop adequate pressure to maintain a uniform flow of the mobile phase; but on the whole such separations are time consuming. Various measures such as vacuum or high pressure pumps were adopted to increase such a flow rate without affecting the plate height in a chromatographic column. Attempts to reduce particle size of the stationary support were not always rewarding. High performance liquid chromatography (HPLC) has solved all the problems of classical methods of separation. In HPLC, narrow columns with an internal diameter of 2-8 mm are employed, with packing particles in 50 μm in size with high inlet pressure flow rate increased to 0.1-5 cm/sec.

It is interesting at this stage to compare gas liquid chromatography (GLC) with high performance liquid chromatography (HPLC). From the point of view of speed and simplicity of equipment GLC is better. For isolation of nonvolatile substances, including inorganic ions and thermally unstable materials HPLC is preferable. Both techniques are, as a matter of fact, complimentary, efficient, highly selective, and need small samples, which can be handled by nondestructive testing and are applicable for quantitative analysis. The special feature of HPLC is that it can accommodate thermally unstable and nonvolatile compounds and inorganic ions.

17.1 PRINCIPLES OF HIGH PERFORMANCE LIQUID CHROMATOGRAPHY

Chromatographic peaks in elution curves are broadened by three processes called eddy diffusion, longitudinal diffusion and nonequilibrium mass transfer. Flow rate, particle size, diffusion rate and thickness of stationary phase affect the above parameters. Van Deemeter put forth equation to chemically relate the efficiency of chromatographic columns. It was tested for GLC. Accordingly it provided

a relationship between the flow rate (U) and plate height (H) by an expression: $H = A + \dfrac{B}{U} + C \times U$ if A

is associated with eddy diffusion, B with longitudinal diffusion and C nonequilibrium mass transfer. The eddy diffusion (A) is related to average particle diameter by $A = 2\lambda\, dR$, where λ is the packing factor. Longitudinal diffusion results from the tendency of a molecule to migrate from the central concentrated part to the dilute region on the side. This longitudinal diffusion $B = 2\psi D_M$, where ψ is the obstruction factor, a measure of hindrance to free molecular diffusion, and D_M is the diffusion coefficient of the solute in the mobile phase. Finally, nonequilibium transfer in the broadening of the band due to fast movement of the mobile phase. The term C in the Van Deemeter equation is represented as

$$C = \frac{qR(1-R)d_f^2}{D_S} + \frac{wd_R^2}{D_M}$$

where d_f is the thickness of a stationary particle having a diameter d_R if R is the retention ratio and D_S and D_M are the diffusion coefficients for solutes in stationary and mobile phases and q and w are constants. This equation of Van Deemeter provides a first approximation of plate height. However, this equation does not provide a relationship between efficiency and velocity of the mobile phase as far as HPLC is concerned. The efficiency of separation is better at low particle diameters of support. Reasonable flow rates can be realised by high pressure pumps. Hence a sophisticated pump with controls is most essential in HPLC.

The chromatography technique involves isolation of the constituents on account of differential migration as may pass through the column. The separation depends upon distribution between mobile and the stationary phases. Type of column, its packing nature of the stationary and mobile phase, column dimensions, flow rate, temperature and size of sample determine the efficiency of separation.

The chromatography can be done in various modes. Such mode depends upon structural properties of solutes to be separated in analysis.

17.2 COMPONENTS OF HIGH PERFORMANCE LIQUID CHROMATOGRAPHY

The important components of HPLC unit include

(a) pump to force mobile phase through a system; suitable pressure gauges *i.e.*, flow meters control such pressure.
(b) sample injection port or valves to inject sample in the system of the mobile phase as head of separation column.
(c) separation column in which sample components are separated into separate peaks before elution and finally.

(d) The detectors and read out device, including computer accessories. First we would consider mobile phase, stationary support and stationary phase, kind of the columns and finally various kinds of detectors employed in analysis.

17.3 MOBILE PHASE

In sample delivery stem, one has to mainly consider kinds of pump used for the dispersation of the mobile phase. Variety of pumps like reciprocating piston pumps, syringe type pumps, constant pressure pumps are commonly used. An ideal pump must operate at 1500 lbs per square inch (i.e., 103 cm or bar) however reciprocating pump is most popular, as its permits continuous solvent delivery. While syringe type of pump operates via positive solvent displacement by variation in voltage. It has capacity of 500 ml to provide pulseless flow rate. The constant pressure pump operates on gas cylinder. A low pressure gas source generates high pressure (400-6000 atm) with the pulseless flow, facilitating large flow rates. In order to protect the detectors, pulse dampers are used. These electronic pulse dampers are most popular. As connecting tubes are made of polymers with average pressure, one has to use stainless steel tubes. Sample valves and loops are calibrated. These valves decide sample injection system. If valve is in load position sample loop (10-50 µl) is filled the degasers and filters are rarely used. A separation involving single solvent system is called as 'isocratic elution' but gradient elution enhances efficiency where eluant concentration is successively increased.

The volume of sample used is 500 µl at pressure of 1500 psi but sampling loops efficiently introduce sample into the mobile phase. It forms integral part of HPLC instrument with interchangeable loops providing choice of sample size (5-500 µL).

17.4 STATIONARY PHASE

It can be liquid, solid or bonded phase. It is coated on solid porous support in form of thin film e.g. silica support is a good example with polymer bonding. Large surface area is provided by the porous support. Pore size of 30 nm is best. Silica degrades beyond pH = 8.0. Microporous polymers have large channels. Beads are unsuitable for work as they do not swell but good for work in nonaqueous media in gel permeation technique.

17.5 STATIONARY SUPPORTS IN HPLC

Some important aspects of stationary support will be also considered here. The stationary support for HPLC must be pressure stable. Inorganic supports are stable and can withstand pressures of 600 atmospheres. A highly porous structure may collapse at high pressure. Ion exchange for instance has a low permeability under pressure. Silica gel and alumina are the best bonded supports; chemically bonded supports are superior because they need not be saturated with liquid phase and do not need a precolumn (PC); no contamination from the stationary phase occurs during sample isolation, the stationary phase cannot be stripped off as it is covalently bound to the solid. Various programming methods including gradient elution can be adopted on such supports and finally column equilibrium is not disturbed by external changes such as temperature and humidity.

Ion exchange supports have a limited utility due to their compressibility, but pressure stable resins are now available which can withstand a pressure of 200 atmospheres and have an exchange capacity of 3-5 meq/g. Pellicular ion exchange resins are now also available.

The problem of stationary support in exclusion chromatography is more complex than ion exchange resins. All types of gels used in columns are not pressure stable. Shrinkage under pressure reduces permeability, consequently reducing efficiency of separation.

The bonded phase supports are commonly used. They are made from silica by chemical bonding of organic substances through (Si-O-Si-C) siloxane bond. They act like immobilised stationary liquid phase and do not stick to walls of the column but are good for solution with pH < 7.5. Polystyrene-DVB matrix being stable at any pH (1-14) are also used as universal stationary supports.

The column length should be within range of 50-300 mm. The best length being 3.9 mm for analytical microcolumn analysis. As particle diameter is reduced interaction between two particles becomes more pronounced rendering inefficient column separation. The rapid compression technology is used to improve their efficiency.

17.6 DETECTORS USED

First, general properties of detectors are examined. The response time should be at the best 10 times less than peak width of solute. The detection limit should be at least 1/5-1/350 of initial sample concentration due to dilution factor. The linearity and the dynamic range is of crucial importance specially in the quantitative analysis.

The microprocessors are used to control operation of HPLC. Such as confirming completion of analysis, controlling stop flow analysis, developing methods, collecting respective peaks in a fraction collector and monitoring of the column performance when carrying out important separations.

The detector generates electrical signal which is proportional to concentration of solute. Similar property of the mobile phase or solute can be measured. It is signal either for mobile phase or solute component. Refractive index detector measures properties of both mobile phase and solute while UV detector measures the property of solute only. The latter detectors are more sensitive and useful for the quantitative analysis. The important detectors are

(i) *UV detector:* This works on Beer's law principle viz. A = abc with usual notions. 70% work is done by these kinds of detectors. Mobile phase is passed in flow colorimeter to measure absorbance of the sample. It has three kinds of fixed wavelength, variable wavelength and photodiode array type detectors. The photodiode variety provides complete spectra of solute in the mobile phase. Any wavelength and not necessarily a maximum absorbance (λ_{max}) wavelength is used for variable wavelength detector. Fixed wavelength detectors are inexpensive, with least noise and sensitive at nanogram (10^{-9}) concentration. Those measuring absorbance of mobile phase, is called bulk detector and are as good as for qualitative analysis.

(ii) *Fluorescence detectors:* The fluorescence of solute is measured. The excitation and emission of radiations are at right angle to each another. They have selectivity as well as sensitivity, due to possibility of varying excitation and emission radiation. Flow cell has capacity of 5 μL. Its use is limited by presence of background light. In working it is more or less similar to UV-visible detector. The phenomena of quenching must be absent.

(iii) *Infrared absorbance detector:* The range is 2.3-14.5 μm or 4000- 600 cm^{-1}. A FT-1R detector is also available. The cells and windows are made of NaCl or CaF$_2$ with optical path of 0.2-1 mm and sample volume of 1.5-10 μL. The greatest disadvantage is, it has low transparency for many solvents e.g. H$_2$O$_1$ or alcohols cannot be easily analysed by IR detectors.

(iv) *Refractive index detectors:* It is bulk detector for mobile phase. It usually measures refractive index of eluant and reference stream of mobile phase. They have universal applications. Small difference in refractive index of mobile phase does not permit precise measurements. RI is also dependent on temperature. Measurement of gradient eluant is difficult. The differential refractive index detector monitors the difference of RI between mobile phase and the column eluant. The sensitivity is not good but it responds to all solutes. They are temperature sensitive.

(v) *Electrochemical detectors:* They measure the current associated with solutes as they are eluted from the column. Redox property of solute is of prime importance. They are sensitive, selective with wide linear range. The common example is that of conductivity detector. Voltammetric properties of solute must be known to analyst. Polar mobile phase measurement is best carried out with such detectors. Flow cell is 5 μL. Such detectors are of amperometric, polarographic, coulometric or conductometric type.

(vi) *Mass spectrometric detectors:* They are most useful in HPLC. If structural and molecular weight information of eluate/solute is known. This permits separation as well as identification in single step provided interferences are absent. Such detector is unfit for nonvolatile and thermally labile products. Thermospray has been recently designed. This allows direct introduction of effluent from column into HPLC network. It is useful for polar substances. The detection limit is 1-10 pg concentration (Table 17.1)

Table 17.1 Relative efficiency of detectors in HPLC

Detector	Mass load limit	Detectors	Mass load limit	Detectors	Mass load limit
Absorbance	100 pg → 1 ηg	Refractive index	100 ηg-1μg	Electrochemical	–
Fluorescence	1-10pg	Conductivity:	100pg-1ηg	Mass spectrometry	100 pg-1ηg
Optical activity	1-10μg	Photoionisation	–	IR detector	1μg

17.7 INSTRUMENTATION FOR HPLC

A typical HPLC unit is shown in Fig. 17.1. The solvent reservoir (A, B, C) and degassing system (D, E, F) is quite important. (A, B, C) can hold 2 litres of solvent. The dissolved gases in the solvent can be eliminated by degassers (D, E, F) which works in conjunction with the vacuum pump. Separation with a single solvent is called isocratic elution, but gradient elution with change in eluant concentration gives better results. A good HPLC unit has provision for more than two solvent reservoirs to release solvents into the mixing chamber (M) at varying rates. Most HPLC pumps have outputs of 1000 psi (lbs/in²) or 1500 psi. with a flow delivery rate of 20 ml/min. Mostly screw driven or reciprocating pumps are employed. Screw driven pumps produce pulse free delivery which can be controlled but one has to change the solvents periodically. Reciprocating pumps (Fig. 17.2) are more common and can produce pulsed flow which is later damped and can be used with unlimited volumes of solvent and permit gradient elution. The pneumatic pumps are also used. The high pressures generated by pumping devices are not dangerous because liquids are not very easily compressible. Many HPLC models are provided with precolumns (PC), which embody stationary supports as in chromatographic columns (CC), except that they make use of a larger particle size. The role of the precolumn is to remove impurities from the solvent and stop contamination. It also reconditions the column with mobile phase and stripping of stationary phase from packing of chromatographic column is thus eliminated completely. The sample injection port is similar to the one used in gas chromatography. The support is made of KelF. A sample is drawn into the valve by means of hyperdermic syringe. Valves can hold 2-100 μl

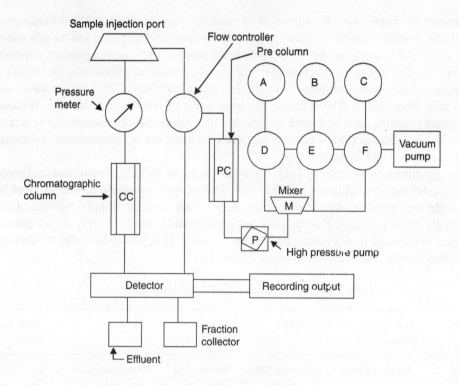

Fig. 17.1 Typical HPLC unit

of the sample. Like GLC, a self sealing septum of silicon or teflon is used in the sample injection port. The chromatograph λ or analytical columns are made in HPLC units of glass or stainless steel. Glass is useful upto 600 psi. The column length is 15-150 cm with an internal diameter of 2-3 mm. Coiled columns like GLC are sometimes employed, but the efficiency of separation is then lost. The packing material is silica gel. alumina and celite (diatomaceous earth) are also used. A column material called 'pellicular particles' is also used. It is made of glass beads coated with layers of porous materials like silica gel, alumina or ion exchange resin. A higher equilibrium is reached because of nonequilibrium mass transfer as shown in the Van Deemeter equation. The only limitation of pellicular particles is limited sample capacity. Most separations are carried out at room temperature. One of the most important parts of the HPLC unit is the detector (Table 17.1).

The choice depends upon the type of sample solution used. The basis of detection is UV, IR absorption, fluorescence, refractive index, conductance, radioactive polarography and mass spectrometric measurements. Except refractive index type detectors, all others are very selective. Only conductometric and polarographic detectors are affected by variation in flow rate. Mass spectrometric and radioactive detectors are not temperature sensitive, while UV, IR and fluorometric detectors have a low temperature sensitivity. Most of them are useful for gradient elution and are easily available in the market. Most commonly used detectors are less accurate than spectrophotometric detectors. For analysis of organic functional groups, UV radiation is most useful. In comparison with absorption detectors, refractive index indicators are not selective and respond to the presence of all solutes.

The chromatogram is recorded with potentiometric recorders, which should be adapted to the particular detector. The response should be 0.5 seconds or better, for full scale, to avoid distortion of rapidly eluting peaks. Chart speed should be adjustable.

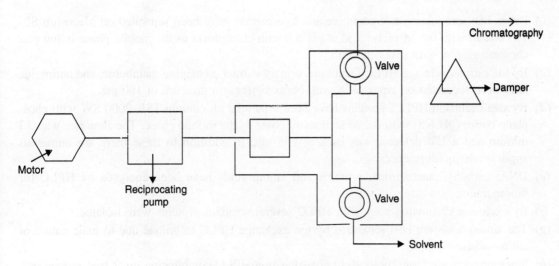

Fig. 17.2 Working of the reciprocating pump

Table 17.2 Characteristics of solvents used in HPLC

S. No.	Solvent	UV maximum (nm)	Refractive index	B.P. °C	Viscosity at R.T.	Density
1.	Hexane	190	1.37	69	0.30	–
2.	Benzene	280	1.50	80	0.60	–
3.	THF	212	1.40	66	0.46	0.87
4.	$CHCl_3$	245	1.44	61	0.53	–
5.	Dioxane	215	1.42	101	0.53	–
6.	CH_3COOH_3	330	1.36	56	0.30	0.78
7.	C_2H_5OH	210	1.36	78	1.08	0.78
8.	CH_3OH	205	1.33	65	0.54	–
9.	CH_3COOH	–	1.37	118	1.1	–
10.	H_2O	–	1.33	100	0.90	–

17.8 APPLICATIONS OF HPLC TECHNIQUES

As far as the applications of HPLC with respect to adsorption chromatography are concerned, they are the same as columnar methods but their main use is in the pharmaceutical and pesticide industries. In liquid-liquid partition chromatography with HPLC techniques, separation of moderately polar to polar substances is possible. As regards a combination of ion exchange and HPLC, good work has been carried out on porous layer beads (PLB). Some important applications of this technique are given:

(*a*) A mixture of sugars containing fructose, glucose, sucrose, maltose and lactose have been separated on column zorbax NH_2 with 70% acetonitrile in aqueous solution as the mobile phase with RI detector.

(b) Drugs like scropolamine, ergotamine and hyoscamine have been separated on Mecosorb SL-100 coated with 0.6 M picric acid at pH 6.0 with chloroform as the mobile phase in ion pair chromatography with a UV-detector.

(c) By the cation exchange HPLC technique, urinary extract containing metabolite, and antipyrine have been separated on zipax SCX with borax buffer at a pressure of 100 psi.

(d) By size exclusion HPLC, proteins have been separated on column TSK 2000 SW with phosphate buffer (pH 6.5) with 0.2 M sodium sulphate as the mobile phase. The flow rate was 0.3 ml/min and a UV detector was used at 205 nm. In addition to these there are numerous applications in Biosciences.

(e) DNA, carbohydrates, lipids, proteins and amino acids have been analysed by HPLC for bioseparation.

(f) By exclusion chromatography with HPLC several standard proteins were isolated.

(g) The amino acids are best separated by ion exchange HPLC technique due to ionic nature of amino acids.

(h) Since proteins are large molecules gel permeation HPLC combination gives best separations.

(i) In environmental pollution analysis ion analysis separate standard cations at trace concentration.

Such ions include lithium, sodium, ammonium, potassium, magnesium and calcium. In ion chromatography eluant used was 0.1 mm EDTA in 3 mm nitric acid. The conductivity detector was used. The flow rate was 1ml/mt.

PROBLEMS

17.1 A mixture containing oxalic, malonic, adipic and palmitic acids, when chromatographed gave areas of 4.3, 3.8, 12.5 and 1.2 cm^2. If the detector response was equal, calculate the percentage composition of the mixture.

17.2 A mixture of acetone and diacetone when analysed by gas chromatography gave the area under the peak as 5.6 and 1.2 cm^2. Since the major peak is due to diacetone, calculate the percentage composition of the mixture.

17.3 A 2 mg sample gave the following peaks: hexane 35 cm^2, toluene 80 cm^2, methylbutyl 1 ketone 15 cm^2 and 1-heptanol 20 cm^2. Calculate the percentage composition of the mixture.

17.4 In a Van Deemeter equation, *viz.*,

$$\text{HETP} = A + B/U + C \times U$$

where A, B, C represent eddy diffusion, longitudinal diffusion, C = mass transfer and U = linear velocity, the values given are A = 0.08 cm, B = 0.13 cm^2/sec, C = 0.03 sec at temp. 150°C. Calculate the gas velocity (U) under these conditions.

17.5 The retention times for air, propane, pentane, acetone and xylene are 43, 90, 140, 150 and 9000 seconds respectively. What is the relative retention time of organic compounds using pentane as the standard?

17.6 For a gas chromatographic column, A, B, C in the Van Deemeter equation were 0.15 cm, 0.36 cm^2 and 4.3×10^{-2} S. Calculate the minimum plate height and the best flow rate.

17.7 If the values of A, B, C are respectively 0.06 cm, 0.057 cm^2/s and 0.13 S, calculate the minimum plate (HETP) height and best flow rate.

17.8 The areas of peaks of methylacetate, methyl propionate and methyl n-butyrate were respectively 18.1, 43.6 and 29.9. The corresponding retention time were respectively 2.12, 4.32 and 8.63 minutes with the retention time of air as 4 minutes. Calculate the percentage composition of the mixture if the relative detector response was respectively 0.60, 0.78 and 0.88.

LITERATURE

1. S. Lindsay, *"High Performance Liquid Chromatography"*, John Wiley, (1992).

2. B.A. Bidlingmeyer, *"Practical HPLC Methodology and Applications"*, John Wiley, (1993).

3. V.R. Meyer, *"Practical High Performance Liquid Chromatography"*, John Wiley (1994).

4. L.R. Synder, J.L. Glajch, J.J. Kirkland, *"Practical HPLC Method Development"*, John Wiley, (1988).

5. R.R. Brown, *"High Pressure Liquid Chromatography–Biochemical and Biomedical Applications"*, Academic Press, New York, (1973).

6. L.R. Snyder, J.J. Kirkland, *"Introduction to Modern Liquid Chromatography"*, 3rd ed. Wiley (1996).

7. F. Settle, Ed., *"Handbook of Instrumental Techniques for Analytical Chemistry"*, Pearson Education (1957).

8. P. Patnaik, Ed., *"Dean's Analytical Chemistry Handbook"*, McGraw Hill Pub. 2nd ed. (2004).

9. D.A. Skoog, F.J. Holler, T.A. Nieman, *"Principles of Instrumental Analysis"*, 5th ed. Harcourt Brace College Publishers (1998).

Chapter **18**

Supercritical Fluid Chromatography and Extraction

A supercritical fluid is defined as any substance that is above its critical temperature (T_c) and critical pressure (P_c). This denotes highest temperature at which gas can be converted into liquid by increasing pressure (P_c). It is highest pressure where liquid can be converted to a conventional gas by increasing temperature of liquid. There is one phase in critical region having properties of both liquid and gases. Above critical temperature (T_c) pressure rises from 0.7-2 times critical pressure. Consequently solvent power of supercritical fluid (SF) increases with density. Such rise in density is not linear with pressure as rate of increase is much greater at critical point.

Alternatively critical temperature can be defined as that temperature above which the liquid phase cannot exist or the temperature above which gas cannot be liquefied by raising pressure. Vapour pressure at critical temperature is critical pressure. At a temperature and pressure above critical point a substance is called 'supercritical fluid'. Its density or viscosity is in between that of liquid and gas phase. Hannay and Hogarth discovered this phenomena in 1879. Their diffusivity rises with rise in temperature. However since 1980 (SFE) supercritical fluid extraction became powerful tool of the separation. SFC has all good qualities of HPLC and GLC. It has ability to interface between these two well known techniques. In 1962, SFC was used as mobile phase in chromatography. It proved a boon for analysis of thermally labile and non volatile substances. SFC could provide large number of theoretical plates in column.

18.1 CHARACTERISTICS OF SUPERCRITICAL FLUIDS

Supercritical fluids possess densities, viscosities and other physical properties in between those of gas and its liquid state. On can compare its properties with liquid and gases (Table 18.1).

Table 18.1 compares properties of SF to liquids or gases while the Table 18.2 provides information on common critical fluids with knowledge of critical temperature and pressure, density, critical point etc. It also summarises important physical properties of supercritical fluids, which can be used as the mobile phase in chromatography. The P_c and T_c are within operational range for HPLC technique. They can dissolve large non-volatile compounds e.g. CO_2 dissolves n-alkanes, diakyl pthalates and

Table 18.1 Comparison of critical fluids

	Density	Viscosity	Diffusion	Remarks
Gas	0.001	0.0001	0.1	Viscosity same as SF
Fluid	0.1–1.0	0.0001–0.001	0.001–0.0001	SF Similar to gas in viscosity
Liquid	1.0	0.01	<0.00001	Density same as gas and liquid

Table 18.2 Typical properties of supercritical fluids

S. No.	Name	T_c °C	T_p atm	Density g/m	Critical (Point)	Remarks
1.	CO_2	72.9	31.3	0.96	0.47	Excellent
2.	N_2O	112.5	36.5	0.94	0.45	V. good
3.	NH_3	71.5	132.6	0.40	0.24	Reasonable
4.	H_2O	217.6	374.1	1.0	–	Unfavourable
5.	Methanol	79.9	239.1	–	–	Average
6.	n-butane	37.5	152.0	0.50	0.23	Reasonable
7.	n-propanol	47.0	235.1	–	–	Tolerable
8.	Acetonitrile	47.7	274.5	–	–	Unfavourable

polycyclic aromatic hydrocarbons. A analyte dissolved in them can be easily recovered by exposing them to atmosphere at low temperatures, specially for thermally unstable molecules. They are inexpensive, nontoxic products. They have solute diffusivities (higher) and viscosity (lower) order of magnitude over than liquid solvents. These are beneficial in chromatography and extraction. The mobile phase is analogous to one in HPLC with large diffusion coefficient than liquids. This favours good speed for separation and resolution for giant molecules. Mobile phase is generally cooled to retain molecules in the liquid state, which in turn is heated till T_c for subsequent analysis.

18.2 MECHANISM OF WORKING OF SFC

This technique is in between GC and HPLC having best features of both the methods. This technique permits separation and determination of group of compounds which are not easily separable by other chromatographic methods; as they may be non-volatile, thermally unstable with absence of any functional group for analysis.

Consider gas as the solvent, with varying vapour pressure. A sample is placed in gas at high temperature when substance dissolves in a gas, as function of their volatility. The enthalpy of dissolution requires strong interaction between analyte and gas as solvent to overcome solute-solute or solute-matrix interactions. The solubility rises due to the interaction between solvent and analyte. In addition to CO_2 one can substitute it with xenon which has also comparable properties. Initially open tubular columns were used at 50°C. Later on columns of 0.25 μm thickness were used with internal diameter of 50-100 mm. Thermostated columns are preferred. Back pressure device is used to control the pressure. A commercial model is shown in Fig. 18.1.

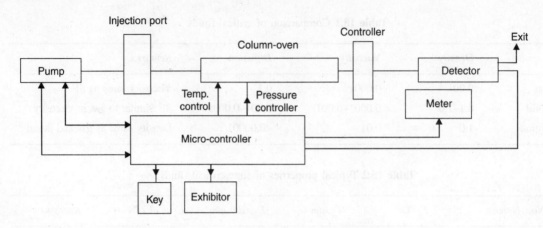

Fig. 18.1 Supercritical fluid chromatograph

In order to understand working of SFC, we must first consider effect of pressure, kind of stationary and mobile phase used and finally the detector used to characterise the compound. One uses thermostated column as in GLC. A restrictor/controller is used to control the pressure enabling conversion of eluant from supercritical fluid to gas to reach detector.

The pressure variation in SFC has remarkable effect on retention capacity of solvent (gas). With rise in pressure mobile phase density also rises and reduces elution of time e.g. CO_2 pressure if varied from 70-90 atm then elution time decreases from 50-5 minutes.

The stationary phase consists of fused silica in open tubular column with internal coatings of bonded and cross linked siloxanes. 10-20 m column is used with d 0.05-0.10 mm with film thickness of 0.05-0.1 μm. The packed columns are also used for the purpose of separations.

The most widely used mobile phase is CO_2 which is good solvent, odourless, nontoxic, easily available and inexpensive. It has $T_c \sim 31°C$ and P_c 72.9 atm providing wide selection of temperature and pressures. Methanol is added in mobile phase as modifier. Also ethane, butane, N_2O, diethyl ether, NH_3 THF also can be used as suitable modifier for polar analytes. CO_2 transmits ultraviolet radiation and is best solvent for organic compounds.

A variety of detectors can be used like Flame Ionisation Detector (FID) which is not frequently used in HPLC. It exhibits general response to organic analyte. It is trouble free and has good sensitivity. Mass spectrometers can be added as detectors. One can also use UV, IR, fluorescence, thermonic and flame photometric detectors.

18.3 COMPARISON WITH OTHER CHROMATOGRAPHY TECHNIQUES

SFC properties are in between those of gas and liquids; so SFC possess the advantages of HPLC as well as of GLC. The method is faster like GLC due to low viscosity permitting higher flow rates. Diffusion rates are also in between those of gas and liquid, so band broadening is large in SFC. The overall separation is faster with lower zone spreading as is seen in GLC. As compared to plate height of 0.039 mm, here plate height is 0.013 mm in SFC influencing the overall efficiency. In GLC, mobile phase facilitates zone movement while in SFC, mobile phase provides transport of solute and also interaction with solutes which influences selectivity. A process of dissolution of molecule in mobile CO_2 phase is analogous to volatisation process in GLC at low temperature than one used in GLC. So high molecular compounds are easily eluted at low temperature in a column. Interaction of the solute molecule and those of SFC must continue to account for solubility in the said media. Solvent power is function of composition and density of the fluids. Thus in general it provides wide scope of applications for SFC methods.

18.4 SELECTION OF SUPERCRITICAL FLUID FOR SFC

One prefers CO_2 or CO_2 mixed with methanol. Polarity, solubility of analyte, nature of matrix decides choice of fluid. CO_2 which is mostly preferred is an excellent solvent for nonpolar species, for aldehydes ketones, alcohol, fats and pesticides, but unsuitable for highly polar solvents. In such case a modifier like methanol is added to CO_2. There are several kinds of modifier like propylene carbonate, methylene chloride or 2 methoxy ethanol. Modifier improves efficiency of isolation e.g. if extraction is 50%, it improves to 90% in presence of methanol for 2, 3, 7, 8 tetrachloro-di-benzo-p-dioxine. It is worthwhile at this stage to compare efficiency of separation by GLC, HPLC, SFC and SFE (to be discussed in next section). If compounds with molecular mass varies from $1-10^9$ daltons, it is noted GLC works upto MW = 10^3, HPLC operates for MW = 10^5 while SFC is good for molecular mass of 10^6 and SFE till 10^8 level.

18.5 APPLICATIONS OF SUPERCRITICAL FLUID CHROMATOGRAPHY

First we would consider general uses. It facilitates separation of both volatile and nonvolatile organic compounds. It can analyse impurities in sample. The isolation is quantitative. It can separate chemically analogous compounds at ultra trace concentration. The methods are nondestructive, and can consume fraction of complex mixture. Any kind of detector is useful for analysis. There are several specific application in field of drugs, pesticides etc. It can also analyse environmental samples, surfactants, polymers. Polar compounds have been analysed by SFC with capillary column. Polysaccharides with 18 glucose units were easily isolated. Azodyes (thermally labile) were also analysed. Surfactants separation is usually carried out on capillary column (*i.e.* 2.1 mm). Lipid, steroids, barbiturates, fatty acids, penicillin were analysed in CO_2 with 0.3% formic acid as the modifier. N_2O can also be used as the mobile phase instead of CO_2 with UV detector. The greater diffusivity and low viscosity facilitates rapid separation. An average column of 10 m × 100 μm is favoured which has film of 0.25 μm of 5% phenyl polysiloxane. Temperature of 150-250°C was good.

18.6 SUPERCRITICAL FLUID EXTRACTION (SFE)

This technique eliminates trouble of sample preparation. It reduces extraction time, is non-polluting, non-toxic, inexpensive. CO_2 is used as the fluid for extraction. SFE has better efficiency and is economically viable method of isolation. In chemical laboratory much time is lost in sample preparation with wastage of chemicals. SFE circumvents all these problems. For instrumental analysis the sample has to be in pure form which can be easily accomplished by extraction. Soxhlet extractor was preferred device for this, but solution becomes dilute in such process.

Supercritical fluid extraction (SFE) has several advantages. The method of separation is very rapid. High diffusivity and low viscosity provides fast separation. The solvent strength of fluid can be varied by altering pressure or the temperature. Many fluids are in gaseous state so recovery of analyte is rendered simple, one does not need dry ashing or evaporation of organic solvents prior to analysis. Fluids are cheap, inert and nontoxic and can be very easily disposed off after extraction is through, by permitting them to evaporate in the atmosphere.

18.7 SELECTION OF SUPERCRITICAL FLUID FOR EXTRACTION (SFE)

CO_2 is used most commonly along with modifier like methanol. The choice is decided from the knowledge of polarity and solubility of analyte in it. A vigorous discussion on this aspect is so far lacking and all conclusions are empirically drawn with several variables. Alkenes, terpenes need CO_2 as solvent. Polycyclic aromatic hydrocarbons, polychlorinated biphenyls are very well extracted in CO_2 although some information on use of N_2O or xenon as fluids is available in the literature. Polar solvents need use of modifier (methanol) to achieve quantitative fluid extraction.

Two kinds of off-line and on-line extractions are popular. In first the analyte is collected by immersing in solvent and permitting the gases and fluid to be released into atmosphere. Analyte can be sorbed on silica followed by elution with suitable solvent. On the contrary in the online extraction process the effluent after depressurisation is transferred in column (GLC, LC) of chromatography. Sample handling is thus avoided before measurement.

18.8 APPLICATIONS OF SUPERCRITICAL FLUID EXTRACTION

Single base gun propellants (with nitrocellulose, TNT) or pesticides and plastisizers are analysed by SFE due to short stability of these products. The instrumentation used is relatively simple. (See Fig. 18.2).

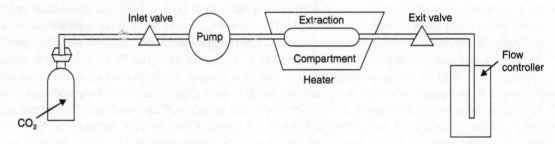

Fig. 18.2 SFE model (online)

The SFE instrument setup is made out of CO_2 storage vessel, valve and pump to control flow and pressure of the fluid (~400 atm) control valve (2 ml/mt) for flow variation, an extraction chamber with heater, then exist valve attached to controller or restrictor. In dynamic extraction process the exit valve is kept open so fresh sample is continuously made available with CO_2 and extracted material flow to collection vessel (not shown in diagram). While in static extractor exit valve is closed and extraction chamber is pressurised. After certain interval exit valve is opened and cell contents are transferred to the collector. However the first dynamic method is most popular.

The problem of coextraction as seen in conventional solvent extraction is absent in SFE. We have high recovery with low standard deviation. The average recovery is just 68.6% with relative standard deviation as 20.6%. In paper industry SFE is of great advantage to extract bleached wood pulp. The time needed for complete characterisation is just three hours.

In conclusion, SFC and SFE are complimentary techniques and are very powerful tools of separation of variety of materials e.g. PAHs from soils samples were analysed in 15-20 mts with offline process. Fats in food products was analysed with (CO_2 + CH_3OH) in 12 mts. The cholesterol in serum was analysed in half hour with CO_2 as mobile phase or extraction of PCB in soil, coal, ash was analysed in 15 mts on online process.

Although flame ionisation detector is commonly used as the detector one can use electron capture detector. Also thermol conductive detector is used for analysis of chlorinated hydrocarbons in SFC. Sulphur containing mercaptans can be similarly analysed. Now-a-days low cost SFC-MS hyphenated instruments are available. With SFC-1R model xenon was used as mobile phase in SFC. Separation of the chiral drugs are best achieved by SFC (Fig. 18.3).

An alternative set up for the supercritical fluid extraction is also shown in Fig. 18.3.

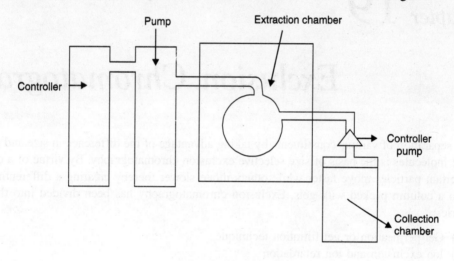

Fig. 18.3 SFE assembly

LITERATURE

1. L. Taylor, *"Supercritical Fluid Extraction"*, Wiley, NY (1996).
2. M.D. Luquede Castro, M.V. Alcarcel, M.T. Tena, *"Analytical Supercritical Fluid Extraction"*, Springer Verlag (1994).
3. S.A Westwood (Ed.), *"Supercritical Fluid Extraction and its use in Chromatographic Sample Preparation"*, Boca Raton CRC Press (1993).
4. R.M. Smith (Ed.), *"Supercritical Fluid Chromatography"*, Royal Society of Chem., London (1988).
5. C.M. White, *"Modern Supercritical Fluid Chromatography"*, Heidlberg FRG Verlag (1988).
6. M.L. Lee, K.E. Morkides (Ed.), *"Analytical Supercritical Fluid Chromatography and Extraction"*, Provo UT Chromatography Conference Inc. (1990).
7. M. McHugh, V. Krukonis, *"Supercritical Fluid Extraction"*, 2nd Ed., Butterworth, Boston (1994).
8. B. Wenclawiak, (Ed.), *"Analysis with Supercritical Fluid Extraction and Chromatography"*, Springer Verlag (1992).
9. F.A. Settle, (Ed.), *"Handbook of Instrumental Techniques for Analytical Chemistry"*, Pearson Education (1997), Indian Reprint (2001).

Chapter 19

Exclusion Chromatography

The separation of various constituents by taking advantage of the difference in size and geometry of the molecules is the basis of size selective exclusion chromatography. By virtue of a difference in size, certain particles move faster while others move slower thereby creating a differential migration front in a column packed with gels. Exclusion chromatography has been divided into the following categories:

    (*a*) Gel permeation or gel filtration technique
    (*b*) Ion exclusion and ion retardation
    (*c*) Inorganic molecular sieves

19.1 GEL PERMEATION CHROMATOGRAPHY

Gel permeation or Gel filtration technique as the name signifies is a technique that fractionates substances largely according to their molecular size. It is based upon inclusion and exclusion of solutes through a stationary phase consisting of a heteroporous crosslinked polymeric gel. It is a special case of liquid solid chromatography. Separation takes place between the liquid phase within the gel particle and the liquid surrounding the particle. The mechanism is based upon the different permeation rates at which each solute molecule enter into the interior of the gel particles. By flow of liquid, molecules diffuse in all parts of the gel, only those molecules having a larger size remain behind and are prevented from entering open regions, *i.e.* voids of gels. Subsequently they pass through the column by way of the interstitial liquid volume; smaller molecules penetrate deep inside. Hence large molecules appear first in the effluent lot followed by entrapped smaller molecules. Thus this separation is possible due to the size barriers that exist within the gel particles. The degree of cross-linking in the gel is important. With lower molecules, polarity, and with larger molecules, size, is the important consideration.

    The stationary supports in such separations are xerogels. They are gels which are organic in nature and are either hydrophilic (e.g. agar, agarose and polyacrylamide cross-linked dextrans) or hydrophobic (based on polystyrene). They are available in the market under the trade names BioGel p-2 (polyacrylamide), Sephadex G-10-200 (Dextrant) also Styrogel (modified polystyrene gels) and agarose. Thus the gels swell in water as hydrophilic gels are water insoluble. Their size varies from 10-300 μ (*i.e.* Biogel, Sephadex etc.). For Styrogel, tetrahydrofuran is preferred as the solvent as the latter has a low viscosity but other organic solvents can also be used e.g. benzene, toluene, chloroform and carbon tetrachloride, cyclohexane and perchloroethylene. Agarose is sold under the trade names e.g. Sepharose and BioGel A which contain varying amounts (~10.2%) of agarose; Sepharose 2B contains 2% agarose.

Such gel contain H-bonding. They are stable in the temperature range 0-30°C and between pH 5-8. No borate buffers can be used as they react with it.

A conventional column is used and the eluant is allowed to flow under gravity. The flow rate increases with an increase in particle size as in ion exchange. The smaller the particle size the lower will be the flow rate. The flow rate is further affected by varying the porosity of the gel. In non rigid gel a pressure device is useful as it will decrease interstitial volume. The total volume of eluant handled is around 20-100 ml. As in other chromatographic methods, here also a suitable physical property of the effluent is measured to ascertain its concentration. Physical properties include refractive index, absorbance, fluorescence intensity or any electrical property.

The columnar behaviour is given mathematically as:

$$V_b = V_o + V_i + V_r = V_o \, V_s$$

$$V_i = m \, S_2 / \rho_s$$

$$V_i = \frac{S_r \rho_r}{\rho_s \left(1 + S_r / \rho_s\right)} = V_0 + K_d V_i$$

$$\frac{V_i}{V_o} = 1 + K_d \, V_i / V_o$$

$$K_{av} = \left(V_C - V_o\right) / \left(V_b - V_o\right), \ i.e., \ K_{av} K_s = K_d V_i$$

In the above expression V_i = inner volume, V_r = volume of gel matrix, V_s = total volume of stationary gel phase, m = wt. of gel, S_r = solvent uptake, ρ_s = density of solvent, K_d = dist. coefficient, P_r = density of gel with subscript r, V = volume, b = bed, O = outside gel. Those solutes which have K_{av} = K_d = 0 pass outside the gel particles and appear in the effluent lot. They are large in size. The volume which separates the fronts of the elution curve of one substance from that of another substances, *i.e.* the difference between their elution volumes is called the 'separation volume'. The eluants should have a low viscosity.

The choice of gel largely depends upon the molecular size and the chemical properties of the substances to be separated e.g. Bio Gel 0-10 operates for substances with a molecular weight between 5000-17000 units. Molecules having a molecular weight above this upper limit, *i.e.* exclusion limit are excluded from the gel. Below this molecular weight they are eluted at an elution volume equal to the total bed volume. For working in nonaqueous media use of lipophilic gels like Sephadex LH-20 is preferred.

19.2 APPLICATIONS OF GEL PERMEATION CHROMATOGRAPHY

It is primarily used for the analysis of mixtures of molecules of different molecular weight. e.g. separation of raffinose, maltose, glucose and KCl at pH 7.0 on Sephadex G15 at flow rate of 5ml/hr with H_2O as the eluant. It is also possible to separate molecules of the same molecular weight by proper selection of gel type and column height. Desalting is another application which involves removal of salts and low molecular weight compounds from macromolecules. Gels with a small size are suitable for such separations which can be also carried out by dialysis. It is possible to estimate the molecular weight by molecular exclusion techniques. Proteins, dextrans and oligomers of ethylene glycol have been separated by this method. It is also used for copolymerisation studies to study whether one has obtained a pure copolymer or it is a mixture of homopolymers. Nucleic acids are separated from nucleotides by gel filtration and similarly, polysaccharides have been separated from sugar, and proteins from low molecular weight substances by gel permeation techniques. This is possible because substances with a molecular weight above the exclusion limit for the gel in question can be concentrated by adding dry gel beads until a thick suspension is obtained. The components in the external solution can be recovered by centrifugation. Low lactose milk can be purified by this technique.

19.3 ION EXCLUSION

Ion exclusion is a process for separating ionic materials from nonionic materials based on inherent differences in the distribution of two types of solutes between an ion exchange resin phase and true aqueous solution. When an ion exchange resin is placed in dilute aqueous solution of an electrolyte, the equilibrium electrolyte concentration is less in the resin phase than in the surrounding solution, *i.e.* the external phase. However, when resin is placed in an aqueous solution of nonelectrolyte, the latter tries to establish equivalent concentration in the two phases. Thus, ion exchange is not involved in absorption of electrolytes or nonelectrolytes as both types of solute can be extracted from the resin by diluting the external solution with water. Now if a solution containing both ionic and nonionic materials is introduced to the top of a resin bed and washed with water, the ionic material flows around the resin particles while nonionic material diffuses into the resin particles as well as flowing into the void spaces. The net forward migration rate of the nonionic material through the column is less than that of the ionic component, hence the ionic component will appear first in the effluent and separations are feasible. Ion exchange resin is used as the immobile phase, *i.e.* stationary support. On passing the sample solution and water through the column, alternate fractions of electrolytes and nonelectrolytes can be collected in the effluent. Since there is no exchange taking place, the resin is never exhausted and need not be regenerated.

19.4 MECHANISM OF THE ION EXCLUSION PROCESS

An ion exchange resin is quite unlike most common absorbents. Only one ion of the electrolyte is excluded by the fixed charge network of the resin; the total amount of the electrolyte diffusing into the resin is limited by the law of electroneutrality. More strongly ionised high capacity resins are most efficient for ion exclusion. The proportion of salt that diffuses into the resin decreases as the exchange capacity of the resin increases. Most low molecular weight water soluble nonelectrolytes freely diffuse in and out of the resin phase regardless of the capacity of the resin and tend to exit at the same concentration in both phases at equilibrium. So long as there is difference in distribution coefficient for different solutes, separations are possible and one can predict the order of elution from the column. The actual position of the solute in the effluent is determined by the resin bed characteristics. Since ion exclusion depends on a Donnan equilibrium, separation will be more efficient at low ionic concentration. The concentration of nonelectrolyte however is less important in considering separation efficiency.

19.5 MERITS AND DEMERITS OF THE ION EXCLUSION TECHNIQUE

Ion exclusion is cheaper than ion exchange as no regeneration is required. Thus the operating cost is restricted to the circulation of water and sample components through the system. The method is applicable to a wide range of materials. The process is simple and adaptable to continuous and automatic operations. Good separations have been achieved in solutions containing nonionics in concentrations up to 40%. The upper concentration of ionic materials is 8%. The principal demerits are that the resin must be in the same ionic form as the electrolyte in solution e.g. for passing NaCl we must use cation exchanger resin in the sodium form. The sample composition is restricted to a single cationic or anionic species. Exclusion works in the wrong way by removing the major component in separations of ionics from nonelectrolyte solutions. The volume of the sample solution is limited by the volume of solution adsorbed by the resin and in practice is found to be less than the occluded volume. On account of rinsing, dilution of the nonionic component occurs; recycling can minimise this demerit. It can thus produce a greater solute concentration in the effluent than in the influent.

19.6 APPLICATIONS OF ION EXCLUSION TECHNIQUES

Sugar and salt cannot be easily separated by other methods but ion exclusion. It permits such separation because of the size and relative inability of sucrose and glucose molecules to diffuse rapidly into the resin phase. The separation of a strong electrolyte from neutral molecules can be best achieved by ion exclusion e.g. purification of commercial glycerine when high ionic concentrations are overcome. The separation of NaCl and ethylene glycol, HCl and CH_3COOH, CH_3COOH and $CH_2ClCOOH$; NaCl and ethanol. Reichenberg applied this technique for the separation of members of a homologous series. Acetic acid and n-butyric acid were thus separated. The strongly acidic sulphonic groups of an cation exchange resin exert a marked salting out effect on the organic material and tend to counteract the adsorptive effect of the resin matrix. Reimann used salt instead of H_2O as an eluant and separated methanol, ethanol and propanol. The beneficial effect of salt in the eluant is probably the selective salting out of nonionics from the solution phase.

19.7 ION RETARDATION

Ion retardation is a column operation very similar to ion exclusion but employs ion exchange resin containing both cationic and anionic functional groups within the resin matrix. Hence resin will absorb anions and cations from the sample solution. However, the absorption affinity is so weak that water can be used as the regenerant for the said ions. Thus ion retardation differs from ion exclusion in that the ions are also retarded as they flow through the column. Ion retardation can effect a separation of electrolytes from large nonelectrolytes and also from other electrolytes if the distribution constants are sufficiently different e.g. $NH_4OH+NaOH$, $NH_4Cl+ZnCl_2$; $FeSO_4+ZnSO_4$ etc. can be easily separated by the process of ion retardation.

19.8 INORGANIC MOLECULAR SIEVES

The natural and synthetic zeolites form molecular sieves for the separation of gases and smaller organic molecules. A mass of zeolite consists of a large number of cavities connected by channels. Screening and sieve action of these channels combined with surface adsorption activity of the crystal matrix makes it possible to separate molecules smaller than the size of channels, from those which are larger. The separation is feasible due to surface activity and also molecular geometry.

Structurally, zeolites are tetrahedral skeletons forming a honeycomb structure with large cavities connected to one another by small apertures. The maximum cross section of an aperture determines the size of the molecules which can penetrate into large cavities. The size and position of metal ions (e.g. Na, Ca) in the crystal and this particular type of aluminosilicates control the effective diameter of the interconnecting channels.

There are various types of sieves, for example, Type 4A is $[Na_{12}(AlO_2)_{12}(SiO_2)_{12}]$. Molecules smaller than $4A°$ in diameter are sorbed by it e.g. H_2O, CO_2, H_2S, SO_2 and hydrocarbons containing 1-2 carbon atoms. Propane is physically excluded. Ethane passes through type 5A which is produced from 4A by ion exchange wherein Na is replaced by Ca and K, thereby changing the diameter of the entrance pores to about 5A°. It allows straight chain paraffins, olefins and alcohols up to C_4 but cyclopropane is excluded, so also naphthenic acid and aromatic molecules are excluded. Type 10 X and 13 X are similar [Na_8 $(AlO_2)_{80}(SiO_2)_{106}$]. They absorb up to 6-10A° diameter of molecules.

There are several applications for molecular sieve techniques. They show affinity for polar molecules and polarizable compounds in preference to nonpolar molecules inspite of having the same size. They are tightly held within the crystals. Unsaturated materials can be removed from refinery and synthetic gases. Crystalline sieves are used as chemical compound carriers. The compound can be released from them by heavier vacuum or by displacement with more adsorbed material such as water. These sieves can be reactivated at 200-350°C. They are also used as the stationary phase in gas solid chromatography.

SOLVED PROBLEMS

19.1 A column of Sephadex G 75 with 2.5 × 36 cm^2 dimensions gave the following results with symbols having same notations in this chapter:

Ribonuclease A (M = 13,700, V_e = 109.5 ml), Aldolase (M = 158,000, V_e = 55 ml), blue dextran (M = 2 × 106; V_e = 53.1 ml) and ovalbumin (M = 4.5 × 103, V_e = 71.8 ml). Estimate the molecular weight of the protein with V_e = 76.5 ml.

Answer: Gel permeation chromatography permits determination of the molecular weight of proteins by plotting a graph of K_d *versus* log (mol. wt). Here K_d is calculated from the knowledge of V_o, V_e and V_i, where V_o is measured with a large molecule like blue dextran (m.wt = 2 × 10^6) and V_g is assumed negligible while calculating V_i and V_t.

19.2 From the following data calculate the molecular weight of the unknown substance: Lysozyme (M = 14,400, relative mobility = 0.90) Typsin (M = 23,300, RM = 0.73) ovalbumin (M = 43,000, RM = 0.54) Catalase (M = 62,000, RM = 0.38) unknown substance (relative mobility 0.68 M = x).

Answer: Electrophoresis in a polyacrylamide gel containing sodium dodecysulphate (SDS) is used. SDS denatures proteins so that they adopt a coil conformation, hence electromobility depends wholly on molecular weight. The relative electrophoretic mobility of a protein is measured by comparing its distance of migration to that of a low molecular weight ionic dye. A graph of log (mol. wt.) *vs* relative mobility is linear and helps to determine the molecular weight of the unknown protein molecule.

LITERATURE

1. E.W. Berg, *"Physical and Chemical Methods of Separation"*, McGraw-Hill (1963).
2. H. Determann, *"Gel Chromatography"*, Springer-Verlag, New York (1968).
3. J.A. Dean, *"Chemical Separation Methods"*, Van Nostrand Reinhold, London (1970).
4. T. Kremman and G.L. Boross, *"Gel Chromatography"*, John Wiley (1979).
5. J.C. Giddings, *"Unified Separation Science"*, John Wiley (1991).
6. W.N. Yau, J.J. Kirkland, D.D. Bly, *"Modern Size Exclusion Liquid Chromatography–Practice of Gel permeation and Gel Filtration Chromatography"*, John Wiley (1979),

Chapter 20

Electrochromatography

The technique of separation of constituents under the influence of an electric current by taking advantage of differences in the migration rate of ions is called electrochromatography. Broadly this technique has been classified into three main areas: free boundary electrochromatography, zone electrochromatography and continuous electrochromatography. In recent years exciting discoveries have been made in the field of capillary electrophoresis. It is considered in subsequent chapters.

20.1 PRINCIPLES OF ELECTROCHROMATOGRAPHY

If a foreign phase which is charged and is subjected to a potential gradient, the foreign phase will migrate through the continuous medium to the cathode or anode depending upon the sign of the charge on the particle. This phenomenon is called electromigration. The basis of this technique is the initial establishment of an inhomogeneity or concentration gradient across the system. Colloids, proteins and enzymes show peculiar migration mobilities and isoelectric points which can be used for identification of specific substances. Separations are accomplished only if the separated components do not spontaneously remix by convective currents. The limitations of free boundary migrations are partly eliminated by development prior to migration of a gravitationally-stable density gradient in a vertical tube by the approximate mixing of buffer solution and varying amounts of soluble nonelectrolyte. On the other hand, if such migration is allowed to proceed within a porous or stabilized medium, the migrating zones are not affected by convection currents produced by temperature and density gradients. In such media, flow of liquids in the tube is inversely proportional to the square of the radius of the tube. In this

V - represent rate of flow of liquid of migrating ions

r - stands for radius of tube through which ions flow

A = area of cross section of tube in question as

$V \propto r^2$

$A \propto r^2$ so we have $r^2 = VA$ or

$V = r^2/A$ ...(20.1)

while the cross sectional area of the tube is proportional to the square of the radius. Filter paper, agar, starch, gelatin, foam rubber or glass spheres are used to provide capillary interstices as a stabilising factor in migration. However, filter paper is most common because complete separations are feasible and samples are easy to handle. It is possible to achieve separation horizontally or vertically and boundary anomalies interfere less. Such migrations on paper was introduced in 1931. The term electrochromatography is used with reference to the physical transport of charged solutes through paper under the influence of a potential gradient when paper acts as stationary support.

The migration of charged particles in paper depends on the magnitude and sign of the net charge on the solute, surface, charges, the applied voltage, electrolyte concentration, ionic strength, pH, temperature, viscosity, solute absorptivity, and other physicochemical properties of the migration medium. Ions migrating in one direction also affect the mobility of other ions moving in the opposite direction. The path traversed by a charged particle is given by (*d*) as

$$d = \frac{Ute}{qk} = \frac{U_t V}{L}$$

where V = volts, L = length of channel, k = conductivity, e = current, U = mobility of ion at time t, q = cross sectional area and d = path traversed.

20.2 ELECTROCHROMATOGRAPHY INSTRUMENTATION

An arrangement of sheets of filter paper moistened with an electrolyte and stretched horizontally between two electrode vessels to which a potential is applied is employed. The sample is placed in the centre of the strips or the strip is sandwiched between two glass plates in order to avoid warming and evaporation of an electrolyte. Direct current is applied at 100-300 volts, carbon and platinum electrodes are used. The combination of electrochromatography with the simultaneous flow of solvent to separate ionic substances has also been developed (Fig. 20.1).

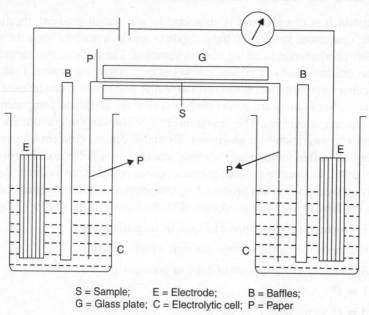

S = Sample; E = Electrode; B = Baffles;
G = Glass plate; C = Electrolytic cell; P = Paper

Fig. 20.1 Electrochromatography assembly

20.3 CURTAIN ELECTROCHROMATOGRAPHY

In this technique electrochromatography is carried out continuously. A curtain of filter paper is employed which is immersed in the buffer solution. The sample is applied through as applicator. The sample is fed to an applicator by a sample reservoir. The two ends of filter paper are immersed in two electrode compartments in the form of the small ends of filter paper (fluted form). On a application of potential difference, the solute particles start moving under the influence of the potential gradient in downward flow of liquid. The components are resolved in different fronts as shown in Fig. 20.2.

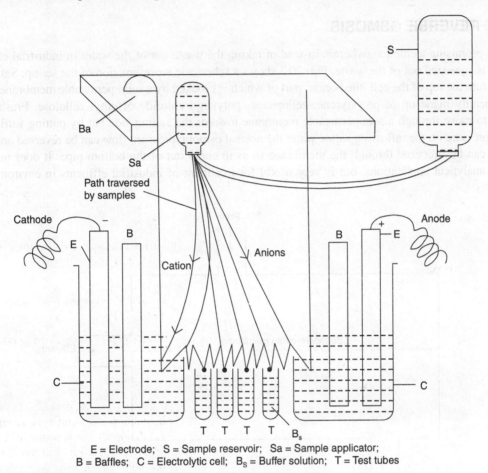

E = Electrode;   S = Sample reservoir;   Sa = Sample applicator;
B = Baffles;   C = Electrolytic cell;   B_s = Buffer solution;   T = Test tubes

Fig. 20.2 Curtain electrochromatography

20.4 APPLICATIONS OF ELECTROCHROMATOGRAPHY

Electrochromatographic homogeneity of zones is an important criterion in the establishment of purity. Electrochromatographic techniques on paper have been applied for the separation of inseparable components. Inorganic separations are most readily achieved in the presence of complexing agents. The latter affect the rate of chromatographic development. The electromigration of complexes have different adsorption affinities for the migration medium and charges different from simple ions e.g. Ag^+, Ni^{2+} migrate to cathode but in the presence of EDTA, the anionic complex of nickel with negative charge migrates towards the anode. The alkali metals and magnesium can also be separated, and alkali and alkaline earth have been similarly separated. Good separations of Pb, Pt, Rh, Ir, Os, Ru and Au have been obtained on paper strips by electrochromatography in different electrolyte solutions. Next to paper the most promising stabilised medium for electromigration experiments is starch. It is used in columns and blocks for the separation of a number of components and is credited to have definite advantages over paper. Electrochromatographic techniques have been used extensively in the separation of inorganic ions, and in clinical diagnosis.

20.5 REVERSE OSMOSIS

It is a promising technique, wherein instead of taking the waste out of the water in industrial effluent, water is squeezed out of the waste. Fig. 20.3 shows a schematic representation of the set-up. Salt water is fed into the top of the cell, the bottom part of which is blocked by a semi permeable membrane which is generally made up of polystyrene, cellophane, polyvinyl chloride or ethyl cellulose. Fresh water tends to move through a semipermeable membrane towards the salting side but by putting sufficiently large pressure on the inflowing saline water the normal osmotic pressure flow can be reversed and fresh water can be squeezed through the membrane so as to come out of the bottom pipe. It does not have many analytical applications, but is very useful for treatment of industrial effluents in environmental analysis.

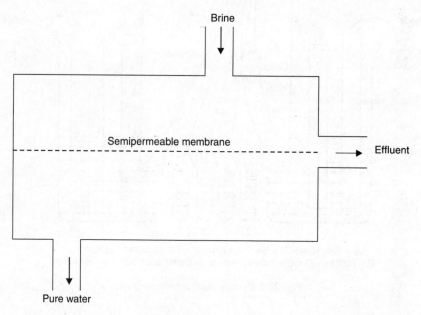

Brine

Semipermeable membrane

Effluent

Pure water

Fig. 20.3 Reverse osmosis

20.6 ELECTRODIALYSIS

This is another promising method where in positive and negative ions are separated out from a flowing current of brackish water by passing it through ion exchange membranes under the influence of electric fields. A typical electrodialysis cell is shown in Fig. 20.4. Salt water is fed into three input pipes at the top of the cell. An electric field having a polarity as shown, tends to make the positive and negative ions drift off in opposite directions. By proper use of ion exchange membranes, it is possible to make a cation membrane permeable to cations and anion exchange membrane permeable to anions. Under proper flow conditions, the water coming out of the middle effluent pipe will be less saline than that coming out of the two outside pipes. This technique has formed the basis for development of the reverse osmosis technique as described in earlier section.

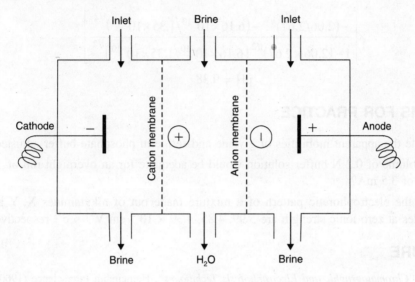

Fig. 20.4 Electrodialysis

SOLVED PROBLEMS

20.1 Calculate the volume of buffer required for electrophoresis if its concentration is 0.05 M and the time required for the separation is 3 hours and current flow is 4 mA.

Answer: The volume of buffer is calculated from the expression

$$V = \frac{0.00373 \times i \times t}{C}$$

If i = current in mA, t = time in hrs and C = concentration in molar

$$\therefore \quad V = \frac{0.00373 \times 4 \times 3}{0.05} = 0.9 \text{ litre } i.e., 900 \text{ ml.}$$

20.2 What is the pH for the separation of the binary mixture of two amino acids A_1 and A_2 if their respective pK values are 9.87 and 9.21 and limiting mobilities are respectively 2.67×10^{-4} and 2.06×10^{-4}?

Answer: The pH is generally calculated from an expression $\mu = \mu K_a/(H^+) + K_a$ if μ = apparent mobility and K_a is ionisation constant. For a two component system

$$[H^+] = (K_1 K_2)^{1/2} \left[\frac{(\mu_1/\mu_2)^{1/2} - (K_1/K_2)^{1/2}}{1 - (C\mu_1/\mu_2)^{1/2} (K_1 K_2)^{1/2}} \right]$$

In the given problem $\mu_1 = 2.67 \times 10^{-4}$, $\mu_2 = 2.06 \times 10^{-4}$ = pK = log K_1 = 9.87 and pK$_2$ = log K_2 = 9.21. On substituting these values, we get

$$[H^+] = (6.16 \times 10^{-10})(1.35 \times 10^{-10})^{1/2}$$

$$\times \left[\frac{-(2.06/2.67)^{1/2} - (6.16 \times 10^{-10}/1.35 \times 10^{-10})^{1/2}}{1 - (2.06 \times 2.67)^{1/2}(6.16 \times 10^{-10}/1.35 \times 10^{-10})^{1/2}} \right]$$

$$pH = 9.38$$

PROBLEMS FOR PRACTICE

20.3 Calculate the apparent mobilities of glycine and serine in phosphate buffer adjusted at pH 9.2.

20.4 What volume of 0.2 N buffer solution would be adequate for an overnight run of 15 hours at a current of 3.5 mA?

20.5 Predict the electrophoretic pattern of a mixture made out of alkylamines X, Y and Z whose mobilities at zero ionic strength are 5.99, 4.85, 4.30 × 10^{-4} cm V^{-1} sec^{-1} respectively.

LITERATURE

1. Z-Smith, *"Chromatographic and Electrophoresis Techniques"*, Heinemann Interscience (1960).
2. G.D. Christian, *"Analytical Chemistry"*, 5th Edition, John Wiley (1995).
3. E.W. Berg, *"Physical & Chemical Methods of Separation"*, McGraw Hill & Co. (1963).
4. J.A. Dean, *"Chemical Separation Methods"*, Van Nostrand Reinhold & Co. (1969).
5. E. Heftman, *"Chromatography"*, 2nd Edition, Van Nostrand (1967).
6. E. Lederer, M. Lederer, *"Chromatography-Principles and Applications"*, Elsevier & Co. (1968).
7. D.A. Skoog, E.J. Holler, T.A. Nieman, *"Principles of Instrumental Analysis"*, Harcourt Brace College Pub., (1998).
8. F. Settle (Ed.), *"Handbook of Instrumental Techniques for Analytical Chemistry"*, Pearson Education Co. (2004).

Chapter 21

Electrophoresis

In 1930, a Swedish chemist named Tiselus introduced the technique of Electrophoresis and he was awarded Nobel Prize in 1948 for this discovery. Electrophoresis is a technique based in part on chromatography, as it is concerned with the separation of the components of the sample. The electrophoresis and electrochromatography are used in combination with each other to gain high resolution in the separation of molecules. This is the most useful method of separation specially in characterisation of complex, biological products such as proteins. A Russian scientist called Tulesett as well as Tiselius separated proteins into four major components as serum albumin, α, β and γ-globulins. In fact analogous to the development of paper chromatography the paper electrophoresis and zone electrophoresis were subsequently discovered. Such invention led to the simplification of equipment; increase in resolving power and direct assay of biological activity on supporting media and extrapolation for prediction. The electrophoresis is defined as the transport of electrically charged particles under the influence of the direct current field or the applied potential on supporting media. The particles under consideration may be ion, complex macromolecules, colloids, or living cells or inert material like oil emulsion or clay. The migration can also occur in non-uniform way by alternating current provided the particles are polarisable. Such transport phenomena is called as "dielectrophoresis". In the gaseous system this is referred to as electrostatic process (ESP). Thus xerography or ESP is the most recent application of such electrostatic procedure in controlling the air pollution.

Electrophoretic separation is based upon the difference in rate of migration of charged species in suitable supporting media like liquid, solid or paper. It is not concerned with the kind of reaction occurring at electrode. Under the influence of an applied electric field positively charged molecules migrate to cathode while negatively charged molecules move towards anode. Neutral molecules or those without charge will not move provided if medium is liquid.

21.1 PRINCIPLES OF ELECTROPHORESIS

The essential requirement of the electrophoresis is that the molecule must possess a charge either positive or negative. The velocity with which the molecules move towards a particular electrode is called as migration velocity (v). It is measured in terms of cms^{-1}. This velocity depends upon applied electrical potential as: $v \propto E$, if E is applied potential field i.e., $v = \mu E$ if μ is some constant called "electrophoretic mobility".

$$\mu = \frac{v}{E} = \frac{\text{migration velocity}}{\text{field strength}\,(E)} = \frac{\text{cm}^{-1}\text{s}}{v\,\text{cm}^{-1}}$$

So μ is expressed as $cm^2v^{-1}s^{-1}$. It has a sign of +ve or −ve while the neutral species will have no mobility. If a molecule has charge 'Q' then in liquid media, force will be = Q × E. This would promote molecule to move towards cathode or anode. The initial drag of ion is called as "viscous drag" which depends upon the viscosity of the liquid media as well as size and shape of the molecule. In order to retain the migration velocity (v) constant, the viscosity of medium (η), magnitude of applied potential (E), charge of the molecule (Q), size and shape of molecule which is measured in terms of radius (r) be constant. Therefore the migration velocity (v) is given as

$$v \propto \frac{1}{\eta} \text{ as } v \propto E, \text{ also } v \propto Q \text{ and } v \propto \frac{1}{r} \text{ so } v = \frac{EQ}{\eta r} \qquad ...(21.1)$$

This technique facilitates separation due to difference in charge or size of molecule. There are various factors that affect electrophoretic mobility, which includes type of (i) charge of molecule, (ii) kind of system used, (iii) type of support used, (iv) the ionic composition of buffer, (v) applied voltage and (vi) the temperature employed. Let us consider each factor as follows:

21.2 PROPERTIES OF CHARGED MOLECULE

(a) Charge
(b) Electrophoretic system
(c) pH of buffer
(d) Ionic concentration
(e) Applied potential
(f) Temperature

(a) The characteristics of molecule charge depends upon net charge, size and molecular mass. The ions with high charge move faster than those ions with small charge i.e., v ∝ Q/r. So charge to size ratio is of vital importance. The term (Q/r) is called "charge to size ratio". For molecules having same (Q/r) value, the separation is feasible on sieving gels e.g. agarose or polyacrylamide as support.

(b) *Kind of electrophoretic system used:* In zones electrophores, the kind of supporting media used is important. For gel permeation pore size is critical. The pH of buffer, the composition of buffer, voltage, temperature are also important. The choice of support media depends upon the size of the sample, cost of support, availability, purpose of separation, time of work and the knowledge of work carrying out such separation.

(c) *pH of the buffer:* For nucleotide and polynucleotides, net charge is negative due to presence of phosphate group which forms acidic backbone of the molecule. At high pH, ions migrate faster, the separations depends upon size as (Q/r) is same. At the isoelectric point (pI) molecules have zero charge because pH < pI molecules will have +ve charge for condition when pH > pI, charge will be negative. For proteins of same size and shape separation depends upon buffer pH, but pH also depends on temperature. Therefore constant temperature must be maintained. The pH depends also on concentration of buffer.

(d) *Nature of ionic concentration of buffer:* The ions in buffer solution should not participate in electrophoretic migration. The concentration of the buffer solution should be as low as possible. The large evaporation of buffer solution during electrophoresis should be minimised.

(e) *Applied potential:* A kind of resistance which is encountered during the electrophoretic studies, may be due to nature of power supply, kind of support medium and wick used. By Ohm's Law (V = i × R) i.e., potential = current × resistance. Current must be kept fixed since current varies with potential. With high voltage high (E) field strength is employed. The total power is (Power = $i^2 \times R$ watts if i = current and R = resistance).

(f) *Temperature:* The density difference due to temperature generates convection currents. So cooling is resorted to avoid density difference. The best results are obtained in a refrigerated chamber. The diffusion should be kept to minimum. There should be minimum loss of solution by evaporation. A change in viscosity of gel affects the electrophoretic components. At 4 °C, water is having maximum density which is best temperature for carrying electrophoresis. Temperature must be kept constant by use of thermostated chamber in order to get reproducible results.

21.3 THEORY OF ELECTRIC DOUBLE LAYER

The electric charge originates from the ionisation of functional group which in turn depends upon the molecular structure. The ion adsorption modifies charge of macromolecule. Adsorption is due to anions but hydrated cations do not exhibit adsorption.

The theory of electrical double layer pertains to structure of boundary between two phases. If ions of one sign are adsorped, ions of opposite sign will be also attracted and get accumulated at phase boundary. A fixed charge of one phase and the mobile charge of second phase of counterions of opposite charge forms a electric double layer. The net charge will be zero with some electrical potential which is defined as the work required to bring unit charge of same sign from infinity to given point in space.

Zeta Potential

(a) Zeta potential is symmetrically distributed over its surface in centre: The solvent molecules are bound to surface which in turn move along with particle during electrophoresis. The potential ψ_0 at surface of particle is given as $\psi_0 - Q/\in_a$ where ($\in$) is dielectric constant and a = radius of the sphere.

It is a different potential from the potential present at the surface of shear. If $x = (\partial + \Delta a)$ if Δa is in statistical average of thickness of hydration shell, the potential which controls electrophoretic mobility is defined as the zeta potential (Z) or electrokinetic potential. This potential decreases with distance (x) as per equation as potential at surface is $\psi_\lambda = \psi_0 \exp(-k_x)$ if k is constant called Debye Huckel constant.

(b) The zeta potential and thickness of double layer decreases rapidly with increase in the concentration. Now this so called Zeta potential cannot be directly measured. A charged particle is subjected to many forces leading to constant electrophoretic migration velocity. Forces have the electrophoretic attraction (K_1) = QE and Stoke's friction $(k_a') = -f_c/v$ if v = electrophoretic velocity and f_c = friction coefficient; usually $f_c = 6\pi\eta a$, if η = velocity of solvent. Electrophoretic retardation (K_a) and relaxation (K_4) are obtained. At steady state electrophoretic velocity

$$(v) = \frac{QE + K_a + K_4}{f_c}.$$ The electrophoretic retardation force $(K_3) = (\in F_a - Q)E$. The electro-

phoretic mobility is $\mu = \dfrac{v}{E} = \dfrac{\in Z}{6\pi\eta} = \left(\dfrac{tf}{6\pi\eta}\right) f(K_a)$, if $f(K_a)$ varies from 1-15. This equation facilitates calculation of zeta potential, if (K_4) relaxation effect is neglected.

21.4 CLASSIFICATION OF ELECTROPHORESIS METHODS

Classification of electrophoretic techniques: There are several ways of classifying electrophoretic methods. The basis of such classification is based on amount of substance separated, kind of procedure used, method of stabilisation, principle of separation and applied potential. The classification based upon the initial and marginal conditions of the separations are follows as:

(a) Moving boundary electrophoresis comparable with free boundary chromatography.

(b) Zone electrophoresis similar to zone electrochromatography

(c) Isotachophoresis is a specialised class of electrophoresis.

(d) Isoelectric focussing which depends upon the mode of focussing.

(e) Immunophoresis is a specific technique.

(a) *Moving boundary electrophoresis:* What Tseilus originally discovered was the moving boundary electrophoresis technique (see Fig. 21.1).

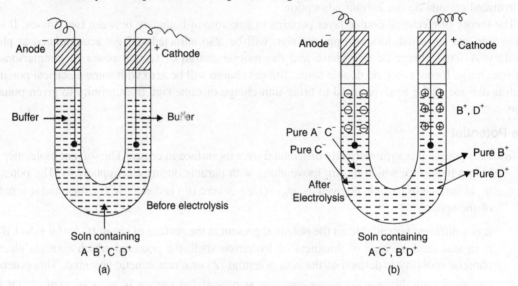

Fig. 21.1 Moving boundary electrophoresis

Till 1950, this method was extensively used. The equipment consisted of U-shaped tube cell fitted with anode and cathode. The protein solution ($A^- B^+$; $C^- D^+$) was put at the bottom and buffer solution was kept above this solution. The electrodes were immersed in buffer solution. On applications of potential anions (A^-) and (C^-) migrated to anode while cations (B^+) (D^+) migrated to cathode (Fig. 21.1a). Their concentrations at electrodes can be measured by refractometer at boundary of two layers in U-tube. The velocity of movement of anions is not identical to cations due to the difference in their corresponding mobility so also there is no similarity in velocity of cations (C^+) and (D^+). Therefore ions with large value of mobility migrates faster in relation to other ions with low mobility. This method is similar to frontal analysis in chromatographic separations.

21.5 ZONE ELECTROPHORESIS

The only difference from free boundary electrophoresis techniques is the fact that the separated ions are likely to mix with each another, defeating the purpose of separation. This is due to the presence of convective current present in U-tube (Fig. 21.1*b*). To avoid this problem of mixing, stable supporting media like filter paper, cellulose acetate, silica gel were used to prevent migration and mixing of electrophoretically separated ions (Fig. 21.2). The cell is filled with buffer solution of specific conductivity. The sample of mixture is fed on centre of support like filter paper. On the application of an electric field, the different component migrate towards electrodes at different velocity due to the difference in mobilities of the individual ions. The schematic representation of separation of molecules A, B, C is given in (Fig. 21.3) order C > A > B.

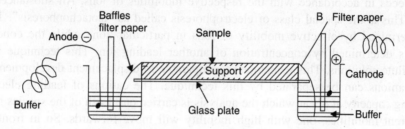

Fig. 21.2 Zone electrophoresis

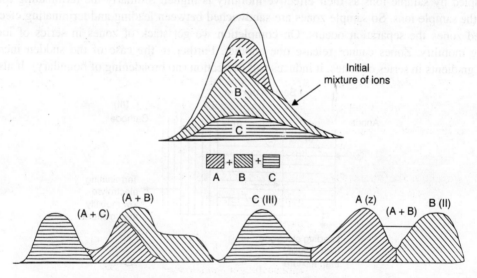

Fig. 21.3 Separation of ions by Zone electrophoresis

The shape of zones is influenced by participation of substance in adsorption equilibria. The time of separation is vital. Zone electrophoresis is analogous to the method of elution in the chromatography. The density gradient stabilisation involves effecting electrophoresis in density gradient column which prevents convection currents thereby preventing mixing of separated ions. Sucrose, glycerol, ethylene glycol are used as density gradient barriers. They dissolve in water, but do not "ionise" hence they do not undergo electrophoresis, as only ionised material can exhibit electrophoretic movement.

21.6 ISOTACHOPHORESIS

In 1897 Kohlrausch developed this method, but since 1970 it became popular with chemists for the separation of charged small molecules. In this method the charged components get stacked one after other in accordance to their electrophoretic mobilities which might vary by 10%. So at times it is not possible to separate cations and anions but it is applicable for separation of one of the ions. The electrolysis compartment is divided into three parts of unequal capacity. The separation column (II) is filled with leading electrolyte (I) while second part of compartment (II) contains mixture of ions to be separated and the third part (III) is a compartment filled with terminating electrolyte. The mobility of these ions is much higher than those of ions which are to be separated, while in terminating electrolytes in third part contains ions with lower mobility in comparison to those ions to be separated. The cations of terminating electrolyte are unimportant. On applying electric field steady state is reached. The separation proceeds in accordance with the respective mobilities of ions. All substances move with same velocity. This is the special class of electrophoresis called as "isotachophoresis". The potential gradient is determined by effective mobility of ions in particular zone while the concentration of separated ion is determined by concentration of another leading ion. This technique is useful for separation of dilute solutions. This technique is comparable to displacement development in chromatography. The anions can be separated by this technique. The cations of leading electrolyte must possess buffering capacity at pH at which the analysis is carried out. Each of the species in the sample will have different mobilities, one with high mobility will move forwards. So in front and behind original sample zone set two series of mixed zones. The ionic species of the leading electrolyte cannot be occupied by sample ions as their effective mobility is higher. Similarly the terminating ion never bypass the sample ions. So sample zones are sandwiched between leading and terminating electrolyte. In mixed zones the separation occurs. On completion we get stack of zones in series of ions with differing mobility. Zones cannot release one another. Further in the case of the sudden increase in voltage gradients in series of zones, it induces self correction (no broadening of boundary). It also leads

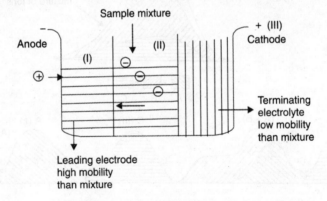

Fig. 21.4 Isotachophoresis

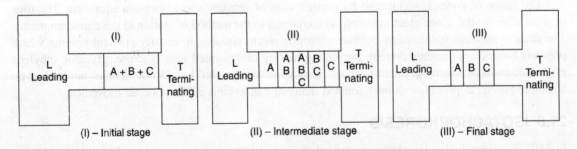

Fig. 21.5 Separation by isotachophoresis in three zones

to increase in temperature in preceding zones, facilitating detection of zones by thermometric detectors. The concentrations always attains a fixed value by composition of leading electrolyte.

21.7 ISOELECTRIC FOCUSSING

This is a kind of steady state electrophoresis technique wherein steady state is attained, in which the widths and the positions of the zones of the separated components do not change with time. In isoelectric focussing, a stable pH gradient is created between the anode and cathode. With applications of potential the charged molecules move through the medium until they reach a position in the pH gradient where their net charge is equal to zero which in turn prevents further migration.

The individual (pI) bands can be easily collected. Isoelectric focussing has better resolving power and can be used in preparative electrophoresis sometimes proteins precipitate at pI as they have zero charge. This method is used for the separation of ampholytes and macromolecular compounds, because $pH_{mix} < pH_{iso} < pH_{max}$. A stable pH gradient in column is must for the isoelectric focussing in order to effect electrophoretic separation. (Fig. 21.6)

21.8 IMMUNOELECTROPHORESIS

This technique exploits the high sensitivity and specific immunochemical precipitation of antigen (*i.e.*, antibodies). If both parts in complex come in contact with each other at optimum concentration, a precipitate line is formed. One uses differently oriented diffusion and migration flows. The mixture of antigen is separated in agarose gel by zone electrophoresis. The slow diffusion flow of antigens against antibodies is replaced with faster migrational flow. It is called "crossed electrophoresis". The precipitation lines after separation can be intensified by staining with dyes. Two or more dyes can be used for two different species in the sample.

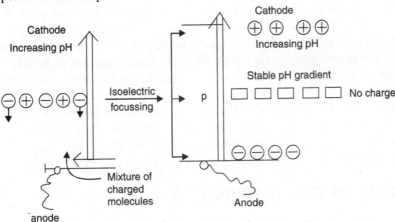

Fig. 21.6 Isoelectric focussing

21.9 TECHNIQUES OF ELECTROPHORESIS

The three important experimental techniques used in electrophoresis are as follows:

(*a*) Free boundary electrophoresis
(*b*) Zone electrophoresis
(*c*) Continuous electrophoresis

(a) *Free boundary electrophoresis:* As discussed earlier in Fig. 21.1, a U-tube is taken and is filled with unstabilized buffer, then the sample is injected and change in electrolytes concentration is measured by knowing change in refractive index of the sample. The technique is largely used to measure electrophoretic mobility of ions. Tiselius had used this kind of model. The U-tube must be kept at constant temperature to get reproducible results.

(b) *Zone electrophoresis:* This is by far the most popular method used in chemistry. It is some times called as electrochromatography. A stable media consists of porous material like gels cellulose acetate, starch gels, polyacrylamide gel, agarose gels. The filter paper is simplest form of support. The support is soaked in buffer and sample is placed in the centre. DC potential is applied. The different migrating molecules move in order of their electrophoretic mobility. On completion separated zones are detected as in chromatography. The convection currents do not distort electrophoretic patterns. One can use various kinds of supporting media in zone electrophoresis. Thus proteins, oligo nucleotides, lipids or carbohydrates are separated on filter paper as support. The paper should not be too thin. It must contain 94% cellulose. The phenomena of electroosmosis must be absent which originates due to migration of H_3^+O ion. The cellulose acetate is another supporting media used in form of membrane. The (OH^-) group are esterified by process of acetylation. Electrophoretic separation are done very rapidly on the membrane as it is homogeneous. On impregnating it with mineral oil, it readily becomes transparent, which can be then directly used after separation and for spectrophotometry. SO_4^{2-} or -COOH groups if present promotes electroosmosis. Methylation arrest electroosmosis. In place of acetylation of OH groups, one can use cellulose acetate membrane. A supporting gel is polymeric material with crosslinked structure with inert property with absence of ionisable groups.

The gels are not usually granulated. They exert sieving effect on passage of molecules in the column. The porosity of gel depends upon number of cross linkage present in gel molecule. The separation proceeds as per pore size and the geometry of the molecule (e.g. exclusion chromatography). The starch gels are prepared from potato starch. They become transparent on heating. The process is cheap, easy to operate and rapid. The isoenzymes are separated by starch gel. The polyacrylamide gels are better as concentration of polyacrylamide can be varied. The adsorption of proteins by polyacrylamide gel is negligible. By cross linking of NN′ methylene bis acrylamide is polymerised with actylmide. They are used as columns. At 1mA current samples gets inside gel. Their chief merit is that they can dissipate heat faster. They can be used horizontally/vertically, in column with better resolution. Finally use of agarose gels in fine as it is linear polymer of D-galactose and its derivative.

21.10 CONTINUOUS ELECTROPHORESIS

This is also called as curtain electrochromatography. It is used when large volume of sample is analysed. In cell, separations occur due to simultaneous flow of electrolyte and electrical migration transversing to the solvent flow. The solution is introduced through wick. The eluting electrolyte flows downward by gravity through supporting media. At right angles electric field is applied to lower corners of support which is dipping in solution. It is called "curtain electrophoresis". During electrophoresis, spilt sample in different fractions. Electrophoretic patterns are similar to fan with fractions emerging from the bed at various points. The neutral species travel downward, but the charged species move to appropriate electrodes. The difference in electrophoretic mobility and zeta potential leads to clean separation of the several ions. (Fig. 21.7).

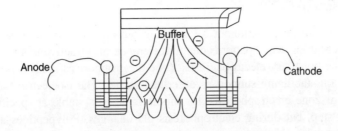

Fig. 21.7 Curtain electrophoresis

21.11 DETECTORS USED IN ELECTROPHORESIS

It depends upon nature of sample, kind of electrophoretic system and purpose of analysis. Many methods like spectral, staining radiochemical or biological assay methods are used for the purpose of detection of molecules. Initially refractive index (Schlieven) method was used for detection. UV was used for proteins (230-280 nm). DNA, RNA showed absorption maxima at 260 nm. Similarly nucleic acid and polynucleotide were best detected by spectroscopy however polyacrylamides cannot be separated. The staining is one of the best method of detection. The separated species is reacted with dye or chromogenic agent but reagent must be selective and the generated products are stable and insoluble. The reaction kinetics should be fast and colored complex must possess high molar absorptivity. The excess staining agents must be washed off before spectroscopic measurement. The destaining is also electrophoretically feasible. The common staining agents used are ninhydrin, amido-black, alizarine blue, acridine, brilliant blue etc. Those giving fluorescence can be also fluorimetrically measured e.g. fluoresenic, acridine, etc. The radiochemical methods were also effective if radioactive isotopes were used. Isotopes like I^{125}, P^{35}, C^{14} are used for measuring γ or β activity. The biological assay method was also efficient as these methods were selective and sensitive. Such methods were based on histochemical localisation of enzymes in cell. They were used to assay enzyme activity. A very brief discussion of four important detectors is given.

(a) *UV absorption detector:* At steady state all compounds move in separate zones. The chief demerit of this kind of detector is need to use small cell volume. The broad neck tubes can also be used. Some nonabsorping material serve better during electrophoresis.

(b) *Conductivity detector:* The sample species are separated according to mobilities. The stepwise increase in field causes an electrical resistance and hence change in the conductivity. The microsensing electrode is used to differential system.

(c) *High frequency detectors:* To give zone same speed, field strength is increased. The current density is same for all. So resistances form front to back. Such method needs good shielding, so zone length of 2 mm can be detected; but reproducibility is poor. The development is relatively interference free.

(d) *Thermometric detector:* Electric field varies from zone to zone and is inversely proportional to the effective mobilities of ionic species, rendering all zones with equal speed. With passage of stabilised current, heat production rises from the front to back stage. With sharp change in temperature, zone boundaries are obtained. By temperature measurement of particular ionic species, species location can be found out.

21.12 INSTRUMENTATION

Tiselius did not use any special instrument for carrying out electrophoresis. The separation was observed as bands of 'Schlieven' waxy patterns due to difference in refractive index. The band resolution is however poor in free boundary electrophoresis. Zone electrophoresis method is commonly utilised. Zones are distributed on stabilising support as described earlier. The isoelectric focusing as discussed earlier is a variation of zone electrophoresis, for dipole ions. It is stable at specific pH (PI) wherein stable pH gradient is step, but during electrophoresis pH changes. Polypeptides and proteins can be separated, when at particular pH the movement stops facilitating separation from moving (other) species. In paper electrophoresis capillary flow from electrolyte apparatus will make for up any loss of liquid by evaporation. Such movement does not interfere in separations. The voltage of 100-300 V (1000 in extreme case) is applied to effect separation . The continuous and discontinuous separations are also possible. Most of the variation in instrument originate from kind of detection system used. One can use just (i) thermometric detector[1], high[2] (ii) frequency detectors, (iii) conductivity detectors, (iv) UV-absorption meter, etc. These detectors are grouped according to the sensing element which is in contact with internal electrolyte. A discussion of some important detectors used is given section 21.11.

21.13 APPLICATIONS IN INORGANIC CHEMISTRY

Amongst the techniques free boundary electrophoresis is not used in practice due to spreading of zones by convection current. In zone electrophoresis, mention must be made for paper electrophoresis, gel electrophoresis or cellulose acetate membrane separation. The purpose of study is to consider, application for inorganic and organic and biological compounds and separation of proteins etc.

(A) In inorganic separations, we have several number of applications. Principal amongst them were the separation of silver and nickel or arsenic, antimony tin, or silver, mercury, lead and iron.

(1) Separation of alkali metals or typical separations of silver, nickel in cationic form. They migrate to same cathode but on adding EDTA (disodium salt) anionic complex formed for nickel migrates to anode and silver to cathode, as negatively charged complex of nickel was formed with EDTA.

(2) Tin, arsenic and antimony were separated, with stationary supporting electrolyte as the paper in 0.02M lactic acid, 0.02 M tartaric acid and 0.04 M DL-alanine. We first see tin which migrates, then arsenic and finally antimony with arsenic in middle-fraction.

(3) Magnesium and alkali metals were separated on paper as stationary support. Ammonia was the supporting electrolyte and voltage of 200-500 V applied. A complete separation was accomplished.

(4) Similarly Cs, Rb, K, Na were separated on paper consisting of ammonium formate and trichloroacetic acid as buffer solution. Ag, Pb, Hg, Tl were separated in 0.1 M sodium chloride, potassium chloride and potassium cyanide in just 180 mts.

(B) The separation of organic compounds, such as aliphatic amines, dyes, hydrolysed products of nucleic acid, organic acids amino acids, and proteins were separated by electrophoresis technique.

(a) In separation of typical amino acids by proper selection of pH; it was seen that the uppermost electrograph showed a case where the pH of the buffer was between the isoelectric points of two amino acids, which in turn migrated in opposite direction. The fractions of acidic, basic and neutral species was obtained e.g. Glutamic acid, Leucine and Glycine.

(b) The protein analysis by electrophoresis was most significant. Such separations were carried out either at high or low pH or in neutral media. (e.g. proteins) The time of analysis was as short as 15 minutes.

(c) Large number of separation of nucleotides, polypeptides, as well as those of the proteins were carried out.

(d) In the separation of nucleic acid containing nucleotides or bases, a simple paper electrophoresis was used at pH 3-4 when negatively charged particles moved to anode. The gel electrophoresis is best for such separation.

20.14 SEPARATION OF BIOLOGICAL PRODUCTS

A large number of biological products such as peptides, proteins, glycerol, lipoproteins, enzymes, nucleic acid, nucleotides, nucleosides, were separated. Similarly organic compounds like alcohols, aldehydes, carboxylic acid, steroids, amines, aminoacids, alkaloids, vitamins, antibiotics, dyes, and pigments were separated by electrophoresis. The principal use of electrophoresis is in the separation of biological molecules, both of low molecular mass like amino acids, peptides, nucleotides, or those with high molecular weight compounds like DNA and RNA molecules. The highest resolution was possible if discontinuity is introduced in liquid media by varying pH, or use of high density gels. Even membrane barriers could be also incorporated to effect clean cut separation.

The use of electrophoresis technique in separation of biological products like proteins, serum or body fluids is well known. The biological activity of enzymes can be traced with specific substrates. The technique of the immunoelectrophoresis is extended for resolution of proteins. In clinical medicine, electrophoresis is used for the diagnosis of few diseases e.g. sickle cell anemia. For non protein mixtures high voltage electrophoresis was used.

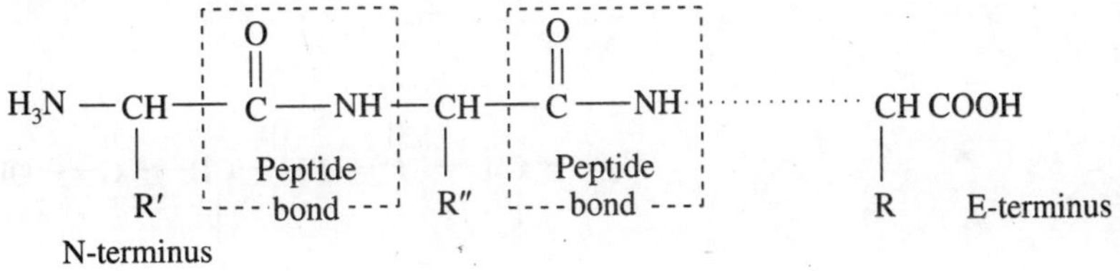

Fig. 21.8 Peptide link

A significant contribution of electrophoresis technique was made for the separation of proteins and polynucleotides. Proteins are linear polymer of aminoacids, macromolecular in nature. These molecules are composed of peptide bonds with side chain of every aminoacid residues (Fig. 21.8). The size and net charge of protein is of relevance in electrophoretic separation. At low pH, free carboxylic acid group will not be ionised while amine groups will be protonated. So protein will have +ve charge. On the contrary at high pH, the free carboxylic acid groups will be easily ionised, but the amino group would not be protonated so protein will have net negative charge as shown in Fig. 21.9. (a & b).

While at intermediate pH some amine groups will be protonated and some will not be protonated. At such pH, amino acid will have no charge. Such point is called "isoelectric point". The amino acid exists as dipolar ion with ionised carboxyl group and protonated amine group. This is called as "pI". So if pH is lower than pI protein will be positively charged but if pH is higher than "pI", protein will have 'negative charge'. Therefore pH < pI, proteins will migrate towards cathode but when pH > pI it will migrate to anode. Thus electrophoresis is quite useful to evaluate pI values for proteins by watching pattern of migration of proteins during electrophoresis at varying pH (Table 21.1).

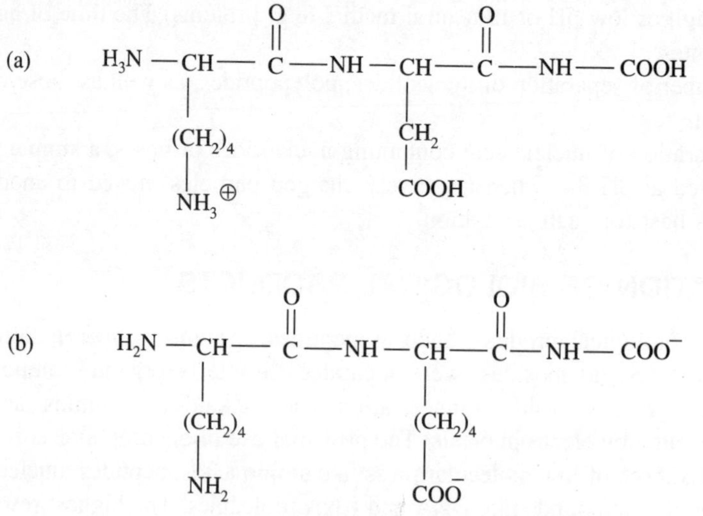

(a)

H_3N —CH— C —NH —CH— C —NH — COOH

(CH$_2$)$_4$

$\overset{\oplus}{NH_3}$

CH$_2$

COOH

(b)

H_2N —CH— C —NH —CH— C —NH — COO$^-$

(CH$_2$)$_4$

NH$_2$

(CH$_2$)$_4$

COO$^-$

Fig. 21.9 (*a*) Cationic species (*b*) Anionic species

The proteins are polymers obtained from aminoacid monomers. The polymerisation of amino group belonging to acid and carboxyl group form peptide bond with release of water molecules (Fig. 21.10).

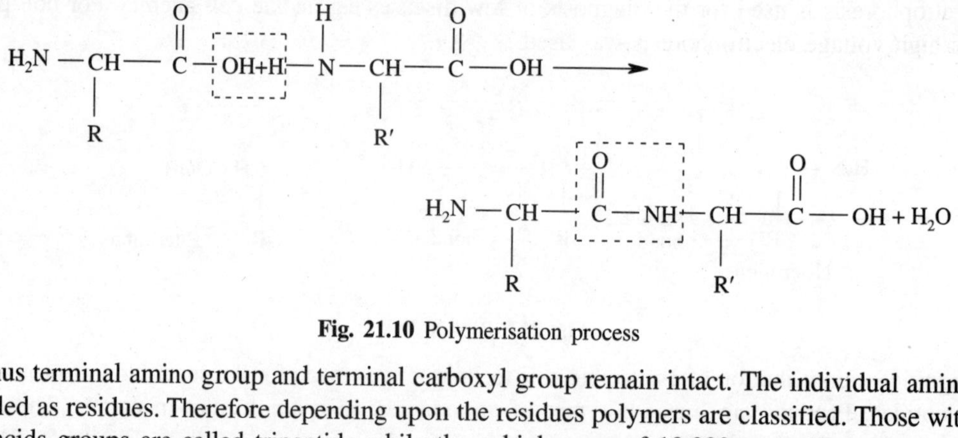

H_2N —CH— C –⌐OH+H¬— N —CH— C —OH ⟶

R

R′

H_2N —CH– ⌐C¬–NH– CH— C —OH + H_2O

R

R′

Fig. 21.10 Polymerisation process

Thus terminal amino group and terminal carboxyl group remain intact. The individual amino acids are called as residues. Therefore depending upon the residues polymers are classified. Those with three aminoacids groups are called tripeptide while those high mass of 10,000 are called polypeptide and those above mass > 10,000 (M.W.) are called proteins. The fibrous protein are not soluble in water but globular proteins are soluble in aqueous solution. The simple structure of aminoacid is shown in Fig. 21.10.

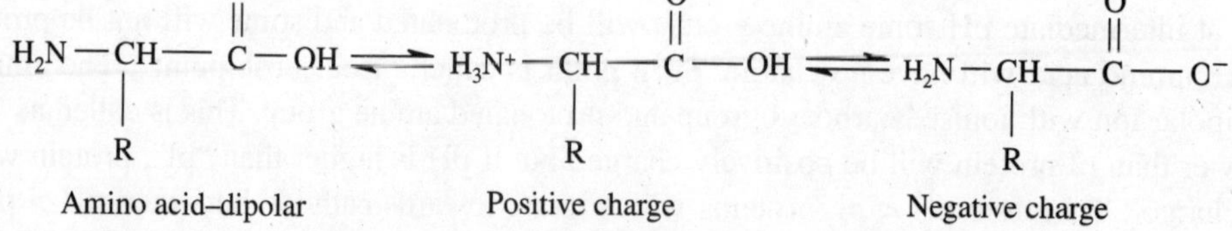

H_2N —CH— C —OH ⇌ H_3N^+—CH— C —OH ⇌ H_2N —CH— C —O$^-$

R

Amino acid–dipolar

R

Positive charge

R

Negative charge

Fig. 21.11 Amino acids.

Table 21.1 pI values of proteins and their charge

S. No.	Name of Proteins	pH = 3.0	pH = 7.4	pH = 10.0	Isoelectric point (pI)
1.	Serum albumin	+	−	−	4.8
2.	Haemoglobin	+	−	−	7.1
3.	Insulin	+	−	+	5.4
4.	Pepsin	−	−	−	1.0
5.	Cytochrome C	+	+	−	10.0

Polynucleotides are similar to proteins but they differ in their structure. Like nucleic acids they are basis of heredity. They are polymeric molecules containing long chains with varying length. Such polyester chain forms backbone. Those with sugar base are called nucleoside while base with sugar and phosphoric acid are called as nucleotide. The general structure of polynucleotide is given in (Fig. 21.12).

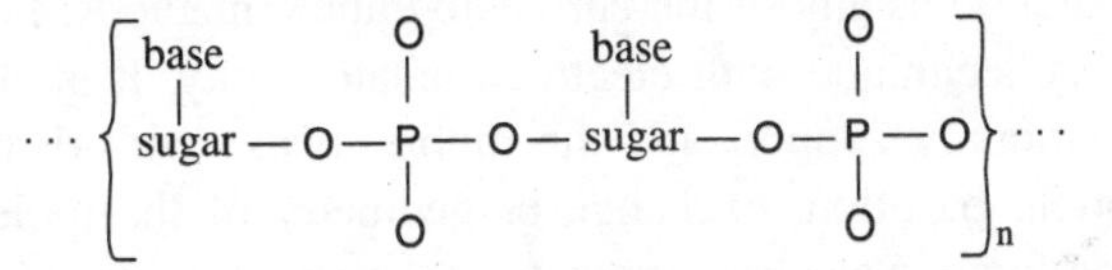

Fig. 21.12 Polynucleotide

The RNA, DNA are also polynucleotides. They have the following structures. See Fig. 21.13 (*a*) and 21.13 (*b*).

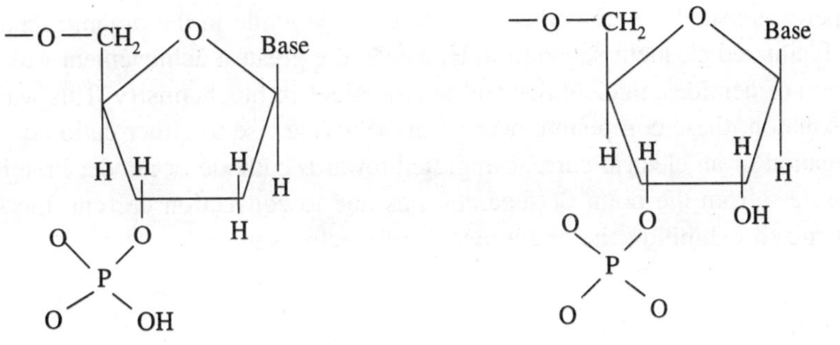

DNA – Double strand

(a)

RNA – Single strand

(b)

Fig.21.13 Polynucleotides - DNA and RNA

Further on dissolution in water polynucleotides form anion which are negatively charged. Thus phosphate (PO_4^{3-}) group forms a negatively charged backbone of polynucleotide. The single stranded polynucleotides are soluble in neutral or basic solution with negative charge due to dissociation of phosphate. This is true for RNA while DNA has two polynucleotide strands with H-bond. So the structure is "double helix". Therefore ionisable groups on purine and pyramidine bases are unavailable for ionisation. DNA molecules are soluble in basic solutions; with negative charge and migrate towards anode. Electrophoresis is commonly used for their separation. A small quantity of material is required for analysis. Their properties can be studied on gel electrophoretic technique which is also used on preparative scale. The generated heat creates problem as the separated components lose resolution and isolation of the components becomes problematic.

21.15 PREPARATIVE ELECTROPHORESIS

The volume of supporting media must be large with large column size (e.g. 20 × 10 × 1.5 cm optimum size). The blocks in form of trough are prepared from starch or polyvinyl chloride (PVC). Foam rubber in form of strips can be also used. The sample volume is 50-500 ml. The body fluid, tissue culture and other biological material is fractioned by the preparative mode. In clinical diagnosis zone electrophoresis is best for analysing serum, urine, spinal fluid, gastric juice and body fluids. The electrophoresis is carried out in slabs of porous material. Milligram level separations were achieved. Thick Whatman 3 MM filter paper is also useful. It is essential to cool (~10°C) the supporting media on both sides. After separation the zones are cut, eluted with solvent and then analysed. Staining agent can be used for visualisation. As much as 10 gm of material can be separated with about 2 W current.

CONCLUSION

From the foregoing discussion, we have come to the conclusion that electrophoresis method is one of the powerful tools of separation as any of the chromatographic methods. In fact electrophoresis can be used as the complementary technique with electrochromatography. It is clear in chromatography, we need stationary phase, stationary support and the mobile phase with driving force originating from difference in the adsorption, partition, exchange or geometry of the molecule, while in the case of electrophoresis the driving force is electrical potential to create difference in mobility leading to differential migration front. The buffer solution serves as the stationary phase while solvent molecule itself acts as the mobile phase. The overall technique is simple and rapid. Within 15 minutes several separations can be carried out. The process is economical and results are just easily reproducible. In inorganic separations paper was usually used as the supporting media while in the organic separations, cellulose acetate or gels facilitated cleancut separation. However, the greatest achievement was significant in the ease of separation of peptides, nucleotides and polypeptides in biochemistry. This was mainly possible as in aqueous solution, these compounds were ionised giving rise to either cationic or anionic species which on application of an electric current migrated towards cathode or anode. Free boundary electrophoresis was useless from the point of separation as due to convection current, the species separated once again got mixed exhibiting the phenomenon of electroosmosis.

LITERATURE

1. M.A.L. Melvin, *"Electrophoresis"*, John Wiley & Co. (1987).
2. J. Bruno, *"Chromatography & Electrophoresis Methods"*, Prentice Hall & Co. (1994).
3. R. Block, E. Durram, G. Zeweig, *"Paper Chromatography and Paper Electrophoresis"*, Academic Press Ltd. (1958).
4. F.M. Everaerts, J.L. Beckers, T.P.E.M. Verheggen, *"Isotachophoresis—Theory, Instrumentation and Applications"*, Elsevier Sci. Publ Co. (1976).
5. Z. Deyl (Ed.), *"Electrophoresis—A Survey of Technique and Applications Part A– Theory, Part B–Applications"*, Elsevier Sci. Publication (1983).
6. E. Lederer, M. Lederer, *"Chromatography—Principles and Applications"*, Elsevier Pub. House (1968).
7. A.T. Andrews, *"Electrophoresis - Theory, Technique and Biochemical and Clinical Applications"*, Clarendon Press, Oxford (1986).
8. O. Gaal, G.A. Medgyesi, I. Kereczkey, *"Electrophoresis in the Separation of Biological Macromolecules"*, John Wiley & Sons (1980).
9. M. Lederer, *"Paper Electrophoresis"*, Elsevier Co. (1955).
10. M. Bier, *"Electrophoresis"*, Academic Press (1959).

Chapter 22

Capillary Electrophoresis

On the introduction of electric field in a tube containing conductive solution, an electrolyte undergoes migration of the charged solutes. The rate and direction of flow depend upon sign and value of the charges of solute particles. This is the basis of electrophoresis. The electroosmotic flow (EOF) starts movement of solute in a tube. Those with high charge migrate faster while one with lower charge flow with slow speed. Neutral molecules do not migrate, but they can be separated by miscellor electrokinetic capillary chromatography (MECC). Charge electrolytes move in different zones depending upon the magnitude of the charge. Initially glass tubes were used as they promoted diffusion of various zones on the supporting media by process of thermal convective diffusion preventing thereby segregation of the ions. Joule heat is generated by passage of current. Broad tubes produce more heat and hence large diffusion. Molecules at centre of the tube migrate faster. The temperature difference between end and middle of tube is

$$\Delta t = \frac{0.239\,W}{4K}r^2 \qquad\qquad ...(22.1)$$

where W = heat in watts, K = thermal conductivity, r = radius of tube.

So reduction in tube diameter reduces diffusion. Addition of non conductive stabilizing medium (e.g. gel) reduces diffusion. It is called 'slab gel electrophoresis. Paper gel polyacronamide can be used as the supporting media. Reduction in diameter of tube reduces the convective diffusion (~200 μ mid) or (pyrex 75 μm id). These small capillary tubes have reduced diffusion, heat generated is released due to high area/volume ratio. One can use as high as 30 kV voltage. It reduces time of separation. The most popular detector is UV/visible absorbance measurement. For UV region glass tubes are useless as they are unsuitable in UV region of the spectra. So fused silica tubes are preferred. The technique has high speed and resolution separations in simple volume (0.1-10 ηl). The separated species is easily eluted from end of tube. The beauty of separation lies in fact that one can isolate the substance at $(10^{-9} - 10^{-15})$ concentrations.

22.1 THEORY OF CAPILLARY ELECTROPHORESIS

Since single phase is involved only longitudinal diffusion occurs. The plate count (N) is given by equation

$$N = \frac{\mu_e V}{2D} \qquad\qquad ...(22.2)$$

If D = difference coefficient, V = voltage, μ_e = charge of solute. Plate count does not increase. One can use 20,000 to 60,000 V potential. For this N = 3,000,000 in capillary electrophoresis, N = 10,000,000 in gel support. One encounters electroosmotic flow, when solvent migrates to one of the electrode. This leads to bulk migration. The electroosmotic flow velocity is given by

$$V = \mu_{e_0} E \qquad \qquad ...(22.3)$$

In presence of electroosmosis velocity

$$(V) = \left(\mu_e + \mu_{e_0}\right) E \qquad \qquad ...(22.4)$$

First the movement of cations takes place. Then neutral molecules and finally anions migrate. Addition of cationic surfactant reduces osmotic flow as well as it becomes positively charged in capillary tube. Coating of trimethylchlorosilane slows the flow to eliminate surface silicol groups. The adjoining Fig. 22.1 shows the assembly for capillary electrophoresis. The cathode (−ve) attracts cations while anode (+) attracts anions at the end of these capillary tubes. The solution is placed in the capillary and high voltage potential is imposed. The −ve charged particles move to cathode and +ve charged ions migrate to anode thereby accomplishing separation of cations and anions. By difference in rate of migration various cations are separated sequentially.

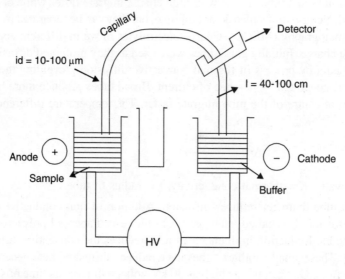

Fig. 22.1 Assembly for work

22.2 INSTRUMENTATION

Since volumes of the sample are small, its introduction in tube poses problem. The introduction can be effected by electrokinetic injection and pressure injection. Sample ends by ionic migration and electroosmotic flow at one end of capillary tube. Mobile ions are readily introduced. A pressure difference is used in pressure devices, by applying vacuum at one end of tube. It is useful for gel filled tube. About 5-50 ηl sample is introduced. Thus hydrodynamic and electrokinetic pressure facilitates introduction of the sample in a tube. The injection volume is calculated by equation as

$$V_i = 2.84 \times 10^{-8} \frac{Htd^4}{l_i} \quad \text{(Hydrodynamic method)} \qquad \qquad ...(22.5)$$

$$V_i = \frac{Pd^4 \pi l}{128 \eta \, l_i} \qquad \text{(Electrokinetic method)} \qquad \qquad ...(22.6)$$

wherein V_1 = sample volume, H = no. of sample, t = time, d = inner diameter of tube, l is total capillary length in cms. This is true for hydro dynamic method. In pressure injection, P = pressure, h = viscosity, injection volume is η.

Finally
$$Q = V_{\pi c} tr \frac{2\mu_{EP} + \mu_{EO}}{L} \qquad \qquad ...(22.7)$$

The total quantity injected is given by (Q) if V = voltage, C = sample concentration, t = time, r = id, μ_{EP} = electrophoretic mobility, μ_{EO} = electroosmotic mobility. l = length = (50-75 cm), id = 375 μ.m, V = 30 kV
$$...(22.8)$$

22.3 SAMPLE SEPARATION

In time of 10 minutes, 19 anions can be separated. If migration velocity of solute is V_{EP} then
$$V_{EP} = \mu_{EP} E \qquad \qquad ...(22.9)$$
where μ_{EP} is the electrophoretic mobility, E = potential applied. Since solutes migrate at different speed they are separated if electrophoretic mobility is

$$\mu_{EP} = \frac{q}{6\pi \eta r} \qquad \qquad ...(22.10)$$

if q = charge, η = velocity of electrolyte, r = internal diameter of tube. If q/r is high μ_{EP} is also high. High charged, small molecules have high velocity. The solute molecules are separated due to difference in their electrophoretic mobility (μ_{EP}) and are moved by electroosmostic force (EOF). This promotes charged solutes, but not neutral molecules with usual notions as described above, we have

$$V_{EOF} = \frac{E_\varepsilon \eta}{4\pi \eta} \qquad ...(22.11)$$

$$\mu_{EOF} = \frac{\varepsilon f_2}{4\pi \eta} \qquad ...(22.12)$$

$$V_{EOF} = \mu_{EOF} E \qquad ...(22.13)$$

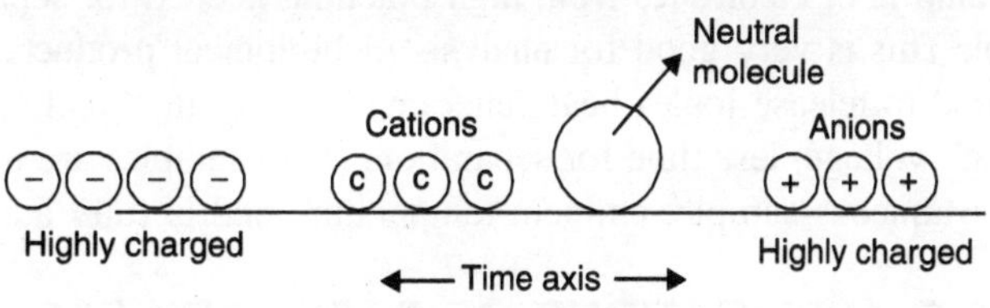

Fig. 22.2 Electropherogram

If ε = dielectric constant of electrolyte, J = zeta potential, E = voltage, η = viscosity of electrolyte. The overall migration can be depicted pictorially in Fig. 22.2.

Various factors influence (EOF) electroosmostic pressure or force. (*a*) increase in E, (*b*) increase in pH of electrolyte, (*c*) increase in concentration, (*d*) increase in temp., (*e*) addition of organic solvent (*f*) modification of wall of capillary tube.

$$R = 0.177 (\mu_2 / \mu_1) \left[V/D (\mu_{av} + \mu_{EOF}) \right]^{1/2} \qquad \qquad ...(22.14)$$

if $\mu_2 - \mu_1$ = difference in electrophoretic mobility, μ_{av} = average electrophoretic mobility.

22.4 SAMPLE DETECTIONS

There are several kinds of detectors used in analysis. They include absorbance, indirect, electrochemical and mass spectral detectors. The ions move due to difference in electrophoretic mobility, therefore the peaks are dependent on retention time. The following Table 22.1 summarises the kind of detectors used and minimum detection limit of analysis.

Table 22.1 Detectors in capillary electrophoresis

S.No.	Name	Detection Property	Detection Limit (moles)	Remark
1.	Spectrophotometry	– absorbance	$10^{-15} - 10^{-13}$	Not good for low absorbance
2.	Fluorimetry	– Fluorescence	$10^{-17} - 10^{-20}$	Dependable, most sensitive
3.	Raman spectroscopy	– Transmittance	2×10^{-15}	Not common
4.	Thermal methods	– thermallense	4×10^{-11}	Uncommon
5.	Mass Spectrometry	– Mass	1×10^{-17}	For biological products
6.	Conductivity	– Conductance	1×10^{-16}	Need separation of electrode
7.	Amperometry	– Diff. Current	7×10^{-19}	Graphite joint used
8.	Radiometry	– Charged isotope	1×10^{-19}	Limit 1 mg/L to 1 µg/L

The first two methods are widely used in analysis i.e. absorbance and the fluorescence measurement. On column studies they are preferred. 50-100 µm is path length; but attempts to increase it are made to get reliable results. Those substances having lower absorbance are measured by indirect measurement with external chromophore as chromation indicates slow decrease in absorbance as we go down the column (λ max. = 254 nm).

Laser fluorescence detector has better sensitivity (~10 zeptomoles). Electrochemical detectors need isolation of electrodes from high potential needed for separation. Mass detector permits flow of 1 µL/ml. This is very good for analysis of biological products e.g. DNA, proteins etc. Small capillary are good to release joules heat generated by warming. Technique that has high separation efficiency need high voltage, less time for separation, small samples are needed, small volume is better. It can be used for aqueous sample; ambient temperature and is fully automated technique in analysis.

22.5 APPLICATIONS OF CAPILLARY ELECTROPHORESIS

It provides both qualitative and quantitative determination of polycyclic amines and hydrocarbons, chiral compounds, vitamins, amino acids, inorganic ions, environmental pollutants, and peptides in biological investigation. There are minor variation in techniques as capillary zone electrophoresis (CZE), capillary gel electrophoresis (CGE), capillary Isotachophoresis (CITP) and capillary isoelectric focussing (CIEF) e.g. CZE is used for separation of small ions, as well as separation of anions. This method is comparable with AAS, ICP-AES and IC. In 3 minutes 30 anions can be separated. Protein, carbohydrates are best separated. CGE is carried out on porous gel polymer, with buffer mixtures. They act as

molecular seives. Polyacrylamide is the best gel for microanalysis (It is $CH_2 = CH - \underset{\underset{O}{\|}}{C} - NH_2$).

Agarose, polysaccharides can be also used.

6

CITP is used for bands moving with same speed. Finally CIEF is used for separation of amphiprotic species e.g. amino acids, proteins, say, analysis of glycine, which contain "zwitter ion". The pH at which no net migration occurs is designated as 'isoelectric point'. (pI) which can characterise amino acids. In CITP sample is sandwiched between the leading and terminating or trailing buffer in capillary and by application of the electric field in constant current mode. Both cations and anions can be simultaneously separated by CITP. In CIEF species are separated due to the difference in the isoelectric point (pz). It is a pH at which protein is neutral. Below pI cations and above pI anion exist in mixture. In CGE gel is crosslinked polymer. They have pores, through which solute migrates, small molecules migrate faster facilitating separation from large molecule. Electrophoretic mobility and size difference facilitates such separations. Large biomolecules can be thus separated by slab gel electrophoresis. It can work on on-line system. In CZE, simultaneous separation of cations and anions is possible. Separations are due to the electrophoretic mobility which in turn is related to charge to size ratio of anion. EOF promotes movements of ions. All ionic compounds can be separated by CZE.

22.6 CAPILLARY ELECTROCHROMATOGRAPHY

It is in between capillary electrophoresis and HPLC. In 1980 two techniques were developed as packed column and miscellar electrokinetic capillary (MEC) techniques. Capillary electrochromatography (CEC) has many merits. It is useful for the separation of uncharged species. It furnishes high efficient separations at micro-volume sample. The mobile phase in CEC transported through a stationary phase by electroosmotic flow (EOF) pumping. It leads to plug profile flow as opposed to hydrodynamic mechanical pressure with greater efficiency. The technique is based upon packed column. Electroosmotic flow (EOF) pushes polar solvents in column. Separation depends upon partitioning of solute species in the stationary as well as mobile phase. Several polyaromatic hydrocarbons were isolated in 30 cm tube (id 75 μm) with acetonitrile (4mM) as mobile phase. The stationary phase was octadecyl silica particles acting as stationary support (3μm).

22.7 MISCELLAR ELECTROKINETIC CAPILLARY CHROMATOGRAPHY (MECC)

It provides a good method for neutral molecules, where basis of separation is the actual difference in distribution of neutral solutes between miscelles and an electrolyte. Sodium dodecyl sulphate (SDS) is used as electrolyte. It is $[CH_3(CH_2)_{11}O-SO_3]$. It has hydrophobic negative charge on one end of the molecule, SO_4^{2-} being neutral hydrophobic 12 carbon hydrocarbons. Miscelles are formed with nonpolar hydrophobic hydrocarbon ends of the SDS (detergent) in centre of sphere away from aqueous electrolyte. Outside sphere are polar ionic ends. The miscell may be cationic, anionic, neutral (i.e., nonionic) and zwitter ion. On introduction of sample in capillary, it spreads between aq. electrolyte and interior of miscelles. Water solubility or polarity determines such distribution. Water insoluble solute is solubalised by miscelles. As told earlier such separation is based upon the difference of distribution of solute between miscelles and electrolyte. Anionic miscelles migrate to anode but EOF diverts them towards cathode instead of anode to render it with less velocity.

Insoluble solutes moves through capillary and are first eluted, while those which are completely solubalised miscell travel last. In actual practice, the miscelles are not stationary but are non polar pseudo-phase. MECC can separate inorganic anions. MECC facilitates separation of low molecular weight aromatic phenols and nitrocompounds. Capillary electrophoresis carried out in the presence of miscelle is MECC or MEKC. Aqueous phases moves fast but not misceller phase. The technique is analogous to LLPC or Liquid-liquid partition chromatography. The difference in partition coefficient between mobile phase and hydrocarbon pseudostationary phase is exploited. It is true chromatography technique. The column efficiency is excellent (N = 100,000 plates) and changing second phase in MECC is simple to change in concentration of miscellar portion.

LITERATURE

1. D.R. Baker, *"Capillary Electrophoresis"*, Wiley NY (1995).
2. S.Y. Li, *"Capillary Electrophoresis"*, Elsevier NY (1992).
3. P. Bocek, *"Capillary Zone Electrophoresis"*, VCH, NY (1993).
4. D.N. Heiger, *"High Performance Capillary Electrophoresis—An Introduction"*, Hewlett Packward Co. (1992).
5. R. Weinberger, *"Practical Capillary Electrophoresis"*, Academic Press, New York, (1993).
6. D.R. Baker, *"Handbook of Instrumental Techniques for Analytical Chemistry"*, (Ed.) F. Settle, Pearson Education Co. (1997).
7. D.A. Skoog, F.J. Holler, T.A. Nieman, *"Principles of Instrumental Analysis"*, Harcourt Brace College, Publishers, p. 778 (1998).
8. P. Patnaik, *"Dean's Analytical Chemistry Handbook–Second Edition"*, McGraw Hill Co. (2004), p. 5.110.

Chapter 23

Analytical Spectroscopy

Chemical analysis consists of two parts, the first part involves separation, while the second part pertains to quantitative determination. Separation involving different kinds of chromatographic methods have been thoroughly discussed in earlier chapters. In subsequent chapters, we propose to dwell on methods for quantitative determination. Such analysis is carried out either by spectral or optical methods or alternatively by electroanalytical techniques. No doubt, in addition to these, radio-chemical and thermal methods are of significance but are less commonly utilised. Optical methods involving spectral techniques are most popular with the chemists as they are simple.

23.1 ABSORPTION SPECTROPHOTOMETRIC METHODS

These are most commonly used in analytical laboratories. The principal reason for their popularity is the easy availability of photometers, colorimeters or simple single beam spectrophotometers. Further, fortunately various chromogenic ligands have been discovered by the organic chemists which have in turn facilitated spectral analysis. Most of such methods involve molecular absorption spectroscopy in the ultraviolet and visible region. Infrared spectral methods which involve molecular vibrations leading to molecular absorption have largely been used for the elucidation of the structure of compounds. The classical examples of spectrophotometric determination of iron with thiocyanate or uranium with 8-hydroxyquinoline or nickel with dimethlglyoxime or cobalt with 1-nitro 2-napthol are very well known. Assignment of the structure of an organic compound or structure of coordination complex of a metal with a ligand is best carried out by infrared spectroscopy, e.g. assignment of the carbonyl group in aldehyde or ketone.

In comparison with molecular absorption, atomic absorption spectroscopic methods have become popular as a tool of analysis only in recent years. The principal advantage of the AAS technique is the analysis of one metal in the presence of another without involving separation. It not only permits rapid determination of multiple samples but multielement analysis is also possible. It has a few demerits also, which are considered later.

23.2 EMISSION SPECTROSCOPY METHODS

Atomic or molecular emission spectral methods have come to the forefront only in recent years. Flame photometry is a simple example of atomic emission spectroscopy. However, it has limitations as it is applicable only for analysis of s-block metals or metals like copper. Further, such analysis is done in the visible region. Subsequent discovery of arc or spark atomic emission spectroscopy had very limited practical applications. Only with the discovery of inductively coupled plasma atomic emission spectroscopy (ICP-AES) such methods become very popular. ICP-AES is extensively used by analytical chemists as it ensures complete atomisation and permits analysis at a low level. Molecular emission methods have become popular with the discovery of the phenomena of fluorescence and phosphorescence. The former proved to be beneficial for analysis of metals like beryllium with morine or aluminium with quinanzarine or zinc with oxine. In comparison, phosphorescence methods were not so commonly used for quantitative chemical analysis.

23.3 MAGNETIC RESONANCE SPECTROSCOPY

Under this category, we have nuclear magnetic resonance spectroscopy and electron spin resonance spectroscopy. NMR is principally used for the elucidation of structure and assignment of hydrogen in chemical structure. ESR or EPR methods are mainly used for characterisation of free radicals. Much progress has since been made in the area of NMR spectroscopy with the discovery of P^{31} NMR or C^{13} NMR in addition to the usual proton H^1 NMR. The greatest contribution has been in the field of chemical shift. Recently two and three dimensional 2D-NMR and 3D-NMR have come to stay. Fourier techniques in IR and NMR have given more refined hyperfine structure of spectra.

23.4 ELECTRON SPECTROSCOPY

This is a technique of recent origin principally used for surface analysis. This covers important methods like Aüger electron spectroscopy (AES method) and Electron spectroscopy for chemical analysis (ESCA), also called X-ray photoelectron spectroscopy. The two methods differ in the way that electrons impinge on a surface and different kind of electrons are available for emission and subsequent measurements. These methods are useful to chemists in the study of polymeric materials, catalysts and smooth surfaces. SIMS is the latest addition.

23.5 SCATTERING METHODS

This constitutes a separate class of analysis. It originated with the discovery of the Tyndall phenomenon. Scattered light may be absorbed as in turbidimetry or (e.g. analysis of barium as barium chromate) or the scattered light may be transmitted and the extent of scattering is measured in Nephelometry *i.e.* analysis of ammonia with Nessler's reagent. When there is a change in frequency of radiation between the incident and emitted radiation, it gives rise to Raman spectroscopy. These methods are very useful in environmental monitoring e.g. analysis of $NO-NO_2$ from air or SO_2 from the atmosphere is done best by fluorescence techniques. Raman spectroscopy can be used for the study of water soluble complexes.

23.6 PHOTOACOUSTIC SPECTROSCOPY AND RELATED METHODS

This is a technique which does not fall strictly into any of the above categories but is extremely useful for the analytical chemist. The change in radiation causes a change in temperature which in turn causes a change in pressure leading to a change in sound pressure which is directly proportional to the concentration. Finally mass spectrometry in the true sense is not a spectroscopic method, with the exception of fragmentation patterns which resemble spectra in spectroscopy. While optical rotary dispersion (ORD) and circular dichromism (CD) coupled to polarimetry are once again not pure spectroscopic methods, they correlate the spectral and refractive index and the constitution or structure of complex or organic compounds. In subsequent chapters, we shall consider these spectroscopic methods from the view point of quantitative chemical analysis.

23.7 ORIGIN OF SPECTRA

Colour is an important criterion for the identification of objects. In spectrochemical analysis, we make use of the electromagnetic spectrum to analyse chemical species and study their interaction with electromagnetic radiation. Planck's equation states: $E = hv$ where E is energy of the photon, v is the frequency, h is Planck's constant (6.624×10^{-27} ergs sec). Thus a photon has a certain energy, and can cause transitions in energy states in an atom or molecule. Since the energy state differs in chemical species, energy change in transition will also differ. A spectrum obtained by plotting some function of frequency (v) against the frequency of electromagnetic radiation is typical of a particular chemical species, and is useful for its identification. In spectrochemical analysis frequencies from (10-10,000 Hz) *i.e.* audio waves to γ-rays (10^{22} Hz) can be used for characterisation. Energy changes are due to transitions like rotational, vibrational, electronic, nuclear etc.. Figure 23.1 shows the range of frequencies used in spectroscopy. The basis for spectrochemical analysis is the interaction of radiation with a chemical species. Assume a sinusoidally oscillating electromagnetic wave, consisting of an electric (E) and magnetic (M) field (3×10^{10} cm/sec in vacuum) with the wave travelling with a velocity (C) and constant wave frequency (v) then spacing between maxima is the wavelength (Fig. 23.2, say point 1-2). This is the distance travelled during one period (l/v) of its oscillation so that the distance (λ) =

velocity (C) $\times$ time $\left(\dfrac{1}{v}\right) = \dfrac{C}{v} = \lambda$ wavelength of spectral region. The interrelation is: 1 cm = 10^{-2} m,

1 nm = 10^{-9} m and 1 Å = 10^{-10} m. Wavelength is inversely proportional to frequency *i.e.* $\lambda = \dfrac{1}{v}$ and

hence inversely proportional to energy. Wave number (cm^{-1}) is defined by an expression

$$\overline{V} = \frac{1}{l} = \frac{v}{C}$$

During spectrochemical analysis it is essential to use light of one wavelength *i.e.* monochromatic radiation. The bandwidth refers to the frequency range and represents the wavelength interval. As shown in Fig. 23.2 the electric and magnetic field are orthogonal to each other, and have orientation within a plane perpendicular to the direction of travel of the wave. If E and M are in the same plane the wave is plane polarised, but if the plane rotates with a definite velocity around the axis of propagation

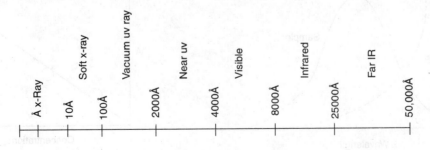

Fig. 23.1 Electromagnetic spectrum

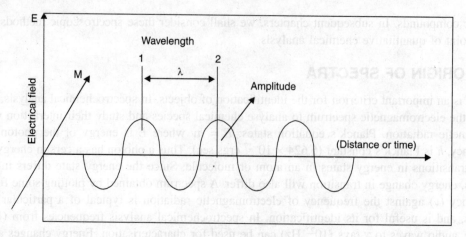

Fig. 23.2 Electromagnetic wave

then it is called circularly polarised. The details of these will be considered later. Radiant power (*P*) is related to the energy (*E*) of photon by $P = E\phi$ = where ϕ is the photon flux or number of photons per unit time. Radiant power is expressed in terms of intensity.

23.8 INTERACTION OF RADIATION WITH MATTER

Radiation interacts with a chemical species to provide information about the species. These interactions are termed as reflection, refraction and diffraction or scattering. The ways in which radiation interacts with chemical samples are absorption, luminescence, emission and scattering which thus depend upon the properties of matter, without involving a long term change in energy.

(*i*) *Absorption:* When a beam of electromagnetic radiation is passed through a chemical sample, part of it is absorbed. Electromagnetic energy is transferred to the atoms or molecules in the sample, and therefore the particles are promoted from lower to higher energy states, *i.e.* excited states. At room temperature they are at ground state. Thus absorption involves transition from the ground state to higher energy states. The study of frequency of absorbed radiation by the species provides a means for identification and analysis of a sample e.g. plot of absorbance *vs* wavelength gives an absorption spectra, (Fig. 23.3 and 23.4). Such spectra may be due to atomic or molecular absorption and is dependent upon the physical state, environment of absorbing species and other related factors.

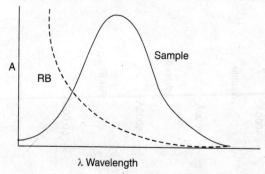

Fig. 23.3 Absorption spectra

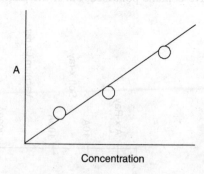

Fig. 23.4 Calibration curve

In atomic absorption spectroscopy, atoms are raised to higher energy states. UV and visible radiations cause transition of bonding electrons in elements (while X-rays interact with electrons closest to the nucleus). The atomic absorption spectra consist of limited narrow peaks. Contrary to this, in molecular absorption for polyatomic molecules in a condensed state, energy is greatly increased. Thus total energy change is due to a change in the electronic, vibrational and rotational states of the molecule. Molecular spectra in the UV-visible region are characterised by absorption bands in a definite wavelength range. They sometimes involve electronic transitions. Pure vibrational absorption can be seen in the IR region where the energy of radiation is insufficient to cause an electronic transition. It is also possible to have absorption due to an induced magnetic field. Thus on subjecting electrons or nuclei to a magnetic field, energy levels are produced and electromagnetic radiation of a suitable frequency cause transitions between these levels. Such absorption is studied by NMR or ESR.

Table 23.1 Interaction of radiation

S.No.	Wave-length (λ), nm $3.3 \times$	Wave-number (cm^{-1}) $3.3 \times$	Fre-quency Hz	Energy eV $4.1 \times$	Kind of radiation	Spectroscopic technique
1.	10^{-11}	10^{10}	10^{21}	10^6	γ – ray	γ – ray emission
2.	10^{-9}	10^8	10^{19}	10^4	X – ray	X – ray absorption/ or emission
3.	10^{-7}	10^6	10^{17}	10^2	Ultraviolet	UV absorption
4.	10^{-5}	10^4	10^{15}	10	Visible	Absorption or emission or fluorescence
5.	10^{-3}	10^2	10^{13}	10^{-2}	Infrared	IR absorption Raman
6.	10^{-1}	10	10^{11}	10^{-4}	Microwave	Microwave absorption also ESR
7.	10^1	10^{-2}	10^9	10^{-5}	Radio	NMR
8.	10^3	10^{-4}	10^7	10^{-8}		

(ii) *Emission of radiation:* Electromagnetic radiation is produced when excited ions, atoms or molecules return to lower energy levels or the ground state. Excitation is possible by several means such as a flame, arc or spark. Radiating particles produce radiations of a definite wavelength producing discontinuous or line spectra. A continuous spectrum consists of wavelengths which are closely spaced. Such spectra are due to the excitation of solids or liquids where atoms are closely packed. Both spectra are useful in analytical chemistry. Sources such as thermal, chemical, electrical and other forms can be used to excite atoms, molecules or ions to higher energy states.

(iii) *Fluorescence and phosphorescence* also called luminescence, is a special kind of emission process. The atoms or molecules are excited by absorption of electromagnetic radiation, and emission takes place after a lapse of time when the excited species return to the ground state. Fluorescence occurs more rapidly than phosphorescence and is complete in after about 10^{-5} (or even less) seconds from the time of excitation. Phosphorescence emission takes place over periods longer than 10^{-2} –100 seconds and can continue for minutes or even hours after the irradiation has ceased. Resonance fluorescence describes a process in which the emitted radiation is identical in frequency to the radiation employed for excitation. Phosphorescence occurs when an excited molecule relaxes to a metastable excited electronic state which has an average life time $> 10^{-2}$ seconds. In all these cases, the excited species undergoes deactivation by emission of radiation.

(*iv*) *Scattering:* Like the processes of absorption, emission and luminescence, the scattering of electromagnetic radiation need not include transition of energy. Scattering involves randomization of directions of a beam of radiation. When a beam of electromagnetic radiation falls upon a small particle, the particle will experience a disturbance due to rotating electric and magnetic fields of passing radiation. The temporary retention of radiant energy by the particle causes polarisation of ions, atoms or molecules. This is followed by re-emission of radiation in all direction as the particle returns to the ground state. A small portion of radiation is transmitted at angles from the original path and the intensity of the scattered radiation increases with particle size. With colloidal particles, scattering can be seen with the naked eye. Scattering of molecules smaller than the wavelength of radiation is called Rayleigh scattering where its intensity is related to wavelength (an inverse fourth power dependence), the dimension of particles and polarizability. The measurement of scattered radiation can be used to determine the size and shape of a polymeric molecule such as a colloidal particle e.g. in nephelometry. However the Raman effect differs from ordinary scattering in that, part of the scattered radiation suffers a quantized frequency change due to vibrational energy level transitions occurring in the molecule as an effect of the process of polarisation.

If a constituent is present in amounts greater than 0.01% it is usually termed a 'macro' constituent whereas one less than 0.01% in the sample is generally termed a 'micro' constituent. These micro amounts can also be termed as 'trace' amounts. In earlier days, it was thought that it would be very difficult to quantitatively determine trace concentrations. However, with the introduction of colorimetric, spectrofluorimetric and spectrographic methods it has become quite simple to analyse even at microlevels. If the constituent is capable of producing colour of its own accord or in combination with some reagent in solution, then the measurement of light absorbed by the coloured solution is measured; such a technique is called a colorimetry. On the other hand, if the constituent is capable of forming a suspension, the light absorbed by the suspension is measured and it is termed 'turbidimetry'. When the amount of light scattered by a suspension is measured it is called 'nephelometry'. The amount of delayed light emitted by a solution under an excited source of radiation like a mercury or xenon lamp is measured in 'fluorescence' methods of analysis.

In visual colorimetry, natural or artificial white light is generally used as the light source and determinations are carried out with simple instruments like the visual colorimeter. In a few cases, the eye is replaced by a photoelectric cell, thereby eliminating error due to personal factors. These instruments are called 'photoelectric photometers'. They make use of light contained within a narrow range of wavelengths which is furnished by passing white light through filters. Such devices are called photoelectric filter photometers. However, in spectrophotometry, light of a definite wavelength at 180-400 nm is UV, 400-750 nm is visible and 750-1500 nm is absorbed near the infrared region. A normal spectrophotometer is composed of a spectrometer to produce light of a selected wavelength, *i.e.* monochromatic, and a photometer to measure the intensity of the monochromatic beam.

23.9 FUNDAMENTAL LAWS OF SPECTROSCOPY

When a beam of light impinges upon a homogeneous medium, a portion of the incident light (P_0) is absorbed as (P_a), a negligible portion of it is reflected (P_r) and the reminder is transmitted (P_t) with the net effect of intensities

$$P_0 = P_a + P_t + P_r \qquad \qquad ...(23.1)$$

wherein P_0 = intensity of incident radiation, P_a = intensity of absorbed light, P_r = intensity of the reflected light, and P_t = intensity of transmitted light. However, in actual practice it has been observed that the amount of reflected light (P_r) is so small (~4%) that for all practical purposes it can be neglected for the air–glass interface with glass cuvetts

$$\therefore \ P_0 = P_a + P_t \qquad \qquad ...(23.2)$$

Table 23.2 Colours of the visible region

Wavelength nm	Transmitted colour	Complementary colour	Wavelength (λ) nm	Colour (transmitted)	Color (complimentary)
< 380	Ultraviolet	–	560-580	Yellow, green	Blue, Violet
380-435	Violet	Yellow, Green	580-595	Yellow	Blue
435-480	Blue	Yellow	595-650	Orange	Blue, green
480-490	Green, blue	Orange	650-780	Red	Blue, green
490-500	Blue, green	Red	650-780	Near IR	–
500-560	Green	Purple	–	–	–

Lambert (1760) and Beers (1852) and also Bouger formulated the law as

$$T = \frac{P_t}{P_0} = 10^{-abc} \qquad \qquad ...(23.3)$$

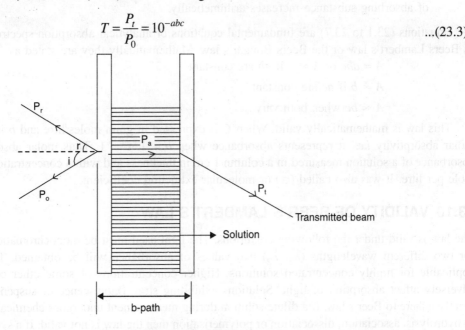

Fig. 23.5 Beer's-Lambert's law

if b = optical path, c = concentration

$$\log(T) = \log\left(\frac{P_t}{P_0}\right) = -abc \qquad \text{...(23.4)}$$

a = constant, T = transmittancy

$$\log\left(\frac{1}{T}\right) = \log\left(\frac{P_0}{P_t}\right) = abc = A \qquad \text{...(23.5)}$$

A = absorbance,

$$-\log(T) \ i.e., \ A = abc \qquad \text{...(23.6)}$$

$$\left(\frac{1}{T}\right) = T^{-1} \text{ is opacity}$$

So $\qquad\qquad\qquad A = abc \qquad \text{...(23.7)}$

$\qquad\qquad\qquad a$ = absorptivity

The above law can be easily stated in words as:

(a) When a beam of parallel monochromatic radiation enters an absorbing medium at right angles to the plane parallel to surface of the medium, each infinitesimally small layer of the medium decreases the intensity of the beam entering that layer by a constant fraction.

(b) When monochromatic light passes through a transparent medium the rate of decrease in the intensity with reference to the thickness of medium is proportional to intensity of light.

(c) The intensity of a beam of monochromatic light decreases exponentially as the concentration of absorbing substance increases arithmetically.

Equations (23.1 to 23.7) are fundamental equations of analytical absorption spectroscopy, known as Beers Lambert's law or the Beers Bougar's law. Mathematically they are stated as

$\qquad\qquad A = abc$ or $A \propto c$ if ab are constant

$\qquad\qquad A \propto b$ if ac are constant

$\qquad\qquad A \propto bc$ when both vary.

This law is mathematically valid. When C is expressed in gram moles/litre and b in cms, a = the molar absorptivity *i.e.*, it represents absorbance when $b = 1$, $c = 1$. Thus molar absorptivity is the absorbance of a solution measured in a column 1 cm in thickness and with a concentration of one gram mole per litre. It was also called $(\in)$ or molecular extinction coefficient.

23.10 VALIDITY OF BEER'S LAMBERT'S LAW

The law is valid under the following conditions. The light used must be monochromatic, or otherwise for two different wavelengths (λ_1, λ_2) two values of absorbance will be obtained. The law is not applicable for highly concentrated solutions. Higher concentrations of some other colourless salts adversely affect absorption of light. Solutions exhibiting stray fluorescence or suspensions may not strictly adhere to Beer's law. If a dilute solution during measurement undergoes chemical reaction such as hydrolysis, association, dissociation or polymerisation then the law is not valid. If a system conforms to Beer's law, a graph of absorbance against concentration (Fig. 23.6) should be a straight line passing through the origin. In case Beer's law is strictly adhered to, we may get straight line graph or calibration curve. The slope of this graph is equal to ab which in turn may help to evaluate the value of molar absorptivity (ϵ).

23.11 DEVIATION FROM BEER'S LAW

If the law is obeyed, we get a straight line graph (A) as shown in Fig. 23.6. Otherwise we may not

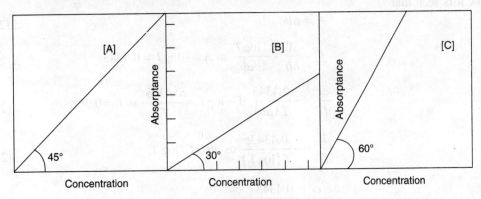

Fig. 23.6 Various calibration curves

get a straight line graph. Consider the following reaction.

$$2CrO_4^{2-} + 2H^+ \rightleftharpoons Cr_2O_7^{2-} + H_2O \qquad ...(23.8)$$

$$(\lambda_{max}\ 375\ nm) \qquad (\lambda_{max}\ 350\ nm,\ 450\ nm)$$

In dilute solution, the law is not obeyed so also in highly concentrated solution. Similarly in 4M HCl for the following reaction:

$$4Cl^- + \left[Co(H_2O)_4\right]^{2+} \rightleftharpoons \left[CoCl_4\right]^{2-} + 4H_2O \qquad ...(23.9)$$

The equilibrium is shifted from left to right as the concentration of cobalt increases. Thus in both the examples mentioned above, Beer's law is obeyed if the actual absorbance of the individual species is considered. The law is valid if a monochromatic beam is considered. If the bandwidth of radiation is greater, *i.e.* light is polychromatic, there will be a greater deviation. The deviation becomes apparent at higher concentrations because in an absorbance *versus* concentration plot the curve bends towards concentration axis (B), hence Beer's law is invalid. With a photometer, the radiant power of the component wavelengths are additive but the law requires the logarithm to be additive. If the absorption curve is flat over the spectral bandwidth employed, then Beer's law is expected to apply. Lack of adherence to Beer's law in a negative direction is always undesirable as a large increase in relative error in concentration is reported. Stray light which enters the detection housing is also a source of error. However, double monochromators minimise such error. Kortum and Seiler state the Beer's law is only a limiting law and should be expected to apply at a low concentration. It is not α or ϵ which is constant and independent of the concentration but $\left[\dfrac{an}{\left(n^2 + 2\right)^2} \right]$ wherein n = refractive index of the solutions at low concentration. Thus n is constant, but at a higher concentration, it needs correction in order to get a correct value of absorbance.

23.12 RELATIVE CONCENTRATION ERROR AND RINGBOM'S PLOT

The error T in the transmittance may cause an error in the concentration (C) measurement. The cause may be ascribed either to mechanical or electrical imperfections of the instrument or nonuniformity of the potentiometer, nonlinearity of the meter scale or detector and variation in the instrument. Rewriting Beer's law it is seen that

$$A = abc \qquad \qquad ...(23.10)$$

$$C = \frac{A}{ab} = -\frac{\log T}{ab} \text{ as } A = -\log T = 0.4343 \qquad ...(23.11)$$

$$\frac{dc}{dT} = -\frac{0.4343}{T(ab)} \text{ if } -\log T = \frac{0.4343}{T} \text{ as } T = 0.368 \qquad ...(23.12)$$

$$\frac{dc}{dT} = -\frac{0.4343c}{T(\log T)} \text{ as } ab = C/\log T \qquad ...(23.13)$$

$$\frac{\Delta c/c}{\Delta T} = \frac{0.4343}{T(\log T)} \qquad ...(23.14)$$

This shows that the relative concentration error depends inversely upon the product [($T \times \log T$) i.e. $T \times$ absorbance] or absorbance and transmittancy. If we differentiate equation (23.10) w.r.t. T by setting the function equal to zero it gives a point of minimum error when $T = \frac{1}{\in} = 0.368 \ (A = 0.4343)$. The variable quantities such as cell path, sample size and volume of the solution should be adjusted so as to yield solutions whose transmittance is 0.368, i.e. % $T = 36.8$. Thus this evaluation shows that sample transmittance can lie in the range $0.200 - 0.650$ (i.e. 0.700 to 0.200 absorbance) for a minimum and nearly constant error. The relative error increases at the extremes of the transmittance scale. The slope of a plot of C versus T, i.e. Ringbom plot gives the relative error coefficient (Fig. 23.7).

i.e. $$\frac{\Delta c/c}{\Delta T} = \frac{2.303}{T/(\Delta \log C)} = \frac{2.303}{\text{slope of Ringbom plot}}$$

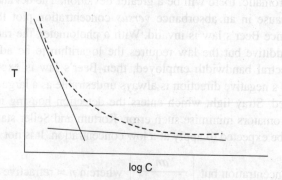

Fig. 23.7 Ringbom plot

The main limitation of the Ringbom plot (*i.e.* the plot of log C *versus* T) is that it provides no information concerning the concentration range of good precision unless it is combined with ΔT vs T relation. The above expression is valid irrespective of Beer's law's validity. A Ringbom plot does not provide the concentration range as stated earlier.

SOLVED PROBLEMS

23.1 If the molar absorptivity of a coloured complex is 3.20×10^3 at 240 nm, calculate the absorbance of a 5.0×10^{-5} M solution in a 50 mm cell when measured at a wavelength of 240 nm.

Answer: $A = abc$. In the given example

$a = 3.20 \times 10^3$, $b = 5$ cm, $c = 5 \times 10^{-5}$ M, hence we have

$$A = (3.20 \times 10^3) \times (5) \times (5 \times 10^{-5}) = 0.800$$

23.2 Calculate the absorptivity of a compound with a molecular weight = 144 if concentration is 1×10^{-5} g/ml solution exhibits an absorbance of 0.400 when the optical path is unity.

Answer: In the expression $A = abc$ we have

$$A = 0.400, \ b = 1.0, \ c = \frac{10 \times 10^{-6}}{144} = 6.9 \times 10^{-5}$$

$$\because \ 0.400 = a \times \frac{1 \times 10 \times 10^{-6}}{144}$$

$$\in = a \ \frac{0.400 \times 144}{10^{-5}} = 5.76 \times 10^7 \ \text{g mol.}/\text{cm}^2/\text{ml}$$

PROBLEMS FOR PRACTICE

23.3 If a given solution transmits 20% of light at certain wavelength what dilution will reduce the absorbance to half? What will double the transmittance?

23.4 A column of a solution of potassium dichromate has a transmittance of 0.340 for light of a wavelength of 480 nm. What is the transmittance of light through a column which is 1/5 of original column?

23.5 A solution of molybdenum thiocyanate complex has a concentration of (x) with a transmittance of 80%. What would be its % transmittance if its concentration is trebled?

23.6 How much would an absorbance of 1.176 correspond to on the transmittancy scale?

23.7 Convert 0.523 absorbance to % transmittance and 75% transmittance to the absorbance scales.

23.8 A coloured complex has molar absorptivity of 6.74×10^3 litre/cm/mol. If the optical path is 25 mm and % transmittance of is 7.77%, what will be the concentration of the solution?

23.9 Calculate the optical path of a coloured complex which has a molar absorptivity of 1480 and shows an absorbance of 0.750 when its concentration is 1.4×10^{-3} M.

23.10 What is the concentration of a coloured complex if it shows an absorbance of 0.250 in glass cuvetts with an optical path of 10 mm and has a molar absorptivity of 5890?

LITERATURE

1. R.J. Abraham, J. Fischer, P. Loftus, *"Introduction to NMR Spectroscopy"*, John Wiley (1988).

2. D. Briggs and M.P. Seah—Practical Surface Analysis Vol-1, *"Auger and X-Ray Photoelectron Spectroscopy"*, John Wiley (1990).

3. R.J.H. Clark, R.E. Hester, *"Advances in Infrared and Raman Spectroscopy"*, John Wiley (1985).

4. R.C. Denney and R. Sinclair, *"Visible and Ultraviolet Spectroscopy"*, John Wiley (1987).

5. R. Davies, M. Frearson, *"Mass Spectrometry"*, John Wiley (1987).

6. D. Rendell, D. Mowthrope, *"Fluorescence and Phosphorescence Spectroscopy"*, John Wiley (1987).

7. E. Metcalfe, F.E. Prichard, *"Atomic Absorption and Emission Spectroscopy"*, John Wiley (1987).

8. R.H. Clark, R.E. Hester, *"Spectroscopy of Surfaces"*, John Wiley (1988).

9. P.W.J.M. Boumans, *"Inductively Coupled Plasma Emission Spectroscopy"*, John Wiley (1987).

10. D.A. Skoog, J.J. Leroy, *"Principles of Instrumental Analysis"*, Harcourt Brace (1994).

11. R.M. Silverstein, G.C. Bassler, T.C. Morit, *"Spectrophotometric Identification of Organic Compounds"*, 5th Ed., Wiley (1991).

12. E.B. Sandell, *"Colorimetric Determination of Traces of Metal"*, Inter Science, Wiley (1958).

13. G.D. Christian, *"Analytical Chemistry"* 5th Ed. John Wiley & Co. (1997).

14. R.D. Braun, *"Introduction to Instrumental Analysis"*. Indian Edition, McGraw Hill & Co. (2004).

Chapter 24

Ultraviolet and Visible Spectrophotometry

The measurement of absorbance or transmittance of radiation in the ultraviolet and visible region of the spectrum is used for the qualitative and quantitative analysis of chemical species. The absorption of such species proceeds in two stages. In the first stage $M + h\nu = M^*$ involves the excitation of the species by absorption of photons ($h\nu$) with limited lifetime (10^{-8}–10^{-9} seconds). The second stage is of relaxation by converting M^* to new species by a photochemical reaction. Absorption in the UV or visible region leads to excitation of bonding electrons. Thus (λ_{max}) or the absorption peak can be correlated with the kind of bonds existing in the species. UV and IR spectroscopy is valuable for characterisation of the functional groups in the molecule and for quantitative analysis.

24.1 ORIGIN OF SPECTRA AND ELECTRONIC TRANSITIONS

The absorbing species may include transitions involving (a) π, σ, n-electrons (b) d and f electrons (c) charge transfer electrons. Such transitions are discussed below.

(a) *Absorbing species involving π, σ and n-electrons:* This class mainly includes organic molecules, ions and a few inorganic anions. Such molecules absorb electromagnetic radiation due to the presence of valence electrons which are generally excited to higher energy levels. Such absorptions are in the UV (<185 mm) region; while chromophores with low excitation energies have an absorption maximum beyond 180 nm. The electrons contributing to absorption in organic molecules are those involving bond formation between atoms, and nonbonding or unshared electrons which are generally localised. Electronic transitions in energy levels are possible by the absorption of radiation due to $\sigma - \sigma^*$ $n - \sigma^*$, $n - \pi^*$ and $\pi - \pi^*$ transitions where σ^* and π^* are antibonding atomic orbitals, while n-involves nonbonding orbital having an energy in between bonding and antibonding orbitals. Thus in $\sigma \rightarrow \sigma^*$ transitions, an electron in a bonding σ-orbital is excited to an antibonding σ^* orbital by absorption of radiation e.g. CH_4 has λ_{max} at 125 nm due to $\sigma - \sigma^*$ transitions. Such absorptions are rarely seen in the accessible UV region. Secondly $n \rightarrow \sigma^*$ transitions (*i.e.* from a nonbonding to an antibonding orbital) are seen in saturated compounds with unshared electron pairs requiring less energy in the region of 150-250 nm when $\epsilon = 1 \times 10^2 - 3 \times 10^3$. The absorption maxima (λ_{max}) for formation of the $n - \sigma^*$ state tend to shift to shorter wavelengths in polar solvents like ethyl alcohol and H_2O. $n - \pi^*$ as well as $\pi \rightarrow \pi^*$ transitions cover most organic compounds, as the energy needed for transitions to give an absorption maxima is in the region 200–700 nm. To have π orbitals an unsaturated functional group is required. Thus for $n - \pi^*$ transitions when $\in = 10 - 10^2$ L/cm/mol while for $n - \pi^*$ transitions $\in = 1 \times 10^3 - 10 \times 10^4$ L/cm/mol. In the former case, peaks are shifted to shorter wavelength (hypsochromic or blue shift) with increasing polarity of solvent. While a reverse trend (bathochromic or red shift) is observed for $n - \pi^*$ transitions, the blue shift is due to increased solvation of the electron pair by lowering of energy. The red shift proceeds with increased solvent polarity (~5 nm). The polarisation force between solvent and species, lowers the energy levels of excited and unexcited states.

Table 24.1 gives some common organic chromophores and aromatic compounds with an absorption peak (λ_{max}) and values of molar absorptivity ($\in$) and possible electronic transitions. This helps us to identify functional groups in spectral analysis. The absorption of multi chromophores are additive and conjugation has a significant effect upon spectral properties. Finally, the UV spectra of aromatic hydrocarbons are characterised by three sets of bonds originating from $n - \pi^*$ transitions. C_6H_6 has as absorption peaks at 184, 204 and 256 nm of which the long wavelength bands have sharp peaks. Polar

Table 24.1 Absorption of chromophores and of aromatic compounds.

Chromophore compound	λ_{max} (nm)	$\in_{max}$ L/cm/mol.	Electronic transitions
Alkene	177	1.3×10^4	$n - \pi^*$
Alkyne	178 – 225	$10 \times 10^3 - 150$	$n - \pi^*$
Carbonyl	186 – 280	$1.0 \times 10^3 - 16$	$n - \pi^*$
Carboxyl	204	41	$n \rightarrow \pi^*$
Amide	214	60	$n \rightarrow \pi^*$
Azo	339	5	$n \rightarrow \pi^*$
Nitro	280	22	$n \rightarrow \pi^*$
Nitrate	270	12	$n \rightarrow \pi^*$
Olefin	184	1.0×10^4	delocalise n^*
Triolefin	250	–	delocalise
Diolefin	217	2.1×10^4	delocalise
Ketone	282	27	$n \rightarrow \pi^*$
Ketone (unsat)	278	30	$n \rightarrow \pi^*$
Ketone (satd)	324	24	$n \rightarrow \pi^*$
H_2O	167	1.48×10^3	$n \rightarrow \sigma^*$
Methanol	184	1.5×10	$n \rightarrow \sigma^*$
Methyl chloride	173	2×10^2	$n \rightarrow \sigma^*$
Dimethyl ether	184	2.5×10^3	$n \rightarrow \sigma^*$
Methylamine	215	9×10^2	$n \rightarrow \sigma^*$
Benzene	204	7.9×10^3	$n \rightarrow \sigma^*$
Toluene	207	7×10^3	$\pi \rightarrow \pi^*$
Phenol	211	6.2×10^3	$\pi \rightarrow \pi^*$
Aniline	230	8.6×10^3	$\pi \rightarrow \pi^*$
Napthalene	286	9.3×10^3	$\pi \rightarrow \pi^*$
Styrene	244	1.2×10^4	$\pi \rightarrow \pi^*$

solvents eliminate fine structure; these bands are affected by ring substitution. An auxochrome is a functional group that has no absorption in the UV region but shows the effect of shifting the chromophore peak to longer wavelengths (e.g. OH, NH_2) Table 24.1 has auxochrome effect shift from 211 and 230 nm to 270 and 254 nm). This is specially dominant for the phenolate anion.

A number of inorganic anions exhibit λ_{max} in the UV region due to $n \rightarrow \sigma^*$ transitions e.g. NO_3^- (313 nm) CO_3^{2-} (217nm) NO_2^- (360 and 280 nm) azido (230 nm) and trithiocarbamate-2m (500 nm) ions.

(b) *Absorption involving d and f electrons:* The *d* block elements absorb in the UV-visible region. For *f* block metals transitions are possible due to *f* orbital electrons. The inner transition elements have narrow peaks due to the interaction of 4*f* or 5*f* electrons e.g. in lanthanides or actinides. Bands are narrow due to the screening of inner orbitals from external influence. On the other hand, in the case of the first (3*d*) and second (4*d*) transition series of elements, broad peaks are generally seen in the visible region; and such peaks are influenced by the surrounding environment. Spectral properties of transition metals involve electronic transitions among different energy levels of *d*-orbitals. From the knowledge of crystal field theory the t_{2g} (*dxy, dyz, dzx*) and e_g (*dx*^{2} - *y*^{2}, *dz*^{2}) are split by an amount Δ in the presence of ligands. Depending upon splitting capacity *i.e.* anionic ligands are arranged in the spectro-chemical series where Δ represents the energy of splitting as:

$$I^- < Br^- < Cl^- < F^- < OH^- < COOH^- < H_2O < SCN^- < NH_3 < en < NO_2^- < CN^-$$

This series helps to predict the position of absorption peaks for various complexes with these ligands. As Δ increases in the above series with increasing field strength, λ_{max} goes on decreasing e.g. $[NiCl_6]^{4-}$ (1370 nm), $[Ni(H_2O)_6]^{2+}$ (1279 nm); $[Ni(NH_3)_6]^{2+}$ (925 nm), $[Ni(en)_3]^{2+}$ (863 nm).

(c) *Charge transfer spectral absorption:* In such species $\in > 1 \times 10^4$, this provides sensitive mean for determining nature of absorbing species.

e.g. $$\left[Fe(SCN)_6\right]^{3-} ; \left[Fe(o\text{-}phen)_3\right]^{-3}, \left[Fe^{2+}Fe^{3+}(CN)_6\right]^-$$

are charge transfer complexes. Its component should be both electron donor and electron acceptor which in turn involves transfer of electrons to give absorption of radiation. The complex generally absorbs at longer wavelengths due to electron transfer requiring less radiant energy. In all charge transfer complexes the metal serves as an electron acceptor.

The concentration of a coloured solution is measured by finding out the absorption of light. In the visible region of the spectrum it is possible to determine the concentration by utilising three techniques namely colorimetry also called visual colorimetry, photometric photometry or spectrophotometry. The latter technique is useful for determination of absorbance in both the ultraviolet and visible region while the first two techniques are generally restricted to measurements in the visible region.

24.2 VISUAL COLORIMETRY

In colorimetry, a matching of the colour is effected in two solutions containing the same amount of coloured substance in the column having the same cross section normal to the direction of examination. White light is used and conditions of equal transmission of light by standard and sample solution are found by the eye of an observer. Since Beers Lambert's law states $A = abc$ with the usual notations, for two solutions $A_x = ab_x c_x$ and $A_y = ab_y c_y$ if they have the same colour producing chromogenic agents. Now, for two solutions optical balance is possible if $A_x = A_y$ i.e. $a\,b_x\,c_x = a\,b_y\,c_y$. In other words, we get a reduced equation $\dfrac{b_x}{b_y} = \dfrac{c_y}{c_x}$ provided a is constant as chromogenic group is same. One can verify this expression experimentally under the following conditions namely (a) $c_x b_x$ constant and both $c_y b_y$ vary, (b) $c_x b_x$ constant and b_y varies (c) $c_x b_x$ are constant and c_y varies. This can be done by the following methods:

1. Standard series method in Nessler's tubes (where $c_y b_y$ vary)
2. Dilution method in Hehner's cylinder (where b_x varies)
3. Balancing method in Dubosque colorimeter (where c_y varies)

In a standard series method, colourless glass tubes of uniform cross section and flat bottom, called Nessler's tubes, are used for holding colored solution of a definite volume. The colour is then compared with that of a series of standards prepared similarly from a known amount of the components being analysed. The Nessler meter, and colour comparators work on this principle of matching colour. In the dilution methods, sample and standard solutions with say concentration c_x and c_y are placed in glass tubes of identical dimensions. The more concentrated solution is diluted until the colours are identical in intensity. Alternatively in tube-I with known concentrations c_x and height b_x fixed, the colour is compared with tube-II with fixed (c_y) by varying the colour height (b_x) by opening the cock so as to have equal intensities of light seen through the tubes.

i.e., we have $c_x b_x = c_y b_y$ i.e., $c_y = \dfrac{c_x b_x}{b_y}$

Thus we can evaluate c_y provided the balancing is done as shown in Fig. 24.1, i.e. ($A_x = A_y$). This is done by using Hehner's cylinders.

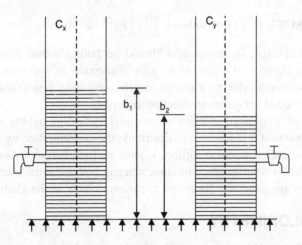

Fig. 24.1 Hehner's cylinders

The balancing method is the most commonly utilised method in visual colorimetry. The instruments using this technique are called 'Dubosque colorimeters' or plunger colorimeters. In this model $c_x b_x$ are kept constant and for a solution of concentration c_y, the path is varied till the colour intensities in both the tubes are equal. Intensity is observed vertically. By knowing b_x, we can evaluate c_y as

$c_y = c_x \left(\dfrac{b_x}{b_y} \right)$. Thus is Fig. 24.2 one sees a light from a even source of illumination pass through the

windows in the top of the base, through solutions to be tested and through plungers b_x and b_y.

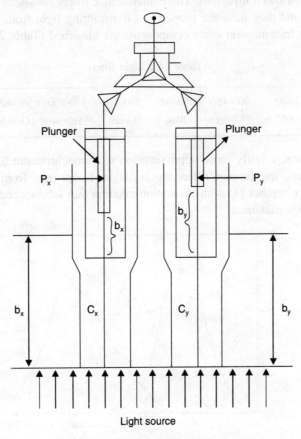

Fig. 24.2 Dubosque colorimeter

Some light is absorbed in passing through the liquids, the amount of absorption being dependent upon the concentration and depth of the solution. Two beams of light from the plunger are brought to a common axis by a prism system. Through the eyepiece a wide circular field is visible, the two halves of the circle being illuminated by light from two different cups. The depths of the liquid columns are adjusted until the two halves of the field are identical in luminosity. When this condition holds, Beer's law is obeyed and the concentrations of the two solutions are inversely proportional to their depths. As a precaution, cups and plungers should be kept clean and the reading should be zero when the plungers touch the bottom of the cup. Fig. 24.2 shows a model where cups are fixed and plungers are allowed to move up and down. There are some models where both plungers are fixed and the cups are moved up or down to adjust the heights to get equal illumination.

24.3 PHOTOELECTRIC CELLS AND LIGHT FILTERS

In visual colorimetry, we see the colour intensities with the naked eye. However, due to personal error, it may not be possible to get reproducible results. In order to eliminate this personal error, intensity is measured by a photocell. Similarly, in the visual method we use either daylight or an even source of illumination which may not be necessarily monochromatic. Hence in photometry we use interference filters which make the light almost monochromatic. Thus in order to get accurate results light filters are used for isolating desired spectral regions. These interference filters consist of coloured glass or gelatin impregnated with dye and they have the property of transmitting light from a specified region of the spectrum. During such transmission other components are absorbed (Table 24.2).

Table 24.2 Light filters

Wavelength (nm)	400	400-450	450-500	500-570	570-590	590-620	620-750	>750
Colour of filter	UV	Violet	Blue	Green	Yellow	Orange	Red	Near IR

By using such filters, a fairly close approximation to monochromatic light may be obtained. In addition to simple filters, interference filters are available. It can seen from Fig. 24.3 that the main principle is a glass filter number (1) with absorption maxima (λ_1) sandwiched with glass filter number (2) with λ_2 as absorption maximum.

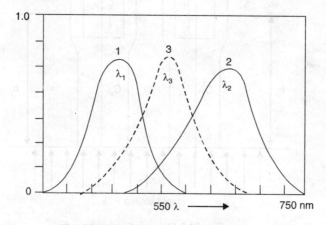

Fig. 24.3 Principle of interference filter

In between the two filters a transparent dielectric spacer such as magnesium bromide ($MgBr_2$) is placed. In transmission, the light passing will have an absorption maximum λ_3 which is between λ_1 and λ_2. Thus by a combination of filters, it is possible to get a closer approximation in the wavelength to procure monochromatic radiation as $(\lambda_1 + \lambda_2)/2 = \lambda_3$.

As regards photocells, we use three types *viz* photoemissive or phototubes, barrier layer or voltaic cell, and photomultiplier tubes. Of this the first two are generally used in the photometer while the last one is used in the spectrophotometer. The photoemissive cell (Fig. 24.4) consists of a glass bulb coated internally with a thin sensitive layer of K_2O Cs_2O AgO which are capable of emitting electrons when illuminated and acts as a cathode. A metal ring inserted inside, near the centre of the bulb acts as an anode which is maintained at a high voltage. The bulb is filled with an inert gas. When light falls upon the sensitive layer, electrons are emitted which in turn cause a current to flow through an outside circuit which can be amplified by electronic means and is taken as a measure of the amount of light striking the photosensitive surface. At any time, emission of electrons leads to a fall in potential

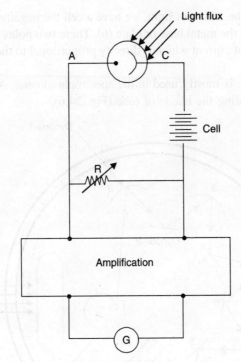

Fig. 24.4 Phototube

across a high resistance in the circuit which can be measured by a suitable potentiometer and is related to the amount of light flux falling on the cathode. The photovoltaic cell or barrier type cell Fig. 24.5 operates on a battery where a metal base plate (1) is deposited with a thin layer of selenium (2), this in turn is covered by a thin transparent metal layer (4) and under the side of the metal surface is covered with a layer of non-oxidising metal. When light falls upon the selenium surface, electrons are emitted which in turn penetrate a hypothetical barrier layer (3) and impart a negative charge to a transparent

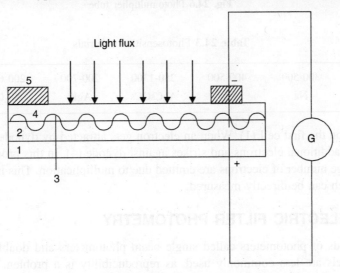

Fig. 24.5 Voltaic cell

metal layer (4). Thus under the action of light, we have a cell the negative pole of which is a collecting ring (5) and a positive pole is the metal base or plate (6). These two poles are connected to a galvanometer which indicates the flow of current which is directly proportional to the intensity of the incident light (Fig. 24.5) falling on plate.

The photomultiplier tube is mostly used in the spectrophotometer. Any kind of material as shown in Table 24.3 is used for coating the inside of tube (Fig. 24.6).

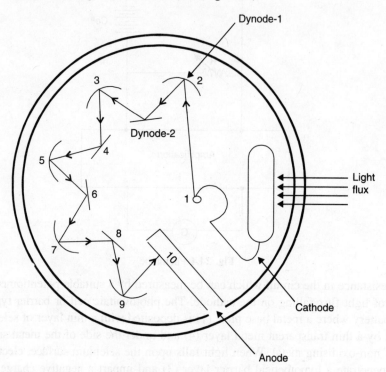

Fig. 24.6 Photomultiplier tube

Table 24.3 Photosensitive materials

Wavelength (nm)	300-500	400-500	250-1300	200-700	200-600	200-750
Coating material	Na	K	AgCsOCs	AgK	Sb-Cs	Bio Ag-Cs

Light falls upon the first cell (1). When an electron gets attracted by dynode (2) it strikes a photo cathode (3) ejects additional electrons and strikes against dynode (4). In this process at the conclusion of 10 strokes a large number of electrons are emitted due to multiplication. This imparts a large current to the anode, which can be directly measured.

24.4 PHOTOELECTRIC FILTER PHOTOMETRY

There are two kinds of photometers called single beam photometers and double beam photometers. Single beam models are less commonly used, as reproducibility is a problem if there is too much current fluctuation. Now let us consider the working of a single beam model. A light beam from a constant source passes through a colliminating lens and then a filter to become monochromatic and the beam then passes through the solution, and impinges on the photocell which in turn generates a current in the circuit and the galvanometer (*G*) shows a deflection. This current is measured by means of a sensitive meter which is fitted with two scales, one log and the other percentage transmittance. A variable resistance is connected in parallel with the galvanometer so that it controls the total photoelectric current passing through and is used to obtain full scale deflection with the blank solution. It is possible to adjust the iris diaphragm in such a way that the galvanometer shows zero deflection when light passes through the blank solution (such zero adjustment can also be done by a variable resistance). Now on placing the analyte sample in the beam of light, the light passes through the sample and falls on the photocell showing a deflection which is proportional to the concentration of the solution. If the response of the photocell is linear then the photo current gives the transmittancy (*T*). The chief requirement of this technique is constancy of the light source during the interval between two photocurrent measurements (Fig. 24.7).

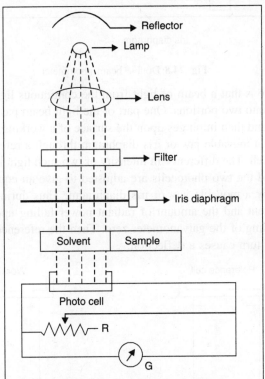

Fig. 24.7 Single beam photometer

In the double beam filter photometer, there are two types of models. In the first, both photocells are stationary and the variation in intensity of a light beam is carried by means of a variable resistance or iris diaphragm (Fig. 24.8) while in the second category of instrument (Fig. 24.9 and 24.10) one of the two photocells is mobile and can be rotated to the light beam falling upon this cell. In both the models the basic idea of utilising two photocells is to allow the light flux to impinge on simultaneously through reference and analyte solution under identical conditions and to eliminate any operational error. In reality, we measure the difference in intensity of two light beams, one passing through the reference solution and another passing through the sample solution.

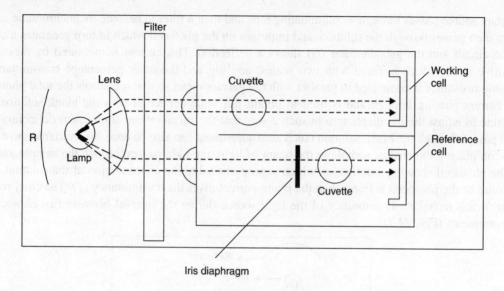

Filter

Lens

Cuvette

Working cell

R

Lamp

Reference cell

Cuvette

Iris diaphragm

Fig. 24.8 Double beam photometer

The working principle is that a beam of light from the continuous light source passes through the filter and is then divided into two portions. One part of the light beam passes through cuvettes holding analyte sample solutions and then impinges upon the surface of a working photo cell. The other part of the light beam passes via a movable jaw or iris diaphragm through a reference solution and impinges directly on the reference cell. The difference in intensity of two such light beams should give a measure of the absorption provided the two photocells are adjusted to give an equal response initially. This is generally accomplished by a movable jaw or iris diaphragm. Thus during operation the cuvettes are filled with reference solvent and the amount of radiant power falling upon the reference cell is first adjusted to make the reading of the galvanometer zero. Then the reference solution is replaced by the sample solution which in turn causes a deflection.

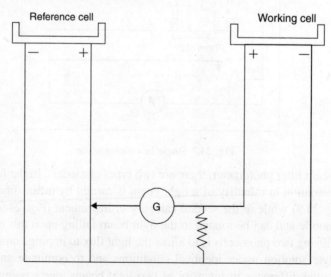

Reference cell

Working cell

− +

+ −

G

Fig. 24.9 Electronic circuit

The deflection due to reference solution is made zero on the galvanometer by means of a voltage divider. The dial setting thus reads absorbance. In another model of a double beam photometer, a light beam is allowed to pass through the lens and filter and is then divided into two portions (Fig. 24.10). The first part of the beam passes through the cuvette and then falls upon a reference photocell. The net response of the cell gives a measure of the light absorbed by the sample solution provided both photo-cells are initially adjusted for equal response. This is generally done in the second type of model by rotation of the photocell itself about a vertical axis. Hence during operation of such an instrument a beam of light is allowed to pass through a reference solution. The galvanometer is zeroed by means of a zero adjuster or neutral wedge. Then the beam of light is allowed to pass through the sample solution when the galvanometer shows a deflection. Such a deflection is nullified by the dial setting or by movement of the photocell which in turn is connected to the dial scale of absorbance as well as transmittancy (Fig. 24.10).

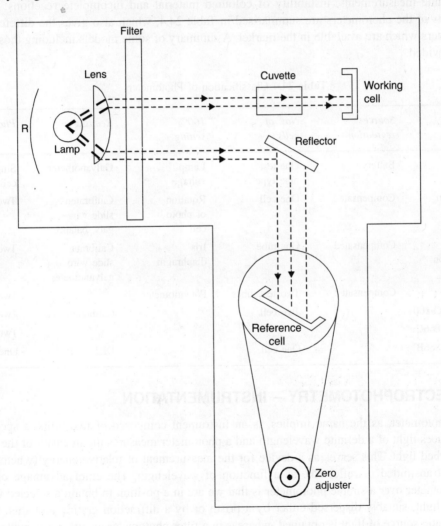

Fig. 24.10 Double beam photometer

24.5 ERRORS IN PHOTOELECTRIC PHOTOMETRY

There is a possibility of causing an error during photoelectric measurements. Any nonlinearity of response in the light sensitive device and in the associated measuring circuits, can be overcome by using matching photocells and selecting a good amplifying circuit. If the variation in intensity of the source causes error, use double beam models and control the voltage applied to the source. Stray light striking the cells can give serious error hence one must use light tunnels and properly placed baffles in the instruments. If there is a rise in temperature of the measuring photocell which thus causes errors, heat absorbing filters and smaller cells with good thermal insulation should be used. Dust scratches and imperfections in the optical system can be source of error so instruments should be protected from dust and breakage of optical parts or rough use. Inspite of taking care of all these factors, errors can also be caused due to the presence of optical defects in the system, wrong weighing, improper control of pH, wrong volume measurements, instability of coloured material and incomplete reactions. The main requirements of the photometers are summarised in Table 24.4, which also gives the different models of photometers which are available in the market. A summary of some models including those made in India is provided.

Table 24.4 Classification of Photometers

Name of Model	Source regulation	Nos. of cells	100% setting	Reading	Photocells
Evelyn	Battery	One test tube type	Lamp voltage	Galvanometer	Single cell
Lumentron	Compensated	One cell	Rotation of photo cell	Calibrated slide wire galv, balance	Two
Klett Summerson	Compensated	One tube	Iris diaphragm	Calibrate slide wire galvanometer	Two
Fischer	Compensated	Two test-tubes	Potentiometer	"	Two
SICOSPEC-100	"	Two cell	"	Calibrated dial	Two
ECIL GS 866C	"	"	"	"	Two
ECIL GS 866B	"	One cell	"	Dial	One cell

24.6 SPECTROPHOTOMETRY – INSTRUMENTATION

A spectrophotometer, as the name implies, is an instrument composed of two units, a spectrometer which produces light of a definite wavelength and a photometer measures the intensity of the transmitted or absorbed light. This serves as a device for the measurement of relative energy (whether energy is emitted, transmitted or reflected), as a function of wavelength. The chief advantage of using a spectrophotometer over a simple photometer is that we are in a position to obtain a selected waveband from white light, suitably dispersed either by a prism or by a diffraction grating and slits, thereby a monochromatic source of light is obtained, whereas in a filter photometer an attempt is made to obtain light of known waveband and not necessarily a known wavelength, by passing it through filters of different colours. In a photometer with filters, the light source arrangement is such that it is not possible to get light exactly of a particular wavelength but it varies between 30-40 nm. However in a spectrophotometer it is possible to obtain light of a selected waveband by suitably dispersing it through a prism and slit or diffraction gratings.

A spectrophotometer is composed of a source for the continuous visible spectrum, a device for obtaining monochromatic light, absorption cells for the sample and blank solution and a means of measuring the difference in absorption between the sample and blank or reference (Fig. 24.11).

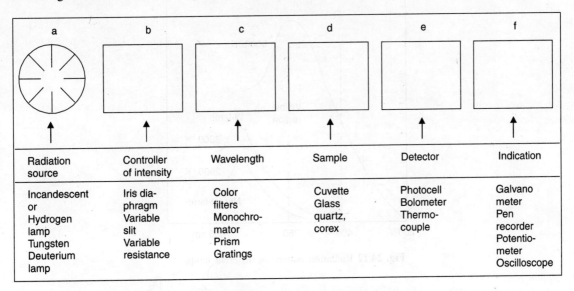

a	b	c	d	e	f
Radiation source	Controller of intensity	Wavelength	Sample	Detector	Indication
Incandescent or Hydrogen lamp Tungsten Deuterium lamp	Iris diaphragm Variable slit Variable resistance	Color filters Monochromator Prism Gratings	Cuvette Glass quartz, corex	Photocell Bolometer Thermocouple	Galvano meter Pen recorder Potentiometer Oscilloscope

Fig. 24.11 Block diagram for typical photometer

1. *Source:* The source most commonly used in UV-visible absorption spectroscopy is a tungsten lamp. The photocurrent depends upon the lamp voltage, $i = K V^n$ if i = photocurrent, V = voltage, n = exponent (3-4 in W lamp); 0.2% variation is reasonable. A DC source such as a storage battery is a constant source of radiation. A hydrogen or deuterium lamp is used as the source to obtain light in the ultraviolet region. An important feature of the tungsten lamp is the constancy of the overall energy output at various wavelengths. This is attained by controlling the voltage for which a constant voltage transformer is utilised. Inspite of controlling these factors we may obtain a varying output. Compensation is usually carried out by measuring the light transmitted by the sample as well as the reference solution.

2. *Monochromator:* This is an accurate source of getting monochromatic light. It may be either a prism or a grating. A slit may be used to pick out the desired wavelength from the dispersion system. If the slit is in a fixed position, rotation of the prism or grating is used to obtain the proper wavelength. There are two ways of mounting a prism (as shown in Fig. 24.13) called Cornu and Littrow mountings.

A Cornu type mounting uses a prism with the angle of 60° while the Littrow type uses a prism which is right angled with an aluminised surface on one side have an optical angle of 30°. Littrow mountings can also be done with prisms having an angle of 60° or 30°. The

$$\frac{dn}{d\lambda} = \text{dispersive power; and } R = \frac{\lambda}{d\lambda} t \left(\frac{dn}{d\lambda} \right) = \text{resolving power of a prism.}$$

Here n = refractive index, λ = wavelength, t = base for which a high value is preferred. In case of gratings which acts like a mirror, the resolving power is given by an expression, $R = mN$ where m = width and N = number of lines per cm^2 area, Hence $m\lambda = d$ (sin i ± sin $\emptyset$) = 0 if $\hat{\imath} = \hat{r}$, $m\lambda = d$ (sin i ± sin $\emptyset$) = 0 where $\hat{l}$ = angle of incidence; $\hat{r}$ = angle of refraction; t = base.

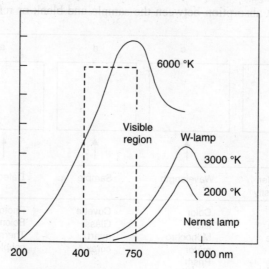

Fig. 24.12 Radiation pattern of different lamps

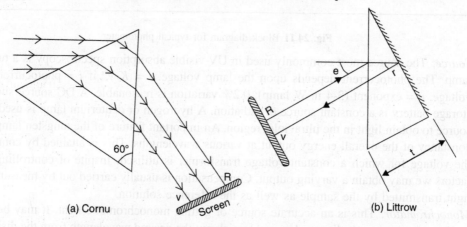

Fig. 24.13 (*a*) Cornu type mounting; and (*b*) Littrow type mounting

3. *Absorption cell:* During measurement in the visible region one can use a glass or corex glass cuvette but for measurement in the ultraviolet region, one must use a quartz cell as glass is opaque in this region. The common optical path used is 10 mm, but cells with a larger or smaller optical path are also available. Cells used are rectangular, but cylindrical cells are also available. One must use cuvettes with lids while handling organic solvents. Finally, good cells are the ones which are fused and not cemented and are matched in pairs.

4. *Receptor:* The important role of the receptor is to respond to change in wavelength. In the spectrophotometer photomultiplier tubes are employed as described earlier in this chapter (see Fig. 24.6).

24.7 FUNCTIONING OF SPECTROPHOTOMETER

Operation of the spectrophotometer is very briefly explained (Fig. 24.14). Keep the reference solution, *i.e.*, blank, in the first cell and keep the analytical solution in the second cell compartment. Then select a proper photocell (200-650 nm or 650-1100 nm) for covering the required wavelength region. With the photocell compartment closed, zero the galvanometer by means of the dark current switch. Select a proper wavelength, open photocell and keep reference bank in beam of light and zero the galvanometer with the sensitivity knob and also slit. Set transmittance dial to 100%. Then keep the cuvette containing the analytical solution in the path of the beam of light and zero the galvanometer by setting to absorbance scale and read absorbance reading directly. The salient features of various instruments of spectrophotometer used are summarised in Table 24.5.

Table 24.5 Various spectrophotometers

Item	Beckman model DU	Hilger UV Spec.	Type SF-4 (Russian instrument)	Spectromom 204 (Hungarian)	ECIL spectrophotometer (Indian)
Light source	H_2/D_2 discharge tube W filament bulb from 6 V storage battery	H_2 lamp W lamp operating on AC	H_2 lamp W lamp from storage battery	H_2/D_2 lamp W lamp from AC supply	H_2 lamp W lamp AC supply
Dispersing device	Quartz littrow prism	Quartz/ glass littrow prism	Quartz littrow prism	Quartz/ littrow mounting	Littrow prism
Slit range	Variable 200-1000 nm	Variable 200-1000 nm	Variable 200-1000 nm	Variable 180-1200 nm	Variable 200-1100 nm
Radiant energy device	Two photo tubes 200-625 nm also 625-1000 nm	Two photo tube 200-625 nm 625-1000 nm	Two photo cells 200-615 nm 650-1000 nm	Two photo cells 180-640 nm 640-1200 nm	Two photo cells 200-625 nm 650-1200 nm
Sample holder (path)	1-50 mm test tube	1-40 mm cuvettes	10 mm cuvettes	10 mm cuvettes	10 mm cuvettes
Control	Potentiometer	Potentiometer	Potentiometer	Potentiometer	Potentiometer
100% standardization	Variable Sensitivity control and slit width	Variable Sensitivity and slit width	Variable Slit width control	Variable Voltage direction	Variable
Attachment	Flame emission Fluorescence Reflectance	Reflectance Fluorescence Flame	Reflectance	Fluorescence Recorder	

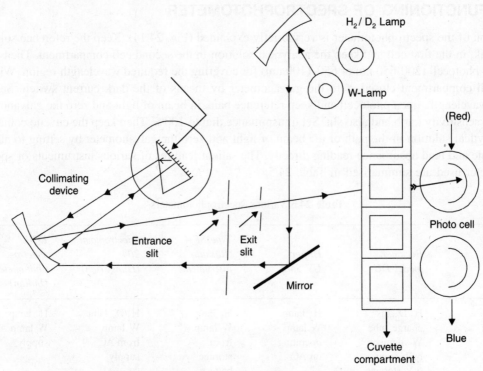

Fig. 24.14 Block diagram of UV-visible spectrophotometer

24.8 SIMULTANEOUS SPECTROPHOTOMETRY

This technique is used to measure the absorbance of a mixture of coloured solutions without separating each component. When no wavelength can be found at which only one component absorbs, it is still possible to determine two components by making the measurements at two wavelengths. If two dissimilar chromophores have different powers of light absorption at some point of the spectrum. If measurements are made on each solution at two points, a simultaneous equation is set at two points from which the concentrations of the individual constituent can be found. The first point should be selected such that the ratio of molar absorptivity is maximum *i.e.*

$$\left(\frac{a_1}{a_2}\right) \text{ at } \lambda_1 \text{ and } \left(\frac{a_2}{a_1}\right) \text{ at } \lambda_2$$

These λ_1 and λ_2 may not necessarily coincide with wavelength of maximum absorbance but should not fall on sharpened portions or a shoulder. Secondly, it is necessary to calculate the molar absorptivity (*a*) for each component using a particular set of sample containers and spectrophotometer. Absorbance is directly proportional to the product of molar absorptivity (*a*) and concentration (*c*) if the light path (*b*) remains constant. Thus using the fundamental equation of absorbance, $A = abc$; for one solution $A_1 = a_1bc_1$ and for the second solution $A_2 = a_2bc_2$. Since *b* is kept constant and A_1 and A_2 are different, we have $A_1 = a_1c_1$ and $A_2 = a_2c_2$. However we measure absorbance at λ_1 and λ_2, since at these wavelengths absorbances are additive and optical path is the same

$$A\lambda_1 = (a_1c_1)\lambda_1 + (a_2c_2)\lambda_1 \qquad \qquad ...(24.1)$$

$$A\lambda_2 = (a_1c_1)\lambda_2 + (a_2c_2)\lambda_2 \qquad \text{...(24.2)}$$

We simplify for c_1 and c_2 as

$$c_1 = \frac{(a_2)\lambda_1 A\lambda_1 - (a_2)\lambda_1 A\lambda_2}{(a_1)\lambda_2 (a_2)\lambda_2 - (a_2)\lambda_1 (a_1)\lambda_2} \qquad \text{...(24.3)}$$

$$c_2 = \frac{(a_1)\lambda_1 A\lambda_1 - (a_1)\lambda_2 A\lambda_2}{(a_1)\lambda_1 (a_2)\lambda_2 - (a_2)\lambda_1 (a_1)\lambda_2} \qquad \text{...(24.4)}$$

The accuracy is poor unless the spectras are quite discrete as shown in Fig. 24.15. Simultaneous determination rests on the important assumption that the substances concerned contribute additively to the total absorbance at an analytical wavelength. The calculations are further simplified if a wavelength

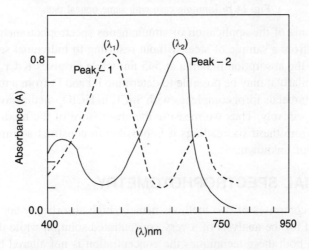

Fig. 24.15 Simultaneous spectrophotometry

is available at which only one of the two components absorbs e.g. as shown in Fig. 24.16. In such a case, the concentration of a single absorber is ascertained from the calibration curve and from the ratio of its molar absorbance at the two wavelengths. Its contribution to the overall absorbance at the second wavelength is readily computed and subtracted from the total observed to give a net absorbance at the second wavelength, which is due to the second component. Mathematically the whole observation can be simplified as:

$$A\lambda_1 = (a_1c_1)\,\lambda_1$$

$$A\lambda_2 = (a_2c_2)\,\lambda_2$$

as

$$(a_2c_2)\,\lambda_1 = 0; \text{ also } (a_1c_1)\,\lambda_2 = 0$$

We have

$$c_1 = \frac{A\lambda_1}{(a_1)_{\lambda_1}} \text{ and } c_2 = \frac{A\lambda_2}{(a_2)\lambda_2}$$

We directly evaluate c_1 and c_2.

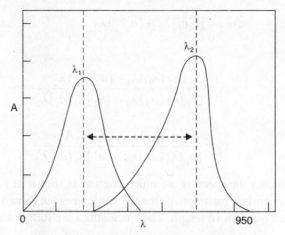

Fig. 24.16 Limiting case with same optical path.

The common example of the application of simultaneous spectrophotometric determination is the analysis of Cr and Mn from a sample of steel without resorting to individual separation. On oxidation Mn gives $KMnO_4$ with the absorption maxima at 545 nm while Cr gives $K_2Cr_2O_7$ with the absorption maxima at 440 nm. Similarly it may be possible to determine Pd and Pt from a mixture by simultaneous spectrophotometry. Both metals form complexes with $SnCl_2$ in $HClO_4$ media having absorption maxima at 635 and 405 nm respectively. Thus we measure the absorbance of the mixture and both individual samples at λ_1 and λ_2: from these six readings it is possible to construct a simultaneous equation and evaluate concentration of unknown.

24.9 DIFFERENTIAL SPECTROPHOTOMETRY

This technique usually comprises of two methods namely, high absorbancy and low absorbancy methods. The first is applied for the analysis of a very concentrated solution while the latter is applied to a very dilute solution. In both these techniques the concentration is not altered by external changes.

In the high absorbancy method, for which dilution is not feasible, the instrument scale reading is set to 100 scale units with the standard solution of finite but lower concentration than that of the least concentrated sample solution. To compensate for the lower amount of transmitted energy reaching the detector, the spectrophotometer is brought to a balance with scale reading at a 100 scale units by increasing the slit width, the source intensity or the amplifier phototube sensitivity. Under these conditions the actual scale is effectively lengthened although there is no change in its physical dimensions. Thus if the actual reading lies in 0-20% T scale by this procedure it is expanded to 0-100% scale units using the standard with changed scale to make 100% setting. A common example will be clear from Table 24.6.

Table 24.6 High and low absorbance method

	I	II	III	IV	V	VI	VII
Concen. µg/ml	0	5	10	40	80	200	280
Absorbance	0	0.025	0.050	0.200	0.400	1.00	1.4

In the above example, ordinarily we use solution number I as the reference blank. However if the absorbance is very large as shown by solution number VII, it may be difficult to measure the absorbance with solution number I as blank. It will be ideal under such circumstance to use solution number VI as the blank with the result that the actual absorbance will be 0.400 instead of 1.4 thereby increasing the accuracy in the region 0.200–0.800 on the absorbance scale as per Ringbom's plot as discussed earlier.

In the low absorbancy technique for trace analysis the reading scale is set to 100% T with the adjustment of instrument for solvent in the usual manner. Then the reading is set to zero transmittance with a standard solution whose concentration is somewhat larger than that of the most concentrated unknown solution that is anticipated. The effective length of the transmittance scale is again increased, this time favouring the low absorbance system *i.e.* if the sample series falls in the 70-100% T range, this range can be covered to 0–100% T units using 70% T to make the zero setting. The error at this end of the scale now becomes finite. In this method, one is forced to construct a calibration curve showing the instrument reading as a function of concentration. The relationship between the logarithm of the instrument reading and concentration is not linear. However this is not considered a serious disadvantage, e.g. in Table 21.6 for measuring solution III we use solution no. IV as the blank and carry out the necessary adjustment (Table 24.6).

In the maximum precision method one can resort to both high absorbance and low absorbance methods to measure the concentration of a solution e.g. in Table 24.6, if we are interested in measuring the absorbance of solution V we can use solution IV as the blank and measure the absorbance of solution by the high absorbancy method. Similarly by utilising solution VI as the blank one can measure the absorbance of solution V by low absorbance. High or low absorbancy techniques must have the same value of absorbance. Thus solution V, with solution IV with an absorbance of 0.200 as the blank, would have an absorbance of 0.200 with net absorbance of 0.400. Similarly the same solution with solution VI as the blank, with an absorbance of 1.00 would show on scale adjustment value of 0.600, giving the actual value as (1.0–0.6) = 0.400. Thus this method can provide the kind of advantages gained with the high absorbance method and low absorbance method which still use an optimum value of 36.8% T.

24.10 REFLECTANCE SPECTROPHOTOMETRY

An attachment for making reflectance measurements is available for standard spectrophotometers. The reflectance curve is the result of a series of such measurement as shown Fig. 24.7. Absorbance measurements are made at different wavelengths. If desired, readings can be converted to world standard CIE trichromatic values. For white light colours of all wavelengths are present in equal intensities, so the spectral reflectance curve would be (XX') a horizontal line with the position on the vertical axis depending upon the brightness of the sample *i.e.* line XX' in Fig 24.17. This is relative to light reflected by pure MgO which is chosen as the standard for comparison. A horizontal line at any point between 100% T and 0% T would indicate equal absorption of all wavelengths of radiation in the visible spectrum resulting in a grey appearance to the eye. Reflectance spectroscopy is useful for the measurement of coloured samples which are ordinarily not soluble in any solvent.

24.11 PHOTOMETRIC TITRATIONS

Volumetric titration techniques are based upon the detection of the end point at the equivalence point by noting the change in colour of a solution. Such a change in colour is usually noted by eye in the visible region. If one uses a photometer for detection of the end point, it is called photometric titration.

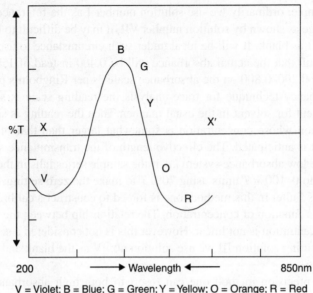

V = Violet; B = Blue; G = Green; Y = Yellow; O = Orange; R = Red

Fig. 24.17 Reflectance spectra of blue colour

If a spectrophotometer is used for the detection of end point it is termed spectrophotometric titration. This is an illustration of a linear titration. In fact it is much more dependable than an amperometric titration where the concentration varies on a logarithmic scale—in photometric titration the concentration varies directly with absorbance on a linear scale.

The basic requirements for a successful photometric titration are the absence of interfering absorbing species in solution at the analytical wavelength of measurement; the change in absorbance should be additive if two solutions have different absorbances. The molar absorptivity of the absorbing analyte solution should be reasonably high and should show a sharp change so that the titration can be performed within a reasonable limit of volume of titrant. It is essential to maintain a concentration of titrant 20 to 50 times higher than that of the analyte solution so as to give a sharp change in colour by addition of a final drop of titrant. It is also necessary to use a dilute solution to facilitate absorbance measurement. However, one is restrained from utilizing the too much dilute solution as one has to make a correction to the volume of the titrant.

If the correction is not made then the graph of absorbance *versus* volume of titrant will bend towards the volume axis and give a wrong intersection and hence an incorrect end point. Therefore we multiply the absorbance by the factor $\left(\dfrac{V+\overline{V}}{V}\right)$ where V represents the initial volume of analyte solution and $\overline{V}$ represents fraction of the volume of added titrant. This correction is necessary to get a straight line graph and a reliable intersection for the equivalence point.

In a photometric titration either the titrant or the analyte solution is coloured or the product of the reaction is coloured. During titration the solution becomes coloured or colourless. There are various possibilities of such colour transformations as indicated in Fig. 24.18. Graph (*a*) indicates the titration curve where the titrant alone absorbs. Graph (*b*) indicates the curve in which the products of the reaction absorb. Further when the coloured analyte is converted to a non absorbing product we get a graph as shown in (*c*). When a coloured analyte is converted to a colourless product by a coloured titrant we get a graph as shown in (d) and finally graphs (e) and (f) represent the addition of a ligand to form two complexes of different molar absorptivity.

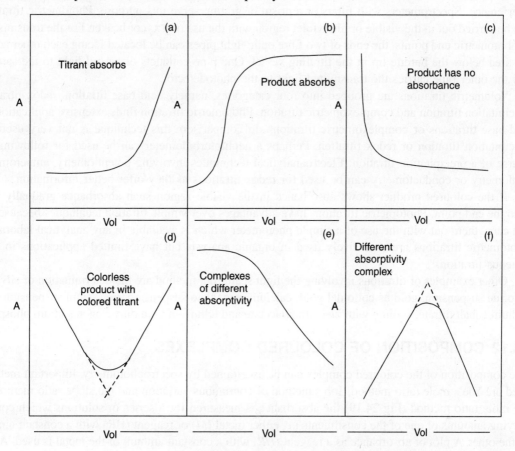

Fig. 24.18 Various titration curves

A photometric titration is carried out at the pH where it is equal to the pKa value at equivalence point. This facilitates exact detection of the end point. It is interesting to note that it is possible to carry out such a titration by making use of an indicator in the case where the titrant or analyte solution are both not coloured. If, inspite of using an indicator in titration, the end point is not sharp and visible, it is impossible to detect the end point. Such titrations are difficult to carry out. For instance in such a titration even if one uses other linear titrations techniques such as amperometry, potentiometry or conductometry titration, one fails to detect the end point. Many of these titrations can be successfully carried out by a photometric titration with judicious selection of indicators. In the event that the analyte solution or titrant are colourless, it is advisable to use a suitable indicator e.g., complexometric titration with metallochromic indicators.

The instrumentation in a photometric titration is simple. A common photometer or spectrophotometer with minor modification can be used for a photometric titration. Specially fabricated instruments are also available. The important requirement for a photometric titration is to have a container which can hold at least 20 to 40 ml of solution in which a absorbance measurement can be made. It is absolutely necessary to maintain the homogeneity of the solution after addition of each increment of the titrant in the container. This can be achieved by making use of a magnetic stirrer. What is most important is that the absorbance measurement must be carried out at a fixed wavelength to eliminate spectral interference. Spectrometers with filters or a prism or grating serve this purpose. Photometric titrations can be carried out in the visible or ultraviolet region with the use of a xycor beaker. For the transmission of photometric end points, the ends of two fibre optic light pipes can be located facing each other across the area below the burette tip in the titrating vessel. One pipe conducts dispersed light to the sample, and the other pipe carries the transmitted light to the photodetector.

Volumetric titrations are grouped into four categories, namely acid base titration, redox titration, precipitation titration and complexometric titration. Photometric titration finds extensive applications in acid-base titrations or complexometric titrations. In comparison, this technique is not very useful in precipitation titration or redox titration. Perhaps a nephloturbidometer can be used for following the course of a precipitation titration. Electroanalytical techniques involving potentiometry, amperometry, coulometry or conductometry can be used for redox titrations as they offer better information.

If the coloured product shows absorbance in the visible region such absorbance gradually rises upto the end point. Photometric titrations have advantages over simple titration techniques because one can carry them out with the use of a simple photometer which is available in any analytical laboratory. Photometric titrations are extensively used in organic analysis but have limited applications in non-aqueous titration.

Other examples of titrations involving the light scattering method are the determination of silver as chloride suspension, gold as colloidal gold, calcium as oxalate suspension, potassium suspension with sodium cobalt nitrate, sodium with zinc uranylacetate and tellurium with dihydrogen sodium phosphite.

24.12 COMPOSITION OF COLOURED COMPLEXES

The composition of the coloured complex can be ascertained by spectrophotometry. Important methods used are Yoe's mole ratio method, Job's method of continuous variation and the slope ratio method. In the mole ratio method (Fig 24.19) the absorbance is measured for a series of solutions which contain varying amounts of one of the constituents *i.e.* either metal (*M*) or reagent (HR) with a constant amount of the other. A plot of absorbance as a reagent (*HR*) with a constant amount of the metal is used. A plot of absorbance as a function of ratio of moles of reagent to moles of metal ion gives a straight line passing through the origin with an inflection at the equivalence point it then becomes horizontal as all the metal cation has been consumed and addition of excess of reagent produces no more colour e.g. the

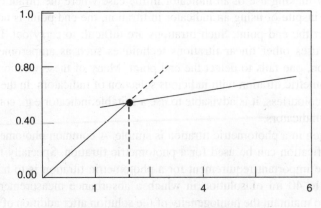

Fig. 24.19 Mole ratio method (Yoe's)

intersection at 3.0 indicates M:R 1:3 in MR_3 complex, $n = 3$ where n represents the number of ligand molecules attached to the molecule (Fig. 24.19).

In Job's method of continuous variation (Fig. 24.20), a series of solutions of varying mole fractions of metal $\left[\dfrac{M}{M+R}\right]$ or reagent $\left[\dfrac{R}{M+R}\right]$ is used wherein their sum is kept constant. They are measured for their absorbance spectrophotometrically. The absorbance, when plotted against the mole fraction of one of the constituents gives a straight line showing a maximum at point Y, then decreasing at a mole fraction corresponding to that of the complex. Finally, in the slope ratio method a quantitative comparison of the slopes of straight line tangents to the two members of the continuous variation plot at their intersection with the axis of zero absorbance gives final solutions. Out of these three methods described, the mole ratio method is most commonly utilised for practical purposes. In comparison, Job's method or slope ratio method have several limitations. The most significant demerit is dilution of solution during variation of mole fraction, or dissociation of coloured complex after a certain concentration limit of the metal or ligand. The mole ratio method is not useful in extractive photometry. It is invalid if the metal forms a dimeric or polymeric complex. It is then necessary to resort to the log D against Log (HR) method to ascertain the composition of the extracted species, where slope gives idea on nature of extracted species.

An application of spectrophotometry is to determine the pH of a solution by using the equation:

$$pH = pKa + \log \frac{[\text{alkaline form}]}{[\text{acid form}]}$$

This is extensively used to determine the pK value of several acid–base indicators:

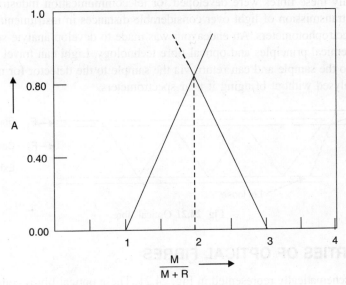

Fig. 24.20 Continuous variation method (Job's)

24.13 SANDELL'S SENSITIVITY

A knowledge of sensitivity is of utmost importance in colorimetric determination of traces of metals.

It is defined as the weight of a substance that can be determined in a column of solution having unit cross section. The weight is expressed in (µg) micrograms and the area in square cms. This is valid only if the system conforms to Beer's law indefinitely at low concentrations. This is true for all reactions. Two factors are involved in sensitivity. The intrinsic sensitivity is proportional to the molar absorptivity of the solution of the coloured product and the ability of the observer directly or indirectly to detect small difference in the light transmission or absorption of the solutions. In spectrophotometry the minimum amount of coloured substance that can be determined usually depends upon the reproducibility of the measurement of transmittancy of a faintly coloured solution. About 0.001 difference in absorbance can be detected on a good instrument (0.2 T). Thus if 0.001 absorbance is given by 0.025 µg of the metal ion, the Sandell's sensitivity is 0.025 µg/ml/cm^2. Also if the molar absorptivity of colored compound is known we can calculate the sensitivity as

$$\text{Sensitivity } (S) = n\left(\frac{M}{\epsilon}\right) = \left[\frac{\text{molecular wt.} \times \text{no. of atoms in the element}}{\text{molar absorptivity of coloured species}}\right]$$

Finally we define sensitivity as the number of micrograms of elements converted to the coloured product which are in a column of solution having a cross section of 1 cm^2 which shows 0.001 absorbance and is expressed as µg/ml/cm^2. Organic reagents with high molecular weights give maximum sensitivity when used as chromogenic agents.

24.14 OPTICAL FIBRES IN SPECTROSCOPY

Optically based sensors are a new development in the field of absorption spectroscopy. These optical fibres behave exactly like electrochemical sensors. They can transmit light along the cable called a wave guide. Initially these fibres were developed for telecommunication industries but later on they were used for the transmission of light over considerable distances in instrumental analysis, transmission of light to spectrophotometers. An endeavour was made to develop analyte-selective sensors by a combination of chemical principles and optical fibre technology. Light can travel over a long distance through the fibre to the sample and can return via the sample to the detector for measurement. Thus a sample can be analysed without bringing it near spectrometers.

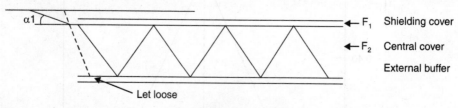

Fig. 24.21 Optical fibre

24.15 PROPERTIES OF OPTICAL FIBRES

A typical fibre in schematically represented in Fig. 24.21. These optical fibres consist of a coaxial core which is like a wavelength guide and is surrounded by shielding material of higher refractive index followed by an external coating of a protective buffer. The transmitted light travels by internal reflection within this coaxial core tube. The angle of acceptance is the higher angle of radiation which will be reflected for a known refractive index of protective coating or shielding material. Light entering at an angle greater than (α) will not be transmitted and (α) defines the fibre numerical aperture (NA) as

$$NA = f_{ext} b \sin a = \sqrt{f_1^2 - f_2^2}$$

where f_2 is the refractive index of cladding, f_1 is refractive index of the core and f_{ext} is refractive index of external medium, NA—numerical aperture and α = angle. NA data for different fibres are provided. The loss of light per unit/length at different wavelengths is equally important. A significant property called "attenuation" in decibel/km is expressed as:

$$db = 10 \log P_0/P$$

where P_0 is the input intensity and P is the output intensity of optical fibre. Thus db = 10 × absorbance

It is worthwhile to consider a few optical fibres which are used in spectrometry. A compound with glass itself acts as the protective coating with core size 15-75 μm which has a NA of 0.5–0.8 with attenuation of 800 db/km in the visible region of spectra which is commonly used, while silica with silicon cladding material with a nylon buffer with a core size of 100–2000 with NA = 0.5–0.6 in the visible region with attenuation of 200 db/km is also useful. These are the optical fibres useful for working in the UV or IR regions. Silica with suitable cladding with fluoropolymer, silicones or doped silica is also used. Such fibres are used for long distance transmission. These optical fibres can be used as probes for usual spectrophotometers and fluorimeters. Such instruments contain bifurcated fibres. Several dozen small fibres are used. The light from the sample falls on a small mirror and is reflected on the fibre before it is transmitted to detectors. The radiation path length is twice the distance between the fibre and the mirror. A mirror is not used in the fluorimeter where optical fibres are used.

24.16 SENSORS WITH OPTICAL FIBRES

Optical fibres can be changed into selective absorbance sensors by immobilising suitable reagents on extremities of the optical fibre. This operates better than electrochemical sensors, since it needs no reference or calomel electrode. For instance, for fluorimetric analysis, an optical fibre was fabricated by coating one end with the dye fluoroscein isothiocyanate which is pH sensitive. Penicillin with a suitable indicator was used for enzyme analysis as biosenser. Recently optical fibres have been developed for oxygen, and s-block metals. Exciting progress is expected in the next decade in this field.

24.17 ERRORS IN UV/VISIBLE SPECTROPHOTOMETRY MEASUREMENTS

(a) *Stray light:* It is unrecognised hindrance in spectral analysis. All wavelength devices produce some intensity of light at different wavelengths than the one which is used for measurement. This may be due to optical imperfection of the components or due to scattered radiation due to dust particles. The stray light errors may be having negative bias for the absorbance values. If $T_{abs} = \left(\dfrac{T_{true} + \rho}{1 + \rho} \right)$, ρ is fraction of light coming from other sources, T_{abs}, T_{true} are the actual and the true transmissions. The stray light seriously affects such measurements. It is worthwhile to consider their influences adding (%) errors as shown in Table 24.7.

The fraction of stray light is wavelength dependent. This is because the amount of energy at particular wavelength depends on the source intensity at that particular wavelength. At long and short wavelength error is large. A false peak is obtained at the lower wavelength limit. The source intensity is decreasing and analyte shows end absorbance. The stray light of longer wavelength has better detector response than one at the shorter wavelength. A negative derivation from Beer's Law is noted as incident radiation is not monochromatic. The manufacturers generally specify stray light properties of their model in their specifications.

Table 24.7 Influence of stray light on absorbance

S. No.	Stray Light (%)	Real Reading	Measured reading	Deviation %	Remarks
1.	1.0	1.0	0.963	3.8	High
2.	0.001	1.0	1.000	0	Medium
3.	0.1	2.0	1.953	2.1	Low
4.	0.01	2.0	1.996	0.2	Low
5.	0.1	3.0	2.995	0.10	High
6.	0.001	3.0	2.996	0.1	High

(b) *Spectral bandwidth:* Apart from stray light, spectral bandwidth is also of importance. There should be finite slit width. A small % of adjacent wavelength creeps in during measurements so it is called spectral band width (SBW). This wavelength spans on the wavelength of measurement containing 75% radiant energy coming from slit. The narrower SBW gives reproducible results. Too narrow slit .eads to reduced signal to noise ratio (S/N). Wide slit is also not dependable. The resolution is also lost; and one notes loss of fine structure of spectra. The optimum SBW is obtained by succession of spectral scans with narrow SBW till peak height is not changing.

(c) *Blank error:* The use of distilled water as blank also leads to error generation. A true blank must be preferred and prepared. Due to difference in the refractive index of analyte and blank is substantial hence must be maintained close with minimum reflective loss.

(d) *Refractive index:* The difference in refractive index promotes reflection of light from cuvettes. If reference solution is used as sample and if it has same refractive index, no reflection from air window is observed. The fraction of the light reflected as an interface varies with wavelength and is given by relation as shown.

$$\rho_{(\lambda_1)} = \frac{(\eta_2 - \eta_1)^2}{(\eta_2 + \eta_1)^2}$$

If ρ = fraction of incident light reflected, η_1 and η_2 are is the refractive index of two solutions. The reflection can also produce multiple reflection errors. So Kortem and Seller gave correction factor.

(e) *Quality of cuvettes:* Finally improper handling of cuvetts leads to errors in analysis so the cuvettes must be very clean. Plaqueing is a phenomena of shaking of colour material to surface of cuvettes. The chemical system producing chromophore may lead to generation of errors. The consideration of the stability of the chromophoric group is also of significance as with change in temperature, atmospheric oxidation, light sensitivity, hydrolysis may influence intensity of chromophore.

Thus we have seen that not only the stray light but also the improper use of the blank or reference solution as well as utilisation of uncleaned cuvettes leads to wrong measurements of absorbance which in turn must be very carefully avoided during analysis.

24.18 SPECTROELECTROCHEMISTRY

The controlled potential coulometry or controlled potential electrogravimetry is coupled with spectrophotometry to have technique of spectroelectrochemistry. The incident beam of light is passed through optically transparent electrode or it is reflected from the electrode surface or alternatively it is passed parallel to the surface of said electrode, to monitor concentration of reactants or products in electrolysis. The optically transparent electrodes are fabricated by coating thin layer of electrode material on substrate which is transparent to light and one which is used to study electrode reaction. The coating must be transparent, with very thin layer (10-500 nm). One can use gold, platinum or tin oxide as electrodes. Glass or plastic is also used as base for coating for study in visible region of the spectra or quartz as the basis for UV radiation and germanium is used for infrared radiation.

The optically transparent electrode (OTE) can be fabricated from fine grid of conducting material or porous piece of glass electrodes. In first case radiation passes through holes in grid (where holes are quite small). Porous glass carbon plate can be used if small pores are present. UV, visible, IR regions are used for spectral investigation. An investigation involving study of absorbance measurement at fixed wavelength or function of time of electrolysis is carried out. The rate constant of chemical reaction is thus measured. Instead of absorbance measurements, one can use fluorescent measurements to study electrochemical reaction. In reflectance spectrophotometry light is reflected from surface of electrode before falling on striking detector. Apart from the method of the UV, visible and IR spectrophotometry one can resort to the technique of fluoro-electrochemistry and specially for the characterisation of the organic compound. In recent years resort is also taken to use nuclear magnetic resonance spectrometry in conjunction with electroanalytical method for the quantitative analysis of chemical compounds.

24.19 HOLOGRAPHIC GRATINGS

Transmission gratings disperses UV, visible and IR radiation. A transmission or reflection grating serves the purpose. The replica gratings are manufactured from master gratings made of metal with grooves. A UV-visible gratings has 300-2000 grooves/mm. The common figure is 1200-1400 grooves/mm. For IR only 10-200 grooves are sufficient. The master grating fabrication needs skill and patience. The special Echellette grating has blazed lines on broadfaces where reflection takes place and it has narrow unused faces. The constructive ($n\lambda$) interference is given as:

$$n\lambda = d\left(\sin\hat{i} + \sin\hat{r}\right)$$

if $\hat{i}$ and $\hat{r}$ are incident and reflected angle, d = spacing in gratings.

The holographic gratings is an outcome of laser technology. They are produced on a large scale as demand for them is rising. They have greater perfection with lines and they provide spectra, free from stray light and double images. In their fabrication beams from pair of similar lasers are fixed at suitable angles on surface coated with photo resistant material. This results in interference fringes (from two beams), sensitive to the photoresistant material such that it can be dissolved away leaving grooved surface which can be coated with aluminium or similar reflecting surface. The spacing of the grooves can be changed by changing angle of two laser beams. A gratings with 6000 lines/mm can be produced at low cost. Replica gratings can be obtained from master gratings, so much so that it is difficult to distinguish between replica and the master gratings after polishing process.

SOLVED PROBLEMS

24.1 A 1.0×10^{-3} M solution $K_2Cr_2O_7$ shows an absorbance of 0.200 at 450 nm and an absorbance of 0.050 at 530 nm. A 1.10×10^{-4} M solution of $KMnO_4$ shows no absorbance at 450 nm and an absorbance of 0.420 at 530 nm. Calculate the concentration of $K_2Cr_2O_7$ and $KMnO_4$ present in a solution which exhibits an absorbance of 0.370 and 0.710 at 450 and 530 nm respectively. Assume the cell width is constant at 10 mm.

Answer: Calculate molar absorptivity (*a*) of $K_2Cr_2O_7$ as well as $KMnO_4$ at 450 and 530 nm. Since $A = abc$ if $b = 1$, $A = ac$, $a = A/c$

For $K_2Cr_2O_7$, at 450 nm, $\in = \dfrac{0.200}{1\times10^{-3}} = 200 = 2\times10^2$

For $K_2Cr_2O_7$ at 530 nm, $\in = \dfrac{0.050}{1\times10^{-3}} = 50 = 5\times10$

For $KMnO_4$ at two wavelengths, $\in$ shall be

$KMnO_4$ at 450 nm, $\in = 0$

$KMnO_4$ at 530 nm, $\in = \dfrac{0.420}{1\times10^{-4}} = 4.2\times10^3$

Now using simultaneous equations,

$A\lambda_2 = (a_1c_1)\,\lambda_1 + (a_2c_2)\,\lambda_1$ if $\lambda_1 = 450$ nm ...(*i*)

$A\lambda_2 = (a_1c_1)\,\lambda_2 + (a_2c_2)\,\lambda_2$ if $\lambda_2 = 530$ nm ...(*ii*)

Substituting the values

$0.370 = 200\,(c_1) + 0\,(c_2)$ at 450 nm

$0.710 = 50\,(c_1) + 4200\,(c_2)$ at 530 nm solving for c_1 and c_2

$c_1 = 1.90 \times 10^{-3}$ M $K_2Cr_2O_7$

$c_2 = 1.46 \times 10^{-4}$ M $KMnO_4$

24.2 Uranyl oxinate in chloroform exhibits an absorption maxima at 430 nm. $\in = 1 \times 10^4$. The absorbance values are greater than unity at a higher concentration; however the complexes are measured by differential spectrophotometry. A 1.0×10^{-4} M solution is used for blank adjustment. With this setting the unknown solution shows a transmittance of 30.2% T. What is the concentration of the unknown solution?

Answer: The standard solution should exhibit 10% T on the usual scale *i.e.*
$A = abc = (1 \times 10^4)\,(1.0)(1 \times 10^{-4}) = 1.0$ *i.e.* 10% T
The scale has thus been expanded 10 times the original value *i.e.* 30.2% T on expanded scale is really 3.02%.

Hence 3.02% $T = -\log\left(\dfrac{3.02}{100}\right)$ absorbance $= -\log 0.0302$

Since $A = abc$ if $A = -\log 0.0302$, $b = 1$, $a = 1 \times 10^4$, $c = ?$

$\because$ $-\log 0.03 = (1 \times 10^4)\,(1.0)\,(c)$

$\therefore$ $c = 1.52 \times 10^{-4}$ M concentration of the diluted solution.

24.3 2-napthylamine has the following absorbances

β - naphthol	3.49	0
β - naphthol anion	3.70	3.47
	log $\in$ (285 nm),	log $\in$ (346 nm)

A solution of buffer containing 1×10^{-4} M β-naphthylamine with pH 9.20 showed an absorbance of 0.373 and 0.098 at 285 and 346 nm. What is the pK value of β-naphthylamine?

Answer: We calculate the concentration of ROH and RO^- anion

ROH $\in_{285} = 3.087 \times 10^3$ (as log $t = 3.40$) $\in_{346} = 0 = \log t = 0$ at 346 nm

RO^- $\in_{285} = 5.010 \times 10^3$ (as log $t = 3.70$) $\in_{346} = 2.95 \times 10^3$ at 285 nm.

$$[RO^-] = \frac{(0.0981)}{(2.950 \times 10^3)} = 3.30 \times 10^{-5} M$$

$0.373 = (RO^-)(5.010 \times 10^3) + (ROH)(3.087 \times 10^3)$

if $RO^- = 3.3 \times 10^{-5}$ m

$(ROH) = (3.3 \times 10^{-5})(5.0 \times 10^3) + (ROH)(3.08 \times 10^3)$

So   ROH $= 6.67 \times 10^{-5} M = pH - \log\left[\dfrac{base}{acid}\right]$ pH $= 9.20 \log(base) = \log 3.3 \times 10^{-5}$, log acid $=$

$\log 6.67 \times 10^{-5}$

So $pKa = pH - \log \dfrac{(base)}{(acid)}$

Substituting these values

$$pKa = 9.20 - \log\left(\frac{3.33 \times 10^{-5}}{6.67 \times 10^{-5}}\right) = 9.20 - \log 0.05 = 9.50$$

PROBLEMS FOR PRACTICE

24.4 A 0.0001 M solution of an organic aromatic amine buffered at pH ~ 3.91 showed an absorbance of 0.440 and 0.080 respectively at 280 and 340 nm. If the molar absorptivity of the amine at 280 nm and 340 nm is 6.607×10^3 and 2.089×10^3 respectively and if an anion of an aromatic amine has no absorptivity at 340 nm but has an absorptivity of 3.02×10^3 at 280 nm what is the pKa of the aromatic amine?

24.5 During simultaneous spectrophotometric determination of complexes of cobalt and nickel with thiodibenzoylmethane at 460 nm and 500 nm, 0.0001 M cobalt solution showed an absorbance of 0.150 and 0.642 at 460 nm and 500 resp. While 0.0002 M nickel solution exhibited an absorbance of 0.684 and 0.088 at 460 nm and 500 nm respectively. If an unknown mixture showed an absorbance of 0.721 and 0.604 at 460 nm and 500 nm, calculate the concentration of cobalt and nickel in the mixture.

24.6 An organic acid has an absorption maxima at 400 nm with a molar absorptivity of 15850. Solutions in the concentration range of 4.1×10^{-5} M are being measured by differential spectrophotometry. Two standard solutions of the above strength have been compared. The solution of unknown concentration showed a percentage transmittance of 82.5%. What is the concentration of the unknown solution?

24.7 In the analysis of a $K_2Cr_2O_7$ on single beam photometer the reading was 75% T. When 1 ml of

0.015 N $K_2Cr_2O_7$ was added to a 5 ml cell, the reading was 65% *T*. What is the concentration of unknown solution?

24.8 What is the sensitivity of determination of lead with cupral at 440 nm, if b = 5 cm, volume of solution is 25 ml and molar absorptivity of coloured complex is 1.28×10^4? (The reading for the absorbance is 0.005).

24.9 The molar absorptivity of chromium-acetylacetone is 3.16×10^4 around 300 nm. In one gram sample dissolved in 5.0 ml solution with an optical path of 50 cm, what is the smallest content of chromium which can be measured? (The minimum reading is A = 0.025 for measurement)

24.10 Find out the composition of a coloured complex between palladium and dimethyl glyoxime with the following data:

[HR]	0.20	0.40	0.60	0.80	1.0	1.50	1.80
A	0.065	0.130	0.200	0.270	0.315	0.365	0.370

(*Hint:* This is the mole ratio method of Yoes).

LITERATURE

1. E.B. Sandell, *"Colorimetric Determination of Traces of Metals"*, General aspects 3rd Ed., Wiley Interscience (1978).

2. G. Charlot, *"Colorimetric Determination of Elements"*, Elsevier Publishing Co. (1964).

3. R.M. Silverstein, G.C. Bassler and T.C. Morrilla, *"Spectrometric Identification of Organic Compounds"*, 5th Ed., Wiley Interscience (1994).

4. A. Babko and A. Pilipenko, *"Photometric Analysis"*, Mir Publishers, Moscow (1971).

5. D.F. Boltz and J.A. Howell, *"Colorimetric Determination of Nonmetals"*, John Wiley (1978).

6. F.D. Snell and Snell, *"Photometric and Fluorometric Methods of Analysis of Metals"*, Wiley Interscience (1980).

7. Z. Marczenko, *"Spectrophotometric Determination of Metals"*, Ellis Horwood and Co. (1980).

8. E.B. Sandell and H. Onish, *"Photometric Determination of Metals"*, 4th Ed. Vol. 1 B, John Wiley (1989).

9. O.S. Wolfbeis, *"Fiber Optics Chemical Sensors & Bio Sensors"*, Boca Raton, FL, CRC Press (1990).

10. B. Headridge, *"Photometric Titration"*, Pergamon Press (1961).

Chapter 25

Infrared Spectroscopy

For the elucidation of the structure of chemical compounds no other spectral method has proved to be versatile as infrared and Raman spectroscopy. These are the finest tools for the characterisations of compounds. Electromagnetic radiation having a wavelength between 0.78 and 100 μm or a wave number between 12.8 and 10 cm^{-1} is called infrared radiation. This is usually subdivided into the near IR region (12,800–4000 cm^{-1}, 3.8–1.2 × 10^{14} Hz, 0.78–2.5 μm), middle IR region (4000–200 cm^{-1}, 0.012–6 × 10^4 Hz, 2.5–50 μm) and the far infrared region (200–10 cm^{-1}, 60–3 × 10^4 Hz, 50–100 μm). The most useful region for all practical purposes is 4000–690 cm^{-1}, (12–2 × 10^{13} Hz; 2.5–15 μm). This is generally called as the middle region. IR spectroscopy is useful for the elucidation of the structures of organic compounds and also for quantitative analysis. Quantitative analysis of atmospheric pollutants like carbon monoxide from air is possible by non-dispersive technique. IR spectra gives a distinct maxima as well as minima at a particular wave number. Absorption spectra are constructed with the wave number on the x-axis and per cent (T), $i.e.$ transmittancy on the y-axis. IR radiation is confined to molecular species where a difference in the energy due to vibrational and rotational states is used to get IR radiation absorption. Thus for absorption, a molecule must have a change in dipole moment as consequence of mode of vibrations. Then a changing electric field would act with the molecule and cause change in amplitude of one of its motions, resulting in different absorption.

There is no change in the dipole moment in vibration or rotation of homonuclear species such as O_2 or N_2 and hence there is no absorbance. The energy needed to effect an alteration in a rotational level is small, corresponding to radiation of < 100 cm^{-1}. Vibrational rotational transitions correspond to the region 13000–675 cm^{-1}. Such rotation is limited to solids and liquids.

25.1 MOLECULAR VIBRATIONS

The relative positions of atoms in a molecule is not fixed but fluctuates steadily due to different kinds of vibrations. For diatomic or triatomic molecules, the nature of vibrations can be ascribed and can be correlated to absorption energies, but for polyatomic molecules such vibrations cannot easily be predicted, due to interaction of many centres of vibration. These vibrations are classified as 'stretching' vibrations and ' bending' vibrations. The former involves vibration along the axis of the bond between the two atoms, interatomic, while the latter vibrations are changes in the angle between two bonds. They are of four types namely scissoring, rocking, wagging and twisting.

These vibrations are possible in molecules containing more than two atoms. A brief description of modes of the vibrations form is given in Fig. 25.1.

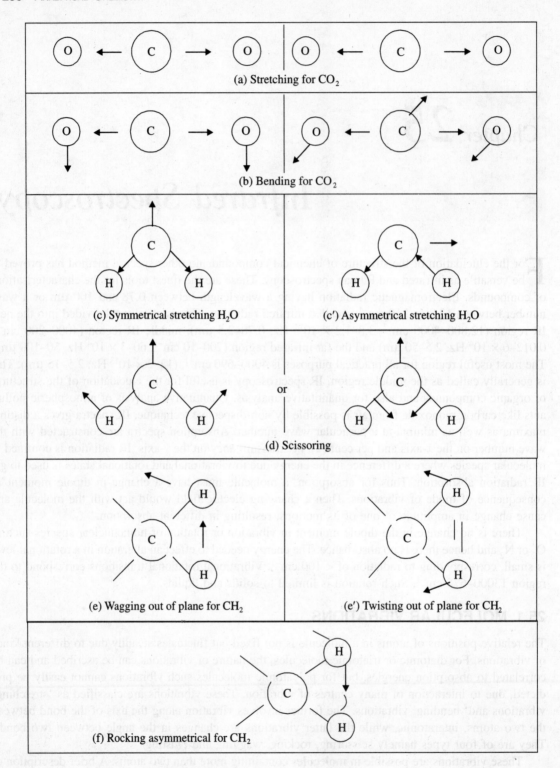

(a) Stretching for CO_2

(b) Bending for CO_2

(c) Symmetrical stretching H_2O

(c') Asymmetrical stretching H_2O

or

(d) Scissoring

(e) Wagging out of plane for CH_2

(e') Twisting out of plane for CH_2

(f) Rocking asymmetrical for CH_2

Fig. 25.1 Modes of molecular vibrations

In Fig. 25.1 (e) wagging, the structural unit swings back and forth in the plane of the molecule, while in (f), rocking, the structural unit swings back and forth out of the plane of the molecule. In (e), twisting, the structural unit rotates about the bond which joins it to the rest of the molecule and finally in scissoring, (d), the two atoms connected to a central atom move toward and away from each other with deformation of the valence angle.

25.2 VIBRATIONAL FREQUENCY

Each normal mode has definite vibrational frequencies. If we assume the bond between a pair of atoms obeys Hooke's law, on displacing one atom a distance x from its equilibrium position w.r.t. the other atom, then the force (F) operating is $F = -kx$ if k = force constant. The negative sign shows F is a restoring force. The simple harmonic motion of atoms is given by the fundamental vibration frequency ν.

$$\nu = \frac{1}{2\pi}\sqrt{K\left(\frac{m_1 + m_2}{m_1 m_2}\right)} = \frac{1}{2\pi}\sqrt{\frac{K}{\mu}} \qquad \qquad ...(25.1)$$

where m_1 and m_2 are masses of atoms, K = force constant and $\mu = \dfrac{m_1 \times m_2}{(m_1 + m_2)}$ is the reduced mass of a diatomic oscillator. The K for single bonds is $2\text{--}8 \times 10^5$ dynes/cm, while for a double bond it is $(8\text{--}12 \times 10^5)$ and for a triple bond it is $(12\text{--}18 \times 10^5)$. K is larger due to higher vibrational frequencies.

In examination of quantized nature of transitions between vibrational levels, (E) the vibrational frequency of the mechanical model is given by an expression

$$E = \left(\nu + \frac{1}{2}\right)h\nu, \qquad \qquad ...(25.2)$$

where E = energy of level having vibrational quantum number (V). When a molecule undergoes transition from one energy (ν_1) to another (ν_2) with respect to quantum numbers,

$$\Delta = \left(E_2 - E_1\right) = \left(\nu_2 + \frac{1}{2}\right)h\nu - \left(\nu_1 + \frac{1}{2}\right)h\nu = (\Delta\nu)h\nu, \qquad ...(25.3)$$

(i.e. A transition between vibrational energy level involves changes equal to $h\nu$ ($\Delta\nu$), where $\Delta\nu$ can be positive or negative. If $\Delta\nu = 1$ then $E = h\nu$.

From the above equation the frequency of normal vibrations of molecules is in the range of 0.6×10^{13} to 12×10^{13} cps, i.e. 4000–200 cm^{-1} (2.5–50 μ). Motions involving H-atoms have higher frequencies than motions with heavier atoms. Bending motions exhibit lower frequencies of absorption than fundamental stretching patterns. Radiation can be expressed in terms of the wave number (σ) as

$$\sigma = \frac{1}{2\pi c}\sqrt{\frac{K}{\mu}} = \frac{1}{2\pi c}\sqrt{\frac{K(m_1 + m_2)}{m_1 m_2}} \qquad \qquad ...(25.4)$$

where $\quad$ c = velocity of light,

$\qquad K$ = force constant, μ = reduced mass, m_1, m_2 = mass of atom.

25.3 VIBRATIONAL MODES

Vibrational modes are simple to calculate for triatomic or diatomic molecules but are somewhat more complicated in polyatomic molecules. In the latter, three degrees of freedom are assumed to fix N-points with three coordinates, each corresponding to one degree of freedom *i.e.* a molecule with N atoms has $3N$ degrees of freedom. We consider the motion of the whole molecule through space, rotation or motion of the molecule around its centre of gravity, motion of individual atoms w.r.t. other atoms. All vibrational modes can be resolved into normal modes with their own spatial properties and vibrational frequencies. Three of these degrees of freedom involve the translational motion of the molecule along a perpendicular axis in space. Then there are three degrees of freedom corresponding to rotation of the molecule about three axes. The remaining $(3N - 6)$ degrees of freedom exhibit normal mode of vibrations of molecule *i.e.* a molecule will possess $(3N - 6)$ independent vibrations. Finally we consider vibrational coupling; here the energy of vibration and hence the absorption peak, is changed by other vibrations in the molecule, e.g. a strong coupling between stretching. A common bond between vibrating groups is needed for interaction between bending vibrations, then coupling between the two takes place. If the individual energies of coupling groups are identical there is maximum interaction. Two or more separated bonds do not exhibit interaction and finally, vibrations of the same symmetry species are required for coupling to take place.

25.4 SPECIAL FEATURES OF IR SPECTROMETERS

These instruments resemble UV–Visible spectrophotometers. Spectrometers for IR are made up of the same basic components as those in the UV–visible range, although sources, detectors and optical components are somewhat different. IR radiation first passes through the sample and then the reference solution and then through a monochromator. This eliminates the effect of stray radiation. The beam is then dispersed through a prism or gratings. On passing through the slit it gets focused on the detector. IR instruments are generally self recording, as point to point manual measurement of absorbance is extremely time consuming. The temperature and humidity of the room must be controlled. The maximum permissible humidity is 50%. If humidity exceeds this limit, prism faces and windows which are made of the alkali halides get fogged. Temperature change also affects the accuracy and wavelength calibration. Both single beam instruments and double beam models are common.

25.5 COMPARISON OF IR AND UV–VISIBLE SPECTROPHOTOMETERS

In IR instruments a Nernst or Glower's lamp is used as the light source, while in UV–Visible models a deuterium tungsten lamp is the radiation source. An IR monochromator gratings or a prism made of alkali halides, whereas in UV–Visible instruments gratings or the prism is made of corex glass or quartz. The detector for IR radiation is a thermocouple or bolometer, while in a regular UV-visible instrument we have a photomultiplier tube. In addition, the sampling technique in IR spectroscopy is altogether different. The monochromator is placed before the sample in a UV–visible instrument whilst, in IR it is kept *before the detector* and after the sample to avoid stray light. Finally, percentage transmittance is measured in IR spectroscopy against wave number whilst in UV–visible measurement of absorbance against wavelength is preferable.

25.6 COMPONENTS OF IR SPECTROPHOTOMETERS

(*a*) *Source:* The most commonly used radiation source is a Nernst or Glower's lamp which is made out of zirconium oxides and hollow yttrium rod 2 mm in diameter and 30 mm in length. It is heated to $1500\text{-}2000°C$ and gives radiation upto 7000 cm^{-1}. A Glower (globar) source is also used in a few instruments with transmittance in the range of 5200 cm^{-1}. This source is satisfactory for the region > 15 µ as its radiant energy output decreases less rapidly.

(b) *Monochromators:* Monochromators used in IR instruments are made out of various kinds of materials, e.g. prism (usually in a Littrow mounting) and windows are made out of glass, fused silica, LiF, CaF$_2$, BaF$_2$, NaCl, AgCl, KBr, CsI. A NaCl prism is used for measurement at $4000\text{-}650$ cm^{-1} and KBr prisms used for measurement at 400 cm^{-1}. Quartz spectrophotometers designed for the UV visible region also extend their coverage into the near IR region. The highest dispersion can be obtained from a reflection grating substituted for the prism. Holographic gratings give the best result as stated earlier.

(c) *Detectors:* The most commonly used detectors in the IR region are bolometers. Amongst thermal detectors, thermocouples are widely used. A bolometer gives an electrical signal as a result of a change in the resistance of the metallic conductor with temperature. A nonthermal detector of greater sensitivity is the photoconductive cell, however, this is rarely used. A PbS or PbTeO$_3$ layer is used as coating material. The long wavelength sensitivity limit seems to be about 3.5 µ for the PbS cell and 6µ for the PbTeO$_3$ cell.

(d) *Recorder:* The two types of instruments, for qualitative and quantitative work, are available. On account of the complexity of IR spectra, a recording model is a must; this is invariably a double beam model. The design is less complicated than the single beam model. All instruments include a low frequency chopper to modulate the source output, these are null type with the beam being attenuated by an absorbing wedge.

(e) *Position of Sample:* The optical design is similar to a UV visible spectrophotometer except that the sample and reference compartment are located between the source and monochromator to eliminate stray radiation generated by the sample and to avoid photochemical decomposition. The radiation from the source splits into two beams, one beam passing through the sample and the other beam through reference solution which in turn passes through an attenuator and chopper. After passing through a prism or gratings the beam falls on a detector which converts it into electrical signals which are directly fed to a recorder. It may be necessary to amplify the signal if it is very weak. The latest models contain a microprocessor for partial automation of measurement.

(f) *Quantitative:* For quantitative work, a double beam model is not always satisfactory, as it is noisy due to complex electronic circuits, and zero adjustments are likely to cause an error from the optical null design. Hence, several mechanical devices are employed so as to get a correct zero adjustment. Such errors are negligible for quantitative work. A nondispersive models or filter models are employed for quantitative work, e.g. CO analysis from atmosphere is carried out by a nondispersive model called an infrared filter photometer with nichrome as the source of radiation which transmits radiation between 3000 and 750 cm^{-1}. The optical path is 0.5 mm. The cell length can be varied. Acryl nitrite, chlorinated hydrocarbons, COCl$_2$, HCN are analysed by such models. In addition, automated instruments are also available for quantitative work.

25.7 SAMPLE HANDLING BEFORE MEASUREMENTS

Sample handling presents a number of problems in IR spectroscopy. Alkali halides are used as window material, e.g. NaCl is transparent upto 626 cm^{-1}, KBr upto 385 cm^{-1}, CsI upto 250 cm^{-1}. On exposure to the atmosphere, these surfaces are fogged by moisture (> 50% in air). The samples to be handled are liquid at room temperature and are scanned in the pure form. The thickness of the film for the purpose of measurement ranges form 0.01 to 0.05 mm. Hypodermic needles are used for filling samples. Various path lengths ranging from 0.002 to 3 mm are employed. Wherever possible, solids are dissolved and examined as a dilute solution, CS_2 and CCl_4 can also be used as solvents. CCl_4 is useful in the region 800–740 cm^{-1} while CS_2 is useful in the region of 2222–1540 cm^{-1}. The material is categorised as transparent if the transmitted radiation is > 75%. All solvents used should be moisture free. Powders and solids are reduced to particles which can be examined as a thick slurry or mull by grinding the pulverised solids in a greasy liquid medium of the same refractive index, to reduce energy losses due to scattering. Paraffin oil or Nujol oil are used for this purpose. For qualitative identification the mull technique is rapid but for quantitative analysis the internal standard method is more useful. The KBr pellet technique involves mixing of the finely divided sample and KBr, and pressing the mixture into a transparent disk in an evacuated die under a hydrolytic or other similar high pressure device. CsI and CsBr can also be used for the measurement of long wavelengths. Good dispersion of samples is most essential in order to procure dependable IR data.

25.8 INFRARED IN STRUCTURE ELUCIDATION

IR spectra have several qualitative as well as quantitative applications. The group frequencies can be calculated from masses of atoms and force constant by using the relationship discussed earlier:

$$\sigma = \frac{1}{2\pi c}\sqrt{\frac{K}{\mu}}$$ with the usual notations. However, one generally prefers to use correlation charts for

identification of the functional group. Table 25.1 summarises some important frequencies. It is useful to classify the entire region into three to four broad regions. One way to categorise is: near IR region (0.7–2.5 μ); fundamental region (2.5–50 μ) and far IR region (50–500 μ). The other way to classify is: hydrogen stretching region (2.7–3.7 μ); triple bond region (3.7–5.4 μ); double bond region (5.1–6.5 μ) and finger print region (6.7–14 μ). Both classifications show that in the second category all regions are from the fundamental region.

(a) The hydrogen stretching region (3700–2700 cm^{-1}) shows a peak due to stretching vibrations of hydrogen and other atoms. The frequency is very large so interaction is negligible. The absorption peaks are due to (3700–3100 cm^{-1}) O–H or N–H stretching vibrations. H–bonding broadens the peak and shifts them to lower wave numbers. While aliphatic C–H vibrations give a peak in the range 3000–2850 cm^{-1}, a structural change in CH bond strength will shift the peak to a maximum. The C $\equiv$ H bond peak is at 3300 cm^{-1}. Hydrogen on the carbonyl group of an aldehyde gives a peak at 2745–2710 cm^{-1}. CH stretching peaks are identified by substitution of deuterium for hydrogen.

(b) In the triple bond region (2700–1850 cm^{-1}), only limited groups absorb e.g. triple bond stretching occurs in the region 2250–2225 cm^{-1} (e.g. $-C \equiv H$ at 2120 cm^{-1}, $-C^+ = N^-$ at 2260 cm^{-1}. A peak for SH is at 2600–2550 cm^{-1} for PH at 2440–2350 cm^{-1} and SiH at 2260–2090 cm^{-1}.

(c) In the double bond region 1950–1550 cm^{-1}, carbonyl stretching vibrations are characteristic, e.g. ketones, aldehydes, acids, aminoles and carbonates—all have peaks at 1700 cm^{-1}. Esters, acid halides, acid anhydrides absorb at 1770–1725 cm^{-1}. Conjugation lowers the absorption peak to 1700 cm^{-1}. The peaks due to $- C = C-$ and $> C = N$ stretching vibrations are located in the range 1690-1600 cm^{-1}. They are useful for qualitative identification of olefins. Aromatic rings show peaks in the region 1650–1450 cm^{-1}, those with a low degree of substitution exhibit peaks at 1600, 1580, 1500 and 1450 cm^{-1}.

Table 25.1 Characteristic IR absorption peaks

Name of region	Group	Environment	Frequency cm^{-1}	Spectral region (cm^{-1})
Normal	OH	Alcohols	3580–3650	3333–3704
		Acid	2500–2700	(2.7 – 3 μ)
	NH	Amines - primary	~3500	
		-secondary	3310–3500	
		Amides	3140–3320	2857–3333
Hydrogen Stretching	CH	Alkyne	3300	(3.0 – 3.3 μ)
		Alkene	3010–3095	
		Aromatic	~3030	
		Alkane	2853–2962	
		Aldehydes	2700–2900	2500–2857
	SH	Sulphur compounds	2500–2700	(4.0 – 4.5 μ)
Triple bond	–C ≡ C–	Alkyne	2190–2260	
	C ≡ N	Alkylnitriles	2240–2260	2222–2500
		Isocyanate	2240–2275	(4.5 – 5.0 μ)
		Arylnitriles	2220–2240	
	–N = C = N	Diimides	2130–2155	2000–2222
				(5.0 – 5.5 μ)
	–N$_2$	Azides	2120–2160	
	> CO	Aldehyde	1720–1740	1818–2000
				(5.5 – 6.0 μ)
		Ketone	1675–1725	
		Carboxylic acid	1700–1725	
		Ester	2000–2300	
Double bond		Acyl halides	1755–1850	1667–1818
		Amides	1670–1700	(6.0 – 6.5 μ)
	> CN	Oximes	1640–1690	
	> CO	β-diketone	1540–1640	1538–1667
	> C = O	Ester	1650	(6.5 – 7.5 μ)
	> C = C <	Alkenes	1620–1680	
	N – H	Amines	1575–1650	1538–1667
Finger print	–N = N –	Azo	1575–1630	
	– C – NO$_2$	Nitro compounds	1550–1570	(7.5 – 9.5 μ)
	– C – NO$_2$	Aromatic nitro	1300–1570	
	C – O – C –	Ethers	1230–1270	1053–1333
	– (CH$_2$) n	Any compounds	~722	666–900
				(11.0 – 15 μ)

(d) The finger print region extends from 1500–700 cm^{-1} where a small difference in the structure and constitution of a molecule makes a marked change in absorption peak distribution. Close matching is necessary, especially in this region absorption bands are due to various interactions, hence exact interpretation is not possible, though sometimes complexity is useful for identification. e.g. C–O–C in ethers and esters absorbs at 1200 cm^{-1} while C–Cl vibrates at 700–800 cm^{-1}, SO_4^{2-}, PO_4^{3-}, NO_3^-, CO_3^{2-} show strong absorbance below 1200 cm^{-1}.

While interpreting an IR spectrum it is essential to examine the entire spectrum rather than an isolated portion. The correlation by Colthup chart, or Table 25.1 serves as a guide for extensive study of the compound. IR in conjunction with NMR and elemental analysis may give definite information on the nature of the compound. A comparison of the experimental spectrum with that of the pure compound is absolutely necessary.

25.9 SPECTRAL INTERPRETATION OF TYPICAL COMPOUNDS

We shall consider a few examples to see how IR spectra are used for characterisation of organic compounds. The compounds discussed are: (i) n-hexane, (ii) benzene, (iii) phenol, (iv) hexanol, (v) n-butylmethyl ether. Nujol CH_3 $(CH_2)_8$ CH_3 is commonly used as a solvent for sample preparation.

(i) The IR spectrum of n-hexane shows vibrational absorption bands characteristic of aliphatic hydrocarbons, e.g. asymmetric and symmetric C–H stretching at 2899 cm^{-1}, –CH$_2$– scissoring at 1471 cm^{-1}, and C–CH$_3$ bending at 1381 cm^{-1}. The weak band near 725 cm^{-1} is caused by bending vibrations of the –(CH$_2$)$_4$ unit.

(ii) In benzene, aromatic stretching C–H vibrations show up at 3030 cm^{-1} while ring skeletal vibrations –C = C– are at 1613–1471 cm^{-1} with a strong peak at 1481 cm^{-1}. C–H out of plane bending occurs at 769 cm^{-1}.

(iii) n-hexanol shows an intense O–H stretching band at 3356 cm^{-1}; the broadness is due to hydrogen bonding. The broad band at 1062 cm^{-1} is peculiar to the C–O stretching vibration of a primary alcohol. When an aliphatic hydrocarbon hydrogen atom is replaced by an OH group, the spectrum changes to include absorption due to OH and CO vibrations in addition to C–H vibrations.

(iv) The effect of hydroxy substitution in benzene is shown in the spectrum of phenol (C$_6$H$_5$ – OH). The OH vibrations appear over a broad band at 3333 cm^{-1} and a shoulder at 3030 cm^{-1} due to aromatic C–H stretch. The bands at 1595 cm^{-1} and 1497 cm^{-1} represent benzene skeletal vibrations. C–OH stretching vibration is shown by a band at 1359-1218 cm^{-1}. Aromatic C–H out of plane bending vibrations at 750 and 685 cm^{-1} show that the benzene ring is monosubstituted.

(v) n-butylmethylether is obtained by the substitution of the methylene group in n-hexane (case (i) above). Absorption due to C–O–C is around 1126 cm^{-1} in addition to the other C–H vibration bands.

Example: If an unknown compound shows sharp absorption peaks at the following frequencies: 710, 1300, 1510, 2700 cm^{-1} and if it is saturated alkane identify the compound. (Fig. 25.2).

We proceed in the following manner to characterise this organic compound. We closely examine the spectra as given in Fig. 25.2 which shows a compound possibly containing C and H in a straight chain. The four significant peaks are due to C–H and C–C linkages at 2940 cm^{-1}; C–H stretching at 1460 cm^{-1}, scissor like bending of H–C–H groups at 1380 cm^{-1}, symmetrical distortion of end methyl groups and at 720 cm^{-1} rocking vibrations of the methylene hydrogen attached to cations along the chain, when the four or more (actually in this case, eight) methylene group are aligned linearly. The absorption peaks at 3000 cm^{-1} – 2890 cm^{-1} are assigned to C–H stretching in the methyl and methylene group respectively.

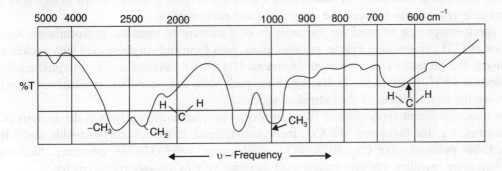

Fig. 25.2 IR spectrum of unknown compound

Further, since unsaturation is absent, C–H stretching frequencies are at a wave number greater than 3000 cm^{-1}. The absence of any aromatic structure shows in the frequency region of 200–670 cm^{-1}. This shows the compound is CH$_3$ (CH$_2$) CH$_3$ *i.e.* Nujal oil.

25.10 QUANTITATIVE ANALYSIS BY INFRARED SPECTROSCOPY

Beer's law can be applied to IR determinations. We can calculate the molar absorptivities (ε) at a particular wavelength where one component absorbs strongly, and the other components have weak or no absorption peaks. The absorbance of the unknown compound is determined at these two wavelengths by simultaneous spectrophotometry. Stray light at times makes the application of Beer's law impossible especially at high values of absorbance. Hence empirical methods are employed. The base line method (Figs. 25.3 and 25.4) involves the selection of an absorption band of an analyte which does

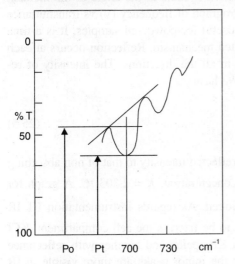

Fig. 25.3 Quantitative analysis

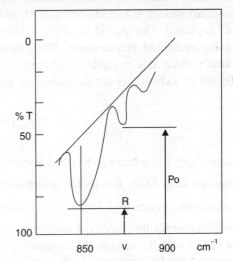

Fig. 25.4 Quantitative analysis of an organic compound

not fall too close to the bands of the analyte. P_0 represents the intensity of incident radiation which is obtained by drawing a straight line tangential to the spectral absorption P_t is measure of the point of maximum absorption. The value of log (P_0/P_t) is plotted against concentration in the usual manner to get a calibration curve. Many possible errors are eliminated by the base line method.

Since IR bands are narrow, a deviation from Beer's law is quite possible, which in turn may lead to a nonlinear relationship between the absorbance and concentration.

IR spectroscopy can be used for the analysis of a mixture of aromatic hydrocarbons, e.g. the resolution of C_8H_{10} (containing xylene, in ortho, meta, para form and ethylbenzene) with cyclohexane as a solvent. We get peaks at a wavelength of around 12-15μ. We calculate molar absorptivities for all compounds at 13.47, 13.01 12.58, 14.36μ which are peak areas and write four simultaneous equations to calculate the concentration of the individual species.

The most significant application of IR spectroscopy in quantitative analysis is the analysis of air contaminants, e.g. for pollutants like CO, methylethylketone, methanol, ethyleneoxide and $CHCl_3$ vapour. Other pollutants like CS_2, HCN, SO_2, nitrobenzene, vinyl chloride, diborane, chloroprene, methylmercaptan, pyridine can also be analysed quantitatively by IR spectrophotometry.

The limitations of the IR technique in quantitative analysis cannot be ignored. The first is non adherence to Beer's law and complexity of spectra leading to an overlap of peaks. The second is narrowness of peaks, and the effect of stray radiation leading to large slit widths. Even narrow cells are also not of much practical use. Therefore IR spectroscopy is less frequently used in quantitative analysis. However it is the best tool for the structure elucidation of organic compounds and complexes.

25.11 IR–REFLECTANCE SPECTROMETRY

This is the best method for the handling of the solid samples like polymeric materials, food, and agricultural materials. Mid IR spectra furnishes valuable information. Reflectance spectra can be used for qualitative and quantitative analysis. An adopter is provided with original IR model for the working of reflectance spectra. There are four kinds of reflections as are (1) spectral reflection, (ii) diffuse reflection (iii) internal reflection and (iv) attenuated total reflection (ATR). The first kind of spectral reflection is useful for smooth surfaces, when angle of reflection is same as the angle of the incident provided surface is non absorbing of IR radiation. The spectrograph of frequency (ν) vs transmittance (T) is plotted. The second (2) diffuse reflectance spectra is useful for powdered samples. It is a time saving method of measurement. The diffusion is a complicated mechanism. Reflection occurs at each plane surface, due to random distribution, reflection occurs in all the directions. The intensity of reflected IR radiation is not dependent on angle of viewing. We have

$$f\left(R'_{ec}\right) = \frac{\left(1 - R'_2\right)^2}{2R'_c} = \frac{K}{S} \qquad \qquad ...(25.5)$$

Here $f\left(R'_{ec}\right)$ is relative reflectance intensity, R'_{ec} is ratio of reflected intensity to that of non absorbing standard (e.g. KCl). K = molar absorptivity (ε), c = molar concentration, $K = 2.303 \ tC$. A graph for reflectance spectra of $f\left(R'_{ec}\right)$ vs (λ) wavelength can be plotted. As regards instrumentation FT IR models provide this facility if adopter is used. An adopter is to be fixed to the cell compartment. KCl is mixed with the sample. A comparison of IR spectra with KBr pellet and another with reflectance spectra with KCl indicates both are similar in nature. Only the minor peaks are more visible in IR reflectance spectra as compared to usual spectra.

Finally let us consider attenuated total reflectance spectrometry. It is a kind of internal reflection spectroscopy useful for solids of limited solubility. When a beam passes from more dense to less dense medium refraction occurs. Such effect increases if angle of incidence is increased. At critical angle reflection is total. During process of reflection, beam penetrates small distance into less medium before reflecting. The depth of the penetration of the beam depends on the factors as wavelength of incident radiation, index of refraction, and angle of the incident beam. The penetrating radiation is termed as 'evanescent wave'. If less dense medium absorbs evanescent radiation, the attenuation of beam takes place at wavelength of absorption band. This phenomena is called as "Attenuated total reflectance" (ATR). The ATR spectra are analogous to conventional absorption spectra with same spectral peaks. However, thickness of absorber and angle of incidence are critical. ATR is easily available. Most striking feature is such spectras are free from any interferences.

Applications: Now let us consider application of near IR reflectance spectroscopy. A finely ground solid can be easily analysed. Protein, starch, lipids, oils and agricultural products can be easily analysed. In Canada and USA large quantity of grains are analysed by this method. Reflectance spectrum depends upon composition of the sample. Diffuse IR reflectance is valuable method of the analysis. $BaSO_4$ or MgO are used as standard reflecting reference materials. The instruments for diffuse reflectance are commercially available. Some instruments with interferences filters furnish narrow absorption peaks. These measurements are made at two wavelengths. For dependable results the instrument must be calibrated. The particle size must be properly controlled and sample should be measured at two wavelengths.

The most significant advantage of near IR reflectance spectrometry is speed simplicity of sample preparation. Once a material is standardised several samples can be analysed. These methods are accurate and reliable in routine analysis.

25.12 ANALYSIS OF AIR POLLUTANTS BY IR–SPECTROSCOPY

IR is considered to the one of the sensitive methods for the analysis of air pollutants (Table 25.2) because NDIR is an established method for the analysis of carbon monoxide in air.

Table 25.2 Characterisation of air pollutants by IR

Air pollutant	Level (ppm) in air	Analysed (ppm) expt.	Error (%)	Source of Pollution
CO	50	49.1	1.8	Incomplete combustion of fuel
$CHCl_3$	100	99.5	0.5	Emission from Latex industries
MEK	100	98.3	1.7	Laboratory disposal of solvent
$CH_2=CH_2$ oxide	50	49.9	0.2	Photochemical smog in air
CH_3OH	100	99.0	1.0	Destructive distillation of coal

In USA, OSHA (Occupational Safety & Health Administration) and also EPA (environmental protection agencies) stipulate stringent standards for tolerance of these pollutants. The peak overlaps are noticed. The IR filter photometer is ideally suited for outdoor environmental for analysis (Table 25.2).

For other environmental pollutants from air or water the IR spectroscopy is unsuitable because it does not always adhere to Beer's law, or displays excessive overlap of spectral curves (Table 25.3). Most of the times the peaks are too narrow. The slit width and wavelength selection is critical task in analysis. In comparison UV–visible spectroscopy is considered reliable technique.

Table 25.3 The analysis of other air pollutants by IR.

Pollutant	Threshold Level TLV (ppm)	Wavelength λ (nm)	Detection limit (ppm)	Source of pollution
SO_2	2.0	860	5×10^{-1}	Combustion of coal contg. sulphur
HCN	4.8	306	4×10^{-1}	Electroplating industry effluent
$C_6H_5NO_2$	1.0	1180	2×10^{-1}	Vulcanisation process in air
B_2H_6	0.1	390	5×10^{-2}	Silicon Industry processing
CS_2	4.0	454	0.5	Rubber Latex emissions
En	10.0	1300	4×10^{-1}	Complexation reactions
C_6H_5N	5	1420	2×10^{-1}	Pesticide products, offensive odour
CH_3SH	0.5	304	4×10^{-1}	Latex polymer industry

25.13 ATTENUATED TOTAL REFLECTANCE

We would consider further details of attenuated total reflectance (ATR) technique used for IR analysis of samples which cannot be examined by conventional UV-visible spectroscopy. It is used for characterisation of the polymeric materials. It is based upon the fact that reflected radiation marginally enters the surface from which it is reflected. The sample is placed or coated on the surface of a prism or opposite side of optically transparent element (Fig. 25.5) IR radiation is passed through analyte and is then reflected. Radiation impinges for small distance in sample and is attenuated by absorption of the sample. The angle of refraction by Snell's law is given as per equation as:

$$n_1 \sin a = n_2 \sin b$$

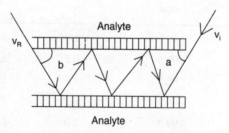

Fig. 25.5 Multiple reflection

if n_1, n_2 are refractive indices of resp. media and $\hat{a}$ is angle of incidence and $\hat{b}$ is angle of refraction of beam. We note that angle of incidence viz. $\hat{a}$ increases so also does angle of refraction viz. $\hat{b}$. Now if $n_2 < n_1$, then angle $\hat{a}$ has angle $\hat{b} = 90°$ for total reflection of the radiation. It is critical angle (θ_c).

If incident radiation strikes at angle greater than the critical angle it is completely reflected from the surface. If $\theta_c = \hat{a}$, $\hat{b} = 90°$, we have $n_1 \sin \theta_c = n_2 \sin 90°$, then

$$\theta_c = \sin^{-1} \frac{n_2}{n_1} \qquad \text{(from above eqn)} \qquad ...(25.6)$$

If $\hat{i} > \hat{\theta}_c$ then reflection takes place. $\hat{\theta}_c$ depends on difference in refractive indices of two media. A substance with high (n) is preferred like AgBr, AgCl, CdTe, Ge, ZnSe. The refractive indexes in the two media are function of the wavelength. $\hat{i} > \hat{\theta}_c$ at all wavelength. $\hat{i}$ is evaluated in terms of $\hat{a}$ as

$$a + 3 = \sin^{-1} \left(\frac{n_2 + 0.2}{n_1} \right) \qquad ...(25.7)$$

The penetration depth is also related to relative refractive indices of prism and sample. To perform, ATM spectrophotometers are usually provided with appropriate attachment.

25.14 FT-IR SPECTROSCOPY AND ITS ADVANTAGES

As compared to the dispersive spectrometer FTIR have several merits in applications as:

(a) They are fast and sensitive as complete spectrum scan can be carried out in few minutes. The (S/N *i.e.*,) signal to noise ratio is proportional to the square roots of total measurements. Multiple spectra can be recorded, and by altering S/N ratio sensitivity of analysis can be improved with repeated scans.

(b) In absence of energy wasting slit the increase of optical throughput is feasible. The circular optical slit is used in FT-IR, so more radiation energy is there. The beam area of FT-IR is hundred times larger than dispersive model.

(c) He-Ne laser when used as internal reference provides built-in facility for calibration with better accuracy and no external calibration of model is necessary.

(d) Mechanical design is simple, with moving parts restricted to moving mirror, so readings are more reliable.

(e) Stray light is totally eliminated. All frequencies are modulated. There is absolutely no contribution by emission phenomena.

(f) Data processing can be accomplished for wide variety of samples like library searching, integration, smoothing baseline correction and spectral subtraction. These operation needs relatively considerable time in usual dispersive models.

(g) FT-IR models have much increased sensitivity as compared to the dispersive IR models. The sampling accessories are easily available.

(h) The complex analysis of petroleum hydrocarbons, oil and greases, multicomponent mixture of sulphur, oxygen anions, heterogeneous catalyst and polymeric films can be readily done by FT-IR technique. As spectral science advances, the use of the Fourier transfer IR or NMR is going to be most popular with analyst.

PROBLEMS

25.1 The IR spectra of zinc cyanide complex at 2100 cm^{-1} showed the following reading for absorbance (metal–ligand ratio):

0.15 (1:2); 0.300 (1:1); 0.450 (3:2); 0.590 (2:1)

0.600 (5:2); 0.600 (3:1); and 0.600 (7:2).

Calculate the composition of the metal ligand.

Answer: We plot the graph of absorbance against (metal/ligand) ratio. We get two straight lines intersecting at points 2:1 around an absorbance of 0.595 thereby indicating that the metal to ligand ratio is 1:2 *i.e.* the complex has the formula: Zn:CN:: 1:2, *i.e.* $Zn(CN)_2$.

25.2 The concentration of an antibiotic sample was to be measured by IR spectroscopy at 2200 cm^{-1}. The unknown sample showed an absorbance of 0.180 while the readings for the calibration curve were as follows for the amount of antibiotic in ppm against absorbance:

0(0.040); 0.8(0.110); 1.4(0.160); 2.2(0.220); and 3.0(0.290)

What is the concentration of antibiotic in the unknown sample?

Answer: We have to plot a graph of concentration of antibiotic in ppm vs absorbance, We get a straight line graph with a slope around 40° of about 0.890. The concentration is 1:70 ppm.

PROBLEMS FOR PRACTICE

25.3 What is the % of unsaturated aliphatic hydrocarbon if at 5,200 nm it shows an absorbance of 38.7%? The readings for the calibration curve are as follows: % hydrocarbon against absorbance

0.30(0.610); 1.0(0.370); 1.5(0.248); 2.0(0.174)

2.5(0.125) and 3.0(0.092)

25.4 Water molecules in an organic solvent are determined by IR spectroscopy at 1430 nm. To 5 ml of sample a known amount of water is added and measured spectrophotometrically in IR region at 1430 nm. The readings are mg of water (absorbance).

0.0(0.210); 1(0.380); 2(0.550); 3(0.720); and 4(0.89)

What is the concentration of water in the sample?

25.5 A chlorobenzene shows no band for absorption between 900-690 cm^{-1}. What is the expected structure?

25.6 The organic compound $C_{13}H_{11}N$ has m.p. of 56°C.

In IR measurement it shows sharp bands at 3060, 2870, 1627, 1593, 1579, 1487, 1452, 1368, 1195, 759 and cm^{-1}. What is the structure?

(*Hint:* The compound belongs to series of substituted idenealine).

25.7 An empty cell showed 8.5 interference peaks in the region of 1000-1250 cm^{-1}. What should be the path length of the cell?

25.8 In an infrared study of 2, 3 dimethylhexane and 2, 3, 4 trimethylhexane the following absorbance values were noted:

ψ	8.0	9.62	10.86	13.02	13.51	cm^{-1}
A	0.024	0.310	0.420	0.050	0.150	

What is the composition of the mixture?

LITERATURE

1. L.J. Bellamy, "*The Infrared Spectra of Complex Molecules*", 2nd Ed., Wiley, New York (1958).

2. J.R. Dyer, *"Applications of Absorption Spectroscopy of Organic Compounds"*, Prentice Hall (1965).

3. R.J. Bell, *"Introductory Fourier Transform Spectroscopy"*, Academic Press, New York (1972).

4. P.S. Kalsi, *"Spectroscopy of Organic compounds"*, Wiley Eastern (1994).

5. R.J.H. Clark, R.E. Hester, *"Advances in Infrared & Raman Spectroscopy"*, John Wiley (1985).

6. C. Socrates, *"Infrared Characteristics Group Frequencies Tables & Charts"*, 2nd Ed., John Wiley (1994).

7. R.M. Silverstein, G.C. Basler, T.C. Morrillo, *"Spectrometric Identification of Organic Compounds"*, 5th Ed., John Wiley (1991).

2. R. Dyer, "Application of Absorption Spectroscopy of Organic Compounds", Prentice Hall (1965).
3. R.J. Bell, Introductory Fourier Transform Spectroscopy, Academic Press, New York (1972).
4. J.R. Dyer, "Spectroscopy of Organic Compounds", 2nd Ed. (1987).
5. K.J.H. Clark, R.J. Hester, Advances in Infrared & Raman Spectroscopy, John Wiley (1985).

Chapter 26

Raman Spectroscopy

In 1931 Sir C.V. Raman was awarded the Nobel prize for the discovery of Raman Spectroscopy in Physics. He had then published a research paper from Indian Association for Cultivation of Science in Calcutta along with K.S. Krishnan about this discovery. The technique is based upon phenomena of vibrations and scattering. It is quite similar to infrared spectroscopy. The principal advantage is one can use aqueous samples during characterisation with the result that the cumbersome sampling practice of use of KBr pellet was abandoned and the use of quartz or glass curvettes was permitted during measurements. On account of lack of proper source of radiation this technique remained unpopular till 1960. However, with the discovery of laser source it has become very popular in the elucidation of chemical structure of the compounds. Laser source is of visible or infrared monochromatic radiation. On account of interaction of the radiation with matter although some light is reflected, part of such radiation is scattered e.g. sky appears blue due to the scattering of radiation.

26.1 RAMAN SPECTROSCOPY

In Raman scattering a change in frequency occurs due to interaction between the incident photons and vibrational energy levels of molecules. It is not just absorption but transfer of a portion of energy of the photons to the molecule or vice versa. It is expressed as $\bar{v}_R = \bar{v}_i + \Delta\bar{v}$ where R represents Raman radiation, $\bar{v}_i$ is the incident radiation and $\Delta\bar{v}$ is the Raman shift corresponding to the energy difference between vibrational levels. Interaction of molecules takes place if the polarizability of the molecule is altered by vibration, $i.e.$, the shape of the molecule is changed without generating a dipole moment. The exciting radiation must be monochromatic if the Raman lines are to be sharp. A highly monochromatic source is thus essential to get small Raman shifts. Raman spectroscopy measurement is carried out in the visible region where high dispersion is easily available with less expensive instruments. Nowadays lasers are used as they produce greater power with a much greater degree of monochromaticity. The He–Ne laser line at $\lambda = 623.8$ nm is generally used. The blue line becomes more sensitive than the red line in such a laser. The Raman spectrum can be used for fingerprint region identification like the infrared spectrum. It is very useful in structure elucidation. Sample preparation requires careful filtration to remove all traces of suspended matter. Fluorescent impurities must be rigidly excluded if a blue source is used. Temperature control is not required but spectra of cooled samples may give an interesting fine structure. Transformation in energy is considered in terms of Stoke's and antistoke's lines.

26.2 MECHANISM OF INTERACTION

A Raman spectra is shown in Fig. 26.1. Source is Argon laser the compound is CCl_4. We get Stokes (scattered), antistokes and Rayleigh scattering which is most intense and has wavelength same as the source of the radiation. The energy of incident photon is larger than vibrational transition energies because the high energy photons are scattered, giving rise to Raman stokelines but antistokes lines have large energy.

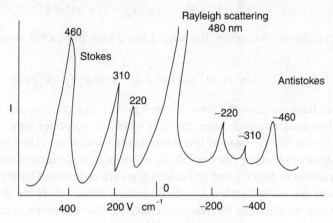

Fig. 26.1 Raman spectra of CCl_4

The difference in the energies of the scattered and incident photons matches vibrational transitions leading to the promotion of the molecules to higher excited vibration state (e.g. IR spectra). Homonuclear diatomic compounds are IR inactive but active in Raman spectra.

Such diatomic molecules may consist of H_2, Cl_2 or X_2. If a band is strong in IR it is invariably weak in Raman and vice versa. Mathematically in the Raman spectra wave number shift $\bar{v}$ which is difference in wavenumber between observed radiation and that the source. In Fig. 26.1 we note stokelines at three points which are smaller than Rayleigh peak while antistokes are at the three points which are greater than wave number of the source. They are less intense. One uses strong stokes lines in Raman spectra interpretation. Only fluorescence can interfere in stoke shifts but not antistokes, so these lines can be then used for analysis. The magnitude of Raman shift (Δr) is independent of excitation wavelength. Raman shifts at longer wavelength are called Stokes shift. We would consider more detailed information on Stokes and antistokes lines.

26.3 STOKE'S AND ANTISTOKE'S LINES

If a molecule M accepts a quantum of energy which is adequate to raise it to an excited state M^*, then the frequency of scattered radiation decreases.

$$V_j^i = \left(V_0 - V_m\right) \qquad \qquad ...(26.1)$$

i.e.,
$$\bar{V}_R = \left(\bar{V}_i - \Delta V\right) \qquad \qquad ...(26.2)$$

Such lines are called Raman Stoke's lines. In the above expression, V_0 or V_i is the incident radiation frequency, V_m is the rotational or vibrational frequency, while V_j^i is the observed scattered radiation frequency. The energy in Stoke's line is given as

$$hv_0 + M \rightarrow M^* + hv_j \text{ with } h(V_0 - V_j) > 0 \qquad \qquad ...(26.3)$$

However, when an excited molecule, M*, gives a quantum of energy to fall back to lower energy state, so that the frequency of the scattered photon V_j^0 is higher than the frequency of the incident radiation, we have

$$V_j^0 = V_j^i + V_m \qquad i.e., \bar{V}_R = \bar{V}_i + \Delta V \text{ if } V_j^i > V_0 \qquad \qquad ...(26.4)$$

Such lines fall on higher frequency Rayleigh lines which are called antistokes lines. We have energy

$$hv_0 + M^{**} \rightarrow hv_j'' + M^* \text{ with } h(V_0 - V_j'') < 0. \qquad \qquad ...(26.5)$$

Thus Stokes lines have wavelengths greater than Rayleigh line wavelengths while antistoke's lines have a wavelength less than Rayleigh lines. In other words if molecules gain rotational energy from photons during a collision, we get lines of low wave number which are called Stoke's lines, while if a molecule loses energy to photons the lines which are on the high wavenumber side are antistoke's lines.

The relative intensities of Stoke's and antistoke's lines are determined by the relative population of electrons in the ground and excited states. As a consequence, Stoke's lines are stronger than antistoke's lines. However, they are weaker than fluorescence emission. The excitation is carried out by radiation at wavelength which is away from the absorption peaks of the analytical sample. Let us see the energy level diagram in Fig. 26.2.

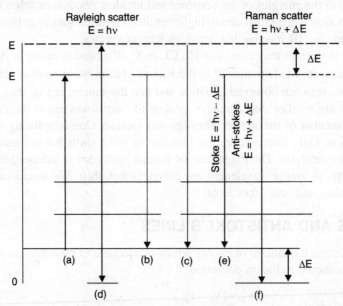

Fig. 26.2 Raman scattering

The line (a) represents energy change in molecule when it interacts with photon from source. Energy increase by hv. The other lines (b) and (c) show types of change occurring on encountering a molecule with photons if it happens to be first vibrational level of ground state. The line (d) depicts the changes that produce Rayleigh scattering, with no loss of energy. Lines (e) and (f) indicate energy changes that generate stokes and antistokes emission. They differ by energy $\pm \Delta E$. The Raman frequency shift and infrared absorption peak frequencies are one and same. The Stokes emission is favoured to antistokes emission. The ratio of stokes and the antistokes will increase with temperature.

26.4 DEPOLARISATION RATIO

It is defined by following equation. $\rho = I_d/I_t$ (26.6). This function has considerable significance. In this equation I_t = intensity of scattered radiation whose plane of polarisation is perpendicular to that of incident radiation while I_d represents the intensity of scattered radiation whose plane of polarisation is parallel to the incident radiation. One can experimentally evaluate 'ρ' by taking reading with polarising pair material inserted between sample and monochromator. Analyser is positioned to pass one type of radiation and then rotated to pass other. In case vibrational motion is symmetrical, $\rho = 0$. If symmetry is decreased by vibration, ratio will be 0.75 of experiment error. ρ is useful for determining shape of molecule and bands with specific vibrational motion. A change in state from trans to gauche form causes change in polarisation of Raman radiation. The polarizibility and polarisation are not synonymous words. The first indicates molecular property while latter describes plane in which radiation vibrates.

26.5 INSTRUMENTATION

One has to consider mainly (*a*) the source of radiation (*b*) sample handling, (*c*) Nature of sample used say liquid or solid, (*d*) spectrometer used.

(*a*) *Source:* Most common source is laser source as it produces a good measure of Raman signal. Most commonly used sources include with range Argon laser (490,520 nm), Krypton laser (530,650 nm), Neon source (630 nm), Diode laser (780 or 830 nm) and Neodymium YAG source (1060 nm). Only Ar and Ky source have practical utility at 488 nm with intense Raman Lines. The last two emitt near IR radiation, producing no fluorescence e.g. Nd/YAG source.

(*b*) *Sample handling:* In comparison to IR it is much simpler as one can use glass or quartz cell. Further laser source can be easily focused. This permits use of small samples either in liquid or solid state. However the most striking feature as compared to IR spectroscopy is one can easily use aqueous sample for measurements.

(*c*) *Nature of Sample:* The sample may be liquid or solid sample. Water is a weak Raman scatter but strong absorber of IR radiation. The aqueous sample is always good for study of inorganic compounds and biological fluids. Solid samples are less common in use e.g. Direct analysis of polymer. The Raman spectra of gel needs special handling. The visible and near IR radiation can be transmitted easily with optical fiber.

(*d*) *Spectrometer:* They are similar to UV-visible spectrophotometer, with double grating system. Transducer is PMT (photomultiplier tube). The new models FT-Raman spectrometers are fitted with germanium transducer. The dispersing spectrometer with fiber optic probes are preferred. Super notch filter eliminates use of double monochromators. For the purpose of detection thermoelectrically cooled photomultiplier tube is used. The FT-Raman instrument use Nd/YAG as source of radiation and Michelson interferometer for wavelength control. For reliable measurement the fluorescent room lights must be eliminated at 1700-3000 cm^{-2}.

26.6 TYPE OF RAMAN RADIATIONS

We have several kinds of Raman radiation as are (*i*) Normal Raman, (*ii*) Resonance Raman (*iii*) FT-Raman (*iv*) SERS Raman radiation which we would now consider.

(i) *Normal Raman:* With monochromator laser beam is focussed to study regions around 2.0 μ. The metals and minerals can be studied around 2917 cms⁻¹. In vibrational mode is used in region on surface where a particular vibrational mode exist. The intensity of normal Raman peak depends upon the polarisibility of the molecule, the intensity of source, and the concentration of the active group. In absence of absorption, Raman emission intensity increase to 4^n power of frequency of source. Such intensities are usually directly proportional to concentration (cf Beer's Lambert's Law). Raman spectroscopy resembles fluorescence spectroscopy with no logarithmic relationship of transmittancy with concentration of the sample.

26.7 RESONANCE RAMAN SCATTERING

(*i*) This refers to the effect in which Raman line intensities are greatly enhanced by excitation with wavelengths which are closer to the electronic absorption peak of a substance. In such cases there is an increase in the Raman peak intensity. This facilitates getting resonance Raman spectra for analytical concentration of 10^{-4}–10^{-5} M. Raman study is confined to concentrations of 0.1 M. Raman bands relate to electronic chromophore which in turn leads to an increase in intensity. An electron goes to an excited state and then relaxation occurs to transfer the electron back to the ground state. Such relaxation occurs in 10^{-14} seconds and as such this phenomenon basically differs from fluorescence effects. Its main applications are in the study of wet biological molecules with a tunable dye laser, e.g. it is possible to ascertain the spin and oxidation state of iron in blood hemoglobin by resonance methods. However, interference by fluorescence is a serious problem.

(*ii*) To prevent thermal decomposition and thermal lensing by laser source it is prevented from irradiating same volume of any significant length of time. Sample must be cooled, to minimise local heating phenomena. Other way is to use flowing liquid sample via peristatic pump. This is phenomena wherein Raman line intensities are increased by excitation with wavelength that closely approaches to electronic absorption peak of the sample. The intensity is jacked up by factor of 10^2 to 10^6. We get spectra at low 10^{-8} M concentration while normal Raman has sensitivity of 10^{-1} M. It has few lines. It is of great use for study of biological samples. Main limitation is interference by fluorescence by analyte sample or impurities present in sample under investigation.

(*iii*) *Surface enhanced Raman Spectroscopy (SERS):* This involves getting Raman spectra on samples which are absorbed on the surface of the colloidal metal particle or on rough surface of metal. Raman lines are obtained of enhanced intensity of 10^3–10^6. Consequently with combination of resonance Raman spectroscopy the detection limit goes down to 10^{-9} to 10^{-12}. One can use colloidal silver or gold particles on glass plate. Thus absorption on metal film and adsorption onto colloids is measured. Latter is good for FT-Raman measurement.

(*iv*) *FT-Raman:* The technique is free of interferences from the fluorescence. Dyes are used at low level. The only drawback is water absorbs in 1 μ region. The advantage of using water solution is eliminated. Polyolefins and morphological studies are made by this method.

(*v*) *Non linear Raman method:* This is possible by laser source of different strength. It is used as coherent antistokes Raman method (CARS). It has better efficiency. Only limitation is that they tend to analyse specific tunable laser. They do not therefore have any widespread applications.

26.8 REMOTE RAMAN SENSING

The light scattering method called as differential absorption i.e. LIDAR (DIAL) is useful to measure the stratospheric molecules in an atmosphere in environmental monitoring. In LIDAR the Rayleigh light (wavelength of incident radiation is not changed) is collected by telescope. The eximer laser supplies pulse of radiation along with it. The return radiation is examined w.r.t. time lapsed and its intensity. It provides concentration vs height profile of the analyte. In DIAL modification of two wavelengths are used of which one is less strongly absorbed. Laser at 308 to 353 nm is generally preferred. DIAL technique is beneficial only for clean atmosphere where entire scattering is only of Rayleigh type. Aerosols scattering is just ignored during measurements, but pollution by gases such as sulphur dioxide leads to variation in particle size concentration. LIDAR was specially designed to differentiate Raman singlet from nitrogen molecules with the resulting scattering at 2331 cm^{-1}. The molecular density decides return of radiation, assuming nothing is contributed by aerosol in the returning radiation with no backscattering. A profile for ozone was carried out in the atmosphere at a distance of 15-50 km. from the surface of the earth.

Apart from remote sensing, Raman spectroscopy is used extensively for the analysis of substances like phosphorus trichloride, turbine fuel and common pharmaceutical products. In turbine fuel for the presence of aromatic compounds narrow peaks are measured but for aliphatic compounds peaks are not very sharp. The fuel additives, antioxidates are recognised by the Raman spectra.

Extensive research is in progress in atmospheric sciences due to increasing interest in study of atmosphere on Mars and Jupiter. Such remote sensing atmospheric investigations were carried out in Lunar sample analysis. It was interesting to note that the remote Raman scattering results almost matched with the results carried out actual Lunar samples in the laboratory.

26.9 ANALOGY OF RAMAN AND INFRARED SPECTROSCOPY

IR is popular because of the availability of a large number of IR spectra of standard materials which is not the case with Raman spectroscopy. For the examination of co-ordination compounds the Raman technique is more useful as it furnishes valuable information in the region of 1000–700 cm^{-1}, especially on metal bonding. Such spectras are easily obtained in aqueous solution by a Raman spectrometer and not by IR spectrophotometer, where interference from moisture is a formidable problem. On account of the availability of cheap instruments this technique should find immense applications in years to come. Raman spectroscopy furnishes similar information to that given by IR or microwave spectroscopy. Unfortunately spectra are somewhat weak with poor resolution. In IR one can go upto a region of 250 cm^{-1}.

26.10 QUANTITATIVE ANALYSIS BY RAMAN SPECTROSCOPY

In the absence of a laser source, quantitative analysis is not simple, but with such a source multicomponent analysis of micro quantities of substances is possible. Mathematically (P_r) i.e., Raman scattered radiation is directly proportional to molar concentration (c) of the scattered constituent in the sample. $P_t = J_c$ if J is the molar intensity of the Raman line. To get good precision an internal standard is used, e.g. in non aqueous samples, CCl_4 is used. On account of their narrow pattern, overlapping of Raman lines is small and base line correction is therefore not needed. The main merit of Raman lines is that they are less complicated than IR-lines. In quantitative analysis the presence of moisture does not pose a serious problem. Raman spectroscopy has been used for the quantitative analysis of mixtures containing benzene and isopropyl benzene with varying number of isopropyl group in benzene. In such analysis carbon tetrachloride is used as the reference solution in order to compare the reference peak which varies in a straight line relationship with volume percentage.

On account of counterfeiting of drugs in the sealed packing in industry many a times it becomes necessary to check such pharmaceutical product from blister packaging specially for the tablets within a short time. FT-Raman helps to sort out this problem. One can measure scattering due to polymer coating which is weak, and fluorescence is almost absent at long wavelength. The spectra of original sample as well as that of drug in blister package more or less is identical. The antihistaminic drugs are tested by FT Raman spectroscopy. Raman spectra is less cluttered with peaks, with no peak overlap. Moisture does not affect spectra with no interferences, and ideal for quantitative analysis but for the cost of expensive laser source and instrument. But availability will be easy if low cost laser diodes are available. Laser microprobes can be used for such purpose. A monochromatic laser beam if focused on object via microscope objective facilitates examination of the sample even at 2μ. The areas are microscopically analysed. When examining surfaces the instrument must be adjusted to provide vibrational mode. One can use quantitative equation for Raman spectra analysis as:

$$I = Kv^4 JC, \qquad \text{Therefore } I \propto C \text{ if } K, V, J \text{ are all constants.} \qquad \dots(26.6)$$

where I = intensity of Raman band, K = constant inclusive of laser power, v = frequency of scattered radiation, J = scattering coefficient and C = concentration. The internal standard method is preferred.

26.11 APPLICATIONS OF RAMAN SPECTROSCOPY

Raman spectroscopy is used for various purposes such as the analysis of inorganic and organic species and it has biological and environmental applications.

    (a) *Inorganic species:* We can use aqueous solutions. Vibration of the metal ligand bond is in the region 700–100 cm^{-1} which cannot ordinarily be studied by IR spectroscopy. Halogen and halogenated compounds have been studied. Spectra of $Sn(OH)_6^{2-}$, $Al(OH)_4^-$ and VO_4^{3-} have been obtained by this technique. The dissociation constants of strong acids have also been derived by Raman spectroscopy. The study of boric acid on dissociation shows species as $B(OH)_4^-$ and not $H_2BO_3^-$.

    (b) *Organic species:* These can be analysed, especially, those involving double bond stretching vibrations such as olefins, cycloparaffin derivatives, and compounds having peaks in the region 1200-700 cm^{-1}. It is also useful for ring size estimation in paraffins and olefinic functional compounds which cannot be analysed by conventional IR-spectroscopy.

    (c) *Biological applications:* The technique can be applied to small samples with less interference from water. Hence it is popular in such analysis. Because peak overlap is low in Raman spectroscopy, mixtures can easily be analysed. As instruments are not attacked by moisture, a small amount of water in the sample to be analysed can be easily neglected.

    (d) *Environmental pollution analysis:* An atmospheric sample can be analysed by Raman spectroscopy. Radiation is scattered from the pollutant molecules and returned to the spectrometer. This permits analysis of pollutants at a low concentration. It can map pollutants as a function of distance. A remote Raman spectrometer permits three-dimensional mapping of air pollutants. The greatest advantage is continuous monitoring of the quality of air. Even lunar samples have been analysed by this method.

    (e) Finally we can cite several examples for the applications as are the analysis of aviation petrol (fuel), PCl$_3$ reactor material and qualitative analysis of pharmaceutical products inside the packing. The fuel additives such as the antioxidants and icing inhibitors (etheyleneglycol derivatives) are easily characterised by Raman technique. In future with easy availability of Laser source like Nd-YAG, Raman spectroscopy will become a common tool in the quantitative analysis of variety of materials and elucidation of structure.

LITERATURE

1. D. Lin-vien, *"Infrared and Raman Characteristics–Raman Frequencies of Organic Molecules"*, Academic Press (1991).

2. F.R. Dolish, W.G. Fateley, F.F. Bentley, *"Characteristic Raman Frequencies of Organic Compounds"*, Wiley, New York (1974).

3. D.J. Gardiner, P.R. Graves, Ed., *"Practical Raman Spectroscopy"*, Springer Verlag (1989).

4. D.A. Long, *"Raman Spectroscopy"*, McGraw Hill Co. (1977).

5. J.G. Grasseli, M.K. Snavely, B.J. Bulkin, *"Chemical applications of Raman Spectroscopy"*, Wiley (1971).

6. K. Nakamoto, *"Infrared and Raman Spectra of Inorganic and Coordination Compounds"*, 4th Ed. Wiley (1986).

7. D.P. Strommen, *"Handbook of Instrumental Techniques for Analytical Chemistry"*, (Ed.) F. Settle, Pearson Indian Edn. (2001), p. 285.

8. P.V. Huong, *"Analytical Raman Spectroscopy"*, Ed. J.G. Grasselli and B.J. Bulkin, Wiley (1991).

9. D.A. Skoog, F.J. Holler, T.A. Nieman, *"Principles of Instrumental Analysis"*, Harcourt Brace Publishers, (1998), p. 429.

Chapter 27

Atomic Absorption Spectroscopy

The first observation of atomic absorption spectra was made by Fraunhöfer while studying dark lines in the solar spectrum. However, in 1955 Alan Walsh from Australia applied the principles of atomic absorption spectroscopy to chemical analysis. In early days, chemists depended on spectrophotometric or spectrographic methods of analysis. Many difficult and time-consuming methods were replaced with the discovery of atomic absorption spectroscopy (AAS). The method is well suited to the analysis of substances at a low concentration, and has several advantages over conventional absorption or emission spectroscopic methods. In the latter, emission depends on the source of excitation. Thermal excitation methods depended upon the temperature of the source. Further, thermal excitation is not always specific, as, in a complex mixture simultaneous excitation of several species is possible. The flame permits the excitation of elements with low excited energy levels. It was necessary to have a reasonable proportion of excited atoms to ground state atoms. Atomic absorption methods rely on measurements connected with ground state atoms and are independent of temperature. These methods are highly specific, hence analysis of a metal from a complex mixture is possible and a high energy source need not be employed.

It is interesting to compare flame photometry with atomic absorption spectroscopy. Elements with low excitation energies can be determined by flame emission while those with high excitation energy are determined by AAS but not by flame photometry. For analysis of elements with resonance spectral lines between 400–800 nm, flame photometry is useful while for those with resonance spectral lines in the range 200–300 nm, AAS methods are more useful. For qualitative analysis, flame photometry is preferable while AAS needs a specific hollow cathode lamp. If AAS is used, a very narrow band pass is necessary. A change in flame temperature will affect the excitation process and hence analysis in flame photometry is variable. AAS is somewhat more expensive than filter flame photometer. Thus flame emission methods and atomic absorption spectroscopy methods are complementary to each other. The greatest advantage of AAS is the analysis of one metal in the presence of another metal.

27.1 PRINCIPLES OF ATOMIC ABSORPTION SPECTROSCOPY

Atomic absorption methods depend upon the absorption of light by atoms. Atoms absorb light at a definite wavelength depending upon the nature of the elements, e.g. sodium absorbs at 589 nm, uranium absorbs at 358.5 nm, and potassium at 766.5 nm. Light at this wavelength has adequate energy to excite to another electronic state. The electronic transition is specific for a particular element. From the ground state an atom is excited to a higher energy state by absorption of energy. There can be several excited states with definite energy. If we take an example of sodium Na (11)–2, 8, 1, or $1s^2$, $2s^2$, $2p^6$, $3s^1$ configuration; the ground state for the valency electron is $3s$ with no energy. The electron on excitation can go to the $3p$ level with energy of 2.2 eV or to $4p$ state with a net energy of 3.6 eV with corresponding absorption wavelength at 589 nm and 330 nm respectively. However, one wavelength is sharply defined with sharp maximum intensity. Atomic spectra are identified by this sharp line which can be distinguished from broadband spectra associated with molecules. The lines arising from the ground state are of utmost importance in atomic absorption spectroscopy. These are called resonance lines. The atomic spectrum peculiar to each element consists of discrete lines including resonance lines. Other lines do not originate from the ground state. Such lines are found in the atomiser of an instrument.

The success of analysis now depends upon the extent of excitation and production of an appropriate resonance line. The temperature of the flame should be extremely high. This can be explained by the Boltzmann distribution equation

$$\frac{N_j}{N_0} = \frac{P_j}{P_0} \exp\left(-\frac{E_j}{KT}\right) \qquad \text{...(27.1)}$$

where N_j and N_0 are the number of atoms in an excited state and the ground state respectively, K is the Boltzmann constant (1.38×10^{-16} ergs/deg), T is the temperature in the absolute or Kelvin scale; E_j is the energy difference in ergs between the excited state and ground state and the quantities P_j and P_0 are statistical factors that are determined by the number of states having equal energy at each quantum level. The fraction of excited atoms in a gas flame is very small. We need close control of temperature for excitation. An increase in temperature increases the efficiency of atomisation. The power of the emitted radiation is proportional to the number of excited atoms according to the relation:

$$P_T = h\nu N_j A_T \qquad \text{...(27.2)}$$

where P_T is total power radiated by the atoms in the flame, h is Planck's constant (6.6×10^{-34} joules/sec) ν is the frequency of peak of the spectral line, A_T is Einstein coefficient (*i.e.* number of transitions each atom undergoes per second) it is approx 10^8 transitions per excited atom. Equation (27.2) is valid for the flame emission method. Table 27.1 gives a comparison of various spectral methods while Table 27.2

Table 27.1 Classification of atomic spectroscopy methods

S.No.	Type	Name	Atomisation	Radiation source	Sample treatment
A	Emission methods	Flame emission	Flame	Sample in flame	Aspirated in flame
		Arc/spark spectroscopy	Arc/spark	Sample in arc/spark	Sample in electrode
		Atomic fluorescence	Flame	Discharge lamp	Aspirated
B	Absorption methods	Flame absorption	Flame	Hollow cathode	Aspirated
		Flameless AAS	Graphite furnace	Hollow cathode	Heated plane
		X-ray absorption	None	X-ray tube	Source beam

gives the temperature of the flame with various fuels with air, oxygen or nitrous oxide as oxidants. This shows the importance of temperature during atomisation.

Since the AAS method (Table 27.2) uses a flame let us consider the type of fuel gases used. Natural gas, propane, butane, hydrogen and acteylene are used as fuels while oxidants used are air, oxygen, nitrous oxide and a mixture of nitrous oxide and acetylene. Table 27.2 lists the maximum temperature of various flames.

Table 27.2 Temperature of flames

Fuels	Air oxidant°C	Oxygen oxidant°C	Nitrous oxide°C
Hydrogen	2100	2780	–
Acetylene	2200	3050	2955
Propane	1950	2800	–

Those metals which are easily vapourised such as Cu, Pb, Zn, and Cd are determined at low temperature. Certain pairs of elements are not easily atomised and hence need a high temperature. Such high temperatures can be attained by using an oxidant in the flame along with the fuel gas, e.g. elements like Al, Ti, Be and rare earths need the use of a oxyacetylene flame or nitrogen oxide, acetylene flame for atomisation while alkaline earths which form refractory oxides need an acetylene–air mixture for atomisation as they are atomised at relatively low temperatures. Table 27.3 indicates importance of temperature for excitation.

Table 27.3 Extent of excitation at different temperatures (K)

Element	Wavelength nm	2000 K	3000 K	4000 K
Cs	852	4×10^{-4}	7×10^{-3}	3×10^{-3}
Na	590	1×10^{-5}	6×10^{-4}	4×10^{-3}
Ca	420	1×10^{-7}	4×10^{-3}	6×10^{-4}
Zn	210	7×10^{-15}	6×10^{-26}	2×10^{-2}

In spite of the best combination it may not be possible to achieve complete atomisation. Hence, nowadays the trend is to use a graphite furnace (Fig. 27.1) which can easily attain temperatures of 2000-3000 K in a few seconds instead of flame. Various optimum conditions for metal to be determined by AAS at various wavelengths are shown in Table 27.4.

As regards the interrelationship between concentration and absorbance, Beer Lambert law is valid if the source of radiation is monochromatic. In AAS, if wavelength of the absorption line is identical to that of the emission line we have a limited choice of lines. For this purpose we must use a narrow slit width, as the peak width is 0.002–0.005 nm. This problem of availability of limited radiation is mitigated by using a source of radiation that emits lines of the same wavelength as the one to be used for absorption analysis. This minimises the broadening effect. Hollow cathode lamp, serves this purpose.

27.2 INSTRUMENTATION FOR ATOMIC ABSORPTION SPECTROMETER

An AAS instrument is composed of the following four components:

(*i*)   Atomisation unit
(*ii*) Source of radiation
(*iii*) Photometric measuring system
(*iv*) Recorder

Atomisation can be carried out either by a flame or furance. Heat energy is utilised to convert the metallic element to atomic dissociated vapour. The temperature should be controlled very carefully to convert it to atomic vapour. At too high or too low temperatures, atoms will be ionised and they will not be absorbed. A typical flame atomiser is shown in Fig. 27.1. Fuel and oxidant gases are fed into a mixing chamber which passes through baffles to the burner head.

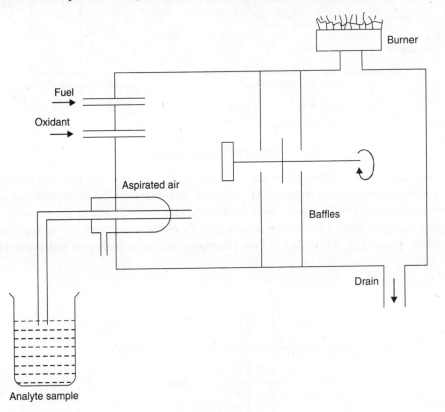

Fig. 27.1 Flame atomiser

A ribbon flame is produced. The sample is aspirated through air into the mixing chamber. Only droplets of a small size pass through the baffles to the burner head. This is not good, however, since if the flame strikes back in the chamber it will cause an explosion, hence a narrow burner is preferred and the gas flow should be carefully adjusted. The maximum temperature attainable is 1200° C. For a high temperature, nitrous oxide is used in the ratio of N : O :: 2 : 1. A flame is not preferred due to interferences and back striking effect. In the case of nonflame atomisers, an electrical heating device is now used. A sample of 1–2μl is placed on a horizontal rod of graphite or a tantalum ribbon. In the graphite furnace, Fig. 27.2 the temperature can be controlled electrically, it is gradually raised in stages, in order to evaporate, dissociate or vapourise the compound.

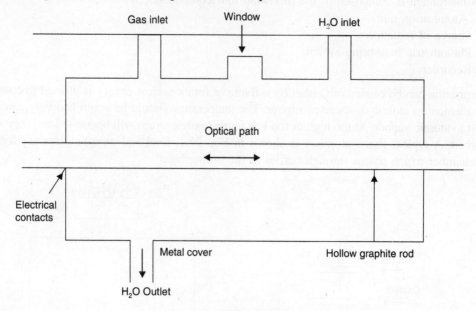

Fig. 27.2 Graphite furnace

Nonflame methods are preferred to the use of a flame, as they are safe. As regards source of radiation, a continuous source is required. The extreme narrowness of the absorption line in the source causes problems. A spectrophotometer with high dispersion optical system is necessary. A series of sources which can give sharp emission lines for a specific element are used. It is called hollow cathode glow discharge lamp (Fig. 27.3). It has two electrodes one is cup-shaped and made of a specific element.

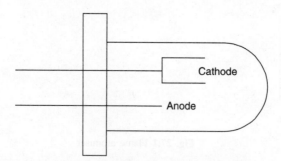

Fig. 27.3 Hollow cathode

The metal used for the cathode is the same as the metal to be analysed. The lamp is filled with noble gas at a low pressure. It will produce a glow discharge with emission from the hollow cathode. Metal atoms are evaporated by sputtering. The atoms accept energy of excitation and emit radiation with lines of the metal. A desired line can be isolated by a narrow band pass monochromator. A typical spectrum of hollow cathode is shown in Fig. (27.4a). The spectra through a monochromator is shown in Fig.

(27.4*b*) and the effect of absorption in the metal vapour is shown in Fig. (27.4*c*). The peak heights are measured as they have absorption and emission lines which nearly equal half width.

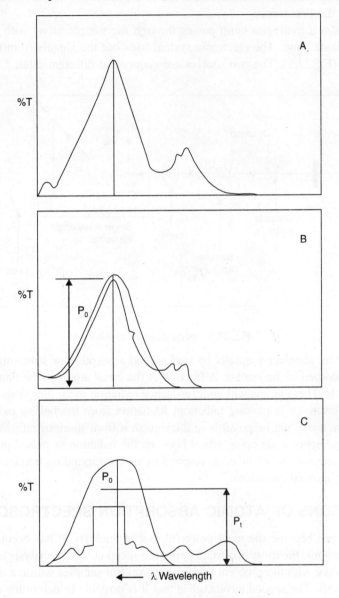

Fig. 27.4 Absorption spectrum

Hollow cathode lamps made out of several elements are available. Such lamps are called multielement lamps, they facilitate determinations without having to change the lamp each time (Ca, Mg, Al), (Fe, Cu, Mn), (Cu, Zn, Pb, Sn) and (Cr, Co, Cu, Fe, Mn and Ni) are several multi-element hollow cathodes but they do not give reproducible results. In Fig. 27.4, (P_0) represents the intensity of incident radiation while (P_t) is the intensity of the transmitted radiation, *i.e.* $\log\left(\dfrac{P_0}{P_t}\right)$ gives the actual absorbance.

As regards instrumentation, flamed and non-flamed or graphite furnace AAS are available. The correction of the background effect involves the use of a continuous source of radiation of hydrogen simultaneously with the line source.

The radiation from a hydrogen lamp passes through the sample along with resonance radiation from the hallow cathode lamp. The electronic system sorts out the signals from the two sources and measures their ratio (Fig.27.5). The two sources are chopped at different rates.

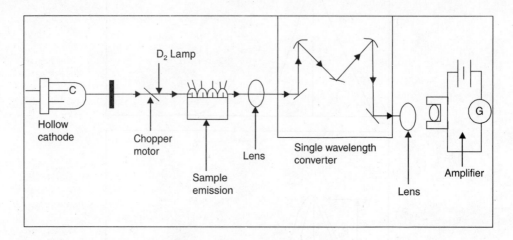

Fig. 27.5 Optical diagram of AAS

Thus the beams are attenuated equally by background absorption or scattering and only the resonance radiation is absorbed in the sample. A fraction of the metal atoms in the flame or vapour will be raised to excited levels. These atoms will emit resonance radiation in all directions at same wavelength at which the monochromator is passing radiation. Radiation from the hollow cathode should not be continuous, otherwise it will not be possible to distinguish it from spurious radiations which are transmitted from the lamp. Hence a chopping wheel between the radiation or pulsed potential is applied to the lamp itself. The detector can be tuned to respond to the corresponding frequency, thus discriminating it from thermally excited emissions.

27.3 APPLICATIONS OF ATOMIC ABSORPTION SPECTROSCOPY

The AAS technique has become the most powerful tool of analysis. It has become popular with the analyst for several reasons, the most important being the speed at which analysis can be carried out. If the analyst is confronted with the problem of analysing several samples within a day. Such analysis is possible only with AAS. The second advantage is that it is possible to determine all elements at trace concentrations. Thirdly, it is not always essential to separate the element in question before analysis because it is possible to determine one element in the *presence* of another, provided we have the requisite hollow cathode for such analysis. AAS can be utilised for the analysis of as many as sixty-seven metals. Typical non-metals which can be analysed are phosphorous and boron. Alkali and alkaline earths can best be determined by flame photometry, *i.e.* emission methods. The elements which can be determined by AAS are listed in Table 27.4. In addition, AAS is used in analysis of elements in a diverse matrix like water. It is also used for isotopic analysis of lithium.

Table 27.4 Optimum conditions for atomic absorption spectroscopic analysis

Elements	Wavelength nm	Type of flame	Sensitivity µg/ml	Working range µg/ml	Detection limit µg/ml
(1)	(2)	(3)	(4)	(5)	(6)
Ag	328.1	AA	0.02	1–5	0.002
Al	309.3	NA	0.75	40–200	0.018
As	193.7	AH	0.60	50–200	0.260
Au	242.8	AA	0.11	5–200	0.009
B	249.8	NA	8.40	400–1600	2.00
Ba	553.6	NA	0.20	10–40	0.020
Be	234.9	NA	0.01	1–5	0.007
Bi	223.1	AA	0.20	10–40	0.046
Ca	422.7	NA	0.01	1–4	0.002
Cd	228.8	AA	0.01	0.5–2	0.0007
Co	240.7	AA	0.05	3–12	0.007
Cr	357.9	AA	0.05	2–8	0.005
Cs	852.1	AP	0.04	5–20	0.004
Cu	324.7	AA	0.04	2–8	0.002
Dy	421.2	NA	0.67	50–200	0.028
Er	400.8	NA	0.46	30–120	0.026
Eu	359.4	NA	0.34	15–60	0.014
Fe	248.3	AA	0.04	2.5–10	0.006
Ga	294.4	AA	0.72	50–200	0.038
Gd	368.4	NA	19.00	900–3600	1.100
Ge	265.2	NA	1.30	70–280	0.110
Hf	307.3	NA	10.00	400–1600	1.400
Hg	253.7	AA	2.20	100–400	0.16
Ho	410.4	NA	0.76	40–160	0.035
In	303.9	AA	0.17	15–60	0.038
Ir	208.9	AA	0.77	40–160	0.400
K	766.5	AP	0.01	0.5–2	0.002
La	550.0	NA	48.00	2500–10000	2.100
Li	670.8	AP	0.01	1–4	0.0002
Lu	336.0	NA	7.90	400–1600	0.040
Mg	285.2	AA	0.003	0.1–0.4	0.0002
Mn	279.5	AA	0.021	1–4	0.002
Mo	313.3	NA	0.28	15–60	0.03
Na	589	AP	0.003	0.15–0.60	0.0002
Nb	334.9	NA	19.00	1000–4000	2.980
Nd	492.5	NA	0.30	350–1400	1.100
Ni	232.0	AA	0.05	3–12	0.008
Os	290.9	NA	1.20	50–200	0.120
Pb	217	AA	0.11	5–20	0.015
Pd	244.8	AA	0.09	4–16	0.016
Pr	495.1	NA	18.00	800–3200	8.300
Pt	265.9	AA	1.20	50–200	0.09
Rb	780	AP	0.03	2–10	2.00

(Contd.)

(1)	(2)	(3)	(4)	(5)	(6)
Re	346.1	NA	9.50	400–1600	0.850
Rh	343.5	AA	0.12	5–25	0.005
Ru	349.9	AA	0.72	30–120	0.087
Sb	217.6	AA	0.29	10–40	0.041
Sc	391.2	NA	0.27	15–60	0.025
Se	196.0	AH	0.47	20–90	0.360
Si	251.6	NA	1.50	70–280	0.200
Sm	429.7	NA	6.60	300–1200	0.075
Sn	224.6	AH	1.00	15–60	0.030
Sr	460.7	NA	0.04	2–10	0.020
Ta	271.5	NA	11.00	500–2000	1.800
Tb	432.7	NA	7.90	500–2000	0.500
Te	214.3	AA	0.26	10–40	0.035
Ti	364.3	NA	1.40	60–240	0.050
Tl	276.8	AA	0.28	10–50	0.013
Tm	371.8	NA	0.27	10–50	0.014
U	358.5	NA	113.00	500–20000	39.000
V	318.5	NA	0.75	40–120	0.050
W	255.5	NA	5.80	250–1000	0.520
Y	410.2	NA	2.30	200–800	0.110
Yb	398.8	NA	0.07	3–12	0.002
Zn	213.9	AA	0.01	0.4–1.6	0.001
Zr	360.1	NA	9.10	400–1600	1.000

Flames: AA – air acetylene; AH – air hydrogen; NA – nitrous oxide acetylene; AP – air propane.

27.4 SENSITIVITY AND DETECTION LIMITS IN ANALYSIS

Atomic absorption spectroscopy is a sensitive means for the determination of several elements (Table 27.4). The two terms, sensitivity and detection limits are commonly used in AAS methods. Sensitivity is defined as that concentration of an element in aqueous solution (µg/ml) which absorbs 1% of the incident radiation intensity passing through a cloud of atoms being determined. A 1% absorbance corresponds to 99% transmittancy or approximately 0.004 absorbance value. Hence another term, detection limit, has been coined. The detection limit is the concentration of an element in solution which gives a signal equal to twice the standard deviation of the series of measurements near blank level or the background signal. Both sensitivity and detection limit vary widely with flame temperature and spectral bandwidth, e.g. on referring to Table 27.4 one can note that for mercury, the sensitivity is 2.2 µg/ml, while the detection limit is 0.16 µg/ml. Interestingly for uranium the values are respectively 113 µg/ml and 39 µg/ml showing a great difference. For this reason it is necessary to mention the flame type as is shown in Table 27.4 as one involving acetylene air (AA), air hydrogen (AH), nitrous oxide acetylene (NA) and air propane (AP). Nonflame atomisation enhances this limit by 10-100 for the ratio of detection and sensitivity limits.

27.5 SPECTRAL INTERFERENCES IN AAS

Interferences can be broadly categorised as involving spectral interferences or chemical interferences. The former are due to the overlap of absorptions of interfering species with the analytical absorption, while in the latter interference results from chemical reactions taking place during atomisation, which in turn change properties of absorption.

Since emission lines of a hollow cathode sources are narrow, interference by atomic spectral lines is uncommon. If two lines are close as for example vanadium at 308.211 nm and aluminium at 308.215 nm, we may have an interference which can be eliminated by selecting another line for Al at, say 309 or 270 nm. The presence of combustion products gives spectral interference. This can be mitigated by using a blank containing the same chemical products and it can be easily corrected with a single beam model. With a double beam model the reference beam does not pass through the flame and poses a problem. A source of absorption or scattering from a sample also creates a positive error in absorbance and hence concentration. As mentioned earlier, scattering by products of atomisation containing refractory oxides of Ti, Zr and W causes spectral interference. This can be avoided by variation of temperature and fuel to oxidant ratio in the flame. One can also employ a background correction. This can be done by the two line correction method, continuous source correction method and the Zeeman effect correction method. Some information is provided at end.

27.6 CHEMICAL INTERFERENCE IN AAS

Chemical interferences are more common than spectral interference. This may be due to formation of compounds of low volatility, dissociation equilibria, ionisation in flames. The anions form compounds of low volatility and reduce the rate of atomisation, e.g. PO_4^{3-} or SO_4^{2-} reduce atomisation of calcium. Cations can also cause such interference, e.g. Al, if present as an impurity in Mg reduces atomisation of the latter element. We can use high temperature flames or releasing agents to eliminate this interference, e.g. addition of Sr or La reduces the PO_4 content. Protective agents like EDTA, Oxine, APPDC (amonium salt of 1-pyrrolidine carbodithioic acid) eliminate interference of several ions like Al^{3+}, Si, PO_4^{3-}, SO_4^{2-} by process of masking.

Table 27.5 Extent of ionisation for alkaline earths

Element	Ionisation energy (eV)	Concentration (µg/ml)	% ionisation	
			AA flame	Non flame
Be	9.32	2×10^{-6}	–	0
Mg	7.64	2×10^{-6}	0	6
Ca	6.11	5×10^{-6}	3	43
Sr	5.69	5.5×10^{-6}	13	84
Ba	5.21	3×10^{-5}	0	88

In this interference with dissociation equilibria, the metallic constituent is converted to elemental form. Ionisation in flames causes serious interferences, especially with flames involving oxygen and nitrous oxide. A noticeable concentration of free electrons exists, e.g. $M \rightleftharpoons M^+ + e^-$ where M is a metal and M^+ a metal ion. Table 27.3 shows the degree of ionization of common elements at different temperatures. Typical cases of Cs, Na, Ca and Zn are considered. The temperature represents approximate conditions that exist in the flame. The degree of ionisation will be influenced by the presence of other ionisable metals in the flame. In the case of alkali metals, increased temperature causes an increase in population of excited atoms as shown in the Boltzmann equation. Counteracting this effect is a decrease in concentration of atoms due to ionisation. Therefore lower excitation temperatures are preferred for alkali metals. A shift in ionisation equilibria can be eliminated by the use of ionisation suppressors providing more electrons to the flame and mitigating ionisation.

Table 27.5 exemplifies the need for free, neutral, ground state atoms as at a higher temperature ionisation causes interference. The table shows the extent of ionisation which is different for different metals. Many transition metals ionise readily in nitrous oxide flames. The percentage of ionisation will change with concentration of an element. This leads to curved calibration graphs during standardisation. The buffering standards and samples with high concentration of easily ionized elements is possible. If the concentration of this element is much greater than the analyte elements and its ionisation potential is lower, complete elimination of ionisation is possible. On account of their low ionisation potential Na, K and Cs are used as ionisation buffers.

However, one must be aware of the exact type of interference in the AAS technique. A chemical reaction taking place in the flame can also give rise to interference. This may be due to incomplete dissociation or formation of refractory compounds, e.g. Ti, Al and V form refractory oxides in the flame. Such interferences can be easily be eliminated by addition of a small quantity of lanthanum salt as it binds Al, Cr, Si and leaves Sr free in the analysis of strontium. In the analysis of Ca, atoms are ionised and hence cannot be analysed. This problem can be solved by adding a more easily ionised element such as Na to keep Ca in a reducible state. Viscosity also makes a difference. With a nonflame atomiser, the problem is more or less similar. Carbon forms refractory compounds, e.g. WC or BC are not decomposible. A coating of graphite with vitreous carbon avoids carbide formation. This also prevents the penetration of samples into porous graphite. Interference due to excessive background emission can be compensated by the use of H_2 or D_2 lamp.

Anions like $C_2O_4^{2-}$, SO_4^{2-}, PO_4^{3-}, interfere in analysis of calcium but one can eliminate such interference by addition of EDTA as a protective agent. As in flame emission methods, we can also use a standard addition method.

The formation of volatile compounds of analyte and loss of analyte in the vapour phase also occurs. The formation of refractory stable carbides, especially in the nonflamed method, decreases the sensitivity and memory effect (i.e. residual signal). The coating of sample cells with pyrolytic graphite avoids carbide formation. Further, the cloud of atoms should be localised in the cell for a longer time to effect complete atomisation and the sample must be vaporised in a constant temperature environment.

Table 27.6 Comparison of limits of analysis by different methods.

Element	Wavelength (nm)	Flame emission (ppm)	AAS Flame Method (ppm)
Al	309	0.01	0.50 (0.03 NF)
Cd	228	0.3	0.004
Cr	357	0.003	0.001 (0.005 NF)
Co	240	0.03	0.003
Cu	324	0.01	0.0005
Fe	248	0.03	0.003 (0.003 NF)

(Contd.)

Pb	217	0.03	0.01
Mg	285	0.003	0.003 (0.00004 NF)
Mo	313	0.03	0.01
Ni	232	0.03	0.003
P	–	1.0	100
Sn	224	0.3	0.02
Ti	364.3	0.5	1.00
W	255.5	4.0	250
Zn	213	3.0	0.0002
Li	670	0.00003	0.005 (0.005 NF)
Na	590	0.0005	0.002 (0.0001 NF)
K	766	0.0005	0.005 (0.0009 NF)
Ca	422	0.005	0.002 (0.0003 NF)

NF = nonflamed AAS.

27.7 ATOMIC LINE BROADENING

The line width in atomic absorption spectroscopy is of great significance. One would prefer narrow lines because they reduce problem of interferences due to overlapping of spectral lines. Several variables influence the atomic lines. Lines are usually distributed very symmetrically around wavelength (λ-max) of maximum absorbance. The line width should be zero but many variables broaden the atomic spectral lines. Line width = $\Delta\lambda_{1/2}$ is defined as width of wavelength units when measured at 1/2 of maximum signal. The line broadening exist due to uncertainty effect or Doppler effect or pressure effects or finally by effect of magnetic and electrical field *i.e.*, Zeeman effect. We would consider all four terms causing line width.

(a) *Uncertainty effect:* Due to uncertainty in transition time of electrons line broadening takes place. It is zero if the life time of two states approach infinity, but ground state electrons have long lifetime while excited atoms have as low as 10^{-7} to 10^{-8} seconds lifetime. Such of the line broadening is also termed as 'natural line width' which are 10^{-4} Å units.

(b) *Doppler broadening:* The wavelength of absorbed or emitted radiation decreases if motion of atoms is forward towards a transducer and it increases if atom is going away from the transducer. This effect is called 'Doppler shift'. It is visible in electromagnetic as well as with sound waves. When a sound generating mobile van approaches a human being, the sound becomes louder and louder, however, it decreases in stages if the sound producing vehicle recedes away from the listener. The magnitude of Doppler shift increases with velocity at which emitting or absorbing species approaches or recedes from the transducer. The relation between Doppler shift or red shift (Δλ) and velocity (v) of the receding galaxy is given by equation as:

$\dfrac{\Delta\lambda}{\lambda_0} = \dfrac{v}{c}$ if λ_0 = wavelength of unshifted or original line, c = velocity of light and $\Delta\lambda$ = Doppler

shift so we have (Δλ) written as

$$\Delta\lambda = \left(\frac{v}{c}\right)\lambda_0 \qquad \qquad ...(27.3)$$

The maximum Doppler shift ($\Delta\lambda$) is shown by atoms with highest velocity of atomic species towards or away from transducer; which is case of sound in human ear. Those moving perpendicular to path of transducer shows no shift in wavelength. Transducer experiences the symmetrical distribution of wavelength with maximum shift corresponding to zero Doppler shift (Fig. 27.6).

In AAS, source of radiation used has many excited atoms which emit radiation while returning back to lower energy level. The radiation originates from excited atoms. The atoms have random motion; such random motion leads to Doppler broadening of the wavelength. The frequency and wavelength of radiation changes on account of the motion of atoms. If (v) is frequency and (λ) is wavelength we have $\lambda v = c/n$ if c = velocity of light. Thus velocity of emitting radiation travelling to the detector increases the observed frequency (v) of radiation. Due to random motion, overlapping of spectral lines are possible.

(c) *Pressure broadening:* It originates from collision of absorbing or emitting species with other ions or atoms in warm atmosphere. They lead to small change in ground state energy levels and hence also change in wavelength. We notice broadening in hollow cathode lamp or discharge lamp due to effect of collisions between two emitting or absorbing atoms. In high pressure xenon or mercury lamp the pressure broadening of this kind is maximum. Also the temperature has pronounced effect on atomic spectra. It is given by Boltzmann equation which we have considered initially, is

$$N_j/N_0 = P_j/P_0 \exp\left(E_j/KT\right) \quad\quad ...(27.4)$$

if N_j and N_0 are numbers of atoms in excited and ground state, $K = 1.28 \times 10^{-23}$ Boltzmann's constant, T = absolute temperature, E_j = energy difference, P_0, P_j are statistical factors. Since atomic absorbance or fluorescence (AAS or AFS) are based upon unexcited atoms, no temperature effect is seen but it directly influences AAS or AFS by increasing atomisation and consequent line broadening. This in turn enhances Doppler effect.

The Doppler effect can be easily depicted by the following diagram.

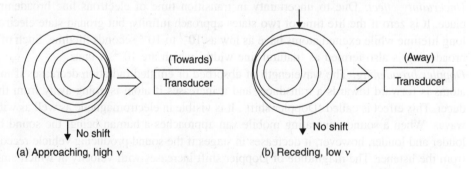

Fig. 27.6 Doppler's effect

(d) *Zeeman's Effect:* A background correction can be given in AAS based upon Zeeman's effect. On exposure of atomic vapour to strong magnetic field (~10 kG) splitting of energy level occurs with the formation of several absorption lines for each electronic transition; where each is separated by 0.01 nm with addition of the absorbances for the energy lines being same as original lines. This phenomena is called as 'Zeeman effect'. It is true for all atomic spectras. Several splitting patterns arise as per electronic transition during absorption in AAS. They lead to formation of π and σ line. The application is based upon differing response of two types of absorption peaks to polarised radiation. Zeeman effect is mainly used for background correction. In instruments magnet surrounds hollow cathode lamp where emission spectrum is split. The Zeeman effect instruments provide a more accurate background correction.

27.8 BACKGROUND CORRECTION

This is also called as nonspecific interferences. This is due to scattering of radiation. Such interferences are large at short wavelength due to scattering. Such background interference can be corrected by measuring absorbance of background. It is then deducted from total absorbance. The background absorbance at nonanalyte line is maximum. It is seen in use of double beam instruments but one can solve this problem. Such correction is not required in visible region for measurements of AAS.

(a) *Zeeman background correction:* It takes advantage of the polarisation of specific spectral line in the magnetic field (*i.e.*, Zeeman effect). On exposure to magnetic field spectral line is splitting to three parts. The central part is *i.e.* (2) π-component. The remaining two are sigma parts (1, 3) (see Figure 27.7). The total integrated area of the σ-component (1, 3) is equal to π-(2) component. The change in energy (ΔE) in presence of magnetic field depends upon the applied field as well as the electronic transitions in spectral lines. Two components can be plane polarised at 90° to each other. π component is parallel to magnetic field and σ-component is at 90° angle to field. Zeeman correction exploits the difference in polarity of the components. Zeeman correction has several merits. The matching of radiative source is not necessary as single lamp is used. Alignment of two optical beam is not required. The spectral interference by narrow lines is absent due to absorbance measurement at wavelength which is same as correction line provided if analyte and interfering line is separated by 0.02 nm such of the correction is effective in reliable measurements.

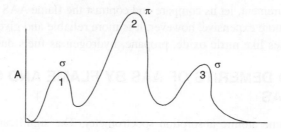

Fig. 27.7 Zeeman effect

(b) *Hieftje correction:* It is also called Smith-Hieftje correction. An advantage is taken of self reversal of hollow cathode lamp operated at high current. Initially absorption of background and analyte is measured with low lamp current. Then absorbance is measured later during short pulse with increased high lamp current. At high current emission peak from lamp is pressure broadened due to increased spluttering. At high current condition self reversal occurs at wavelength of maximum absorbance. The absorbance at high current is due to background. One needs single source. Fairly fine background correction is accomplished when narrow lines are super imposed on broad lines. Such instruments embodying this idea are relatively inexpensive.

It is also called as background correction based upon source self reversal. It is better than Zeeman correction. It is based on self reversal or self absorption behaviour of emitted radiation from lamp. Hollow cathode is operated at high current. It produces nonexcited atoms which can absorb radiation. It also broadens emission band. A band minimum in middle is produced. It is maximum absorption peak (λ max). In this method the lamp has to be successively operated on high and low current. The background absorbance is deducted from total absorbance to give true results.

Thus spectral interference can be present due to atomic line broadening. Such broadening may be due to uncertainty effect or Doppler's effect, or it may also be due to pressure broadening or finally due to Zeeman's effect. In order to get reliable results one has to resort to background correction. AAS instruments are readily available in the market with these corrections. Zeeman correction has certain limitations, however, the Hieftje correction is inexpensive. In relation to these spectral interference the chemical interference due to formation of compounds of low volatility, problems in the dissociation equilibria, or ionisation equilibria can be easily sorted out as discussed earlier.

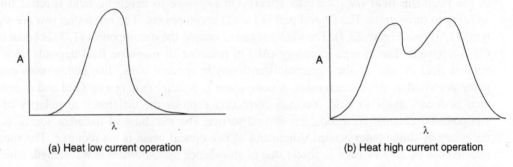

(a) Heat low current operation   (b) Heat high current operation

Fig. 27.8 Hieftje's correction

After the discussion of various kinds of the correction which can be applied before atomic absorption spectroscopic measurement, let us compare and contrast the flame AAS and graphite furnace AAS. As for cost the latter is more expensive, however, it is more reliable and gives reproducible results. The flamed model needs gases like nitric oxide, propane, hydrogen as fuels during analysis.

27.9 MERITS AND DEMERITS OF AAS BY FLAME AND GRAPHITE FURNACE AAS

Let us first consider flame atomic absorption spectrometry. The significant merits are it is a robust method, easy to use, needs little time for analysis and its quite inexpensive. However, the prominent limitations of flamed AAS are one has to use toxic gases as fuel, detection limits are larger (0.1 - 10 µg/L) and hot flame with nitrous oxide acetylene leads to the formation of the refractory oxides of specially transition elements like titanium, zirconium, vanadium.

With graphite furnace, the principal advantages are that microgram samples can be handled, detection limits are 10-100 times lower than flamed model, no gases like propane, hydrogen or acetylene are required for analysis and can be operated by skilled workers. Inspite of these, there are very few restrictions for graphite furnace AAS viz. analysis takes longer time. It may have several interferences as discussed earlier. The instrument maintenance is headache and precision is somewhat lower.

Before concluding this chapter it is worthwhile to compare detection limits of common metals by foresaid technique in Table 27.7.

Table 27.7 Detection limit for AAS (two models)

Element	Wavelength nm	Flame model µg/ml	Graphite ng/ml	Furnace (absolute) ng/ml
Al (III)	310	0.5	0.1	0.0001
Cd (II)	229	0.01	0.2	0.0002
Co (II)	241	0.03	0.3	0.0003
Cr (III)	358	0.03	0.01	0.00001
Fe (III)	248	0.02	0.20	0.0002
Mn (II)	280	0.02	0.001	0.000001
Ni (II)	232	0.05	0.50	0.0005
Pb (II)	217	0.10	0.10	0.0001
Sn (IV)	235	0.10	3.0	0.003
Zn (II)	214	0.01	0.002	0.00002

The perusal of Table 27.7 provides the following significant observations:

(i) For cadmium, cobalt, iron and nickel (tin) the flame atomic absorption spectroscopy limits of detection are quite low (2). While for other elements like aluminium, chromium, manganese, lead and zinc show better sensitivity in the determination of graphite furnace model.

The overall advantages of AAS are that instrumentation is simple and inexpensive. It has low limits of detection (specially for graphite furnace AAS). It provides rapid analysis of multielements needing no previous separations. The main disadvantage is, it is a single element technique, shows multiple interferences and is applicable to liquid samples. No information is provided on chemical form of the analyte sample though it can be used for analysis of metals and metalloids. It is a destructive technique with complex procedure for the sample preparation.

Apart from hyphenated technique (as described elsewhere) one can couple AAS with flow injection analysis. This provides the sample continuously and one can easily analyse as many as 20 samples per hour. The hollow cathodes are available for single element. However, these days hollow cathode for multielements (as many as eight) are available. However, their accuracy sought by multielement hollow cathode is not satisfactory good and at any time single element hollow cathode are preferable. The accuracy of the analysis is 15-20 times the detection limit with accuracy of +1%. The time of analysis may vary from 10 seconds to 2 minutes as it is mainly related to the mode of the sample preparation. As many as 70 metals can be analysed by AAS at level of ppm with little volume of the sample. Future will see use of solids samples. Both the technique have very wide applications in biological, medical, clinical samples analysis. Environmental samples are best analysed by AAS. So also samples from pharmaceutical industry, food industry, metal industry and many more sources are analysed by atomic absorption spectroscopy.

PROBLEMS

27.1 In atomic absorption spectroscopic analysis by cold vapour technique for mercury at 253.7 nm with an air acetylene flame, the following readings were obtained: Amount of mercury ppm (absorbance) are 0(0); 2(0.053); 3.8(0.104); 5.8(0.160); 8(0.220); 9.6(0.260); 11.2(0.310). If the unknown sample shows a concentration (x) with absorbance of 0.179 what is the value of x?

Answer: We should prepare a calibration curve by plotting absorbance (*y*-axis) against amount of mercury (ppm) on the *x*-axis. We get a straight line graph. The concentration corresponding to 0.179 absorbance is 6.4 ppm.

27.2 A sample of hard water was analysed for its calcium content by atomic absorption spectroscopy at 422.7 nm in nitrous oxide acetylene flame. The following readings were obtained. Concentration (absorbance) 0(0); 1(0.090); 3.10(0.273); 3.9(0.340); 5(0.440); and 6(0.525) unknown (0.380). What is the concentration of calcium in the unknown sample?
Answer: As usual prepare a calibration curve by plotting calcium concentration vs absorbance at 423 nm. A straight line graph is obtained.
The concentration corresponding to 0.380 absorbance is 4.3 ppm of calcium.

27.3 A sample of brass (5g) was dissolved in a suitable acid and made upto 500 ml. Atomic absorption analysis of zinc at 213.9 nm with air acetylene flame gave the following readings: Zinc concentration (absorbance): 0(1.0); 5(0.061); 10(0.122); 15(0.185); 20(0.248); 25(0.309); 30(0.370); unknown (0.164). Calculate the concentration of zinc in the unknown sample.

27.4 Soil samples were analysed for the detection of manganese at 279.5 nm in an air–acetylene flame by the atomic absorption method. Manganese (% transmittance) values are: 0(100%); 0.20(88.2%); 0.40(77.6); 0.60(68.2%); 0.80(60.1); 1.0(52.7%); 1.2(46.5); unknown (75%). Calculate the concentration of manganese in the unknown sample of the soil.

27.5 A sample of potash ash gave a reading of 37% T. The second and third solution containing the same quantity of unknown solution plus 40 and 80 µg/ml of added potassium gave a reading of 65% T and 93% T. What is the concentration of potassium in the original unknown sample in µg/ml?

LITERATURE

1. J. Ramirez Munoz, *"Atomic Absorption Spectroscopy"*, Elsevier Publishing Co. (1968).
2. M.D. Amos, *"Basic Atomic Absorption Spectroscopy: A Modern Introduction"*, Varian Techtorn (1975).
3. G.F. Kirkbright, *"Atomic Absorption Spectroscopy And Related Methods"*, Marcel Dekker (1977).
4. G.F. Kirkbright, M. Sargent, *"Atomic Absorption & Fluorescence Spectroscopy"*, Academic Press (1974).
5. A. Varma, *"CRC Handbook of Furnace Atomic Absorption Spectroscopy"*, Boca Raton, FL, CRC Press (1994).
6. D.L. Tsalev, *"Hydride Generation: A Atomic Absorption Spectrometry"*, John Wiley (1995).
7. G.D. Christian, *"Analytical Chemistry, 5th Edition"*, John Wiley (1994).
8. D.A. Skoog, L.J. Leoray, *"Principles of Instrumental Analysis"*, Harper Braun (1994).

Chapter 28

Atomic Emission Spectroscopy

A tomic emission spectroscopy pertains to electronic transitions in atoms which use an excitation source like flames, arcs or sparks and argon plasma sources. Such methods are widely applicable and are specific and sensitive. They need minimum sample preparation, as the sample can be directly introduced into the arc or spark. Inter-element interference at higher excitation temperatures is marginal. Spectra can be taken simultaneously for more than two elements. The limitations are: that such recording is done on a photographic plate which needs some time to develop, print and interpret the results; the radiation intensities are not always reproducible; and relative error exceeds 1-2%. However, the greatest advantage of the method is that it permits non-destructive analysis.

28.1 PRINCIPLES OF EMISSION SPECTROSCOPY

The source of excitation influences the form and intensity of emission in such measurements. The source provides sufficient energy to vaporise the sample and secondly it causes electronic excitation of the elementary particles in the gas. Though we can get band and line spectra, the latter are useful for emission spectroscopic analysis. Excited molecules in the gas phase emit band spectra. Thus a molecule in an excited state of energy (E_2) undergoes a transition to a state of lower energy (E_1) and a photon of energy $h\nu$ is emitted where

$$h\nu = E_2 - E_1 \qquad \qquad ...(28.1)$$

However, in each electronic state a molecule may exist in a number of vibrational and rotational sub states of different energies so radiation from excited molecules comprises a number of frequencies which are grouped into bands; each band matches a transition from one excited state to another electronic state of lower energy. Finally, excited atoms or monoatomic ions in the gaseous state emit line spectra. As vibrational and rotational fine structure is absent from spectra of monoatomic species, so the emission spectrum is composed of a series of individual frequencies matching transitions between various electronic energy levels. A spectral line has a natural width. The emission, absorption or fluorescence spectra of atomic particles consist of well defined narrow lines due to electronic transitions of the outermost electrons. It is possible to consider such transitions for a simple element (e.g., sodium). From consideration of its energy level diagram one can predict the number of lines encountered.

The measurements in emission spectroscopy are feasible because each atom has a definite quantum of energy in which electrons reside. They are in the ground state with the lowest energy. On addition of energy by thermal or electrical means, one or more electrons are removed to a higher energy state.

However, these excited electrons return to the ground state and in this process extra energy in the form of a photon of radiant energy is emitted This results in definite transitions. If the excitation energy is large then the emitted energy is also large, which results in several lines for measurement. Thus the intensity of the spectral line is dependent upon transitions taking place in an atom. Self absorption sometimes decreases the intensity of such a transition. The use of a high energy source is not always helpful as it ionises the gas with the loss of one or more electrons. The spectrum of an ionised ion is radically different from that of a neutral atom. A spectrum of a singly ionised ion, e.g. Mg (I), will be similar to a neutral atom like Na with atomic number less by one unit (Fig. 28.1). The types of emission are depicted in this energy level diagram.

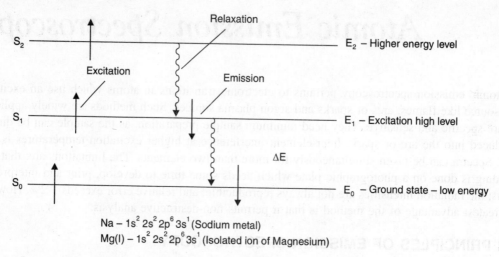

$Na - 1s^2\,2s^2\,2p^6\,3s^1$ (Sodium metal)
$Mg(I) - 1s^2\,2s^2\,2p^6\,3s^1$ (Isolated ion of Magnesium)

Fig. 28.1 Atomic emission phenomena

28.2 SOURCE OF EXCITATIONS

There are various sources for excitation of electrons. They inclued the flame, alternating current arc (AC arc), direct current arc (DC arc) and alternating current spark (AC spark). Each method involves introduction of a sample into the source in a vaporised form leading to excitation of electrons to higher energy levels. Flame excitation will be discussed separately under flame photometry. The DC arc is generated with a potential gradient of 50-300 volts; arc temperature is 4000-8000 K; emission is due to neutral atoms. The current used is about 1-300 amps (Fig. 28.2). A DC arc is struck between carbon and graphite electrodes; sometimes flickering of the arc is observed. Selective vaporisation may also occur. It is useful as a sensitive source to detect very small concentrations of material. An attachment (*i.e.* stallwood) to DC enhances the sensitivity. The AC arc (Fig. 28.3) employs a potential difference of 1000 volts or more. The arc is drawn at a distance of 0.5-3 mm. For reproducible results the separation distance between the two electrodes, potential, and the current must be properly controlled. The AC arc is steadier than the DC arc. The AC spark gives higher excitation energies but poor reproducibility. High voltage transfer 10-50 kV across two electrodes which gives a spark; use of a condenser increases the current. The spark is preferred to an arc when high precision rather than high sensitivity is required. A spark usually excites ionic spectra. It is reproducible and stable; less material is generally consumed; high concentration of solution can be employed; the heating effect is less which is useful for analysis of low melting materials with a low melting point and there is no interference from ions like

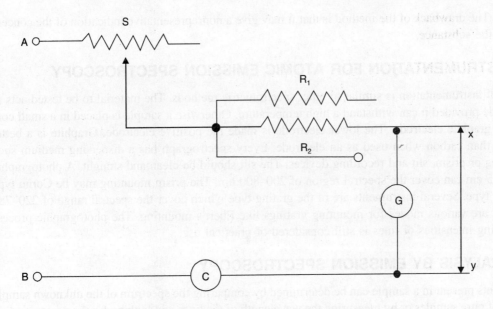

A–B = Potential difference 300 V; x-y = Gap for the arc;
S = Reactor; G = Galvanometer; C = Ammeter;
R_1, R_2 = Variable resistance

Fig. 28.2 DC arc circuit

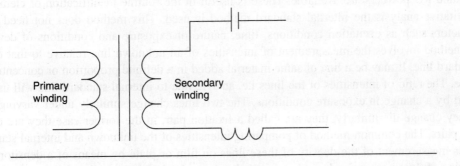

Fig. 28.3 AC arc circuit

cyanogen. The drawback of the method is that it may give a nonrepresentative indication of the concentration of the substance.

28.3 INSTRUMENTATION FOR ATOMIC EMISSION SPECTROSCOPY

The overall instrumentation is similar to other spectrometric methods. The material to be tested acts as an electrode provided it can withstand a high temperature. Otherwise a sample is placed in a small core carbon or graphite electrode. The lower electrode is made the positive electrode. Graphite is a better conductor than carbon when used as an electrode. Every spectrograph has a dispersing medium such as a grating or prism, slit and recording devices. The slit should be clean and straight. A photographic plate of 7.5 cm can cover the spectral region of 200–800 nm. The prism mounting may be Cornu type or Littrow type. Several instruments are of the grating type which cover the spectral range of 220–780 nm. There are various modes for mounting gratings like Ebert's mounting. The photographic process for recording intensities of lines is still considered of practical use.

28.4 ANALYSIS BY EMISSION SPECTROSCOPY

The elements present in a sample can be determined by comparing the spectrum of the unknown sample with that of pure samples or by measuring the wavelength of the lines and looking for the corresponding element in the tables. If three or more lines of the element are identified it is quite adequate for identification. RU lines and RU powders are those lines of each element which are the last lines to disappear as the concentration of the element is gradually decreased. These are persistent lines. Such lines are useful in detecting small concentrations. Powders of 50 elements showing RU (i.e. rares ultimates) called RU powders are available. These is useful in the routine identification of elements.

In quantitative analysis the internal standard method is used. This method does not need close control of factors such as excitation conditions, time, nature of exposure and conditions of development. This method involves the measurement of intensities of an unknown line relative to that of the internal standard line. It may be a line of same material added in a definite proportion or concentration to the sample. The ratio of intensities of the lines i.e. analyte line to internal standard line will usually be unaffected by a change in exposure conditions, The two lines change similarly under varying conditions; if they change dissimilarly, they are called a fixation pair, in the former case they are called homologous pairs. The common method of comparing intensities of the unknown and internal standard lines involves measurement of the density of these lines on film or plate by means of a densitometer. A plot of ratios of densities versus log of concentration is useful for analysis.

Finally let us consider what is meant by the log and step sector methods. These methods are now outdated. It involves running a log or step sector in front of the slit during exposure of emitting radiations so that the resulting lines are of different lengths. Thus, the stronger line is longer and weaker line is shorter in length due to less exposure. Thus if C = concentration, D = density, P = intensity of particular line, h = height of line image, then we have $\log C \propto \log P$. If sector is logarithm $h \propto \log P$ because the height of the line is proportional to exposure intensity between $h \propto \log C$. We plot a graph of difference in height of internal and unknown standard vs. log concentration which gives a straight line graph.

28.5 FLAME EMISSION SPECTROSCOPY

Flame emission or flame photometry involves excitation by spraying the sample into a flame. Radiation from the flame enters a dispersing device to isolate the desired region of the spectrum. The intensity of the isolated radiation is then measured. The use of a solution spray permits uniform distribution of the sample throughout the body of the flame and difficulties associated with an arc or spark are avoided.

A flame photometer is composed of a pressure regulator and flow meter for fuel gases, an atomiser, a burner, optical system, photosensitive detector and recording output of the detector (Fig. 28.4).

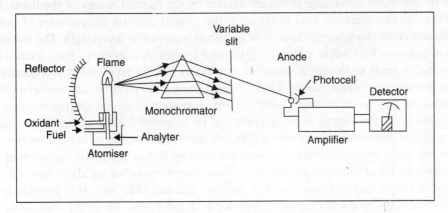

Fig. 28.4 Flame photometry assembly

(i) *Accessories:* The pressure regulator and flowmeter, are used for the proper adjustment of pressure and flow of gases. A 10 lb gauge for fuel and 25 lb gauge for oxygen are needed. Double diaphragm and needle valves are used to control pressure. To control gas flow, a rotameter should be inserted in the gas line. A flow rate of 2–10 ft hr is best for getting good results.

(ii) *Atomisers:* The atomizer is used to introduce a liquid sample into the flame at stable and reproducible rates. They are classified as those which introduce the spray into a condensing chamber for removing large droplets and those which introduce the spray directly into the flame. One needs about 4–25 ml of sample per minute of which 5% reaches the flame. In another method, a slurry of powder material in isopropanol is employed for spraying. Glycerine acts as solvent.

(iii) *Flame:* The burner should produce a steady flame. A Meker burner is good for working at low temperature. A deep metal grid across the mouth of the burner prevents the flame striking back. A combination of burner and aspirator introduces the sample directly into the flames and gives a uniform flame.

(iv) *Monochromator:* The optical system functions as a collector and monochromator of light and focuses it on a photosensitive detector. The light is focussed on the entire detector by a concave mirror by adjustment of the flame. Use of absorption or interference filters isolate the characteristic radiations used, but better isolation can be done by a monochromator. A good slit with a narrow opening is necessary.

(v) *Detector:* Photosensitive detectors such as a barrier layer cell are not useful because the response is not amplifiable. However, filter flame photometers use only these devices but temperature should be uniformly controlled. Flame photometers that restrict the bandwidth of the radiant energy reaching the detectors employ phototubes and amplifier units to boost the output. However, photo electron multiplier tubes are best.

28.6 PRINCIPLES OF FLAME PHOTOMETRY

The flames transform the solid or liquid into the vapour state and decomposes it to simpler molecules or atoms. They finally excite these electrons to higher energy states and return them to the ground state by emission of radiation. On dispersing material in a flame, water or solvent is evaporated and dry salt is left in the flame. On further heating at a higher temperature the dry salt is vaporised and the molecule is dissociated to neutral atoms which are responsible for emission phenomena. The vapours of netural metal atoms or molecules containing atoms are excited by the thermal energy of the flame. From the excited levels of atoms, electrons tend to return to the ground state by the emission of radiation. A particular element emits characteristic spectra of its own at a particular wavelength. The line spectra are obtained from atoms or ions while molecules give band spectra. A continuous band formation is also possible. Excitation leads to raising of electrons to a higher energy level and is accompanied by the return of electrons to the ground state by loss of radiant energy. An electron may not return directly to the ground state but may return in stages leading to the generation of several spectral lines e.g. E_2 to E_1 to E_0 *i.e.* ground state. An energy level diagram can be prepared for all atoms. Such energy level diagrams are simple to construct for mono or diatomic molecules like Na or Mg but become complex for transition and main group elements. The most outstanding line is due to the transition of electrons between the lowest excited state and ground state. These transitions occur for alkali metals like lithium (671 nm), sodium (590 nm), potassium (766.5 nm) and calcium (423 nm). It is possible to regulate energy level transitions by controlling the temperature of the flame. We prefer transitions from the lowest energy state of an excited ion or atom to the ground state. Neutral atoms emit characteristic sharp lines but for second group s-block elements lines are possible from ionised atoms at a higher tempera-ture. The spectrum of an ion is not the same as that for the neutral atom. In such a case the spectrum is similar to one emitted by the element of the preceding atomic number, e.g. for the Al ion the spectrum will resemble that of Mg element. Molecules give band spectra as they have energy of internal rotation and vibration and electronic excitation. This leads to distribution of excited radiation in a spectrum in the form of bands instead of lines.

The hydrogen flame gives the best metal to background signal. The measurement of intensities of spectral lines is dependent upon the amount of salt impregnated in the flame, the amount of salt dissociated, the extent it is ionised, the number of atoms excited, the changes of transition from the excited to ground state and self absorption. After dissociation the variation in emission intensity with temperature is regulated by excitation. Ionisation diminishes the concentration of the neutral atom in the flame leading to a decrease in emission intensity. The energy needed for the dissociation of metal into atoms is closer to the ionisation potential or ionisation energy of a metal atom.

There are many factors contributing to variation in emission intensity. One such variation is due to the formation of metal hydroxides of alkali metals in the flame. Oxyacetylene provides an atmosphere more favourable for the existence of free atoms of those elements which can form stable monoxide molecules. The material is dissolved in a hydrocarbon solvent and the emission intensity is increased by use of an organic solvent. A number of organic solvent–water mixtures are used. Solvent extraction techniques can be exploited to achieve analytical separation and then the organic phase can be aspirated directly into the flame to take advantage of enhancement of emission intensity. Simultaneous extraction and direct flame photometry gives excellent results.

28.7 INTERFERENCES IN FLAME PHOTOMETRY

There are various kinds of interference such as spectral interference, background emission, self absorption, ionisation, interference due to anions and solution properties. Spectral interference is encountered while isolating the desired radiant energy. Such interference is eliminated with monochromators. However, if interference cannot be totally obviated by increased resolution it can be eliminated by alternatively selecting other spectral lines or by prior removal of foreign elements by solvent extraction. Background emission may be due to certain constituents present in the flame which can be eliminated by using a blank solution. In self absorption, the process of excitation is followed by the loss of a discrete amount of energy in the form of radiation as the electron falls back to its original position or to a lower energy level. If some of the radiant energy is self absorbed, the strength of the spectral line is weakened. The effect is more pronounced for resonance lines particularly those arising from the lowest excited level. At low concentrations such phenomena are insignificant. Ionisation causes serious interference. An oxygen gas flame possesses sufficient energy to ionise the alkali and alkaline earth metals. The influence of other ions, if large amounts of salts and acids are present, the emission intensity is lowered (< 0.1 M) H_2SO_4, H_3PO_4 and HNO_3 show a marked effect in lowering metallic emission. The use of a releasing agent or protective chelating agents circumvents this type of interference. Solution properties like vapour pressure and surface tension influences droplet size, use of LiCl eliminates this effect. Non ionic surfactants are also useful in increasing the surface tension of the solution.

28.8 EVALUATION METHODS IN FLAME PHOTOMETRY

There are two methods, one is the standard addition method while the other is called the internal standard method.

$(L_1 - H) a = K X$ found $\qquad$ If L_1 emission of unknown a, $\qquad$ ...(28.2)

$(L_2 - H) b = K (X+S)$ $\qquad$ L_2 emission with adding standard b $\qquad$ ...(28.3)

$(L_2 - L_1) = KS$ found $\qquad$ H background emission $\qquad$ ...(28.4)

X found $\dfrac{S. \text{ added}}{S. \text{ found}} = X$ actual present $\qquad$ ...(28.5)

S = amount of standard added

X = conc. of unknown compound.

The net emission readings are obtained on two solutions: first, containing an aliquot of the unknown solution and second containing the same quantity of unknown solution plus a measured amount of standard solution of the element. The quantity of test element in each of these solutions is then determined from the measurement of emission intensities and by constructing the standard calibration curve (Fig. 28.5).

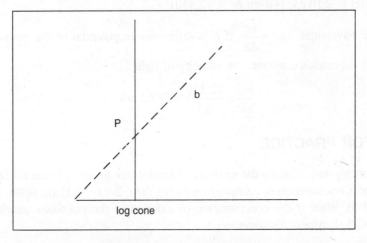

Fig. 28.5 Standard addition method

In the internal standard method, a fixed quantity of the internal standard element is added to samples and standard solution. The radiant energy of emission is measured for two solutions. The ratio of emission intensity of analyte line to that of internal standard line is plotted against concentration of analysis on a log–log graph. The log of emission vs log of C will give straight line with intercept on *y*-axis. Li is used as the internal standard (Fig. 28.5). This eliminates several interferences.

PROBLEMS

28.1 A flame photometric analysis of lithium was carried out with sodium as the internal standard. The readings were:

Li in ppm	:	1.0	2.0	5.0	10.0	20.0	50	(unknown)
Li intensity	:	10.0	15.3	22.2	35.4	56.4	77.5	38.5
Na intensity	:	10.0	10.5	9.5	10.0	11.0	10.0	10.5

Calculate the concentration of lithium if an unknown sample gave readings of 38.5 and 10.5 for sodium and lithium intensity.

Answer: A log of (Li/Na) is plotted against log [Li]. We thus have new table for logs :

log (Li/Na)	:	1.0	0.163	0.368	0.549	0.709	0.889	0.564
log (Li)	:	0	0.30	0.69	1.0	1.30	1.69	log [x]

By plotting these values on log-log paper corresponding to 0.564 we get 1.08, *i.e.* antilog 1.08 =12.02 ppm of Li.

28.2 In a flame photometric determination of potassium with standard addition method the following observations were made: [K] (intensity), *i.e.* 0(18); 1.0(19.5); 2(21.0); 5.0(25.5); 10.0(33.0), 20.0(48.0); 50(93.0). For the unknown sample, the reading is 18.0. What is its concentration? *Answer:* Here the blank reads 18.0. The addition of K in an amount equal to the unknown should increase the intensity by 18.0 units, *i.e.* for 36 units. If we plot graph of [K] against intensity reading for 36 intensity unit we shall have about 12 ppm of potassium. Hence 12.0 ppm is the answer.

28.3 Calculate the wavelength of the resonance line of the sodium atom if the excitation energy of the resonance level is 2.10 eV. (Given *hc* = 12,330).

Answer: The wavelength $(\lambda) = \dfrac{hc}{\Delta E}$ if *E* is difference in potential of the ground and excited of state in eV, *h* = Planck's constant, *c* = velocity of light

$$\therefore \qquad \frac{hc}{\Delta E} = \frac{12,330}{2.1} = 587.2 \text{ nm}$$

PROBLEMS FOR PRACTICE

28.4 Flame photometry was used for the analysis of hard water for its calcium content. The readings obtained were: Concentration of calcium (intensity): 0(4); 2(9.6); 4(15.0); 6(20); 8(26.5); 10(31.9) unknown (11.5). What is the concentration of calcium in the unknown sample? Express your results in $CaCO_3$ hardness for water.

28.5 From a sample of milk, strontium was determined by flame emission methods. The results obtained for calibration were: 0(4.5); 4(21); 8(38.5); 12(56.5); 16(73); 20(90); unknown (43.5).

What is the concentration of strontium in the unknown sample?

28.6 A sample of stainless steel is to be analysed for its chromium content by emission spectroscopy. Three standard samples containing 4.7×10^{-2}, 1.04×10^{-1} and 2.54×10^{-1} chromium gave corresponding intensity ratios of I_{Cr}/I_{Fe} of 1.182, 0.531 and 1.980 respectively. If an unknown sample gives a ratio of 0.685 what is the concentration of chromium in the unknown sample of stainless steel?

28.7 From plant matter, zinc is estimated with cadmium as the internal standard. With log sector attachment heights of various lines were measured on photoplates. The data obtained were as follows:

% Zn	:	1×10^{-3}	1×10^{-2}	1×10^{-1}	1.0	Unknown
Zn ht (mm)	:	7.6	9.0	9.9	10.7	9.0
Cd ht (mm)	:	4.2	4.5	4.4	9.0	4.3

What is the concentration of zinc in the unknown sample?

28.8 Between iron ($\lambda_1 = 304.2$ nm) and ($\lambda_2 = 304.508$ nm) there is one line. What is the wavelength of this line (λ_x) if on screen it projects 1.5 mm away from the first line and 2.3 mm away from the second line?

(*Hint:* Use relation $\lambda_2 = \lambda_1 + (\lambda_2 - \lambda_1)\dfrac{I_x}{I}$ where I_x = distance between first iron and unknown line and I = distance between known line and the second line.)

LITERATURE

1. W.West, "*Chemical Applications of Spectroscopy*", Wiley Interscience, New York (1956).
2. J.A. Dean, "*Flame Photometry*", McGraw-Hill, New York (1960).
3. H.H. Willard, L.L. Meerrit, J.A. Dean, F. Stross " *Instrumental Methods of Analysis*", Van Nostrand (2001) 2nd Ed.
4. G.W. Ewing, "*Instrumental Methods of Chemical Analysis*", 4th Ed., McGraw Hill (1978).
5. D.A. Skoog, D.M. West, "*Principles of Instrumental Analysis*", 2nd Ed., Holt Saunders (1980).
6. R.M. Barnes, "*Emission Spectroscopy*", Stroudsburg Hutchinsond Ross (1977).
7. J.A. Dean, T.C. Rains, "*Flame Emission & Atomic Absorption Spectrometry*", Marcel Dekker (1969).

Chapter 29

Inductively Coupled Plasma–
Atomic Emission Spectroscopy

It is amazing to note that atomic emission spectroscopy has not become as popular in analytical chemistry in comparison to atomic or molecular absorption spectroscopy. One exception is flame emission spectroscopy. Flame photometry is commonly used along with the more usual UV visible and IR spectroscopic methods. The sensitivity of emission methods is much better than that of absorption methods. One reason for this anomaly is the lack of easily available instruments. Of late, atomic absorption spectroscopy has become common only because simple and cheap instruments are now available. Emission spectroscopy by and large remains a tool for qualitative identification of elements with limited applications in quantitative analysis. AC arc or spark emission spectroscopy or DC arc spectroscopy are available. However, flame emission spectroscopy has remained a widely accepted tool of analysis. Unfortunately, the technique has its own limitations.

29.1 LIMITATIONS OF FLAME EMISSION SPECTROSCOPY

This technique has by and large been used for the analysis of alkali and alkaline earths. The primary reasons are that the atomisation temperatures or the s-block metals is comparatively low (say less than < 2500°C) and the elements can be easily measured in the visible region e.g. sodium at 589 nm, caesium at 840 nm and lithium at 620 nm. It is quite easy to have interference filters which can be used as monochromators in flame photometers, thereby rendering them relatively inexpensive. However, this technique suffers from one serious drawback. The volume of sample which is actually atomised and one which reaches the flame is just 15% while 85% of the sample is lost during atomisation. This factor is of great importance when one is handling small volumes of samples during analysis. The flame photometer permits analysis of one sample at a time while other methods like AAS not only allow multiple samples but also facilitate multielement analysis. Very few flame photometers have been provided with a recording facility and most of the time the analyst has to rely on manual measurements. The extent of ionisation depends upon the flame temperature, as a knowledge of the ionisation potential facilitates quick prediction of atomisation temperature.

It will be observed from Table 29.1 that good ionisation occurs at temperatures greater than 3000°C. The temperature of the flame varies. For instance with air as the oxidant and hydrogen as the fuel the temperature is 2100°C while with air and acetylene the temperature is 2200°C, with air and propane 1950°C and with oxygen as oxidant with hydrogen as fuel gas it is 2700°C but with air-acetylene

$3050°C$ and with air-propane it is $2800°C$. As shown in Table 29.1 the extent of excitation of an atom also varies with temperature. For instance for calcium it is 6×10^4 % of ionisation at 4000 K. Thus temperature plays a significant role in atomisation. Most flame photometers are provided with liquid propane (LPG) gas as the fuel with air as oxidant which cannot attain a temperature more than $2500°C$, limiting the analysis of several transition elements. Hence it was essential to look for alternative source where sample loss is minimum and temperature of atomisation is very high, permitting complete atomisation of the element to be analysed.

Table 29.1 Degree of ionisation

Element	Ionisation potential eV	% Ionisation at		Concentration μg/ml.
		2000°C	3000°C	
Be	9.32	—	0	2.0
Mg	7.64	0	6	2.0
Ca	6.11	3	43	5.0
Sr	5.69	13	84	5.5
Ba	5.21	0	88	30.0

29.2 PRINCIPLES OF PLASMA SPECTROSCOPY

Let us consider the process of atomisation (Fig. 29.1). In actual practice one observes that an aqueous solution when put in a flame gets desolvated and converts it into the solid form. This in turn gets vaporised which on subsequent heating gets atomised. An electron gets excited and moves to the next orbit having higher energy during excitation. However, since this electron cannot remain in the excited state all the time, it releases the energy when it returns to the ground state by emitting energy.

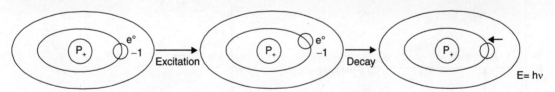

Fig. 29.1 Energy level for electrons

The whole process can be expressed by equations:

1. $M(H_2O)_n^+ X^- \xrightarrow{\text{desolvation}} (MX)_n \text{ (solid)}$

2. $(MX)_n \xrightarrow{\text{vaporisation}} MX. \text{ (gas)}$

3. $MX \xrightarrow{\text{atomisation}} M \text{ (elemental)}$

4. $M \xrightarrow{\text{ionisation}} M^+ \text{ (ion formation)}$

5. $e_{-1}^0 + M^+ \xrightarrow{\text{excited}} M^+ \text{ (additional electron)}$

6. $M^+ \xrightarrow{\text{decay}} e_{-1}^0 + M^+ \xrightarrow{h\nu} \text{[atomic emission]}$

The above transformation can be depicted in energy level diagrams as shown in Fig. 29.2.

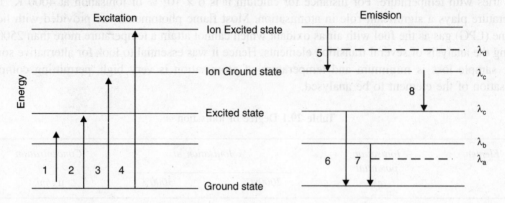

Fig. 29.2 Energy level diagram of atomic transitions

1, 2 excitation	5 – ion emission
3 – ionisation	6, 7, 8, – atom emission
4 – ionisation/excitation	

Such a transition is given by the equation

$$E = h\nu \text{ or } E = hc/\lambda$$

where, as described earlier E = energy, h = Planck's constant, ν = frequency, λ = wavelength and c = velocity of light. In Fig. 29.2 the difference in energy for (5) is large in comparison to (6), hence wavelength for (5) is short (λ_d) as compared to one for (6) λ_a. During such multiple transitions we prefer the first resonance line for measurement.

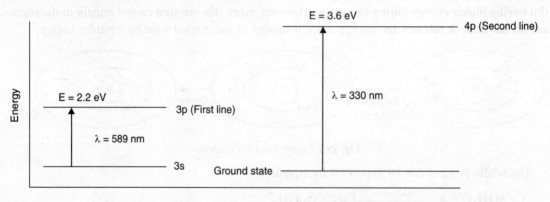

Fig. 29.3 Energy level diagram for sodium

Figure 29.3 gives the energy level diagram for sodium with atomic number 23 and electronic configuration as $1s^2\ 2s^2\ 2p^6$. During the process of excitation an electron from the $3s$-orbit goes to $3p$ by absorbing energy of 2.2eV but when it emits to ground state, it does at λ = 589 nm. The electron on further excitation can go to the $4p$-level and return to the 3s ground state at λ = 330 nm, but this second wavelength is not preferred for analysis. This property makes atomic spectroscopy useful for specific analysis of elements.

29.3 PROCESS OF ATOMISATION AND EXCITATION

In AES a sample is subjected to a temperature high enough not only to cause dissociation into atoms but also to cause significant collisional excitation (and ionisation) of the sample atoms. Such excited atoms or ions decay to lower states through thermal or radiative emission or energy transitions and the intensity of light emitted at a particular wavelength is measured.

For the purpose of atomisation and subsequent excitation, three kinds of thermal sources are used to dissociate sample molecules into free atoms. They are (*a*) flames, (*b*) furnaces and (*c*) electric discharge. In flame photometry we have seen the use of a flame as the source of excitation. In atomic absorption spectroscopy we use both flames and graphite furnaces as the source of atomisation. Unfortunately flames or furnaces are not always sufficiently hot enough to dissociate many molecule into free atoms e.g. refractory carbides or oxides remain in molecules at 4000 K. Most free atoms are in the ground state especially for the p-block elements. One can easily use atomic absorption spectroscopy for analysis. The simplest example is that of alkali and alkaline earths. In the AES technique an arc or spark is used for atomisation. In such electrical discharges the arc is created by an applied potential across electrodes to produce a temperature of 7500 K. However, often such electrical discharges do not give the reproducible results. It is therefore necessary to go in for a plasma source.

29.4 PLASMA AS AN EXCITATION SOURCE

A plasma contains a remarkable fraction of electrons and positive ions in equal numbers along with neutral molecules. Such plasmas can conduct electricity and are affected by magnetic fields. They are composed of highly energetic and ionised gases, and are produced in inert gases like argon. They are useful not only for dissociation of atoms but also for excitation and ionisation to give atomic and ionic emissions. Argon supported inductively coupled plasma (ICP) is the most common. DCP i.e. direct current plasma or MIP *i.e.* microwave induced plasma are less common. A plasma source has high stability and can overcome depression or interference effects caused by formation of stable compounds and give increased sensitivity of detection. It is simple to operate especially for liquids which cannot be handled by an arc/spark source. One does not need an electrode for discharge, and contamination with low background emission is avoided.

29.5 INDUCTIVELY COUPLED PLASMA SOURCE

This uses argon gas. A typical ICP source is indicated in Fig. 29.4.

Various cross sections are shown in Fig. 29.4. At the base of the discharge tube the sample punches a hole through the centre of the discharge at the induction region (IR) where the inductive energy transfers from load coil to a plasma. At this point argon light is emitted liquid samples are nebulised into an aerosol, and a fine mist of sample in the form of droplets is introduced into the ICP. The sample is carried to the centre of the plasma by nebuliser argon flow. Figure 29.5 shows various stages of the process taking place when a sample droplet is introduced into the ICP. As shown in Fig. 29.5 argon is circulated throughout the coaxial tube (I) as shown by arrows.

In the second stage (II) when a radio frequency field is applied, argon gas starts circulating around radio frequency signal. In stage (III), argon gas gets charged and produces electrons and charged (V) argon ions. A spark is applied at the mouth of the coaxial tube to produce a plasma as shown in (IV). Finally in the last stage (V) a sample of liquid to be analysed is injected into the plasma to puncture the plasma and encapsulate the sample at a very high temperature within it. A radio frequency field causes ionisation and formation of plasma in region (IV) and subsequently the sample aerosol carrying

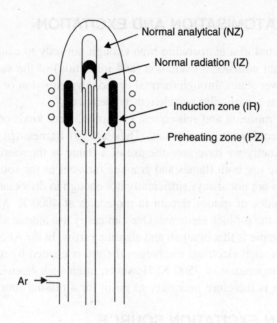

Normal analytical (NZ)

Normal radiation (IZ)

Induction zone (IR)

Preheating zone (PZ)

Ar →

Fig. 29.4 ICP zones

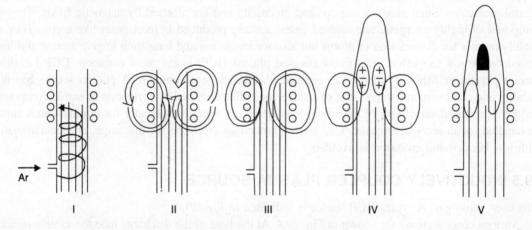

Ar →

I   II   III   IV   V

Fig. 29.5 Cross section of ICP torch

nebulizer flow punches a hole in the plasma. The sample of aerosol is carried into the centre of the plasma by nebulizing argon flow. The sample then undergoes various transformations like desolvation, vaporisation, atomisation, ionisation and excitation. During excitation, outer electrons are excited to higher energy states. The excitation and ionisation occurs due to collisions of the analytical atoms with high energy electrons. We get very high temperature in plasma. Various temperatures zones are shown in Fig. 29.6.

The gas temperature in ICP is around 6000 °C, it helps to eliminate chemical interference. The sample inside the plasma remains for two milliseconds. Such long residence reduces matrix interference. The sample in the ICP discharge does not travel around the outside of the discharge and experiences a uniform temperature. Emission from plasma is usually polychromatic. Separation is done by a monochromator and detection by a photomultiplier tube. On account of the large number of emission lines, spectral interferences can easily be avoided by use of a monochromator. The detection limit is µg/l (ppb) range. The sample concentration should be five times the detection limit. The upper limit is called the linear dynamic range (LDR). A large LDR needs less dilution.

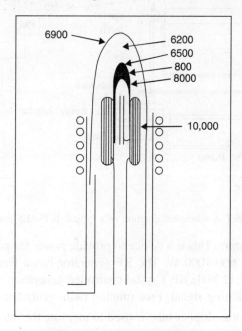

Fig. 29.6 Temperature zones of ICP

29.6 ICP-AES INSTRUMENTATION

It will be seen that in the flame photometer, atomisation is effected by a flame, in atomic emission technique by an arc or spark discharge, in atomic absorption largely by a graphite furnace (GF–AAS) while in ICP–AES it is done by an ICP torch. With the exception of this difference the rest of the instrumentation is more or less the same as used in the AAS or FES technique. A schematic diagram of a typical ICP–AES assembly is shown in Fig. 29.7.

The instrument is composed of important components such as (a) nebulizer (b) pumps (c) spray chamber, (d) sample injector (e) ICP torch, (f) RF generator and (g) detector. A very brief description of each component is as follows:

Nebulizers are devices to convert liquid into aerosols for onward transmission to the plasma. Pneumatic or ultrasonic forces are used for breaking up droplets of sample.

Pumps are usually used to pump liquids into nebulizers. The peristatic pumps are used which do not come in contact with the liquid. Tubing should be dependable and made from appropriate material.

Sprayer—the spraying chamber is used to transport liquid droplets into the plasma. It is located between the nebulizer and torch and removes large droplets. Droplets of 10 µm pass through it whilst it allows extra solution to drain off.

Sample injector—the sample is kept in it and via a pump, it goes to the nebuliser. A hydride generator is used for Hg, Sb, As, Bi, Ge, Pb, and Se, Te, Sn. Solid samples are introduced *via* an arc/spark source. 1PC torch - has mostly been described earlier.

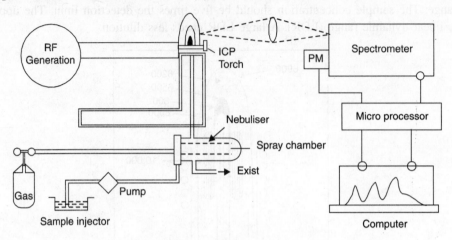

Fig. 29.7 A schematic diagram of a typical ICP-AES assembly

Radio frequency generator: This is a device to provide power for generation of plasma discharge. Its power ranges from 600-1000 W. The RF generator has a frequency of 27-56 MHz. Most instruments work on 27.12 MHz RF. Crystal controlled generation uses piezoelectric quartz crystals to produce an oscillating signal. Free running radio generators are also available.

Detector: usually a photomultiplier tube is used to measure the intensity of the emitted radiation. It is capable of amplifying an electrical signal by a combination of dynodes. About 9-10 dynode stages are useful for amplification of current.

29.7 APPLICATIONS OF PLASMA SPECTROSCOPY

Analysis is considerably simplified with ICP-AES techniques. On an average, RF works on 1-1.5 kW while plasma argon flows at a rate of 12-15 l/min in the auxiliary, and nebulizer unit, plasma flow is 1-2 l/min. The wavelength selected should be free of spectral interference. ICP-AES techniques can be used for analysis of agricultural products and foods. The traces of metals like Ca, Cu, Fe, Mn, Mg, P, K and Zn from beer or wine are best analysed by ICP-AES. Similarly Al from blood, Cu in brain tissue, Se in liver, Na in breast milk can easily be analysed. In earth sciences, lanthanides in rock can be analysed, while in environmental analysis Fe, Cd, Cu, Mo, Ni, V, and Zn from sea water are easily analysed. Trace metals from alloys and steel are also analysed. Metals like Cu Fe Ni and Si from lubricating oils or gasoline can be analysed at tracer concentration.

29.8 COMPARISON OF ICP–AES WITH AAS

In ICP-AES one needs to be highly skilled, while average skill is required in operating AAS. Accuracy is better in AAS than in ICP-AES. There are also less interferences in AAS in comparison with ICP-AES. The detection limit is better in ICP-AES while in AAS it depends upon the kind of metal. e.g. for analysis of alkali metals, flame emission or ICP-AES is better. Table 29.2 summarises various spectroscopic methods.

Table 29.2 Comparison of detection limit in different atomic spectroscopy (μg/ml)

Element	λ nm	AAS	FES	ICP-AES
Al	309	0.005	0.05	2
As	193	0.02	0.0005	40
Co	422	0.02	0.1	0.02
Cd	329	0.0001	800	2
Cu	325	0.002	10	0.1
Fe	248	0.005	30	0.3
Hg	253	–	0.0004	1
Mn	279	0.0002	5	0.06
Na	589	0.0002	0.1	0.2
Ni	232	0.002	20	0.4
Pb	217	0.002	100	2
Sn	220	0.1	300	0.30
V	318	0.1	10	0.2
Zr	360	0.0005	0.0005	2

It will be observed that the technique of analysis depends upon several parameters such as the matrix, its history, interferences present, level of analysis, time and number of analyte samples to be handled. No generalized formula can be offered for selection of a particular method.

29.9 COMPARISON OF AFS, AAS AND ICP-AES

So far we have considered two atomic spectroscopy techniques. At this stage it is worthwhile to compare and contrast all the spectral modes of analysis including atomic fluorescence from the viewpoint of detection limit. For analysis of mercury with atomic fluorescence (AFS) detection is 20 times lower than by AAS using cold vapour method. AFS is best for analysis of ppt level of metals. AFS is the limits of detection which is 2-15 times lower than those by AAS methods. See Table 29.3.

Table 29.3 Detection limits of AFS, AAS, ICP-AES

Periodic Group	Element	AAS (fg) (1×10^{-15})	ICP-AES (fg) (1×10^{-15})	AFS (absolute) (1×10^{-5})	AFS (ηgl) (1×10^{-9})
VII	Mn (II)	1000	6	100	5
VIII	Fe (III)	2000	580	70	3.5
IB	Ag (I)	500	5	10	0.5
IB	Au (III)	4000	15	10	5
IIB	Cd (II)	300	12	0.5	0.01
IIB	Hg (II)	25000	9	90	9
IIIA	Al (III)	4000	500	100	5

(Contd.)

Periodic Group	Element	AAS (fg) (1×10^{-15})	ICP-AES (fg) (1×10^{-15})	AFS (absolute) (1×10^{-5}) (ηg)	AFS ($\eta\delta$) (1×10^{-9})
IIIA	Tl (I)	10000	3	0.1	0.005
IVB	Ge (IV)	40000	4	1	0.05
IVB	Sn (IV)	20000	10	30	1.5
IVB	Pb (II)	5000	50	0.2	0.01
VB	Sb (III)	20000	12	10	0.5
VIB	Te (VI)	10000	10	20	1

The high sensitivity of AFS is obvious from Table 29.3. The AFS method is more sensitive than AAS method by 2-4 order of magnitude. AFS detection limit is considerably lower than one by ICP-AES technique. However, ICP-AES has significant advantage that it permits multielemental analysis. Thus AFS inspite of being sensitive method has restricted applications to few specific elements. Elements like mercury, arsenic, antimony, selenium and tellurium are best determined by AFS and the detection limit (ηg/L) varies from 1-50 ηg/litre.

LITERATURE

1. C.B. Boss, K.J. Freedom, *"Concepts, Instrumentation and Techniques of Inductively Coupled Plasma Atomic Emission Spectroscopy"*, Perkin Elmer Co. (1989).

2. A Montaser, D.W. Golightly, *"Inductivity Coupled Plasmas in Analytical Atomic Spectroscopy"*, VCH Publisher, NY (1992).

3. M. Thompson, J.N. Walsh, *"A Handbook of Inductivity Coupled Plasma Spectroscopy"*, Blackie Glasgow (1983).

4. G.F. Wallace, P. Barrett, *"Analytical Methods Development for Inductivity Coupled Plasmas Spectrometry"*, Perkin Elmer Co. (1981).

5. A. Varma, *"CRC Handbook of Inductivity Coupled Plasma Emission Spectroscopy"*, Boca Raton, FL, CRC Press (1991).

6. D.A. Skoog, J.J. Leory, *"Principles of Instrumental Analysis"*, Harcourt Brace Pub., London (1992).

7. G.D. Christian, *"Analytical Chemistry"*, 5th Edn., John Wiley and Co., New York (1994).

8. R.D. Braun, *"Introduction to Instrumental Analysis"*, McGraw Hill, C.B.S. Indian Edition (2004).

Chapter 30

Chemiluminescence, Atomic Fluorescence and Ionisation Spectroscopy

The spectrochemical methods can be broadly divided into four categories viz. one involving emission (atomic emission) or absorption (atomic and molecular spectroscopy) or scattering (Raman spectroscopy, turbidimetry and nephelometry) and finally those involving Luminescence (Atomic and molecular fluorescence, phosphorescence and the chemiluminescence).

Although fluorescence and phosphorescence are known since ages in molecular spectroscopy they differ from chemiluminescence and electrochemiluminescence. This is because in fluorescence or phosphorescence after the excitation of electrons energy generated is first absorbed and subsequently emitted on relaxation at lower energy level after sufficient time lag. While in chemiluminescence energy is *not absorbed* in higher energy level, it is dissipated in form of chemiluminesence during the process when electron returns to ground state.

The spectra generated by chemical reaction is called chemiluminescence while one generated by electrochemical reaction is termed as the electrochemical luminescence phenomena. For analysis of atmospheric pollutants like SO_2, $NO-NO_x$, O_3, chemilumiscence is used for quantitative monitoring. In few instances it is also used in titrations with indicator showing the chemiluminescence. It is used for measurement in gases as well as in liquids. Atomic fluorescence spectroscopy is of recent origin. It needs very high energy of excitation of electrons. Atoms after absorption of photons emit out radiation while returning to ground state. We have resonance as well as Stoke's fluorescence. The technique is sensitive for metals like Mg, Cd, Zn for quantitative analysis. The AFS instruments are not easily available and are very expensive. Therefore in comparison to AAS, the AFS methods did not find much applications in quantitative chemical analysis. Recent addition in these techniques is Resonance Ionisation Spectroscopy (RIS) and Laser Enhanced Ionisation Spectroscopy (LEIS). On account of limitation to use for the gaseous sample in absence of inexpensive instrument it is not commonly used.

30.1 LUMINESCENCE

It occurs when an electron which is excited falls back to lower energy electron level with emission of radiation with wavelength much *longer* than that of the wavelength of original excited electron. The energy corresponds to (ΔE), the difference of energy between two levels. It occurs usually in UV, visible or X-ray regions. Electron occupying lower level has most stable configuration in ground state. For the appearance of luminescence electron from ground state is excited to higher energy level. If excitation is by chemical reaction, it is called chemiluminescence but if extraction is by electrochemical reaction it is termed as electrochemiluminescence. These methods are hardly used in quantitative chemical analysis. However, the one originating after absorption of electromagnetic radiation is used in chemical analysis. If a molecule possesses all paired electrons we call it singlet state with zero spin. When unpaired electrons are present, it is called doublet state. Finally a system with two unpaired electron is called as triplet state. The absorbed energy of electron can be released by collisional deactivation, or by radiationless loss of energy. The energy loss by vibration is vibrational relaxation. The process of 'internal conversion' occurs when a molecule loses energy in small steps by transforming from high to low vibration level. However, after absorption if higher energy level electron reverses its spin orientation the state is called as Triplet state by release of some energy. Such conversion process is called 'intersystem conversion'. Fluorescence occurs during electron transition from excited singlet state to lower energy singlet state while phosphorescence occurs due to transition of the electron from excited triplet state to lower energy by singlet state. If the wavelength of excitation and emission is same in fluorescence it is called resonance fluorescence phenomena. One with longer wavelength for emission is normal fluorescence. If the excitation and emission takes place in 10^{-6}–10^{-10} seconds it is called fluorescence but if time of emission is between 10^{-2} – 100 seconds it is termed as phosphorescence, as spin reversal in intersystem conversion needs longer time. As molecules have the vibrational and rotational energy it leads to broadband of spectra, while atoms do not possess such energy. The measured intensity of Luminescence (P_i) is proportional to incident radiation intensity (P_0) and the concentration (C). So we have: $P_i = KP_0 C$ if K is some constant; while incident radiation P_0 = constant.

30.2 CHEMILUMINESCENCE

The origin is related to the chemical reaction. It occurs in UV, visible or IR region. Bioluminescence is occurring in biological transformation. The process occurs by one of the following transformations viz.,

$$R \rightarrow P^* \rightarrow P + h\nu, \text{ or } P + \text{heat or } A^* \rightarrow A + h\nu \text{ or } A^x \rightarrow A + \text{heat.} \qquad ...(30.1)$$

Here 'R' undergoes chemical reaction to form (P^*) in the excited state, which in turn loses energy by emission of radiation by release of heat or just transferring energy to chemical acceptor 'A' which also emits radiation or heat. Since reactions are slow chemiluminescence is also slow. For organic compounds lowest excited state is triplet state. Longed lived triplet state promotes quenching by arresting chemilumiscence due to radiationless relaxation. In chemiluminescence or bioluminesence, finally the electron transfer and fragmentation reaction occurs.

30.3 MEASUREMENT OF CHEMILUMINESCENCE

The instrument measuring these luminescences are calibrated with lamp or fluorescent standards. The standard tungsten or quartz iodide lamps are used for calibration for the UV or visible region respectively. Fluorescent standard are very useful. Quinine sulphate satisfies the need for calibration. Luminol reaction is also good. For quantitative analysis standard addition or working curve is used.

30.4 QUANTITATIVE ANALYSIS

It is used for the analysis of air pollutants from atmosphere or record end point in titration. Trace metal analysis is also possible. In environment, for monitoring of SO_2 or NO-NO_x, the most dependable method is the chemiluminescence technique.

The details of such monitoring method is furnished in following section. Apart from SO_2 and NO-NO_x even O_3 can be analysed by chemiluminescence. The atmospheric SO_2, H_2S or similar compounds are analysed by this method. In hydrogen rich air-hydrogen flame, sulphur (atomic) reacts to form S_2 molecule which can have chemiluminescence at 394 nm. The resulting material is analysed by gas chromatography.

The most important work is on NO-NO_x analysis by chemiluminescence. NO–NO_2 can be analysed by reaction with ozone. NO_2 is dissociated to NO

$$NO_2 \rightarrow NO + \frac{1}{2}O_2 \; ; \; NO + O_3 \rightarrow NO_2 + O_2 + h\nu$$

From last equation chemiluminescence is measured at 600 nm. As small as 0.002-1000 ppm of NO_2 can be analysed by this method.

The analysis of NO is directly done by reaction with O_3 to give NO_2 e.g.

$$NO + O_3 \rightarrow NO_2 + O_2 (g) \uparrow$$

The NO_2 from auto emission and cigarette smoke is easily analysed by this method. Also ammonia with oxygen gets converted to NO_2 at high temperature.

$$2NH_3 + 2O_2 \rightarrow 2NO_2 + 3H_2 (g) \uparrow$$

The chemiluminescence is also used to monitor air flow in upper atmosphere (> 90 km). This is done by tracing rocket trajectory.

30.5 CHEMILUMINESCENCE TITRATIONS

Apart from gas-phase chemiluminescence analysis, this technique is used for liquid as well as gas phase titration as is done in potentiometric or photometric linear titrations. The plot of chemiluminescence intensity vs volume of titrant is prepared and the line of intersection after extrapolation if necessary is considered as the end point of titration e.g. with solution of iodine as the titrant, arsenic (III), SO_2, sulphite is carried out with chemiluminescence indicator like Luminol. The luminescence is observed only after end point is reached. Excess titrant give more intense chemiluminescence in the solutions.

Apart from liquids one can titrate gases also. O_2, N_2, H_2 by chemiluminescence methods e.g. Atomic O is titrated with NO_2 in following manner.

$$NO_2 + O \rightarrow NO + O_2, \; O + NO + M \rightarrow NO_2 + M + h\nu$$

where 'M' is metallic catalyst. Such titration is done for flowing gases. For H we have

$$H + NO_2 \rightarrow OH + NO; \; H + NO + M \rightarrow HNO + M + h\nu.$$

Such chemiluminescence occurs at 600-800 nm. In similar manner nitrogen gas is titrated e.g.

$$N + NO = N_2 + O_2, \; N + O + M \rightarrow NO + M + h\nu.$$

Both in UV or IR region chemiluminescence can be measured.

30.6 CHEMILUMINESCENCE FOR LIQUIDS

In biochemistry, AMP and ADP (*i.e.*, adenosine mono phosphate and adenosine diphosphate respectively) are analysed by chemiluminescence technique; a typical reaction is that of fireflies. Many nitrogen containing compounds have been analysed by this method using Luminol as an indicator. It shows following reaction as:

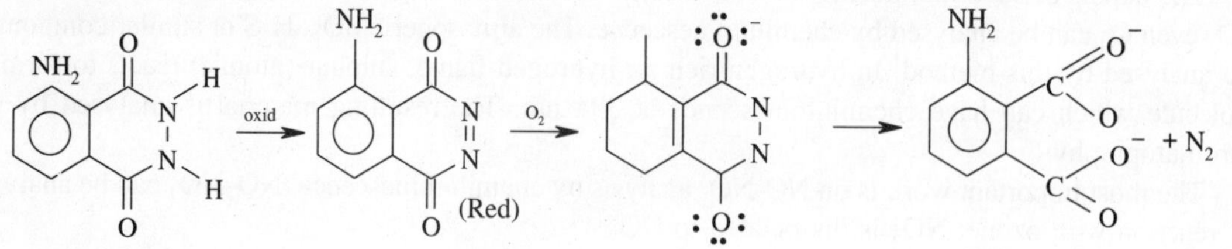

Fig. 30.1 Chemical reactions

We can analyse Cl^-, HOCl, OCl^-, H_2O_2, O_2, $\left[Fe(CN)_6^{-2}\right]$; NO_2 with this indicator. The optimum pH is 8-10 and S^{2-} is analysed with hypobromite solution.

Trace metal analysis in presence of hydrogen peroxide and Luminol is possible. Quantum yield is large, but pH control is critical. Luminol (1 mM) is preferred. Metals like Co^{2+}, Cr^{3+}, Cu^{2+}, Fe^{3+}, Mn^{2+}, Ni^{2+}, V^{+4} can be analysed by it. Most of metals change oxidation state during conversion. In many reactions peroxide is not required e.g. MnO_4^-, OCl^-. Instead of luminol, galic acid with peroxide is also used for determination of Co^{2+}, Ag^{2+}. Lophine *i.e.*, 2,4,5 triphenyl imdazole is also used for analysis of Co^{2+}, Cr^{+6}, Cu^{2+}, Fe^{3+}, Ir^{4+}, OS^{4+} and Mn^{7+}.

30.7 ELECTROCHEMILUMINESCENCE

The electron transfer reaction in electrogenerated cation and anion are used in these kinds of luminescence studies. Three electrode assembly is essential. Both electrogenerated cation as well as electrogenerated anions interact with each other to provide neutral molecule which actually exhibits electrochemiluminescence.

This technique has been used mainly for study of electron transfer reactions between electrogenerated radical cation and radical anions. Two or three electrodes are placed in solution of the percursors to the cations and anions. The potential applied causes generation of an ion in electron transfer reaction to form excited cation or anions. The generated cation reacts with anion to form an excited neutral molecule exhibiting phenomena of chemiluminescence. The radical cation is formed if applied potential is sufficient to detach electron from highest filled molecular orbital of the said molecule, so R $\rightarrow$ $R^+ + e^-$. The radical anion is also formed when potential is applied to electrode which is capable of the addition of electron to the lowest unoccupied molecular orbital of the molecule, so R + $e^- \rightarrow R^-$. Chemiluminescence occurs if cation reacts with anion. One prefers platinum electrode for successful completion of the reaction. The combination to form molecule proceeds in the following reactions.

$$\underset{\text{anion}}{R^-} + \underset{\text{cation}}{R^+} \rightarrow \underset{\substack{\text{Excited}\\\text{singlet}}}{R^*} + \underset{\substack{\text{Ground}\\\text{State}}}{R_t} \qquad \qquad ...(30.2)$$

$$R^* \rightarrow R_t + h\nu \qquad \qquad ...(30.3)$$

This exhibits chemiluminescence. All reaction cannot lead to chemiluminescence as energy is lost by radiationless process. The singlet route leads to exhibition of the chemiluminescence. The excited triplet state can also show luminescence during the formation of molecule by process of relaxation in triplet state to ground state or by triplet-triplet annihilation. The reaction can proceed as follows:

$$2R^* \rightarrow R^* + R_t \qquad \qquad ...(30.4)$$

$$R^* \rightarrow R_t + h\nu \qquad \qquad ...(30.5)$$

The compound 9-phenyldiamine exhibits chemiluminescence by singlet route while compound NNN′N′–tetramethyl-p-phenyldiamine (TMPD) shows chemiluminescence by triplet route. The aromatics, heterocyclic, metallic and metalloporphyrines show chemiluminescence. Acetonitrile is best solvent to dissolve luminating matter. Propylene carbonate, methylchloride, THF are the best solvents. Rotating ring disk electrode is suitable. The reaction mechanism is studied by this technique, but this technique is less frequently used for quantitative chemical analysis by analyst.

The chemiluminescence originates due to relaxation in triplet state. Both cationic and anionic radicals of 9,10-diphenylanthracene exhibit the electroluminescence in triplet state. Organic solvents are preferred for the reaction to occur. Metalloporphorine and metal complex show the phenomena. Solvents like acetonitrile, DMSO, methylene chloride, propylene carbonate are commonly used. The platinum or rotating ring disk electrodes are used. Appropriate potential is applied to form cation at cathode and anion at anode; these ions in turn are reduced or oxidised in subsequent operations respectively. On contact with appropriate ions they interact to show chemiluminescence. These methods are used to study reaction mechanism. Bioluminescence is another area which would come to forefront in future years to come.

30.8 ATOMIC FLUORESCENCE

This technique uses radiation from the continuous source of excited atoms to lift electrons to higher energy level. Fluoroscence originates when excited electron returns to ground state. The absorption of energy by atom leads to excited electronic level. Such atomic fluorescence is feasible when excited atoms emit radiation after earlier being excited by absorption of photons. We can have four types of fluourescence e.g. Resonance fluorescence, direct line phenomena; stepwise fluorescence and thermally helped fluorescence as is observed in atomic absorption spectroscopy. The ground state absorption is a prerequisite for fluorescence leading to decrease in intensity of excited radiation. In resonance fluorescence wavelength of absorption is same as that of emission. They are depicted in Figures 30.2 to 30.5. In these figures, A = absorption, F = fluorescence. 0, 1, 2 indicates ground or first or second energy level for the atom. Wavy line with arrow indicates fluorescence.

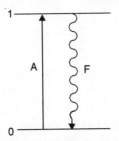

Fig. 30.2 Resonance fluorescence

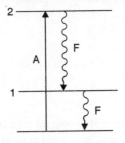

Fig. 30.3 Direct line stoke fluorescence

In direct line fluorescence electrons returns to lower energy level which is above ground state, so wavelength of absorption is *shorter* than wavelength of emission. This is called also 'Stroke fluorescence'. In stepwise fluorescence a photon is emitted while returning to ground stage. (Fig. 30.4). This is also an example of stoke fluorescence. In thermal process the electron after excitation to one level gets further excited to higher level by the radiationless process to higher state when fluorescence occurs from both energy levels. Thus emission is thermally assisted fluorescence. The wavelength of absorption and emission may be different. Finally in antistokes fluorescence, emitted radiation has *shorter* wavelength than that of the absorption radiation. In sensitised fluorescence the excitation of the radiation is transferred to neighbouring atom leading to its fluorescence.

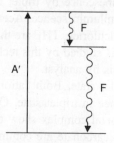

Fig. 30.4 Stepwise

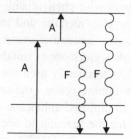

Fig. 30.5 Thermal

30.9 PRINCIPLES OF ATOMIC FLUORESCENCE

The equation representing the intensity of emitted radiation varies with experimental conditions. The source of radiation and analytical concentration are of prime importance. The equation for singlet state for an atom showing fluorescence is simple. Fluorescence usually occurs during return journey of an electron at constant temperature. An element absorbs energy initially from source. So fluorescent radiation (X_F) is proportional to (P_2) the absorbed radiation. Later some atoms of this show fluorescence. So fluorescence power efficiency (F_2) is the ratio of power of fluorescent radiation during electronic transitions from level (2) to the level (1) of energy. Fluorescence is proportional to P_a and F_2 so we have

$$X_1 \propto F_2 P_a \qquad \qquad ...(30.6)$$

In order to get integrated absorption coefficient, we have

$$P_a = E_r \int K_v dv \qquad \qquad ...(30.7)$$

E_v is spectral radiation from source and 'K' is absorption coefficient when v is frequency of absorption. By putting proportionality constant, equation is modified to

$$X_v = \frac{1}{4n} F_2 E v_1 \int K_v dv \qquad \qquad ...(30.8)$$

$$\int K_v dv = \frac{nhv_1' X_1}{c} \frac{E^0 v_1}{E^* v + E v_1} \qquad \qquad ...(30.9)$$

With proportionality constant $l/4\pi$ if l is path length integrated coefficient of absorption. At saturation numbers of atoms in ground state is equal to same number of atoms in excited state. Therefore X_r will be

$$X_r = \frac{lF_2 E \nu_1 nhr_1 X_1}{4\pi c} \frac{E^* \nu_2}{E^* \nu_1 + E \nu_1} \qquad \qquad ...(30.10)$$

Finally we have an equation.

$$X_F = \frac{lF_2 E \nu_1 nhr_1 X_1}{4\pi c} \qquad \qquad ...(30.11)$$

The integrated absorption coefficient is shown in equation (30.9), where n = number of analytical atoms, ν_1 = is frequency of absorbed radiation, X_1 is speed of electromagnetic radiation in vacuum $E^* \nu_1$ is saturated spectral irradiance. It is the irradiance from source leading to generation of the fluorescence radiance. At saturation numbers of atoms in groundstate is same as number of atoms in the excited state so we get X_F as per equation number (30.11).

This equation shows that fluorescence radiance is directly proportional to the concentration of analyte element. The product of $K\nu l$ is 0.05 if l is length tube or container for sample. This leads to the formation of linear curve. Finally at high concentration fluorescence is proportional to square root of concentration.

30.10 APPLICATIONS OF ATOMIC FLUORESCENCE

Inorganic quantitative analysis is carried out with calibration curve, which are not always straight line or linear. Therefore standard addition method cannot be used. As many as 58-60 elements can be analysed by atomic fluorescence spectral methods (AFS) using resonance fluorescence methods. Few elements cannot be done by resonance fluorescence spectroscopy as are Ag, Ba, Co, Cu, Fe, Mn, Mo, Ni, Pb, Zr, while Al, Bi, Ga, Pd, Sn, Ti are analysed by non-resonance fluorescence spectroscopy. For

Table 30.1 Atomic fluorescence analysis of elements (AFS)
(Detection limit ηg/ml) (1×10^{-9} g/ml)

Element	AFS by ICP	AFS (Flame)	Emission Wavelength (λ)	Kind of fluorescence
Al	2	5	309.2	Direct line
As	40	100	193.2	–
Ca	0.02	0.01	422	–
Cd	2	0.01	228	–
Cr	0.3	4.0	357	–
Cu	0.1	1.0	324	Direct line
Fe	0.3	8	373.5	–
Hg	1	20	185.0	–
Mg	0.05	1.0	285.0	–
Mn	0.06	2	279	–
Mo	0.2	60	313	–
Na	0.2	–	589	–
Ni	0.4	3.0	232	–

few metals limits of detection are better than one obtained by AAS. In AFS, we encounter similar kinds of interference as are encountered in AAS methods, like say, chemical or ionisation interferences. With continuous source the spectral interferences are observed. (AsCd), (CoIn), (CoHg), (FeMn), (NiSn) show fluorescence at same wavelength. Scattering is another problem and hindrance which must be controlled.

30.11 INSTRUMENTATION FOR ATOMIC FLUORESCENCE SPECTROSCOPY

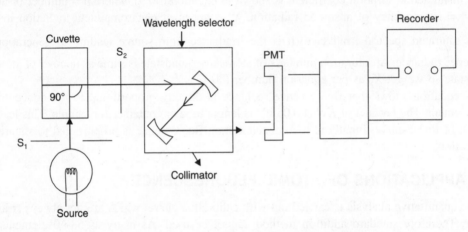

Fig. 30.6 Atomic fluorescence spectrometer

The important components of AFS are source of radiation, sample holder, wavelengths selector and detector (Fig. 30.6). The incident radiation and emitted radiation are at angle of 90°. In line measurements are rarely made, however, mechanical chopper is used if line measurement is used to select particular radiation. A strong intensity source like xenon lamp is used as the source of radiation. For line source hollow cathode lamp is used. Inductively coupled plasma (ICP) is the best source. The electrodeless discharge lamp is also a good source. Laser (light amplification by stimulated emission of radiation) is dependable as the source of radiation as it provides narrow line bandwidth. The ruby laser is also used. Tunable dye laser is most useful for analytical investigations. The dyes used are coumarine, Rhodamine, Nile blue etc. in region from 400 to 780 nm. For sample holder, cell or cuvettes are used and they are kept at low temperature permitting smooth atomisation. Similar to AAS we have air acetylene, $N_2O-C_2H_2$, or argon flame are used. Graphite furnace is usually preferred. ICP-AFS with argon plasma is best. Induction coil generates argon plasma. ICAP is used as intensive source of excitation radiation in AFS. Temperature of plasma is quite high. Low resolution monochromator is used for choice of line source. Filters can also be used. For detection of radiation PMT *i.e.* photo multiplier tube is used but the readings are best obtained by what is called as resonate detector.

Atomic fluorescence methods were first discovered in 1964 for the quantitative analysis of elements. However, in comparison to AAS which is popular due to its simplicity, this technique has not become popular. The commercial AFS instruments are not easily available in the market. The method has no special advantage over AES or AAS methods. Further, AFS needs continuous strong flow from source of radiation. Initially the hollow cathode was employed as the source; however, less electrode discharge lamp was more useful for the purpose of source. Lasers are ideal for the job. The nondispersive instruments are simple, inexpensive, facilitate the multielement analysis; they have better sensitivity and high energy output. The background emission can be eliminated by use of electrothermal devices. The interference encountered in AAS such as spectral interference, background emission (Zeeman effect); chemical interferences, inclusive of the low voltality compound formation, shifting of dissociation equilibria as well as the ionisation equilibria are also similarly encountered in Atomic fluorescence method. The presence of metals in sea water, lubricating oils, biological products and agricultural samples are best analysed by AFS. The detection limits of some elements is shown in parenthesis. The elements in question are as follows: Cd (0.0001) Cu (0.002), Fe (0.005), Mg (0.00002), Mn (0.0002), Mo (0.005), Na (0.0002), Pb (0.002), Zn (0.00005) ηg/ml. This shows the sensitivity is excellent for magnesium and zinc while the sensitivity of AFS analysis is not satisfactory for mercury (0.1), tin and vanadium (0.1 ηg/ml).

However, we expect in future exponential growth for the use of atomic fluorescence methods for the analysis of common metals like magnesium and zinc specially in biological samples. Finally, let us consider the principles of resonance ionisation spectroscopy (RIS) and laser enhanced ionisation spectroscopy (LEIS). In both methods the electrons or atoms are excited to high energy state to form ion which in turn is used for spectral analysis.

30.12 RESONANCE IONISATION SPECTROSCOPY

This was the outcome of the fast development in laser technology. This method facilitates micro or trace level analysis of the elements. In abbreviated form we call them as RIS and LEIS. The latter pertains to Laser Enhanced Ionisation Spectroscopy (LEIS). In 1974 it was first discovered. In RIS pulse laser is used to excite atoms from ground to higher excited state of orbits. The excitation of atom from ground state leads to resonance spectroscopy. Before returning to ground level electrons absorb energy from the second laser to remove out electron from the atom leading to the formation of ion. Such ion is detected by suitable device or means. The laser operates at such wavelength which is corresponding to difference in the ground and excited state thereby rendering the method selective. RIS is more sensitive than AAS, AFS and AES as it deals with all ions or electrons not few as in AAS or AES. They are formed during process of ionisation and all atoms and not part of them within the beam of laser are ionised. Many ionisation reactions facilitate applications of RIS to assay all seventy elements (exceptions He and Ne) due to absence of appropriate laser needed for excitations. One can study the ionisation energy of each atom. Ionisation takes place only if applied energy of the laser exceeds ionisation energy or ionisation potential. There are several pathways to achieve this effect. In first pathway same laser is used to excite and as well as for ionisation. In second pathway single laser with energy is doubled to enable atom to go to excited state. In third and fourth pathways, two lasers are used. In third case one laser lowers atom to lower excited state while for the other it rises from lower to higher excited state leading to process of ionisation. While in fourth pathway doubled frequency from one laser raises atom to the first excited state while second laser adjusted at different wavelengths is used to further raise atom to second excited state. This leads to its ionisation. In another pathway two photons from the same or different lasers are simultaneously absorbed and lead to excitation and subsequent ionisation. Many elements can be analysed by known mechanism of RIS. Important elements which are analysed are: Al^{3+}, Sb^{3+}, As^{3+}, Ba^{2+}, Cd^{2+}, Ca^{3+}, Cr^{3+}, Co^{2+}, Cu^{3+}, Ga^{3+}, Ge^{4+}, Ag^+, Au^{3+}, Na^+, Tl^+, Sn^{2+}, W^{3+}, U^{6+}, V^{5+}, Zr^{4+}, Zn^{2+}, etc.

Merits: The example of each type of pathway for excitation and ionisation with kind of laser is listed in Table 30.2. Al, Ca, W, Cd and As respectively exhibit 1, 2, 3, 4, 4 pathways of excitation as well as ionisation. The detectors in RIS measure electric charged particles, which may be an electron or even an ion. The detectors are like GM counter, proportional counter, electron multiplier. The gaseous atoms are used for sample and laser is fired in chamber. Pulsed ionisation chamber is one place of firing, but it cannot amplify current but proportional counter can do so. In GM counter output from detector is independent of the number of electrons caused to pulse during process of excitation and ionisation.

Table 30.2 Mode of lasers and ionisation process (RIS, LEIS)

Pathway	First Laser	Second Laser	Example	Effect
1.	Same laser	Excitation	Al	Ionisation
2.	Single laser	Double energy	Ca	Excitation and ionisation
3.	Two lasers at diff. wavelength	One laser raises atoms to lower level	W	Other raise lower to higher level.
4.	Two lasers	double frequency raise v to excited state	Cd	Second laser diff. raises atom to second excited state.
5.	Same or different laser used	Two photons simultaneously absorbed.	As	Ionisation

Limitations: The principal demerits of RIS is the sample must be present in gaseous state and it is imperative to convert sample first into gaseous state. For analysis of sodium, laser enhanced resonance ionisation spectroscopy was used by Mayo et al. from solid sample. For sodium second pathway is followed with use of single laser. However, RIS is an excellent technique for the analysis of the element at ultratrace concentrations. Apart from metals some non metallic elements (gases like Ar, Br^-, Cl^-, F^-, H^+, Kr, O^{2-}, S^{2-} can be easily analysed by the technique of Resonance ionisation spectroscopy. In absence of suitable prototype instrument so far this technique is not fully utilised in chemical analysis.

30.13 LASER ENHANCED IONISATION SPECTROSCOPY

This technique makes use of flame to atomise the analyte, while laser excites the atom to the high electron state as in resonance ionisation spectroscopy. The excited atoms are ionised by heat and these ions are electrically studied with electrodes placed across the flame. The laser selectively excites sample to atoms as in RIS. Since excited atoms need less energy to be ionised than the ground state atoms, the excited atom are readily ionised by flame. Then ions and atoms of sample are studied with two electrodes near flame. As with RIS, LEIS can exhibit any kind of mechanism as shown in Fig. 30.7 by mechanism as A, B, C, D. In this figure kind (A) is material is in resonant laser excitation and subsequent thermal ionisation. In kind (B) non-resonating laser excitation is followed by ionisation. In kind (C) resonant laser excitation is followed by non-resonant laser excitation and ionisation. It is also called 'stepwise resonance laser excitation'. Finally kind (D) involves sequential thermal excitation, step-wise laser excitation and thermal ionisation. Kind (D) exhibits non-resonant laser excitation.

The elements like Al (A), Ca (B), Co (C), Ni (D) can be analysed by LEIS with kind of transition shown in brackets. As many as 70 elements can be analysed by LEIS. Like AAS same flame gases are used for excitation along with oxidant. A cool flame facilitates atomisation. Acetylene air flame is best. We use electrodes to study electron charge migration phenomena with potential drop of 1000-2000 V between two electrodes. Tuneable dye lasers are most useful due to availability of several wavelengths. The quantitative analysis is done with working curve. Lower limits is one part per billion. The effective working range is 10^4-10^5 for LEIS. Like AAS there are many interferences as are seen—one causing ionisation from ground state is serious e.g. alkali metals. However, this technique like RIS is also in development stage and it is expected that in future it may be used for ultratrace analysis of elements.

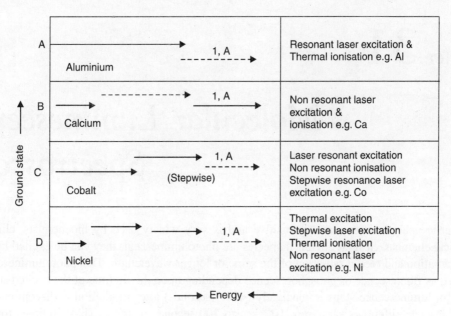

Fig. 30.7 Energy transition

LITERATURE

1. R.D. Braun, "*Introduction to Instrumental Analysis*", McGraw Hill Book Co. (1987).

2. D.A. Skoog, F.J. Holler, T.A. Nieman, "*Principles of Instrumental Analysis*", 5th Ed. Harcourt Brace. Col. Publ. (1998).

3. R.Smith, (Ed.) J.D. Winefordner, "*Spectrochemical Methods of Analysis: Advances in Anal. Chem. and Instrumentation*", Wiley (1971).

4. E.B. Sandell , H. Onish, "*Photometric Determination of Traces of Element*", Interscience, Wiley, (1978).

5. D. Kealy, P.J. Haines, "*Instant Notes in Analytical Chemistry*", Viva Publications, New Delhi, (2002).

6. H.A. Strobel, "*Chemical Instrumentation*", John Wiley and Co. (1973).

electrons are at a higher energy level. Like SₐSₒ there are many more pairs of states like excited singlet states. Transitions occur from one excited state to another. However, the distinction of S₁, S₂ in excited states need not be explained but that transitions can be seen by emission of electromagnetic radiation.

Chapter 31

Molecular Luminescence
Spectroscopy

Fluorimetric analysis is extremely sensitive and is very widely used by biochemists, clinical and analytical chemists. Several chemical species are photoluminescent; they can be excited by electromagnetic radiation and reemit radiation of the same or longer wavelength. This photoluminescence can be classified as fluorescence or phosphorescence depending upon the lifetime of the excited state. With fluorescence, luminescence stops immediately (<10^{-6} seconds) after irradiation is discontinued, while in phosphorescence lifetimes vary from 10^{-2} sec to 100 seconds or more. Thus lifetimes for fluorescence are very short: viz. 10^{-6}–10^{-9} seconds. Exciting radiations and reemitted radiation are either in the visible or ultraviolet regions. They may sometimes be in high energy X-ray regions.

31.1 PRINCIPLES OF FLUORESCENCE AND PHOSPHORESCENCE

Let us consider the molecular events leading to a luminescence by transition of electrons. Electrons in the ground state are stated to be in the singlet state with all electrons paired since they are occupying the lower orbital. However, if electrons have the same spin then they are unpaired and the

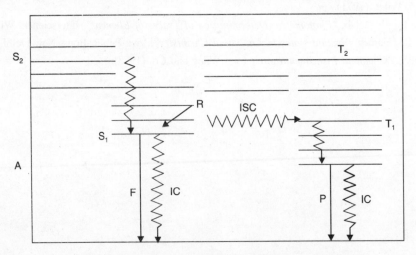

Fig. 31.1 Energy level diagram for diatomic molecules

molecule is in the triplet state. These two states denote the multiplicity of the molecule. During electronic transitions, if the transition is higher than the S_1-state a process of *internal conversion* occurs. The process involves removal of excess energy from the higher level to cause what is called vibrational relaxation.

It occurs in 10^{-12} seconds. Another possibility is that the molecule is in the excited state and the electron reverses its spin thereby transferring the molecule to a lower energy level. This is the triplet state. Such a process of conversion is called *inter system conversion* (ISC). By a process of vibrational relaxation the molecule reaches its first excited triplet (T_1) and finally returns to ground state in 10^{-2} seconds. Transitions between states of different multiplicity are prohibited or forbidden. T_1 has a long life compared to S_1.

In this energy level diagram (for, say, a diatomic molecule) Fig. 31.1, S represents a *singlet state*— a state where all the electrons in a molecule *are paired* and T-represents a *triplet state*—a state where two electrons have *unpaired spin*. A ground state is a singlet state. The *unpaired electrons* have a lower energy than paired electrons, hence a *triplet state* has a lower energy level than a singlet state. Molecules release energy which has been absorbed during a transition from the ground to singlet (S_2) state. Such dissipation can occur by radiating a photon equal in energy to the difference between the exciting (S_2) and ground state. The competing process of relaxation by vibrations is also present (shown as R) which involves the transfer of vibrational energy to neighbouring molecules and is a rapid process especially in solids and liquids. If photon emission energy is equal to that absorbed, we call the process resonance fluorescence. However, in solution an excited molecule will rapidly relax to the lowest vibrational level of electronic state S_1. The internal conversion (IC) would shift the molecule from S_2 to S_1 involving a small energy difference. Now the molecule is quickly deactivated by vibrational relaxation to the lowest vibrational level, S_1. Thus most excited molecules in solution will decay to the lowest vibrational level of the lowest excited singlet state before any useful radiation is emitted. After reaching this state the molecule can return to the ground state say by emission of radiation. This loss of radiation is termed as fluorescence (*i.e.* from S_1 to S_0) as shown by (F) in Fig. 31.1. Fluorescence quantum efficiency (ϕ) represents the fraction of excited molecules that fluoresce. The lifetime of an excited singlet state is usually 10^{-6}–10^{-9} seconds. The wavelength of fluorescent radiation is longer than that of the absorbed radiation.

As an alternative to internal conversion and fluorescence, a molecule in the singlet state (S_1) can undergo intersystem crossing (ISC) which involves electron spin flip to put the molecule into triplet state (T_1). Vibrational relaxation is rapid and since ISC is a spin forbidden process it cannot compare with fluorescence. Now any transition from the triplet state (T_1) to the ground state (S_0) is also a spin forbidden phenomenon, therefore the lifetime (T) of the triplet state is long *viz.* 10^{-4} to 10–100 seconds. The entire process is termed as phosphorescence (P) in Fig. 31.1. Phosphorescence is a luminescence process in which a molecule undergoes a transition from the triplet to the ground state. Since the triplet state is at lower energy than the singlet state, the former is closer to the ground state. Hence internal conversion to the ground state is more efficient than phosphorescence and is common at room temperature for loss of triplet state energy. Phosphorescence quantum efficiency can be increased by cooling the solution to a lower temperature ($\sim$ 77 K).

31.2 FLUORIMETRY IN CHEMICAL ANALYSIS

The intensity of fluorescence depends upon concentration. According to Beer's law the fraction of light transmitted will be

$$\frac{P}{P_0} = e^{-\in bc} \qquad \qquad ...(31.1)$$

where P_0 is the intensity of incident radiation, P is the intensity of the transmitted radiation, b = optical pathlength, c = concentration, $\in$ = molar absorbivity.

The fraction of light absorbed will then be

$$\left(1-\frac{P}{P_0}\right) = \left(1-e^{-\in bc}\right). \qquad \qquad ...(31.2)$$

The equation can be rewritten as

$$(P_0 - P) = P_0\left(1-e^{-\in bc}\right) \qquad \qquad ...(31.3)$$

which is equal to the total fluorescence intensity. If it is multiplied by quantum efficiency of fluorescence ϕ, we then get

$$F = (P_0 - P)\phi = \phi \times P_0\left(1-e^{-\in bc}\right) = \phi P_0\left(1-10^{-abc}\right) = 2.303\phi P_0 bc \qquad ...(31.4)$$

However, for dilute solutions a small fraction of light is absorbed so $\in bc \geq 0.05$ so we can presume $F = \phi P_0 (2.3\in bc)$, if K is the instrument constant. For all practical purposes, for very dilute solutions intensity of fluorescence varies directly with concentration. At higher concentrations a straight line relationship is not valid.

Errors: The possible errors in fluorimetry as well as phosphorimetry. The quantum efficiency of a luminescent process must be the same and reproducible. If such quantum efficiency decreases, it leads to the phenomenon of quenching which is discussed later in this chapter. Heavy atoms and paramagnetic species affect ISC *i.e.* intersystem crossing (Fig. 31.1) and quantum efficiency, especially in fluorimetry. Even O_2 contributes its paramagnetism and leads to quenching. A drift or change in source intensity and position of the cell leads to wrong measurement. Finally the inner filter effect which originates due to difference in intensity of fluorescence will lead to wrong measurements.

Merits: Advantages of fluorimetry as well as phosphorimetry in quantitative analysis are many. The methods are selective and have no spectral interference; if such interference is present it can be mitigated by the right choice of both excitation and luminescence wavelength. The method is sensitive. Time resolution is of importance in phosphorimetry, due to its long lifetime. This also helps to eliminate scattering from the sample. In comparison to fluorimetry, phosphorimetry is however not commonly used in chemical analysis due to the complexity of instrumentation and the need to cool the sample to get reproducible results. EPA, 5 : 2 : 2 mixture of diethylether, isopentane and ethanol is used as a special solvent in phosphorimetric analysis.

31.3 LUMINESCENCE INSTRUMENTATION

A typical system employed in the method is shown in Fig. 31.2. Since fluorescence is an isotropic property, *i.e.* because there is emission in all directions, it is possible to detect the emitted radiation in any desired direction from the sample.

To facilitate elimination of interferences of excitation radiation, luminescence is measured at right angles to the exciting radiation. This 90° orientation is common in all instruments. The exciting radiation is generally provided by an intense ultraviolet radiation source such as a xenon arc or mercury lamp. A monochromator is used to select the proper wavelength from excitation source for analysis. After excitation the molecule shows luminescence which may be in the form of fluorescence or phosphorescence which in turn is measured utilising a photosensitive detector (Fig. 31.2), Those models using filters (for monochromatisation of light) are called fluorimeters or phosphorimeters. Those using prisms or gratings are called spectrofluorometers or spectrophosphorimeters. The latter are more sensitive and versatile and provide a complete excitation spectrum. Filter fluorimeters are less expensive but very useful; though they are not very useful for obtaining structural information. With a few minor modifications a photoelectric colorimeter can be used for fluorimetric analysis.

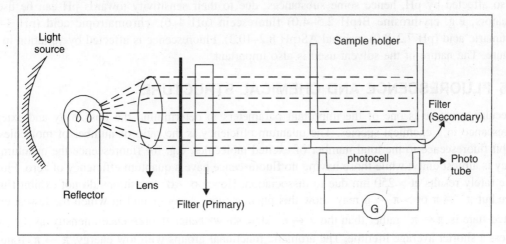

Fig. 31.2 Optical diagram of fluorimeter

31.4 APPLICATIONS OF FLUORIMETRY

This technique has varied applications in medicine, forensics, and environmental science in addition to inorganic and organic analysis, e.g. drugs like quinine can be analysed at nanogram levels. LSD *i.e.* lysergic acid diethylamide can be analysed fluorimetrically especially from a sample of blood or urine. The excitation and fluorescence wavelengths are respectively 335 and 435 nm. Other metabolites do not interfere in such determinations. In environmental chemistry, atmospheric carcinogens belonging to polynuclear aromatic hydrocarbons like 3-4 benzopyrine which are formed by incomplete combustion of fuels like petroleum and which is present in autoexhaust and cigarette tar can be analysed by fluorimetry. Such analysis is carried out in sulphuric acid medium, with an excitation wavelength of 545 nm. Reproducible results are obtained at −190°C. It is interesting to note that one cigarette contains 10 ηg of benzopyrine. It can be determined with accuracy to nanogram concentration by fluorimetry.

Finally inorganic analysis of metals like Al, Be, Ca, Cd, Cu, Ga, Ge, Hg, Mg, Nb, Sn, Ta, Th, W, Zn and Zr can be done by fluorimetry. Reagents like 8-hydroxyquinoline, 2, 2-dihydroxyazobenzene, dibenzoylmethane, flavonol, benzoin, alizarine and quinalzarine can be used as complexing ligands.

The measurement of the intensity of fluorescence is a suitable and rapid method for the determination of a large number of specific substances, e.g. riboflavin, thiamine hydrochloride and vitamins. Many inorganic materials also exhibit fluorescence in aqueous solution or with organic reagents, e.g. uranium complexed with NaF exhibits fluorescence and hence can be determined fluorimetrically. Zn, U, W and Mo also exhibit fluorescence on complexation. Al, Ga, Zn, Mg also show this phenomenon when complexed with 8-hydroxyquinoline at tracer levels. Al and Be complexes with Morin (pentahydroxy flavanol) also exhibit fluorescence. Similarly complexes of quinalizarine with metals like thorium show fluorescence. This fluorescence intensity can easily be measured and the element in question quantitatively determined.

Precautions: However, certain precautions should be taken during such measurements. Since vegetable and animal matter also shows fluorescence it should be removed before measurement. Sometimes the total intensity of fluorescence is measured. Then the reagent producing fluorescence is destroyed and fluorescence of the solution is once again measured and the difference between the two readings gives the fluorescence due to cations which show fluorescence with the ligand. Fluorescence is also affected by pH, hence some substances, due to their sensitivity towards pH can be used as indicators, e.g. erythrosine B(pH 2.5–4.0) fluorescein (pH 4–6), chromatropic acid (pH 3–4.5), o-coumaric acid (pH 7.2–9.0), napthal AS(pH 8.2–10.3). Fluorescence is affected by variation in temperature. The nature of the solvent used is also important.

31.5 FLUORESCENCE AND CHEMICAL STRUCTURE

Molecular structure is one of the important parameters in determining the possibility and extent of fluorescence in a chemical species. The quantum efficiency is the ratio of number of molecules that exhibit fluorescence to the total number of excited molecules, e.g. for fluorescence the quantum efficiency is almost unity; while those having no fluorescence have a quantum efficiency of zero. Fluorescence rarely results at > 250 nm due to dissociation. Hence $\sigma - \sigma^*$ transitions do not exhibit fluorescence but $\pi^* \to \pi$ or $> \pi^* \to n$ may show this phenomenon. A compound in which the lowest energy excited state is $\pi \leftarrow \pi^*$ rather than the $n \leftarrow n^*$ state shows better fluorescence intensity, as the former possess a shorter average lifetime. The aromatic functional groups with low energy, $\pi \to \pi^*$ transition levels exhibit fluorescence. Aliphatic and alicyclic carbonyl compounds or highly conjugated double bond structures also show this effect. The unsubstituted aromatic hydrocarbons show fluorescence in solution. Heterocyclics like pyridine, furan, thiophene and pyrrole do not fluoresce; but fused ring structures do. Quinoline, isoquinoline and indole show fluorescence. Substitution in the benzene ring shifts the absorption maxima and also changes the fluorescence peak, e.g. both benzene and toluene exhibit fluorescence maxima at a wavelength of 270-310 nm but halobenzenes show maxima at a wavelength of 275-380 nm and benzoic acid shows a shift to a wavelength of 310-390 nm. Thus the effect of halogen substitution is significant. The substitution of a carboxylic acid or carbonyl group of

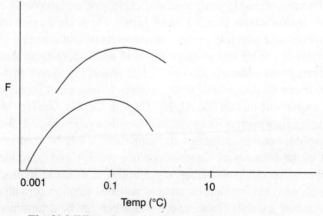

Fig. 31.3 Effect of temperature on fluorescence of arenes

a aromatic ring inhibits fluorescence. Fluorescence is favoured in molecules having a rigid structure. The quantum efficiency of fluorescence is lowered with a rise in temperature, due to a higher frequency of collisions. From Figs. 31.3 and 31.4 it will be noted that temperature as well as pH influence the fluorescence intensity hence close control of pH is essential.

Further the molecules containing multiple conjugated double bonds with a high degree of resonance stability exhibit fluorescence. Thus electron donating groups e.g., $-NH_2$ or $-OH$ and OCH_3 should be on a resonating nucleus. Electron withdrawing groups such as $> COOH$, $-NO_2$, $-N=N-$ and $-Br$, $-I$ and CH_2COOH decrease or destroy fluorescence. Ring closure favours an increase in the fluorescence of aromatic compounds. Formation of metal chelates promotes exhibition of fluorescence. Such structures are coplanar with respect to the chromophore. Dissociation can enhance fluorescence, substances fluoresce more in a rigid state or in glycerol as compared to one in alcohol or water solutions.

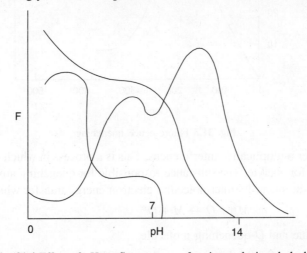

Fig. 31.4 Effect of pH on fluorescence of various substituted alcohols

31.6 FLUORESCENCE QUENCHING AND INNER FILTER EFFECT

This represents a reduction in the intensity of fluorescence due to the specific effects of constituents of the solution. Quenching may occur due to excessive absorption of either primary or fluorescent radiation by the solution. It is called concentration quenching or inner filter effect. Coloured species in solution with the fluorinating species will cause serious interference by absorbing fluorescent radiation. This phenomenon is termed as "inner filter effect." It can arise from too high a concentration of the fluorescence producing material. Such molecules reabsorb the emitted radiation leading to the inner filter effect (Figure 31.5). If the effect of quenching is produced by the fluorescent substance itself it is known as self quenching; e.g., the decline in fluorescence of phenol with concentration. Quenching can also be effected by nonradiative loss of energy from the excited molecules. Chemical quenching is sometimes applied to the reduction in emission due to actual changes in the chemical nature of the fluorescent substance. A common observation is pH sensitivity. As described earlier, fluorescence intensity depends upon temperature. A hazard in fluorimetry is contamination with fluorescent substances in amounts undetectable in the solution (see Figs. 31.3 and 31.4).

Fluorescence can be totally suppressed by some constituents of the sample. It is a multiplicative interference and is called as 'Inner filter effect' as defined earlier. An interference absorbs at same wavelength as analyte sample to decrease radiant power available to excite analyte sample. Alternatively, interferent absorbs at the wavelength at which analyte also show fluorescence leading to escape of number of emitted photons in sample and reach to detector to show and permit instrumental correction.

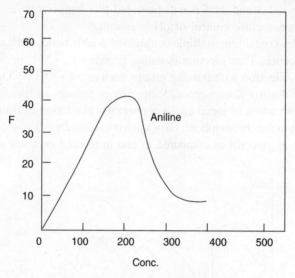

Fig. 31.5 Fluorescence quenching

Quenching is another multiplicative interferences. This is a process in which sample shows decrease in fluoresence intensity for analyte. The substance responsible for quenching are many e.g. O_2. It must be removed before measurement. Its intermolecular electron energy transfer which follows the path as:

$$M^* + Q \rightarrow M + Q^* \qquad \qquad ...(31.5)$$

If M^* is excited molecule and Q-quenching molecule.

Thus Q-acts as multiplicative/M deactivator molecule, Q^* excited quenching molecule and additive interferences. We give quenching mathematically as

$$\phi_F^0 / \phi_F = \left[1 + K_{\delta v}(Q) \right] \text{if } \phi_F^0 \text{ and } \phi_F^0 \qquad \qquad ...(31.6)$$

are fluorescence quantum yield for analyte in absence and presence of quencher at concentration (Q) and K = quenching constant. The analyte fluorescence depends upon Q-concentration. The effect of quenching decreases as the solution is diluted. So quenching can be eliminated by mere dilution with solvent which does not itself quench. In Figure 31.6 note that the original quenched intensities is shown

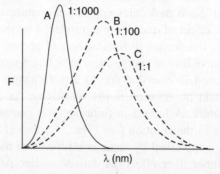

Fig. 31.6 Effect of dilution

by curve (C) but its dilution to 1:100 shows curve (B) and 1:1000 dilution gives curve A with high intensity for fluorescence of analyte. The potential quenchers must be removed before measurement.

The miscelle forming surfactact should be added to analyte so that quenching molecule is segregated from analyte molecule by surfactant which in fact protects analyte from quencher. Triplet state of organic molecule quench more. The phosphorosence would show more quenching. Cyclodextrins is used to protect analyte from quenching effect.

31.7 PHOSPHORESCENCE SPECTROSCOPY

These methods are based upon the nature and intensity of light emitted by molecules in the triplet state. On account of their long life they are subjected to a deactivation process and when the substance is dissolved in a rigid medium, phosphorescence emission can usually be observed. It is unique in regard to frequency, lifetime, quantum yield and vibrational pattern. Such properties are used for quantitative identification. The correlation of intensity with concentration serves as a basis for quantitative measurements.

Phosphorescence emissions arise from transitions between the first excited triplet state and the singlet state. Such states are of different multiplicities. When a phosphor absorbs radiant energy of a frequency within the normal absorption bands, the electrons of one of the molecules in the ground state are elevated to the first or higher excited singlet electronic state. The internal conversion and deactivation causes a drop in the potential energy of the excited molecule from higher excited singlet state to the first excited single state. The triplet state persists for a relatively long period of time and phosphorescence emission will continue over a relatively long period of time. It even continues after the excitations source is turned off, leading to afterglow. The best results are obtained at a 196°C.

31.8 PHOSPHORESCENCE AND CHEMICAL STRUCTURE

The introduction of paramagnetic metal ions such as copper and nickel gives rise to phosphorescence but not fluorescence. However Mg and Zn compounds show strong fluorescence. Phosphorescence is affected by the molecular structure such as unsubstituted cyclic and polycyclic hydrocarbons and those containing $-CH_3$, $-NH_2$, $-OH$, $-COOH$, $-OCH_3$ substituents which have lifetimes in the range of 5–10 seconds for benzene derivatives and 1-4 seconds for naphthalene derivatives. The introduction of a nitro group in a structure diminishes the intensity of phosphorescence, as does the introduction of aldehyde and ketonic carbonyl groups. The emission life time (τ) is in seconds in rigid media and is 10^{-2} –100 seconds in fluid media. The frequency of the vibrationless transition ($\Delta v = 0$) will be the highest frequency band of the phosphorescence spectrum. The vibrational pattern represents the spacings of the vibrational levels of the ground state. The quantum efficiency is the ratio of the quanta of light phosphorescence to the quanta observed.

31.9 PHOSPHORIMETRY IN QUANTITATIVE ANALYSIS

The expression relating phosphorescence intensity to concentration is similar to the one derived for fluorescence. With the usual notations it is derived as follows:

$$\frac{P}{P_0} = e^{-\in bc}$$

$$...(31.7)$$

$$\left(1-\frac{P}{P_0}\right)=\left(1-e^{-\epsilon bc}\right) \qquad ...(31.8)$$

$$\left(P_0-P\right)=P_0\left(1-e^{-\epsilon bc}\right) \qquad ...(31.9)$$

$$I=\phi P_0\left(1-10^{-abc}\right)=2.303\,\phi\,P_0\,2bc \qquad ...(31.10)$$

Since $I=(P_0-P)\,\phi$, where ϕ = quantum efficiency of phosphorescence.

$$I=K\,\phi\,P_0\,(2.3\in bc) \qquad ...(31.11)$$

hence $\qquad I=(2.303\,\phi\,P_0)\,2bc \qquad ...(31.12)$

$K\,\phi\,P_0\in b\,2.3$ are all constant for dilute solutions. Radiation power of the excitation source should be as large as possible and the length of time of excitation should be as long as possible to fully populate the triplet state. The interval of time following cessation of excitation is an important factor in phosphorescence. The influence of impurity molecules leading to deactivating collisions is significant, hence a need for a rigid medium is essential.

The phosphorescence spectrometer is similar to the fluorescence spectrometer with the addition of a resolution phosphoroscope as described earlier. The details of instrumentation are not given because they are exactly the same as that of a UV-visible spectrometer with the source of radiation, monochromator and detector. As stated earlier the detector is at right angles to the beam of incident radiation.

PROBLEMS

31.1 An organic compound was measured fluorimetrically. A standard solution of 1 µg/ml showed a fluorescence intensity of 73.0 units. If the unknown solution shows a fluorescence intensity of 62 units what is its concentration?

Answer: A standard solution of 1 ug/ml of solution has a fluorescence of 73.0 Hence the unknown solution should have concentration:

$$\frac{62}{73}\times1.0=0.8492\ \mu g/ml.$$

31.2 A beryllium-morin complex of strength of 0.2 mg/ml, *i.e.* 200 µg/ml showed an intensity of 78.8 with a 2.0 blank value. What will be the concentration of beryllium if the unknown solution shows 63.5 units with the same blank?

31.3 A solution of zinc with 8-hydroxyquinoline shows fluorescence. The calibration graph shows zinc concentration/fluorescence intensity as: 0/0, 0.10/7.50; 0.20/14.8; 0.30/22.6; 0.40/30; 0.50/37.7. What is the concentration of zinc in the unknown solution if the fluorescence intensity is 18.7 with a 2.4 blank?

31.4 Aluminium with oxine shows a strong fluorescence. A standard solution of 5 µg/ml of aluminium shows a fluorescence of 0.450. What is the concentration of aluminium in the unknown solution which exhibits a fluorescence of 0.990 units?

31.5 A 0.150 g of selenium shows a fluorescence of 0.250. If 50 µg of gallium are added, the fluorescence rises to 0.750. Determine the percentage of gallium in selenium if a blank shows a fluorescence of 0.05 unit.

31.6 The fluorescence of uranium-TBP is given by $C = 2 \times 10^{-5} + 5 \times 10^{-4} \log F$, where C is the concentration and F = fluorescence intensity. Determine the concentration of uranium (g/ml) if its fluorescence intensity is 50 units.

LITERATURE

1. D.M. Hercules, *"Fluorescence and Phosphorescence Analysis"*, Wiley Interscience, New York (1966).

2. F.D. Snell, *"Photometric and Fluorometric Methods of Analysis of Metals"*, Wiley Interscience (1980).

3. G.G. Guibault, *"Practical Fluoresence"*, 2nd Ed., Marcel Dekker (1990).

4. R.J. Hurtubise, *"Phosphorimetry—Theory, Instrumentation and Applications"*, VCH Publisher, New York (1990).

5. S. Udenfriend, *"Fluoresence Assay in Biology and Medicine"*, Academic Press (1979).

6. E.I. Wehry, *"Modern Fluorescence Spectroscopy"*, Plenum Publishing Co. (1981).

7. C.E. White, R.J. Argaaer, *"Fluorescence Analysis–A Practical Approach"*, Marcel Dekker (1970).

8. G.D. Christian, *"Analytical Chemistry"*, 5th Edition, John Wiley (1994).

Chapter 32

Nephelometry and Turbidimetry

The total phenomena of scattering of radiation is radiative scattering, which occurs due to the change in direction of radiation. There are several kinds of the scattering as (*i*) Tyndall scattering, (*ii*) Rayleigh scattering and (*iii*) Raman scattering. The Tyndall scattering occurs when diameter of scattering particle is equal or greater than that of incident radiation wavelength. This is true for colloidal solution. The Rayleigh as well as Raman scattering have the diameter of scattering particle much smaller in comparison to wavelength of incident radiation. In the Rayleigh scattering wavelength of incident and scattered radiation is same. However, in Raman scattering the wavelength of scattering radiation is greater or smaller than wavelength of incident radiation. Of the scattering phenomena, Tyndall and Raman scattering are useful in chemical analysis. Mathematically intensity (P) is proportional to the fourth power of radiation frequency *i.e.*

$$P = kv^4 \qquad \qquad ...(32.1)$$

if v = radiation frequency. The techniques of turbidimetry or nepholometry utilises this phenomena of radiation. Let us consider each kind of scattering from the viewpoint of analysis.

32.1 TYNDALL SCATTERING

The incident radiation frequency is larger than the radiative frequency. It is composed of reflection, internal reflection, diffraction and refraction of radiation. Due to similarity in Tyndall scattering and radiative scattering as far as wavelength is considered it is not used in qualitative analysis but turbidimetry or nephelometry can be used with Tyndall scattering. One opts out for nephelometry at low concentration of the analyte (~concentration 0.005). One can have small scattered radiation. The technique operates invariably in visible region. The intensity of Tyndall scattered radiation is proportional to second power of the frequency (v^2) if $\hat{i}$ is incident radiation. A incident wavelength chosen is such that there is no absorbance or fluorescence of radiative source. Both absorption and turbidance are proportional to analyte concentration. For nephelometric work fluorescence instruments are preferred. The preparation of working curve is a specialised technique. In environment analysis nephelometry is used to ascertain suspended particle matter in water and air.

32.2 RAYLEIGH SCATTERING

Both Rayleigh and Raman scattering occurs when small particles are analysed with size less than wavelength of the incident radiation. If size and refractive index is small for particles, they behave as radiative dipole if electromagnetic radiation passes through. On passage of electromagnetic radiation, the particle varies at the frequency of incident radiation. Oscillating dipole acts as point source which emits radiation in all direction. Since incident and scattering radiation frequencies leads to Rayleigh scattering, polarised particle shows great Rayleigh scattering. This is depicted in Fig. 32.1. The dipole moment (μ) which is induced by the electric field is proportional to field strength (E) and polarizability (a) of the particle. So $\mu = aE$. This further changes to

$$P = \frac{16\pi^4 v^4 \mu^2}{3c^3} \qquad \qquad ...(32.2)$$

(More polarisation more Rayleigh scattering), where ρ = intensity of scattered radiation, v = frequency scattered, c = concentration, μ = dipole moment.

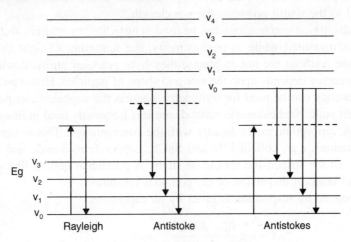

Fig. 32.1 Energy level diagram

Rayleigh scattering should not be confused with phenomena of fluorescence.

32.3 RAMAN SCATTERING

It takes place when frequency of scattered radiation is not same as incident radiation. (See Fig. 32.1) Such difference in frequency can be used for two mechanisms. If scattering species is in vibrational level above ground level, scattering occurs by return of species to vibrational level (it is lower energetically than the level in the beginning). This leads to antistokes Raman scattering which has higher frequency than that of incident radiation. There may be situation when in the Raman scattering radiation of low energy than incident radiation shows phenomena of stokes. It is used for quantitative analysis. The intensity of Rayleigh scattering is greater than that of Raman scattering. It is used for qualitative analysis.

We have

$$\Delta v = (v_1 - v_s) \qquad \qquad ...(32.3)$$

is change is frequency (Δv) between the incident radiation (v'_1) and scattered radiation (v_s). This corresponds to difference is energy between vibrational level of scattering species. If Δv is positive, scattered lines are stokes but if Δv is negative they are antistokes lines.

32.4 TURBIDIMETRY

Turbidimetry refers to the peculiar spectral properties of dispersion and may be taken as the ratio of light reflected to the light incident upon a dispersion. The intensity of light reflected by a suspension is a function of concentration when the other conditions are constant. The method of measurement of turbidity fall under three classes: measurement of the ratio of the intensity of scattered light to the intensity of incident light; measurement of the intensity of transmitted light to that of the incident light and measurement of the extinction effect *i.e.* the depth at which a target disappears beneath the layer of a turbid medium. The instruments measuring the Tyndall ratio are called Tyndall meters. In such instruments the intensities are measured directly. Turbidimetry involves the measurement of transmitted light, and is directly proportional to the concentration and depth of the dispersion. It also depends upon the colour. For smaller particles the Tyndall ratio is proportional to the cube of the particle size and is inversely proportional to the fourth power of the wavelength.

The principle of absorption spectroscopy can be used in turbidimetry wherein absorption due to the suspended particles is measured while in nephelometry, the scattering of light by a suspension is measured. Although the methods are not so precise they have practical utility. It will be seen that in much measurement accuracy depends upon the size and shape of particles. Hence any instrument used for absorption spectroscopy can be used for turbimetry whereas the nephelometer needs a receptor at right angles to the light path. Nephelometric methods are less frequently used in inorganic analysis. At a higher concentration, absorption varies directly with the concentration. This however does not hold good at low concentrations, e.g., colloidal Te and $SnCl_2$, copper ferrocyanide and heavy metal sulphides. The solubility of the suspension should be slight. A protective colloid like gelatin is used to obtain a nonstable and uniform dispersion of the colloidal substance.

The turbidity produced by suspension is given by an expression

$$S = \log \frac{P_0}{P} = \frac{Kbcd^3}{\delta^4 \alpha \lambda^4} \qquad \qquad ...(32.4)$$

where S = turbidance; P_0 = intensity of incident radiation; λ = wavelength; P = intensity of transmitted radiation; c = concentration; b = thickness of a layer of sample; d = average diameter of a particle and δ, K are both constant. This expression holds good for a dilute solution. In a monochromatic radiation, α, K, d, λ are constant hence the above equation reduces to

$$S \propto bc \text{ or } S = Kbc \qquad \qquad ...(32.5)$$

This expression is similar to Beer's law.

32.5 NEPHELOMETRY

All the laws governing turbimetric determinations are also applicable to nephelometric measurements. The only point of difference in nephelometry is that the light source and receptor are at right angles. In both turbidimetry and nephelometry one must use a uniform suspension. It is possible to obtain particles with a uniform physical character. It is necessary to control the concentration of the two ions which produce the precipitate. The ratio of concentrations of the two solutions to be mixed should be controlled. The manner and order of mixing should be observed and the time of mixing should be standardised. The amount of protective colloids present should be properly controlled and the temperature carefully noted. For best results during nephelometric measurements, turbidity should be homogeneous with a low density, turbidities should have the same dispersion. A constant degree of turbidity should be maintained through the entire process of measurement and the results should be definitely reproducible.

The applications of both turbidimetric and nephelometric techniques are numerous e.g. in water pollution studies the amount of sulphate in water is measured by nephelometry as barium sulphate. Previously, the turbidity of water was measured by a Jackson candle turbidimeter in JU-units but nowadays it is measured by a turbidimeter in NTU units. Ammonia from water can be determined with Nessler's reagent by turbidimetry. Similarly phosphate can be determined as phosphomolybdate. The amount of selenium or tellurium in a sample can be determined by nephelometry by reduction with stannous chloride as the blue sols. Other determinations include colloidal sulphur, acetone, pepsin, proteins and trypsin. Other examples of the use of light scattering methods are determination of silver as chloride suspension, gold as colloidal gold, calcium as oxalate suspension, potassium suspension with sodium cobaltnitrite, sodium with zinc uranylacetate and tellurium with dihydrogen sodium phosphite.

When a beam of light is passed through a transparent material, a small fraction of radiation energy is scattered. Further, if monochromatic radiation with a narrow frequency band is used, scattered energy will consist of radiation of the incident frequency, *i.e.*, Rayleigh scattering, but certain frequencies below and above that of the incident beam will be scattered which are strong and are termed as Raman scattering.

How it is used in chemical analysis for the elucidation of structure as well as in quantitative analysis was already discussed in an earlier chapter.

PROBLEMS

32.1 In turbidimetric analysis of lead as sulphate as standard solution contained 2.5 mg/ml of Pb $(NO_3)_2$. The calibration curve was prepared in a total volume of 50 ml by adding the requisite amount of solution. The corresponding values for absorbances were as follows:

Volume of stand. solution (ml)	2	4	6	8	10
Absorbance	0.15	0.25	0.32	0.39	0.43

A 50 ml of the solution was diluted to 250 ml and 5 ml of this solution showed an absorbance of 0.35. What is its concentration?

Answer: We should plot a graph of absorbance *vs* concentration. In the above samples the amount of Pb $(NO_3)_2$ is 2.5 mg/ml so we have:

5 (0.150); 10 (0.25); 20 (0.390); 25 (0.480)

as the amount of Pb $(NO_3)_2$ *vs* absorbance. The unknown solution has an absorbance of 0.350 which corresponds to 17.9 mg/ml in 50 ml. It is 179 mg or 250 ml ≡ 11.19 mg for 50 ml *i.e.*, 17.9 mg/ml of Pb $(NO_3)_2$ = 11.19 mg/ml of Pb as 331.0 Pb $(NO_3)_2$ corresponds to 207 Pb.

32.2 100 ml of HNO_3 (density = 1.50) was made upto 250 ml and to 25 ml of this diluted solution $AgNO_3$ was added to precipitate Cl^- ions. The volume was made up to 50 ml and the absorbance was 0.350. For the calibration curve the following readings were obtained.

Cl^- (Absorbance) as: 0.05 (0.05); 0.1 (0.1); 0.14 (0.2); 0.19 (0.3) and 0.24 (0.4). What is the chloride content of the unknown sample?

Answer: We plot graph of $[Cl^-]$ *vs* absorbance which is a straight line. The unknown solution showing 0.35 corresponds to 0.21 ng/ml of Cl^- ion. The whole lot of nitric acid will contain 0.21 × 10 = 2.1 mg/ml of Cl^- ion.

PROBLEMS FOR PRACTICE

32.3 0.556 g of a sample of sulphide mineral was made into a solution of 250 ml and an aliquot of 5 ml was made up with $BaSO_4$ to 50 ml to give an absorbance of 0.450. The readings for the calibration curve are: (mg/ml) absorbance as 0.05 (0.1); 0.1 (0.2); 0.16 (0.03); 0.22 (0.4); 0.27 (0.5). What is the percent of sulphur in the mineral?

32.4 A standard solution containing 7.59 mg/250 ml of $BaSO_4$ was prepared and gave a reading of 5.0 while the unknown solution gave a reading of 4.26. What is the concentration of sulphate in the unknown sample?

32.5 0.75 g of a photosensitive material containing selenium was dissolved in 10 ml of water then 1 ml of the solution gave 25 ml of suspension with intensity of 35 units. Determine the concentration of selenium by using a calibration curve as (μg/5 ml Se) Absorbance: 2(25); 4 (50); 6 (745); 8 (100).

LITERATURE

1. H.C. Van de Hulst, *"Light Scattering by Small Particles"*, Wiley (1957).
2. J.H. Harley, S.E. Wiberley, *"Instrumental Analysis"*, John Wiley (1962).
3. C.N. Bonaveli, *"Fundamentals of Molecular Spectroscopy"*, 2nd Ed., McGraw Hill (1980).
4. P.S. Sindhu, *"Molecular Spectroscopy"*, Tata McGraw Hill (1985).
5. G. Herzbers, *"Molecular Spectra Molecular Structure"*, Van Nostrand (1950).
6. D.A. Skoog, F.J. Holler, T.A. Nieman, *"Principles of Instrumental Analysis"*, Harcourt Brace Publ. (1998).

Chapter 33

Refractometry and Interferometry

Refractometry is the branch of sciences which dwells on the measurement of refractive index while interferometry concerns measurement with fraction of light during transmission. In earlier days in chemical analysis, refractive index of material was considered as the unique physical property of substance like melting point or boiling point. But it was hardly used for the qualitative or quantitative analysis of substances. However, with the advent of sophisticated refractometer like one designed by Abeg or immersion refractometer or advanced version of interferometer, the measurement of refractive index became simple. This was specially useful for the characterisation of organic compounds with reference to liquids or solids; specially for identification and ascertaining the exact composition of biological products like purines and amino acids. The recent applications of the refractive index measurement is in high performance liquid chromatography wherein RI detector *i.e.* refractive index detector has been extensively utilised for the quantification of the number of organic compounds. Most of the organic liquids have their refractive index between 1.25-1.80. Accurate analysis was most essential. This index varies with temperature. Although better instrumental techniques like infrared spectroscopy or nuclear magnetic resonance spectroscopy have come to forefront. Relatively more reliable and simple, measurement of the refractive index provides a rapid method in industrial analysis or while handling bulk of chemicals. It was used for the analysis of gases like carbon dioxide or alcohol in blood, sea water, marine effluents, synthetic mixtures and heavy water samples. The measurement of refractive index in gases needed more care and better equipment.

33.1 PRINCIPLES OF REFRACTOMETRY

Refractive index is defined as the ratio of velocity of electromagnetic radiation (c) in vacuum to the velocity of electromagnetic radiation in a medium (v). Therefore we get the equation

$$n = \frac{c}{v} \qquad\qquad ...(33.1)$$

The compounds have varying refractive indices. The difference between two or more compound's refractive index is very small ranging from $1.80 - 1.25$. We need to measure it to third place of decimal as 0.001. The actual measurement of 'c' is rather difficult in practise, hence a change in the direction of the beam of light as it enters sample from medium with different refractive index is measured. The change in direction invariably occurs (see Fig. 33.1) on refraction. Thus when a beam of radiation strikes against a medium (denser) we have angle of incidence radiation as "$\hat{i}$" while when it passes from denser to rarer media we have angle of refraction as "$\hat{r}$",

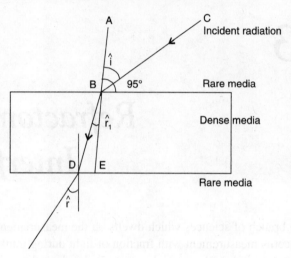

Fig. 33.1 Refraction in dense/rare media

As angle $\hat{i}$ increases, the ratio of $\left(\dfrac{\sin i}{\sin r}\right)$ remains constant. The total reflection can occur only when light passes from denser to rarer atmosphere. The value of refractive index (n) depends upon temperature and pressure e.g.

$$\text{specific refraction } (r_D) = \frac{n^2-1}{n^2+2}\times\frac{1}{\rho} \qquad ...(33.2)$$

where ρ is the density. The mole fraction is obtained by multiplying specific rotation (r_D) by molecular weight. It is an additive property. According to Snell's law we have

$$\frac{\sin i}{\sin r} = \frac{v_1}{v_2} = \frac{n_2}{n_1} \qquad ...(33.3)$$

wherein n_1, n_2 are refractive indices while v_1, v_2 are velocities of radiation of two compounds. If (n_1) is known, (n_2) can be easily evaluated. The angle $\hat{i}$ increases if radiation is passing from optically thick or dense to rare medium, but if the radiation passes from rare to optically dense medium, radiation is refracted to perpendicular line (BE in Fig. 33.1). The angle of refraction approaches 90° with increase in $\hat{i}$ (dense $\rightarrow$ rare travel media). The transformation are depicted in Fig. 33.2 and Fig. 33.3.

In Figure 33.2, θ represents "critical angle" obtained by extending refracted beam to first dense medium. The refractive index (n) can be calculated from knowledge of critical angle ($\hat{\theta}$). If $\hat{r} = 90$, $\hat{i} = \theta$ we have

$$\sin\theta = \frac{n^2}{n_1} = n_{21} \qquad ...(33.4)$$

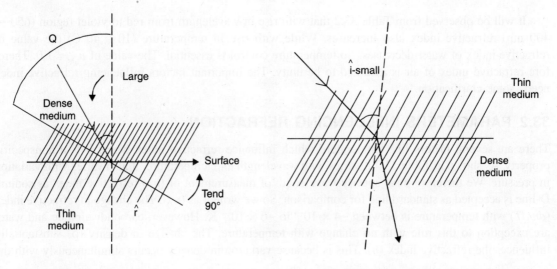

Fig. 33.2 Dense to thin media ($n = 1.5$)   **Fig. 33.3** Thin to dense media ($n = 1$)

Now in Fig. 33.3 the radiation travels from rare to dense media both $\hat{i}$ and $\hat{r}$ increase, however, $\hat{i}$ is always greater than $\hat{r}$. It is worthwhile to examine (Table 33.1). The atomic refraction of organic groups from which we can calculate mole fraction (Mr_D).

Table 33.1 Refractions of atomic groups (n_r)

Atom	Ref. index	Atoms	Ref. index
H 1.10		Cl 5.96	
C 2.41		Br 8.86	
C = C 1.73		I 13.90	
C ≡ C 2.40		–C ≡ N	5.45
> C = O 2.11		F 1.60	
O—H 1.52		C—O—C 1.64	

The mole refraction (M_r) is property dependent on the structure of the arrangements of atom within a molecule e.g. CH_3COOH, $M_r = 13.09$. The temperature influences the refractive index (n) as shown for H_2O in Table 33.2.

Table 33.2 Variation of refractive index with wavelength and temperature

Effect of λ		Effect of temperature °C	
λ (nm)	n (increases)	temp. °C	n (decreases)
650	1.3312	10	1.3337
590	1.3330	20	1.3330
480	1.3371	40	1.3306
430	1.3404	50	1.3290

It will be observed from Table 33.2 that with rise in wavelength from red to violet region (650 → 400 nm) refractive index also increases. While with rise in temperature (10 – 40°C) the value of refractive index of water decreases. So temperature control is essential. The value of n_{air} = 1.0. Therefore refractive index of air is assumed to be unity. The important factors influencing refractive index needs some clarification.

33.2 PARAMETERS INFLUENCING REFRACTION

There are several factors like temperature which influence refractive index hence it is a nonspecific property. As shown in Table 33.2 it varies with wavelength and temperature. It also varies with variation in pressure. We therefore must specify wavelength of measurement of (n_D) refractive index as sodium D-line is accepted as standard line for comparison. So we write it as (n_D). The change in refractive index (dn/dT) with temperature is between -4×10^{-4} to -6×10^{-4} K. However, carbondisulphide and water are exception to this rule with no change with temperature. The change in density (ρ) substantially influences the refractive index (n). This is because variation in density occurs simultaneously with the variation of temperature. So when we write (n_D^{20}) it means the measurement is carried out at 20°C with sodium light source (D). The change in pressure results in change in density and consequently refractive index also changes. Usually (dn/dp) change is around 3×10^{-5}/atm. For solid and liquid change is insignificant but is significant for gases. The concentration of the solution also affects refractive index, more the concentration, more change will be in $\hat{r}$ angle of refraction thereby influencing refraction index (n). Apart from sodium vapour lamp (D) one can use mercury lamp with absorption peak at 580, 540, 430 and 400 nm or cadmium lamp (~640 nm). However, both need particular filter to allow restricted band of light to pass through. Even xenon lamp can be used but natural while light is best for all measurement.

33.3 SIGNIFICANCE OF CRITICAL ANGLE DURING MEASUREMENTS

This principle is commonly used in refractometer for measurement of refractive index. The critical angle is defined as the angle of the refraction in a medium when incident radiation ($\hat{i}$) is at the grazing angle of 90°. So we get simplification of equation if ($\hat{\theta}$) is critical angle

$$\frac{n_2}{n_1} = \frac{1}{\sin\theta} \qquad ...(33.5)$$

and since sin θ = sin 90° = 1.0. If we have value of n_1 we can find n_2 if we know the critical angle $\hat{\theta}$ at fixed wavelength. This is the basis of (n) measurement in most of the refractometers.

33.4 REFRACTOMETERS

There are several kinds of refractometers for the measurement of refractive index. They include (*i*) Abbe's refractometer (*ii*) Immersion refractometer (*iii*) Pulfrich refractometer (*iv*) Differential refractometer (*v*) Interferometric of refractometer. Although it is not worthwhile going into details of discussion on each model a salient feature from the viewpoint of comparison in terms of significant applications and principle are covered in Table 33.3.

Table 33.3 Various refractometers—salient features

Sl.No.	Property	Abbe's Model	Immersion Type	Pulfrich model	Differential model	Interferometer type
1.	Range (n)	1.3 –1.70	1.32 –1.60	1.30 –1.60	1.30 –1.90	–
2.	Precision	$\pm 1 \times 10^{-4}$	$\pm 3 \times 10^{-5}$	$\pm 1 \times 10^{-4}$	$\pm 2 \times 10^{-3}$	0.002
3.	Sample size	0.05 – 0.10 ml	5 – 30 ml	0.5 – 5 ml	5 ml	0.10 ml
4.	Application	Routine	Routine	Precision	Qualitative	High precision
5.	Source of light	White light or Sodium D lamp	White light commonly used	Monochromatic source of light	Sodium Lamp D	Monochromatic light used.
6.	Optics	Refracting prism	Prism lens collimator	Prism disk with Vernier scale	Filter and hollow prism	Prism chopper Balance meter.

Some of salient features of these instruments are considered very briefly as

  (i) Abbe's model is most popularly used in the laboratories. At least 0.03 ml sample is required for analysis. It is useful for analysis of carbohydrates.
 (ii) Immersion type model is also called dipping model wherein prism immerses in solution. With combination of prisms one can measure n = 1.67.
(iii) *Pulfrich refractometer:* It uses monochromatic radiation to measure the critical angle using relation

$$n = \left(n_p^2 - \sin^2 \theta_2 \right)^{1/2} \qquad \qquad ...(33.6)$$

Large volume of the sample is needed for measurement. If ρ = density, θ = angle of incidence, n = refractive index.

(iv) *Differential model:* A difference between 'n' of sample and reference is measured. The readings obtained are very accurate due to less difference in the temperature coefficient of two liquids namely analyte and reference samples.
 (v) Finally interferometer type instruments have highest precision and accuracy, One needs reference and analytical solution. In IR region measurement is possible. The principle involves the characterisation of the interference to fringes.

33.5 QUALITATIVE AND QUANTITATIVE ANALYSIS

The substance must be present in pure form. A refractive index of known compound is compared with unknown by measurement of 'n' under identical conditions. By comparison with Lorentz-Lorenze molar refraction (M_r) is used.

$$M_r = \left(\frac{n_2 - 1}{n^2 + 2} \right) \frac{M}{\rho}, \qquad \qquad ...(33.7)$$

if M = molecular weight, n = refractive index and ρ-density of sample expressed as gm/cm^3. The mole fraction of atoms if added gives molar refraction of compound.

This is possible by constructing a calibration curve as done in UV-visible spectrophotometry. A straight line plot is obtained but in few places calibration curve is also used. One expresses the quantitative figure in terms of refractive increment (K') which is given as

$$K' = \left(\frac{n - n_0}{c} \right) \qquad \qquad ...(33.8)$$

if c = concentration, n = refractive index of sample, n_0 = refractive index of the solvent. The differential instruments are preferred for such analysis.

33.6 ANALYTICAL APPLICATIONS

The concentration of protein, quality of urine and analysis of sugar is invariably carried out by this technique where the scale directly gives concentration figure. The mixture of (NaCl + KCl) can be best analysed by the refractive index measurements.

The measurement of refractive index is used on very extensive scale in liquid chromatographic work. It is also used in studies involving electrophoresis. The degree of separation by distillation can be ascertained by refractometry. For the analysis of refractive indices of gases a interferometer type model is preferred. The permeability of ballon for hydrogen or helium is ascertained by this method. The analysis of sea water is done by interferometer. The study of the colloidal solution, potable water, sewage, fermented broth can also be done by refractometer. Serum, carbon dioxide in blood, alcohol in blood (rash driver's blood) heavy water composition can be estimated by it. The index of refraction is useful in qualitative characterisation. A plot of 'n' against 'ρ' i.e., refractive index vs density of compounds helps us to identify the organic substances (Fig 33.4). The instrumental precision should be reasonable. The composition of sample must be uniform. A curve of n gives quantity of analysed sample. One needs extremely high precision for the analysis of gaseous compound as the refractive index varies linearly with pressure and temperature. On molecular or molar basis refraction is an additive property. The refraction per mole or molar refraction (M_r or R) is

$$M_r \text{ or } R = \left(\frac{n_2 - 1}{n^2 + 2} \right) \frac{M}{\rho} \qquad \qquad ...(33.9)$$

if ρ = density, M = molecular weight. This value is temperature independent. This value throws light on external interactions, intramolecular association and other related parameters.

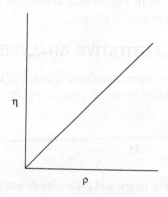

Fig. 33.4 Relation of refractive index with density of substance

The refractive index is one of the dependable physical property which is dependent upon temperature as well as pressure. It is very useful for both qualitative and quantitative analysis. In recent years it has attained an importance as the detectors used in high performance liquid chromatography and in electrophoresis. We expect its extensive applications in the biological sciences in future.

33.7 INTERFEROMETER PRINCIPLES

The refractometer depends upon the refraction of light which passes from one medium to another media. The intereference of light decides the finest measurement. In interferometer, there is no diffraction. The light enters and leaves the analyte at an angle of 90° (Fig. 33.5).

A beam of parallel light enters through the apertures A_1 and A_2 (if A_1T and A_2T are at the same distance). Two light beam reach at point (T) to indicate bright image. At all other points viz. S_1S_2 two beams of light are not of the same length.

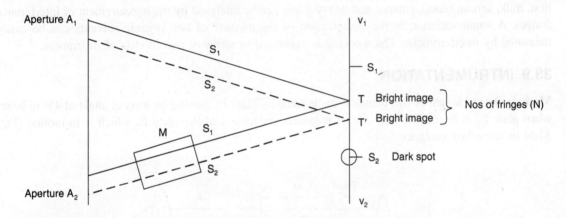

Fig. 33.5 Interferometer

A_1S_1 and A_2S_2 are not same and at point S_1 or S_2 two beams differ by half a wavelength. Interferences of two beam produces dark spot. At point S_2 the difference is one wavelength and a bright spot is formed. With monochromatic radiation the succession of light and dark spot continues all along. Now if a substance is having slightly greater refractive index is placed in cell M, the length of beam A_2T is increased by an amount Δx as the velocity of light through 'M' is decreased. The extent of increase depends upon thickness of 'M' and refractive index 'n'

$$\Delta x = x\,(n - n_0) \qquad \qquad ...(33.10)$$

if x = thickness, n = refractive index, n_0 = refractive index of media. The velocity of light in two different media is proportional to their refractive indices. Two beams will not arrive as 'T' in phase but at other points 'T' which is at optically equal distant from A_1 and A_2. If λ is wavelength the distance between T and T′ is measured in number of fringes N (composed of dark and light band).

$$N = \frac{\Delta x}{\lambda} \quad \text{or} \quad N = \frac{x(n - n_0)}{\lambda} \qquad \qquad ...(33.11)$$

Thus if $N > 1.0$ all bands are similar. White light rather than monochromatic light produces dependable results. It is blue in centre and red on outer edge with substance 'M' in path of beam has been shifted when two plates of equal thickness are placed in two beams, the number of bands that the central band shifts provides a method for calculating refractive index of one plate. Intereferometer is used mainly for measuring concentration of solutions and gases.

The interferometers are commercially available. They can be used as the direct reading instruments. As used a calibration plot is constructed by plotting concentration against reading by interferometer, to provide a smooth straight linear curve. In a similar manner unknown sample is measured and from plot of reading (N) vs. concentration (C) the unknown concentration can be calculated.

33.8 APPLICATIONS

Interferometer is used for analysis of gases. The permeability of balloon fabrics to hydrogen or helium can be measured. Methane or carbon dioxide can be analysed at ppm level. The sea water can be characterised by interferometer. The analysis of alkali metals can be easily carried out. Water properties like adsorption can also be measured by interferometer. The sewage, fermentated broth, colloidal solution, milk, serum, blood, ethanol and heavy water can be analysed by the measurement of interference fringes. A small variation in the composition of the mixture of two organic solvents can be easily measured by interferometer. The working is expressed in terms of two models of instrument.

33.9 INTRUMENTATION

Mode I: Optical length of two beams is made same by plate E_2 moving on lever at angle of 45° to beam when plate E_1 is fixed. The micrometer measures movement of the plate E_2 which is in motion (Fig. 33.6) in secondary category.

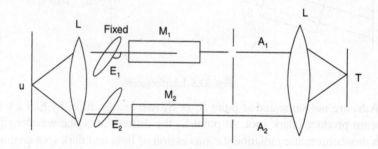

Fig. 33.6 Extended interferometer

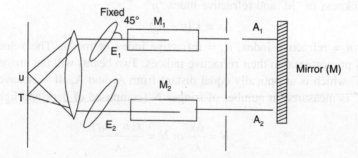

Fig. 33.7 Folded type interferometer

Mode II: This is portable model. The light is reflected by mirror (M) so that it passes twice through chambers and bands are visualised (Fig. 33.7). A precision can be obtained with half length of beam. A_1 and A_2 are slits to focus beam. The scale reading is proportional to thickness of solvent with cell of 1-4 cm length. However on account of availability of efficient spectroscopic methods involving atomic or molecular absorption or emission, these methods involving refraction or interferometry have gone out of date and obsolete.

LITERATURE

1. H.A. Strobel, *"Chemical Instrumentation: A Systematic Approach"*, Addison Wesley Publication Co. In. (1973).
2. R.D. Braun, *"Introduction to Instrumental Analysis"*, McGraw Hill and Co. (1987). Indian edition (2004).
3. S.S. Batsenoy, *"Refractory and Chemical Structure"*, New York Consultant Bureau (1902), cited in ref. No. 1.
4. G.W. Ewing, *"Instrumental Methods of Analysis"* 5th Ed., McGraw Hill and Co. (1960).
5. H.H. Willard, J.J. Merit, J.A. Dean, S. Newman, *"Instrumental Methods in Chemical Analysis"*, D. Vanstrad and Co. (1958), C.B.S. Publishers, Indian edition (2001).
6. D.A. Skoog, F.J. Holler, T.A. Nieman, *"Principles of Instrumental Analysis"*, Fifth edition, Harcourt Brace College Publ. (1998).

Mode II: This A portable model. The light is reflected by mirror (M) so that it passes twice through the sample and bands are visualised (Fig. 35.7). A precision can be obtained with half length of beam A, and A₁ are slits to focus beam. The scale reading is proportional to the ratio of solvent with that of path length. However on account of availability an efficient directness other methods become more common.

Chapter 34

Polarimetry and Related Methods

Many transparent non-symmetrical structures rotate the plane of polarised radiation. Such materials are called optically active compounds. e.g. quartz, sugar etc. The rotation is dextro-rotatory (+) if it is in a clockwise direction and it is laevo-rotatory (−) if it is in a counterclockwise direction to an observer looking toward the light source. The extent of rotation depends upon various parameters such as the number of molecules in the path of the radiation, concentration, length of the containing vessel, wavelength of radiation and finally the temperature. The specific rotation is defined as $[\alpha]t = \dfrac{\alpha}{dc}$ where α is the angle through which the plane of polarised light is rotated by a solution of concentration 'c' grams of the solute per ml of the solution, when contained in a cell of 'd' (decimeter) in length. The wavelength commonly specified is that of sodium vapour of 5893 Å or 590 nm. Some typical values of specific rotations for optically active compounds are listed in Table 34.1.

Table 34.1 Specific rotations of some compounds

Compound	$[\alpha]_D^{20}$	Compound	$[\alpha]_D^{20}$
d-Glucose	+ 52.7	Sucrose	+ 66.5
d-Fructose	−92.4	L-tartaric acid	+ 14.1
Maltose	+ 130.4	(All compounds measured in water)	

34.1 POLARISED LIGHT

Monochromatic radiation from a sodium lamp is made parallel by a collimater and polarised by a calcite prism. Then one half of the beam is intercepted through a half shade nicol prism. The radiation then passes through a protective window, the sample and then through a nicol analyser to the eyepiece for visual observation. The light is a bundle of rays going in all directions perpendicular to the plane of paper. In Fig. 34.1 light on passing through the polariser AOA′ resolves into BOB′ and COC′ of which COC′ is eliminated and BOB′ is transmitted. The resulting beam is plane polarised light. If polaride sheets (A and B) are placed as shown in Fig. 34.1, light will pass through, but no light will pass through sheet C, which is at right angles to the plane. We say that two planes are crossed if their axes are mutually perpendicular (e.g. A and C). Thus a beam of radiation may pass any of the plane polari-sation from zero with complete symmetry to 100% with complete polarisation. Thus some crystals or solutions and liquids rotate the plane of polarized light through them. This is optical activity. Its variation with wavelength is termed rotatory dispersion (ORD).

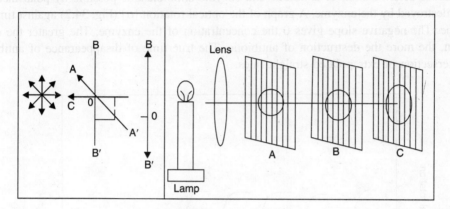

Fig. 34.1 Plane polarised radiation

A polarimeter can work without a small auxiliary prism. The angle through which the analyser is turned from the position initially crossed without the sample and then the position with sample present is measured to know the angle of rotation. The prism in half beam is oriented with its polarising axis at an angle of a few degrees from that of a polariser. At a particular position in the analyser, radiation passes in two halves of beam of equal power. This can also be done with photoelectric polarimeter.

34.2 APPLICATIONS OF POLARIMETRY

Sucrose can be analysed by direct application of the equation:

$$C = \frac{\hat{\alpha}}{d[\alpha]_D^{20}} = \frac{\hat{\alpha}}{2 \times 66.5} = \frac{\hat{\alpha}}{133.0} \qquad \text{...(34.1)}$$

if $[\alpha]_D^{20}$ for sucrose is 66.5 (d = decimeter). Since organic compounds have an expansion of about 0.1% per °C, the temperature must be rigorously controlled.

The equation is valid in the absence of other optically active substances.

$$\underset{\substack{\text{sucrose} \\ +66.5°}}{C_{12}H_{22}O_{11}} + H_2O \xrightarrow[\text{inversion}]{\text{acid}} \underset{\substack{\text{glucose} \\ +52.7°}}{C_6H_{12}O_6} + \underset{\substack{\text{fructose} \\ -92.4°}}{C_6H_{12}O_6} \qquad \text{...(34.2)}$$

if $[\alpha]_D^{20}$ is known. During inversion $[\alpha]_D^{20}$ changes from 66.5 (for sucrose) to (−92.4; + 52.7), *i.e.* −19.8° for glucose and fructose corresponding to an equimolar mixture of the products. By measuring rotation before and after rotational inversion it is possible to calculate the percentage of sucrose. If $\Delta\alpha$ is the angle of rotation it gives

$$\Delta\alpha = \frac{360 W_S [\alpha]_{D_1}}{342(V+10)} + \frac{W_X [\alpha]_{D_X}}{(V+10)} - \frac{W_S [\alpha]_{D_S}}{V} - \frac{W_X [\alpha]_{D_X}}{V} \qquad \text{...(34.3)}$$

where $[\alpha]_D$, $[\alpha]_{DS}$, $[\alpha]_{DS}$, $[\alpha]_{DX}$, are the specific rotations of sugar, sucrose and other active materials respectively. W_S = weight of sucrose, W_X = weight of active materials, V = volume of solution (100 ml).

The weight of sucrose will be $W_S = -1.17\Delta\alpha - 0.00105 \times (\alpha)_{D_X} W_X$.

Simultaneous determination of antibiotic and enzyme antibiotase is possible by polarimetry. The antibiotic is destroyed by the enzyme. A graph of the optical rotation [α] (Fig. 34.2) against time gives a straight line. The negative slope gives θ the concentration of the enzyme. The greater the enzyme concentration, the more the destruction of antibiotic. The true time of disappearance of antibiotic is found by intersection of extrapolated straight lines.

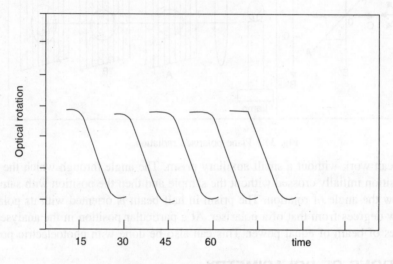

Fig. 34.2 Destruction of antibiotic with enzyme

34.3 OPTICAL ROTATORY DISPERSION AND CIRCULAR DICHROMISM

The wavelength dependence of optical activity is an important source of structural information about asymmetric compounds in comparison with the specific rotation at a single wavelength. It is closely related to the phenomenon of circular dichromism (CD). The change in optical activity with respect to wavelength is called optical rotatory dispersion (i.e. ORD). Certain crystalline materials show the effect of double refraction or birefrigence when a beam of light passing into a crystal spilts into two beams of equal power, which diverge from one another at a small angle. The two beams are seen to be plane polarised at right angles to each other. Thus plane polarised radiation can be resolved into two beams said to be circularly polarised in opposite sides (Fig. 34.3).

The refractive index and also the absorptivity for a medium for left and right hand circular components may not be the same and will be designated as n_L and n_R and $\in_L$ and $\in_R$ resp. For an isotropic medium such as glass or water $n_L = n_R$ and $\in_L = \in_R$, i.e. the refractive indices and absorptivities are independent of the state of polarisation. However, if $n_L \neq n_R$ a phase difference appears between the two components, i.e. the plane of polarisation is being rotated hence $\alpha \neq 0$ and if $\in_L \neq \in_R$. Thus one component is absorbed more strongly than the other, $(\in_L - \in_R) \neq 0$. The vector diagram showing polarisation then turns out to be elliptical with the eccentricity φ (Fig. 34.4) a measure of $(\in_L - \in_R)$. It can be shown that the angle of rotation and the eccentricity are given by equations:

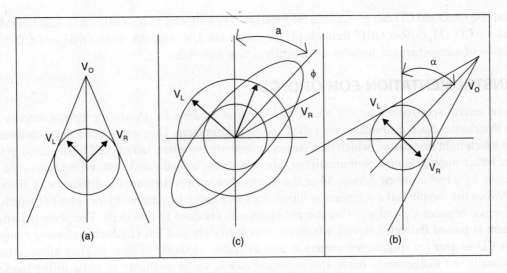

Fig. 34.3 Circularly polarised light

$$\angle\alpha = \frac{\pi}{\lambda}(n_L - n_R) \quad (\text{circular birefringence is } n_L - n_R) \qquad ...(34.4)$$

$$\phi = \frac{1}{4}(\in_L - \in_R) \quad (\text{circular dichromism is } \in_L - \in_R) \qquad ...(34.5)$$

In the above expression the quantity $(n_L - n_R)$ is called *circular birefringence* and $(\in_L - \in_R)$ the *circular dichromism* of the medium. In actual practice (α) is measured by polarimeter, while ϕ is determined indirectly through circular dichromism.

It has been shown mathematically that the ORD curve and CD are not independent. The relationship between the two is the "Cotton effect" which is schematically represented in Figs. 34.4 and 34.5.

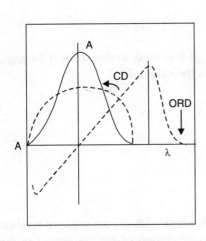

Fig. 34.4 Cotton effect

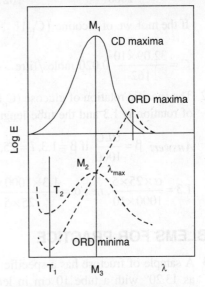

Fig. 34.5 ORD and UV spectral relation

Both the ORD and CD can give useful information. The midpoint of the ORD curve matches (M_2), the peak of CD (M_1). Also ORD through (T_1) matches the UV maxima. Both ORD and CD give information of stereochemical features of optically active materials.

34.4 INSTRUMENTATION FOR ORD, CD

There are several spectrophotometers which provide arrangements for measuring optical activity as well as absorbance. Such instruments are called spectropolarimeters. They consist of a monochromator through which light first passes, which then passes through the polariser, sample and finally through the analyser before impinging on a photomultiplier tube. The beam is modulated with respect to its state of polarisation by a motor driven device. Since the spectrophotometers measure the difference of absorbance between the sample and a reference solution, they can easily be adapted to measure CD which is the difference between $\in_L$ and $\in_R$. Circular polarisation is obtained in two steps. The plane polarised light beam is passed through a device which resolves it to right and left circularly polarised components. A CD adapter for spectrophotometers is also available. However in spite of great advancement internationally, no indigenously made spectropolarimeter is so far available in India in the market, although models with appropriate filters are fabricated.

PROBLEMS

34.1 The specific rotation of nicotine is 162° at 589 mm. What is the concentration of the solution (moles/1) if the rotation is 0.52° with a tube of length 10 cms?

Answer: $\beta = \dfrac{\alpha L C}{1000}$, where α is specific rotation, β is angle of rotation,

L – length of tube, C – concentration in g/100 ml. We have

$$0.52 = \frac{162 \times 10 \times C}{1000}, \text{ i.e., } C = \frac{0.52 \times 1000}{162 \times 10} = 32.09 \text{ g}/100$$

If the mol. wt. of nicotine $\left(C_{10}H_{14}N_2 = 162 \right)$, then

$$C = \frac{32.09 \times 10}{162} \ 0.020 \text{ moles/litre}.$$

34.2 The specific rotation of glucose ($C_6H_{12}O_6$) containing 5 g/l is to be determined when the angle of rotation is 1.3 and the tube length is 25 cms. What is the value of α?

Answer: $\beta = \dfrac{\alpha L C}{1000}$ if $\beta = 1.3$, $l = 25$, $C = 5$ g/1, we have

$$1.3 = \frac{\alpha \times 25 \times 5}{1000 \times 10}, \text{ i.e., } \alpha = \frac{1.3 \times 1000 \times 10}{25 \times 5} \text{ at } 104°$$

PROBLEMS FOR PRACTICE

34.3 A sample of fructose has a specific rotation of 45.20°. A solution shows the angle of rotation as 13.20° with a tube 10 cm in length. What is the concentration of the compound in the unknown solution?

34.4 A sample of raffinose with a specific rotation of 45.20 has a rotation of 12.12° with the length of the tube as 100 mm. What is its concentration?

34.5 What is the specific rotation if α = −3.50 when 9.88 g of solute is dissolved in 94.6 g of solvent? (tube length = 5 cm).

34.6 What is the specific rotation of morphine (0.45 g/30 ml methanol) if its angle of rotation is 4.92° and tube length is 25 cm?

34.7 If the specific rotation for glucose at different wavelengths is

λ (nm)	447	479	508	535	656
α	+96.62	+83.88	+73.61	+65.35	+41.89

Determine the specific rotation at 589 nm, 486 nm and 434 nm.

34.8 The specific rotation of strychnine hydrochloride is −104°. What would be its concentration if the angle of rotation is −1.56° and the tube length is 25 cm?

34.9 (+) malic acid rotates the plane of polarisation to the right while (−) malic acid rotates it to the left. If the specific rotation (β) is 2.3° for both acids, what is the concentration of (+) malic acid if it has an angle of rotation + 0.8 and if the total concentration of acids is taken as unity?

LITERATURE

1. C. Dejerass, *"Optical Rotatory Dispersion; Applications to Organic Chemistry"*, McGraw-Hill, New York (1960).

2. M. Velluz, M. Legrand, M. Grosjean, *"Optical Circular Dichromism"*, Academic, New York (1965).

3. D.C. Walker, *"Origins of Optical Activity in Nature"*, Elsevier (1979).

4. P. Crabbe, *"Optical Rotatory Dispersion and Circular Dichromism"*, Holden Day (1965)

5. J.C. Bailer, D.H. Bush, *"The Chemistry of Coordination Compounds"*, Reinhold (1956).

6. D.A. Skoog, D.M. West, *"Principles of Instrumental Analysis"*, 2nd Ed., Holt Saunders (1980).

14.4 A sample of raffinose with a specific rotation of 45.20 has a rotation of 12.12° with the length of the tube as 100 mm. What is its concentration?

14.5 What is the specific rotation if α = +1.50 when 9.55 g of solute is dissolved in 91.6 g of solvent? Tube length = 5 cm).

24.6 What is the specific rotation of morphine (0.45 g/10 ml methanol if the angle of rotation is It it has an angle of rotation + 0.8 and if the total concentration of acids is taken as unity?

Chapter 35

Nuclear Magnetic Resonance Spectroscopy

An interesting kind of interaction between matter and electromagnetic radiation can be seen by exposing a sample simultaneously to two magnetic fields, one stationary and the other varying at the same radio frequency. At optimum field, energy is absorbed by the sample and absorption can be noted as a change in the signal developed by a radio frequency detector and amplifier. The energy of absorption can be related to the magnetic dipolar nature of the spinning nuclei. If I represents the spin quantum number its values are $n/2$ where n can be 0, 1, 2... When $I = 0$ the nucleus does not spin, e.g. ^{12}C, ^{16}O, ^{32}S. Maximum sharpness of absorption peak occurs with nuclei for $I = 1/2$ e.g. ^{1}H, ^{19}F, ^{31}P, ^{13}C and ^{29}Si. Spinning nuclei simulate tiny magnets and so interact with the externally impressed magnetic field 'H'. Their rotatory motion causes them to precess. There are $(2I + 1)$ possible orientations and hence energy levels; thus the proton has two levels. The energy difference between them is given by

$$\Delta E = \frac{\mu H}{I} \qquad \qquad ...(35.1)$$

where μ is the magnetic moment of the spinning nucleus. Energy will be absorbed from the radio frequency field at a frequency where the condition $h\nu = E$ is fulfilled. This is called the Larmov frequency. We have therefore

$$\frac{\omega}{H} = \frac{2\pi\mu}{hI} \qquad \qquad ...(35.2)$$

where ω is the angular frequency of precession and is set equal to $\pi\upsilon$. The ratio $\omega/4$ is a constant peculiar to any nuclear species that has a finite value of I. This is called gyromagnetic ratio (ν). At 100 MHz frequency, the energy difference ΔE is about 10^{-2} cal mol^{-1}. Purcell (1946) and Bloch stated that most nuclei (including protons) and electrons possess inherent magnetic fields though too small to be observed in the ambient magnetic field of the earth. In an intense magnetic field the nuclei can assume specific orientations with corresponding potential energy levels. A minute amount of energy is absorbed or emitted as the nuclei jump from one energy level to another. This gave rise to a new tool in spectroscopy called nuclear magnetic resonance or NMR. Thus as a result of the magnetic properties of nuclei arising from their axial spin, radio frequency radiation is absorbed in the magnetic field. The method is also called nuclear spin resonance spectroscopy or NMR. For any nucleus NMR consists of one to several groups of absorption lines in the radio frequency region of the electromagnetic spectrum.

35.1 PRINCIPLES OF NMR SPECTROSCOPY

In 1924 Pauli suggested that certain atomic nuclei have properties of spin and magnetic moment. Exposure of these nuclei to a magnetic field leads to splitting of their energy levels. Bloch and Purcell showed that nuclei absorb electromagnetic radiation in a strong magnetic field as a result of energy level splitting induced by the magnetic force. They were awarded the Nobel prize (1932) for this discovery. It is possible to correlate this finding with molecular structure.

Table 35.1 Nuclear spin quantum numbers

Sl No.	No. of neutrons	No. of protons	Spin number	Example
i	even	even	0	S^{32} Si^{28} C^{12} O^{16}
ii	even	odd	1/2	$H^1 N^{15}$ F^{19} P^{31}
iii	odd	even	1/2	C^{13}
iv	odd	odd	1	$H^2 N^{14}$
v	even	odd	3/2	Br^{29} B^{11}
vi	odd	even	5/2	$_{52}^{127}$

Nuclei which have either an odd number of protons or an odd number of neutrons but not both, exhibit half integral quantum numbers, e.g. H^1, N^{15}, F^{19} (case ii) P^{31}. These nuclei behave like spinning spheres. The circulating charge generates a magnetic field much like an electric current in a coil of wire. There is an associated nuclear magnetic moment μ along the axis of spin. For nuclei in which the number of neutrons and protons are both odd the charge is distributed nonsymmetrically, e.g. H^2, N^{14} ($I = 1$) (case iv). Nuclei which have the number of protons and number of neutrons both even have an angular momentum (I = 0) and exhibit no magnetic properties (case i) e.g. C^{12} O^{16} S^{32}, are magnetically inert and hence are not detected in NMR. The magnetic nuclei (I > 0) interact with an external magnetic field by assuming discrete orientations with corresponding energy levels. The energy levels are $I, I - 1$, $I - 2, - I$ for a given nucleus there are $(2I + 1)$ possible orientations. For a proton in a particular $I = $ 1/2, two orientations are possible.

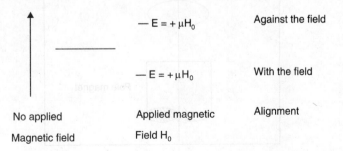

Fig. 35.1 Proton spin states under the magnetic field

Figure 35.1 shows separation into two levels with an energy of μH_0 or $-\mu H_0$, where H_0 is the intensity of the applied magnetic field. The difference in energy ΔE is equal to $2\mu H_0$. It is a special case of the relationship

$$\Delta E = \mu H_0 / I .\qquad\qquad\qquad ...(35.3)$$

Thus three classes of nuclei can be distinguished. (1) Those with zero spin in which both the number of neutrons and protons is even. They do not give NMR signals (e.g. case (i), Table 35.1, C^{12} O^{16}, S^{32}; (2) those with half integral spin, with either the number of protons or number of neutrons odd (e.g. H^1, B^{11}, F^{19}, P^{31}, Cl^{35}, Br^{79} case (ii), (iii), (v), (vi) of Table 35.1; and (3) those with integral spin in which both the number of neutrons and the number of protons is odd (e.g. H^2 and N^{14} case (iv), Table 35.1). Nuclei with spin values of 1/2 act as though they were spherical bodies having a uniform charge distribution over all surfaces.

If a probing electrical charge approaches the nucleus, it experiences an electrostatic field whose magnitude is independent of the direction of approach. The electrical quadruple moment is zero, e.g. H^1, F^{19}, P^{31}. Other spinning nuclei possess nonspherical distributions over their surface, therefore the electrical quadruple moment affects their relaxation time and their coupling with neighboring nuclei. When an alternating radio frequency is applied in the x direction (Fig. 35.2) and superimposed over a stationary magnetic field, nuclei will have enough energy to undergo a transition from a lower energy

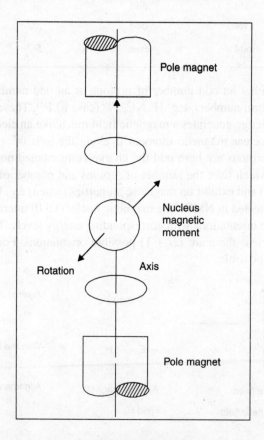

Fig. 35.2 Spinning nucleus

to a higher energy state. The frequency ν of radiation that will give a transition between energy levels is given by an expression:

$$\Delta E = h\nu = \mu H / I \qquad ...(35.4)$$

In the above expression ΔE is directly proportional to the intensity of the applied magnetic field. In NMR 14,000 gauss is the usual field. For protons in this field E is 5.7×10^3 cal/mole (very small energy)

$$E = h\nu = 2\mu H_0 \text{ for } I = \frac{1}{2} \text{ if } \nu = 60 \times 10^6 \text{ cycles/pls} \qquad ...(35.5)$$

i.e., 60 megacycles (MC) which is the radio frequency range with Hz = 5 nm. We can supply energy of this frequency. A proton in a lower energy level may absorb this energy and jump to an upper level, by magnetic resonance.

Table 35.2 Magnetic moment of some elements

Element	μ gauss	Spin No.	gauss/sec	ν at 14,000 gauss MC
H^1	1.4×10^{23}	1/2	2.7	60
B^{11}	1.3×10^{23}	3/2	2.6	19.2
C^{13}	0.35×10^{23}	1/2	0.68	15.1
F^{19}	1.32×10^{23}	1/2	2.5	56.4
P^{31}	0.37×10^{23}	1/2	1.1	24.3

So

$$2\pi r = \frac{2\pi \mu H_0}{hI} \gamma H_0 \qquad ...(35.6)$$

where 2π is introduced to convert the linear frequency to angular units of frequency and a new parameter γ is introduced called the gyromagnetic ratio. This

$$\gamma = \frac{2\pi}{H_0} \qquad ...(35.7)$$

is of fundamental importance in NMR. With a fixed field H_0 (*e.g.* 14,000 gauss) we may vary the frequency until we get resonance conditions. Each type of nucleus will resonate at a different frequency. Some instruments are built to maintain a fixed frequency and vary the magnetic field. We have the following conclusions:

A nucleus ($I > 0$) in an intense magnetic field may find itself in one of $2I + 1$ equally spaced energy levels. The energy difference between levels is small, though nuclei prefer to be in a lower energy level, thermal motion equalizes populations. The small excess number of nuclei in the lower level can be promoted to higher levels by absorption of radiant energy in the radio frequency region.

In NMR the absorption signal may reach zero as soon as the populations are equalized. It is possible to have nuclear relaxation. This is of two types called spin–lattice or longitudinal relaxation and second spin–spin or transverse relaxation. In spin–lattice relaxation, the excited nucleus transfers energy to another nucleus. Such energy is conserved within the system. Such relaxation is characterised by a relaxation time (T_1) which is a measure of the half life required for the system in the excited state to return to an equilibrium state. In contrast, in spin-spin relaxation, the nucleus in the upper level can transfer energy to a neighbouring nucleus by an exchange spin. There is an associated spin-spin relaxation time (T_2) which affects the natural line width of the NMR absorption peak.

The narrow lines are associated with long relaxation times and broad lines with short relaxation time. Both T_1 and T_2 enter into relaxation efficiency. In viscous liquids and dilute solutions relaxation times are longer with sharp lines. However magnetic field in homogeneity contributes more to line width than T_1 and T_2 in most high resolution work. The various mechanisms for radiationless energy transfer are termed nuclear relaxation processes.

35.2 NMR SPECTRA MEASUREMENT

NMR spectra can be produced by two methods. The first is similar to obtaining optical spectra, which involves measurement of the absorption signal as the electromagnetic frequency is varied. No dispersing prisms or gratings exist for radio frequency. Variable radio frequency oscillators produce radio frequencies which can be varied from 1–10 KHz. An alternate way is to use a constant frequency radio oscillator and sweep the magnetic field (H_0) continuously. Old instruments had field sweep methods for production of spectra. The linear sweep oscillator which is commonly employed now, has the advantage of being more efficient in producing spin decoupled spectra.

NMR instruments are either high resolution or wide line models. High resolution NMR can only resolve the fine structure associated with absorption peaks. Such instruments use a magnetic field of 7000 G. Such wide line instruments are used for quantitative elemental analysis and studying the physical environment of the nucleus. Wide line instruments which employ magnets with field strengths of a few thousand gauss are simpler and less expensive than high resolution NMR. For high resolution, NMR spectroscopy needs sophisticated models.

35.3 NMR INSTRUMENTATION

NMR instruments consist of the following important components (Fig. 35.3)

- (*i*) Magnet
- (*ii*) Field sweep generator
- (*iii*) Radio frequency source
- (*iv*) Signal detector
- (*v*) Recorder output
- (*vi*) Sample holder and probe

- (*i*) *Magnet:* The accuracy and quality of an NMR instrument depends upon the strength of the magnet. Resolution increases with an increase in field strength provided the field is homogeneous and reproducible. The magnets may be conventional type, permanent magnets or electromagnets and superconducting solenoids. Permanent magnets have G a field strength of 7046–14002 corresponding oscillator frequencies for proton studies are 30–60 MHz. Good thermostats are needed as the magnets are temperature sensitive. Electromagnets need a cooling system; commercial electromagnets have 60, 90, 100 MHz frequency for protons. High resolution NMR has superconducting magnets with a proton frequency of 470 MHz. The effect of field fluctuations is offset by the use of a frequency lock system, which may be either external or internal. The former employs a separate sample container for the reference substance with an analytical container while in the latter the reference substance is dissolved in a solution containing the sample. The reference compound is tetramethylsilane (TMS).

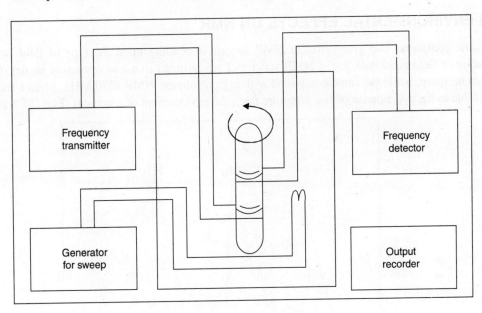

Fig. 35.3 Typical NMR set-up

(ii) *Field sweep generator: A* pair of coils located parallel to the magnet faces, permits alteration of the applied field over a small range. By varying a direct current through these coils the effective field can be changed by milligauss. Such a field is automatically changed linearly with time. For a 60 MHz proton instrument the sweep range is 235 milligauss. For F^{19} and C^{13}, sweeps as great as 10 KHz are required.

(iii) *Radio frequency source:* The signal radio frequency oscillator (*i.e.* transmitter) is fed into a pair of coils mounted at 90° to the path of the field. A fixed oscillator with a frequency of 60, 90 or 100 MHz is used in high resolution NMR; the power output is 100 which is plane polarized.

(iv) *Signal detector:* The radio frequency signal produced by resonating nuclei is detected by means of a coil that surrounds the sample. The electrical signal is small and is amplified before recording.

(v) *Recorder output:* The abscissa drive of an NMR recorder is synchronised with the spectrum sweep: the recorder controls the spectrum sweep rate. Peak areas permit estimation of the relative number of absorbing nuclei in a chemical environment. They are also useful for analytical determination, as peak heights are not useful for measuring concentration.

(vi) *Sample holder and probe:* The NMR sample cell consists of a 5 mm outer diameter glass tube containing about 0.4 ml of liquid. The sample probe is a device for holding the sample tube in a fixed spot in the field. The probe contains a sample holder, sweep source and detector coils, with a reference cell. The detector and receiver coils are oriented at 90° to each other. The sample probe rotates the sample tube at a hundred rotation per minute on the longitudinal axis.

For high resolution NMR work, the sample must be in a nonviscous liquid state. Mostly 2–15% solutions are used. The best solvents for proton NMR should contain no protons, CS_2, and CCl_4 are quite ideal but many compounds are not soluble in them. $CDCl_3$, deuterated chloroform or deuterated benzene (C_6D_6) are the most commonly employed solvents.

35.4 ENVIRONMENTAL EFFECTS ON NMR

The same compound can give different NMR spectra depending upon the type of field strength or resolution of field employed, e.g. a NMR spectra of say, ethanol in a low resolution model (60 MHz) will not be sharp, while the same compound with high resolution NMR (270 MHz) gives a sharp peak. This is due to the dependence of fine structure upon the environment of a nucleus (Figs. 35.4 and 35.5).

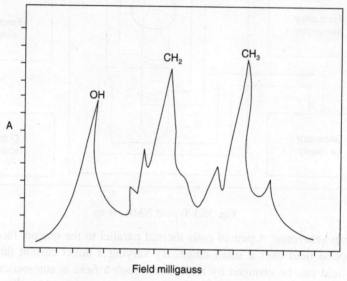

Fig. 35.4 Low resolution NMR (60 MHz)

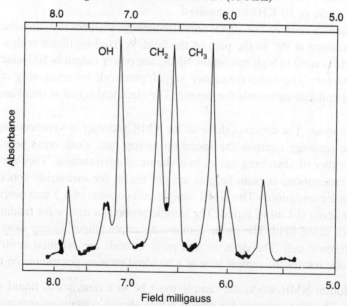

Fig. 35.5 High resolution NMR (270 MHz)

In Fig. 35.4 we have three peaks of OH : CH_2 : CH_3 1 : 2 : 3. If H of OH is replaced by OD then the first peak disappears. A small difference occurs in the absorption frequency of the proton. Difference in position of the peaks depend upon the group to which the H atom is bound. This effect is called 'Chemical shift'. In Fig. 35.5 with high resolution NMR, 2 of the 3 proton peaks are split into additional peaks. The secondary environmental effect which is superimposed upon the chemical shift has a different cause. It is termed 'spin spin splitting'. Both are of importance in structural analysis. Peak separations due to chemical shift are proportional to the field strength or oscillator frequency distance between any pair of peaks. It increases in proportion to an increase in the field, e.g. if there is a change from 60–100 MHz then the shift will be increased by a factor of 100/60, but the distance between fine structure peaks within a group would not be altered.

Chemical shift arises from circulation of the electrons surrounding the nucleus under the influence of the magnetic field. Movement of electrons creates a magnetic field. The splitting of chemical shift peaks can be explained by assuming that the effective field around one nucleus is further enhanced or reduced by local fields generated by H-nuclei bonded to adjacent atoms. As an internal standard, $(CH_3)_4$ Si, *i.e.* TMS or tetramethyl silane, is used being a symmetrical molecule.

35.5 CHEMICAL SHIFT

Every nucleus is surrounded by a cloud of electrons in constant motion. Under the influence of the magnetic field these electrons are compelled to circulate in such a manner as to oppose the imposed magnetic field. This has the effect of partially shielding the nucleus from experiencing the full value of the external field. Hence, either the frequency or the field will have to be altered to bring the shielded nucleus into resonance. This is done by adjusting the magnetic field through an auxiliary winding carrying a direct current which sweeps the field over a narrow span (< 100 μT per test loop field). The value of the added field is converted to its frequency equivalent for presentation on the recorder. Now the value of the shift depends on the chemical environment of the proton. Since this is the source of variation in shielding by electrons it is called the 'chemical shift'. Chemical shift is measured as a field or frequency. In actual practice, the ratio of the necessary change in the field to the applied field or of the necessary change in frequency to the standard frequency is a dimensionless constant designated (δ) in ppm. As we cannot observe the resonance of protons without any shielding electrons, there is no absolute standard with which to compare shifts. For NMR work for organic compounds CCl_4 is used as a solvent, with $(CH_3)_4$ Si, (TMS), as internal standard. TMS is selected because not only are all its H-atoms identical with respect to the environment but they are more strongly shielded than protons of any organic compound. The position of TMS in chemical shift has $\delta = 0$ as the symmetrical and shielded molecule. Some workers give $\delta = 0$ for TMS and denote the shift by τ where $\tau = (10 - \delta)$ to δ shifts a positive value for all samples. Large shielding corresponds to an unshield chemical shift. The field must be increased to compensate for the shielding. (δ) decreases with increased shielding and an increase in the coupling constant (J) increases. It is possible to plot figures of approximate ranges of δ or J for protons in different chemical environments. However, the exact magnitude of δ depends upon the nature of the substituent, solvent concentration and H-bonding.

Thus the nucleus finds an effective field which is smaller than the applied field H_0, hence

$$H_{eff} = (H - \sigma H_0) = H_0(1 - \sigma) \qquad \qquad ...(35.8)$$

if σ is the shielding parameter or

$$\delta = \frac{H_{sample} - H_{TMS}}{H_1} \times 10^6 \quad ...(35.9)$$

if $\quad\quad H_1$ = radio frequency of signal used.

Methanol, CH_3OH has two protons CH_3 and OH. The oxygen atom is more electronegative than the carbon atom hence the electron density around the CH_3 proton is higher than that around the proton of

Table 35.3 Proton shift (δ) or τ

Compound	δ ppm	τ	Group	δ ppm	τ
Si $(CH_3)_4$	0	10	COOH	–14	–3.2
C—CH_2—C	2	8	CHO	10	+2.0
CH_3–C	2	8	RN=CH	8.0	–3.0
NH_2	2	8	C=CH	7.0	–3.8
$CHCl_3$	– 7.25	2.75	OH (phenol)	11.0	5.0
C_6H_6	– 7.27	2.73	CH_3—N	3.5	8.0
H_2O	– 5.1	4.9	CH_3—C=	2.0	9.0
Dioxane	– 3.57	6.43	S—H	2.0	8.5
Acetone	– 2.09	7.91	O—H(alcohol)	3.5	7.0
Cyclohexane	– 1.49	8.51	CH_3S	3.5	8.0

OH, *i.e.* the shielding parameter (s) is greater for CH_3 than for OH, *i.e.* s CH_3 > s OH. Additional shielding and deshielding effects arise from electronic circulations induced within molecules, e.g.

$$— CH_3 — CH_3 — — CH_2 = CH_2 — — CH \equiv \bullet CH — RCHO \quad\quad C_6H_6$$
$$\sigma \to 1 \quad\quad 0.9 \quad\quad\quad 5.8 \quad\quad\quad 2.9 \quad\quad\quad 10.0 \quad\quad\quad 7.3$$

The order of δ values would be wrongly stated from local diamagnetic effects because we have to consider the behavior of π-electrons in unsaturated bonds. In CH $\equiv$ CH the shielding effect of the triple bond overshadows the deshielding effect of electron withdrawal by the carbon atom which occurs in the saturated ethane molecule. Similar explanations can be given for other compounds.

Chemical shift is used for identification of functional groups and as an aid in determining structural arrangements of groups. These applications are based upon the correlation between structure and shift. The exact value of (δ) depends upon the nature of the solvent and solute.

35.6 SPIN-SPIN SPLITTING

The second type of structure seen in NMR is due to the interaction of the spin of a proton with that of another proton or protons usually attached to an adjoining carbon. The interaction involves the spins of the bonding electrons of all three bonds (H—C, C—C, C—H). If protons are in an equivalent environment the interaction will not be manifested but otherwise the maxima at each chemical shift position will be split into a close multiplet, e.g. consider the NMR spectra of C_2H_4I in $CDCl_3$ with TMS as an internal standard. The methylene (CH_2) and methyl (CH_3) protons give δ = 3.2 and 1.8 in resp. the ratio of 2:3. The maximum at $\delta = 0$ is the TMS peak. Two methylene protons should be expected as a result of splitting of the methyl resonance in a 1:2:1 ratio. The spacing between the components of both multiplets are all equal and called coupling constants (J). J has a range of 1.20 Hz. The effects of other protons within the molecule is like a spinning magnet which is oriented with or against the field, thus the resonance position of a proton is affected by the spin of nearby protons. With high resolution NMR we thus observe splitting of lines or a coupling of protons which occurs through bonds by means of a slight unpairing of bonding electrons. We enumerate certain rules about spin–spin coupling and bond multiplicities.

(a) Spin–spin interactions are independent of the strength of the applied field. (By changing the field the coupling constant does not change though there is a change in chemical shift).

(b) Equivalent nuclei do not interact with each other to show splitting.

(c) Multiplicity is determined by the number of neighbouring groups of equivalent nuclei (e.g. for a proton it is $n + a$).

(d) Intensities of multiplets are symmetric with relative intensities given by the coefficients of terms in the expansion of $(a + b)^n$, where n = number of neighbours causing splitting.

(e) Coupling constants decrease with distance.

(f) Coupling constants are the same in both multiplets of a pair which are interacting.

(g) Coupling constant rarely exceed 20 cps while chemical shifts are over 1000 cps.

(h) In undistorted multiplets, chemical shift is measured at the centre of the system of peaks.

35.7 APPLICATIONS OF NMR SPECTROSCOPY

NMR can be used in qualitative identification *i.e.* characterisation of organic compounds. It is possible to elucidate chemical structure by noting the values of chemical shift, spin–spin splitting and decoupling constant. First it is necessary to establish how many different environments of H-atoms exist in the molecule. From the study of the line structure of peaks (multiplicity) one can determine which hydrogen types are closest to each other.

As a quantitative tool NMR has the advantage that a pure substance is not required. Though a pure reference sample is required as internal standard, it can be any compound that has a characterisable spectrum and which is not overlapping with the sample. In conjunction with NMR one uses IR, UV, elemental analysis and mass spectra for characterisation of a pure compound, e.g. the spectrum of a dilute solution of a sample in $CDCl_3$ can be examined. In alcohol, proton exchange on the OH group is so rapid that these protons do not spin-spin couple with neighbouring protons and show a singlet resonance line at $\delta = 2.58$ ppm. Methylene protons split into methyl protons in triplet at $\delta = 1.22$ ppm. Methylene proton does not split but the methyl proton does at $\delta = 3.70$ ppm. In ethanol —OH is more electronegative than carbon and the electron density is shifted away from the CH_2 group, and diamagnetic deshielding results ($\delta \sim 1.20$ to 1.35 ppm), e.g. in the spectrum of ethylacetate a proton shows great amount of diamagnetic shielding and appears as a triplet at $\delta = 1.25$ ppm then the protons are less shielded than the proton due to neighbouring carbonyl group as a result the protons resonate as a shield at $\delta = 2.03$ ppm. The methylene protons appear as a quartet $\delta = 4.12$ ppm.

In quantitative analysis one can use the fact that peak areas and the number of nuclei are directly related. In the absence of overlap, the area of the peak can be employed to establish the concentration of the species if the signal area per proton is known. The internal standard can be used to find the signal peak area. Organic silicon derivatives are unique for calibration purposes. The saturation effect, however, is a big problem in quantitative work, but this can be eliminated by control of variables such as the relaxation time, source and rate of scanning spectrum. NMR is used for the determination of water in food products, paper pulp and agricultural materials. NMR can also be used for elemental analysis.

NMR shift reagents are important in interpretation of proton NMR spectra. They disperse the absorption peak, separate overlapping peaks and permit easier interpretation. They are complexes of Eu or Pr, e.g. dipivalomethane (DPM) complex Pr, Pr (DPM)$_3$. The Pr ion in this neutral complex is capable of increasing its coordination by interaction with lone pairs of electrons. DPM is employed in non-polar solvents such as CCl_4, $CDCl_3$, C_6D_6 to avoid solvent competition with the analysis for electron receiver sites on the metal ion. Eu (DPM)$_3$ gives shifts in lower fields (Fig. 35.6).

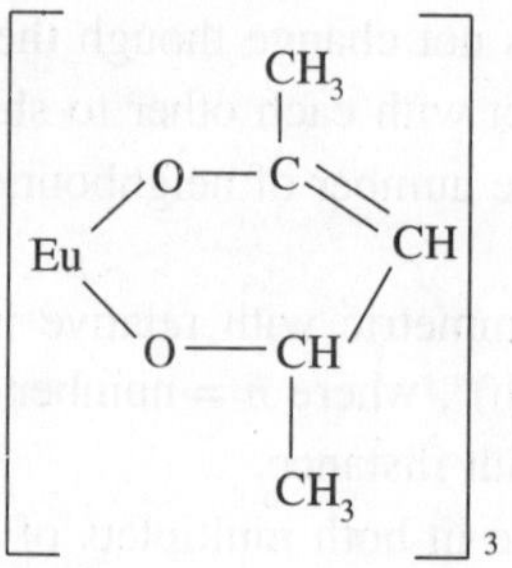

Fig. 35.6 Europium dipivalo methanato complex

35.8 NMR OF OTHER ELEMENTS LIKE C¹³, F¹⁹ AND P³¹

In addition to protons some of the other elements used include fluorine, phosphorus and carbon.

Carbon-13 NMR has advantages over proton NMR in biochemical organic structure elucidation as the shift is > 200 ppm. Compounds with molecular weights between 200 and 400 can be examined. Spin–spin coupling is not encountered. Homo-nuclear spin (*i.e.* ^{13}C and ^{12}C) will also not occur as the spin quantum of ^{12}C is zero. ^{13}C NMR provides information about the structure of molecules and not just the surrounding groups. There are very good methods for decoupling the interaction between ^{13}C atoms and protons. The NMR spectra of solids with ^{13}C NMR are not very useful for solids owing to the limitations of line broadening which removes sharp peaks in the spectrum.

Phosphorus-31 with a spin number 1/2 exhibits NMR spectra with chemical shifts extending over a range of 700 ppm. The resonance frequency of ^{31}P NMR at 14,092G is 24–3 Hz. This has been of immense use in the study of the chemical structure of phosphorus containing compounds, such as condensed phosphates, thio phosphates and hydroxy methyl phosphine.

Fluorine also has a spin number of 1/2 with a magnetic moment of 2.685 nuclear magnetics. The resonance frequency is slightly lower than that for the proton, being 56.4 Hz while the proton has a resonance frequency of 60 MHz at 14.092G. Since the absorption of fluorine is sensitive, the chemical shift extends to 300 ppm compared to 20 ppm for proton and 700 ppm for ^{31}P NMR. This has proved very useful for the study of organic fluorine compounds.

Finally apart from ^1H, ^{31}P or ^{19}F one can use other kinds of nuclei. They include ^{11}B, ^{23}Na, ^{15}N, ^{29}Si ^{109}Ag, ^{199}Hg, ^{113}Cd, and ^{207}Pb. These isotopes have been tried experimentally but did not offer very exciting information.

35.9 FOURIER TRANSFORM (FT) NMR

Fourier transform NMR has resulted in a dramatic increase in the sensitivity of NMR measurements. This is possible by exciting all possible resonances simultaneously. The sample is flooded with radiation of all possible frequencies, followed by time to frequency conversion. A pulse of radio frequency energy is used to get a wide range frequencies. Following each pulse, the excited nuclei continue to emit signals at Larmov frequencies as they relax back to their ground states. The emission following the excitation pulse is called free induction decay (FID) which can be picked up by a probe coil with multiple resources if the FID pattern becomes complex. An interferogram is thus obtained. FT NMR can also be used with ^{13}C.

This technique has widespread applications especially for ^{13}C–NMR, ^{31}P–NMR and ^{19}F–NMR giving high signal to noise ratio facilitating rapid scanning. The optics of these instruments provide a large energy output. FT–NMR instruments are not as good as FT-IR models if one has no good quality grating spectrophotometers. FT instruments, permit high resolution work, study of samples with high absorbance or possessing weak absorption bands, fast scanning work and in a few cases emission (TR) studies.

In comparison to FT–IR the FT–NMR instruments have not become popular with research workers. The kind of refinement and accuracy as can be obtained by FT–IR over simple IR instrument is not at all comparable with the FT-NMR and NMR instruments. However a greater progress has been made in recent years on the application of three dimensional NMR spectroscopy. Further with the discovery of MRI technique (*i.e.*, magnetic resonance imaging), it has made spectacular progress in diagnosis of diseases of brain or other organs of body. It has opened new avenue for appropriate treatment of patients in hospitals. Some details of such techniques are also covered at the end of this chapter.

35.10 TWO DIMENSIONAL FT–NMR

2D–NMR is a new technique which helps to decode complex spectra. A data is procured as a function of time vs. intensity of NMR. Before FID signal procurement, system is perturbed by pulse for period (time FT of FID as function of (t_2) when (t_1) is fixed gives spectrum similar to obtained by usual pulse experiments. (Fig. 35.7) The process is repeated for different values of t_1 furnishing two dimensional spectrum with variable frequencies as v_1 and v_2 with chemical shift of δ_1 and δ_2 (Fig. 35.8). The nature and timing of the pulses have been used in 2D-NMR and more than two repetetive pulses are also used. We would get kinds of spectra as shown in Figs. 35.7 and 35.8 for an organic compound CH_3–CH (OH)–CH_2–CH_2OH. It is 1, 3, butanediol.

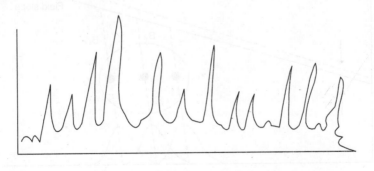

Fig. 35.7 C^{13} NMR of 1, 3 Butanediol

When proton broadband decouple is turned off, a 90° pulse is applied to analyte. After time (t_1) decoupler is turned on and new pulse is applied thereby resulting FID is digitilised. The process is then repeated for other values of (t_1) leading to generation of series of spectra. The spectrum is made up of quartret, two triplets and a doublet. Such information can be procured only from two dimension NMR. However, on account of the prohibitive cost the technique is not popular for conventional analysis.

Finally we would consider one of the important techniques in medical diagnostic work. It is called MRI *i.e.*, magnetic resonance imaging. Cost-wise it is very expensive diagnostic tool, however, from the viewpoint of accuracy and exact diagnosis of the disease this is one of the most powerful tools of analysis.

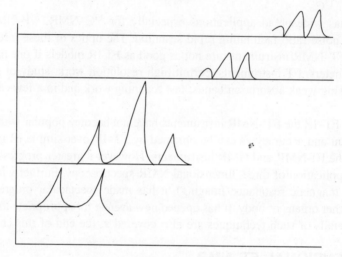

Fig. 35.8 2D-NMR

35.11 MAGNETIC RESONANCE IMAGING (MRI)

NMR technology is extensively used not only in chemistry but also in medicine. The MRI data from radio-frequency excitation of solids or liquids are subjected to Fourier transform spectra and converted to three dimensional images of the interiors of the object. The images are formed non invasively. Like X-ray there is no damage to human or living organisms by exposure to magnetic field.

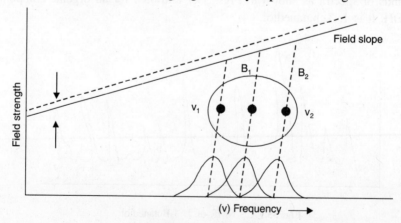

Fig. 35.9 Concept of MRI

The fundamental concept of MRI is described in Figure 35.9. In this the magnetic field is varied so as to get a profile as is indicated in the figure. The linear variation in B_1, B_2 is created by auxilary coils in magnet controlled by computer. Protons in different points or locations experience varying magnetic field strength viz. B_1 and B_2 at two regions.

The Larmov equations states that these nuclei show different resonance frequencies viz. v_1 and v_2 which is given as $v_1 = \dfrac{vB_1}{2\pi}$ and $v_2 = \dfrac{vB_2}{2\pi}$ if B_1 or B_2 are magnetic fields. Thus if magnetic field gradient (1×10^{-5}) is applied along z-axis of MRI magnet, resonance frequency range $(2.7 \times 10^8)(1 \times 10^{-5})$ $(2\pi$ radians) = 425 Hertz results *i.e.*, Proton (H_1^1) one cm apart along field gradient have resonance frequencies by difference of 425 Hz *i.e.* by changing the centre frequency of NMR problem pulse increment of 425 Hz. It is feasible to probe successively one cm positions in the varying direction of the field gradient. When FIDs are subjected to FT-NMR transformation. The information on concentration is generated as examplified by heights of the peak (as shown in Fig. 35.9). The position of the slice on z-axis can be altered by adding a DC offset to auxilary coils by dotted lines for field gradient. Each portion is probed by changing the pulse width of RF pulse to tune to frequency of precision frequency of protons in the said portion. The middle peak represents centre circle indicating the signal from protons in the central position of the object. If field is large, thickness of portion is made smaller.

The position selection provides spatial information in one direction along two axis by flipping the spins of protons within the selected portion. All information within each portion along z-axis is then procured. A selected portion is pictured as two dimensional grid lying on the x-y plane with different rows and columns. One can have grid of 3×3 to 256×256 or more elements with each element or pixel have definition $(1 \times 1$ mm) area. In such three dimensional imaging the z-axis field gradients is selected to give a portion of 1 mm. At this point of measurement sequence all protons have their spins flipped. A second gradient on application on axis perpendicular to z-axis by using second set of y-axis coil oriented at $90°$ to the axis of magnet. Nuclei at different position (on y-axis) precess at different frequencies depending upon their position in the field gradient. If y-axis field is turned off all the nuclei have precessed through fixed phase angle whose magnitude is related to size of section. The distance on y-axis is encoded to the phase angle of precessing proton.

As y-axis gradient is closed, a third set of magnet coils ($90°$ to ban z and y-axis is then activated to produce a field gradient in third dimension along x-axis. A judicious and timely application of various RF pulse sequences and magnetic field gradients produce three dimensional images. The possibility of producing high resolution images from three dimension data is an unique strength of MRI technique. MRI is a non-invasive technique reproducible, rapid and with high spatial resolution. It is thus a popular tool in diagnostic medical sciences. But it is very expensive.

It is also used progressively in food industry, water analysis or to know lipid contents of sunflower oil, structure of food and cookies, study of polymer etc. However, the application of magnetic resonance imaging is a boon in the field of medicine specially as a powerful diagnostic tool for identifying abnormal anamolies in brain or related organs.

PROBLEMS

35.1 β-alanine and alanine can be analysed in D_2O by NMR. A 5% solution of alanine and β-alanine in D_2O shows readings of 980 at 1.48 ppm and 650 at 2.54 ppm resp. The first peak is due to α-methyl and the second peak is due to α-methylene group. An unknown sample mixture shows a reading of 377 at 1.48 ppm and 400 at 2.54 ppm level. What is the % of alanine in the mixture?

Answer: Since the unknown is measured under identical conditions to the standard solution. We evaluate

$$\text{Amount of alanine} = \frac{377 \times 100}{980} = 38.46\% \text{ at } 1.48 \text{ ppm level}$$

$$\text{Amount to } \beta\text{-alanine} = \frac{400 \times 100}{650} = 61.53\% \text{ at } 2.54 \text{ ppm level}$$

Thus 30.46% and 61.53% of alanine and (β–alanine are present.

35.2 Various derivatives of butanol were analysed by NMR and the following results were obtained

5%	2,2-dimethyl-2-butanol	47	4.7 ppm	(65)
8%	2,2-dimethyl-3-butanol	75	4.3 ppm	(55)
6%	2,3-dimethyl-2-butanol	57	4.0 ppm	(20)

A solution mixture exhibited integrated intensities of 65 at 4.7 ppm, 55 at 4.3 ppm and 20 at 4 ppm. What is the composition of the mixture?

Answer: Since the mixture is measured under the same conditions as mixture they are comparable.

$$\text{For 2, 2-dimethyl-1-butanol} = \frac{5 \times 65}{47} = 6.91\%$$

$$\text{2, 2-dimethyl-3-butanol} = \frac{8 \times 55}{75} = 5.86\%$$

$$\text{2,2-dimethyl-2-butanol} = \frac{20 \times 6}{57} = 2.10\%$$

Hence   6.91% 1-butanol,
5.86% 3-butanol
2.10% 2-butanol are present in mixture.

PROBLEMS FOR PRACTICE

35.3 With fluorobenzene as an internal standard and CCl_4 as the solvent, the following results were obtained. Amount of fluorine = 1.94 g; weight of unknown = 0.57 g, volume of CCl_4 = 5 ml; reading due to fluorobenzene = 784; reading of unknown = 387. Calculate the percentage by weight of fluorine in the unknown compound.

35.4 Thebaine shows three peaks in NMR spectra due to protons in 4.8–5.75 ppm and 5.75–7.15 ppm but crytopine has no peak in the first region (*viz.* 4.8–5.75 ppm) and has absorption in the second region due to six protons. Calculate the concentration of cryptopine in thebaine.

35.5 For a diketone there are five isomeric structures — proton NMR shows two signals of equal intensity. If the empirical formula is $C_4H_4O_2$ what will be its spatial configuration?

35.6 The NMR spectrum for the compound C_3H_6O has a triplet centered at 84.73 ($\tau = 5.27$) and quartet at $\delta = 2.72$ (($\tau = 7.28$). Deduce a structure.

35.7 A sample containing hydrogen was prepared in 0.320 gm/ml of n-decane in CCl_4 with peak at 922. An unknown sample containing 0.285 gm/ml had peak of 325. What is the % of hydrogen in the unknown sample? (*Hint:* Total area = KWH, if K = constant, W = weight of sample in standard volume, H = % hydrogen in sample.)

LITERATURE

1. J.A. Pople, W.G. Schnieder, H.J. Beinstain, *"High Resolution Nuclear Magnetic Resonance"*, McGraw Hill (1959).
2. W.N. Paudler, *"Nuclear Magnetic Resonance"*, Allyn & Bacon, Boston (1971).
3. R.K. Harris, *"Nuclear Magnetic Resonance Spectroscopy"*, John Wiley (1986).
4. H. Gunther, *"NMR Spectroscopy, Basic Principles, Concepts & Applications in Chemistry"*, 2nd Ed. John Wiley (1995):
5. R.J. Abraham, J. Fischer, P. Loftus, *"Introduction to NMR Spectroscopy"*, John Wiley (1988).
6. H.O. Kalinowsky, S. Berger, S. Braun, *"Carbon 13 NMR Spectroscopy"*, John Wiley (1988).
7. L.D. Field, S. Sternhell, *"Anaytical NMR"*. John Wiley (1989).
8. J. Schraml, J.M. Bellama, *"Two Dimensional NMR Spectroscopy"*, John Wiley (1988).

Chapter 36

Electron Spin Resonance Spectroscopy

Another name for this technique is Electron Magnetic Resonance (EMR) or Electron Paramagnetic Resonance (EPR) spectroscopy. As electrons have spin, they possess a magnetic moment. Therefore, magnetic resonance theory applies to electrons in a similar manner to the way it is applied to spinning nuclei. The magnetic ratio of electrons is 1.76×10^7 rad G^{-1} sec^{-1}; protons have a comparable value of 2.6 rad G^{-1} sec^{-1}.

The greatest application of ESR spectroscopy has been in the study of free radicals, even at a very low concentration. It has also proved useful for determination of trace amounts of paramagnetic ions in biological samples. Similarly, odd electron molecules, transition metal complexes, lanthanide ions, and triple state molecules have been studied by ESR techniques. ESR also uses what are called spin label reagents, which have stable molecules containing odd electrons having the capacity to react readily with amino acids. ESR of such compounds furnishes valuable information about structure, polarity, viscosity, phase transformations and chemical reactivity. The basic theory, methodology and instrumentation used is comparable to those used in nuclear magnetic resonance spectroscopy (NMR). The only exception is the absence of chemical shift phenomena in ESR spectroscopy.

36.1 ELECTRON SPIN RESONANCE SPECTROSCOPY

Electron spin resonance (ESR) also called electron paramagnetic resonance (EPR) is based on the fact that atoms, ions, molecules or molecular fragments which have an odd number of electrons exhibit characteristic magnetic properties. Such a display of magnetic characteristics is due to arbitrary action, spinning action, or both, of unpaired electrons about the nucleus. An electron has a spin and due to spin there is magnetic moment. Absorption of microwave radiation by an unpaired electron when it is exposed to a strong magnetic field is possible.

The principles of ESR spectroscopy are similar to those of NMR spectroscopy. When a strong constant magnetic field H is applied to the unpaired spins of electrons a torque acts to make the electron dipoles line up either parallel or antiparallel to the direction of the magnetic field. For electron spins which are parallel in a field, $E = \mu H$, whereas, for those which are aligned antiparallel, $E = -\mu H$ as shown in Fig. 36.1.

Since an electron bound to an atom has two distinct motions associated with it — spin about its own axis, and an orbital motion about a nucleus — the electron magnetic moment is better replaced by the products $g\mu_g H$ in *BM*. (Bohr magneton) and is a real constant equal to $eh/4\pi m_e C$ where g is a variable known as the splitting factor (g = 2.0023 MC/gauss) but it varies when strong coupling occurs (Fig. 36.1). Now by the Boltzmann's law of distribution the relative number of electrons in the upper and lower energy states is given by

$$n_{upper}/n_{lower} = e^{-2}\mu H/KT \qquad ...(36.1)$$

where K = Boltzmann's constant.

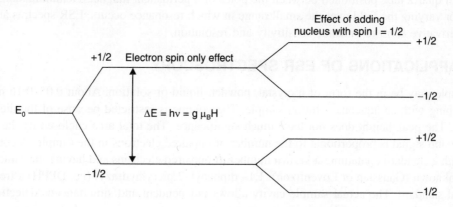

Fig. 36.1 Splitting pattern by field

Therefore, absorption and hence sensitivity of measurement increases steeply with field strength. Spectroscopic measurements become possible when a sample is simultaneously subjected to a constant magnetic field and irradiated by a much weaker radio frequency field applied perpendicular to the fixed magnetic field and held at the frequency of precession. At resonance, the absorption of energy from the rotating field causes the spins of electrons to flip from the lower energy level to the higher field. The two levels are separated by the energy difference.

$$E = 2\ \mu H/h\nu \qquad ...(36.2)$$

with the usual notations and for a free electron the frequency of absorption is given by

$$\nu = 2\ \mu H/h = (28026 \times 106)\ H \text{ megacycles per gauss} \qquad ...(36.3)$$

In a field of 3400 gauss the precession frequency is 9500 MC/ sec.

In other words, the electron, like the proton has a spin quantum number of 1/2 and possess two energy levels which differ in energy content under the influence of a magnetic field. The lower energy level $m = -1/2$ and higher energy $m = + 1/2$. Hence

$$E = h\nu = \mu B_N \frac{H_0}{I} g\beta_N H_0 \qquad ...(36.4)$$

where g = splitting factor and (β_N is Bohr's magneton which is 9.7×10^{-21} erg G^{-1}. For free electron, it is 2,0023. Now ESR spectrometers often employ fields of 3400 G, where

$$\nu = I\beta\frac{H_0}{h} = 9.51 \times 10^9 \text{ Hz} \qquad ...(36.5)$$

or 9500 MHz. This lies in the microwave region or 340 KT. ESR involves the absorption of radiation in the microwave region of the electromagnetic spectrum.

36.2 INSTRUMENTATION FOR ELECTRON SPIN RESONANCE

The source of microwave radiation is a Klystron tube which is operated to produce monochromatic radiation having a frequency of about 9500 MHz. This tube is an electronic oscillator in which a beam of electrons is pulsed between a cathode and a reflector. The oscillating output is transmitted to a waveguide by a loop of wire which sets up a fluctuating magnetic field in the guide. The waveguide is a rectangular metal tube, and transmits the microwave radiation to the sample which is generally held in a small quartz tube positioned between the poles of a permanent magnet. Helmholtz coils provide a means for varying the field over the small range in which resonance occurs. ESR spectras are recorded in the derivative form to enhance sensitivity and resolution.

36.3 APPLICATIONS OF ESR SPECTROSCOPY

The sample may be in the form of a crystal, powder, liquid or solution. About 0.05–0.16 ml is put in pyrex tubing with an aqueous solution sample. The volume is restricted because of the dielectric loss of water. The peak height does not have much significance. The total area enclosed by the absorption or dispersion signal is proportional to the number of unpaired electrons in the sample. A comparison is made with a standard containing a known number of unpaired electrons and having the same line shape as the unknown (Gaussian or Lovenlizon). 1,1– diphenyl–2 picrylhydrazyl (i.e. DPPH) a free radical is the usual standard. The actual sample cavity allows independent and simultaneous detection or resonance of sample and standard.

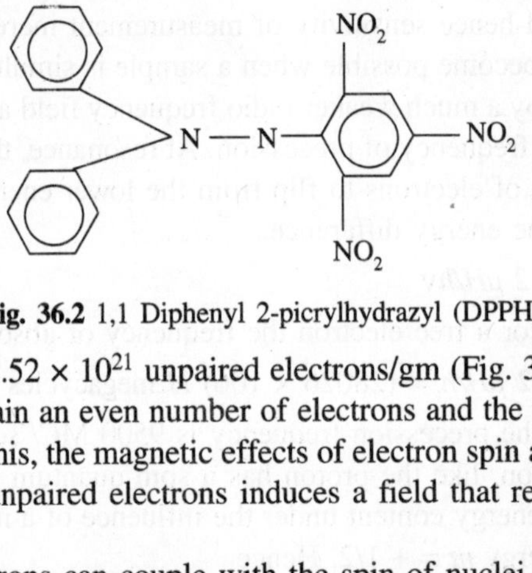

Fig. 36.2 1,1 Diphenyl 2-picrylhydrazyl (DPPH)

DPPH generally contains 1.52×10^{21} unpaired electrons/gm (Fig. 36.2). Molecules do not exhibit ESR spectra because they contain an even number of electrons and the number in the two spin state is identical (spin paired). Due to this, the magnetic effects of electron spin are cancelled. These substances are diamagnetic. The spin of unpaired electrons induces a field that reinforces the applied field; this paramagnetic effect is larger.

The spin of unpaired electrons can couple with the spin of nuclei in the species to give splitting patterns similar to those observed for nuclear spin-spin coupling. Thus when an electron interacts with n-equivalent nuclei its resonance peak is split into *(2n + I)*, where I is the spin quantum number of the nuclei and n is the number equivalent nuclei. This process is termed as hyperfine splitting, e.g. in the case of hydrogen the spin of the hydrogen nucleus couples with that of the unpaired electron to produce

$\left(2 \times \dfrac{1}{2} \times 1 + 1\right)$ or 2 peaks (Fig. 36.3).

ESR has been applied to the study of chemical, photochemical and electrochemical reactions which proceed via free radial mechanisms. The ESR method provides structural information about intermediate radicals. Transition metal complexes are also studied by ESR at ppb levels. Spin label reagents which contain odd electrons can react selectively with aminoacids, the ESR of such compounds provides information on structural features, polarity, viscosity, phase changes and chemical reactivity.

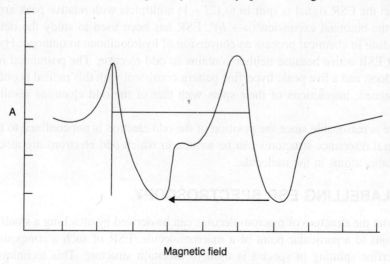

Fig. 36.3 ESR spectra of hydrogen

No counterpart of chemical shifts in NMR spectra exist in ESR since the g-value of the free radical is ~2.0023. Further, electrons available for spin resonance are usually near. Spectra are analysed for interaction between electron spin and nuclear spins of adjacent magnetic nuclei from the number and relative intensities of multiplet structures. Definite means, for ascertaining the presence of free radicals beyond any doubt exist in ESR spectroscopy provided, of course, unpaired electrons are present. The effective magnetic field experienced by a given unpaired electron is $H_{eff} = H_0 \pm H_{Local}$ ..., where H_0 is the applied magnetic field, H_{Local} = magnetic field generated by magnetically active neighbouring

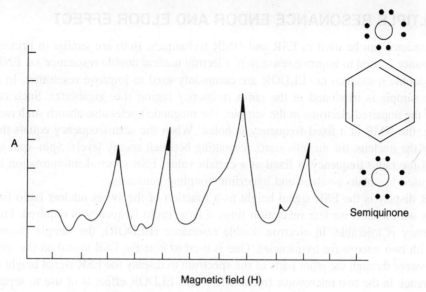

Fig. 36.4 ESR spectra of semiquinone radical

nuclei. The number of splittings caused by 1 electron-1 nucleus interaction will equal $2I + 1$, where I is the spin quantum number of the nucleus. When an electron intervenes magnetically between $n =$ equivalent nuclei the ESR signal is split in to $(2I + 1)$ multiplets with relative peak areas equal to the coefficients of the binomial expansion $(a + b)^n$. ESR has been used to study the detection of a free radical intermediate in chemical process as conversion of hydroquinone to quinone. Hydroquinone and quinone are not ESR active because neither contains an odd electron. The postulated intermediate (i.e. semi quinone) does, and a five peak, hyperfine pattern consistent with this radical is obtained. Four ring protons are assumed; interactions of their spins with that of the odd electrons results in five peaks. (Fig. 36.4)

This picture is reasonable since the position of the odd electron is not confined to the oxygen atom as shown. Several resonance structures can be written in which odd electrons are associated with any of the several other atoms in the molecule.

36.4 SPIN LABELLING ESR SPECTROSCOPY

Information about the structure of macromolecules can be derived by attaching a small compound with unpaired electrons to a particular point of a macromolecule. ESR of such a compound can easily be obtained. Hyperfine splitting in spectra is used to ascertain structure. This technique is called spin labelling. The compound containing the unpaired electron is called the spin label. Proteins and cell membranes have been studied by this method. The spin label compound contains a nitricoxide ring or a transition metal complex providing unpaired electrons. The label must have chemical activity and a group of atoms to form bridges between the free radical and the macromolecule. Selected spin labels and the group on the macromolecule to which each label is attached (in brackets) are given below e.g. $RC{\equiv}N$ (Heme), ATP–R (myosin), $CH_3CO\,NHR$ (Lysozyme), $BrCH_2CO{-}NHR$ (Methionine), $I{-}CH_2$ $CO\,NHR$ (Cystein imidazole). Thus five or six membered heterocyclic molecules containing a nitrosyl group whose nitrogen atom is bonded to a tertiary carbon are the best spin labelling compounds for macromolecules.

36.5 MULTIPLE RESONANCE ENDOR AND ELDOR EFFECT

Multiple resonance can be used in ESR and NMR techniques. Both are similar in principle. In ESR, double resonance is used to improve resolution. Electron nuclear double resonance *i.e.* ENDOR as well as electron double resonance *i.e.* ELDOR are commonly used to improve resolution. In the ENDOR method, the sample is irradiated in the radio frequency region (i.e. gigahertz). Such radiations are absorbed by the unpaired electrons in the sample. The magnetic nuclei also absorb such radiation. Then on scanning, the NMR at a fixed frequency is noted. When the scan frequency equals the absorptive frequency of the nucleus, the nucleus starts resonating between energy levels. Spin–spin splitting does not occur. If the radio frequency is fixed at a certain value, ESR spectral interpretation is simplified. ENDOR is used to obtain g-values and hyperfine coupling constants.

ENDOR display is the ESR signal height as a function of the swept nuclear radio frequency. For free radicals with a short nuclear relaxation time, a large radio frequency is required. For solids, 5W radio frequency is tolerable. In electron double resonance (ELDOR), the sample is simultaneously irradiated with two microwave frequencies. One is used to note the ESR signal on the spectrum while the other is swept through the other parts of the spectrum to display the ESR signal height as a function of the difference in the two microwave frequencies. This ELDOR effect is of use to separate overlapping multiradical spectra and study the relaxation effect. It also aids in assigning peaks.

36.6 CHARACTERISATION OF METAL COMPLEXES BY ESR

ESR can be used to study transition metal complexes. The unpaired electron is associated with the metal in the complex. The ligands have definite electromagnetic fields due to the change in polarity of the atoms. The fields are quite strong near the metal and unpaired electrons. These complexes can couple with these electrons to cause splitting of the ESR spectral line into multiplets. This leads to fine splitting or hyperfine splitting. We get fine structure due to the spectrum of multiplet lines. The extent of such splitting depends upon the energy level of these unpaired electrons. Such splitting can occur only if the electric field is unsymmetrical around these electrons. However, if the ligand contains a magnetic nucleus and if the unpaired electron is delocalised in the complex, more splitting is noted due to spin–spin coupling with the magnetic nucleus. This is called super hyperfine splitting, resulting in a complicated ESR spectrum.

36.7 OTHER APPLICATIONS OF ESR

Apart from the elucidation of structure ESR can be used for other purposes e.g. to show the presence of unpaired electrons, or for location of an unpaired electron in an intermediate. The quantitative analysis of constituents containing metallic ions with unpaired electrons can be carried out e.g. Cr (III.), Cu (II), Fe (III), Mn (II) and Ti (III) have been analysed in this way. ESR is used to measure surface area provided the paramagnetic compound used in such studies has a known number of unpaired electrons.

LITERATURE

1. R.S. Alger, *"Electron Paramagnetic Resonance Technique and Applications"*, Interscience *(1968).*
2. J.E. Wertz, J.R. Bolton, *"Electron Spin Resonance Elementary Theory and Practical Application"*, McGraw Hill (1972).
3. P.B. Ayscough, *"Electron Spin Resonance in Chemistry"*, Methuen (1967).
4. M.M. Dono, J.H. Freed (Ed.), *"Multiple Electron Resonance Spectroscopy"*, Plenum (1979).
5. R.D. Braun, *"Introduction to Instrumental Analysis"*, McGraw Hill (1987).
6. R.M. Silverstein, G.C. Bassler, T.C. Morill, *"Spectrometric Identification of Organic Compounds"*, 5th Edn., John Wiley (1991).
7. J.R. Dyer, *"Applications of Absorption Spectroscopy of Organic Compounds"*, Prentice Hall (1965).
8. R. Beer, *"Remote Sensing by Fourier Transform Spectrometry"*, John Wiley (1992).

Chapter 37

X-Ray Spectroscopy

X-ray spectroscopy is gradually getting importance because it belongs to a category of non-destructive method of analysis. X-ray spectroscopy is a broad based area comprising of four techniques viz. X-ray absorption; X-ray emission, X-fluorescence and X-ray diffraction. Of these four methods of quantitative chemical analysis, X-fluorescence technique is most useful while for the elucidation of crystal structure of compounds, X-ray diffraction is most promising. Although X-ray absorption is analogous to optical method like UV, or visible spectroscopy there is critical problem confronted with preparation of sample in form of target. While X-ray emission has largely ramification around fluorescence methods. X-ray diffraction is scarcely used for quantitative analysis although qualitative analysis is possible if one possesses standard samples with known composition containing same elements which are being analysed.

The significant aspects of X-rays is that they have wavelength much shorter < 0.01- 10 nm than UV-visible radiation. It is generated by bombardment of target with high density particles like α-rays, electrons or protons as in proto induced X-ray emission (PIXE). The electrons from innermost shell like 'K' shell is pushed into nucleus to form new element with atomic number one unit less. The vacancy created in inner shell is filled by the outer shell electrons (including valency shell). Such process of filling of electrons generates radiation called as X-rays. The target material is copper, tungsten or chromium. The windows in apparatus are made out of beryllium. The sampling of solids is not really a problem, but liquids or gases have to be embodied in mylar film plates. Radioactive elements can also produce X-rays by what is known as 'K-capture' or K-electron capture.

We would consider source, kind of instrumentation, sampling, detection and analysis by all four methods as mentioned earlier.

37.1 BASIC PRINCIPLES

In 1901 Roentgen received a Nobel prize for the discovery of X-ray. These rays exist in the region of 0.01–10 nm but 0.08–0.2 nm is productive in chemical analysis. The electronic levels of an atom are designated by shell as K, L, M, N, etc. lines of which K being innermost shell. The electrons migrate from the vacant outer orbital to inside vacant orbit to fill up the vacant slot. The kind of energy generated by this kind of migration leads to origin of the X-rays. The time scale is approximately 10^{-12}–10^{-4} seconds for the transition. The electron vacancy in inner shell is created when an atom is bombarded with a particle which possess adequate energy to knock out electrons from the innermost orbit to the centre of nucleus. Thus e_{-1}^{0}, H_1^1, He_2^4 are used as the projectiles. Nuclear capture also leads to generation of X-rays in the radioactive elements, leading to formation of new element with the atomic number one unit less than original element.

e.g.

$$M_N^Z \rightarrow e_{-1}^0 + M_{(N-1)}^Z$$

This creates a vacancy in innershell. The major X-ray generated are used for the purpose of absorption, emission, fluorescence or diffraction in the chemical analysis. X-rays are generated by (*a*) bombardment of metal target with beam of high energy electrons or (*b*) exposure of matter to primary X-rays beam to generate secondary X-rays showing fluorescence (*c*) use radioactive element which on disintegration leads to X-ray formation (*d*) from synchronisation of radiation source, but the last is most expensive process. The continuum and line spectra is produced. The first is called white or Bremsstrahlung radiation which is of continuum type.

The characteristics of white radiation is noteworthy. In an X-ray tube electrons generated at hot cathode are directed to anode when potential difference is 100 kV, which leads to X-ray generation. White radiation or continuum spectra (Fig. 37.1) is produced.

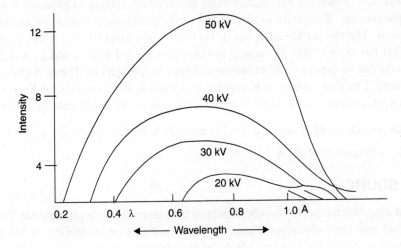

Fig. 37.1 Continuum radiation

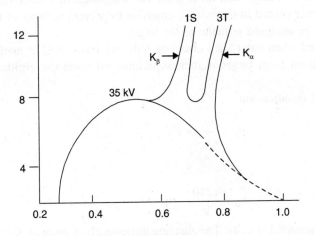

Fig. 37.2 Line spectra

A line spectra (Fig. 37.2) is obtained under other conditions of generation. A photon produced at collision of electron has an energy equivalent to difference of energy of electron before and after collision. A maximum photon energy matches zero kinetic energy during collision. Therefore

$$h\nu_0 = \frac{hc}{\lambda_0} = V_e \text{ (This is a product of accelerating voltage and charge)} \qquad ...(37.1)$$

because V_e is accelerating voltage, h = Planck's constant, c = velocity of light, ν_0 = maximum frequency, λ_0 = wavelength of radiation, we rewrite it as

$$\lambda_0 = 12.398/V = \frac{hc}{eV} \qquad ...(37.2)$$

This is called as Duane-Hunt Law.

As regards line spectra (Fig. 37.2) intense emission lines are seen at 0.67–0.71Å. Such X-ray lines are of two types having shorter and longer wavelength. If N < 23 shorter (K) lines are produced while (L) are longer lines. Every element has characteristic accelerating voltage to generate X-rays. Moseley discovered this phenomena. X-rays are produced due to the electronic transition of electrons in inner-most atomic orbitals. The shorter K-series has α_1 α_2 level while longer L-series has β_1 β_2 levels (e.g. Tungsten (N = 74) has α_1 = 0.209 Å, α_2 = 0.184 Å while β_1 = 1.475 Å and β_2 = 1.282 Å as the wavelength). X-rays can be generated from radioactive rays by γ-emission. The α, β emission leave the atom in excited state. Electron capture or K-capture also leads to the formation of X-rays. On account of K-electron capture a vacancy in K-shell is created, leading to electronic transition from outer shell

with simultaneous generation of X-rays e.g. Fe(55) exhibits K capture as $Fe_{26}^{53} + e_{-1}^0 \longrightarrow Mn_{25}^{53} + h\nu$, where K_d line has a frequency of 2.1Å.

37.2 X-RAY SOURCE

It is an evacuated tube which involves bombardment of the electrons on target material. The thermionic emission from coil generates electrons. Temperature of coil can be controlled in order to regulate generation of electrons. The potential barricade drives the electrons on target (anodic). Tubes are of two types, one with demountable target and other fixed vacuum tubes. In former vacuum is changed every time while target is water cooled in thermionic emission to prevent melting of target. In dismountable tubes material tube to be analysed is made as the target.

X-rays are produced when outer shell electrons falls into inner shell or nucleus. The line spectrum due to electron expulsion from target is then superimposed upon the emitted continnum or white spectra.

From Duane-Hunt equation viz.

$$\lambda_0 = \frac{hc}{eV} \text{ or } eV = \frac{hc}{\lambda_0} \qquad ...(37.3)$$

We have

$$\lambda_0 = \frac{1240}{V} \qquad ...(37.4)$$

if λ is in nm and V is measured in volts. The absolute intensity (I) of emitted X-rays at particular voltage rises with atomic number (N) of an element.

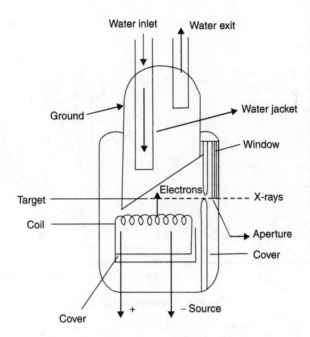

Fig. 37.3 X-ray tube

So

$$I = aN(E_V - E) + bN^2 \qquad \qquad ...(37.5)$$

if a, b are constants, E_V = energy at peak while E = energy of emitted X-rays. The area under curve (Fig. 37.2) is given as

$$A = K_f N t^{1/2} = K_f N V^2 \qquad \qquad ...(37.6)$$

if K = constant. The metals used as target are having high melting point, good thermal conductivity and large atomic number (N). Such metals consist of Ag, Co, Cr, Cu, Fe, Mn, Ni, Pt, Rh, W and Y. However, the most popular metal is Tungsten (W). Next to W, Cr is preferred for analysis of elements with atomic number Cr (N = 22). It contains high intensity line.

As (v) the frequency of K_α line is given as

$$v = 2.48 \times 10^{15} (N-1)^2 \qquad \qquad ...(37.7)$$

37.3 X-RAYS ABSORPTION

On passing X-ray through film its power is reduced due to either absorption or diffraction by film material. A typical absorption spectrum for silver and lead is shown in Fig. 37.4. It has well defined absorption peaks. A sharp discontinuous portion called as the 'absorption edge' after maxima is observed. The maximum for absorption rises when quantum energy is equivalent to the energy required to remove electron out of atom. In Fig. 37.4 leading spectra has four peaks viz. 0.14 Å, 0.75 Å, 0.80 Å and 1.0 Å with definite significance. For the silver also absorption peaks are seen in the figure at 0.485 Å.

The mass absorption coefficient is of importance in chemical analysis. It is analogous to Beer's law in the visible-ultraviolet spectroscopy. As

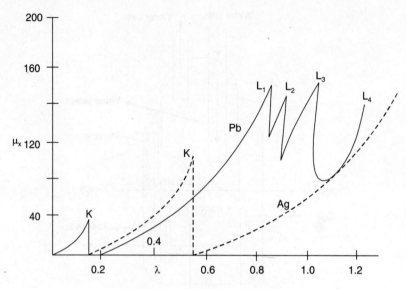

Fig. 37.4 X-absorption spectras

$$\ln \frac{P_0}{P} = \mu_x$$

...(37.8)

if x = thickness (cms) of sample, P, P_0 are intensities of transmitted and incident beam, μ = linear absorption coefficient which is characteristic of an element. The law can be further modified as in terms of density of sample as

$$\ln \frac{P_0}{P} = \mu_M \text{ as } \mu = \mu_M \acute{P}$$

$$\ln \frac{P_0}{P} = \mu_M \rho_x$$

...(37.9)

if ρ density of sample; μ_M = mass absorption coefficient. It is independent of the state of element. For μ_M units are cm²/g. It is an additive function of weight fraction of elements in the sample.

$$\mu_M = W_A \mu_A + W_B \mu_B + W_C \mu_C$$

...(37.10)

if W_A, W_B, W_C are mass fraction of elements A, B, C and μ_A, μ_B, μ_C are mass absorption coefficient of corresponding elements.

It is

$$\mu_M = \Sigma \mu m_1 W_1 = \Sigma \mu_n W_n$$

...(37.11)

for n-number of the elements. The absorption edges are used for the qualitative characterisation of elements. The mass absorption coefficient (μ_M) is proportional to fourth power of atomic number of an element at fixed wavelength

i.e.

$$\mu_M \propto [N]^4 \text{ at fixed wavelength}$$

...(37.12)

It rises with wavelength from 2.5–3.0. This coefficient is inversely proportional to the atomic weight (N) of the absorbing element. Absorption and scattering are always additive. If J is absorption (photoelectric) coefficient, σ = scattering coefficient then we have mass absorption coefficient,

$$\mu = (J + \sigma) \qquad \qquad ...(37.13)$$

as both absorption and scattering coefficient are additive.

37.4 INSTRUMENTATION

Before undertaking the discussion on application of X-rays in three modes viz. absorption, fluorescence and diffraction it is worthwhile to undertake a discussion on instrumentation which is more or less common for all three techniques with minor variation. Amongst important components we have (a) source, (b) collimator, (c) sampler (d) dispersive device and (e) detectors. The last two forms consists of variety of variations. Schematically, the working instrument can be shown in Fig. 37.5.

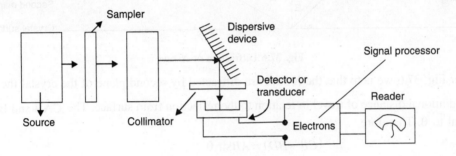

Fig. 37.5 X-ray unit

(a) *Source:* Apart from X-ray tube already described earlier we have the radioisotopes or fluorescent source. X-ray tubes are also called as the Coolidge tube. Tungsten is cathode and copper is anode. The process is efficient in X-ray tubes. Due to heating by thermionic emission the anode is cooled with water. With transducer cooling is not essential. Radio isotopes encapsuled and emitting of X-ray radiation in region of 0.30 – 0.45 Å– are good for fluorescence and diffractions studies.

Fluorescent source gives K_α and K_β lines from molybdenum or tungsten electrode. The process involves electron capture or γ or β-rays emission, but all radiations have low intensity. Many elements like Am^{241}, Cd^{109}, Co^{52}, Fe^{51}, H^3, I^{125} and Pm^{147} can be used as the radioactive source to demonstrate electron capture or γ-emission, but the continuously emitted X-rays cannot be stopped like in X-ray tube in radioactive source.

(b) *Collimator:* This is equivalent to monochromator in the optical spectroscopy. For using Bragg's equation ($n \lambda = 2d \sin\theta$) we need to know angle of incidence. The metallic plates collimator the generated rays; where only parallel radiation are allowed to pass through, cylindrical hollow tubes can also be used to collimate radiation. Two collimators are placed between sample holder and dispersive device as well as in disperser device and detector. Filters were also used to segregate K_β radiation. Thin metal sheets like Zr act as filters for X-rays.

(c) *Sampler:* The analytical sample must be reasonably thin to prevent all absorption of X-rays. The solutions are kept in cells with transparent windows. Mylar or with low atomic number metals like beryllium are used as windows.

(d) *Analytical Crystal:* This is related to monochromator where in the radiation is obtained by reflection of X-rays from crystal faces. The relation between (λ) wavelength and angle of diffraction (θ) and distance between each set of atomic planes of crystal lattice (d') is given by

Bragg's equation as

$$m\lambda = 2d \sin\theta \qquad \ldots(37.14)$$

where m = order of diffraction (Fig. 37.6).

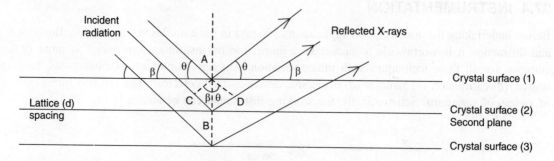

Fig. 37.6 Diffraction of X-rays

As per Fig. 37.6 we note that the X-rays are diffracted by second plane of the crystal the distance $\overline{CBD}$ is additional distance of travel in relation to the reflection from surface. The $C\hat{A}B$ and $B\hat{A}D$ i.e., β are equal to θ. Therefore

$$(CB - BD) = AB \sin\theta \qquad \ldots(37.15)$$

and

$$\overline{CBD} = 2(AB) \sin\theta \qquad \ldots(37.16)$$

wherein $\overline{AB}$ is interplanar spacing 'd'. The angle between direction of the incident beam and that of diffracted beam is $2\hat{\theta}$. The analysing crystal is mounted on 'Gonimeter' and rotated through required angle. d-value must be small to make angle $2\hat{\theta} > 8°$. The small d-spacing is good in order to get high dispersion

$$d\theta/d\lambda = m(2d \cos\theta) \qquad \ldots(37.17)$$

by differentiating Bragg's equation. We get

$$\lambda = 2d \qquad \ldots(37.18)$$

angle $2\hat{\theta}$ becomes 180°. Gonimeter can be moved for $\angle 150°$. Variety of materials are used as analyser crystals. They include LiF (3.8 – 0.218 Å) Al (4.5 – 0.32 Å), NaCl (5.4 – 0.3 Å), CaF_2 (6.1 – 0.4 Å), Mica (19.2 – 1.39 Å). The figures in the bracket indicate useful range of wave region in Å. The lattice spacing varies from 2.0–9.9 Å while the reflecting plane varies from 200–002. Beyond 26Å multiple monolayer soap film crystal is used.

The X-ray filters with their corresponding K-edges are consisting of Co (160.8) Fe (174), Mn (189), Ni (148) Pd (51), V (226) and Zr (69). Use of one single crystal cannot cover entire effective region of spectrum. Therefore two or more crystals in combination are preferred. The angle $\hat{\theta}$ should be adjusted between 5–80° angle. Large spectral range crystals have poor dispersion ($d\theta/d\lambda$).

The dispersion is obtained by differentiating Bragg's equation.

$$\frac{d\theta}{d\lambda} = \frac{n}{2d \cos\theta} \qquad \ldots(37.19)$$

This shows dispersion is inversely proportional to crystalline spacing and is function of angle of incidence.

(*e*) *Detectors:* These are also termed as transducer and signal processors. There are several variations including those like photon counting, gas filled d-electrodes, GM tube, proportional counter, ionisation chamber, semiconductor and scintillation counters. It is out of scope for us to undertake discussion of all kinds of detectors; we would only consider those which are most commonly used in X-ray spectral analysis. Since photon counter, GM tubes and proportional counters were considered earlier in the chapter on radiochemical methods we shall restrict discussion to ionisation chamber, scintillation counters and the semi conductor detectors. Photon counting is useless for high intensity beam of radiation, but it is good in absence of monochromator. A gas filled transducer is more efficient. GM tube lacks the large counting time.

(*f*) *Gas ionisation detector:* They are operated in voltage range of 0-400 volts when current is $10^{-13} - 10^{-16}$. However, their use is restricted due to lack of sensitivity but find more use in radiochemical analysis. Inert gas with two electrodes are used. As seen in Fig. 37.7, windows are made of mica or beryllium. X-rays reacts with inert gas like Argon in chamber to form positive ions and electrons (*i.e.*, $+e^-$). These electrons ionise the gas which in turn form charged ion pair of +ion and electron. +ions are neutralised while electrons generate a measurable current at (*c*). The limitation of this detector is lack of facility for amplification of current generated. However, it has the advantage that the output pulse is function of energy of incident photon.

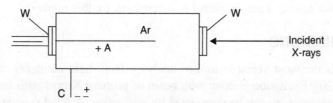

Fig. 37.7 Ionisation chamber

(*g*) *Scintillation counters:* With availability of good photo tubes this method has become popular. It consists of transparent (NaI) crystal containing trace of thallium. It is a 3.4″ dimension crystal with plane surfaces out of which one faces to cathode of photo multiplier tube. The energy of incoming radiation is lost to the scintillation and is in turn subsequently released as fluorescence radiation. The dead time is 0.25 μS. Flashes produced are transmitted to photo cathode of PMT and are converted to electrical pulses. Each flash generated is in proportion to energy of incoming photon. Organic liquid scintillation are better because of less self absorption. Thallium doped iodide crystal is covered with reflecting coating of aluminium which prevents loss of the radiation, further it does not permit external radiation and it directs X-rays to PMT in centre. The atmospheric moisture is prevented from deposition on crystal surface. They have low dead time (~0.25 μS) with better sensitivity.

(*h*) *Semiconductor Transducers:* They are best in X-ray detection process. They are called "lithium drifted silicon or germanium detectors". Such detector is formed by permitting lithium to diffuse into silicon semiconductor. Lithium is deposited on the surface of p-doped silicon. On heating at 400°C lithium diffused to silicon. This lithium can donate electrons to crystal thereby changing p-type to n-type silicon. A 'np' junction is formed. A DC potential is applied, which causes the electrons to withdraw from lithium-region and positive holes to be withdrawn from p-type silicon region. After drifting of this crystal it is cooled, rendering lithium immobile. The crystal has now three regions: n type region has lithium atoms in silicon than p-type region is analogous to silicon diode and central region with lithium in silicon with high resistance. The outer surface of p-type is coated with film of gold for good contact. The signal output is taken from aluminium layer which is coated to n-type silicon. It is then fed to amplifier for measurement. These devices must be cooled in liquid nitrogen as lithium atoms would diffuse in all directions in silicon reducing efficiency of detector. Use of germanium in place of silicon can detect short radiation of 0.3 Å. These intrinsic germanium detectors are to be cooled during their use.

37.5 X-RAY ABSORPTION IN CHEMICAL ANALYSIS

These applications are limited in comparison to X-ray emission and X-ray fluorescence processes. The only advantage of this method has minimum interferences. These methods are similar to optical spectroscopic methods. Only one needs monochromatic radiation. The single element with high atomic number is to be analysed in a sample of only light elements. The method can be used both for qualitative and quantitative analysis. For qualitative work absorptive edges are compared with standard when $N = 12 - 100$. However, quantitative analysis heavily depends upon the concept of absorbtive edges. The transmitted X-ray intensity is measured at two wavelengths which are shorter than edge and two wavelengths which are longer than edge. In comparison to X-ray fluorescence method, this technique is less frequently used. It is used for ascertaining thickness of metallic plate. One needs portable X-ray model with radioactive source. Lead in petrol is determined by this method.

37.6 X-RAY EMISSIONS

X-ray fluorescence is the most versatile tool of analysis. Only light elements cannot be analysed by these methods ($N < 20$). Excitation is done with beam of primary X-rays with large energy. (Tungsten target tube). Various wavelength can be segregated by diffraction. A good separating crystal must have good dispersion. Light alloy steels can be best analysed by this method. Nickel from aluminium alloy can be also analysed. Lead and bromine in petroleum can be also analysed. Water causes excessive scattering of X-rays hence solutions cannot be used. Thus the emission spectra is useful for quantitative analysis of heavier elements. The sample is placed on copper holder as target material. Metallic target can be also machined. The emitted spectrum is dispersed and used for qualitative analysis. However, to prepare sample as target material is a cumbersome process. Further in such investigation hydrochloric acid provides reproducible results while other acids like nitric or sulphuric acids are not preferable when dealing with solution chemistry.

37.7 X-RAY FLUORESCENCE METHODS

In generation of X-rays the photon from source is absorbed by sample producing positive ion with vacant space in inner electronic orbit. An electron from outer orbit falls into vacant place, when X-ray photon is emitted. It is X-ray fluorescence if initial excitation occurs by absorption of X-ray photons. This incident photon has very large energy but fluorescence photon has comparatively very small energy. Therefore radiation is fluoresenced at wavelength which is greater than the absorbed radiation. Incident X-ray radiation must be strong to expel electrons from inner orbital. Bandwidth of emitted radiation is small. The sample must be thin to allow X-rays to pass through with minimum absorption. If there is partial penetration of X-rays in sample then we have:

$$1 = \frac{kW}{\mu_i + \mu_f}$$ ...(37.20)

if k = constant, W = weight of fraction of sample and μ_i and μ_f are the linear absorptive coefficient for sample at incident wavelength as well as at fluorescent wavelength. A change in fluorescence radiation is caused by sample than analyte sample. This is matrix effect or interferences. However, excitation is brought out by irradiation of sample with X-ray beam from tube or radioactive source. Primary beam excites absorption and emits fluorescence. XRF is best used for qualitative analysis if N > 8. The method is popular as it permits non-destructive analysis.

37.8 FLUORESCENCE INSTRUMENTS

It is same as in X-ray absorption.

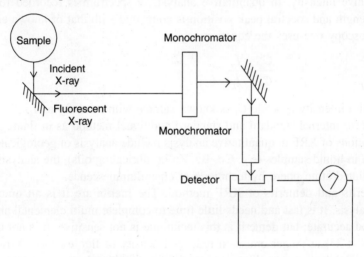

Fig. 37.8 X-ray on wavelength dispersion.

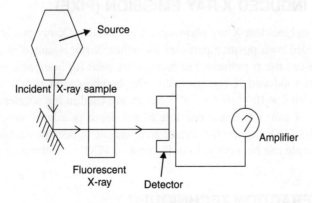

Fig. 37.9 X-ray on energy dispersion

Two models viz. wavelength dispersive and energy dispersive instruments are available and are shown respectively in Figs. 37.8 and 37.9.

In former impinging beam is not monochromatic. The energy dispersive models have in comparison less resolution (Figs. 37.7 and 37.8). Solids, liquids, gases can be analysed by X-ray fluorescence. Powders are used in form of pellets or disks made with sodium tetraborate. The solutions are put in aluminium cells. Beryllium windows are common. We can have instruments with single or multichannel models. Energy dispersive instrument are relatively cheaper than wavelength dispersing instruments. They are good at shorter wavelength. Nondispersive instruments are used for specialised job as analysis of sulphur in oil.

37.9 X-RAY FLUORESCENCE IN ANALYSIS

The qualitative analysis is carried out by comparison of peak of sample with known element, and they can be identified. Quantitative analysis is analogous to UV-visible technique. One can approximately evaluate quantitative data using following equation viz. $P_x = P_y W_x$ if P_x- line intensity, W_x is weight of sample, P_y is relative intensity. In quantitative analysis, a spectrum is recorded for intensity as the function of wavelength and spectral peak position is compared with that of known element. Similar to UV-visible spectroscopy one uses the equation

$$\frac{C_x}{C_k} = \frac{I_x}{I_k} \qquad \qquad ...(37.21)$$

if C_x unknown, with intensity I_x while C_k is known sample with intensity as I_k.

One can use also internal standard and standard additional method as in flame photometry. Some significant applications of XRF in quantitative analysis include analysis of geological materials, quality control operations in liquid samples (e.g. Co, Ba, Zn in lubricating oils), the analysis of SO_2 and NO-NO_2 from the polluted atmosphere by technique of chemiluminescence.

There are merits and demerits of XRF method. The merits are it is an unique technique for nondestructive analysis. It is fast and needs little time to complete multi elemental analysis. The methods are precise and accurate; but demerit is this technique is not sensitive. It is not useful for analysis of light metals (N < 23) as Auger emission reduces intensity of fluorescence. X-ray instruments are inexpensive (Rs. 250,000) or for sophisticated model Rs. 500,000. Latter is a automated model.

37.10 PARTICLE INDUCED X-RAY EMISSION (PIXE)

The incident electrons and incident X-ray photons cause secondary X-ray emission. A technique where in if sample is bombarded with positive particles to induce X-ray is called as particle induced X-ray emission or PIXE. One can use α-particles but protons are used for bombardment of the sample hence they are called as proton induced X-ray emission. The bombardment results in generation of X-ray fluorescence. Protons with 2×10^{-14} to 3×10^{-12} energy are used in the accelerator. It permits analysis of sample from 1nm to 1 μm. It is quite possible to get depth profile of sample. Proton microprobe facilitates PIXE use. This PIXE is good for surface analysis due to deep penetration and high sensitivity. As small as 1 ρg of sample can be analysed but the cost of PIXE is prohibitive for common use in the laboratory.

37.11 X-RAY DIFFRACTION TECHNIQUES

Since 1912 Laue used this method primarily for elucidation of structure of crystalline solid matter. The physical properties of metals were better understood by this method. The elucidation of structure of variety of organic compounds like natural products, steroids, vitamins, antibiotics was carried out by X-ray diffraction. A qualitative identification of crystal is also possible. Powdered samples are best analysed by XRD. The composition of samples can be easily arrived at. Since X-ray diffraction pattern is unique for each crystal like fingerprints identification was possible. In crystal the angle at which radiation is diffracted from crystal is related to atomic spacing between two atoms in a crystal.

The crystalline substance is ground to fine powder. It contains small tiny crystals. On exposure to X-rays they orient for reflection from interplannar spacing. Sample is placed in a glass or cellophane tube or it is moulded with noncrystalline binder. One uses automatic diffractometer or photographic recording device.

The diffraction pattern can be obtained, with X-ray source with filter. A powder sample replaces crystal on mount. However, photographic plates are most commonly used, specially when sample amount is too small. A powder camera (Debye-Scherrer) is used. A monochromatic beam is collimated (From Cu or W as K_α Lines) and is passed through narrow orifice. Undiffracted radiation passes out. Camera is cylindrical and can wrap strip of film around its inside wall. The sample is held in centre of beam. We get a pattern as shown in Figures 37.10 and 37.11.

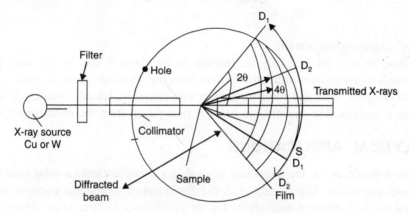

Fig. 37.10 X-ray Powder Goniometer

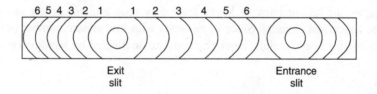

Fig. 37.11 Developed photograph.

Above figures show representation of diffraction lines from plane of crystal. Bragg angle θ for each line is indicated. A 1 mm³ portion of the powder contains 10^{12} crystals. Each crystal occupies 10^{-2} m³ volume. X-ray film is wrapped around inner circular casting that surrounds sample.

The developed film shows impression as in Fig. 37.11. The angle between the beam of X-rays and the line that connects the sample with arc is 2θ. By using equation

$$n\lambda = 2d\sin\theta \quad \text{(Bragg's Equation)} \qquad \qquad ...(37.22)$$

we can then calculate interatomic spacing within a crystal. The distance 's' can be measured between two arcs of cone on opposite of exit orifice. The angle between two lines connecting arcs with sample is $4\hat{\theta}$. It can be also calculated as

$$4\theta = \frac{s}{r} \qquad \qquad ...(37.23)$$

in radians. However, in Bragg's equation, angle $2\hat{\theta}$ must be measured in degrees as 1 rad $= 57.9°$. So we do conversion as

$$\theta_{rad} \times 57.3°/rad = \theta_{deg} \qquad \qquad ...(37.24)$$

We have then

$$\frac{4\theta_{deg}}{57.3} = \frac{s}{r} \qquad \qquad ...(37.25)$$

or finally

$$\theta_{deg} = \frac{57.35}{4r} \qquad \qquad ...(37.26)$$

is used in Bragg's equation calculation.

The diffraction pattern can be easily interpreted. If we know θ and 2θ and also their relative intensities, one needs to use powder diffraction pattern from file. One can carry out quantitative analysis by measuring intensities of lines and comparing them with known standards. Powder camera is mostly used for qualitative analysis. The quantitative analysis rarely used from the diffraction measurements.

37.12 ANALYTICAL APPLICATIONS

X-rays constitute a fine tool for the elemental analysis for metallurgical samples containing silver, copper nickel and aluminium. The facility of X-ray fluorescence to analyse complex solid samples without involving any wet chemical analysis is of great significance in metallurgy, mineral and cement analysis. It is a boon in analysis of metals like nickel in petrochemical products. The entire technique is based upon simple procedure viz. the excitation of inner electron (e.g. K-shell) of atoms promotes some to high energies. In falling back to lower level they may emit radiation in the X-ray region characteristic of the concerned element. Since in elements with low atomic number (<12) the electrons are involved in bonding for heavy metals, inner electrons are unaffected by bonding or valency changes.

LITERATURE

1. J. Wandkovic, "*Instrumentation in Analytical Chemistry*", Part I, Ed. J. Zyka, Ellis Harwood Publishers (1991).

2. H.A. Lichafsky, H.G. Pfeiffer, E.H. Winslow, P.D. Zemany, "*X-Ray Absorption and Emission in Analytical Chemistry*", John Wiley and Sons (1960).

3. H.D. Klug, L.E. Alexander, "*X-Ray Diffraction Procedures*", Wiley (1954).

4. J.G. Brown, "*X-Rays and Their Applications*", Plenum (1966).

5. T.G. Dzubay (Ed.), "*X-Ray Fluorescence Analysis of Environmental Samples*", Ann Arbor Science (1977).

6. L.Bragg, "*The Development of X-Ray Analysis*", Haffner Press NY (1975).

7. D. Kealey, P.J. Haines, "*Analytical Chemistry – Instant Notes*", p.214, Viva Books (P) Ltd. New Delhi. (2002).

Chapter 38

Mössbauer Spectroscopy

Mössbauer spectroscopy is not used extensively as are other spectroscopic methods of quantitative analysis. This is because we do not have inexpensive instruments for such investigations and many elements to be studied do not have suitable isotopes. In fact Fe^{57} and Sn^{118} are the only two isotopes which have been extensively used in Mössbauer spectral measurements.

As far as its definition is concerned, we can state that "it is the recoil free gamma ray resonance absorption or alternatively state that it is a phenomena of gamma resonance fluorescence spectroscopy. Fe^{57} and Sn^{118} are two isotopes which posses the narrow resonance lines for study of hyperfine interactions of nucleii. While investigating this phenomena, an account for isomeric shift as well as the quadrupole and magnetic splitting is also necessary. Out of 104 elements only 46 elements can be easily analysed as in all they have more than 88 isotopes for characterisation. Other isotopes which are commonly used are Zn^{69}, Sm^{147}, Sn^{119}, Sb^{121}, I^{127}, Ni^{61} and Ir^{191}.

The Mössbauer spectroscopy has immense applications not only in analytical chemistry but also in other physical, chemical and the biological sciences. The elucidation of nuclear structure is best carried out by this method. In biological sciences the movements of microorganism under influence of magnetic field, or study of proteins or hemoproteins or enzyme is carried out by this technique. The application in chemical sciences are numerous. They include analysis of lunar soil, alloys, corrosion products, organometallic compounds and chemical dynamics. One can get more reliable and best results if this technique is coupled along with other spectral methods such as XRF, ESR and NMR.

38.1 HISTORICAL SIGNIFICANCE

Mössbauer spectroscopy is the recoil free γ-ray resonance absorption. It is also termed as nuclear gamma resonance fluorescence. The technique has wide uses in various physical, chemical and biological sciences. Its extensive rapid use in science has proved its importance and utility. In fact in 1958 Mössbauer, a german scientist, discovered this nuclear resonance absorption of gamma radiation region of particular isotope. He only pointed out that ^{57}Fe and ^{119}Sn isotopes possess the narrow resonance lines permitting study of hyperfine interactions of nucleii and its effects. For his discovery, Mössbauer was awarded Nobel prize in 1961. This technique later on became very popular for the elucidation of the structure. It was mainly due to easy availability of isotopes of iron in natural and synthetic resources. It has several applications like those of analysis of the lunar samples, study of oxidation products of iron, analysis of alloys and steel and biological products like hemoproteins and analogous substances.

38.2 PRINCIPLES OF MÖSSBAUER SPECTROSCOPY

It can be considered under broad areas of

(a) Nuclear γ-ray resonance fluorescence
(b) Hyperfine spectra
(c) Modulation of energy
(d) Isomeric shift
(e) Quadrupole splitting
(f) Magnetic splitting.

The phenomena involves the transition of the nucleus from excited state (E_0) to ground state of energy (E_g) with the emission of the γ-ray of energy (E_r) as:

$$E_r = (E_0 - E_g) - E_R \qquad ...(38.1)$$

wherein E_R is recoil energy of the nucleus.

$$E_R = E_r^2 / 2mc^2 \qquad ...(38.2)$$

wherein 'm' is mass of the recoil atom. As E_R is positive number, E_r will be less being the difference of $(E_0 - E_g)$. The absorption of γ-radiation by the nucleus leads to pumping of the absorper nucleus from the ground state to the excited state. The newly excited nucleus remains in higher energy for a time 'J' and returns to ground state by reemission of γ-radiation, which occurs equally in all directions.

(a) *Nuclear γ-ray resonance fluoroscence:* Let us consider all factors responsible for this effect. As per nuclear gamma resonance fluorescence effect, the gamma radiation with energy E transfers during absorption or emission by the nuclei, as

$$p = E/C \qquad ...(38.3)$$

related with energy change of

$$E = \frac{p^2}{2m_1} = \frac{E^2}{2m_1c^2} = \Delta E \qquad ...(38.4)$$

here 'm_1 is the nuclear mass and 'c' is the velocity of light. Now if we consider resonance in transition of nuclei between excited state with energy (E_0) with lifetime of 'J' and the ground state, the resonance energy of γ-photon during absorption or emission is

$$E = E_0 \pm \Delta E = E_0 + \frac{E_0^2}{2m_1c^2} \qquad ...(38.5)$$

This is because with small natural 'J' lifetime is

$$J = \frac{h}{r} \ll E \quad \text{(Heisenberg principle)} \qquad ...(38.6)$$

Now for crystal lattice, the recoil impulse is transferred to crystal instead of nucleus. Consequently equation (38.4) and (38.5) contains mass of entire system as

$$M = N_A m \qquad ...(38.7)$$

(if N_A is Avogadro's number)

if
$$\Delta E = E_0 \qquad \qquad ...(38.8)$$

This results in photon less absorption and emission of γ-photons with the probability:

$$f = \exp\left(-p^2\left(x^2/h^2\right)\right) \qquad ...(38.9)$$

if x^2 = mean quadaric amplitude of the lattice vibrations. Consequently, photon less radiative transition *i.e.* the Mössbauer effect is observed with non-zero probability. We have eq. (38.9) as

$$f = \exp\left(-\frac{3E_g^2}{4m_1 c^2 K\theta_n}\right) \qquad ...(38.10)$$

if K = Boltzmann's constant and θ_n is Debye temperature. So at high temperature

$$f = \exp\left(-\frac{3E_0^2 T}{2m_1 c^2 K\theta_D}\right) \qquad ...(38.11)$$

With decrease in temperature it decreases. The important conclusion is that the Mössbauer effect can occur for low energy levels of heavy nuclei in solid state at low temperature.

(b) *The hyperfine structure* of spectra as per eq. (38.6) natural width of lines is exceedingly small e.g. ^{57}Fe has energy

$$E_0 = 14.4 \text{ keV} \qquad ...(38.12)$$

and lifetime

$$J = 1.41 \times 10^{-7} \text{ sec} \qquad ...(38.13)$$

to

$$J = 4.7 \times 10^{-9} \text{ eV} \qquad ...(38.14)$$

and width

$$J/E_0 = 3.2 \times 10^{-13} \qquad ...(38.15)$$

It permits noting of changes in photo energy in earth's gravitational field, and quadratic Doppler vibrations. All Mössbauer applications are based on hyperfine structure.

(c) *Modulation of an energy:* Resonance fluorescence is one of the interesting spectroscopic technique if you understand the energy of modulation of γ-ray emitted in the decay process. It can be alternatively arrived from Heisenberg uncertainty principle which states that

$$J = \frac{h}{2\pi r} \qquad ...(38.16)$$

(if h = Planck's constant; J = lifetime of excited state for ^{57}Fe)

$$J = 4.6 \times 10^{-2} \text{ keV} \qquad ...(38.17)$$

To modulate γ photon which is 14.41 keV can take use of Doppler effect which enunciates that if a radiation source has velocity relative to observer as 'v' the energy will be shifted by an amount

$$E = \left(\frac{v}{C}\right)E \qquad ...(38.18)$$

which leads to calculation

$$v = C \cdot \frac{J}{E_r} = 3 \times 10^{-10} \times \frac{4.6 \times 10^{-12}}{14.4} = 0.0096 \text{ cm/s} = 0.0036 \text{ in/s} \qquad ...(38.19)$$

(*d*) *Isomeric Shift:* The variation in transition energy for the free nucleus by interaction with electron shell is given by expression no. (38.21) as

$$\Delta E = K \frac{\Delta R}{R} |\psi(0)|^2 \qquad ...(38.20)$$

if $\Delta R/R$ = relative change in nuclear radius in ground and excited state, K = nuclear constant and $|\psi(0)|^2$ is the electron density. Since difference between electron densities of the emitting and absorbing nuclei shift is

$$\Delta E_1 = K \frac{\Delta R}{R} \Delta |\psi(0)|^2 \qquad ...(38.21)$$

The shift represents electronic states in absorption and emission of the γ-photons *i.e.* chemical states of resonating ions or atoms. It is analogous to chemical shift in NMR spectroscopy; facilitating evaluation of electron density changes. Similar temperature shift can be also considered.

(*e*) *Splitting Quadrapole:* A shift is caused in the nuclear magnetic spin number m_1 due to the interactions of electric field, gradient with unsymmetric spherical nucleus. For ^{57}Fe the split into two levels are

$$2\Delta E_q = eQ r_{i/2} \qquad ...(38.22)$$

if e = charge of an electron. This leads to the quadrupole doublet formation. The crystal field of the lattice is

$$v_i(0) = (1 - R_0) v^0(0) + (1 - R_i) v^S(0) \qquad ...(38.23)$$

In equation (38.15) '*a*' and '*c*' designate atomic and crystal contributions respectively and R_s = Sternheimer (antishielding) factors. The quadrupole splitting is typical of electronic structure of the resonating atom.

(*f*) *Magnetic splitting:* Let us consider magnetic splitting which occurs due to the action of magnetic field with induction 'B'_{ef}' on nucleus *i.e.* the nuclear 'Zeeman effect'. It depends on nuclear magnets 'm_l' quantum number and is expressed as

$$E_m = -\mu B_{ef} m_l / l \qquad ...(38.24)$$

Here '*n*' represents magnetic moment of the nucleus e.g. Fe^{57} with $l = 1/2$ split is

$$2\Delta E_m = \mu B_{ef} \text{ (see Fig. 38.1)} \qquad ...(38.25)$$

As averaging of magnetic field occurs in lifetime of nucleus, (J) does contain term for effective magnetic field as shown by vector 'B_{ef}'. In field it decreases with temperature for para and diamagnetic material it is zero. So magnitude of magnetic splitting and distribution of transition intensities speaks for magnetic state of an atom in lattice or crystal. The Mössbauer spectra can be related to the composition and structure of the substance. Number of examples are provided at the end of chapter to illustrate as to how the Mössbauer effect spectroscopy for elucidation of the structure of an atom, electronic configuration and chemical characteristics.

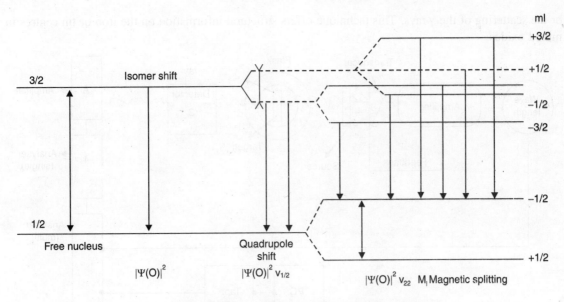

Fig. 38.1 Magnetic splitting and shift of Fe^{57}

38.3 MÖSSBAUER INSTRUMENTATION

Before consideration of the instrumentation, it is worthwhile to consider Mössbauer isotopes. There is need for low energy excited state and the heavy nuclei. Only 46 elements are included. They are as follows. These elements have in all as many as 88 isotopes.

Table 35.1 Mössbauer isotopes

Group I (alkali metals) & IB	K^{40}, Cs^{133}, Ag^{107}, Au^{197}
Group II (alkaline earths) & IIB	Ba^{133}, Zn^{67}, Hg$^{199-201}$
Group III (lanthanides) & IIIB	La$^{137-139}$, Sm$^{147-154}$, Eu$^{151-152}$, Gd$^{154-160}$, Dy$^{160-164}$, Ho^{165}, Er$^{164-171}$, Yb$^{170-176}$, Lu^{175}.
Group IV & IVB	Hf^{126}, Ge^{73}, Sn^{119}, Th^{232}
Group V & VB	Ta^{181}, Sb^{121}
Group VI & VIB	W$^{180-186}$, Te^{125}, U^{234}
Group VII & VIIB	Tc^{99}, Re^{187}, I$^{127-129}$
Group VIII	Fe^{57}, Ni^{61}, Ru$^{99-100}$, Os$^{186-190}$, Ir$^{191-193}$, Pt^{195}

We need suitable excited state for the decay of artificial radioisotopes where in the Coulomb's excitation of nuclei can be used. A large 'f' factor (e.g. eqns. 38.10 and 38.11) is essential. The contents of Mössbauer isotope is also vital in natural mixture. Therefore Fe^{57} and Sn^{113} are commonly used as Mössbauer isotopes. These metals which have high 'f' factor and have $I = 3/2 \rightarrow 1/2$ transitions in both the metals.

Mössbauer spectroscopy technique is limited to relatively few nuclei. These methods include Fe^{57} therefore it finds use in analysis of iron related products like corrosion studies and enrichment of iron. Its use is confined to the solid state system. The sample must be reasonably thin to prevent absorption or the scattering of the γ-rays. This technique offers structural information on the iron or tin centres in mixed oxides.

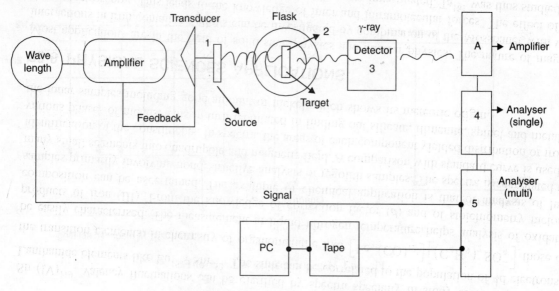

Fig. 38.2 Mössbauer spectrometer

An active source is mounted on Transducer (1) which imparts motion relative to absorber. The γ-rays strike sample being analysed (absorber) (2) is then transmitted to detectors (3). These in turn are stored in analysers (4, 5). Then after amplification (A), few γ-rays are absorbed by sample and again re-emitted in multidirections. The remaining γ-rays pass through multichannel analyser (5). It will be difficult to carry out measurements with 10^{-13} resolution. The velocity spectrometers record intensity of resonance γ-radiation as function of absorber velocity. When source is radioactive, a hyperfine structure of spectra is obtained for absorber. A radiation passing through absorber by transmission is measured by the detector or analyser. Two working modes are employed. The velocity spectra is recorded with increasing velocity. A method of parabolic motion is recorded by analyser.

38.4 INTERPRETATION OF SPECTRA

The majority of results obtained from Mössbauer spectroscopy are in the form of velocity spectrum; where intensity (I) is recorded as a function of velocity (v). In ideal cases resonance line has Lorentzian shape. The real spectra of absorber with many resonating atoms are super position of number of curves corresponding to hyperfine structure of each component. A common method for analysis is the fitting spectras by least square method. Let us consider Fig. 38.3.

For spectra (*a*) is result of many repetitive scans through velocity range. The spectra is character-ised by position of 'δ' (resonance maximum and minimum intensity) of transmitted line where 'A' represents total area under curve.

For spectra (*b*) active nuclides like Fe^{57} or Sn^{119} are used. A quadrupole coupling is observed. Thus Δ is equal to $e^2 q_0/2$, if e = charge, q = gradient of field and O = nuclear quadrupole moment, where S, Δ, J, A, H_0 are all dependent upon temperature.

For spectra (*c*) can give information on the magnetic field (H). Only single resonance line is seen in absence of magnetic interaction. So far, more than 100 Mössbauer structure transitions with 43 elements have been observed.

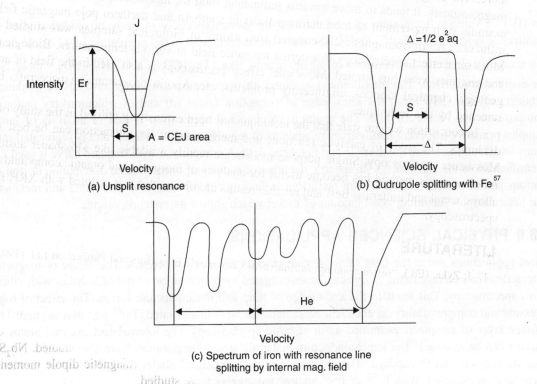

Fig. 38.3 Various spectras

(a) Unsplit resonance

(b) Qudrupole splitting with Fe⁵⁷

(c) Spectrum of iron with resonance line splitting by internal mag. field

38.5 MÖSSBAUER SPECTROSCOPY APPLICATIONS

These applications can be considered in three groups as those in eqn. (38.8) physical science involving mainly nuclear and solid state physics, eqn. (38.7) chemical science applications pertaining to the structural chemistry involving analysis of lunar soil, oxidation product of corrosion, analysis of alloys etc. and eqn. (38.9) biological science applications involving analysis of protein and hemoproteins etc.

38.6 ANALYTICAL SCIENCE APPLICATIONS

Analytical science applications involves qualitative and quantitative analysis. Mössbauer spectra is characteristic of the electron structure. Such interpretation of spectra can be done in two ways. One involves the segregation of individual spectras and subsequent classification by their hyperfine structure. In symmetrical cubical field, unsplit line spectra is obtained while in unsymmetrical cubical structure, quadrupole doublet is observed. Fe^{52} and Sn^{119} are used as standards. Other is fingerprint region wherein is unknown spectra is compared with spectra of known compound. It is called 'stripping method'. While in quantitative analysis the number of resonating atoms is directly proportional to the intensity of the component *i.e.* area 'A' analysis of mixture is a little complicated.

38.7 CHEMICAL SCIENCES APPLICATIONS

It is largely used for the elucidation of the structure of compound. Fe^{52} and Sn^{119} which are used as standards but I^{129}, Ru^{99}, Sb^{121} are also equally effective. Resonance maxima (δ) or chemical shift is related to electronic configuration of atom. The shifts for $Sn(II)^{(119)}$ are large as compared to those for

Sn (IV)[119]. Valency fluctuations can be clarified by spectra specially in study of oxidation state of Lanthanide elements like Eu[151], Sm[152]. The shift can be correlated to the population of 4d electrons in the transition elements. In chemistry of organometallic like $\left[Fe_3\left(CO\right)_{12}\right]\left[\left(C_4H_9\right)_2 SO_4^-\right]$ those can be easily characterised. The measurement at liquid nitrogen temperature helps analysis of oxidation products of iron (III). From the knowledge of correction factor (d) and of stoichiometry factor(s) composition can be ascertained. The example of chemical application is that of analysis of lunar samples primarily involving nondestructive analysis of regolith samples. The spectra is segregated into many small segments into quadrupole and magnetic field. A comparison with standard curve is used for identification of the constituents. In spectra, the areas of each component yielded distribution of iron in various phases of metals. This in turn facilitated in finding out silicate, illmenite, spinel and metals in the lunar samples including good amount of nickel which shows its meteoric origin.

38.8 PHYSICAL SCIENCES APPLICATIONS

Most applications are in the field of solid state physics and nuclear physics. The nature of magnetic interactions in iron containing alloys can be investigated by combination of the Mössbauer with vibrations spectroscopy. This leads to the knowledge of inter and intramolecular forces. The effect of high pressure and compressibility on chemical characteristics were investigated. Ta[181] was thus studied. The surface layer of sample is examined as in electron spectroscopy The internal and external atoms of surface can be assessed. The superconducting materials at low temperature were also studied. Nb_2Sn was studied with Sn[119] standard. The electromagnetic moments of nuclei (magnetic dipole moment) were also investigated. With Ir[191] or Ir[192] nuclear parameters were studied.

38.9 BIOLOGICAL SCIENCES APPLICATIONS

The proteins and the hemoproteins were characterised by Mössbauer spectroscopy. From 1961 such studies were undertaken by biochemists for the iron containing compounds. Paramagnetic iron compounds can be studied at Neel temperature for compounds with molecular weight of 50,000 or more. The magnetic orientation of microorganism under the influence of the magnetic field with magnetosomes. It tends to move towards sedimental food for its consumption. Fe[57] isotope assisted to examine such movement of the microorganisms in southern and northern pole magnetic field and its influence. Antiferromagnetically coupled iron atoms in biological samples was studied by using Mössbauer effect spectroscopy in external magnetic field at various temperatures. Biological catalyst and enzymes were thus studied. Mössbauer effect spectroscopy is also used in the field of archeology, geology, chemical kinetics, engineering problems, etc. Apollo programmes (spacecraft) have been screened by this technique.

In conclusion we can state that the technique has been extensively utilised in the study of chemical dynamics. The study of catalytic reactions and mechanism of the interaction can be best studied by Mössbauer spectroscopy. Single purpose models are readily available like Mössbauer austenitometer. As of now this method has become useful for analysis of inorganic and organic compounds including alloys, metal and minerals. Best and reliable results are obtained if it is coupled with XRF, ESR, NMR spectroscopy.

LITERATURE

1. J. Zyka, (Ed.), "*Instrumentation in Analytical Chemistry*", Ellis Horwood Publication Ltd. (1991), p. 299.

2. P. Kealy, P.J. Haines, "*Instant Notes in Analytical Chemistry*", Viva Books (P) Ltd. New Delhi (2002), p. 189.

3. T.E. Cranshaw, "*Mössbauer Spectroscopy and Its Applications*", (1986).

4. R.H. Herber (Ed)., "*Chemical Mössbauer Spectroscopy*", (1984).

5. G.J. Long (Ed.), "*Mössbauer Spectroscopy Applied to Inorganic Chemistry*", (1984).

6. G.M. Bancroft, "*Mössbauer Spectroscopy*", (1973).

7. T.C. Gibb, "*Principles of Mössbauer Spectroscopy*", (1976).

8. U. Gonser (Ed.), "*Mössbauer Spectroscopy*", (1975).

2. P. Kealy, P.J. Haines, "Instant Notes in Analytical Chemistry", Viva Books (P) Ltd, New Delhi (2002), p. 189.
3. T.E. Cranshaw, "Mössbauer Spectroscopy and Its Applications", (1986).
4. R.H. Herber (Ed)., "Chemical Mössbauer Spectroscopy", (1984).
5. G.J. Long (Ed.), "Mössbauer Spectroscopy Applied to Inorganic Chemistry", (1984).
6. G.M. Bancroft, "Mössbauer Spectroscopy", (1974).
7. T.C. Gibb, "Principles of Mössbauer Spectroscopy", (1975).
8. B. Gonser (Ed.), "Mössbauer Spectroscopy", (1975).

Chapter 39

Electron Spectroscopy

Surface analysis has come into prominence only recently. It involves the study of concentration differences at a surface and in bulk, layer by layer. The composition of the surface of a solid differs from the interior or bulk of the solid. Surface analysis is of great importance in the study of catalysis, semiconductors, thin film technology and corrosion studies. The study of the functions of biological membranes is also possible. Electron beams are best suited for such studies because electrons can penetrate the outermost layer of the surface. However, the probe beam may also include photons or ions. Energy exchange interaction is possible with electromagnetic radiation in electron spectroscopy for chemical analysis (ESCA), with an incident electron beam as in Auger Electron Spectroscopy (AES) or with a beam of incident ions as in secondary ion mass spectrometry (SIMS). Excitation by electromagnetic radiation provides good information but lacks sensitivity, though it is nondestructive in nature while an incidental beam gives better sensitivity with restricted resolution with no chemical information. Finally, an electron beam gives better spatial resolution as well as sensitivity. The surface must be flat and clean. Surfaces can be cleaned by a process of sputtering with argon ions. Thus ESCA and AES are examples of electron spectroscopy while SIMS is ion spectroscopy.

39.1 PRINCIPLES OF ELECTRON SPECTROSCOPY

The signal produced by excitation of the analyte consists of a beam of electrons and measurements are made of the power of this beam as a function of electron energy. Excitation of the analyte is possible by irradiation with electrons. The first technique involves irradiation with monochromatic X-radiation for ESCA. This is also called X-ray photoelectron spectroscopy (XPS). In the second method, auger electron spectroscopy (AES), a beam of electrons is used. Electron spectroscopy can identify any metal or element except helium and hydrogen.

These techniques permit the determination of the oxidation state of an element. The kind of bonding and electronic structure can be well ascertained by ESCA or AES. The ionisation is:

$$S + h\nu_1 (\text{X rays}) \rightarrow S^{+*} + \overline{e} \qquad ...(39.1)$$

The inner shell electrons are ejected to give $[S]^{+*}$ in the excited state, creating a vacancy in the inner shell. We then monitor the kinetic energy of the excited state viz. $[S]^{+*}$. ESCA can study electrons in the core and valence orbitals. After excitation, electrons in the outer shell fall into vacancies in the inner shell with simultaneous emission of X-ray photons as $S^{+*} \rightarrow S^+ + h\nu_2$, (X-ray fluorescence). It is also possible to lose energy when the outer shell electron is absorbed by a second electron from a vacant orbital. Such a second electron can be ejected from the sample giving a doubly charged ion causing the "Auger effect":

$$S^{+*} \rightarrow S^{2+} + \overline{e} \quad \text{(Auger electron)} \qquad ...(39.2)$$

In ESCA the entire energy of the incident photon is absorbed and the emitted electron possesses kinetic energy. The process of ionisation proceeds as

$$S + \overline{e} \rightarrow S^{+*} + 2\overline{e} \quad \text{(continuum)}. \qquad ...(39.3)$$

Since the kinetic energy of each electron emitted during ESCA or AES is related to the energy of the orbital from which the electron is ejected and since orbital energies are peculiar to atoms or molecules, electron spectroscopy can be used for qualitative analysis. Further, as the number of emitted electrons is proportional to the concentration of the emitter, it can also be used for quantitative analysis. The transformation is shown in Fig. 39.1.

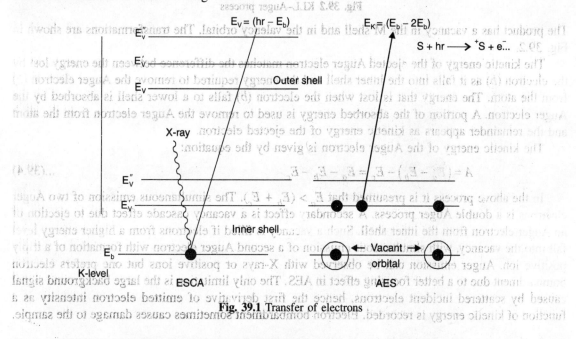

Fig. 39.1 Transfer of electrons

39.2 AUGER EMISSION SPECTROSCOPY

After initial ejection of inner shell electrons, an outer shell electron falls into the vacancy i.e. from the K-level to the L-level and from there once again to the L-level giving the transition KLL. The energy lost when the electron falls into the vacancy is used to eject a second outer shell (i.e. L) electron from the atom. Since, in all two electrons are ejected, a doubly charged positive ion is created. Thus the KLL auger process is one involving initial ejection from the 1s-level followed by relaxation of 2s electrons into the vacancy 1s and subsequent emission of a 2p electron. If V is used to indicate a valancy orbital similar to a KLL transition, we may have MMV transitions also, wherein the electron initially ejected from the M shell, falls to a vacancy with simultaneous emission of an electron from a valence orbital.

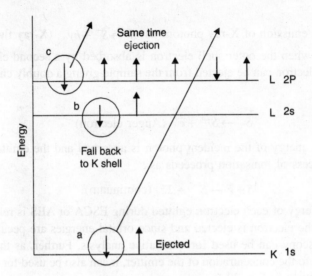

Fig. 39.2 KLL–Auger process

The product has a vacancy in the M shell and in the valency orbital. The transformations are shown in Fig. 39.2.

The kinetic energy of the ejected Auger electron matches the difference between the energy lost by the electron (b) as it falls into the inner shell and the energy required to remove the Auger electron (c) from the atom. The energy that is lost when the electron (b) falls to a lower shell is absorbed by the Auger electron. A portion of the absorbed energy is used to remove the Auger electron from the atom and the remainder appears as kinetic energy of the ejected electron.

The kinetic energy of the Auger electron is given by the equation:

$$A = (E_a - E_b) - E_c = E_a - E_b - E_c \qquad \qquad ...(39.4)$$

In the above process it is presumed that $E_a > (E_b + E_c)$. The simultaneous emission of two Auger electrons is a double Auger process. A secondary effect is a vacancy cascade effect due to ejection of an Auger electron from the inner shell. Such a vacancy is filled if electrons from a higher energy level fall into the vacancy, with simultaneous emission of a second Auger electron with formation of a triply positive ion. Auger emission can be observed with X-rays or positive ions but one prefers electron bombardment due to a better focusing effect in AES. The only limitation is the large background signal caused by scattered incident electrons, hence the first derivative of emitted electron intensity as a function of kinetic energy is recorded. Electron bombardment sometimes causes damage to the sample.

39.3 INSTRUMENTATION FOR ELECTRON SPECTROSCOPY

Most instruments permit both ESCA and AES measurement and are made up of a source, sample holder or container and analyser analogous to monochromator, detector and signal processing unit. They need a high vacuum ($10^{-8} - 10^{-10}$ torr).

Source and sample holder: The electrons are directed through a slit to the electron analyzer. The source consists of an X-ray tube (in ESCA) or an electron gun or gas discharge tube. Electron guns produce a beam of electrons with an energy 1-10 keV for producing Auger electrons. A beam of 500-5μm is used with microprobes. While in ESCA we use an X-ray source, the K_{ex} lines for two elements have a narrow bandwidth; viz. 0.8–0.9 eV to give a better resolution. Solid samples are mounted in a fixed position close to the electron source. The vacuum is used to avoid attenuation of the electron beam. The sample is freed from moisture and oxygen and is cleaned by Argon spluttering.

Electron Analysers: These are of two kinds viz. retarding field and dispersion type. In the former, electrons pass from the sample to a cylindrical collector through metallic grids, giving 70% transmission. The potential across the grid is gradually raised to retard the flow of electrons. The signals so collected are amplified, which in turn are successively decreasing. Such instruments lack resolution of dispersion instruments. In the dispersion type model, the electron beam is deflected by an electrostatic field when such beams travel in a circular path. In such a case the radius of the circle depends upon the intensity of field. We can thus focus the electron. From cylindrical plates V_1 (plate voltage of V_1 and V_2) the energy of the electron E is given as $V_2 - V_1 = 2E_K R \log(R_1/R_L)$ where R_1 and R_L are radii of two plates and R is average radius.

Detectors: Multichannel photodetectors are available. Resolution elements of electron spectra are monitored simultaneously. The path of electrons in the analyzer is affected by the earth's magnetic field. For better resolution, external fields are reduced to 0.1 mG by ferromagnetic shielding.

39.4 APPLICATIONS OF AUGER ELECTRON SPECTROSCOPY

Chemical shifts can be seen by Auger spectral peaks. This also throws light on the oxidation state. Quantitative analysis by AES is restricted to elemental analysis. All elements except H_2 and He can produce Auger spectra, in which the area under the peak is measured to give a quantitative content. It is also used for depth profiling analysis, wherein a change in the peak shape of differentiated spectra affects the usual peak height estimates of concentration. The X-ray technique can offer the same information as AES and serves as a complementary technique. AES is good for elements with a low atomic number (N < 10) and high spatial resolution. AES is largely used for qualitative analysis of solid surfaces. Lowering Auger electrons (20–1000 eV) can penetrate few atomic layers (3–20 Å) of solid. It is also used for depth profiling of surfaces. This involves determination of elemental composition of a surface as it is being etched or sputtered by argon ions. In addition, the line scanning technique is used to characterise the surface composition of the solid as a function of distance along a straight line of 100 μm or less. An Auger microprobe is used to produce a beam that can be moved across a surface.

39.5 ELECTRON SPECTROSCOPY FOR CHEMICAL ANALYSIS (ESCA)

This procedure was discovered in 1981 by K. Siegbalin who was later awarded a Nobel prize for his discovery. Figure 39.3 depicts ESCA transitions. The electrons from the inner K shell (lower lines) and electrons from outer or valency shells (upper lines) undergo transitions. The reaction can be represented as:

$$A + hr \rightarrow \overset{+*}{A} + \bar{e} \qquad \qquad ...(39.5)$$

if A represents the atom/molecule/ion and $\overset{+*}{A}$ is the excited ion with a higher positive charge *i.e.* A* > A as far as charge is concerned. The (E_K) kinetic energy of the emitted electron is measured. The binding energy of the electron, E_h, is obtained which is termed as the core energy of binding.

$$E_h = hr - E_k - W, \text{ where } W \text{ is the work function.} \qquad ...(39.6)$$

More than one peak can be observed due to 1s electron can be seen due to 2s, 2p electrons. ESCA is more useful for characterisation rather than for determination as it is concerned with core electron binding energy. X-rays are the only source of excitation of electrons commonly used. The electrons whose binding energy is less than the X-ray energy are ejected. What we measure is E_K i.e. kinetic energy by an analyser. ESCA furnishes information on the outer 2 nm of a surface layer.

39.6 CHEMICAL SHIFTS IN ESCA

If W is kept constant in the above equation we can calculate the chemical shift of say, a metal oxide by the equation:

$$E_{oxide} = E_k \text{ (metal)} - E_k \text{ (oxide)}$$

This chemical shift (Fig. 39.3) indicates transitions of electrons in ESCA measurement.

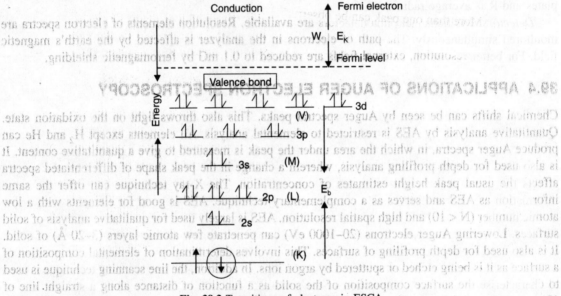

Fig. 39.3 Transitions of electrons in ESCA

This is used for surface analysis of solids. In Fig. 39.4, the Fermi level is the point of zero potential energy on the energy level scale. The chemical groups which are bound to an atom can change electron density and cause a chemical shift in ESCA peak position. As in NMR spectroscopy, chemical shift can be used for qualitative analysis. Chemical shifts for various compounds have been investigated, especially the chemical shift corresponding to the ejection of a 1s electron from carbon has been studied in depth, however, carbon compounds easily get contaminated. Chemical shifts can be measured relative to a reference compound e.g. N_2 for nitrogenous compounds. Chemical shifts can be measured for all elements which have an inner shell electron in a chemical compound which can be ejected (except H_2 and He).

39.7 ANALYTICAL APPLICATIONS OF ESCA

As the name implies, this technique is extensively used for chemical analysis. It can be used for estimating elemental composition of a sample if it is present in amounts more than 10:1%. The identification of an oxidation state of inorganic compound can easily be carried out. ESCA can be used for the elucidation of chemical structure, but has not been commonly used for quantitative analysis. Only in rare cases are peak overlaps encountered e.g. O (1s); Sb (3d); Al (2s); Cu (3s 3p) etc. This can be eliminated by measuring spectra at an additional peak such as the Auger peak which as a rule remains unchanged on the kinetic energy scale, while photoelectric peaks are displaced. Table 39.1 shows various electron spectroscopic methods while Table 39.2 shows chemical shifts with respect to the oxidation state.

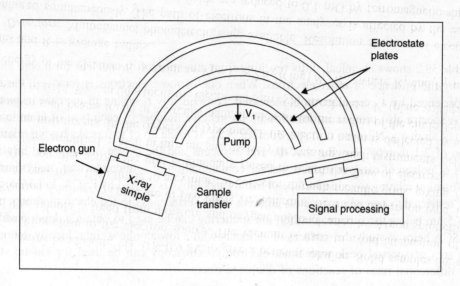

Fig. 39.4 Electron spectroscopy for chemical analysis

Table 39.1 Various types of electron spectroscopy

Sr. No.	Name	Abbreviated (Short form)	Mode of electron generation from sample
1.	Electron Spectroscopy for Chemical Analysis	(ESCA/XPS)	Monochromatic X-rays
2.	Auger Spectroscopy	(AES)	Bombardment with electrons
3.	Ultraviolet Photo-electric Spectroscopy	(UPS or PES)	Vacuum UV radiation
4.	Electron-Impact Spectroscopy	(EIP)	Monoenergetic electrons
5.	Penning Ionisation Spectroscopy	(PIS)	Collision of the gas phase excited atom
6.	Ion Neutralisation Spectroscopy	(INS)	Bombardment with low energy ions

In this chapter we have so far considered only ESCA, AES. These have real time applications in surface analysis.

Table 39.2 Chemical shift with respect to the oxidation state

Element	−1	0	+1	+2	+3	+4	+5
N_2 (1s)	0	−	+4.5	−	+5.1	−	8.0
S (1s)	−	0	−	−	−	+4.5	−
Cl_2 (2p)	0	−	−	−	+3.8	−	+7.1
I_2 (4s)	0	−	−	−	−	−	+5.3

In Table 39.2 shows chemical shifts resulting from attraction of the nucleus for a core electron is decreased by the presence of outer electrons. When one of these electrons is removed, the effective charge experienced by a core electron is increased, consequently resulting in increase in binding energy. ESCA spectra also provide information on the relative number of each type of atom in addition to the kind of atom e.g. Nitrogen (1s) spectrum for NaN_3 is made up of two peaks having relative areas in the ratio of 2:1 thereby showing that the end nitrogens and the centre nitrogen are different. The active sites of poison of catalyst surface, surface contamination of semiconductors, analysis of human skin, investigation relating to oxide surface of metals and alloys is carried by ESCA. In ESCA chemical analysis is restricted to several nanometers adjacent to the surface, as ejected electrons cannot penetrate more than 5 nm below the surface. Further, ion sputtering can be used to obtain a depth profile of the sample. ESCA helps to study the surfaces of elements with a low atomic weight. Finally, by noting the change in area of spectral peak with time, electron spectroscopy can be used for kinetic studies to determine differential rates of reactions of double surfaces.

39.8 SECONDARY ION MASS SPECTROMETRY

It is used mainly for the identification of surfaces of materials, molecular species, depth distribution and ascertain isotope ratios. Similarly, surface adsorbed gases, erosion or corrosion, polymer surfaces and analysis of impurities in semiconductor devices is done by SIMS. All solids can be analysed, in sample size of 1-4 cm^2. Ultra high vacuum is employed during analysis; specially for the analysis of nonvolatile substances. Time of analysis varies from 0.1–5 minutes, but depth profile studies need more time of 5-20 mts. The volatile samples and insulating samples need special handling process. Accuracy is 20-50 Å. It is sensitive for the monolayer surfaces. Sensitivity is 1-ppb to 1 ppm. It compares favourably well with AFS or XPS techniques of surface analysis.

SIMS is thus most challenging method for surface analysis. It is mass analysis of +ve and −ve ions on the bombarded surface by the process of sputtering. Ion beam is used for such sputtering. This is primary ion beam with energy of 0.5-25 kV. The removal of ions from surface depends upon the intensity, mass, energy, angle of incidence of primary beam and as well as the properties of the sample. The sputtered surface provides secondary ions where chemical bond is ruptured. Ion and neutral particles are ejected. If current density of primary beam is raised the sputtered rate also goes up and peak is reached ultimately at 60-70° angle of incidence of beam. The monolayer provides sputtered ions. The secondary ions are detected in SIMS. Primary beam species have O_2^+ or Cs^+ ions. They enhance positive or negative ion beam respectively. O^{2-}, N^{3-}, Cl^-, F^- ions reduce secondary ion yield if they are present on surfaces of the sample.

The dynamic SIMS has better depth resolution. The widening of delta layer is related to change in width of thin layer which is measured by transmission electron microscopy (TEM) (See next chapter). SIMS is performed at high or low primary ion intensities, but high rate provides better results. For thin films lower primary ion intensity is used. The static SIMS provides an information of molecular species (ionic) while dynamic SIMS is used for the analysis of organic and polymer surfaces.

Instrumentation: SIMS instrument is made up of UHV chamber, ion source, specimen holder, secondary ion extraction device, mass spectrometer, secondary ion detector and data recording or processing system. In order to increase negative secondary ions, Cs^+ ion gun is used. A filter is some times used. The energy generated is 10-30 keV. The sample holder holds 1.5×1.5 cm surface piece. The double focusing mass analyser is used in mass spectrometer. Such choice depends on mass resolution and range, transmission efficiency, imaging capability, resolution and cost. The quadrupole ion analysis system is also useful as only particular m/e ratio ions pass through it. It permits rapid switching from the +ve to −ve ion detection system. With the time of flight analysis (TOF-MS), there is better mass resolution with no compromise with the sensitivity, permitting the simultaneous collection and analysis of anions. Transmission efficiency is 20% with mass range of 0-10,000 AMU. Since sputtering process is destructive the surface gets eroded. The uniform bombardment provides accurate depth profile. SIMS can be used in an imaging mode also.

Surface analysis needs low intensity ion beams so that the surface layer is not totally removed. TOF-SIMS provides best results. The atomic or molecular species are identified at ppb to ppm level by SIMS. A calibration curve can be constructed. The concentration of species is quantitatively analysed. The quantitative analysis of static SIMS or TOF-SIMS is of great importance. The duoplasmatron ions guns generating O_2^+ is best to get good results. The Cs^+ source needs replenishment of caesium every year. Gallium liquid metal ion gun also gives reproducible results. SIMS instrument with double focusing magnetic sector has resolution of 10,000 (M/δM) with 35% efficiency while quadrupole model can work for 0-1000 AMU with resolution of 25 (M/δM) with 10% efficiency. TOF model is best. We would also consider electron microprobe analysis (EMA), UV photo-electron spectroscopy (UPS) and comparison of all techniques with each other in electron spectroscopy.

39.9 ELECTRON MICROPROBE ANALYSIS (EMA)

One uses X-ray emission for focussing on solid surfaces. They are then analysed by energy disperse spectrometer. The instrumentation for electron microprobe is shown in Fig. 39.5. Three beams are used. They are electron beam, light beam and X-rays. A vacuum of 10^{-5} torr is provided. The electron beam is generated by heating W-cathode fitted with suitable accelerating anode. The beams (two) (1) are focused on surface of sample. The optical microscope (2) locates area for bombardment. The generated X-rays are dispersed and located by gas filled transducer. The three beams do not interfere with each other. In Fig. 39.5 the electron probe is located in (1) while viewing system is in (2), while (3) represents X-ray analyser and (4) depicts specimen stage.

There is a mechanism to rotate the sample in *x* and *y* directions and also rotate around itself also. There are numerous applications of electron microprobe. It can provide enormous data relating to chemical and physical quality of the surfaces. It is used in material science for phase studies, the grain boundary investigations, diffusion dynamics in semiconductors and occulsion phenomena in solids. It is useful tool for the characterisation of heterogenous catalysis and in polymer field. As a matter of fact the lunar rocks were analysed mainly by electron microphobe technique. The intensity of four elements characteristics were examined in rocks.

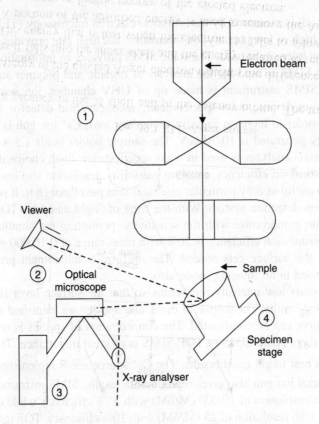

Fig. 39.5 Electron microprobe

39.10 ULTRAVIOLET PHOTOELECTRON SPECTROSCOPY

We have already seen in ESCA wherein one uses X-rays for bombardment of surfaces while in the Auger spectroscopy one uses electrons for bombarding surface of sample. Now if one uses ionising radiation in ultraviolet region it is called as ultraviolet photoelectron spectroscopy or photoelectric spectroscopy. Depending on the kind of shell used it is called as PES (OS) in UPS. It is called as the PES (OS) *i.e.*, photoelectric spectroscopy of outershell of electrons in ultraviolet electron spectroscopy (UPS). Technically UPS is analogous to ESCA as in both cases photons (*i.e.*, X rays) are bombarded on the solid surfaces. After absorption of energy by photons, the kinetic energy of ejected electron is measured. UPS is useful for analysis of gaseous species. The energy of photons in UPS method is inadequate to eject inner shell electrons but is of sufficient energy to eject electrons from outer shell (*i.e.*, valency electrons). The kinetic energy of electrons is equal to the difference between incident photon and ionisation energy of the sample. The ionisation energy is actually the difference between the energy of ion after ejection of electron and energy of the electron at ground state molecule. So UPS is useless for inner shell studies, but good for valence shell studies. This permits study of vibrational and rotational mode of transitions, which in turn throws light on bonding as well as oxidation state of an element. UPS is very useful for qualitative work and for the study of electrons in valency shell. The ejected electrons originate from molecular orbits and not from usual atomic orbitals. Such electrons are delocalised. The study of ionisation energy spin-orbit coupling in polyatomic molecule, instability in poly atoms ions can be carried out extensively by ultraviolet photoelectron spectroscopy (UPS).

Instrumentation: The instrumentation for UPS is more or less same as was used in ESCA. Microwave lamps are used as source. They primarily consist of Ar (106–104), He (58–24 nm); Kr (123), Ne (74) and Xe (146). It is caused by emission of photon from the gas. A monochromatic radiation is absolutely necessary for the spectral characterisation of solid surface. The resonant radiation is preferable if it has high energy. In addition to inert gases one can use or take resort to hydrogen, nitrogen and oxygen as filler gases.

Uses: As regards its application, UPS is used for analysis of gases or volatile solvents and UPS is used in gas phase investigations. The mixture of liquids can be analysed by UPS. Technique with high vapour pressure provides sharp peaks. UPS is used for qualitative analysis by comparison with standard spectra of the sample like in IR.

39.11 COMPARISON OF ESCA, UPS AND AUGER SPECTROSCOPY

As discussed earlier ESCA confines to electrons in inner shell, while UPS is concerned with the valence electrons and finally Auger spectroscopy relates to secondary electrons. All of them undergo process of ionisation. During ionisation electrons are expelled from the sample. The kinetic energy of expelled electrons and the total number of electrons in process provides information for identification and quantitative analysis. Thus electron spectroscopy characteristics electrons emitted from the sample. Ionisation is initiated by bombardment of sample by X-rays, UV radiation or electrons or ions. ESCA is also called as XPS i.e. X-ray photoelectron spectroscopy. In UPS ionisation radiation is used in UV region. Also UPS is called PES i.e., photoelectron spectroscopy. On bombarding sample with electron and emission of other secondary electrons is studied in Auger spectroscopy (AES) or (OHJAY). ESCA, UPS and AES are most commonly used.

In the ionisation process transformation any one kind of four transformations as

(1) $S + h\nu (\text{x-ray}) \rightarrow S^{+*} + e^-$ (ESCA) or ...(39.7)

For surface studies

(2) $S^{+*} \rightarrow S^+ + h\nu^2 \left(X\text{-ray fluorescence} \right)$ ...(39.8)

(3) $S^{+*} \rightarrow S^{2+} + e^- \left(\text{Auger} \right)$ ...(39.9)

(4) $S + e^- \rightarrow S^{+*} + 2e^- \left(\text{continuum} \right)$ ...(39.10)

The inner shell electron is ejected and S^{+*} ion is given in excited state. In UPS only the valence electron is ejected leading to ionisation. The energy of incident radiation is in UV region. It is insufficient to eject inner shell electrons. However, ESCA is used for study of both inner as well as valence shell electrons and the transformation in their kinetic energy. The resolution is better in UPS than ESCA and the former resolves vibration and rotational levels. On ejection of electrons, excited ion relaxes either by showing X-ray fluorescence or electron from outer falls in inner shell with emission of x-ray photon. The potential energy lost by outer shell electron is absorbed by second electron and in turn eject out of the electron in sample to show Auger effect. XRF and Auger competes with each other. For small atomic number elements Auger effective is prominent. The ionisation is also possible by bombardment of photon e.g. SIMS. Auger electrons have sufficient kinetic energy. The depth of 1-5 nm electrons are emitted hence all are called as surface spectroscopy methods. Finally the kinetic energy of electron emitted in ESCA, UPS, AES is related to energy of orbital from where electron is ejected. They are characteristic of atom or molecule electron. This permits both qualitative and quantitative analysis.

39.12 COMPARISON OF AUGER, ESCA AND SIMS TECHNIQUES

Auger spectroscopy is used for quantitative analysis of element on the surfaces for study of depth profile, study of desorption absorption phenomena and chemical reactivity. It is used in finger printing, spectral analysis, characterising chemical state of elements, interface analysis in situ. Any solid can be analysed, with no special sample preparation required as they are analysed in high vacuum. Time of analysis varies from 5.25 mts. It cannot analyse conducting or semiconducting samples. Only solids can be analysed, decomposing samples cannot be also analysed. However, its accuracy in terms of resolution is 0.2 μm with sensitivity of 0.3%.

X-ray photoelectron spectroscopy also called as ESCA is used for identification of elements, knowing binding energies, quantitative analysis of surface elements, depth profiling to know the chemical states of elements. Is mainly used in fingerprinting spectras, finding atomic concentration, differentiation of conducting and insulating material. Polymers are best analysed by it. So also absorption, chemisorption of gases can be investigated on metal surfaces by it. A sample of (0.2 × 0.2 m) to (1 cm × 1 cm) is required after degreasing. Time needed is 5–25 minutes. Not useful for nonconducting materials, applicable to solids and nondecomposing sample. The accuracy is 0.2 mm with sensitivity of 0.3% carried in high vacuum.

The secondary ion mass spectrometry (SIMS) is used for identification, elemental surface distribution profile studies, study of molecular species, trace analysis and study of isotopes. It is used in identification of surface contamination, or organic residues or polymer surfaces via molecular fragmentation, study of diffused dopants in semiconductors and analysis of impurities which affect electrical properties of semiconductor devices. A sample size needed is (0.1 × 4 cms²). Analysis is done in (UHV) ultra high vacuum. Time of the analysis is 0.1-5 mts. For TOF-MS time needed is 0.5–2 mts. Solid or liquid samples is analysed. Accuracy is 20-50 Å or range of ppb to ppm for bulk specimens.

All three are compositional analysis methods. XPS or ESCA is most commonly used. (AFS and STM are imaging techniques). SIMS is most sensitive. Standard preparation is difficult. SIMS is destructive technique. Space resolution is poor in XPS. Table 39.3 summarises the properties, merits and demerits of these techniques.

Table 39.3 Comparison of Auger, ESCA and SIMS techniques

S.No.	Property	Auger	ESCA or XPS	SIMS
1.	Depth profile	5 Å	5 Å	20 Å
2.	Resolution	50 nm	150 nm	50 nm
3.	Sensitivity	50	50	10,000
4.	Detection order	0.1%	0.5%	0.0001%
5.	Elements analysed	Z >3	Z >1	Z >1
6.	Source-excitation	e_{-1}^0	X-rays	ion
7.	Emission seen	e_{-1}^0	ion	e_{-1}^0
8.	Sample destruction	high	low	low
9.	Information	Spatial resolution	Sensitivity	Chemical in form
10.	Matrix effect	negligible	negligible	Significant

LITERATURE

1. F. Settle, (Ed.), *"Handbook of Instrumental Techniques for Anal. Chemistry"*, Pearce Education, Indian Edition (2004).
2. D. Briggs, M.P. Seah (Ed.), *"Practical Surface Analysis by Auger and X-photoelectron Spectroscopy"*, Wiley NY (1983).
3. J.F. Moulder, (Ed.), K. Chastion, *"Handbook of X-ray Photoelectrons Spectroscopy"*, Eden Praire MN, Physical Electronic Publishers (1992).
4. R.G. Wilson, F.A. Stevie, C.W. Magee, *"Secondary Ion Mass Spectroscopy"*, Wiley, NY (1989).
5. T.A. Carlson, *"Photoelectron and Auger Spectroscopy"*, Plenum, NY (1975).
6. J.W. Rabalais, *"Principles of Ultraviolet Electron Spectroscopy"*, Wiley (1977).
7. D. Betteridge, A.D. Baker, *"Analytical Potential of Photoelectrospectroscopy"* Acs (1978).
8. C.R. Brundle, A.D. Baker, *"Electron Spectroscopy–Theory, Techniques and Applications"*, Academic Press (1977).

Chapter 40

Electron Microscopy

Electron microscopy technique plays a significant role in characterisation of materials in material science, heterogeneous catalysis and polymer technology. This is a non-destructive method of analysis as their characterisation of solid surfaces is so valuable. There are electron spectroscopic methods for surface analysis as well as electron microscopy techniques. Many times chemical properties of surface are not identical with the bulk properties of inside material. There are as many as 5–6 electron spectroscopic methods for surface characterisation which we have considered in previous chapter. They include ESCA *i.e.*, electron spectroscopy in chemical analysis, Auger Electron Spectroscopy (AES), UV-photoelectron spectroscopy (UPS), secondary ion mass spectrometry (SIMS) and the electron microprobe (EM). In all these methods, the primary beam is composed of X-ray photons, UV photons, electrons, ions and photons while the secondary beam invariably consists of electrons except in secondary ionisation mass spectroscopy, *i.e.* SIMS. The primary beam is used to irradiate surfaces while secondary beam originates from the surface of sample due to scanning, sputtering or emission. The photon beam is inferior to electron beams. The microscopic methods for imaging surfaces and getting morphology or physical feature is of prime value in analysis. The surface sampling in electron spectroscopy is carried out by focussing primary beam on surface and measuring secondary beam intensity. In other method viz. electron microscopy the surface is mapped by moving primary beam in a "raster pattern" of the measured increments and noting changes in secondary beam while in last method of electron microscope involves depth profiling. The adsorption causes contamination of surfaces. It occurs in vacuum also, hence the sample surface needs good cleaning. This is done by heating or sputtering with rare gas ions; or polishing or by ultrasonic cleaning. ESCA, AES, UPS are called electron spectroscopy methods. They provide information on oxidation state of metal with layer thickness of 20–50 Å. They are useful in their own form for surface analysis.

40.1 CLASSIFICATION OF ELECTRON MICROSCOPY METHODS

In earlier chapter we had briefly reviewed electron spectroscopy methods which have the merit of facilitating qualitative as well as quantitative analysis of compounds. Similarly, the electron microscopy methods are broadly grouped into following class as:

(a) Scanning electron microscopy (SEM)
(b) Scanning probe microscopes (SPM) including scanning tunnelling microscopy (STM)
(c) Atomic force microscopy (AFM) which is part of SPM method mentioned earlier.

The scanning electron microscopy provides the morphologic and topographic information of the sample while scanning probe microproby provides insight to structure of material with good resolution. In semiconductor studies SPM methods are used for the analysis of silicon surfaces and location of defects. They are also used for imaging of magnetic domains on magnetic substances. DNA imaging is done by the atomic force microscopy as it permits imaging of aquatic biological samples. The optical microscopy provides information on physical nature of material. However, its resolution has limitations; but all methods viz. SEM, STM, AFM provide better resolution with more details of the surfaces. The STM and AFM together are termed as SPM *i.e.*, scanning probe microscopy (SPM). The surface is cleaned in raster pattern by beam of electrons in *x* and *y* direction. In a raster pattern electron beam is focussed in straight line on surface in both directions and is returned to starting location and it is shifted downwards. The process is repeated several times to get better scan of material. We would consider in next few sections each technique in greater detail with special reference to principles, instrumentation, sampling and important applications.

40.2 SCANNING ELECTRON MICROSCOPY (SEM)

In such measurements the surface is invariably solid, so raster pattern for scanning is possible. Several signals e.g. backscattering, secondary and Auger electrons, X-ray fluorescence photons are originated. The first two viz. back scattering and secondary emissions are most useful for analysis. They serve as basis for scanning electron microscope (SEM).

40.3 WORKING OF SEM INSTRUMENT

The Figure 40.1 presents a schematic representation of both scanning electron micropscope/microprobe involving the use of either electron detector (SEM) or X-ray detector (SPM) with common electron gun source. The focusing system comprises of condenser (1) and objective lens (2) to facilitate reduction of an image (~200 nm). Condenser lenses (1) which permits electron beam to reach objective lens (2) and ultimately reaching to sample (3). Now SEM is achieved by two coils. These coil (4) deflects beam in x < y direction. Scanning is controlled by electronic signal on one coil (4) in the line scan.

Left coil (5) is used to deflect beam. This helps to irradiate sample with electrons beam. The signal (digital or analog) from specimen is stored in the digital form (6). Same signal permits movement of cathode ray tube as displayed (6). The (7) magnification (M) is given by formula M = W/w if W = width of display, w = width of signal and M = magnification. W is made constant by just decreasing (w). So we get greater (M) *i.e.* magnification from 10× to 100,000× range.

The sample scanning is an important process in procuring reproducible results. The electricity conducting samples are easier to work but biological samples do not conduct either heat or electricity. Then in such instance non-conducting sample is coated on the conducting plate but its thickness must be carefully controlled. Apart from sampler, transducer is an important component of SEM assembly. There are scintillation devices. There are also semiconductor transducers available.

The electron-hole pairs produce change in electrical conductivity. Their devices are cheap and inexpensive. The induction of electron beams with solid sample is another way of measurement. The back scattered electrons, secondary electrons and the X-ray emission interact readily with solids. The interaction may be elastic or inelastic in nature. In former, in elastic interaction, there is no change in energy of the trajectory electrons while in latter *i.e.*, in elastic interactions change in energy is due to transfer. The secondary electrons originate from excited solid which emits secondary electrons. Elastic electrons do change direction of travel but not velocity of movement, so kinetic energy is constant. The angle of diffraction varies as beam has 20–25 keV energy which penetrates 1–1.5 µcm. The secondary electron production is one half to one fifth or less of numbers of background electrons. These secondary electrons arise due to interactions of beams of electron and weakly bound conduction of electrons in solid leading to release of conduction electrons which in turn facilitates measurements.

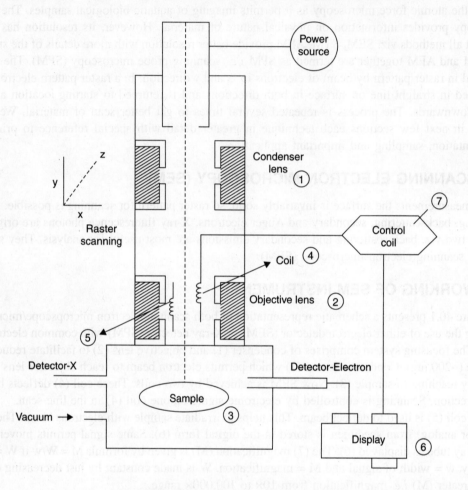

Fig. 40.1 Scanning electron microscope/probe

Applications: Scanning electron microscopy provides topographic as well as morphological information about surface solids. Electron microscopy is a sure test to throw light on the composition of the material.

40.4 COMPARISON OF SCANNING PROBE MICROSCOPY (SPM OR STM)

This technique also includes scanning transmission microscopy. This technique provides details of surface characteristics. The scanning tunneling microscopy was first invented in 1982 while in 1986 its discoverers viz. Ruher and Binning were awarded Nobel prize in Physics. The technique provides details in x, y and z axis dispositions. It has 20 Å as its resolution but can go down to 1 Å. This technique of SPM largely has ramifications on STM *i.e.* scanning tunneling microscope and AFM atomic force micropscopy. Both the methods provide atomic size and scale details of the sample.

40.5 SCANNING TUNNELING MICROSCOPY (STM)

It could be used for conducting solids. In fact conducting electricity is basic requirement of their applications in surface examinations while for atomic force microscopy, conductivity is not the prerequisite for analyst.

40.6 BASIC PRINCIPLES OF STM

The raster pattern (x, y direction) is employed for scanning with metal tip. The vertical movement of the metal tip provides picture of overall topography. The tip is kept at the constant or fixed distance from the surface on all x, y, z axes. The tunneling current is also kept constant flowing between tip and sample. The current is detected when two conductors come in proximity with each other. The current originates in medium which has no electrons e.g. vacuum, nonpolar solvent or aqueous solution. If (I_t) is tunneling current, then: $I_t = V_e - C_d = (V_e = C_d)$, if V_e = voltage applied in two conductors, C = constant, while d = distance between two conducting surfaces viz. lowest and highest atom on the sample. The piezoelectric transducer controls the to and fro movement of the metallic tip. On distance variation, the current changes. This in turn provides high resolution (Fig. 40.2). For three directions (x, y, z) three piezoelectric transducers are provided. The transducer can vary distance of length of transducer with change in DC potential. This in turn influences the sensitivity of the measurement. Some instruments are provided with hollow type piezoelectric device with metal coating on external surface. They are divided into four parts. By variation of potential in two strips which are bent into cylinder can provide means to vary length of the tube (on z-axis). On account of the scanning process the method is designated as scanning probe microscopy. The length, diameter, wall thickness of so formed cylinder influences the scan size. The computer control is main part of STM. The output signal is sensitive to distance between sample and metal tip. Tunneling tip is vital part of STM. Good images are procured if tunnel bending is confined to one metal. The gap and current change is in direct proportion. Pt/Ir wires can be used to manufacture tip with single atom. With decrease in gap, current increases to provide better magnification.

40.7 ATOMIC FORCE MICROSCOPY (AFM)

One of the special features of the atomic force microscope is the application for the characterisation of the surfaces of nonconducting or the insulating materials. As opposed to this for SPM/STM one needed essentially a conducting material. A force detecting cantilever stylus (1) (Fig. 40.3) is used for scanning. Such force causes deflection of cantilever (3) detectable by optical sensors. A piezoelectric tube (4) is employed to control movement of this stylus. Force is kept fixed to get reproducible results while scanning is to provide uniform topographs.

Figure 40.3 shows AFM scanning assembly. It has piezoelectric tube (4) (pz) which permits scanning of sample in all three directions (viz. x, y and z) under the tip (2) adjacent to the sample (6). A laser beam signal (5) is fed to sample pz transducer. The sample moves with constant force in all the three directions. The cantilever and tip, control the atomic force. The cantilever is fabricated from metal sheet while tip is made from diamond or equivalent hard material. Tip is fixed on cantilever bar by sealing e.g. the silicon compounds can be used as semiconductor for fixing tip on cantilever surface since both the components are delicate and fragile. Tips may have cone shapes. They are in micrometer dimensions e.g. cantilever bar is just ten micrometers.

The principal limitations of atomic force microscope is that the tip is constantly under downward pressure in contact with the sample. This is turn damages surface of the sample. This is a serious

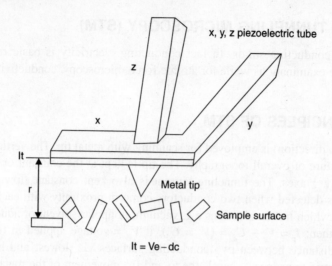

Fig. 40.2 STM scanning

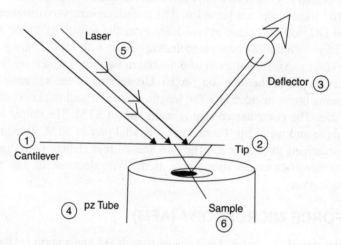

Fig. 40.3 AFM scanning

limitation while handling the biological sample which are quite soft. It faces similar problem while handling polymeric materials. To avoid this problem the tip is allowed to come in contact with sample surface at regular time period. This process is termed as the 'taping mode operation'. This is accomplished by to and fro motion of cantilever bar oscillations. In this fashion, the tip is perfectly protected.

40.8 APPLICATIONS OF ATOMIC FORCE MICROSCOPY

We get excellent resolution of the nonconducting surfaces. The silicon surfaces and their defects if any in semiconductor devices can be located by atomic force microscopy. The magnetic materials can be best studied by this method. In bioscience and bioengineering field one can examine DNA, proteins, enzymes interactions and the membrane viruses. The substances in aquatic environment and specially below water surfaces can be examined easily by AFM with no interferences of other micro organisms during process of imaging. It is specially useful for soft material examination. The microscopic examination under water gives disturbances in imaging process but as capillary forces below and above tip balances it prohibits distortion of the images. The solid surfaces at the atomic level can be critically examined by scanning tunneling microscopy. The detection of the functional groups can be carried out by AFM e.g. the carboxyl group coated on tip and sample coated with > COOH as well as CH_3 group. This method of the characterisation is termed as "chemical force microscopy" (CFM). This technique can furnish information on spatial configuration as well as on specific qualitative analysis of the desired sample.

40.9 COMPARISON OF ELECTRON MICROSCOPY WITH ELECTRON SPECTROSCOPY

It is interesting to compare electron spectroscopy with electron microscopy. The quality of surface is many times not same with materials inside. Electron spectroscopy is rapid provided if emitted electrons from incident beam are regular in the absence of photon. The spectral measurement relates to assessing of this beam as a function of energy of the electrons. The electron spectroscopy facilitates identification of the oxidation state of metal and its electronic structure. On account of poor penetration capacity of electrons technique, it is largely used for analysis of solid surfaces and not liquids or gases. It permits qualitative but not always quantitative analysis. The electron microscopy is reliable but is very slow method of the characterisation of the surfaces. Further, its resolution is limited by diffraction effects restricted to specific wavelength. The surface of sample is screened in raster pattern. It is a pattern analogus to one shown by cathode ray tube or television set when electron beam scans in x-direction and then shifts to y-direction. The proces is repeated. The back scattered and secondary electrons provides basis for SEM. While SPM provides minor details of surfaces in all three directions viz. (x, y, z) axis which is not the case in SEM methods. It has good resolution (~1Å). Both in STM and AFM surface of sample is moved in x/y direction in raster pattern and feeded to computer. It is quite difficult to adopt SEM to soft samples like biological products. The structure of DNA was thoroughly studied by AFM technique.

Overall these techniques have come to forefront only after 1960. Before that the sophisticated techniques of measurement of surfaces were lacking. However, with the growth of knowledge in biological science, material science and in the field of heterogenous catalysis it is expected to make great revolution in elucidation of surface structure of these materials. This holds good in future because better spectroscopic methods as well as engineering technology is available and it has progressed with reference to fabrication of metal or alloy tips and the support of three or more piezoelectric tubes or bars.

LITERATURE

1. O.C. Wells, "*Scanning Electron Microscopy*", McGraw Hill (1974).
2. R.S. Howland, "*Atomic Force Scanning Travelling Microscopy*", Eds. S.H. Cohen, M.T. Brey, M.L. Lighttbedy, Plenum Press (1994).
3. S. Park, R.C. Barret, "*Scanning Tunneling Microscopy*", Ed. J.A. Strosciv, W.J. Kalser, Academic Press (1993).
4. T.R. Weisendanger, "*Scanning Probe Microanalysis and Spectroscopy*", Vol.1 to 4, Cambridge University Press (1994).
5. A. Goldstein, "*Scanning Electron Microscopy and X-ray Microanalysis*", Plennum Press (1981).
6. P.K. Ghosh, "*Introduction to Photoelectron Spectroscopy*", Wiley, New York (1983).

Chapter 41

Photoacoustic Spectroscopy

The processes resulting from excitation of molecules involve either absorption or emission of radiation. These are termed as radiative processes. In nonradiative processes, transitions from higher vibrational levels to a lower ground state is involved. If a substance in the gaseous or liquid state is undergoing a transition, the energy of transformation results in an increase in thermal movements of all molecules of the sample. However, if the substance is in the solid state, the energy is converted to lattice vibrations and transferred to media in contact with such a solid. Such processes results in a rise of temperature which can be measured by photoacoustic spectroscopy or optoacoustic spectroscopy.

If modulated UV–visible, IR, or microwave radiation falls upon a sample it excites it during absorption. Some of the molecules return to the ground state by radiationless or non-radiative processes. The released thermal energy causes expansion of gases or liquids. Since the incident radiation is controlled, the surrounding gas periodically contracts and expands depending upon the process of control or modulation. These pressure waves resulting in sound waves are measured by a detector. The sound waves, which are emitted when the sample is successively cooled or heated due to the absorption of electromagnetic radiation are measured. Opaque substances can also be analysed by photoacoustic spectroscopy.

41.1 PRINCIPLES OF PHOTOACOUSTIC SPECTROSCOPY

UV–visible, IR, or microwave rays cause excitation of the molecule. The electrons from the excited state return to the ground state by a process of relaxation in the form of fluorescence, phosphorescence or a nonradiative process. In the nonradiative process, energy is dissipated as heat (H). The heat so generated is directly proportional to the absorptivity (A) of the sample and the intensity (I) of the radiative source. It is expressed by the equation, $H = AI$. The absorptivity (A) is a measure of the quantum of radiation absorbed by the sample. This means a high intensity source will produce a large quantum of heat and give a strong signal. In IR and microwave excitation, relaxation occurs by emission of thermal energy. Such energy causes a pressure wave which can easily be detected. Modulation or control of the exciting radiation alters such pressure waves. This results in a signal. The modulation frequency is between 20–100 Hz. An acoustically resonant frequency is hence used (λ), which depends on cell length e.g. $\lambda = 2L/n$ where n is an integer and L is the cell length. The process for conversion of heat to a pressure wave is determined by the nature of the sample. In a gas or liquid, a temperature rise is encountered, which in turn causes a rise in pressure for a fixed volume in the cell. The pressure wave results in sound waves. A microphone or piezoelectric transducer is used as the detector. However, if the sample is solid, then an absorbent filter gas, like air or inert gas, is used. The gas absorbs energy resulting in a pressure wave which in turn is measured by an acoustic detector. This dual process results in the formation of a wave in a gas. The intensity of the photoacoustic signal is proportional to absorptivity (A) and thermal diffusion length (L). With an increase in thermal diffusion length, more radiation is absorbed in the sample, consequently warming the gas. If $L = 0.1$ μm when the modulation frequency is 10–100 kHz or if $L = 10$–100 μm then the modulation frequency will be 10-20Hz. Finally, in optically opaque solids, all incident radiation is absorbed at the surface e.g. for carbon black, the signal is dependent upon the modulation frequency, hence the intensity of incident radiation serves as a reference for the normalisation of photoacoustic spectra. For thick samples, the spectra is proportional to the variables shown in the given equations.

41.2 INSTRUMENTATION FOR PHOTOACOUSTIC SPECTROMETRY

Photoacoustic instruments usually consist of a source, monochromator, sample cell, detector and recorder. A brief description of each is given. Figure 41.1 gives a schematic representation of a photoacoustic instrument.

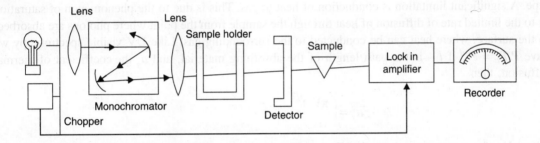

Fig. 41.1 Single beam photoacoustic spectrometer

The optical diagram is more or less similar to the usual UV-visible spectrophotometer with the exception of the chopper which modulates the intensity radiation. For a pulsed source, the chopper is not needed. The detector is not a photomultiplier tube but an acoustic detector. The signal is hence preamplifed and fed to a lock-in amplifier which is connected to a read-out system.

Source: An intense source is required to give a good output signal. The most commonly used source is a xenon arc lamp to give a pressure of 50-70 atmospheres and is operated at wavelengths from 250-2500 nm. Hydrogen, tungsten, krypton lamps can also be used at lower wavelengths. Lasers are also used e.g. argon ion laser. A tunable dye laser is also used in the 350-1200 nm region.

Sample holder and monochromator: A xenon arc lamp is used as monochromatic source with a diffraction grating monochromator. Lasers or filters can also be used. Line sources are used in nondispersive models. The shape and size of the sample holder depends upon the physical state of the sample. One side of the holder has a transparent window. External signals cannot be allowed to interfere with the sample. The construction material of the cell has thermal mass. The materials commonly used are steel or aluminium, silver or gold; the latter are best but costly. Quartz windows are in common use. Single and multiple pass cells are used for gas cell holders. A differential cell holder possesses two cells (Fig. 41.2).

Detector and recorder: A microphone or piezoelectric transducer is used as a detector, but the former is most popular. The condenser microphone detector is the best. For liquids one has to use a piezoelectric detector.

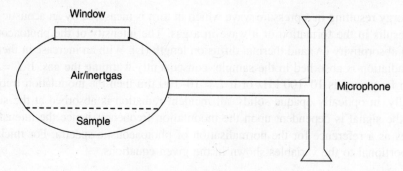

Fig. 41.2 Differential cell holder

The detector is a solid material such as $PbZrO_2$, $PbTe : O_3$, whose structure is altered under pressure. Such a change causes a change in charge and hence a difference in potential. It changes the pressure wave which in turn can easily be amplified. The detector can be used in the form of a tube or a disk. Most instruments have an *XY*-recorder. Scanning instruments are usually of the double beam type. A significant limitation is conduction of heat to gas. This is due to the phenomenon of saturation or to the limited rate of diffusion of heat through the sample from the point where photons are absorbed, to the surface, where heat can be conducted to the surrounding gas. Like UV visible spectroscopy we have $P/P_0 = e^{-al}$ if l – is the path length of the absorbing material, and a_s the coefficient of thermal diffusion, then

$$a_s = \left(\frac{\pi V}{\alpha s} \right)^{1/2}$$

where V = frequency, α_s = thermal diffusion. If $a < a_s$ a photoacoustic signal will emerge but when a increases or exceeds a_s the signal becomes saturated. However, the saturation effect can be minimised

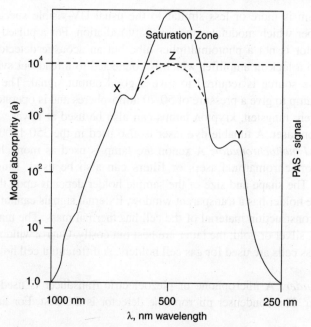

Fig. 41.3 Spectra with saturation zone

by grinding the sample along with a transparent diluent like potassium bromide or alumina (Al_2O_3) (Fig. 41.3).

41.3 ANALYTICAL APPLICATIONS OF PHOTOACOUSTIC SPECTROMETRY

These spectra are similar to absorption spectra. In the absence of fluorescence or phosphorescence, these methods are very useful. They provide good spectra for opaque materials. It has a high sensitivity and is useful for the measurement of radiation by transparent media, because it does not suffer from interference by scattered radiation. It is useful for the analysis of biological samples, and in flowing liquid analysis like liquid chromatography. Qualitative analysis is thus facilitated by it. Since the signal increases linearly with concentration of the analyte, it is used for quantitative analysis, and is specially useful for quantitative analysis of single absorpting species. Like visible spectrophotometry, one can also resort to simultaneous spectrophotometry. The method can be made more sensitive by increasing the intensity of the source. It is comparable to fluorescence spectroscopic measurement.

It is interesting to compare the PAS spectra and absorption spectra (Fig. 41.4). Similarly a saturation zone can be visualised in Fig. 41.3 when a > 10^4 with jamming effect.

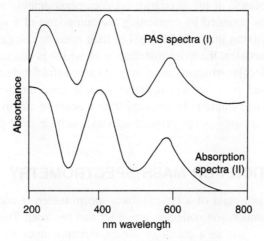

Fig. 41.4 PAS (I) and absorption (II) spectra

When the sample loses energy by luminescence a small PAS signal results while those analyte samples which do not show good fluorescence or phosphorescence spectra are bound to show good PAS signals. Thus the methods are complementary to each other.

LITERATURE

1. R.D. Braun, "*Introduction to Instrumental Analysis*", McGraw Hill Book Co., London (1987), Indian ed. (2001).
2. G.M. Ewing, "*Instrumental Methods of Chemical Analysis*", 5th Ed., McGraw Hill Co. (1985).
3. Y. You-han Pao, "*Optoacoustic Spectroscopy and Detection*", Academic Press, New York (1980).
4. A. Rosencwaig, "*Photoacoustics and Photoacoustic Spectroscopy*", Wiley, New York (1980).
5. F.P. Schafer, "*Dye Lasers*", Springer Verlag, Berlin (1973).

Chapter 42

Mass Spectrometry

The mass spectrometer is an instrument that sorts out charged gas molecules or ions according to their weight or mass. This technique has nothing to do with spectroscopy; however, the name spectrometer was chosen because of the similarity between photographic records and optical lines in spectra. As mass spectrum is obtained by converting the compound of a sample into rapidly moving ions (+ve in nature) and resolving them on the basis of their mass to charge (m/e) ratio. The ionisation process produces positive particles, the mass distribution of which is characteristic of the parent compound. In elucidation of molecular structure, mass spectras are useful for determining molecular weight. Since the ion current at various mass settings is proportional to concentration, quantitative analysis can also be easily carried out. The sample is bombarded with a beam of electrons which in turn produces an ion molecule or ionic fragments of the original species; such charged fragments can be separated according to their mass.

42.1 INSTRUMENTATION FOR MASS SPECTROMETRY

Figure 42.1 indicates the main parts of a typical mass spectrometer. It consists of a sample injector, ionisation chamber, ion separator, ion collector, amplifier and recorder. The function of the mass analyser is to disperse particles just as a prism or monochromator does in a spectrophotometer. It is absolutely essential to maintain all the components travelling to the detector at a low pressure ($10^{-4} - 10^{-8}$ torr), hence, a good vacuum system is indispensible. A sample is volatilised and pushed into the ionisation chamber (10^{-5} torr pressure). Then molecules of the sample are ionised directly or indirectly by a stream of electrons to produce positive ions. A collimated beam of positive ions enters the separation area of the mass analyser (10^{-7} torr), and the fast moving particles are subjected to a strong magnetic field which causes them to describe a curved path; the radius of the path depends upon their velocity, mass and field strength. Particles of different mass are focussed on the exit by varying the accelerating potential or field strength. The ions passing through the exit slit fall upon a collector electrode; the ion current that results is amplified and recorded as a function of field strength or accelerating potential.

Sample handling system: This introduces a sample into an ion source as a gas at a low and reproducible pressure (~0.01 torr). For less volatile samples, heating or vaporisation is necessary, provided the compound is thermally stable. Nonvolatile and thermally unstable compounds are often introduced into the ion source by a sample probe which is equipped with a heater to volatize the sample at low pressure.

Ion source: This converts molecules into gaseous ions. The most common way of producing ions involves bombarding the sample with a beam of energetic electrons from an ion gun. The electron impact source causes fragmentation of the molecules leading to a large number of positive ions of various masses. In a chemical ionisation chamber or source, methane is introduced into the electron beam source (~p = 1 torr). Interaction of the electron beam with the reagent gas produces a host of ions such as CH_5^+, CH_4^+, CH_2^+. The CH_5^+ is a strong proton donor and further reacts: $CH_5^+ + MH \rightarrow MH_2^+$ + CH_4, where MH is the analyte sample. However, this method gives less fragmentation of the analyte. In a spark source, ion formation from non volatile inorganic samples is possible. A field ionisation source with a metallic anode and cathode is also used for fragmentation.

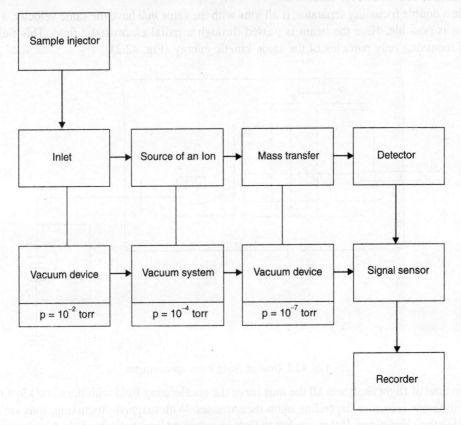

Fig. 42.1 Mass spectrometer

Mass analyser: This is a device resolve ions with different mass to charge ratios. A mass analyser should differentiate between minute mass differences and produce a high level of ion current. There are various types of analysers:

(a) Single focussing analysers with magnetic deflection.

(b) Double focussing analyser.

(c) Time of flight analysers.

(d) Quadrupole analyser.

(a) In a single focussing analyser the separator has a circular beam path of 180°, 90° or 60°. The centripetal force is $F_m = H_{ev}$. If H is the magnetic field; v = particle velocity, e = charge on ion, balancing the centripetal force (F_m) against the centrifugal (F_e) force, we have $H_{ev} = mv^2/r$ where r = radius of circular path. By combining these equations we have $\dfrac{m}{e} = \dfrac{H^2 r^2}{2V}$, where V is the accelerating voltage applied in the ionisation chamber.

In order to traverse a circular path, F_m and F_e should be equal i.e., $H_{ev} = \dfrac{mv^2}{r}$. In most instruments H and v are fixed hence e/m of the particle is inversely proportional to the acceleration voltage. The type of separation is shown in Fig. 42.2.

(b) In a double focussing separator, if all ions with the same m/e have the same velocity, a very high resolution is possible. Here the beam is passed through a radial electrostatic field. This field has the effect of focussing only particles of the same kinetic energy (Fig. 42.2).

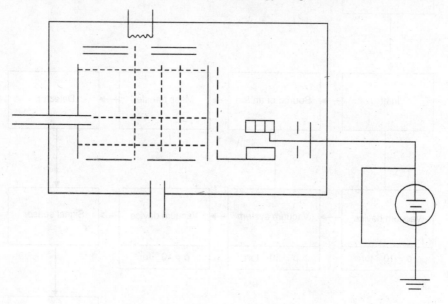

Fig. 42.2 Time of flight mass spectrometer

(c) In time of flight analysers all the ions leave the accelerating field with the same kinetic energies but with different velocities depending upon their masses. With magnetic focussing, ions are separated by changing their directions. If they are left to float in a straight line through a field-free region they will take different times to travel a given distance.

(d) The measurement of this time of flight is the basis of the non-magnetic separator as shown in Fig. 42.3.

If ions are allowed to enter the drift tube continuously there is no possibility of measuring the time required for a given ion. This is sorted out by pulsing a control grid in to an electron beam so that ions are produced in pulses lasting about 0.25 μ sec at a frequency of 10,000 times per second. The time of flight (t) is given as $t = k\sqrt{m/e}$ if k is a constant which is dependent on length of flight time. Thus this is a non-magnetic means of separation. The detector used is an electron multiplier tube.

A quadrupole analyser, employs short parallel metal rods arranged around the beam. The combined field causes the particles to oscillate about their central axis of travel (Fig. 42.4). The oppositely placed

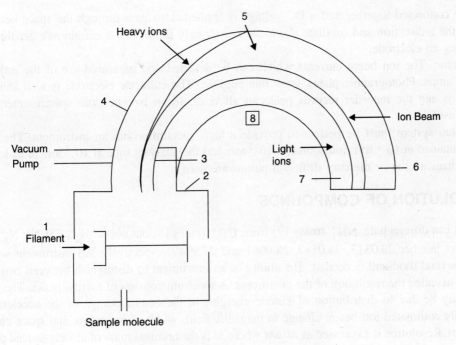

Fig. 42.3 Single deflection mass spectrometer

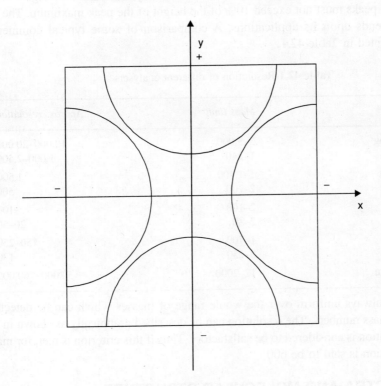

Fig. 42.4 Quadrupole mass spectrometer

electrodes are connected together and a DC voltage is applied. Ions pass through the space between electrodes in the z-direction and oscillate about the z-axis; only those with a certain m/e get through without striking an electrode.

Ion collector: The ion beam currents which can be detected and measured are of the order of 10^{-15} to 10^{-5} amps. Photographic plates serve this purpose. The collector electrode is well shielded from stray ions and the recorder records peaks of all sizes. In an isotope ratio spectrometer, two detectors are used.

The vacuum system must be perfect to provide a high vacuum within an instrument. The inlet system is maintained at 10^{-2} torr, ion source at 10^{-4} torr and the analyser tube at 10^{-7} torr or as low as possible. Mechanical, ion or mercury diffusion pumps are used.

42.2 RESOLUTION OF COMPOUNDS

An instrument can differentiate NH_3^+ (mass 17) from CH_4^+ (mass 18) but not $(C_2H_4^+, CH_2N^+, N_2^+$ and CO^+ with mass number 28.0313, 28.0187, 28.0061 and 27.9949 respectively. An instrument with a resolution of several thousand is needed. The ability of an instrument to distinguish between two ions of equal mass in called the resolution of the instrument. A spectrum consists of narrow peaks. The peak broadening may be due to distribution of kinetic energies in the beam, change in the accelerating voltage, a badly collimated ion beam, change in magnetic field, width of ion beam, and space charge of the ion beam. Resolution is expressed as $M/\Delta M$ where M is the nominal mass of closely spaced peaks of equal height, one at M and the other at $(M + \Delta M)$, with the stipulation that the signal at the valley minimum between the peaks must not exceed 10% of the height of the peak maximum. The resolution of an instrument depends upon its applications. A comparison of some typical commercial mass spectrometers is presented in Table 42.1.

Table 42.1 Resolution of different analysers

Type	Mass limit	Approx. resolution
Double focussing	2–5,000	10,000–20,000
	1–240	1,000–2,500
Signal focussing	1–1,400	1,500
	2–700	500
Quadrupole	2–100	100
	2–80	20–50
Time of flight	1–700	150–250
	0–250	130
Fourier transform	12–2000	50,000–760,000

Resolution is not always uniform over the whole range of masses which can be detected. It can become poor at high mass numbers. The resolution can be described graphically as shown in Fig. 42.5.

If $M > 1$ the resolution is considered to be satisfactory. Thus if this criterion is met, for masses upto 600 to 601, the resolution is said to be 600.

42.3 MASS SPECTRA AND MOLECULAR STRUCTURE

A mass spectrum consists of an array of peaks of different heights. The nature of the spectrum is dependent upon properties of the molecule, ionisation potential, sample pressure and instrument design. A minimum electron beam (7–15 mV) is used for generation of mass spectra in the ionisation process:

$$M + e^- \rightarrow M^+ + 2e^-$$

If M is the molecule and M^+ is the molecular ion or parent ion. A large increase in beam energy decreases the molecular ion beam, as extra energy breaks bonds to give fragments with a smaller mass. For molecules containing a large number of atoms the number of positive ions produced is also large. For a beam with 50–70 eV energy, a reproducible pattern is obtained.

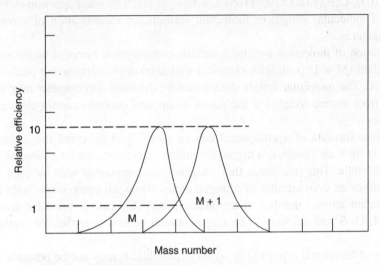

Fig. 42.5 Resolution of spectra

The mass of an ion is useful for identification of a compound. Ion–molecule collision produces peaks of a higher mass number than the molecular ion peak. The intensity of the molecular ion peak depends upon the stability of the ionised particle. The stability of the ion is influenced by structure and shape of molecular peaks. The largest peak in a mass spectrum is termed the 'base peak'. The peak height is reported as a fraction of base peak height. Electron impact spectra contain peaks of small fragments of the parent molecule. With an electron beam of energy of 70 eV the molecular ion splits into many fragments. The ion abundances are given in terms of percentage of base peak or percentage of total amount of ions produced.

42.4 FRAGMENTATION PATTERNS IN MASS SPECTRA

It is worthwhile observing the sequence of events within the electron impact ion source as the electron energy is increased. The first ionisation potential is between 8–12 eV when ions are formed. The first ion appearing is called the molecular ion or '*parent ion*'; with a rise in potential, bonds break at the appearance potential. Most mass spectra are run at 70 eV which is energetic enough to break any kind of bond. Each compound yields a characteristic series of fragments called the 'fragmentation pattern' or cracking pattern. Such patterns are shown usually in a bar graph. The most intense peak is called the base peak. A small peak in the natural abundance of an ion is called the "isotope satellite peak". High resolution instruments give an idea of mass defects *i.e.* difference between true atomic or molecular weight and nominal whole number value. A few broader than normal maxima are observed superimposed upon the usual spectrum. These are caused by metastable ions formed in the ion source that spontaneously decompose during their passage through the instrument. The new pattern must include one positive ion with a mass less than the previous ion. The observation of metastable states can be of use in study of organic mechanism.

42.5 QUALITATIVE ANALYSIS WITH MASS SPECTROMETRY

The identification of an unknown compound is possible by mass spectrometry. This is possible by calibration with a known compound like mercury vapour $(m/e = 198–204)$ or perfluorokerosine (PFK) with peaks $CF_3(69)$, $C_3F_3(93)$, CC_4F_3 (105) and C_3F_5 (131). The mass spectrometer is thus useful for determination of molecular weight or molecular formula, or identification of a compound from the fragmentation patterns.

In determination of molecular weight, a volatile compound is essential as the molecular ion peak has to be identified (M + 1) peak with chemical ionisation with methane but peaks due to impurities should be ignored. The molecular weight determined by the mass spectrometer may not be the same as that calculated from atomic weights if the parent compound contains certain elements with high isotopic abundances.

The molecular formula of a compound can be established provided the molecular ion peak is known but mostly for such a purpose a higher resolution mass spectrometer is needed. The nitrogen rule also gives this formula. This rule states that, 'All organic compounds with an even molecular weight must contain zero or an even number of nitrogen atoms while, all compounds, with an odd molecular weight must contain an odd number of nitrogen atoms.' This rule is valid for covalent compounds containing C, H, O, S and halogens. The fragmentation pattern is useful for identification of compounds.

Recognition of functional groups (Fig. 42.6) is possible. It may not be possible to account for all the peaks in a spectrum but characteristic patterns of fragmentation are sought. Alcohols have weak peaks but when they lose water, they give a strong peak at (M = 18). The cleavage of the C–C bond next to oxygen is also common. The primary alcohols have a strong peak at mass 31 due to the presence of the ion $CH_2 = OH^+$ as shown in Fig. 42.6 for pentanol.

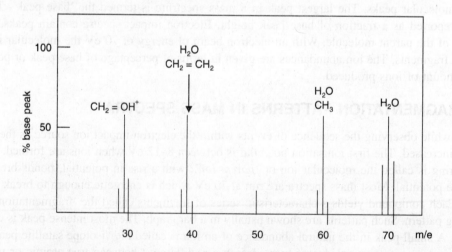

Fig. 42.6 Mass spectrum of pentanol

42.6 QUANTITATIVE ANALYSIS WITH MASS SPECTROSCOPY

Mass spectroscopy is useful for quantitative analysis of mixtures containing closely related compounds. Nowadays it is used for the analysis of both inorganic and organic mixtures of low volatility. Since fragmentation patterns of components of a mixture are additive, a mixture can be analysed if the spectrum of a compound run under similar conditions is available. Mass spectra in combination with GC gives the best possible results. GC and a quadrupole mass spectrometer are used in combination. This is called the hyphenated technique discussed in detail in last sections. The basic essentials of analysis are that each component must show at least one peak differing substantially from another, peak contribution must be additive and sensitivity should be reproducible with availability of suitable standards. By this technique high molecular weight polymeric compounds have been analysed by pyrolysis. It is also useful for the analysis of complex petroleum products, e.g. parafins give a peak at $m = 43–85$; cycloparafins at $m = 41–97$, cycloolefins at $m = 67–96$ and alkyl benzenes at $m = 77–134$. Such analysis is used to characterise properties of fuels, lubricating oils, asphalts etc.

In inorganic trace analysis, mass spectrometry can be used to great advantage, especially utilising a spark source. It is also useful for trace analysis of elements in alloys and minerals and in superconductors. The spark model has high sensitivity, permits determination at ppb levels, it is simple due to presence of one single major line and other weak line, and line intensities are equal. However, the main demerit of spark source mass spectrometry is the erratic nature of the source and lack of reproducibility which can be set right by the use of a photographic detection system. For quantitative analysis such an instrument is based on the use of photographic lines with densitometer with suitable standards.

42.7 APPLICATIONS OF MASS SPECTROMETRY

The mechanism of a reaction or the structure of a complex compound can be determined by the use of a labelled atom to find the exact location in the product. The mass spectrometer can locate the position of the labelled isotope by the change it causes in the spectrum by detecting which fragment is heavy or determining the ratio of a heavy to normal isotope.

High polymers and natural products have low vapour pressure but on pyrolysis the decomposition products can be analysed by a mass spectrometer.

Exact mass determination is possible by mass spectrometry with an accuracy of one part in 10^8. The relative abundances of isotopes have been determined by mass spectrometry Table (42.2).

Table 42.2 Abundances of isotopes of some elements

Sl. No.	Element	Common isotope	Other isotope mass number	Relative % of most abundant
1	Bromine	^{79}Br	81	–
2	Carbon	^{12}C	13	1.08
3	Chlorine	^{35}Cl	37	32.5
4	Hydrogen	^{1}H	2	0.015
5	Nitrogen	^{14}N	15	0.37
6	Oxygen	^{16}O	17, 18	0.04–0.20
7	Silicon	^{28}Si	29, 30	3.4–5.1
8	Sulphur	^{32}S	33, 34	0.8–4.40

39.8 – 39.14

The mass spectrometer is very useful for preparation of pure isotopes.

In the final section of this chapter we shall consider in greater detail the following:

1. Laser mass spectrometry (MALDI)
2. Fast Bombardment and Liquid Secondary Mass Spectrometer (FAB-LSI-MS).
3. Electron Spray Ionisation Mass Spectrometer and (LC-ESI-MS).
4. Various hyphenated methods coupled with mass spectrometer. Such coupling occurs with mainly gas-chromatograph but also IR spectrometer, liquid chromatograph, capillary electrophoresis. The chapter is finally concluded with introduction to Tandem-mass spectrometry.

So far we have considered conventional mass spectrometer with all its accessories and applications. However, in last decade newer techniques have come to forefront.

We would look into each aspect of their working, what vital information they provide and what are their important applications in chemical analysis.

42.8 LASER MASS SPECTROMETRY

This technique is useful for molecules with 200-1,000,000 Dalton mass. It is largely a qualitative method but accuracy is fine and is useful for analysis of nonvolatile samples, proteins in solid or liquid state. Millimole or nanomoles substances are used for analysis. Liquids are deposited on sample probe which may be laser probe. Time required is only ten minutes for complete analysis. Easily ionisable oxygen or nitrogen must be part of sample. Time of flight analyser is used (TOF). Laser desorption ionisation mass spectrometry is of recent origin which is most useful for nonvolatile components (more than 5 mts. for the decomposition). We detect m/e ratios. Laser desorption is most vital. It was developed by Japanese workers in 1989. MALDI stands for "matrix assisted laser desorption/ionisation". SID *i.e.* surface induced deposition can occur by application of voltage to FT-MS trapping plates. On collision and fragmentation they form product ions which can be analysed.

Mass spectra of macromolecule is measured as the function of m/e ratio. MALDI combines chemical ionisation reaction with LD. Laser is fired and mixture desorbed into gas phase and analyte ions. Picolinic acid, gentisic acid, sinnapinic acid or substituted cinnamic acid were used as matrices for biological product analysis and polymers. The kind of laser used for laser desorption are nanosecond pulses. Nitrogen UV laser is best at 337 nm; it is cheap, compact and good. Neodymium/YAG-laser provides best results at 1064 nm. For IR region work, carbon dioxide laser is used but YAG laser is most useful for analysis time of flight analyser; it is ideally suited as it is economical, easy to operate and also has low maintenance, is cost sensitive, useful at high mass number limit, but it had low mass resolution and is applicable for nonvolatile macromolecules. They work at 5-30 kV. The single laser probe analyses entire range of m/e ratios. Ion multiplier is used for mass detection. The measurement takes place in micro seconds time as seen in attached oscilloscope. Flight time is correlated with mass. Ion reflector if used improves resolution; it has optimum m/e range for operation. Since 1991, MALDI-FI MS are available in market, but only drawback is the instrument requires frequent calibration to get reliable results. FT-MS is not effective for higher mass at low pressure (10^{-8} torr). Thus outstanding limitation is it has low limit of mass. Since polymers are clean and many contain ionisable oxygen or nitrogen they are easily analysed. Purely hydrocarbon matrix are not analysed by this method involving soft ionisation method. The sequence of information for polymers can be obtained. With it qualitative analysis is most dependable. By using internal standard, quantitative analysis can be carried out. Mass accuracy depends upon the mode of calibration of the instrument with range of 0.1 to 0.03%. The detection limit is 1-10 pmol. Attempts were made to reach at attomole limits (10^{-18}). Although MALDI has higher mass limit, results are not quantitative. Mass sample < 500 Da are rarely analysed by this device. It has numerous applications, e.g., analysis of hydrocarbon dendrimer. The substituted fullerene are analysed by MALDIN UV region. For the peptides the sequence information is obtained by it. Borin insulin is analysed by UV MALDIFT-MS with accuracy of 70 ppm. Polymer mass distribution can be carried out. Polymer polyethylene glycol (PEG) with mass of 6207-6250 are analysed by proposed method. PEG 6000-4000 are best analysed by FT-MS with IR laser desorption. By and large instruments are quite expensive due to exorbitant cost of nitrogen laser, YAG laser, TOF, FT-MS, the total exceeding $ 1350K. Alkali metals from biological samples are best analysed by it.

42.9 FAST ATOM BOMBARDMENT AND LIQUID SECONDARY ION MASS SUBSTITUTED SPECTROMETRY

Molecular weight of many compounds can be obtained by FAB-LSIMS if they are nonvolatile. The upper limit of analysis is 3000 Da. Trace analysis is possible. The quantitative analysis is facilitated by isotopically labelled internal standards. It is coupled with HPLC. Peptide sequence can be easily obtained. The high boiling liquids are analysed by it. Time of analysis varies from 0.25-2 hrs. The sample must contain one polar group. Fast atom bombardment facilitates ionisation. Most of the products analysed were from biochemistry laboratories. FAB-MS confirms presence of newly synthesised product. Electron or chemical ionisation needs vaporised sample in gaseous state at 300°C and 10^{-6} torr pressure. At high temperature, more vaporisation results in chemical ionisation. The bombarding of surface with high energy ions (~5kV) produces energy spikes (i.e., high temperature region) and surface material is ejected and ionised to give secondary ions with no sample damage. Barber (1981) generated protonated molecular ions $(M + H)^+$ by bombarding solution of sample in liquid media like glycerol. FAB could be replaced with ion beam to improve sensitivity to give SIMS. Both FAB and SIMS give identical results. An ion of high mass can be thus produced from 10,000–15,000 Dalton at 8-10 keV accelerating voltage. FAB analyses is better carried out at lower resolution due to difficulty in the identification of all C^{12} peaks and hence molecular weight. The sample preparation is important in polar low volatility liquid. A beam of energetic power (e.g. Xe) is used. For LIMS Cs^+ is used. The energy is 5-10 keV for LIMS and 30-100 µA for FAB. This led to ejection of sample in all directions. The sample contains either positive ions or negative ions which are easily detected. Neutral sample molecules are converted to + or −ve ions. The positive ions form protonation of sample $(M + H)^+$. With loss of proton, negative ion $(M - H)^-$ are produced. If beam is massive more secondary ions are generated. They are cooled on expansion. The inorganic ions if present in the sample would attach to sample (Na^+K^+ Cl etc.) and detected as cationic species as MNa^+ or MK^+ ions. Even adducts like (M matrix $+H)^+$ or $(M_2 + H)^+$ are formed. Ions pass to mass analyser and m/e is measured. For mass greater than 1000 detection efficiency for m/e measurement is reduced. FAB-MS generates high ion currents. The sample preparation is critical for FAB-MS and LSMS. In 15C5 trimethylsilyl derivative gives MH^+ but nothing in glycerol. The sample matrix should show low volatility, be fluid and unreactive with sample. Such matrix like glycerol contains hydroxyl group or thiol (-SH) group present. Methanol, dichloroethane are used as the liquid solvents. Other liquid matrix used in FAB-MS are thioglycerol, diethanolamine, triethanolamine and 15 crown 5 (for organometallics). The sample concentration be 10^{-3} M. The exact confirmation of positive or negative ions leading to knowledge of molecular weight is of primary concern of FAB-MS. High background spectra must be less. One can encounter cluster of compositions e.g. glycerol forms such cluster. Alkali metals if present reduce MH^+ peak due to formation of adducts e.g., carbohydrates form such adducts. FAB spectra has less fragmentation. The analysis of mixture is cumbersome due to excessive fragmentation. By inserting 50–75 µM fused silica capillary, permits to carry analyte in continuous flow of water or appropriate solvent flow. FAB reduces background, improves sensitivity, eliminates suppression effect and can be directly coupled to GC or capillary electrophoresis. A 10% solution of CSI is used for calibration. One can carry out exact mass measurement. High molecular weight compounds or thermally labile ions can be analysed by MALDI — As far as applications of FAB-MS and LSIMS for elucidation of the structures of biopolymers, proteins, peptides (MW > 25,000 Da), these techniques with Tandem Mass Spectrometer is used for analysis of aminoacids or the oligosaccharides. Organometallic complexes, oligonucleotides and several inorganic complexes are analysed in the presence of crown ether. The various solvent matrices employed in FAB-MS are shown in Table 42.3.

Table 42.3 Selection of solvent matrices in FAB-MS

S.No.	Solvent used	Mass number	m.p./b.p.°C	Analysis of	Remarks
1.	15C5	220	100-130°C	Organometallic	Nonhydroxylic reaction.
2.	Glycerol	92	182°C	Peptides	Analyte can be reduced.
3.	Thioglycerol	108	118°C	Proteins	Mixture with glycerol preferred
4.	2 nitrophenyl-octylether	251	198°C	Organometallics	Nonhydroxylic, sequester sodium ion.
5.	Diethnolamine	105	217°C	Saccharides and fatty acids	Negative ion detection possible
6.	Triethnolamine	149	190°C	Fatty acids, saccharides	Used as negative ion.

Table 42.4 Kinds of column used in HPLC-MS

S.No.	Specification of column	Dimensions (cms)	Total volume injected (ml/mt)	Flow rate ml/mt
1.	Narrow bore	1.5 × 0.21	0.05 – 5.0	0.0005
2.	Microbore	1.5 × 0.1	0.01 – 1.0	0.001
3.	Capillary	3.0 × 0.03	0.001 – 1.0	0.0001
4.	Nanoscale	5.0 × 0.005	0.0005 – 0.05	0.0001
5.	Wide scale	1.5 × 0.5	0.5 – 2.0	0.5

42.10 ELECTRON SPRAY IONISATION MASS SPECTROMETRY

This is also called as HPLC electron spray ionisation-mass spectrometry. It is useful for separation of nonvolatile and thermally labile compounds. One can determine molecular weight of peptide or determination of their primary structure and for kinetic studies and quantitative work. It is used extensively for identification of metabolites, drugs, and toxins. One can analyse both solids and liquids at atomole to picomole level (*i.e.* 1×10^{-18} to 1×10^{-11} M). Time of analysis is 2–10 mts. Only limitation is sample must dissolve in electrospray solvent. Further, liquids with high conductivity or large surface tension can not be analysed. Accuracy is better than 0.01%. The combination of the HPLC and MS is beneficial for resolving complex mixtures. It is useful for polar and thermally labile compounds. LC-MS as it is called has many merits as low-cost low-resolution MS can be used. Molecular specifity is improved. The sensitivity is improved, with less sample. Since 1970 attempts were made to combine LC with MS with no success due to difference in operating conditions. MS works at $10^{-5} - 10^{-8}$ torr vacuum with HPLC flow rate of 0.5-1.5 ml/mt. LC-MS instruments contain LC, an interface and MS. We have already familiarised with HPLC in earlier chapter. However, reverse phase HPLC is better suited for electron spray ionisation mass spectrometry technique. As regards electron spray ionisation (ESI) had solved several problems in biochemical analysis of high mass products. ESI produces ions quickly. A high voltage is applied to tip of capillary to produce an electrostatic field to disperse the emerging solution into fine mist or spray of charged drops. The drops shrink in size as they pass down the column. Then repulsive forces cause droplets to explode into smaller charged particles. Fusion process is repeated. The ions are sampled through nozzle which then enter in mass analyser. The basis of high mass analysis is the formation of series of charged ions of the molecules reducing *m/e* ratio. Spray must be stable, electric current be appropriate for electrolyte concentration.

Polar solvents including methanol, ethanol, propanol, acetonitrile are good for electron spray operation. Nonpolar solvents are not easily dispersed. The ion formation occurs by spraying which facilitate LC, MS operations. LS-ESI-MS is used for identification of thermally labile compounds. The basis of identification is the retention time and molecular weight and structure specific fragment ions. Due to high detection, sensitivity is used in all quantitative analysis. Use of internal standards are quite common. If two compounds have same molecular weight then tandem MS-MS is used. As regards applications it is used in several areas of research e.g. sequencing of the proteins, identification of mixtures of compounds, structure determination, drug analysis, pesticides, toxins characterisation. Is used mainly in peptide mapping. The selective detection of compounds from the complex mixtures e.g. several peptides or phosphorylated peptides or glycopeptides.

It is used in solving many biomedical problems. Is also used for the identification of metabolites, biological matrix can be quantitatively analysed. The cost of instrument varies from model to model due to MS-MS capability. Magnetic sector instruments are more expensive.

Online LC-ESI-MS is used with ion chromatograph also. The sample is dissolved in solvent and applied to column. On elution the MS is scanned (within certain mass range) to give mass spectra of LC effluents. The computer base data system base is most useful.

One significant advantage of ESI-MS is its use for molecular weight determination of macro molecules by multiple charging phenomena from m/e ratio. The kind of columns, their flow rate, dimension etc. in RP-HPLC-ESI is presented in Table 42.2.

42.11 HYPHENATED TECHNIQUES IN CONJUNCTION WITH MASS SPECTROMETER

The coupling of two techniques results into hyphenated method. This combination of methods had become popular with the invention of Mass Spectrometry. Mass spectrometer can be coupled with say Inductively Coupled Plasma Spectrometer (ICP-MS), direct current plasma (DC-MS), microwave induced plasma (MIP-MS), spark source mass spectrometer (SP-MS), secondary ion mass spectrometer (SIMS-MS). The thermal ionisation and mass spectra models were first discovered for elemental analysis. The first three methods are really hyphenated methods. The combination of two techniques provide better results than those obtained by single instrument. We shall consider most common type of four hyphenated methods which are:

(a) Gas chromatography – Mass Spectrometry (GC-MS)

(b) Gas chromatography – IR spectrometry (GC-IR)

(c) Capillary Electrophoresis – Mass Spectrometry (CE-MS)

(d) Chromatography – Mass Spectrometry (IC-MS)

(a) Gas Chromatography–Mass Spectrometry (GC-MS)

A gas chromatograph forms good combination with MS or IR spectroscopy. These are excellent hyphenated methods. The previous effluent from exit of GC was analysed by suitable method. However, if same effluent is led into another device e.g. mass spectrometer, one could get good resolution and qualitative determination was possible. The interphasing of two instruments is quite common and popular with chemists. The sample from GC is fed to ionisation chambers. A carrier gas is removed by injet separator. Quadrupole model gives best results. The speed of the separation and sensitivity of analysis is of paramount importance. Ion trap detector is used in gas chromatography. Trapped ions are ejected to electron multiplier detector. Ejection is controlled in proportion to mass/charge ratio. Real time or computer based MS are common. GC-MS is used for analysis of several mixtures with odour and flavour compounds; in water pollution; food analysis; diagnosis in medicine and study of drugs. This provides very rapid and simple tool for characterisation. Alkali metals are also analysed from biological samples.

(b) Gas Chromatogram–Infrared Spectroscopy (GC–IR)

Just coupling of the capillary column of GC with IR spectrometer furnishes a means of separating and detecting components in one single operation. During interface, column and detector are of vital importance. Dead volume is reduced to considerable extent. Radiation detector is sensitive. Sample is allowed to travel from detector region to IR cell. The spectra is printed with help of computer. The spectra of gaseous effluent sometimes does not match with library spectra as former is of fine structure (rotational). A band for intermolecular interaction remains absent e.g. H-bonding in alcohols. For this MS–FT–IR gives lot of data.

(c) Chromatography–Mass Spectrometry (IC–MS)

GC–MS was most popular for analysis of complex mixtures. In any chromatography–MS technique spectra are collected as the compounds emerges from column. LC/MS is a useful combination in hyphenated method. The sample is diluted, so some efforts were made to concentrate it before passing in MS assembly.

(d) Capillary Electrophoresis–Mass Spectrometry (CE–MS)

In 1987 first effort to couple these two versatile techniques was made. It provided the best method for analysis of biopolymers, proteins, polypeptides and DNA. The capillary effluent is passed directly into electrospray ionisation chamber and the product so formed is routed in quadrupole mass analyser.

42.12 TANDEM MASS SPECTROMETER

In comparison to GC very limited applications are indicated for GS-IC gas solid chromatography coupled up with some suitable device. Unlike GC-MS, the ICP-MS is growing very popular in inorganic chemical analysis. Tandem mass spectrometry is another example of hyphenated method where two mass spectrometers (MS-MS) are coupled together to procure better results. The first MS resolves sample while second MS analyses the product. It is also called daughter-ion MS/MS method. The parent ion MS-MS is also common. Dibutyl phthalate and sulphamethizine is analysed by daughter MS-MS while alkyl phenyl in solvent coal is resoluted by parent MS-MS technique. The separations are faster than one in GC or LC-MS methods.

Tandem mass spectrometry is used for elucidation of structure even if LC peak contains more than one component. Two stage MS_1-MS_2 is used with appropriate fragmentation mechanism. A triple stage quadrupole mass spectrometer is good for work with hyphenated method of LC-MS-MS. Tandem mass will become most popular in future.

PROBLEMS

42.1 The mass spectrum of nitrogen produced from air is characterised by the presence of peaks of isotopic forms of nitrogen $^{14}N^{14}N$ and $^{14}N\,^{15}N$ equal to 680 and 5 mm resp. What is the % of nitrogen ^{15}N in the sample?

Answer: The amount of nitrogen is obtained from relation

$$x = \frac{100}{2R+1}$$

where R = ratio of peak heights of two isotopes. Hence on substituting values

$$R = \frac{h14N}{h15N} = \frac{680}{5} = 136$$ if H 14 N is height for 14N and H15 N is height for 15N

Hence amount of nitrogen ^{15}N is

$$x = \frac{100}{(2 \times 136)+1} = \frac{100}{272+1} = \frac{100}{273} = 0.366\%$$

42.2 A mixture of compound A and its saturated analogic compound B are mass spectrometrically analysed. Calculate the mole percent of each compound in a mixture from following data.

Sl. No.	m/e	Compound A	Compound B	Unknown
1	81	0.3	49	48.9
2	95	0.2	43	43
3	137	0.1	100	100
4	141	100	0.1	9.3
5	338	48.0	0	4.4
6	348	0	8.4	8.4
7	Sensitivity	0.797	0.910	–

Answer: For compound A, the peak at *m/e* at 338 (S.N.:5) is useful. For compound B, the peak at 137 *i.e.* (S.N.:3) is useful. The uncorrected partial pressure is determined as

100 = 0.1 (XA) + 100 (XB), 4.4 = 48 (XA) + 0.0 (XB)

Solving the simultaneous equation

$$X\,A = 0.08,\ X\,B = 0.92$$

Now $\quad X\,A/\text{Sens.} = 0.08/0.797 = 0.10\ P\,A$

$\qquad\quad X\,B/\text{Sens} = 0.92/0.916 = 1.01\ P\,B$

The partial pressure can be converted to %

$$\frac{0.10 \times 100}{1.11} = 9\ \text{mole};\ \%\,A\quad \frac{0.01 \times 100}{1.11} = 91\%\ \text{mole}\ B$$

PROBLEMS FOR PRACTICE

42.3 A mass spectrometric analysis of an inert gas gave the following results:

Sl. No.	m/e	% Abundance	Mass
1	36	0.337	35.968
2	38	0.063	37.963
3	40	99.60	39.962

Calculate atomic weight of the inert gas.

42.4 The following data were obtained for unknown compounds

Sl. No.	m/e	A	B	C	D	Unknown
1	67	100	16.0	17.0	12.3	100
2	84	0	25.0	0	0	14.9
3	97	0	7.0	0	44	80.8
4	105	5.4	0	42	8.5	26.1
5	109	4.8	4.2	100	1.5	26.6
6	Sens.	1.42	0.82	0.94	1.08	-

Calculate the mole percent of each species in the mixture.

42.5 In the spectrum of ($CO_2 + CH_4$) the lines seen are at m/e 8, 13, 14, 15, 16, 28, 44. What neutral particles and positive ions might these lines correspond to?

42.6 The spectrum of amyl acetate includes lines corresponding to mass number (m/e) 39, 50, 51, 77, 78, 104. What ions might these lines correspond to?

LITERATURE

1. F.A. Settle (Ed.), "*Handbook of Instrumental Techniques for Analytical Chemistry*", Pearson Education India (2004).

2. D.M. Desiderio (Ed.), "*Mass Spectrometry–Clinical and Biomedical Applications*", Vol.2, Plenum Press (1994).

3. I.R. Chapman, "*Practical Organic Mass Spectrometry—A Guide for Chemical and Biochemical Analysis*", Wiley (1993).

4. R.M. Caprioli (Ed.), "*Continuous Flow Fast Atom Bombardment Mass Spectrometry*", Wiley, N.Y. (1990).

5. T. Matuso, *"Biological Mass Spectrometry: Present and Future"*, Wiley N.Y. (1994).

6. A.L. Vergfy, *"Liquid Chromatography-Mass Spectrometry Techniques and Applications"*, Plenum Press, N.Y. (1990).

7. F.A. White, G.M. Wood, *"Mass Spectrometry Applications in Science and Engineering"*, Wiley Interscience (1986).

8. J.T. Watson, *"Introduction to Mass Spectrometry"*, 3rd Ed., Lippincott Raven (1997).

9. K.L. Busch, G.L. Glitsh, S.A. McLuckey, *"Mass Spectrometry—Mass Spectrometry Technique and Application of Tandem Mass Spectrometry"*, VCH Pub (1988).

10. B.I. Millard, *"Quantitative Mass Spectrometry"*, Heyden, London (1978).

11. F.W. McLafferly, *"Tandem Mass Spectroscopy"*, (1983).

12. I. Howe, D.H. William, R.D. Bowen, *"Mass Spectrometry—Principles and Applications"*, McGraw Hill (1981).

5. T. Monose "Modern Mass Spectrometry, Present and Future", Wiley, N.Y. (1994).

6. A.L. Murfy "Biand Chromatography Mass Spectrometry Techniques and Applications", Plenum Press, N.Y. (1990).

7. F.A. White, G.M. Wood, "Mass Spectrometry Applications in Science and Engineering", Wiley Interscience, (1986).

Chapter **43**

Thermoanalytical Methods

Thermoanalytical methods involve techniques such as thermogravimetric analysis (TGA) involving change in weight with respect to temperature, differential thermal analysis (DTA) with change in heat contents with reference to temperature, thermometric titration, and also differential scanning calorimetry and thermomechanical analysis. The data obtained are used to plot continuously recorded curves which may be considered as thermal spectra *i.e.* a thermogram. These thermograms characterise a system in terms of temperature dependence of its thermodynamic properties and physical chemical reaction kinetics. TGA involves measurements of a change in weight of a system as the temperature is increased at a predetermined rate, while DTA consists of measuring changes in the heat content as a function of difference in temperature between the sample and reference sample under investigation where the two materials are simultaneously heated to an elevated temperature at predetermined rates. The thermometric titrations involve plotting of a change in temperature against volume of titrant. A thermogram provides a means of following a gas absorbing or gas process reaction by continuously recording the change in the volume of gas consumed or evolved as the material is heated to elevated temperatures at a constant rate, while differential scanning calorimetry measure heat flow in analyte and standard. They are classified in Table 43.1.

Table 43.1 Classification of thermal methods

S. No.	Name	What is measured?	Kind of instrument used
1.	TGA	Change in mass w.r.t. temp	Thermobalance
2.	DTA	Difference in heat contents with temp.	DTA assembly
3.	Calorimetric DTA	Heat gained or lost w.r.t temp.	Differential calorimeter
4.	Thermal Titration	Change in temp. w.r.t. volume	Special titration assembly
5.	Derivative TGA	Rate of change in mass w.r.t. temp	Thermobalance
6.	DSC	Difference in heat flow	DSC apparatus
7.	TMA	Change in dimensions	LVDT model

43.1 THERMOGRAVIMETRIC ANALYSIS

It has become common on account of the ease of availability of automatic recording thermobalances. A dynamic and static thermogravimetry is there. In dynamic thermogravimetry, a sample is subjected to conditions of continuous increase in temperature, which is linear with time; in static or iso-thermal thermogravimetry. The sample is maintained at constant temperature for a period of time during which any change in weight is noted.

With regard to apparatus and methodology, three basic requirements are a precision balance, a furnace and recorder. A furnace should have a linear rise of temperature with time and should be capable of carrying out work in inert, oxidising or reducing atmospheres. A continuous record of weight and temperature ensures that no features of thermograms are overlooked. In the diagram (Fig. 43.1) of the thermogram of calcium oxalate, the plateaus on the decomposition curve are indicative of constant weight and represent a stable phase over the particular temperature interval (viz. a, b, c, d) the product being hydrated calcium oxalate (*a*), calcium oxalate (*b*), then calcium carbonate (*c*), and finally calcium oxide (*d*).

The 'abcd' regions correspond to the following transformation

$$\underset{(a)}{CaC_2O_4, 2H_2O} \xrightarrow[226°C]{\Delta} \underset{(b)}{CaC_2O_4} \xrightarrow[478°C]{\Delta} \underset{(c)}{CaCO_3} \xrightarrow[1025°C]{\Delta} \underset{(d)}{CaO}$$

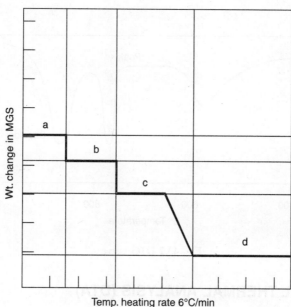

Fig. 43.1 Thermogram of calcium oxalate

The shape of the thermogravimetric curve is influenced for a particular compound by the heating rate of the sample in the crucible and the atmosphere surrounding the sample.

TGA is used for investigation of suitable weighing forms of material. It is useful in testing materials which are analytically standard and indirect application of the technique for analytical determination. It is also used to determine the range of temperature of drying. This can be studied by varying the rate of heating and noting changes in weight. Thermogravimetric data can be used to evaluate kinetic parameters of weight changes in reactions. The order of reaction (*x*) and energy of activation (E_A) can be computed from the relationship:

$$\frac{E_A / 2.303\,R\Delta T^{-1}}{\Delta W_r} = -x + \frac{\Delta \log(dw/dT)}{\Delta \log W_r}$$

wherein R = gas constant, T = absolute temperature, $W_r = (W_c - W)$ weight loss at time t, W_c = weight

loss on completion of reaction and W = initial weight. By plotting a graph of $\left[\dfrac{\Delta \log (dw/dT)}{\Delta \log W_r} \right]$ against

$\left[\dfrac{\Delta T^{-1}}{\Delta \log W_r} \right]$ it is possible to find the values of energy of activation (E_A) and as well as order of

reaction (x).

43.2 DERIVATIVE THERMOGRAVIMETRIC ANALYSIS (DTG)

It is useful to compare a thermogram with its first derivative. In such thermograms a plateau at a particular temperature is derived but its shoulder at high temperature cannot be easily ascertained without a derivative curve. Many TGA apparatus can automatically measure derivatives. Figure 43.2 depicts such a curve.

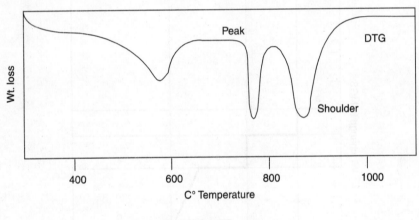

Fig. 43.2 DTG curve

43.3 DIFFERENTIAL THERMAL ANALYSIS (DTA)

In this technique the temperature difference between the analyte sample and reference material is continuously recorded as a function of the furnace temperature. DTA and TGA are complementary techniques, however the range of measurement in DTA is much larger.

The apparatus and methodology of DTA unit comprises a sample holder with a thermocouple assembly, furnace assembly, flow control system, a preamplifier and recorder and furnace power programmer and controller (Fig. 43.3). In operating DTA, one inserts a very thin thermocouple into a disposal sample tube (2 mm diameter which can hold 0.1–10 mg of sample). An identical tube may contain a reference material such as alumina, alumdun powder, quartz sand or it may be empty. Two tubes are put in a sample block, side by side and heated or cooled at a uniform programmed rate. DTA is a dynamic method. All aspects of the technique should be standardised to obtain reproducible results.

The material should be fine (100 mesh). The plot of ΔT as function of T is an indication of energy gain or loss in the specimen under investigation. If an endothermic change occurs, the specimen

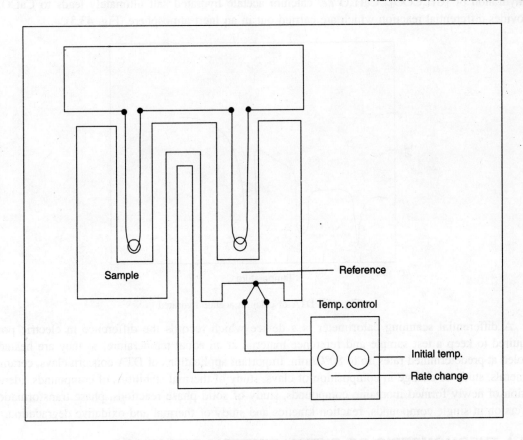

Fig. 43.3 DTA apparatus

A differential scanning calorimeter is a device which records the difference in electric power, required to keep the sample and reference temperature at zero respectively as they are heated or cooled at preset rate. DTA. DTA. Important application of DTA concerns studies of inorganic materials, studies of composition of curves show the thermal stability of compounds having combined newly formed compounds. Compounds, study of solid phase reactions, phase transformations, inorganic single compounds, fraction kinetics and study of thermal and oxidative degradation.

43.4 THERMOMETRIC OR ENTHALPY TITRATIONS

In such titrations heat of reaction are used for estimation the one temperature is plotted against volume of titrant. The titrant is delivered from a motor driven syringe or solution reservoir a fairly dry insulated vessel and the temperature change of solution are recorded which is of minor results. The measure thermometric titration is especially regarded. A sharp break in the curve marks the end point. They are used in all types of titrations including those involving precipitation reactions.

The main apparatus and instrumentation involved in thermometric titrations is made simple. It consists of anion burettes, an isolation titration chamber, thermistor bridge circuit and a recorder. The temperature changes are recorded with respect to those. Thermometric titration curves represents the measure of the quantity enthalpy change which involves the entity ΔG which is the free energy involved. The electrodes used for the titration of some salt solution e.g. barium hydroxide in sodium and acetic acid. The work of neutralisation. A titration of titration acid and sodium hydroxide is shown in Fig. 43.4.

43.5 THERMOMETRIC TITRATION—APPLICATIONS

Some of chemical reactions proceed with heat change or the loss in gain on one observe the course of reaction by noting the resulting change. As the these interactions will not complete.

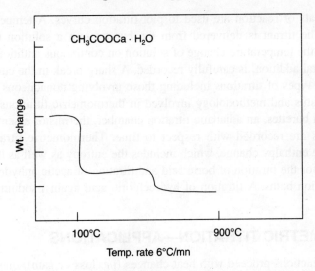

Fig. 43.4 TGA of calcium acetate

temperature lags behind the reference temperature. The two curves shown in Fig. 43.4. It shows thermogravimetric analysis and complimentary relationship of differential thermal analysis. The dehydration of $[CH_3COO]_2$ CaH_2O *i.e.* calcium acetate hydrated salt ultimately leads to $CaCO_3$. It provides differential reaction which are carried out in an inert atmosphere (Fig. 43.5),

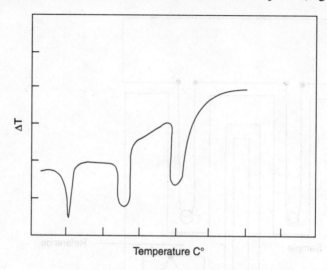

Fig. 43.5 DTA of calcium acetate hydrated

A differential scanning calorimeter is a device which records the difference in electric power required to keep a test sample and reference material at an equal temperature, as they are heated or cooled at predetermined rates up to 80°C/min. Important applications of DTA concern clays, ceramics, minerals, study of change in composition of clays, study of thermal stabilities of compounds, identification of newly formed inorganic compounds, study of solid phase reactions, phase transformations, inversion in single compounds, reaction kinetics and study of thermal and oxidative degradation.

43.4 THERMOMETRIC OR ENTHALPIMETRIC TITRATIONS

In such titrations heats of reaction are used to plot titration curves. A temperature is plotted against volume of titrant. The titrant is delivered from the burette into a solution kept within a thermally insulated vessel and the temperature change of solution on continuous addition of titrant, or after each successive incremental addition, is carefully recorded. A sharp break in the curve marks the end point. They are used in all types of titrations including those involving nonaqueous media.

The main apparatus and methodology involved in thermometric titrations is quite simple. It consists of motor driven burettes, an adiabatic titration chamber, thermistor assembly and a recorder. The temperature changes are recorded with respect to time. Thermometric titration curves represent a measure of the entire enthalpy change, which includes the entropy as well as the free energy involved. The method is used for the titration of boric acid and titration of acetic anhydride in acetic acid and in the study of acetylation baths. A titration of hydrochloric acid against sodium hydroxide is shown in Fig. 43.6.

43.5 THERMOMETRIC TITRATION—APPLICATIONS

Since all chemical reactions proceed with heat changes (*i.e.* loss or gain), one can observe the course of reaction by noting the resulting change in temperature. A simple thermometer and well insulated reaction vessel like a thermos flask with a good magnetic stirrer will suffice. One can use thermistor detectors for accurate measurements. All kinds of titrations viz. acid–base, redox, complexometric or precipitation can be carried out by this technique. In this method we do not measure G, the free energy, but H, the heat contents. So we have $H = G + T\Delta S$ wherein G is the free energy, T is the temperature, ΔS is the entropy change and H is the enthalpy or heat of reaction. Now if G is zero and $T\Delta S$ is large and negative, one can still realise the end point e.g. for H_3BO_3 (boric acid) G is small, while H is -10.2 kcals/mol comparable to HCl ($H = -13.5$ kcals/mol). The change in temperature during titration is given by an expression.

$$T = \frac{nH}{K} = \frac{Q}{K}$$

where n = number of moles of reactants and K = effective heat capacity of system, Q = total amount of heat evolved.

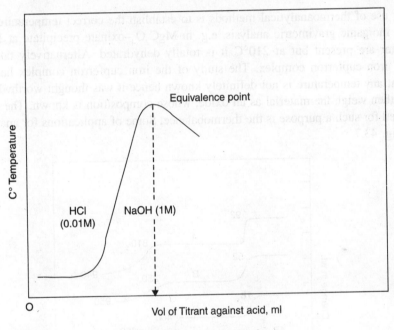

Fig. 43.6 Thermometric titrations

Thus total change in temperature is directly proportional to the number of moles of analyte. One can give several examples of thermometric titrations e.g. Ti (III) with Ce (IV); [Fe (CN)$_6$]$^{4-}$ with Ce(IV); Fe (II) with $Cr_2O_7^{2-}$; Ca with $H_2C_2O_4$ or Ag with HCl, Ca^{2+} with EDTA etc. In all these cases the minimum titrable concentration ranges from 0.1–0.006 when change in H varies from −30 to + 5 kcal/mol.

43.6 ENTHALPIMETRIC TITRATIONS

They are of two kinds of thermometric titrations which we have just discussed and also direct injection enthalpy methods. Thermometric titrations are also called enthalpimetric titrations. We measure temperature vs volume of titrant. In the indirect method, excess reagent is added and the resulting temperature is then measured. The principal merits of enthalpimetric methods are that it has wide utility and is useful for nonaqueous systems, gases, molten salts and furnishes direct information on H (enthalpy), G (free energy) and S (entropy change). The readings are less precise than thermometric titrations, but the methods are rapid and do not require standardisation. It is a one-time measurement technique where we measure temperature T directly.

Enthalpimetric titrations are useful for determining species in a biochemical system such as enzyme catalysed reactions, enzyme activity, measurement and study of proteins and lipids. Several enthalpimetric methods include reactions of acids or bases with strong counterparts. Many systems included in thermometric titrations like that of Ag^+, Ca^{2+}, Fe^{3+} and Tl^{3+} can also be carried out by enthalpimetric methods.

43.7 THERMAL METHODS IN QUANTITATIVE ANALYSIS

An important use of thermoanalytical methods is to establish the correct temperature for drying of a precipitate in inorganic gravimetric analysis, e.g. in MgC_2O_4–oxinate precipitate at 110°C, two molecules of water are present but at 210°C it is totally dehydrated. Alternatively take the case of a precipitate of iron cupferron complex. The study of the iron cupferron complex has shown that its composition at any temperature is not definitely known hence it was thought worthwhile to ignite the complex and then weigh the material as an oxide, whose composition is known. The main instrument commonly used for such a purpose is the thermobalance. Some of applications for analysis of alloys is depicted in Fig. 43.7.

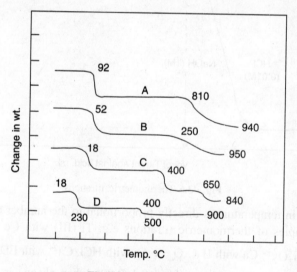

Fig. 43.7 Alloy analysis

In addition typical TGA curves for $AgCrO_4$, $HgCrO_4$, CaC_2O_4, MgC_2O_4 are shown in Fig. 43.8. Similarly an analysis of an alloy containing Ag and Cu is shown in Fig. 43.8. The thermal transformations are:

$$2Ag_2CrO_4 \xrightarrow{\Delta} 2O_2 \uparrow + 2Ag + 2AgCrO_4\,;$$

$$Hg_2CrO_4 \xrightarrow{\Delta} Hg_2O + CrO_3\,;$$

$$Cu(NO_3)_2 \longrightarrow CuO + 2NO_2 + O_2$$

$$MgC_2O_4 \xrightarrow{\Delta} MgCO_3 \xrightarrow{\Delta} MgO.$$

The chemical transformations (Fig. 43.8) for the analysis of an alloy containing Cu and Ag are:

$$AgNO_3 \longrightarrow NO_2 + O_2 + Ag$$

$$Cu(NO_3)_2 \longrightarrow CuO$$

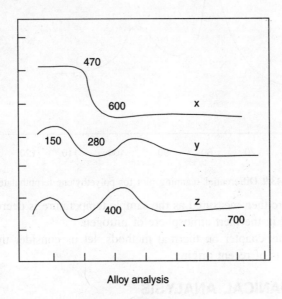

Fig. 43.8 Alloy analysis

The mixture is weighed between 400° and 700°C.

43.8 DIFFERENTIAL SCANNING CALORIMETRY

This is one of the thermal techniques wherein difference in heat flow into substance and reference material is measured as function of sample temperature when both substances are exposed to controlled temperature. This is a calorimetric method separate from the differential thermal analysis wherein here we measure energy difference in DSC.

In DSC, temperature of sample and reference are controlled at same values and the amount of heat which is added to one substance in excess of that added to the other substance is measured. The sample and reference have same temperature. A plot of added heat vs temperature is monitored for isothermal process. DSC is used for study on enthalpic changes. It is used to measure specific heats, heat of fusion and enthalpic changes in transitions. Qualitative analysis can be carried out and also the quantitative work (see Fig. 43.9). In drugs analysis DSC is used to test its purity. A Van't Hoff plot is used for such purposes.

DSC can measure activation energy and rate constant for transitions. Glass transition temperature can be measured by differential scanning calorimeter (DSC).

Instrument: Two kinds of instrument of DSC viz. power compensated DSC and heat flux DSC are known. In latter sample and reference temperature are not same. Figure 43.9 shows a DSC curve for polyethelene terphthalate. Two peaks are due to micro crystal formation and melting. Isothermal measurements are most commonly used in the analysis. The curve in Fig. 43.9 is for exothermic reaction. A crystalline behaviour can be studied. In power compensated DSC model sample and reference material are separately heated by keeping same temperature for both (It can be raised or lowered).

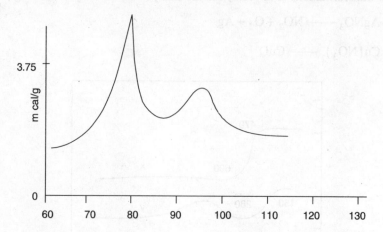

Fig. 43.9 Differential scanning plot for polyethylene terphthalate

The heat flow in two are then measured as the sample temperature is decreased or increased. The experiments are carried out in the inert atmosphere of nitrogen.

Before we conclude this chapter on thermal methods, let us consider the technique of thermo mechanical analysis which is of recent origin.

43.9 THERMOMECHANICAL ANALYSIS

It is a thermal method in which a mechanical property of sample is measured when the sample is subjected to the stress. It is a static or dynamic mechanical analysis (DMA). The method of thermomechanical analysis (DTMA) is that method of analysis wherein the dimension of the sample is measured as the function of the temperature. Thermocouple is used to measure the temperature. The purge gas flows on sample. A linear variable differential transformer (LVDT) is used to change the dimensions variations to electrical transducer or signal. In fact it is movable transformer with mobile core. A change in LVDT is related to displacement of core. TMA is generally used to measure expansion coefficient of the core which is in contact with sample. As sample shrinks or expands, the core moves (in transformer) and this movement is fed to the recorder and measured.

The dynamic mechanical analysis is the method wherein mechanical property of sample is measured as function of temperature; when periodic stress is imposed on sample. The sample is clamped in two arms; wherein one arm moves in relation to other in periodic fashion thereby placing stress on the sample. The frequency of motion is variable. The sample is deformed due to motion. The amplitude of deformation is monitored with LVDT core. The output of LVDT represents frequency and amplitude of motion of area. The system is subjected to oscillate. The energy required to promote the oscillation is related to modulus of sample. Any property can be plotted as the function of temperature. The tensile strength can be thus measured.

However, thermomechanical analysers involving either TMA or LVDT has not become very popular in the chemical analysis for want of proper instrumentation. In comparison the TGA, DTA or thermometric titrations have become most popular due to availability of sophisticated instruments.

PROBLEMS

43.1 In the thermogravimetric analysis of 0.250 g of $Ca(OH)_2$, the loss in weight at different temperatures was :

(a) 0.018 g at 100-150°C (loss of hydroscopic water)

(b) 0.038 g at 500-560°C (dehydration)

(c) 0.0229 at 900-950°C (dissociation)

Determine the composition of calcium hydroxide.

Answer: We determine the composition assuming the reaction:

$$Ca(OH)_2 \, H_2O \xrightarrow[500°C]{\Delta} Ca(OH)_2 \xrightarrow[900°C]{\Delta} CaO + H_2O$$

Amount of hygroscopic water $= \dfrac{0.018 \times 100}{0.250} = 7.2\%$

From mol.wt. $Ca(OH)_2 = 74$, $H_2O = 18$

Amount of $Ca(OH)_2$ dehydrated $= \dfrac{74 \times 0.038}{0.18} = 0.1562$ grams of $Ca(OH)_2$ in 0.250 g of technical sample.

Percentage weight shall be $= \dfrac{0.1562 \times 100}{0.250}$

$= 62.40\%$ of $Ca(OH)_2$ in technical sample.

PROBLEMS FOR PRACTICE

43.2 The thermal decomposition of CaC_2O_4 and MgC_2O_4 proceeds between 160 and 870°C in the following manner :

(i) 0.125 g at 160-250°C (loss of water molecules)

(ii) 0.100 g at 450-600°C (decomposition of oxalate)

(iii) 0.057 g at 600-700°C (decomposition of carbonate to oxide)

(iv) 0.041 g at 750-870°C (formation of SiO_2)

The product undergoes a reaction of losing water molecules, decomposition of oxalate to carbonate, then carbonate to oxide and finally production of SiO_2. Find the % composition of CaC_2O_4, $MgC_2O_4H_2O$ and SiO_2 in the mixture.

43.3 Gypsum, dolomite and calcite, when heated, undergo the following transformation. At 180-200°C there is dehydration; at 720-740°C dolamite dissociates, and at 880-920°C calcite dissociates. The losses at different temperatures were 5.2% (180-200°C), 5.5% (720-740°C) and 16.3% (880-920°C) what is % composition of mixture ?

43.4 For 0.100 g of lime chloride weight losses observed were : 0.012 g at 100°C for elimination of hygroscopic water, 0.0163 decomposition of lime chloride with liberation of oxygen at 300°C and 0.035 mg of $CaCO_3$ formation at 500°C. What is the composition of lime chloride if it contains certain amount of silica?

43.5 0.200 g of precipitate of $MgNH_4PO_4$ was heated with activated carbon to eliminate NH_3 and CaH_2. The product was reacted with H_2O to give H_2. The volumes of H_2 generated were 12.45

ml at 80°C due to removal of H_2O; 102 ml at 105°C to remove H_2O of crystalisation and 8.5 ml at 220°C to remove chemically bound water. Calculate the % composition of the mixture.

43.6 An oxide of potassium is formed by the reaction

$$2KO_3 \rightarrow 2KO_2 + O_2 \uparrow (g)$$

If 0.254 g sample was heated at 70°C for 60 minutes, the weight was found to be 0.212 g. Calculate % of potassium ozonide in the sample.

LITERATURE

1. L. Erdey, *Gravimetric Analysis,* Pergamon Press Ltd. (1963).
2. C. Duval, *Inorganic Thermogravimetric Analysis,* 2nd Ed., Elsevier (1963).
3. W.W. Wendlant,. *Thermal Methods of Analysis,* 3rd Ed., Interscience, New York (1986).
4. W. Wendlandt, L.W. Collins, *Thermal Analysis,* Wiley (1977)
5. R.C. Mackenzie, *Differential Thermal Analysis,* Vol 1 & 2, Academic Press (1970).
6. G. A. Vaughn, *Thermometric & Enthalpimetric Titrimetry,* Van Nostrand (1973).
7. B. Miller, *Thermal Analysis,* Wiley (1982).

Chapter 44

Radioanalytical Techniques

On account of the ease of production of radioactive isotopes of the most common elements, the measurement of radioactivity has become a useful tool in analytical chemistry. Radio isotopes can be used as a means of following an analytical process or we can measure the activity of samples of radioactive materials. In this chapter we consider two important techniques which are commonly used in quantitative analysis, namely neutron activation analysis and isotopic dilution method and substoichiometric extractions.

44.1 RADIOCHEMICAL TECHNIQUES

Most quantitative analysis can be carried out by spectroscopy methods or electroanalytical techniques. In spite of versatility and ease of operation, radiochemical methods have not become popular for want of availability of isotopes of elements with reasonable half life periods. The instruments required for measurement are simple like the γ-ray spectrometer. These isotopes are under direct control of government agencies such as Department of Atomic Energy and Nuclear Power Corporation of India.

Artificial transmutations can be effected by making N/P *i.e.* the neutron, proton ratio unstable by increasing the number of neutrons. The projectiles generally used in such transmutations are protons (H_1^1), deutrons (H_1^2), α-particles (He_2^4) and neutrons (n_0^1). The last named missile has become very popular on account of its capacity for deep penetration. A slow neutron bombardment yields better results than fast neutrons. This property is used in neutron activation analysis. The isotope dilution methods have come to the forefront for ease of operation. The technique of substoichiometric extractions has come to stay. In addition, in the study of columnar behaviour of metals on ion exchange columns, radioactive short lived isotopes have proved to be beneficial for characterisation and evaluation of column parameters like the elution constant, peak elution volume in chromatography.

44.2 SOURCE OF NEUTRONS

Free neutrons are unstable with an average half life of 12.5 min. They decay to give protons and electrons. They have no free existence due to their reactivity. Slow neutrons have low kinetic energy while fast neutrons have an energy of 0.5 MeV. We use thermal neutrons with an energy of 0.025 eV. Nuclear reactors like Zero energy of Canada or Indian (Cirus) reactor provides continuous supply of neutrons with high flux of 10^{11} to 10^{14} n/cm/sec and permit the analysis of metals at level of 10^{-3}-10^{-6} g. Portable neutron sources are also available with a flux of 10^5–10^8 n/cm^2 sec. They are produced by the following reaction

$$\begin{aligned}^{9}_{4}\text{Be} + {}^{4}_{2}\text{He} &\rightarrow {}^{12}_{6}\text{C} + {}^{1}_{0}\text{n} + 5.7\,\text{MeV}\end{aligned}$$

or

$$\begin{aligned}^{2}_{1}\text{H} + {}^{3}_{1}\text{H} &\rightarrow {}^{4}_{2}\text{He} + {}^{1}_{0}\text{n}\end{aligned}$$

One can use an accelerator for generation of beams of neutrons or nuclear reactor.

44.3 NEUTRON CAPTURE AND γ-EMISSION

The usual reaction following transformation is ${}^{23}_{11}\text{Na} + {}^{1}_{0}\text{n} \rightarrow {}^{24}_{11}\text{Na} + \gamma$. This follows emission of γ-rays as shown in the reaction of sodium with neutrons. Other transmutations are:

$$\begin{aligned}^{27}_{13}\text{Al} + {}^{1}_{0}\text{n} &\rightarrow {}^{1}_{1}\text{H} + {}^{27}_{12}\text{Mg}\end{aligned}$$

The elements ${}^{24}_{11}\text{Na}$ or ${}^{27}_{12}\text{Mg}$ can decay by emission of γ or β-rays. The slow neutrons yield better results than fast neutrons. Further, in addition to the measurement of λ activity of ${}^{23}_{11}\text{Na}$ or ${}^{27}_{13}\text{Al}$ one can easily measure their β activity but this is not common practice.

44.4 PRINCIPLES OF NEUTRON ACTIVATION ANALYSIS

The method of activation analysis depends upon the formation of radio nuclides from elements in the sample when they are subjected to bombardment with neutrons of suitable charged particles. The bombarding neutrons may be slow or fast neutrons. It has been experimentally observed that slow thermal neutrons are more powerful projectile materials than fast neutrons. Slow neutrons give radio-active species of the same atomic number but different mass number. After such bombardment the mass number is increased by one unit. The isotope thus produced have typical radiations and modes of decay, thus making radio-activation a useful technique for qualitative identification and quantitative analysis.

During the course of neutron activation, the number of radioactive atoms (N^*) present in the sample is increased by the flux of neutrons. But it is decreased by neutral radioactivity of the radionuclides produced. The rate of production (R) follows an expression:

$$R = N \phi \sigma$$

where N = number of nuclei available; ϕ = slow neutron flux measured as neutrons $\text{cm}^{-2}\,\text{sec}^{-1}$ and σ = slow neutron capture cross section, cm^{-2}.

It is then possible to derive an expression for N^* in terms of R and the time t:

$$N^* = \left(\frac{R t^{1/2}}{0.693} \right) \left[1 - \exp\left(-0.693\, t/t_{1/2} \right) \right]$$

$$= \frac{0.693}{t_{1/2}} \text{ decay constant.}$$

where $t^{1/2}$ = is half-life period of an isotope.

When a sample is composed of more than one element and is subjected to activation analysis different radionuclides will be produced. The different particles or rays emitted can be detected using various counters. It is possible to restore the decay curve obtained by measuring the total radioactivity produced by the sample as a function of time. The logarithm of activity is plotted against time and the slopes are determined mathematically. One important demerit of the latter method is that only those isotopes of moderately long half life can be identified positively. The use of activation analysis for quantitative work usually requires comparison with a known standard in order to know the strength of the neutron beam and to minimize errors in the counting, in chemical treatment and in geometrical considerations.

We can consider the capture rate which is proportional to the slow neutron flux (ϕ) expressed in neutron cm^{-1} sec^{-1} and to the number of target nuclei (N) available. The proportionality constant is called the capture cross section (σ) expressed in barns which are units for an atomic cross section. A correction for fractional abundance 'f' of the target nuclide is required. The activity 'A' will then be expressed as

$$A = N \phi \, \sigma \, fs$$

$$A = N \sigma \phi \left[1 - \exp\left(-\frac{0.693t}{t_{1/2}} \right) \right]$$

where s is the saturation factor which represents the ratio of the amount of activity produced in time 't' to that produced in an infinite time as:

$$S = 1 - e^{-\lambda t} = 1 - 10^{-\lambda t/2.3}$$

If λ = decay constant for the particular radioisotope.

In terms of the weight (W) of the element, assuming that the rate of production of radioactivity, given by N σ ϕ, is constant during irradiation, we have

$$W = \frac{A \times M}{6.02 \times 10^{23} \, \phi \sigma \, fs}$$

if M = molecular weight and 6.02×10^{23} is Avogadro's number.

If two samples of the same element, *i.e.* with the same atomic and mass numbers are bombarded under identical geometries for the same duration of time and if they differ in actual weight taken, the activity displayed by them will be directly proportional to their weight, e.g.,

$$W_1 = \frac{A_1 \times M}{6.02 \times 10^{23} \, \phi \sigma \, fs}, \; W_2 = \frac{A_2 \times M}{6.02 \times 10^{23} \, \phi \sigma \, fs}$$

Since these are the same elements with the same isotopes but different weight we have ϕ, σ_1, f_1, s_1, M all constants by taking the ratio of the two

$$\frac{W_1}{W_2} = \frac{A_1}{A_2}$$

In actual analysis, W_1 may be known, A_1, A_2 can both be counted and W_2 can be found as

$$W_2 = W_1 \left(\frac{A_2}{A_1} \right)$$

This is the fundamental equation of neutron activation analysis. For the neutron source one can use reactors with a flux of 10^{13}–10^{14} or use a portable Pu-Be neutron source with a flux of 10^7 and Sb-Be neutron source with a flux 10^3–10^4. One gram of radio beryllium source provides a yield of approximately 10^7 neutrons/sect and Sb^{124} ($t_{1/2}$ = 60 days) beryllium source provides a yield of 3×10^6 neutrons. Thus in quantitative determinations the comparative method is used. The accuracy of the method is better than 10% at nanogram levels of metal ions. The method is particularly valuable when only minute amounts of samples are available.

44.5 APPLICATIONS OF ACTIVATION ANALYSIS

Neutron activation analysis can be used for analysis of as many as 69 metals and four inert gases. O_2, N_2 and Y can also be activated with fast neutrons. It can be used for analysis of elements in alloys, metals, semi-conductors and environmental samples. It has become popular in forensic science due to its accuracy and speed. The elements which cannot be analysed are Tc, Tl, Po, Li, Be and Pm. The main sources of error are specific shielding, unequal neutron flux, counting uncertainty and variation in geometry of the sample. The method has remarkable sensitivity. On an average 10^{-5} µg of element can be detected in a short time.

44.6 PRINCIPLES OF ISOTOPE DILUTION METHOD

The application of radioactive tracers to analytical problems involves a wide variety of techniques. Both organic and inorganic tracers are used in radiometric titrations, in isotope dilution experiments and for development and evaluation of other analytical methods. A molecule or ion containing a radioactive isotope is present during the course of the reaction or chemical change. By taking measurements of activity before and after the reaction it is feasible to determine the end point of a titration, the course of reaction, or the completeness of separation. In a radiometric titration, the titrant is a radioactive species, e.g. if titration involves precipitation, measurements of the activity of the solution during the course of the titration will remain constant and then increase after the end point has been reached. It is also possible to measure the solubility of a particular compound by incorporating a radioactive isotope in the species being precipitated and then measuring the activity of the precipitate itself or measuring the residual activity of the solution.

Isotope dilution methods are frequently used where it is difficult to measure the yield of a specific reaction or process, or to separate a specific compound from the reaction mixture. However, if it is possible to isolate a portion of the material in a pure form, then a small amount of the material under investigation is taken, in which a radioactive isotope has been incorporated, a separation procedure is followed and the activity of the purified material is compared with the activity of the material added. A comparison permits one to calculate the concentration of the material in the original mixture.

Thus this technique measures the yield of a nonquantitative process or it enables an analysis to be performed where no quantitative isolation procedure is known. To the unknown mixture containing a compound with an inactive element P, is added known weight W_1 of the same compound tagged with the radioactive element P^* whose activity is known, (A_1), the mixture is chemically indistinguishable. A small amount of compound is isolated from the mixture and its specific activity (A) is then measured. The extent of dilution of the radiotracer shows the amount (W) of inactive element present given by an expression

$$W = W_1 \left(\frac{A}{A_1} - 1 \right)$$

This method has proven valuable in the analysis of complex biochemical mixtures and in radiocarbon dating of archeological and anthropological specimens. Thus in the above equation, (W_1) – weight of active element, (A) – activity of diluted sample and (A_1) – activity of known compound can all be measured and it is possible to evaluate the value of W. We use stable, radioactive isotopes.

44.7 APPLICATIONS OF ISOTOPE DILUTION TECHNIQUE

The isotope dilution technique is used for the determination of as many as 30 elements in complex mixtures. It is used for compounds useful in organic and biochemistry. Vitamin D, B_{12}, insulin, and penicillin have been analysed by this method. Its main use is substoichiometric analysis. It is used in inorganic chemistry for the determination of mercury in a sample using ^{203}Hg as the material for tagging with specification of 2,400 cpm/g. Similarly chloride content in the tissue of different animals has been determined using the isotope ^{36}Cl, with a specific activity of 185 cpm/g.

44.8 SUBSTOICHIOMETRIC METHODS OF ANALYSIS

These techniques are beneficial both in activation analysis and isotope dilution. In such methods the mass of the collector greatly exceeds that of the radio isotope. The main requirement is that the same amount of species be taken in measurement. The analytical reagent is deliberately added in an insufficient amount (*i.e.*, substoichiometry) for complete removal of the species from the sample as well as the standard. If the amounts of reagent used for both W and W_1 are identical the amount is determined by added reagent while in the substoichiometry isotope dilution method, identical amounts of tracer are added to two identical solutions (W_0) with only one containing the sample. A suitable reagent is then added to isolate a fixed quantity (W'_r) then we have

$$A'_r = \frac{A_0 W'_r}{W_0} \text{ or } W_x = W_0\left(\frac{A'_r}{A_r} - 1\right)$$

where, in the above equations, W_r is the weight of the species, with activity A'_r after reagent addition, W_0 is radioactive species in grams with activity A_0. This has several practical applications.

PROBLEMS

44.1 A 0.005 g sample containing 20×10^{-1}% Cr is irradiated for 3600 minutes at a neutron flux of 20×10^{10} neutrons/cm^2 sec. The half life of Cr^{51} is 667.2 hrs and its cross section of thermal neutron absorption is 17 barns for Cr^{50}. What is the activity produced in an iron alloy sample containing traces of chromium?

Answer: If $A = \lambda N^*$, $R = N\phi\sigma$; $N^* = \left(\frac{Rt_{1/2}}{0.693}\right)\{1 - \exp(-0.693t_{1/2})\}$

where N = number of nuclei; ϕ = slow neutron flux; σ = slow neutron cross section; λ = decay constant; A = activity; $t_{1/2}$ = half life period; N^* = number of radioactive atoms present in the sample; R = rate of production.
We however use an expression

$$A = N\phi\sigma\{1 - \exp(-0.693t_{1/2})\}$$

Substituting the values

$$\phi = 2 \times 10^{11}; \sigma = 17 \times 10^{-24} \, \text{cm}^2; t_{1/2} = 27.8 \text{ days } (667.2 \text{ hrs.}); t = 60 \text{ hrs.}$$

If Avogadro's number = 6.023×10^{23} then

$$A = \frac{(5 \times 10^{-3})(2.0 \times 10^{-4})(6.023 \times 10^{23})}{52} \times \frac{(17 \times 10^{-24})(1 - e^{-0.693(60/667)})}{52}$$

$A = 2.28 \times 10^3$ dps *i.e.* disintegration per second.

44.2 The technique of isotope dilution was used for analysis of mercury in a catalyst sample. To a 1.0 gram sample of catalyst was added 1 g of a mixture containing 1.0% Hg^{203} with a specific activity of 2400 cpm/g. Then 0.1 g of mercury was separated which showed an activity of 30 cpm. What is the percentage of mercury in the catalyst?

Answer: In isotope dilution we use

$$W = W' \left(\frac{A'}{A} - 1 \right),$$

where W is the weight of compound; W' is weight of compound containing a radioactive isotope of activity A' and A is the specific activity. In this problem.

$W' = 0.01$ gram; $A' = 2400$ cpm/g; $A = 30$ cpm/0.1 = 300 cpm/g
W = weight of mercury which is unknown. We have

$$W = 0.01 \left(\frac{2400}{300} - 1 \right) = 0.01(8 - 1) = 0.01 \times 7 = 0.07, \ i.e. \text{ in 100 gram it is 0.7%}$$

PROBLEMS FOR PRACTICE

44.3 In neutron activation analysis of selenium in copper, a standard sample with 0.75 µg/g of selenium and the unknown sample were bombarded with slow neutrons for 600 hours with γ - radiation at 0.05–0.2 meV. The data obtained were:

E meV	0.04	0.06	0.08	0.10	0.12	0.16	0.20
dpm standard	400	400	500	500	1600	800	400
dpm unknown	600	600	900	1700	2800	1200	600

If the background intensity is 400 and the unknown sample intensity is 600 dpm, what is the concentration of selenium?

44.4 An isotope dilution method is used to determine Cl in plants with Cl^{36}. To 0.89 g sample of plant 1 ml of 0.001 M of $NaCl^{36}$ with an activity of 185 cpm is added. If 0.1 g of HCl was separated and counted to 400 cpm. Calculate the chloride content in the plant.

44.5 0.55 g of tissue contains 6.2% P^{31} when the sample is irradiated with neutrons with a flux (ϕ) of 6×10^{10} neutrons/cm^2 sec for 1200 mts. If (σ), *i.e.* cross section is 0.23 barn and if P^{32} has ($t_{1/2}$) half life as 340.8 hrs, calculate the specific activity of P in the sample.

44.6 A 1.40 gram sample of Devadra's alloy containing 1.24% Al is irradiated with neutrons with flux (ϕ) of 1.0×10^8 neutrons/cm^2/sec. If active Al^{28} has $t_{1/2}$ of 2.27 and after half hour irradiation, what amount of aluminium is Al^{28} (given, $\sigma = 0.21$ barns).

LITERATURE

1. G.R. Choppin, *Experimental Nuclear Chemistry,* Prentice Hall, New York (1961).
2. J.M. Harley and S.E. Wiberley, *Instrumental Analysis,* John Wiley (1962).
3. G. Friendlander, J.W. Kennedy and J.M. Miller, E.S. Macias, *Nuclear and Radiochemistry,* 3rd Ed, Wiley, New York (1981).
4. W.S. Lyon, *Guide to Activation Analysis,* Van Nostrand (1964).
5. H.J. Arnikar, *Essentials of Nuclear Chemistry,* 2nd Ed., Wiley Eastern (1994).
6. W.D. Ehmann, D.E. Vance, *Radio Chemistry and Nuclear Methods of Analysis,* John Wiley (1991).
7. Z.B. Alfassi, *Chemical Analysis by Nuclear Methods,* John Wiley (1994).
8. S.J. Amis, *Non-destructive Activation Analysis,* Elsevier, New York (1981).
9. D. Desoete. R Gybles, J. House, *Neutron Activation Analysis,* Wiley (1972).
10. Ruzicka, J Stary, *Substoichiometry in Radio Chemical Analysis,* Pergamon Press (1968) .

Chapter *45*

Electroanalytical Methods:
Potentiometry

These techniques are as popular as spectrophotometric methods of quantitative analysis. Primarily these techniques are based upon measurement of potential (potentiometry) or conductance (conductimetry), diffusion current (polarography or voltammetry) and evaluation of material generated at electrodes (coulometry). All these measurements are directly related to analytical concentration. These properties have also been used to detect the end point in titrations (e.g. potentiometry, conductometry, amperometry or coulometric titrations). One can also carry out gravimetric analysis by deposition. These techniques have become common because of the easy availability of simple instruments.

45.1 ELECTROCHEMICAL CELLS

There are variety of electrochemical cells. They are called galvanic cells if they produce electrical energy (Fig. 45.2) and if they use electricity they are called electrolytic cells (Fig 45.1). Electricity is conducted in an electrochemical cell if two electrodes are connected externally by a metal conductor and an electolyte solution favours mobility of ions from one electrode to another. A cathode is the electrode at which reduction occurs while an anode is one where oxidation takes place. The reactions at electrodes are called half reactions e.g.

$$Cu^{2+} + 2e^- = Cu(S) \text{ or } 2H^+ + 2e^- = H_2(g)$$

Liquid junctions are used to eliminate direct reaction between two components of two half cells. The cells are represented in the following way

$$Zn \mid ZnSO_4(M) \parallel CuSO_4(M) \mid Cu$$

The anode is put on the left in the cell diagram. All electrochemical systems obey Ohm's law which states that $C = E/R$, where C = current, E = potential, R = resistance. We use direct current to study electrochemical reactions. Alternating current is generally employed in the study of conductance measurements.

A major series of analytical methods are based on the electrochemical properties of solutions. Suppose in electrolyte two electrodes are immersed (Fig. 45.1) and if externally connected, either current will flow through the solution if a battery is put in the circuit or the cell itself may act as a source of electrical energy and produce a current through external connections. This effect will depend upon the nature and composition of the solution, the material of the electrodes, their size and spacing, stirring effect, temperature, properties of external circuits and direction of current flow in the circuit.

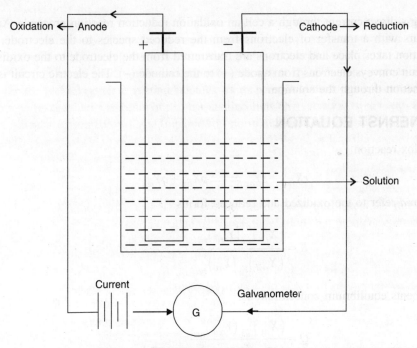

Fig. 45.1 Electrolytic cell

An inert electrode is one which does not enter into the chemical reaction, (Fig. 45.2) Pt, Ag, Au, C are used to make inert electrodes. An active electrode is an electrode which is made of an element which will enter into chemical equilibrium with ions of the same element in solution, e.g. Ag, Hg, and hydrogen form active electrodes. A gas electrode (e.g. a hydrogen electrode) consists of platinum wire

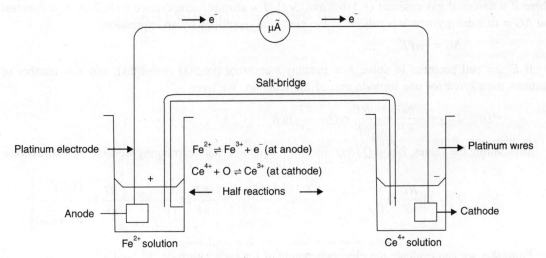

Fig. 45.2 Galvanic cell with junction

foil to conduct electricity with gas bubbling over the surface. The more active elements are seldom used for construction of electrodes due to direct chemical interaction, e.g. Cr, Fe. Calomel and glass electrodes are combinations of inert or active electrodes with appropriate compounds. They are called 'half cells'.

On passing a direct current through a cell an oxidation reduction reaction proceeds. At the anode, oxidation occurs with a transfer of electrons from the reduced species to the electrode, and at the cathode, reduction takes place and electrons are transferred from the electrode to the oxidised species. A external circuit conveys electrons from anode (+) to the cathode (−). The electric circuit is completed by ionic conduction through the solution.

45.2 THE NERNST EQUATION

Consider a redox reaction

$$rX_{red} + sY_{ox} \rightleftharpoons pX_{ox} + qY_{red}$$

where *ox* and *red* refer to the oxidized and reduced forms

$$K = \frac{\left(X_{ox}\right)_{eq}^{p}\left(Y_{red}\right)_{eq}^{q}}{\left(X_{red}\right)_{eq}^{r}\left(Y_{ox}\right)_{eq}^{s}}$$

and '*eq*' represents equilibrium concentrations:

$$Q = \frac{\left(X_{ox}\right)_{act}^{p}\left(Y_{red}\right)_{act}^{q}}{\left(X_{red}\right)_{act}^{r}\left(Y_{ox}\right)_{act}^{s}}$$

If Q = activity coefficient wherein 'act' is the actual value from experiments and not necessarily equilibrium values. Then thermodynamically, a change in free energy is given by the equation

$$\Delta G = RT \ln Q - RT \ln K$$

where R = universal gas constant (8.316 J/mol deg), T = absolute temperature in Kelvin, K = constant and ΔG = free energy which is related to the electrical quantities by an expression:

$$\Delta G = -nFE_{cell}$$

If E_{cell} = cell potential in volts; F = Faraday's constant (96,500 coulombs); and n = number of electrons transferred for one formula unit of the reaction. We have

$$E_{cell} = -\frac{\Delta G}{nF} = -\frac{RT}{nF} \ln Q + \frac{RT}{nF} \ln K$$

Substituting the values, K and Q from the above equations and rearranging we have the expressions:

$$E_{cell} = \left[\frac{RT}{nF} \ln \frac{\left(X_{ox}\right)_{eq}^{p}}{\left(X_{red}\right)_{eq}^{r}} - \frac{RT}{nF} \ln \frac{\left(X_{ox}\right)_{act}^{p}}{\left(X_{red}\right)_{act}^{r}} \right] - \left[\frac{RT}{nF} \ln \frac{\left(Y_{ox}\right)_{eq}^{s}}{\left(Y_{red}\right)_{eq}^{q}} - \frac{RT}{nF} \ln \frac{\left(Y_{ox}\right)_{act}^{s}}{\left(Y_{red}\right)_{act}^{q}} \right]$$

From this we can evaluate the electrode potential for each electrode E_X^0 and E_Y^0

$$E_X^0 = \frac{RT}{nF} \ln \frac{(X_{red})_{eq}^r}{(X_{ox})_{eq}^p} \text{ and } E_Y^0 = \frac{RT}{nF} \ln \frac{(X_{red})_{eq}^r}{(X_{ox})_{eq}^p}$$

Hence

$$E_{cell} = \left[E_Y^0 - \frac{RT}{nF} \ln \frac{(Y_{red})_{act}^q}{(Y_{ox})_{act}^s} \right] - \left[E_X^0 - \frac{RT}{nF} \ln \frac{(X_{red})_{act}^r}{(X_{ox})_{act}^p} \right]$$

By this procedure we have separated the expression into two terms: the effects of the two substances X and Y on the potential of the cell. We define the half cell potential as:

$$E_X = E_X^0 - \frac{RT}{nF} \ln \frac{(X_{red})^r}{(X_{ox})^p}$$

Then from these two equations also E_γ is

$$E_Y = E_Y^0 - \frac{RT}{nF} \ln \frac{(Y_{red})^q}{(Y_{ox})^s}$$

We have the actual value of potential of the cell as

$$E_{cell} = (E_Y \sim E_X)$$

In the same manner the chemical equation can be split as

$$rX_{red} = pX_{ox} + ne^-, \ sY_{ox} + ne^- = qY_{red}$$

where e^- = electron. All half cell reactions with electrons are written as:

$$rX_{red} = pX_{ox} + ne^-,$$

$$sY_{red} = qX_{ox} + ne^-,$$

In the above expression E_X^0 and E_Y^0 are called *standard electrode potentials,* for the half reaction. It is difficult to determine the absolute value of potential of a single electrode as a second electrode is required for measurement. The normal hydrogen electrode (NHE) is usually assigned a zero position on the scale of potentials. In the NHE, H_2 gas at a partial pressure of one atmosphere is bubbled over a platinised platinum foil immersed in aqueous solution, in which the activity of hydrogen is assumed to be unity. So the NHE has a zero potential at all temperatures. In transition elements, formal potentials are often used. Formal potential is defined as:

$$E^{0'} = E^0 - \frac{RT}{nF} \ln \frac{r^{r_{red}}}{r^{p_{ox}}} = \text{Formal potential}$$

45.3 STANDARD ELECTRODE POTENTIAL

In many places two electrodes of a cell are separated while maintaining electrolytic contact between them. This contact may be a porous cup or glass barrier or a salt bridge (refer to Fig 45.2 of an electrolytic cell with liquid junction potential).

The E^0 values are negative w.r.t. NHE for metals, which are more powerful reducing agents than H, and are positive for those which are less powerful reducing agents. This is illustrated in Table 45.1.

Table 45.1 Some standard electrode potentials of metals

Metal	Chemical reaction	Potential E at 25°C (Volts)
Ag	$Ag^+ + e^- = Ag$	+0.799
Al	$Al^{+3} + 3e^- = Al.$	−1.662
Ba^{2+}	$Ba^{2+} + 2e^- = Ba$	−2.906
BiO^+	$BiO^+ + 2H^+ + 3e^- = Bi + H_2O$	+0.320
Ca	$Ca^{2+} + 2e^- = Ca$	+1.44
Cd	$Cd^{2+} + 2e^- = Cd$	−0.403
Co	$Co^{2+} + 2e^- = Co$	−0.277
Cu	$Cu^{2+} + e^- \longrightarrow Cu^+, Cu^+ + e^- = Cu$	+0.153 V; + 0.5221
Fe	$Fe^{2+} + 2e^- \longrightarrow Fe, Fe^{3+} + e^- = Fe^{2+}$	−0.440 V; + 0.771
Hg	$2Hg^{2+} + 2e^- = Hg_2^{2+},$	+0.920
K	$K^+ + e^- = K$	−2.925
Mg	$Mg^{2+} + 2e^- = Mg$	−2.363
Mn	$Mn^{2+} + 2e^- = Mn$	−1.180
Na	$Na^+ + e^- = Na$	−2.714
Ni	$Ni^{2+} + 2e^- = Ni$	−0.250
Pb	$Pb^{2+} + 2e^- = Pb$	−0.126
Pd	$Pd^2 + 2e^- = Pd$	+0.987
Sn	$Sn^{2+} + 2e^- = Sn, Sn^{4+} + 2e^- = Sn^{2+}$	−0.136 V; + 0.154
Ti	$Ti^{3+} + e^- = Ti^{2+}$	−0.369
Tl	$Tl^+ + e^- = Tl$	−0.336
UO_2^{2+}	$UO_2^{2+} + 4H^+ + 2e^- \rightleftharpoons U^{4+} + 2H_2O$	+0.334
V	$V^{3+} + e^- = V^{2+}$	−0.256
Zn	$Zn^{2+} + 2e^- = Zn$	−0.763

In following reactions

$$2H^+ + 2e^- = H_2(g); \ E_H^0 = 0 \, V$$

$$Fe^{3+} + e^- = Fe^{2+}; \ E_{Fe}^0 = -0.771 \, V$$

E^0 = standard electrode potential. This is consistent with experimental values of a galvanic cell with suitable electrodes in contact with their ions. By convention the unit potential of a hydrogen electrode is assigned the value of exactly zero volts at all temperatures. As per IUPAC, the electrode potential is to be used exclusively for half reactions. The sign of the electrode potential will indicate whether or not reduction is instantaneous with respect to the standard hydrogen electrode. For instance it is negative for Zn but positive for Cu and it gives an idea of the relative driving force of a half cell reaction. It is a relative quantity. Some typical standard electrode potentials are listed in Table 45.1.

45.4 EFFECT OF CURRENT ON CELL POTENTIAL

We now consider the concepts of a reversible cell, over voltage and polarisation phenomena. A reaction is irreversible if its products do not react with each another. A cell with ZnAg-AgCl will undergo two reactions:

$$2\,AgCl + Zn \longrightarrow 2\,Ag + Zn^{2+} + 2Cl^- \qquad \text{(if externally short circuited)}$$

$$2\,Ag + 2H^+ + 2Cl^- \longrightarrow 2\,AgCl + H_2 \qquad \text{(connected to source of electricity at high voltage)}$$

Thus Ag–AgCl cell is reversible but the Zn–HCl half cell is not reversible. Thermodynamically a reaction is said to be reversible if an infinitesimally small change in the driving force causes a change in direction. The system is said to be in thermodynamic equilibrium.

An electrode (and hence a half cell) is said to be polarized if its potential shows any departure from the value which would be predicted from the Nernst equation. This will happen if a good current is drawn through it. Changes in potential due to actual changes in the concentrations of ions at the electrode surfaces are called *concentration polarizations*. The activities concerned should always be those of the ions at the electrode surface.

Any half cell is operating irreversibly if a good amount of current is flowing through it. The actual potential of a half cell under such conditions cannot be calculated but it is always greater than the corresponding reversible potential as calculated from the Nernst equation. The difference between the equilibrium potential and the actual potential is known as 'over voltage', over potential' or 'ohmic drop'. It is the extra driving force necessary to cause the reaction to take place at an appreciable rate.

Over voltage increases with current density but decreases with temperature. It varies with the chemical nature of the electrode and is noticeable for electrode processes which produce gaseous products like H_2 and O_2, but is of insignificant magnitude when metals are deposited. Its magnitude cannot be forecast exactly due to a number of variables. A typical values for H_2 over voltage with electrodes made of zinc for instance is -0.716 V at 0.001 A/cm^2 current density and -1.22 V at a current density of 2 A/cm^2. For oxygen with a copper electrode it is -0.582 V at a current density of 0.02 A/cm^2 and -0.793V at current density of 1 A/cm^2.

45.5 CLASSIFICATION OF ELECTROANALYTICAL METHODS

The electroanalytical methods are classified as follows:

Potentiometry: It is the direct application of the Nernst equation through measurements of potentials of two nonpolarized electrodes under conditions of zero current.

Voltammetry and polarography: These are methods of studying the composition of a dilute electrolyte solution by plotting graph of current–voltage curves. In these, the concentration of reducible or oxidisable species at the electrode can be determined. Voltage is applied to a small polarizable electrode (relative to a reference electrode) and increased to a negative value over a span of 1–2 V; the resulting change in current through the solution is noted. Voltammetry is a general name, while polarography is especially applied to applications of the dropping mercury electrode. However, this classification is not strictly followed. In amperometry both electrodes are polarizable.

Coulometry: This is a method of analysis which involves the application of Faraday's laws of electrolysis relating the equivalence between the quantity of electricity passed and the amount of chemicals generated at the corresponding electrodes.

Conductimetry: The two identical, inert electrodes are employed and the conductance (*i.e.* reciprocal of resistance) between the two electrodes is measured, with an alternating current powered Wheatstone bridge. Specific effects of electrodes are eliminated. Technique involving titrations is termed as conductometry.

Oscillometry: This permits one to observe changes in conductance, dielectric constant or both by use of high frequency alternating current. Electrodes are not kept in contact with the solution directly.

Chronopotentiometry: A known constant current is passed through the solution and the potential appearing across the electrodes is observed as a function of time. The related measurement of current changes following application of a constant potential is called chronoamperometry.

Controlled potential electrolysis: It is possible to quantitatively separate species by means of electrolytic oxidation or reduction at an electrode, the potential of which is carefully controlled. The quantity of separated species may be measured coulometrically or gravimetrically.

45.6 POTENTIOMETRY FOR QUANTITATIVE ANALYSIS

The Nernst equation gives a simple relationship between the relative potential of an electrode and the concentration of the corresponding ionic species in solution. Thus, measurement of the potential of a reversible electrode permits calculations of the activity or concentration of a component of the solution.

Consider a galvanic cell as shown in Fig. 45.3. To calculate $[Ag^+]$ consider the cell where the beaker on the left contains a solution of unknown content. The central beaker connects the KCl solution of the saturated calomel electrode with a salt bridge from the silver solution. If an instrument measuring potential is connected to two electrodes Ag will be +ve, SCE –ve. Now if the difference in voltage is

0.400 V, $E_{Ag}^0 = 0.799$ V w.r.t. NHE. By applying the Nernst equation the two half reactions are:

$$E_{Ag} = E_{Ag}^0 + \frac{RT}{nF} \ln(Ag^+); E_{SCE} = +0.246\,V$$

$$E_{cell} = (E_{Ag} - E_{SCE}) = E_{Ag}^0 - E_{SCE} + \frac{RT}{nF}\ln(Ag^+) \text{ if } n=1$$

$$\log(Ag^+) = \frac{E_{cell} - E_{Ag}^0 + E_{SCE}}{2.303(RT/nF)} = 2.57\times10^{-3}$$

We always subtract the more negative value from the more positive value.

The term $2.303\left(\frac{RT}{F}\right) = 0.0591$ at 25°C which should be remembered in all calculations.

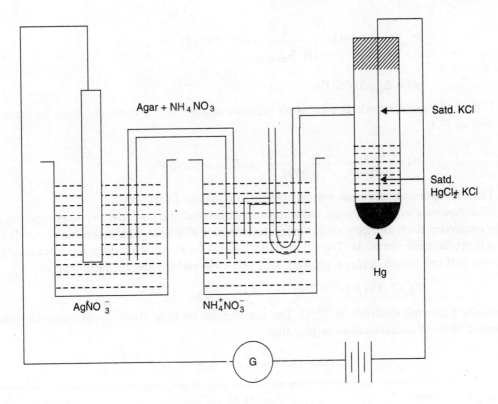

Fig. 45.3 Potentiometry assembly

45.7 VARIOUS ELECTRODES IN POTENTIOMETRY

(a) Hydrogen electrode: Two identical electrodes are placed in beakers containing identical solutions in every respect (but not in concentration) and are connected by a salt bridge, the potential between electrodes is related to the ratio of the two concentrations. The concentration cell provides a highly sensitive tool, because the sources of error arising within the cell cancels out. The overall precision is limited by measuring instruments.

The calculation of $[H^+]$ is simplified if we adopt pH as a measure of acidity or basicity, e.g. pH = $-\log [H^+]$. The pH of 0.05 M potassium hydrogen phthalate which is used as a buffer is defined by the relation as:

$$pH = 4.000 + \frac{1}{2}\left(\frac{t-15}{100}\right)^2 \text{ if } t = 0 \text{ to } 55°C \text{ where } t = \text{temperature}$$

For the measurement of pH, an electrode is required which will respond reversibly to the concentration of hydrogen ions. A hydrogen electrode dipping into the unknown solution is most useful.

$$E_{cell} = -0.0591 \log \frac{[H^+]_{standard}}{[H^+]_{unknown}} \text{ at } 25°C \left(\text{since } \frac{RT}{F} = 0.0591\right).$$

if the standard electrode = 1.0 (*i.e.* normal hydrogen electrode),

$$E_{cell} = -0.0591 \log \frac{1}{[H^+]_{unknown}} = 0.0591 \, pH$$

or
$$pH = E_{cell}/0.0591$$

If the hydrogen electrode is used as an indicator against SCE a correction must be made for the potential of SCE

$$pH = \frac{E_{cell}}{0.0591} = \frac{(E_{cell} - 0.246)}{0.0591} \text{ as } E^0_{SCE} = 0.246 \, V$$

Thus the hydrogen electrode remains the defined standard reference.

(b) *Reference electrode:* During measurement of potential, the potential should be constant for one of the electrodes. Such an electrode is termed the reference electrode. This reversible electrode behaves like a nonpolarizable electrode. The calomel half cell (SCE) is one such reference electrode. It represents the half cell shown. ‖ $Hg_2 Cl_2$ (satd), KCl (M) ‖ Hg and has the chemical reaction

$$Hg_2Cl_2 \, (S) + 2e^- = 2Hg(I) + 2Cl^-; \quad E^0_{SCE} = 0.246 \, V$$

for standard calomel electrode at 25°C. The liquid must be kept above the solution in which it is immersed to avoid contamination or plugging.

Table 45.2 Potential of SCE vs. NHE

Temp °C	0.1 M Calomel	Saturated Calomel
15	0.3362	–
20	0.3359	0.2479
25	0.3356	0.2444
30	0.3354	0.2411
35	0.3344	0.2376
40	0.3335	0.2350

The potential of the saturated calomel electrode and of 0.1M calomel electrode against the hydrogen electrode is listed in Table 45.2. These values are at different temperatures.

(c) *Quinhydrone electrode :* An equimolar mixture of quinone and hydroquinone solution establishes an equilibrium which involves both electrons and hydrogen ions as:

$$C_6H_4O_2 + 2H^+ + 2e^- \longrightarrow C_6H_4O_2H_2 \longrightarrow C_6H_6O_2$$

i.e.
$$Q + H^+ + 2e^- \longrightarrow H_2Q$$

$$E_{QH} = E^0_{QH} + \frac{0.0591}{2} \log \frac{[Q][H^+]^2}{[H_2Q]} \text{ if } E^0_{QH} = 0.700 \, V$$

where Q = quinone, H_2Q = hydroquinone and QH = quinhydrone

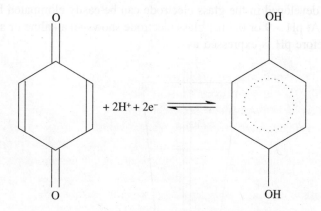

Fig. 45.4 Quinone (Q) and hydroquinone (H_2Q)

$$E_1 = E_{QH} = 0.700 + 0.0591 \log[H^+] = 0.700 - 0.0591 \text{ pH and } E_2 = 0.246 \text{ V}$$

$$E = (E_1 - E_2)\text{pH} = \frac{0.454 - E_{cell}}{0.0591} \text{ w.r.t. Saturated Calomel electrode}$$

The quinhydrone electrode cannot be used for a solution with pH > 9.0, as hydroquinone is neutralised by a base. Strongly oxidising and reducing solutions should be avoided. It is a less expensive electrode.

(*d*) *Metallic electrode* : An electrode of antimony metal inserted in aqueous solution becomes coated with its oxide and responds to changes in hydrogen ion concentration

$$2Sb + 3H_2O \rightleftharpoons Sb_2O_3 + 6H^+ + 6e^-$$

$$\frac{1}{2}Sb_2O_3 + 3H^+ + 3e^- \longrightarrow Sb + \frac{3}{2}H_2O$$

This is useful in industrial applications but is not very much used in analytical laboratories.

$$E = 0.145 + 0.0591 \log(H^+) = 0.145 - 0.0591 \text{ pH}$$

Mn, W, Mo, Hg, Ge can be similarly employed for construction of electrodes.

$$E = E_M^0 - 0.0591 \log \frac{[M^{n+}]_1}{[M^{n+1}]}, \text{ if M = metal with valency as '}n\text{'}.$$

(*e*) *Glass electrode* : This is the most common pH sensitive electrode which consists of a thin membrane of certain varieties of glass which are responsive to changes in [H$^+$] ions. For two solutions separated by a thin membrane we have potentials:

$$E_g = \frac{RT}{F} \ln \frac{(H^+)_1}{(H^+)_2} = (0.0591)(pH_2 - pH_1) \text{ at } 25°C \text{ as } n = 1$$

A typical glass electrode is shown in Fig. 45.5.

The glass electrode against any suitable electrode gives a potential which is related to the pH. The asymmetric potential developed in the glass electrode can be easily eliminated by frequent calibration of the said electrode. At pH > 9 or 0.5 the glass electrode shows an alkaline or acid error but the value is insignificant. Therefore pH is expressed as

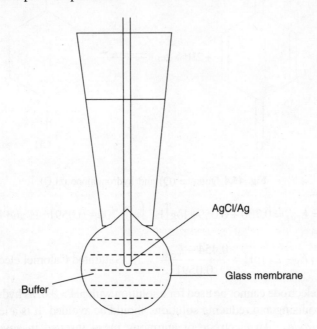

Fig. 45.5 Glass electrode

$$pH = \frac{E_{cell} - E_g}{0.0591} \text{ where } E_g = \text{potential of reference electrode}$$

This is the potential due to unequal strain in the glass. E_g = a constant for a particular electrode assembly formed by measuring the potential produced when the electrodes are dipped into a standard buffer. If it has an internal solution with pH 7.0 and if it is dipped in a solution of pH 7.0, naturally no voltage will be developed.

$$E_g = E_{cell} - 0.0591 \text{ pH}$$

$$E_g = 0 - (0.0591)(7) = -0.4137 \text{ V}$$

$$pH = \frac{E_{cell} - [-(0.0591)(7)]}{0.0591} = \frac{E_{cell}}{0.0591} + 7.0$$

Thus if a glass electrode if filled with an internal solution of pH 4.0 or pH 7.0 resp., then

$$pH = \frac{E_{cell}}{0.0591} + 4 \text{ or } pH = \frac{E_{cell}}{0.0581} + 7.$$

Glass electrodes are most popular and useful for pH measurements but not effective in a pH region greater than 10.0 or in the presence of a high Na^+ ion concentration. Further it causes errors in fluoride solutions or in solutions with pH > 6.0. A combination of glass and reference electrode in a single electrodes assembly is fabricated by many companies. The glass electrode itself is a membrane electrode. At 25°C the SCE is +0.285V for 1M KCl and SCE = + 0.246 for saturated KCl (Table 45.1).

(*f*) *Liquid membrane electrodes:* They are obtained from immiscible liquids that selectively bind certain ions. They permit direct potentiometric determination of activities of cations or anions. Liquid ion exchangers have been used for such purposes on porous inert solid supports. A polyvinyl chloride membrane is also used as a support. Nowadays, liquid cation, liquid anion and macrocylic crown ether membranes are available in the market. The internal solution is the same as the ions to be analysed.

Specially fabricated liquid membrane electrodes for Ca^{2+}, NO_3^-, K^+ and water hardness are available. A recent development is gas sensing probes and enzyme electrodes.

45.8 POTENTIOMETRIC TITRATIONS

A great variety of titrations can be carried out by potentiometric technique. The reaction should involve addition or removal of some ion for which an electrode is available. The potential is measured after successive additions of small volumes of titrant, or continuously, with an automatic recording device. The precision can be improved by use of a concentration cell.

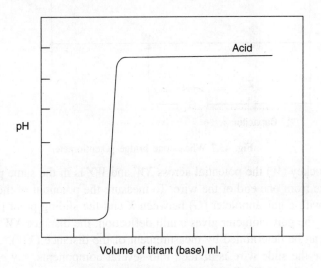

Fig. 45.6 Potentiometric titration

(*a*) *Neutralisation reactions:* Acid–base titrations can be performed with a glass indicator electrode measured against a reference electrode. Figure 45.6 shows a typical titration curve. K should be less than 10^{-8} (*i.e.* ionisation constant). They are useful for mixtures of acid and base titrations.

(*b*) *Precipitation titrations:* A silver indicator electrode is used. Since calomel reacts with silver electrode an agar bridge is used. The formation of precipitate will remove simple hydrated ions from solution.

(*c*) *Complexmetric titrations:* Silver-mercury electrodes are used. Many metals can be titrated with EDTA.

(*d*) *Redox reactions:* Platinum or other inert electrodes can be used in redox titrations. Powerful oxidising agents as ($KMnO_4$, $K_2Cr_2O_7$, Co $(NO_3)_3$) form a metal oxide coat which is removed by cathodic reduction in acid solution. The electrode potential for the half reaction must be the same.

(e) *Non aqueous titration:* The method can be used for titration in nonaqueous media, e.g. 2, 4. dinitrophenyl hydrazine acts as an acid if dissolved in pyridine. Also tetrabutyl ammonium hydroxide in benzene or alcohol can be used as titrant. Glass and SCE are used as electrodes.

45.9 POTENTIAL MEASURING INSTRUMENTS

Although potentials are measured in volts, the potentials of electrodes cannot be determined by connecting a voltmeter across the cell and observing the deflection of the needle, as the voltmeter draws some current from the cell being measured. Hence a potentiometer or electronic voltmeter is used. The principle of the potentiometer is shown in Fig. 45.7. A storage battery sends a current through the wire XY and the variable resistance R. The value of (R) is so adjusted that the potential across terminals XY is about 2 V.

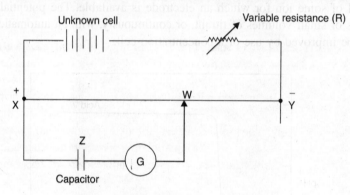

Fig. 45.7 Wheatstone bridge potentiometer

On sliding the jockey (W) the potential across XW and WY is in the same proportion as the length of portion or distance from one end of the wire. To measure the potential of the unknown cell (C) it is connected in series with a galvanometer (G) between X and the sliding point (W). Now (W) is moved to the point at which the galvanometer gives a null deflection; the distance XW represents the potential of the cell (C) and can be determined by measurement of the distance (XW).

In actual practice the slide wire is divided into several components, say eleven separate resistors with equal resistances (Fig.45.8). The first ten wired portions are connected to a multiple tap switch with 10 positions, while R_{11} is a uniform wire wound on a drum and provided with a continuously movable contact. R_1 and R_{11} are provided with calibrations reading directly in volts. R is adjusted until the current flowing is about 1.2 mA (K_1 and K_2 are switches). R is adjusted until the galvanometer shows no deflection. Then the unknown cell is connected. (K_1) is thrown to the left, (K_2) closed and the two main dials are adjusted until the galvanometer shows no deflection. The setting on the dial gives the potential of the cell. The chief advantage of a potentiometer is that at the moment when the potential of the cell is read, current drawn from it is zero (residual errors due to the resistance of the cell or polarisation effects are negligible). The standard cell used to validate calibrations of the potentiometer is the saturated weston cell (potential 1.01864 V at 20°C, decrease by 4×10^{-5} V at 1°C rise in temperature (Fig. 45.9).

This device involves measurement of the potentials of electrodes without flow of current but depends upon the power amplifying properties of an electronic vacuum tube or resistors. An electronic

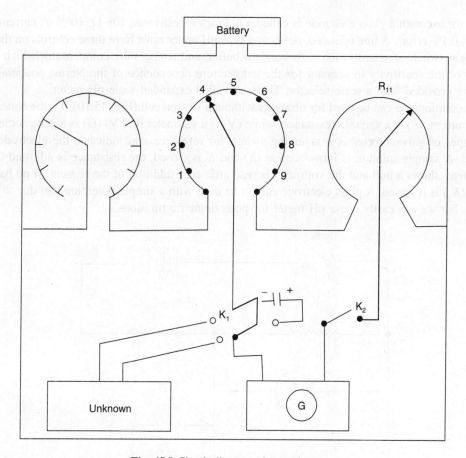

Fig. 45.8 Circuit diagram of potentiometer

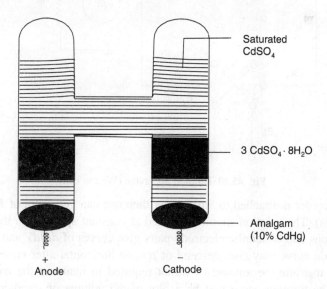

Fig. 45.9 Weston cell

amplifier for use with a glass electrode is called a pH meter (resistance 108 Ω, 10^{-11}. A current causes 10^{-3} drop, 0.1% error). A line operated, deflection type pH meter must have three controls on the panel, an operation switch, standardisation with standard buffer, and temperature compensator which permits alteration of the sensitivity to account for the temperature dependence of the Nernst potential. Some models are provided with a scale selector. They are called expanded scale pH meter.

The potentiometer can be used for titrations without a weston cell (Fig. 45.10) B – the battery gives working current, (R) is a variable resistance, while (V) is a voltmeter (0.2 V) (G) is a galvanometer, and (R_D) damper of galvanometer is a reversing switch for reference and indicates the electrodes to be connected. A sample solution is introduced at (X) and K_1 is closed, the resistance is adjusted until the galvanometer shows a null and the voltmeter is read after each addition of the reagent. If no balance is obtained (K_2) is reversed. A glass electrode cannot be used with a simple potentiometer due to its high resistance but we can easily use a pH meter for potentiometric titrations.

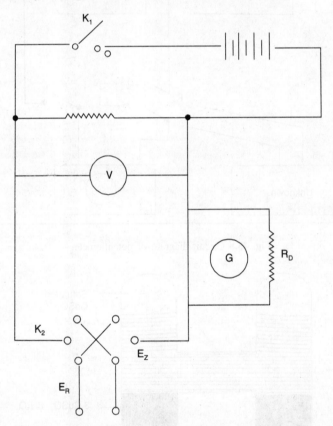

Fig. 45.10 Titration apparatus (Weston cell)

If a stripchart recorder is attached to a pH meter then one can get a plot of E and V (*i.e.* potentiometric titration curves). The solution should be added at constant speed from the burette.

Derivative titrations and bimetallic electrode pairs give curves of dE/dV and d^2E/dV^2 respectively. The bimetallic titration curve may rise, descend or remain horizontal after equivalence is passed.

It is possible to measure the amount of titrant required to maintain the indicator electrode at a constant potential. The titration curve will be a plot of the volume of standard solution added as a function of time. This is extensively used in the field of enzymology.

45.10 ION SELECTIVE ELECTRODES

Electrodes for the measurement of the activity or concentration of specific anions were developed by Pungor in Hungary. They prepared an electrode for the measurement of the activity of the iodide ion in water. Monomeric silicon rubber was polymerised in a dispersion of silver iodide particles. The new polymer so formed was used as a membrane to seal off the bottom of glass tubing. KI solution was filled in the tube and the Ag-electrode was put at the other end of the glass tubing which was then sealed. On dipping this electrode in the solution containing iodide ions, it was seen that the latter were preferentially adsorbed at the solutions membrane interface. The magnitude of such adsorption was a function of interface activity of the iodide ions in two solutions, *i.e.* internal and external solutions with respect to the membrane. Boundary potentials were developed over the membrane's inner and outer interfaces. The electrode potential represented the boundary potential across the membrane and through the Nernst equation was associated with the activity of iodide ions in the solution in contact with the outer membrane interface. When a reference electrode was used with an ion selective electrode, a calibration procedure involving E cell and a known activity or concentration solution of iodide ion was applied. Membrane electrodes have been used for analysis of bromide, chloride and fluoride ions. The classification is shown in Table 45.3.

Table 45.3 Classification of ion selective electrodes (ISE)

Crystalline Single Crystal	Membrane Polycrystal	Non Crystalline Glasses	Extractant Liquids	Membrane Immobilised Liquid	Gas Sensing Electrode	Enzyme Electrodes
Fluoride Electrodes (LaF_3)	Sulphide (Ag_2S) Electrode Silver	Na^+ H^+	High molecular weight amines for anions or Liquid cation exchanger for Ca^2, K^+	LAE LCE in THF for NO_3^- or Ca^{2+} resp.	Hydrophobic for CO_2, NH_3 Environmental probe	Blood Urea with with urea membrane

A recent development is the use of liquid ion exchange membranes in ion selective electrodes. These liquid ion exchange membranes are non-volatile, water immiscible and have functional groups to form chelates or coordination complexes.

A typical assembly of an ion selective electrode involving a liquid ion exchange membrane is shown in Fig. 45.11. It can be used for potentiometric determination of several anions and polyvalent cations. The liquid ion exchange membrane is held between two porous disks. These disks allow contact but at the same time do not allow the outer aqueous solution to mix with the ion exchange liquid. The inner solution has a known concentration of ions for which the electrode is selective. Ag/AgCl is the reference electrode which is immersed in the inner solution and sealed, and is connected to the external circuit. It is absolutely essential to immerse the electrode in the same solution of ions which are to be measured for some time before commencement of measurements. Boundary potentials are established across the two solution membrane interfaces, the outer interface is functionally dependent on the ion activity in the external solution and it is fixed at a constant temperature. The developed potential (E) is $E_{\text{cell}} = \text{Constant} + \dfrac{0.059}{n} \log\left[M^{n+} \right]$ if M^{n+} is an ion which is being measured with a valency 'n'. The cell potential is given by an expression

$$E_V = E_{\text{ref}} + E_j - E_A^0 + \frac{0.059}{n} \log \frac{1}{a}$$
$$\text{Cell}$$

wherein E_{ref} is the potential of the reference electrode, E_j is the sum of the two junction potentials, one arising at each end of the salt bridge. This equation is further simplified as:

$$E_V = L + \frac{0.0591}{n} pA$$

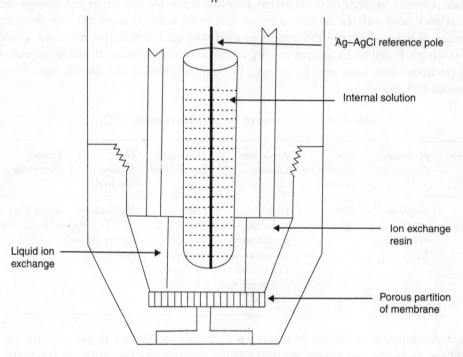

Ag–AgCl reference pole

Internal solution

Ion exchange resin

Liquid ion exchange

Porous partition of membrane

Fig. 45.11 Liquid membrane electrode

where L is the sum of ($E_{\text{ref}} + E_j$) and E_A^0, pA is the negative logarithm of the activity of the analyte and E_V is cell potential.

An ion selective electrode is independent of the solution pH within a certain range of pH only. Thus we see that an ion selective electrode consists of a membrane that responds more or less selectively to a single species of ion in contact with a solution of the ion to be determined on one side and with a solution having a fixed activity of ion in contact with a suitable reference electrode on the other side. Ion selectivity is accomplished through selective extraction of ions into the membrane phase as well as by difference in mobility of the ions within the membrane. Liquid anion exchangers such as Trioctyl amine, Amberlite LA-1, LA-2, Aliquat 336s or a liquid cation exchanger such as (HDEHP) bis-2-ethylhexyphosphoric acid or (DNS) dinonylnapthalene sulphonic acid are used as membrane materials. The ion of interest is selectively exchanged by such membranes.

45.11 APPLICATIONS OF ION SELECTIVE ELECTRODES

Finally let us consider applications of ion selective electrodes. The electrodes can be used as an end point detector in a titration. The concentration of the species can be determined directly from measurement of the potential. Ion selective electrodes are dependable in titrations. Potential measurements of low precision can yield accurate titration or concentration values. Lack of Nernst equation response, may, however, lead to measurement of ion activity. Errors can be minimised by standardising the electrode with two solutions, one which is slightly lower and another slightly higher than the solution to be measured. In addition to liquid membrane electrodes, other solid membranes, enzyme electrodes and specially fabricated electrodes are now available. A practical application of the liquid membrane electrode is the calcium liquid ion exchange electrode where the calcium salt of dodecyl phosphoric acid is the active component in the solvent di-n-acetylphenyl phosphonate. An anion selective liquid membrane is obtained by using bulky cationic species that will selectively extract anions into the organic solvent, e.g. substituted phenanthroline complex of iron (II) which selectively extracts perchlorate, and complexes of nickel which selectively extract nitrate. Solid state membranes are prepared from AgCl or AgBr acting as a membrane to Ag ion activity due to high ionic conductivity of the latter ions. Ag_2S is also used for other metal sulphides like CuS, PbS and CdS giving a membrane responsive to the second metal. Glucose and urea have been measured by enzyme electrodes. Such electrodes have been used for the determination of enzymes and substances which interact with them, e.g. an amygdoline sensitive electrode can be made by retaining β-glucosidase in a gel layer coupled to a cyanide sensitive membrane electrode. A CO_2 sensitive electrode consists of a pH sensitive glass electrode around which a gas permeable layer retains a layer of Na_2CO_3 solution. This is very useful in environmental chemistry for the analysis of dissolved carbon dioxide from water samples.

45.12 ADVANTAGES AND LIMITATIONS OF ION SELECTIVE ELECTRODES

Ion selective electrodes measure activity rather than concentration of an ion. Interference from ionic strength effect is possible. They measure free ions. On complexation or protonation interference can take place. Ion selective electrodes are rather selective but not always specific although the name implies this. They are pH limited. Their great advantage is that they work in coloured or turbid solutions where photometric methods do not work. They have a logarithmic response, hence give a broad range of operation. However, errors caused are also proportionately large during measurement. The response is not rapid for dilute solutions but is otherwise good and can be employed for on-line analysers. The response depends upon the temperature of the solution. We can use portable instruments. The sample does not get fouled or destroyed after measurement. The working range is 10^{-5}–10^{-9}. To get good results periodic calibration is a must. Impurities in solutions cause serious errors.

PROBLEMS

For values of standard electrode potentials refer to Table 45.1 or Appendix-2.

45.1 What is the emf of the following cell?

(a) $Zn \mid Zn^{2+}(1 \times 10^{-6}) \parallel Cu^{2+}(0.010M) \mid Cu$

(b) $Cu \mid Cu^{2+}(2M) \parallel Cu^{2+}(0.010)M \mid Cu$

(Given $E_{Zn}^0 = -0.763$ V and $E_{Cu}^0 = +0.337$ V).

Answer: (a) We use the Nernst equation for evaluation of potential,

$$E_1 = E_{Zn}^0 + \frac{RT}{nF}\log(Zn^{2+})$$

Generally $\qquad \frac{RT}{F} = 0.0591$ and if $n = 2$

Hence $\qquad E_1 = -0.763 + \frac{0.0591}{2}\log(1\times10^{-6}) = 0.940$

Similarly, $E_2 = E_{Cu}^0 + \frac{RT}{nF}\log(Cu^{2+}) = 0.337 + \frac{0.0591}{2}\log(0.010) = 0.028$

Hence potential $\qquad E = (E_1 - E_2)$

$\therefore \qquad E_{cell} = (-0.94) - (+0.28) = 1.22$ volts

(b) This is an example of the concentration cell.

$$E_1 = E_{Cu}^0 + \frac{RT}{nF}\log[Cu^{2+}]. \text{ Since } RT/F = 0.0591; n = 2$$

$$E_1 = +0.337 + \frac{0.0591}{2}\log(2) = 0.3458$$

$$E_2 = +0.337 + \frac{0.0591}{2}\log(0.010) = 0.2779$$

$$E = E_1 - E_2 = (0.3458 - 0.2779) = 0.0679 \text{ volts}.$$

45.2 What is the half cell potential at the end point in the titration of Fe(II) with KMnO$_4$ in a H$_2$SO$_4$ media if at the end point pH = 2.3 ?

(Given $E_{Fe_3}^0+ = 0.771$, $E_{MnO_4}^0 = -1.51$ V)

Answer: The possible reaction is

$$5FeSO_4 + KMnO_4 + 4H_2SO_4 \rightarrow 5Fe(SO_4)_3 + MnSO_4 + 4H_2O$$

or $\qquad 5Fe^{2+} + MnO_4^- + 8H^+ \rightarrow 5Fe^{3+} + Mn^{2+} + 4H_2O$

Since $\qquad E_{Fe}^0 = 0.771$ and $E_{MnO_4}^0 = +1.51$ when $[H^+] = 1.0$

We have $\qquad E_1 = E_{Fe}^0 + \frac{RT}{nF}\log\left(\frac{Fe^{3+}}{Fe^{2+}}\right) = 0.771 + \frac{0.0591}{1}\log\frac{[Fe^{3+}]}{[Fe^{2+}]}$

$$E_2 = E_{MnO_4}^0 + \frac{RT}{nF}\log\frac{[MnO_4^-][H^+]^8}{[Mn^{2+}]}$$

$$= +1.51 + \frac{0.0591}{5} \log \frac{\left[MnO_4^-\right]\left[H^-\right]^8}{\left[Mn^{2+}\right]} \text{ as } n = 5$$

Therefore by adding the two equations after equalising

$$6E = E_1 + 5E_2 = 0.771 + 5(1.51) + 0.0591 \log \frac{\left[Fe^2\right]\left[MnO_4^-\right]\left[H^+\right]}{\left[Fe^{2+}\right]\left[Mn^{2+}\right]}$$

Since $\quad \left[Fe^{3+}\right] = 5Mn^{2+} \text{ and } Fe^{2+} = 5\left[MnO_4^-\right]$

$$E = \frac{0.771 + 5(1.51)}{6} + \frac{0.0591}{6} \log\left[H^+\right]^8$$

but $\quad$ pH $= 2.3 = -\log\left[H^+\right]$

$$E = \frac{0.771 + 7.56}{6} - \frac{0.0591 \times 8}{6} pH$$

$$E = 2.031 + 0.0788\,(2.3)$$

$$= 2.031 - 0.1812 = +1.8498 \text{ volts}.$$

45.3 If 30 ml of 0.1 N $FeSO_4$ is diluted to 100 ml and titrated with 0.1 N $Ce(SO_4)_2$ with SCE, what is the emf of the cell when 20 ml of ceric solution is added? ($E^\circ_{Fe} = +0.771$, $E_{SCE} = +0.285$)

Answer : The redox reaction is

$$Fe^{2+} + Ce^{4+} \longrightarrow Fe^{3+} + Ce^{3+}$$

when 30 ml of a solution of an iron salt neutralises 20 ml of cerric solution, *i.e.* ratio of concentration of ferric to ferrous is 2:1

$$E_1 = E^0_{Fe} + \frac{RT}{nF} \log \frac{\left[Fe^{3+}\right]}{\left[Fe^{2+}\right]}$$

if $\quad E^0_{Fe} = 0.771, \dfrac{RT}{F} = 0.0591\, n = 1$

We substitute these values to have

$$E_1 = 0.771 + \frac{0.0591}{1} \log \frac{[2]}{[1]} \text{ as ratio } Fe^{3+}/Fe^{2+} = 2/1$$

$$E_1 = 0.771 + 0.0177 = +0.7887 \text{ volts}$$

$$E_2 = 0.285 \text{ due to saturated calomel electrode}$$

So $\quad E = E_1 - E_2 = 0.7887 - 0.285 = 0.5037 \text{ volts}.$

PROBLEMS FOR PRACTICE

45.4 Calculate: (*i*) emf of the concentration cell of silver having the cell concentration in two compartments as 0.40 and 0.001 M, $E_{Ag}^0 = 0.799$ V and (ii) emf of the concentration cell of mercury with respect to the concentration of mercury as 0.10 M and 0.0071 M if $E_{Hg}^0 = + 0.920$ V.

45.5 The hydrogen ion concentration of an acid is 3.60×10^{-3} M. What is the pH and OH value? What should be the emf with hydrogen gas and calomel electrodes? ($E_{SCE}^0 = 0.285$, $E_H = 0$ V)

45.6 What is the emf of the following cell?

$\mid$ Pt $\mid$ Sn^{4+} (0.001 M) $\parallel$ Sn^{2+} (0.06) M $\mid$ with calomel half cell.

(Given $E_{SCE} = 0.285$, $E_{Sn^{2+}}^0 = -0.136$ V)

45.7 A solution of alkali hydroxide has a concentration of 5.20×10^{-4} M in (OH$^+$). With a hydrogen gas electrode and calomel half cell what will be the emf of the cell? ($E_H^0 = 0$ and $E_{SCE} = 0.285$ at 25°C)

45.8 In the potentiometric titration of ferrous sulphate with cerric sulphate, 60 ml of 0.1 N ferrous salt (250 ml) was titrated with 0.2 N cerric salt. If the end point was after addition of 10 ml of cerric sulphate what would be the emf i.e. voltage reading?

(Given $E_{Fe}^0 = + 0.771$ V, $E_{Ce}^0 = + 1.61$ V.)

45.9 What is the voltage reading at the end point in the potentiometric titration of Sn^{2+} with KMnO$_4$ with SCE?

(The pH at the end point is 2.0, $E_{SCE} = 0.285$ V and $E_{Sn}^0 = -0.136$ V; $E_{Mn}^0 = + 1.51$ V.).

45.10 Calculate the solubility product K_{sp} of AgCl at 18°C if the potential is 0.518 V vs. H-electrode.

($E_{Ag}^0 = + 0.800$)

LITERATURE

1. P. Delahay, "*New Instrumental Methods in Electrochemistry*", Interscience (1954).
2. L. Meites, "*Handbook of Analytical Chemistry*", McGraw Hill & Co. (1962).
3. J.A. Dean, "*Handbook of Analytical Chemistry*", Wiley (2001).
4. H. Rossoti, "*Chemical Applications of Potentiometry*", Van Nostrand, (1969).
5. R.C. Kapoor, B.S. Agarwal, "*Principles of Polarography*", Wiley Eastern (1991).
6. Y. Umezawa, "*CRC Handbook of Ion Selective Electrodes*", Boca Raton Press (1990).
7. J. Koryta, J. Dyorak, L. Kayon, "*Principles of Electrochemistry*", 2nd Ed., John Wiley, (1993).
8. W. M. Laltimen, "*The Oxidation State of the Elements and Their Potentials in Solutions*", 2nd Ed., Prentice (1952).
9. I. M. Kolthoff, H.A. Lattinen, "*pH and Electrotitration*", 2nd Ed., Wiley (1941).
10. S. M. Khopkar, "*Analytical Chemistry – Problems & Solutions*", New Age International Pub. (P), Ltd. (2003).

Chapter 46

Polarography and Voltammetric Methods

In voltammetry we consider an electrolytic cell having one polarizable electrode. Such a system is studied by means of current–voltage diagrams. These methods are used mainly for quantitative analysis of solutions. The term polarography is used where the polarizable electrode consists of dropping mercury called the dropping mercury electrode (DME).

The nonpolarizable electrode is the saturated calomel electrode, which acts as the reference electrode. A mercury pool can be used as the nonpolarizable reference electrode. Since the polarisable electrode used is much the smaller it is referred to as the microelectrode. It is made of a noble metal such as mercury or platinum.

46.1 DIFFUSION LIMITING CURRENT

In Fig. 46.1, as the microelectrode and (SCE), *i.e.*, reference electrode are connected with a micrometer to a source of known variable potential, the current may vary and is measurable (µa) while the voltage is varied from 0 to -3 V, *i.e.* 0 to -3 to $+3$ V. In Fig. 46.1 (*b*) the potential is measured directly by a voltmeter. The choice of microelectrode depends upon the range of potentials needed (potential is +ve to SCE, when a Pt microelectrode is used). Mercury can be employed due to its high over voltage which is -1.8 V in acid and -2.3 V in bases. It cannot be made more positive than $+0.25$ V against SCE because it undergoes anodic dissolution. If a plane platinum microelectrode in 10^{-1} M KCl containing also 10^{-3} M lead chloride is connected *via* a salt bridge to the SCE and 0.5 V voltage is applied, it will reduce Pb^{2+} ions to Pb-metal. At the same time a number of changes occur; K^{+} and Pb^{2+} ions will start migrating towards the microelectrode and chloride anion will move in the opposite direction. Since K^{+} ions are not easily reducible they form a sheath around the electrode; Pb^{2+} will not form any sheath as they are reducible and they migrate towards the surface of the cathode. However, since the electrode is covered with a sheath of K^{+} ions, Pb will have to migrate in the vicinity of the electrodes by a process of diffusion, in order to maintain electrical neutrality.

The rate of movement of any species by diffusion is proportional to the convection gradient, the difference between the concentrations at two points, divided by the distance between the electrodes. Therefore mathematically

$$\frac{dC}{dt} = D\frac{dC}{dx}$$

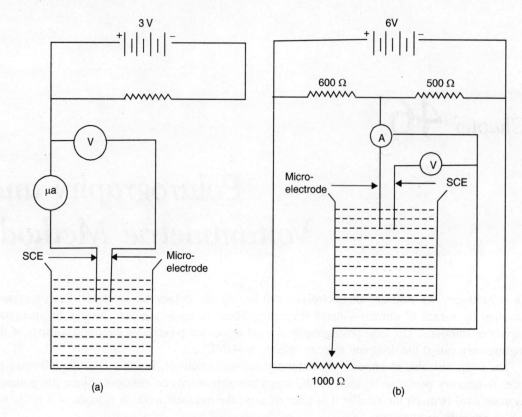

Fig. 46.1 Polarograph circuit

where C = concentration of diffusing species and D = diffusion coefficient and x is distance between two electrodes.

By application of Fick's law to electrolysis we have

$$i_d = nFA\left(\frac{D}{\pi T}\right)^{1/2} C$$

where i_d = electrolysis diffusion current (in microamperes) flowing at a time T (seconds); n = number of electrons involved in the electrode reaction; F = Faraday's constant; A = area of cross section of electrode; D = Fick's diffusion coefficient (sq.cm per second); C = bulk concentration (millimoles/litre). The initial concentration is assumed to be zero at the electrode surface. Both i_d and 'n' are +ve for cathodic reduction and −ve for anodic oxidation. It is important to note the proportionality relationship between the current and concentration. The analysis of current-time curves at constant potential is called 'chronoamperometry'. It is useful in the study of electric transfer kinetics in irreversible systems while a curve of potential *versus* time at constant current is called chronopotentiometry. This is discussed at the end of chapter.

46.2 POLARISABLE DROPPING MERCURY ELECTRODE (DME)

The most common microelectrode is mercury in the form of a succession of droplets emerging from a very fine pore of glass capillary (Fig. 46.2). It has several advantages e.g. high hydrogen formation over voltage of a mercury cathode from H^+ ions which facilitates reduction in acid media without interference. The electrode surface is continually being removed and cannot become fouled or poisoned giving a reproducible current voltage curve. The increasing area of the electrode during the lifetime of a drop more than offsets the decreasing current observed with an electrode of fixed size. It makes quantitative analysis predictable and achieves an immediately reproducible average current at any given potential. One minor limitation is the ease with which mercury is oxidised and hence it cannot be used as an anode. Thus it can only be used for analysis of reducible or very easily oxidisable substances. 10 cm capillary gives a drop time of 3–6 seconds with a mercury head of 50 cm. A significant advantage is the possibility of increasing the head of mercury to give a constant flow through the tip thus facilitating reproducible results.

Metals like zinc and cadmium can be deposited from acid media because of high over voltage of reduction of the hydrogen ion. The behaviour of the electrode is independent of its previous history. Solid electrodes show erratic behaviour on account of deposited impurities. A reproducible average current is instantaneously produced when voltage is brought from a lower to a higher value or vice versa. The main drawback is that mercury gets oxidised hence it can be used as an anode (at +0.4 V). Initially one can see an anodic wave due to the transformation $2Hg + 2Cl \rightarrow Hg\,Cl_2(S) + 2e^-$. It can be used for analysis of chloride ions. It has a nonfaradaic residual or charging current. At very low concentrations DME can easily be clogged due to the small capillary size. A typical DME is shown in Fig. 46.2.

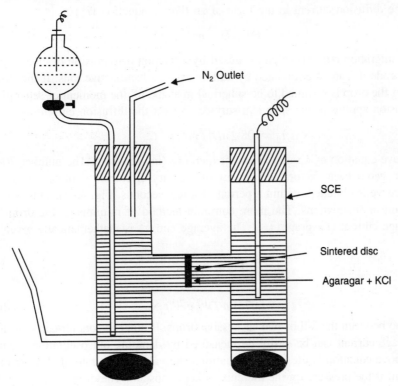

N₂ Outlet

SCE

Sintered disc

Agaragar + KCl

Fig. 46.2 Dropping mercury electrode (DME)

A beaker with a DME immersed in it can be used but N_2 gas must be bubbled to remove the O_2.

The capillary tubes have an internal diameter of 0.03–0.05 mm. Water or aqueous solution should not enter the capillary as it cannot easily be cleaned. The capillary should not be left immersed in water.

46.3 PRINCIPLES OF POLAROGRAPHY

Polarography is exclusively carried out with a DME. The voltage ramp is linear if the drop time is suitably controlled. The range of a DME is 0–2V against a SCE. Over potential is caused by the electrochemical reaction. Anodic currents are negative. The current flowing through the cell, which is limited by the rate of diffusion to the indicator electrode, is thus called a "diffusion current (i_d)". Fast polarography is a variation of polarography in which the current is measured only during the last fraction of a second before the drop falls from the capillary. The current oscillates throughout the polarogram. In the absence of an electroactive species a small current known as the residual current flows. This is due to reduction of impurities inside the solution and the capacitive current or charging current. The diffusion current is the portion of the current controlled by the rate of diffusion of the electroactive species to the surface. It originates due to the concentration gradient. The diffusion current is a difference between limiting current and the residual current, and is an instantaneous current. We measure average diffusion current to the centre of oscillations in polarographic wave.

46.4 THE ILKOVIC EQUATION

We express the diffusion current in the form of an Ilkovic equation *viz.*:

$$i_d = 607 n D^{1/2} m^{2/3} t^{1/6} C$$

We get a migration current which is caused by attraction or repulsion of electroactive ions by the indicator electrode. It can be eliminated by adding excess electroactive ions. If the flow of mercury is constant and if the drop is assumed to be spherical in shape till the moment of separation, then the law of linear diffusion applies to the spherical surface. We get the diffusion current as:

$$i_d = 708.2 n D^{1/2} m^{2/3} t^{1/6} C$$

In the above equation m = rate of flow of mercury (mg/seconds). The number 708.2 includes the Faraday factor, geometrical factors and density of mercury. If current is plotted as a function of time, a fluctuating curve is obtained, within a period of a few seconds (Fig. 46.3). This is not convenient to read on account of fluctuations. The more common method of minimising the drop fluctuation is to measure average current (i average) (i_d). The average current is mathematically given as

$$i_d = \frac{1}{t_d} \int_0^{t_d} i_d \, dt = \frac{6}{7} i_d$$

$$i_d = 607.0 \ n \ D^{1/2} m^{2/3} t_d \ C \qquad \qquad \text{...(Ilkovic equation)}$$

where t_d = time between the fall of two successive drops. If other factors remain constant, then $i_d \propto C$. Thus the average current can be easily be measured by damping out oscillations in the meter or recorder. The above equation is the Ilkovic equation. The value of the rate of flow of mercury (m) will remain constant if the pressure on the mercury is kept constant. The drop time t_d is proportional to V, the pressure of Hg and the applied potential. It is maximum at a potential of −0.5 V measured *versus* a SCE. This is the potential at which the mercury-water interface shows a maximum interfacial tension known as the 'electrocapillary maximum'. Temperature does not enter directly into the Ilkovic equation although each factor (except n) is indirectly dependent upon temperature.

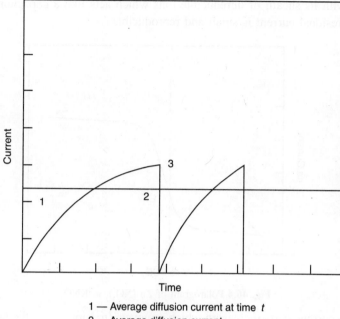

Fig. 46.3 Diffusion current drop time

Fig. 46.3 Diffusion current drop time

The value of i_d (diffusion current) increases at a rate of 1–2% per °C. Hence the electrolysis cell should be kept at a constant temperature. The effect due to the stirring action of fast falling drops which disturb the diffusion layer geometry makes the Ilkovic equation invalid. In such a case the current becomes large especially if the drop time becomes less than 3–4 seconds. The diffusion coefficient (D) varies with the viscosity of the medium, which in turn affects (i_d). The diffusion current (i_d) also varies if the concentration of the supporting electrolyte is less than 25–30 times that of the reducible substance. This is due to the fact that reducible ions carry a fraction of the current called the migration current. Such a current is due to electrostatic attraction or repulsion between the DME and the ions. This can be eliminated by using an excess electroactive sample solution.

46.5 GENERATION OF POLAROGRAPHIC WAVES

So far we have considered current–time curves at a constant potential. Now let us consider the situation when the potential is changed, either stepwise or continuously. The former is a manual procedure. The operator sets the potential, waits for the steady state and records the average current (i_d) then moves to the next desired potential and again records the diffusion current (i_d). The instrument determining the entire current-voltage curves is called a 'polarograph'. The potential applied to the cell is continuously increased in the negative direction (*i.e.* the DME is negative with respect to the SCE) and the current is successively recorded. Since potential is linear with time and the recording paper moves with a constant speed, the resulting curves are called polarograms. In Fig. 46.4 we have a typical polarogram of $Zn(NO_3)_2$ in KNO_3 solution with a DME and a SCE. In region X, the potential is too low to permit any significant reduction. The small current flowing in this region is called the 'residual current', which is due to the reduction of impurities (e.g. Fe, Cu, O_2) and charging current due to the mercury and solution interface with its sheath of unreducible ions which acts like a capacitor of continuously increasing area. The residual current is small and reproducible.

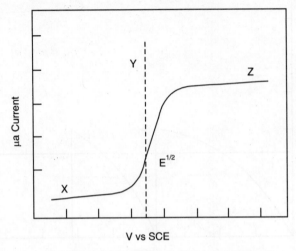

Fig. 46.4 Polarogram of Zn $(NO_3)_2$ in KNO_3

In region Y the current starts increasing which removes Zn^{2+} ions from the layers of electrolyte in contact with the surface of the electrode by reduction of zinc. In the region Z the current shows a saturation effect caused by depletion of Zn^{2+} ions in the vicinity of the DME. More zinc ions reach the electrode by diffusion and are then reduced to the metal. The value of the limiting current at point (Z) is called the 'diffusion current' (i_d) which is directly dependent upon the concentration of the reducible species (after correction for residual current). Subsequently the negative potential may be due to reduction of hydrogen ions or water, by the reaction

$$2H_2O + 2e^- \rightleftharpoons H_2 + OH^- \quad E° = -0.828 \text{ V}$$

46.6 CONCEPT OF HALF WAVE POTENTIAL

The relation for the current as a function of potential can be derived from the Nernst equation provided the reaction at the microelectrode proceeds reversibly. Consider the reduction of a metal cation to a metal soluble in mercury in order to form an amalgam. Since the solutions used are dilute, the activity coefficient of the solution and also in the amalgam is expressed as: $M^{n+} + ne^- \rightarrow M_{(Hg)}$ if M_{Hg} is the metal (M) dissolved in mercury amalgam. The potential at the DME by Nernst equation:

$$E_{DME} = E^0 - \frac{RT}{nF} \ln \left[\frac{[M]_{Hg}}{[M^{+n}]_{aq}} \right]$$

Since the current i is limited by diffusion it follows that

$$i = K \left[[M^{+n}]_{aq} - [M^{+n}]^0_{aq} \right]$$

for limiting current (i_d), $[M^{+n}]^0_{aq}$ is small, hence negligible.

$$\therefore \ i_d = K[M^{+n}]_{aq}$$

From the Illkovic equation, the constant (K) is given as

$$\frac{i_d}{C} = K = 607nD^{1/2}m^{2/3}t_d^{1/6}$$

The concentration of $[M]_{Hg}$ in the amalgam is proportional to the current. Thus we have

$$i = K'[M]_{Hg}$$

where K and K' are identical constants and M is metal ion.

Finally, if $K/K' = \sqrt{D/D'}$. The potential at DME is given by

$$E_{DME} = E^0 - \frac{RT}{nF}\ln\left(\frac{D}{D'}\right) - \frac{RT}{nF}\ln\left(\frac{i}{i_d - i}\right)$$

At the point where $i = (1/2)\, i_d$, last term drops out and we get the expression:

$$E_{1/2} = E^0 - \frac{RT}{2nF}\ln\left(\frac{D}{D'}\right)$$

$E^{1/2}$ is called the 'half wave potential which has great significance. Hence the potential at DME in terms of the half wave potential is

$$E_{DME} = E_{1/2} - \frac{RT}{nF}\ln\left(\frac{i}{i_d - i}\right)$$

The equation for the potential for DME gives the form of the polarographic wave in terms of i_d and $E_{1/2}$ and gives a method to evaluate n which is equal to the slope of the tangent to the curve at the half wave point $\left[(\tan\phi) = n\right]$ by plotting a graph of

$$\log\frac{1}{i_d - i} \ \text{against} - E_{DME} \ \text{if slope} = \frac{0.0591}{n} = \frac{2.303RT}{nF}$$

In the presence of a complex former, the half wave potential is shifted to more negative values with the relation.

$$E_{1/2} = E^0 + \frac{RT}{nF}\ln K_c - p\frac{RT}{nF}\ln(X)$$

where K_c = instability constant of complex; $[X]$ = concentration of complexing agent; p = number of moles of X which combine with one mole of the metal (M). Table 46.1 shows the effect of complexation on halfwave potential.

Since the half wave potential $(E_{1/2})$ is typical of the substance undergoing oxidation or reduction at the microelectrode, it is utilised to identify a substance (Fig. 46.5). The value of $E_{1/2}$ for a species depends upon the nature of the supporting electrolyte, largely because of variation in the tendency to form complex ions. The values of $E_{1/2}$ for Cd are (−0.6 V in KCl); (−0.81 V in NH_4Cl); (−0.78 V in

NaOH); (-1.88 V in KCN). In (1M) NaOH different metals have different values for $E_{1/2}$ as given below:

Cd^{2+} (-0.078 V); Cs^{2+} (-1.46 V); Cu^{2+} (-0.41 V); Fe^{3+} (-1.12 V); Pb^{2+} (-0.76 V); Zn^{2+} (-1.53 V).

Table 46.1 Effect of complexation on half wave potential ($E_{1/2}$) for various metals

S. No.	No Complexation $E_{1/2}$	Complexed as chloride complex	Complexed as cyanide complex	Cation complexed divalent	Nature of complex species
1.	-0.60	-0.64	-1.98	Cd^{2+}	$CdCl_4^{-2}\,\left[Cd(CN)_4\right]^{2-}$
2.	$+0.02$	-0.04	$-$	Cu^{2+}	$CuCl_3^-,\left[Cu(CN)_4\right]^{2-}$
3.	$-$	-1.20	-1.45	Co^{3+}	$CoCl_6^{-3},\left[Co(CN)_6\right]^{3-}$
4.	$-$	-1.20	-1.36	Ni^{+2}	$NiCl_3^-,\left[Ni(CN)_4\right]^{2-}$
5.	-0.40	-0.44	-0.72	Pb^{2+}	$PbCl_4^{2-},\left[Pb(CN)_3\right]^-$
6.	-1.0	-1.00	$-$	Zn^{2+}	$ZnCl_4^{2-},\left[Zn(CN)_4\right]^{2-}$

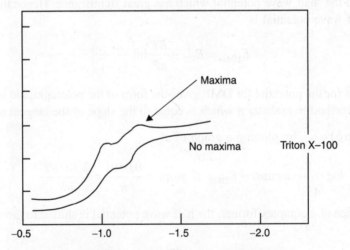

Fig. 46.5 Polarogram of Zn $(NO_3)_2$ in KNO_3

The right selection of electrolyte is thus important. The rapid method envisages plotting a polarogram of known and unknown. The half wave potential can be determined graphically as shown in Fig. 46.6. In this figure AB, DF extended, a tangent is drawn to the curve BD at a point C, GH is bisected and the line *JK* drawn parallel to AB and DF. The abscissa of the point of intersection JK with the curve at C gives the value of $E_{1/2}$. This is a rough graphical estimation of value of $E_{1/2}$.

46.7 POLAROGRAPHIC MAXIMA

If the applied potential is raised, the current, after increasing as the anion is discharged, fails to level off but decreases again, leaving a maximum in the curve (slight hump or sharp peak) (Fig.46.5) Such maxima can be eliminated by the addition of an organic surfactant e.g., gelatin or methyl red dye or a nonionic detergent such as triton X-100 which is a powerful maxima suppressor (0.2% about 1 ml is added to 10 ml of solution). Too much suppressor should not be added. The dissolved oxygen is reducible at the microelectrode in many media. It gives rise to two waves, the first caused by reduction of O_2 to H_2O_2 the second by reduction of O_2 to H_2O. Such interference causes trouble in analytical determinations. In alkaline media O_2 is removed by the addition of sulphite. Flushing of N_2 of course removes oxygen. Freon 12 can also be for flushing.

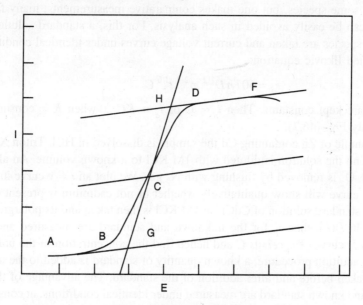

Fig. 46.6 Typical polarogram

A manual polarograph can easily be constructed in the laboratory using electrical circuits as shown in Fig 46.7. The ammeter or galvanometer should have a range of 50 μA, with shunts to measure large currents. Photographic recording instruments are also available, but those instruments with a strip chart recorder are most popular. The current and voltages are standardised with a self contained standard cell; the recorder is scanned at speed of 1 cm/min; maxlength = 25 cms.

The voltage (V) applied to a polarographic cell is $V = E_A - EC + I \Sigma R$ where E_A = potential of anode, E_C = cathode potential; ΣR = ohmic potential drop in the system. The resistance includes that of the solution of each electrode, the column of Hg, the slide wire and current measuring resistor. All three can be controlled except the resistance of the solution. Nonaqueous or mixed solvents require a high resistance for IR to become significant. There are methods to give correction for this potential drop, e.g. the use of a three electrode cell, which has a DME, a SCE and a counter electrode. It also has an operational amplifier circuit.

46.8 APPLICATIONS OF POLAROGRAPHY

Polarography is used for the qualitative identification and quantitative analysis of metals. The applications in inorganic as well as in the organic analysis are well known. Let us consider how it is used for the quantitative analysis of metals.

The magnitude of the diffusion current is related to the concentration of reducible species by the Ilkovic equation.

$$i_d = 607.0\, n\, D^{1/2}\, m^{2/3}\, t_d^{1/6}\, C$$

If several factors of that equation are known or if such factors can be measured, then the concentration of a species can be computed theoretically from the observed current. D (*i.e.* diffusion coefficient) cannot easily be evaluated. It can be easily found by chronopotentiometry or by measurement of conductance of the solution. D is sensitive to the viscosity and temperature of the solution. D is given in the literature for some species, but one makes comparative measurements; many factors which are difficult to handle can be easily avoided in such analysis. For this, a standard solution of the desired species and known species are taken and current voltage curves under identical conditions are plotted. We apply a simplified Ilkovic equation:

$$i_d = 607\, n\, D^{1/2}\, m^{2/3}\, t_d^{1/6}\, C$$

where n, D, m, t_d are kept constants. Then $i_d \propto C$ or $i_d = KC$ ... when K = constant which can be evaluated graphically Fig. (46.7).

To analyse a sample of Zn containing Cd the sample is dissolved in HCl, Triton X–100 is added to eliminate maxima and the solution is diluted with 1M KCl to a known volume. An aliquot of solution is placed in the cell, O_2 is removed by flushing with N_2 gas. We plot an $i \propto v$ curve in the region (−0.4 to 0.8 V). The trial curve will show qualitatively whether or not cadmium is present (wave at −0.64V given if present). A standard solution of $CdCl_2$ in 1M KCl is then taken and its polargram is also plotted as shown in Fig. (46.7)(b), then i_d for the unknown and standard are measured and we construct a calibration (Fig. 46.7) curve of i_d versus C and hence find the concentration of the unknown sample of Cd. In the standard addition procedure, a known quantity of standard is added to the unknown solution and a polarogram taken before and after addition of the standard. The advantage of this method is the unknown solution and known standard are measured under identical conditions, at constant temperature. During quantitative analysis one must take into account the 'residual current'. If two waves are close together, another supporting electrolyte can be used, e.g. Cd^{+2}, Pb^{+2} cannot be distinguished in NaOH but can be differentiated in KCN in analysis.

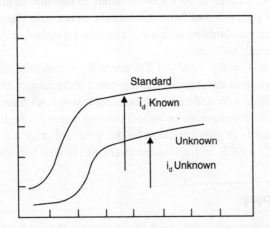
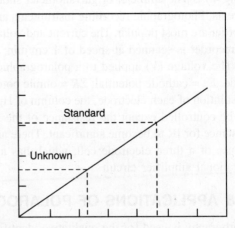

Fig. 46.7 Quantitative analysis

Since D varies from ion to ion there is no uniformity in the height of the waves obtained with equivalent concentrations of the different reducible ions. A plot of current voltage or diffusion current

constant $(I) = \dfrac{i_d}{C m^{2/3} t^{1/6}}$ will be typical. The Ilkovic equation predicts constancy for this ratio for any

given temperature. Its value gives the relation between the diffusion current and the concentration we might determine (m) and (t) for a new capillary. Once a series of constants are determined, it is necessary to repeat only the determination of the constant for one ion to establish those of the entire series for a new capillary. A single standard stock solution can be used. It is called the pilot ion method.

46.9 AC POLAROGRAPHY

The voltage ramp used in AC polarography consists of a small amplitude, sinusoidally alternating potential applied on top of the normal polarographic ramp. It has a value of 5–50 mV with a frequency of 10–100 Hz. An alternating current flows through the cell due to AC potential which in turn is measured and an AC polarogram obtained by plotting the AC current as a functional of DC potential (Fig. 46.8).

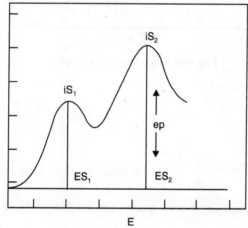

Fig. 46.8 An AC polarographic voltage ramp

For a redox reaction, the AC polarogram is a symmetrical peak. The peak height (*ep*) is related to concentration by

$$ep = \frac{2.5\,n^2\,F^2 D^{1/2}\,\Delta E N^{1/2}\,C}{\Delta RT}$$

In the above equation, ΔE is the amplitude (v) of the superimposed alternating potential; A – area; N – frequency of AC potential while the other terms have the usual significance. One can use a phase sensitive lock-in amplifier to measure the AC current. The detection limit is about 10^{-4} m. AC polarography is useful for kinetic studies of electrode reactions. It has added sensitivity and permits analysis of traces of reducible species in the presence of a large concentration of a substance giving a wave at a lower potential. However, we need complex equipment.

46.10 RAPID SCAN POLAROGRAPHY

Rapid scan polarography involves rapid scanning of a polarogram within the lifetime of a one single drop. The rate of change of voltage required is large and can be sought by an electronic ramp generator. The curve obtained in such a technique is shown in Fig. 46.9. It gives a rapid scan polarogram with the reducible species, as the slow process of diffusion is unable to supply reducible material to electrode fast enough to keep up the rapid potential increase (thus a steady state is not attained.)

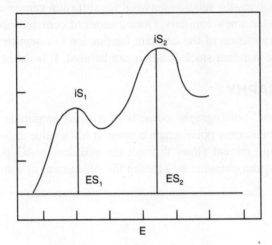

Fig. 46.9 Rapid scan polarography

The summit potential E_S is given as:

$$E_S = E_{1/2} - 1.1 \frac{RT}{nF} \dots \text{ and at } 25°C \text{ it is } E_S = E_{1/2} - \frac{0.028}{n} \text{ volts}$$

$$i_s = K n^{3/2} m^{2/3} D^{1/2} \left(\frac{dE}{dt} \right)^{1/2} C \quad \text{(Randle and Servek equation)}$$

The diffusion current is i_s summit if both species are soluble in Hg or H_2O.

46.11 ORGANIC POLAROGRAPH

This is an important tool in elucidation of structure of organic compounds. Organic products are insoluble in mercury but are soluble in suitable solvents or solvent mixtures. A solvent dissolving the electrolyte is used. Alcohols, ketones, dimethylformamide, urea, ammonium formate are good water soluble solvents. Conjugated unsaturated compounds, carbonyl compounds, quinones, and azo compounds and heterocyclic compounds can be reduced at the DME in LiOH in aqueous solutions. C_2H_5OH gives $E_{1/2}$ for some organic substance as follows: benzaldehyde (-1.51 V); n-propylphenylketone (-1.75 V); isopropylphenyl ketone ($-1,82$ V); tert-butylphenylketone (-1.92 V). The concentration was varied from 0.2 to 2.5 mM while the pH was controlled.

Thus organic polarography is useful for elucidation of structure. Here $E_{1/2}$ is dependent on the pH of the solution as the latter is related to the reaction product. Hence good buffering is necessary to get a good $E_{1/2}$ and reproductible diffusion current. Water soluble solvents are used. Typical functional groups like carbon, produce characteristic polarographic waves. The carbon group gives a wave. A few carboxylic acids are polarographically reduced e.g. dicarboxylic acids like malonic acid. The peroxides and epoxides are reduced. Nitro amine, oxide, nitroso and azo groups also undergo reduction. Organic halogen groups give waves. The $> C = C <$ double bond is reduced. Mercaptans and hydroquinone produce anodic waves. A reactions of organic compounds at microelectrodes are very slow. The reaction shows $R + nH^+ + ne^- = RH_n$ wherein R and RH_n are oxoidized and reduced forms of organic molecules. Electrode reactions using the hydrogen ion will change pH at the electrode surface thereby influencing reduction potential, causing poorly defined waves. This problem can be solved by proper buffering of the solution to give reproducible $E_{1/2}$.

46.12 PULSE POLAROGRAPHY

These are of two types: normal and differential pulse polarography (See Figs. 46.10 and 46.11).

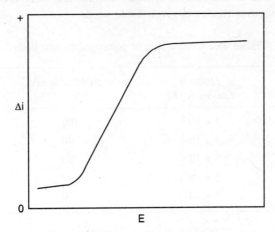

Fig. 46.10 Normal pulse polarogram

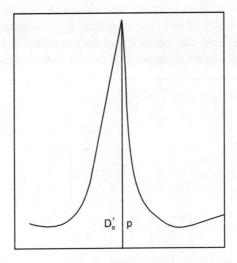

Fig. 46.11 Differential pulse polarogram

In normal pulse polarography the size of the pulse is increased linearly with time. Current is sampled in the last 5–20 msec. of the drop when the increase in current is least. It gives better sensitivity.

Here $i = \dfrac{n\text{FA}D^{1/2}C}{\pi^{1/2}S^{1/2}}$ if S is the time for which the pulse is applied.

In differential pulse polarography a DC potential is increased linearly with time at a rate of increase of 5 mV/sec. A DC pulse of about 20–100 mV is applied for 60 msec before detachment of a drop of Hg. Two current measurements, one prior to the pulse and another after the end of the pulse are made. The difference in current per pulse (Δi) is plotted as function of increasing voltage. The height of this pulse is proportional to concentration (Fig. 46.11). We can monitor ((Δi) for different peak maxima of individual reducible species at intervals of 0.04–0.05 V. It thus increases the sensitivity of method. This is primarily due to increase in faradaic current and decrease in non faradaic charging current. A comparison of various modes is given Table 46.2.

Table 46.2 Comparison of polarographic methods

	Type	Limiting Sensitivity (F)	Resolvable mV	Separatability
1.	Conventional –DC	1×10^{-3}	100	10 : 1
2.	Rapid Scan	3×10^{-7}	40	400 : 1
3.	AC	5×10^{-7}	40	1000 : 1
4.	Square wave	5×10^{-8}	40	20,000 : 1
5.	Pulse	1×10^{-8}	40	10,000 : 1

46.13 SQUARE WAVE POLAROGRAPHY

In this technique the potential ramp is a square wave imposed on a staircase ramp (Fig. 46.17). The height of each step ΔE_S is typically 10 mV. The amplitude E_{sw} is adjusted to 50/nmV E_{sw} is adjusted to 50 mV; for two electrons in 25 mV. The period (τ) of each staircase is twice the time (t_p). The current is measured at the end of each square wave pulse at time (t_1) and before application of the next pulse

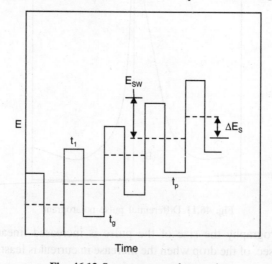

Fig. 46.12 Square wave polarography

at time (t_2). We have a plot Δ_c, difference in $(i_{t1} - I_{t2})$, *versus* applied potential. We get a net polarogram similar to that shown in Fig. 46.11. The square wave frequency is between 10–100 Hz with an average of 206 Hz. The main advantage of the square wave polarogram is speed. As before $\Delta ip \propto C$. One can also use a DME.

46.14 AMPEROMETRIC TITRATION

It is possible to carry out a titration in an electrolytical cell and follow its progress by observing the diffusion current after successive addition of a reagent. Such a titration is called an 'amperometric titration'. Since $\Delta i_d \propto C$, two straight lines are obtained with a point of intersection at the equivalence point. Typical curves are shown in Fig. 46.13. Here in curve (1) results are for the titration of a reducible ion by a reagent which itself gives a polarographic wave, e.g. Th^{4+} with $H_2C_2O_4$ at DME and SCE in KNO_3. The current decreases as Th^{4+} is removed from the solution. The reverse titration of oxalate by thorium gives a curve similar to curve (2) in Fig. 46.13. Thus Th^{4+} gives a diffusion current until all oxalate is precipitated (e.g. Ba against $K_2Cr_2O_7$ at 0 V here against DME and SCE, Ba^{2+} is not reduced but $Cr_2O_7^{2-}$ gives different current). Curve (3) is an instance of titration of Ba with $K_2Cr_2O_7$ at -1.0 V where both Ba^{2+} and $Cr_2O_7^-$ are capable of reducing and giving a diffusion current. Since a diffusion current i_d is decreased by the addition of solvent, any amperometric titration data must be corrected for the dilution effect resulting from reagent addition. It is possible to do so by multiplying the diffusion current by the factor $\left(\dfrac{V + \bar{V}}{V}\right)$ where V = original volume, $\bar{V}$ = volume of reagent added. This can also be avoided by using a reagent solution ten times more concentrated than the solution to be titrated so volume addition will really be negligible compared to the total volume. Otherwise curved lines will be obtained. Too greater solubility of the precipitate will cause a problem at the equivalence point

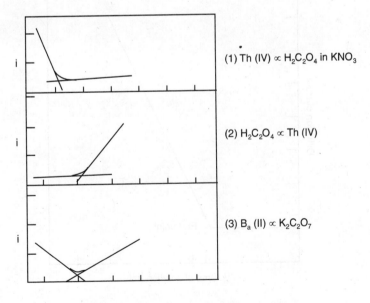

(1) Th (IV) $\propto H_2C_2O_4$ in KNO_3

(2) $H_2C_2O_4 \propto$ Th (IV)

(3) B_a (II) $\propto K_2C_2O_7$

Fig. 46.13 Amperometric titration curves

$(10^{-4}\text{–}10^{-5}$ M tolerable solubility). One can use 5% alcohol to reduce solubility. Table 46.13 describes a typical example of amperometric titrations. Except acid and base, all type of titrations are possible. Selection of the electrode is critical. Amperometric titrations are applicable to precipitation as well as complexation reactions. The results are good and reproducible (Fig. 46.13).

Any polarographic instrument can be used to follow titrations. Temperature should be controlled. Time required is 10 min. A large number of reducible species produce a diffusion current, e.g. Ag^+ and CrO_4^{-2}. The main advantages are: (*a*) calibration of galvanometer is not needed, (*b*) applied voltage is given to nearest tenth of a volt, (*c*) specific characteristics of the capillary are also without effect, (*d*) no polarizing unit need be used, (*e*) we can select the reference electrode such that diffusion potential will appear without any external voltage source, (*f*) a suitable half cell can easily be used for the purpose.

Table 46.3 Various amperometric titrations

S.No.	Type	Electrode	Titrant	Analyte
1	Complexometry	DME	EDTA	Cu^{2+} Cd^{2+}, Ca^{2+}
2	Redox	Rotating Pt electrode	$KBrO_3$, KBr	As^{3+}, Sb^{3+}
3	Precipitation	DME	$K_2Cr_2O_7$	Pb^{2+}, Ba^{2+}
4	Precipitation	Rotating Pt	Ag NO_3	Cl^- Br^- N^- SCN^-
5	Organic precipitate	DME	Oxine	Mg^{2+} Al^{3+}, Zn^{2+}

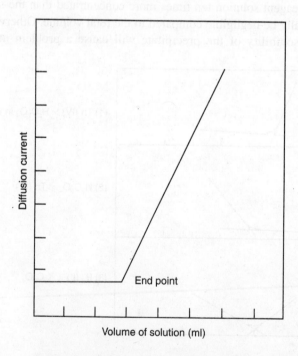

Fig. 46.14 Titration with rotating electrode

The substitution of a rotating platinum electrode for a DME gives increased sensitivity because of the partial disruption of the diffusion layer. A 2–3 mm Pt wire is suspended in a vertical glass tube. This tube is rotated at constant speed. Such an electrode greatly reduces the residual current. In such a titration (Fig. 46.14) we add the reagent 3–4 times to get the current. The extrapolation of the line and point of intersection on the volume axis, furnishes the equivalence or end point.

Amperometric titrations have wide applications compared to potentiometric titrations. In another kind called biamperometric titration one impresses a small potential across two identical inert electrodes, such as platinum electrodes. This is widely used in the Karl Fischer titration. Let us consider biamperiometric titrations in more details.

46.15 BIAMPERIOMETRIC TITRATION

This is technique of generation of dual diffusion current taken and small potential is imposed on them. (approx 50–100 mv). Two platinum electrodes and galvanometer are put in the circuit. No current flows in the absence of oxidation or reduction at electrodes. However, Redox couple will provide two electrons since process are reversible. The transformations are indicated in Table 46.4 where we use Pt-Pt as cathode-anode couple.

Table 46.4 Reversible couples in titrations

S.No.	Cathode (Reduction)	Anode (Oxidation)	Titrant
1.	Fe (III) and Fe (II)	Fe (II) $\rightarrow$ Fe (III)	Ce (IV) soln.
2.	$MnO_4^- - MnO_4^{2-}$	$Mn^{2+} \rightarrow MnO_4^{2-}$	–
3.	I_2	I^0	$Na_2S_7O_3$ soln.
4.	Br_2	Br^0	–
5.	Ce^{4+}	Ce^{3+}	–
6.	$\left[Fe(CN)_6^{3-}\right]$	$\left[Fe(CN)_6^{-4}\right]$	Ce (IV) soln.
7.	Ti^{4+}	Ti^{3+}	–
8.	VO_2^-	VO^{++}	Fe (II) soln.
9.	$H_2O_2 \rightarrow OH^-$	$H_2O_2 \rightarrow O_2$	–

A few titrations can be carried out in this system. In Table 46.4 the reactions at cathode (reduction) and at anode (oxidation) and kind of titration used is indicated. These are all dead stop titration. The current flows till equivalence point is attained with change of as small as 15 mV potential. In absence of free iodine, current cannot flow. The reaction listed in serial no (3) is used in Karl Fischer titration for the purpose of analysis of moisture. In first titration, the current flows after equivalence point as Fe (II)-Fe (III) or Ce (III) – Ce (IV) form reversible couples. The end point in all these titrations are sharply defined since current drops to zero at equivalence point.

46.16 POTENTIOSTAT

This is most essential component of any voltammetric devices. Its function is to monitor current at known fixed potential. It contains integrated circuit operation operational amplifier and digital modules. A simple electronic circuit for potentiostat is shown in Fig. 46.15. The output of Amplifier (1) is contacted to counter electrode with feedback to its own inverting output through the reference electrode. The feedback decreases the difference between the inverting input and noninverting input amplifier and leads to establishment of same potential as $E_{entrance}$ of amplifier I. Since difference in two electrodes is zero, the reference electrode is connected to $E_{entrance}$ through high impedance. In amplifier 3 the current flows in counter electrode. The three electrode setup is preferred (from micro to milliamp scale). With microelectrodes the current is in the range of n (nano) 10^{-9} to p (pico) 10^{-12} ampere range. The operational amplifier acts as a current to voltage converter (amplifier 2) to provide output signal to converter.

Most voltammetric methods are dynamic. Exact control of the potential is critical function of potentiostat. With digital electrons most of them work in digital pattern. Digital fabrication of the applied voltage has given rise to pulsed voltammetry. The latter is fast and sensitive. A commonly used wave form model is used which is of linear scan type. Potentiostat is essential if one prefers to use micro and nanometer size electrodes leading to pico or nano ampere range current. Square wave polarography needs rapid response period from the electronics. Wide range of potentiostats are available in the market.

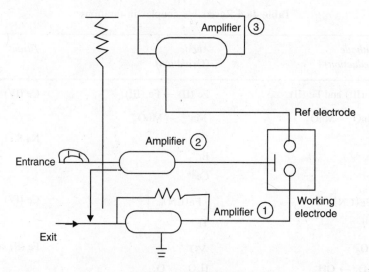

Fig. 46.15 Potentiostat

46.17 HYDRODYNAMIC VOLTAMMETRY

It is carried out in several ways. The solution is vigorously stirred with microelectrode in contact with solution. There is a assembly as shown in Fig. 46.16. We have three electrodes viz. SCE, working electrode and counter electrodes. The analytical solution can be made to pass through tube by flow. During electrolysis the reactant is carried to surface of the electrode by migration, convection and process of diffusion. The efforts are made to reduce migration by adding excess of inactive supporting electrolyte when concentration of supporting electrode rises more than analyte concentration (50-100 fold). Fraction of current carried out by the analyte becomes negligible. The rates of migration of the analyte towards electrode of opposite charge is rendered independent of applied voltage.

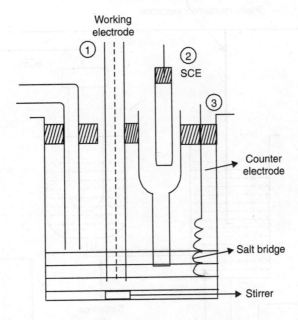

Fig. 46.16 Hydrodynamic model

With advancement in direct polarographic techniques with automated facilities, the hydrodynamic voltammeter does not offer any special advantage and therefore it is not commonly used.

46.18 CHRONOPOTENTIOMETRY

This method is related to polarography in that it is diffusion controlled, but it does not involve a study of current voltage curves. It is actually potentiometry at constant current. Consider a three-electrode cell (Fig. 46.17) with a small constant current passing through the working and counter electrodes, while the potential of the working electrode is controlled against a SCE. The working electrode employed is a pool of mercury; platinum is used as the counter electrode and a SCE is used with the capillary tip close to the working electrode. The solution is not stirred. The solution consists of an inert supporting electrolyte with a low concentration of cations. When the switch is closed at the current source, the potential of the working cathode rises from 0 V to a negative value to permit reduction of metal to take place. It will stay at this value and rise when cation is depleted in the vicinity of the electrode (point Y on Fig 46.18). Then it will again rise sharply until it is negative, to reduce the next reducible ion e.g. H_2O at point Z.

The time needed to deplete the surface layer (XY) is called the transition time (Y)' and it is related to the bulk concentration of the reducible species by the Sand equation

$$T_{1/2} = \frac{\pi^{1/2} \, nF \, A D^{1/2} \, C}{2I}$$

where n = number of electrons transferred; F = Faraday's constant; A = area; C = concentration; D = Fick's diffusion coefficient; I = diffusion current passed; T = transition time.

This permits calculation of the concentration in terms of known or measurable parameters. This extends this method for quantitative analytical applications.

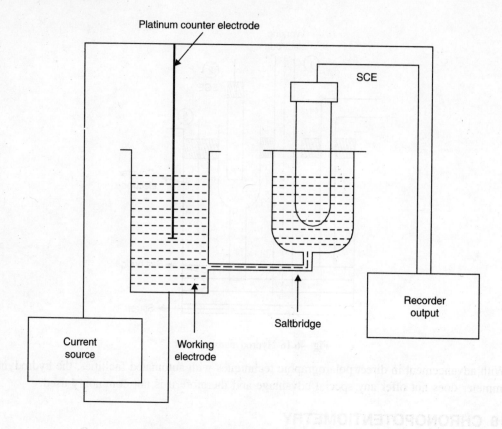

Fig. 46.17 Chronopotentiometry assembly

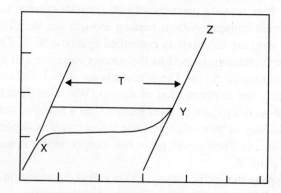

Fig. 46.18 Chronopotentiometry curve

The potential corresponding to the X–Y portion of the curve can be related to the reduction potential of the electroactive species. Karaglonoff equation showed that for a system where oxidised and reduced species were soluble the potential of the working electrode E_{we} was given by an expression:

$$E_{we} = E^0 - \frac{RT}{2nF}\ln\left(\frac{D}{D'}\right) - \frac{RT}{nF}\ln\left(\frac{t_{1/2}}{\tau^{1/2} - t^{1/2}}\right)$$

where R = universal gas constant; T = absolute temperature; F = Faraday's constant; D, D' = Fick's diffusion coefficient; τ = Transition time

$$t^{1/2} = \left(\tau^{1/2} - t^{1/2}\right) \text{ or } t = \tau/4$$

Since $E_{1/2} = E^0 - \left(RT/2nF\right)$ In $(D/D'$ if we write $E\ \tau /4$ for the value of E^0 when $t = \tau/4$ then

$$E_{we} = E\tau/4 - \frac{RT}{nF} \ln\left(\frac{t_{1/2}}{\tau^{1/2} - t^{1/2}}\right)$$

This equation is similar to the polarographic equation.

Finally let us consider the practical aspects of this technique. The main drawback of this method is that the cell cannot be easily changed like the change of cell in polarography because the geometry of the working electrode and the reaction with capillary reference half cell are optimum. Any change in the surface of the electrode due to adsorption or deposition will interfere with the diffusion phenomena and must be carefully watched. This technique is best suited for kinetic studies of chemical reactions. One can ascertain whether a process is diffusion controlled by chronopotentiometry. The process is the same even if the current is changed. Further, the current density (I/A) can be varied by a suitable amount and its impact on the electrode reaction studied. In conclusion one can state that, for quantitative analysis, chronopotentiometry is not very useful when compared to the polarographic technique.

PROBLEMS

46.1 Calculate the diffusion coefficient of thallium (I) if the transition time for a 50×10^{-3} M solution is 22.1 seconds. Area of working electrode (A) is 2.4 cm², current = 144 μ a and number of electrons involved = 2.

Answer: We use the chronopotentiometry equation

$$T_{1/2} = \frac{\pi^{1/2} n F A D^{1/2} C}{2I} \text{ on substituting values}$$

$$22.1 = \frac{\sqrt{(3.14)} \times 2 \times 96,500 \times 2.4 \text{ cm}^2 \times \sqrt{D} \times 5 \times 10^{-2}}{2 \times 144}$$

$$\therefore \quad \sqrt{D} = \frac{22.1 \times 2 \times 144}{\sqrt{3.14} \times 2 \times 96,500 \times 2.4 \times 5 \times 10^{-3}} = 6.5 \times 14^{-5} \text{ cm/sec.}$$

46.2 The diffusion current constant (I) for an zinc is 8.25 when m = 32.5 mg/sec. and t = 4.3 seconds. If the diffusion current for unknown solution of zinc is 4.3 μa what is the concentration of zinc?

Answer: We use an expression

$$I = 607\ n D^{1/2} = \frac{i_d}{C \times m^{2/3} \times t^{1/6}} \text{ with usual notations.}$$

In the given problem, I = 8.25, i_d = 4.3, m = 32.5 and t = 4.3

So $$I = 8.25 = \frac{4.3}{C \times (32.5)^{2/3} \times (4.3)^{1/6}}$$

$$C = \frac{4.3}{8.25 \times (32.5)^{2/3} (4.3)^{1/6}} = \frac{4.3}{8.25 \times 10.18 \times 1.275}$$

Hence $\quad C = \dfrac{4.3}{8.25 \times 10.18 \times 1.275} = 4.02 \times 10^{-2} \, M$

46.3 What is the diffusion current value of cadmium if
$C = 3 \times 10^{-3}$ moles/litre, $D = 0.72 \times 10^{-5} \, cm^2/sec$, $m = 3$ mg/sec. $T = 4$ seconds ?
($n = 2$ for Cd)

Answer: By Ilkovic equation, we have

$$i_d = 607 \, n \, C \, D^{1/2} \, m^{2/3} \, t^{1/6}$$

Substituting the values we have

$$i_d = 607 \times 2 \times 3 \times 10^{-3} \times (0.72 \times 10^{-5})^{1/2} (3)^{2/3} (4)^{1/6}$$

$$i_d = 607 \times 2 \times 3 \times 10 \times (8.4 \times 10^{-3})(9)^{1/3} (4)^{1/6}$$

i.e., $\qquad i_d = 607 \times 2 \times 3 \times 3^{2/3} \times 4^{1/6} \times (0.72 \times 10^{-5})^{1/2}$

$$= 3642 \times 9^{1/3} \times 4^{1/6} \times (0.000072)^{1/2}$$

$$= 3642 \times 2.08 \times 8.4 \times 10^{-3} \times 2.5 \times 10^{-7} = 25.4 \, \mu A$$

PROBLEMS FOR PRACTICE

46.4 In a particular polarographic determination, if $m = 9$ seconds for 2 drops (*i.e.* $m = 4.5$ sec) and if each drop weighs 6.1×10^{-3}, what is the product of the last two terms (*i.e.* $m^{2/3} \, t^{1/6}$) in the Ilkovic equation?

46.5 In a particular polarographic analysis of copper $C = 1 \times 10^{-4} \, M$, $i_{d1} = 17.5 \, \mu a$, and for unknown solution of copper $i_{d2} = 27.9$. What would be its concentration (C_2)?
(Hint: With other factors constant, $id_1/id_2 = C_1/C_2$).

46.6 A certain solution with $C = 1 \times 10^{-3}$ mole/litre has $i = 6.0 \, \mu a$, $I = 4.0$ and $m = 1.25$ mg/sec. What is the drop time?

46.7 A polarographic analysis of mercury with an equal volume of $5.50 \times 10^4 \, M$ solution and unknown gave a diffusion current of $58.5 \, \mu a$, while the unknown solution alone gave a diffusion current of $47.4 \, \mu a$. Calculate the concentration of mercury in the unknown sample.

46.8 In a chronopotentiometric analysis the various parameters had the following values:
$n = 6$, $D = 1.4 \times 10^{-5} \, cm^2/sec$; $\tau = 41$ sec; $i = 1.50$ ma;
$A = 1.60 \, cm^2$, what is the concentration of the solution?

46.9 What is the diffusion current which flows through a cell at -1.0 V if the capillary is 1.92, $C = 2 \times 10^{-3} \, g/l$ and $D = 0.72 \times 10^{-5}$?

46.10 What is the concentration of cadmium in a solution if $D = 0.72 \times 10^{-5} \, cm^2 \, sec^{-1}$, $m = 2.0$ mg/sec; $t = 4.4$ second and $i_d = 10 \, \mu a$?

LITERATURE

1. M. Kolthoff and J.J. Lingane, *"Polarography"*, 2nd Ed., Interscience, New York (1952).
2. L. Meites, *"Polarography Technique"*, 2nd Ed., Interscience, New York (1965).
3. J. Heyrovsky, *"Principles of Polarography"*, Academic Press, New York (1966).
4. R.C. Kapoor, B.S. Agarwal, *"Principles of Polarography"*, Wiley Eastern (1991).
5. M.R. Smyth, J.G. Vos, *"Analytical Voltammetry"*, Elsevier Science Publ. (1992).
6. J.T. Stock, *"Amperometric Titrations"*, John Wiley (1965).
7. P. Zuman, *"Topics in Organic Polarography"*, Pergamon Press (1970).
8. A.M. Bard, *"Modern Polarographic Methods in Analytical Chemistry"*, Marcel Dekker (1980).
9. F. Vydra, K. Stulik, E. Julakova, *"Electrochemical Stripping Analysis"*, Halsted (1977).
10. L. Meites, *"Handbook of Analytical Chemistry"*, McGraw Hill (1976).

LITERATURE

1. M. Robert and J.L. Lauvent, "Polarography," 2nd Edn, Interscience, New York (1975).
2. J. Weber, "Polarography Principles," 2nd Edn, Interscience, New York (1965).
3. L. Bauman, "Techniques of Polarography," Academic Press, New York (1964).

Chapter 47

Conductometric Methods

In previous chapters we have considered potentiometry where both electrodes are inert, and in polaro-graphy where one of the electrodes is polarisable; this chapter will dwell on the situation where both electrodes are polarisable. The electrical properties of solutions which are not dependent on the occurrence of electrode reaction are discussed in this chapter.

47.1 PRINCIPLES OF ANALYSIS

According to Ohm's law $I = E/R$, where I = current in amperes, if E is in volts and R is resistance in ohms. This law is valid if specific electrode reactions and diffusion limitations are eliminated. The conductance, L, is defined as the reciprocal of the resistance so that $I = EL$. The unit of conductance is 'mho', reciprocal of ohms. The conductance of a solution depends inversely upon the distance, d, between the electrodes and directly upon their areas, and the concentration, C_i, ions per unit volume of the solution and on the equivalent conductance of the ions in question λ_i. Therefore

$$L = \frac{a}{d} \sum_i C_i \lambda_i$$

The summation symbol $\sum$ points to the fact that the contributions to the conductance of the various ions present are additive. Since 'a' and 'd' are in cms, C should be in ccs or mls. If the concentration is expressed as normality (equivalents per litre) we must introduce a factor of 1000. In conductometry since the geometry of the vessel is not easily measurable we substitute $\theta = \dfrac{d}{a}$. It has a cell constant value and is called the cell constant. By substituting the above values we have:

$$L = \frac{\sum C_i \lambda_i}{1000\,\theta} = \frac{\sum C_1 \lambda_1}{1000} \frac{a}{d}$$

The summation $\sum_i C_i \lambda_i$ can be replaced by $C\lambda$, where λ is the equivalent conductivity for the case of a single ionised compound in solution. Hence equivalent conductivity is equal to the sum of the equivalent ionic conductivities $\Lambda = \sum C_i \lambda_i$ (Table 47.1).

In order to compare the values of conductance obtained with various electrode assemblies, the conductance may be replaced by the conductivity or specific conductance (K) defined as

$$K = L\left(\frac{d}{a}\right) = L\theta$$

The cell constant can be determined experimentally using the above equation as $\theta = K/L$. Conductance measurements are carried out on a solution of known specific conductance. Potassium chloride is mostly utilised as the reference solution (See Table 47.2).

The values for the specific conductance (K) at 20°C for different concentration (Fig. 47.3) of KCl are given in Table 47.3.

Table 47.1 Equivalent conductance (λ) at 25°C

Cation	$\lambda°_m \ cm^2/mol$	Anion	$\lambda°_m \ cm^2/mol$
Na^+	50.1	OH^-	198.6
K^+	73.5	I^-	76.8
NH_4^+	73.4	Br^-	78.1
Ca^{2+}	119	Cl^-	76.4
Ba^{2+}	127	F^-	55
Fe^{3+}	204	NO_3^-	71.4
Co^{2+}	106	SO_4^{2-}	160
Cu^{2+}	107	PO_4^{3-}	210
Zn^{2+}	106	CO_3^{2-}	139
Pb^{2+}	139	SCN^-	66
Ag^+	61.9	ClO_4^-	67
H^+	344.9	CH_3COO^-	41

Table 47.2 Specific conductance of potassium chloride

Conc. g/kg	Conductance mho/cm
71.135	0.11134
7.419	0.01285
0.745	0.00140

Table 47.3 Specific conductance of potassium chloride at various concentrations

Concentration (M)	Conductance s/cm (K)
1.0	0.111
1.0×10^{-1}	1.28×10^{-2}
1.0×10^{-2}	1.41×10^{-3}
1.0×10^{-3}	1.47×10^{-4}

Electrolytic conductance is temperature-dependent increasing its value by about 2% per °C rise, so all experimental work must be carried out at constant temperature. All measurements are made at 25°C. The equivalent ionic conductivity (λ) is an important property of ions which gives quantitative information concerning their relative contributions to conductance measurements. Thus λ is dependent on total ionic concentration of solution, increasing with increasing dilution. The values of equivalent conductance of cations and anions are summarised in Table 47.1.

47.2 MEASUREMENT OF CONDUCTANCE

To avoid Faradaic current and electrolytic complications, conductance measurements are invariably taken with alternating current (AC). The frequency of the current is about 1000 Hz. Efficient stirring is essential. The circuit for determination of electrolytic conductance (L) is a Wheatstone bridge (Fig. 47.1) modified to operate on AC (E) source energies bridge. R_3 and R_x are ratio arms kept at a ratio of 0.1, 1, or 10 or may be equal (*i.e.* $R_1 = R_2$). R_x is the resistance of the conductance cell. R_3 is the resistance of the standard arm. The capacitance (C_x) is parallel with R_x. It causes a phase shift in the alternating potential across R_x which is balanced by a air capacitor (C) across the balance arm. The alternating potential appearing across the diagonal of the bridge may be observed directly with earphones or an AC galvanometer. A condition of balance in the bridge exists if the output of the amplifier (or sound in the earphones) is zero, at which point

$$R_x = \left(\frac{R_1}{R_2}\right)R_3$$

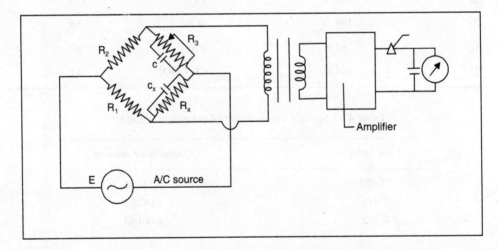

Fig. 47.1 Conductance measurement by AC bridge

The power source (AC) operates in the range 60–1000 Hz at a potential of 6–10 V. Since the ear has good response at 1000 Hz it is preferred to detect the null point with earphones. The Wheatstone bridge is unsatisfactory for conductance experiments with high resistance e.g. in nonaqueous systems or molten salts. In order to calibrate and calculate the cell constant, potassium chloride is preferred (See Table 47.2). Since conductance varies with temperature (2% per °C) it is necessary to keep the temperature constant.

47.3 ANALYTICAL APPLICATIONS OF CONDUCTOMETRY

An aqueous solution is used in conductance measurements (H_2O is a poor conductor, $L_{H_2O} = 5 \times 10^{-8}$ mho per cm at 25°C). Solutions of strong electrolytes show a linear increase in conductance with increasing concentration especially at high concentrations. It reaches a maximum and then diminishes. Fuming nitric acid can be analysed for NO_2: H_2O content by conductivity measurements. Conductance is measured for HNO_3 before and after treatment with KNO_3. A rapid and convenient procedure has been devised for checking the analysis of natural water and brines. It is also useful for the determination of small amounts of ammonia in biological materials. NH_3 is removed by treatment with H_3BO_3 and the specific conductance measured before and after treatment. It is also used to determine a specific ion in the presence of moderate concentrations of other ions if a reagent is available which will selectively remove the desired ion as a precipitate or unionized complex. K is measured before and after addition of the said reagent. However, direct measurements suffer from a lack of selectivity because the charged species add to the value of total conductance. It has maximum applications in ion chromatography. Conductance gives good information on association or dissociation equilibria in aqueous ionic solutions.

47.4 CONDUCTOMETRIC TITRATIONS

Conductometric titrations provide convenient means for locating the end point in titration. The best known example is that of boric acid titration with a strong base when volume changes are noted. The end point is nonspecific especially in redox titrations. For a good reaction an excess of hydrogen ions is necessary to get a visible change in the value of conductance. The conductometry refers to titration procedures while conductimetry refers to non-titrative measurements. The conductance method can be used to follow the course of titration if a great difference in specific conductance exists in the solution before and after addition of the titrant. The cell constant (θ) should be known. The spacing of electrodes should not vary too much during successive measurements. Conductance is proportional to the concentration at a constant temperature, but conductance on addition of a reagent may not vary linearly as dilution of the solution also occurs due to addition of water. Hydrolysis will also cause the similar problem. Hence we need a correction for volume of titrant.

The best example is that of acid–base titrations because OH^- or H^+ show a great difference in the value of conductance. Acid–base titration curves involving (*a*) a strong acid and strong base and (*b*) a weak acid and strong base are shown in Fig. 47.2. The extent of change in the magnitude of conductance does not depend upon the corresponding concentrations. Conductometric measurement involving precipitation and complexation are not very suitable for titration, for a small change in conductance is useless especially in precipitation titrations.

The displacement and precipitation reactions can be considered. Such titrations can be carried out conductometrically. Titration of a weak acid vs. weak base can easily be carried out conductometrically. It is also used in precipitation titrations, e.g. $AgSO_4$ *vs* $BaCl_2$. The conductance drops to a very low value at the equivalence point. $MgSO_4$ *vs* $Ba(OH)_2$ can be similarly titrated conductometrically to the end point.

Conductometric titrations are useful in any reaction where the ionic content is markedly less at the equivalence point than either before or after it. If ionic concentrations encountered are too high, the methods are of no use, e.g. titration of Fe^{+3} by $KMnO_4$, as the change in overall conductance during the titration will be negligible compared to the magnitude of total conductance. Thus conductometric titrations are best suited for acid–base titrations, while potentiometric titrations are best suited for redox titrations.

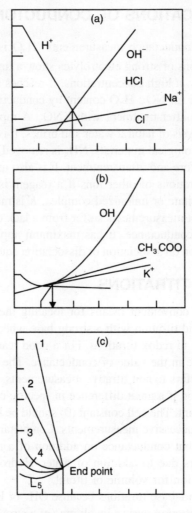

Fig. 47.2 Conductometric titrations

47.5 HIGH FREQUENCY TITRATIONS

Conductance procedures largely depend upon the motion of ions in an electric field. Use of alternating current to avoid electrochemical deposition permits the ions to move a short distance between alternations. However, if the frequency is further increased, a point will eventually be reached beyond which the ions will have no time to acquire their full speed. At such high frequencies the phenomenon of 'molecular polarization' becomes important. Thus if any molecule is subjected to an external electric field the electrons within it are drawn towards the positive electrode while nuclei are attracted towards the cathode in the opposite direction. This leads to actual motion of two kinds of particles relative to each other, giving rise to distortion of the molecule. This effect is temporary and disappears as soon as the field is removed.

Some molecules possess permanent electric dipoles (*i.e.* the centres of +ve and –ve electric charge within the molecule do not coincide) e.g. H_2O, CH_3COCH_3 $CHCl_3$, nitrobenzene etc. have dipole moments but not CH_4, CCl_4, C_6H_6 *p*-dinitrobenzene. Under the influence of an applied electric field such dipolar molecules orient themselves + ve to –ve and *vice versa* and show oriental polarisation in addition to temporary distortion polarisation. Both types, orientation and temporary distortion polarisation results in a flow of electric current for an extremely short period of time, during the application of the electric field. The duration of such currents is very small. At a radio frequency which is measured in millions of hertz (megahertz), the polarisation and conduction current both become of a significant order of magnitude and neither can be neglected. Therefore in high frequency measurements the sample is placed between two plates of a capacitor such that the resonant frequency of a tunable circuit is altered by the absorption of energy in the sample. The apparatus is equivalent to that employed in the measurement of dielectric constants of nonconducting samples as described latter.

Direct high frequency analyses of binary mixtures can be carried out by means of a calibration curve prepared from known solutions. Thus o–p xylene or hexane, have been analysed. The course of titration can be followed by this technique. This is not possible in conventional low frequency conductometric titrations. A significant advantage of the method is that no electrodes are placed in contact with the solution.

47.6 DIELECTRIC CONSTANT

The dielectrics are substances that are non conductors of electricity as they contain no mobile electrons. The dielectrics are optically transparent as opposed to electrically conducting solids as the latter are capable of absorbing radiation or reflecting light very strongly. A typical capacitor consists of a pair of conductors separated by a thin layer of dielectric material which is electric insulator with no free current through conducting charged matter.

The varying solvents have differing dielectric constant. Few solvents with their dielectric constants are shown in Table 47.4.

Table 47.4 Dielectric constants of common solvents

Solvent	Dielectric constant (ε)	Solvent	Dielectric constant (ε)
Acetonitrile	37.5	Formic acid	58.5
Acetone	20.7	Formamide	109.5
Acetic acid	6.15	Ethyl alcohol	24.3
Acetic anhydride	20.7	Ethylene diamine	14.2
Aniline	6.89	Ethylacetate	6.02
Benzene	2.33	Methyl alcohol	32.6
Benzyl alcohol	13.1	Nitrobenzene	34.8
1-Butanol	17.1	1-Propanol	20.1
1-Butylamine	5.3	2-Propanol	18.3
Chloroform	4.81	Phenol	9.78
Carbon tetrachloride	2.24	Water	81.0
1, 4 Dioxone	2.21	–	

It will be noted that apart from water also formamide has highest dielectric constant while 1-propanol, benzene and carbon tetrachloride have relatively low dielectric constant. It is interesting to know as to how this dielectric constant is measured.

47.7 MEASUREMENT OF DIELECTRIC CONSTANTS

Between two electrodes if non-conducting solvent is placed the system behaves as capacitor in series as C_a and C_g (Fig. 47.3). The capacitance of C_a will vary from fixed (C_g) when filled in air. The value of 'D' depends upon dielectric constant of the sample as given by an expression:

$C_a = C_g D$ if C_a = air and C_g = glass capacitors. C_g or C_a depends upon thickness of glass, D = dielectric constant of container and its plate area. The capacitance is not linear function of dielectric constant but will tend towards C_g. A graph of cell response against dielectric constant is shown in Fig. 47.3 as oscillometer. The portion of linear response is extended by inserting variable inductance in measuring cell and the oscillator circuit. Water has highest dielectric constant (See Table 47.4).

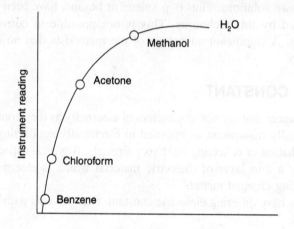

Fig. 47.3 Response (t) for solvents

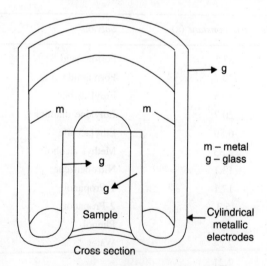

Fig. 47.4 Typical cell of oscillometer

The sample is placed between two plates of capacitor (or within induction coil) such that resonance frequency is altered by absorption of energy in the sample. This circuit is frequency determining element of oscillator and final observation of shift in the frequency caused by sample. Tunable circuit is calibrated with variable capacitor which compensates for changes in reaction of the cell and facilitates it to return to original value. This is used for measurement of dielectric constants of nonconducting samples for liquid due to motion of ions which pertains to absorption energy. This is low frequency conductometry. Standard capacitors are used. The sample cells are capacitance type wherein in the solution is filled in annular space within two concentric cylindrical electrodes joined to the external glass surfaces. No electrodes are in contact with liquid.

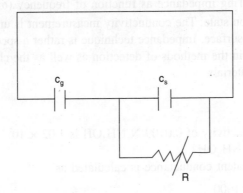

Fig. 47.5 Electrical circuit of cells

47.8 DIRECT CURRENT CONDUCTIVITY ANALYSIS

We use AC but direct current can be also used for measuring high electrolytic conductivity. It is mainly used for measuring conductometry of primary standards. It is restricted to cells of four electrode assembly (Fig. 47.6). All the four electrodes must be reversible. This eliminates polarisation effect. The process is simple in comparison to AC method. It works on Ohm's law viz. $C = E/R$, $E = CR$. A standard resistor is put in series in cell.

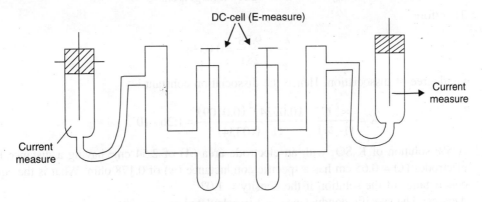

Fig. 47.6 DC conductivity bridge

The current passing via cell is proportional to current passing through standard resistor. Voltameter measures the difference of voltage across standard resistor and by using Ohm's law current is calculated.

Nowadays there is growing tendency to have electrodeless assembly for the measurement of conductivity using inductive coupling or capacitive coupling. The latter is also called by the name as 'Oscillometry'.

47.9 IMPEDANCE

So far we have seen methods to measure the conductivity of the sample irrespective of consideration of capacitance in system. With electrodes, with imposed AC current, voltage of very low amplitude (~5 mV) can detect impedance of the system (Z). This impedance (Z) is mathematically expressed as $Z = R-f/\omega C$ if Z = impedance, R = resistance, C = capacitance, ω = frequency perturbation and $f = \sqrt{(-1)}$. Now by measuring impedance as function of frequency (ω) one can identify the sample quality in solid or in solution state. The conductivity measurement is unreliable when redox reactions takes place at the electrode surface. Impedance technique is rather a specialised mode of measurement which are highly advanced in the methods of detection as well as the characterisation specially of the corrosion of conducting materials.

PROBLEMS

47.1 The electrical conductivity of 0.0109 N NH_4OH is 1.02×10^{-4} ohms/cm^2. What is the dissociation constant of NH_4OH?

Answer: The equivalent conductance is calculated as

$$\lambda = \frac{K \times 1000}{C} \text{ if } K = \text{conductance}, C = \text{concentration}$$

∴ $$\lambda = \frac{1.02 \times 10^{-4} \times 1000}{0.0109} = 9.38 \text{ ohms/geq/cm}^2$$

The equivalent conductance at infinite dilution is calculated from mobilities of NH_4^+ and OH^- ions, *i.e.* $NH_4^+ = 76$, $OH^- \equiv 205$

$$\lambda_\infty = 76 + 205 = 281 \text{ ohms/geq/cm}^2$$

Therefore

$$\propto = \frac{\lambda}{\lambda_\infty} = \frac{9.38}{281} = 0.0334$$

i.e. degree of dissociation. Hence the dissociation constant

$$K = \frac{\propto^2 C}{(1-\propto)} = \frac{(0.0334)^2 (0.0109)}{(1-0.0334)} = 1.26 \times 10^{-5}$$

47.2 A 5% solution of K_2SO_4 with an electrode area (A) of 2.54 cm^2 having a distance between electrodes (L) = 0.65 cm has a specific conductance (w) of 0.178 ohm. What is the equivalent conductance of the solution if the density = 1.0?

Answer: The specific conductance (w) is calculated as

$$w = x\left(\frac{A}{L}\right) \text{ i.e., } x = \frac{WL}{A} = \frac{0.178 \times 0.65}{2.54} = 0.0458 \text{ ohm}$$

Number of gram equivalent of K_2SO_4 in 1 cm^3 of solution of (5%)

$$n = \frac{5}{95 \times 87.13} = 6.039 \times 10^{-4} \text{ geq}$$

Finally,

$$\text{the equivalent conductance} = \frac{4.58 \times 10^{-6}}{6.039 \times 10^{-4}} = 75.8 \text{ ohms/geq/cm}^2$$

PROBLEMS FOR PRACTICE

47.3 In a conductometric titration of NaOH against HCl the following readings were obtained: ml (conductance) 0 (3.15), 1 (2.6), 2 (2.04), 3 (1.40), 4 (1.97), 5 (2.86), 6 (3.66). What is the number of gram eq. of NaOH in solution?

Answer: Here we plot the graph of conductance *(y)* axis vs ml of titrant on *(x)* we get two straight lines intersecting at a point around 2.5 ml for 1 M HCl. The calculations lead to 0.127g of NaOH.

47.4 A solution of 0.01 NaOH when titrated against HCl conductometrically give the following readings:

0(218), 4(230), 9(247), 10(256), 11(269), 12(285), 14(323), 17(380).

Calculate the number of moles of NaCl present in the solution.

47.5 What is the equivalent conductance at infinite dilution of a solution of Ag_2SO_4 at 25°C ?

47.6 If the specific conductance of 0.02M HCl is 7.92×10^{-3} ohms, what is the equivalent conductance of 0.02M HCl solution?

47.7 A 0.02 M H_2SO_4 containing NaI was titrated with 1.1 M $KMnO_4$.

The reading for volume of solution vs. conductance were as follows: 0.2(943), 0.4(872), (0.670), 1.2(622), 1.4(608), 1.6(609), and 2.4(614). Calculate the number of moles of NaI in the original solution.

47.8 A conductance cell at 18°C for 0.1N KCl has a conductance of 0.01315 and one with 0.105 acetic acid has a conductance of 5.58×10^{-3}. What is the dissociation constant of CH_3COOH acid?

LITERATURE

1. H.T.S. Britton, *Conductometric Analysis,* Van Nostrand (1934).
2. P. Delahay, *New Instrumental Methods in Electrochemistry,* Interscience, New York (1954).
3. J.J. Lingane. *Electroanalytical Chemistry,* 2nd ed., Interscience (1958).
4. E. Pungor, *Oscillometry and Conductometry,* Pergamon, New York (1965).
5. G.D. Christian, *Analytical Chemistry,* 5th Ed., John Wiley (1994).
6. D. A. Skoog, D.M. West, *Fundamentals of Analytical Chemistry,* 4th Ed., Sounders (1982).
7. T. Riley, E. Tomlinson, A.M. James, *Principles of Electro Analytical Methods,* John Wiley (1987).
8. D.A. Skoog and J.J. Leory, *Principles of Instrumentation Analysis,* Harcourt Brace (1994).

Chapter 48

Coulometry and Electrodeposition Methods

Coulometric and electrodeposition techniques ensure quantitative conversion of the reactant to a new product by a process of oxidation or reduction. Electrogravimetry is a quantitative method wherein the products of the reaction at the cathode or anode are weighed to ascertain concentration. The methods are rapid, selective and equally sensitive. Interestingly, none of these methods need calibration or standardisation.

48.1 PRINCIPLES OF ELECTROLYSIS

Coulometry involves the study and analytical applications of electrolysis. This involves the passage of currents in such a way that both electrodes are polarized. If we consider a circuit as shown in Fig. 48.1, we will observe that no current flows in the ammeter when the voltage is increased initially. But on further increasing the voltage, as decomposition potential is reached, a current will start flowing with the simultaneous deposition of Cu at the cathode and liberation of O_2 at the anode. The reactions encountered are:

At cathode $\qquad Cu^{2+} + 2e^- \longrightarrow Cu(s), E° = +0.34$ V

At anode $\qquad \frac{1}{2}O_2 + 2H^+ + 2e^- \longrightarrow H_2O; E° = 1.23$ V

Net reaction $\qquad Cu^{2+} + H_2O \longrightarrow Cu(s) + \frac{1}{2}O_2(g) + 2H^+$

The total potential across the cell is given by:

$$V = (E_a + W_a) - (E_c + W_c) + IR$$

where E_a and E_c are reversible half cell emfs at the anode and cathode respectively; W_a and W_c represent the added potential of two electrodes due to over voltage w.r.t. the deposition of copper and oxygen, resp. and in part due to the concentration polarization resulting from the passage of current; and IR is the ohmic drop through the solution itself. Since we are mostly concerned with cathodic deposition of metal and if we consider the reactions occurring at the anode to be the same, then the values of $(E_a + W_a)$ will remain more or less constant and we can determine the deposition on the cathode by the quantity $(E_c + W_c)$. In contrast the polarography uses a dilute solution, coulometric analysis involves the use of a concentrated solution, hence activity corrections cannot be neglected. From the knowledge of the electrode potentials of various metals and hydrogen as a function of ionic activities, it is possible to select the optimum voltage or potential to be applied for the deposition of a particular ion at the cathode (e.g. Sn is deposited at -0.6 V; Cu at -0.2 V; Zn at -1.3 V; Tl at -1.1 V *vs* SCE). The polarization term (W_c) cannot be calculated exactly. If we avoid excess current density and constantly stir the electrolyte, W_c can be kept to the minimum value. On controlling W_c, deposits are formed and there is no local depletion of the solution around the cathode. The increase in H^+ ion concentration due to the anodic reaction does not affect potential (< 0.01V).

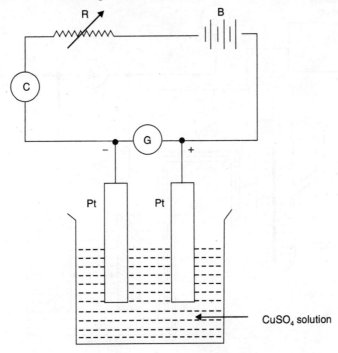

Fig. 48.1 Electrolysis of $CuSO_4$ solution

The cathode is important for electrodeposition but we cannot neglect the anode. Platinum is commonly used as an inert anode but the anodic oxidation product should not interfere with the cathodic process. If the anions present are NO_3^-, SO_4^{2-}, PO_4^{3-} then oxygen will be liberated at the anode. If Cl^- is present Cl_2 gas is formed which will attack the anode, so we add NH_2-NH_2 or use a silver anode to avoid complications. If AgCl is formed at the anode we can get back Ag by putting electrodes overnight in $ZnCl_2$ solution in electrical contact with zinc metal to promote double decomposition.

48.2 ELECTROLYSIS AT CONSTANT POTENTIAL

When the standard electrode potentials $(E°)$ of two ions are close (e.g, copper, tin) one can measure the potential of the cathode by using the circuit as shown in Fig. 48.2. We use a SCE and a potentiometer and we apply the exact potential (E_c). This method is known as constant potential cathode electrolysis. The electrode potential will be $E_{SCE} - (E_c + W_c)$. There are devices to automatically control the cathode potential by what are called 'potentiostats' of chapter 45. We have seen that, this is a device which maintains an exact potential at the cathode to effect exact electrodeposition. The reduction potentials required for deposition of metals may change depending upon the presence of other materials in the solution, e.g. the presence of complex forming substances has an important role in reducing the effective concentration of the free metal ion. In an analogous manner, we can also control the anode potential during anodic oxidations.

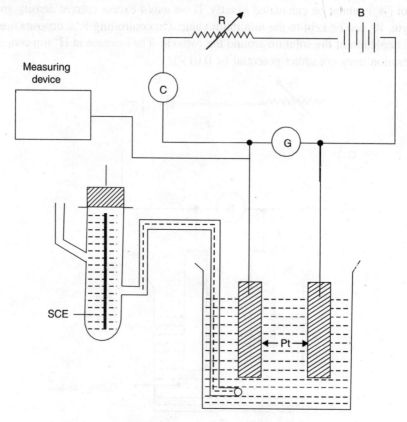

Fig. 48.2 Controlled potential electrolysis

In controlled potential cathode electrolysis, at a potential where only a single species is reducible, the current is limited by diffusion and is proportional to the concentration of reducible species. We have

$$\therefore \quad \frac{I_t}{I_0} = \frac{C_t}{C_0}\left(\text{If } I_t = I_0^{-kt}\right)$$

where C_t = concentration at time 't', C_0 is concentration at time t_0 and I_t and I_0 are currents at time t and t_0 after setting polarization, K = constant

$$K = \frac{DA}{V} \times \frac{25.8}{\delta} = \frac{25.8 DA}{V\delta} = \frac{DAS}{V}$$

D = diffusion coefficient; A = area of surface of cathode; V = volume of solution; δ = thickness of surface layer where concentration gradient is fixed and S = rate of stirring.

From a plot of log I vs t, we can determine the time required to deposit any fraction of the desired species. Controlled potential electrolysis has been found extremely useful in many fields of chemistry (Fig 48.3), e.g. organic and inorganic preparations; separation of Selenium and Tellurium, Tungsten

(III, IV) and NH_2OH. It also provides a valuable method for the separation of radioactive nuclides at microgram level.

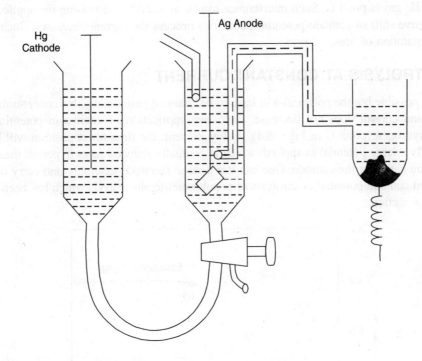

Fig. 48.3 Controlled potential electrolysis with Hg as cathode

A useful method for the separation of metals is electrodeposition on a mercury cathode. Any metal with a deposition potential less than mercury can be deposited on a Hg surface, e.g. Al, Sc, Tl, V, W and U. Alkali and alkaline earths can be deposited if the solution is basic. The process is used for the purification of metals (see Figs. 48.3 and 48.4). It will be seen from Fig. 48.4 that a shift in cathode

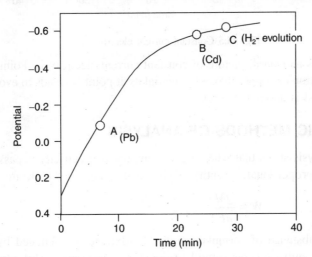

Fig. 48.4 Changes in potential during electrolysis (points for electrolyte deposition)

potential leads to concentration polarization and consequently we notice deposition of other species and loss of selectivity. At point A in Fig 48.4 lead can be deposited, at B cadmium and at point C there is evolution of H_2 gas at point C. Such interference can be avoided by decreasing the applied potential to avoid a negative shift of cathode potential but in this process the current decreases. Such a process is useful for separation of ions.

48.3 ELECTROLYSIS AT CONSTANT CURRENT

This process is possible but the potential will have to be changed periodically. No concentration polarization occurs and current does not decrease. Excessive application of change in potential leads to generation of hydrogen (point C in Fig. 48.4). At low current, the time for deposition will be lengthened. So initially a large potential is applied, which is gradually reduced so as to permit the maximum deposition of the cation at the cathode. One can use a three electrode assembly and carry out what is called controlled cathode potential electrolysis or potentiometric electrolysis which has been discussed in the preceding section.

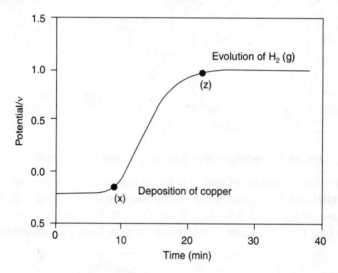

Fig. 48.5 Constant current electrolysis

Figure 48.5 visualises an overall picture of constant current electrolysis. Point (x) shows the optimum potential for deposition of copper. Excess potential as at point (z) leads to evolution of $H_2(g)$. The constant current employed is about 1–5 amp.

48.4 COULOMETRIC METHODS OF ANALYSIS

Faraday's law of electrolysis states that when a given amount of electricity is passed through an electrolyte it will result in a proportionate quantity of chemical change at the electrode, *i.e.*

$$W = \frac{QM}{nF}$$

where W = weight of substance of formula weight 'M' oxidised or reduced by the passage of Q coulombs, n = number of equivalents per formula weight, and F = faraday. This permits the quantitative determination of any substance which can be made to undergo an electrochemical reaction with 100%

current efficiency when no side reactions are present. This method is known as coulometric analysis. It can be carried out in two ways, (*a*) at constant current *i.e.* amperostatic where the amount of material deposited is proportional to the elapsed time and (*b*) at constant potential *i.e.* potentiostatic, where current decreases exponentially from a relatively large value to practically zero. In both processes, the quantity to be measured is the integral of *idt, i.e.* $\int idt$ where i = current in amperes flowing at any instant, t = time in seconds. The relationship between current and time is considered. The classical method of measuring the quantity of electricity is with a chemical coulometer as similar to shown in Fig. 48.6. Weight gain (or loss) after deposition is measured and is used when the primary deposit cannot be measured or weighed directly. A Hg–cathode is used in electrolysis. The water coulometer is useful for measurement.

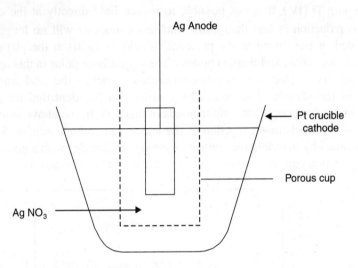

Fig. 48.6 Silver coulometer

Analytical coulometric procedures may be based on any type of electrode reaction including dissolution, electrodeposition, oxidation reduction. If the electrode reaction involves the substance to be determined, it is called a direct process. Others are termed as indirect processes, e.g. when the electrode reaction consumes or liberates a substance which is not itself determined but which can be related on a weight basis to such substances. Direct coulometric determinations substitute electrical measurements for weighing. Coulometric methods are much faster than volumetric and gravimetric methods and need no standardization.

The most important criterion in choosing the best method whether amperostatic or potentiostatic is that there should be 100% current efficiency. Every faraday must theoretically show an equivalent of material deposited. It is not always necessary for the analyte to take part in the process involving electron transfer at the electrode e.g. conversion of Fe (II) to Fe (III) at an inert electrode. Potentiostatic methods are used for deposition of substances with poor physical properties and reactions giving no solid product e.g. H_3AsO_3 to H_3AsO_4.

48.5 APPLICATIONS OF COULOMETRY

As many as fifty elements can be analysed by these methods. Mercury is a popular cathode. Electrolytic determination of organic compounds can be done by coulometry at constant potential. The concentration of a gas or liquid can be monitored by changing current coulometry as in the determination of oxygen. Many dissolved oxygen analysers used in environmental monitoring are based on this principle. A potential drop is measured with silver as the cathode and cadmium as the anode. Amperostatic coulometry is another name for coulometric titrations at constant current.

48.6 COULOMETRIC TITRATIONS

An indirect coulometric analysis consists of the electrolytic generation of a soluble species which is capable of reacting quantitatively with the substance sought. This technique is termed titration, although the titrant is not added as such from a burette, e.g. titration of Fe^{3+} in HBr with Ag anode and Pt cathode. At the anode, the reaction is $Ag + Br = AgBr + e^-$ while at the cathode there is reduction as $Fe^{3+} - 3e^- = Fe$ with Ti (IV). It is not possible to reduce Fe^{+3} directly at the cathode as current efficiency w.r.t. iron reduction is less than 100% and hence analysis will no longer be valid. But if excess Ti(IV) is added, it is reduced in the presence of acid to Ti(III) at the platinum cathode. The current efficiency reaches 100%, and there is no loss. The equivalence point in this reaction is the point in time when the quantity of electricity is just equivalent to generate the total amount of ferric iron present originally in the sample. The equivalence point can be identified by potentiometric or amperometric photometric or conductometric methods of analysis. In the above reaction it is assumed that no interaction of electrochemical reaction at the anode and cathode occurs. Such a problem of interaction can be avoided by shielding the counter electrode (viz. anode) with a glass tube with a fritted tip to discourage convection currents. Ag or agar–agar salt bridges can be also be used with advantage.

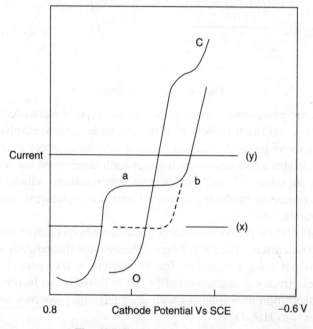

Fig. 48.7 Coulometric titrations

A shield closed by a ion exchange membrane (cation or anion exchanger as the case may be) can also be used instead of a glass fritt. Thus cations or anions are not allowed to escape from the anodic compartment.

48.7 APPLICATIONS OF COULOMETRIC TITRATIONS

Several reagents have been prepared by electric pulses e.g. H^+, OH^-, Ag^+ or oxidants Ce(IV), Mn (III), Ag(II), Br_2, Cl_2, I_2, $Fe(CN)_6^{3-}$ and reductants Fe(II), Fe $(CN)_6^{4-}$, Ti(III), Sn(II). By one or other of these reagents, it has become possible to substitute all volumetric procedures by coulometric techniques. The primary standard is a combination of a constant current source and electric time. The assembly used is shown in Fig. 48.8 with amperostatic arrangement.

Coulometric titration is applicable for samples one or two orders of magnitude smaller than conventional procedures. Some reagents which are unstable and not used in volumetric work can be used e.g. Mn(III), Ag(II), $CuBr_2^-$, Cl_2. The precision is far better than the volumetric method of analysis. The method is less useful for larger concentrations as it is essential to work with a large current and then current efficiency could be reduced to less than 100%. 5–10 mA current should be used for practical work.

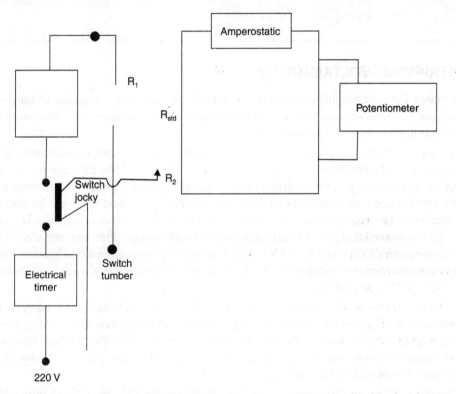

Fig. 48.8 Circuit diagram for coulometric titrations

The technique finds applications in all kinds of volumetric titrations. Some typical titrations are listed in Table 48.1.

The great advantage of coulometric titrations over conventional volumetric titrations is the elimination of standards. Hence, an instable standard need not be stored. Small quantities of reagents are required and microquantities of substance can be handled. It can be used for automatic titration with a constant current source. The possible errors may be due to variation in value of current, lack of 100% current efficiency, wrong measurement of current or time, titration error for departure of equivalence point from end point.

Table 48.1 Coulometric titrations

Kind of titration	Analysis of	Electrode reaction	Chemical reaction
Acid base	Acids	$2H_2O + 2e^- = 2OH^- + H_2$	$OH^+ + H^+ = H_2O$
	Bases	$H_2O = 2H^+ + 1/2\ O_2 + 2e^-$	$H^+ + OH^- = H_2O$
Redox titration	As (III)	$2Cl^- = Cl_2 + 2e^-$	Cl_2 is reagent
	$H_2C_2O_4$	$Mn^{2+} = Mn^{3+}\ e^-$	Mn^{3+} as reagent
Precipitation	Halogens	$Ag = Ag^+ + e^-$	$Ag^+ + Cl^- = AgCl(s).$
Complexations		$CaNH_4Y^{2-} + NH_4^+ + 2e^- =$	$HY^{3-} + Ca^{2+} = CaY^{2-} + H^+$
	$Ca^{2+},$	$Ca + 2NH_3 + HY^{3-}$	

48.8 STRIPPING VOLTAMMETRY

There are three kinds of stripping voltammetry methods as (a) Anodic stripping voltammetry (b) cathodic stripping voltammetry and finally (c) adsorptive stripping voltammetry. We would consider in brief all these three methods of stripping analysis.

(a) *In anodic stripping voltammetry (ASV):* As we have already seen in coulometry it is used for trace analysis with detection limit of ppt (i.e., from 10^{-5} M to 10^{-12} M). One can easily analyse with ASV four to six elements at a time. With mercury electrode, metal ions are concentrated by imposing negative potential. Such amalgamated metals are later stripped (i.e., oxidised) by imposing potential in positive direction. The peak current (i_p) is directly proportional to the concentration (C_m) of metal in analyte with peak potential (E_p). It is different for different metals. With mercury electrode potential range of analysis with SCE is $0 - (-1.2$ V). With mercury film one can reach to low level of metal. By adjusting deposition time or voltage one can analyse more than one metal in analyte. Now we would consider what is CSV and AdSV.

(b) *Cathodic Stripping Voltammetry:* This technique is used primarily for analysis of samples which cannot be readily dissolved in mercury (I). On application of potential in such system an insoluble film is formed on the surface of mercury electrode. A negative direction voltage scan will strip the deposited film into solution. Anions like halides, selenides, sulphides and the oxy anions like $[MoO_4]^{2-}$ or $[VO_3]^{-5}$ can be analysed by this technique.

(c) *Adsorbtive Stripping Voltammetry:* AdSV is analogous to ASV and CSV in principal. The main difference originates at the pre-concentration stage. The analysis is carried out by adsorption on electrode surface or by specific reaction at electrode with no accumulation electrolysis. Several organic compounds like cocaine, codeine, etc. are determined by AdSV at lower concentration. The technique is resorted to for quantification with $i_p \propto C$ if i_p = current, C = concentration.

Several organic compounds like nucleic acid bases and several other biochemical products have been analysed quantitatively, by cathodic stripping voltammetry. The detection limits for ASV in mercury film is 10^{-11} M but with square wave ASV it is 10^{-12} M. The level for the differential ASV is however 10^{-10} M.

48.9 CONVENTIONAL STRIPPING ANALYSIS

In stripping analysis, the analyte is collected by electrode position at a mercury or solid electrode. It is then redissolved or stripped from the electrode to produce a more concentrated solution than was originally present. Quantitative analysis is based either on electrical analytical measurements during stripping or on concentration of solution. Such methods are useful in trace analysis work as concentration of electrolysis allows analysis at trace levels with the greatest accuracy, e.g. 10^{-6}–10^{-9}, or microgram to nanogram concentration analysis is feasible. A micro mercury electrode is used for the deposition process followed by anodic voltametric analysis of the sample to be analysed. Let us consider two steps during such analysis.

(*a*) Electrodeposition step: Part of the analyte is deposited during this step, hence quantitative results depend on control of the electrode potential and electrode size, time of deposition, stirring rate for sample and standard solution. Mercury drop with a platinum tip is used as the electrode (Fig. 48.9). The concentration of the mercury is kept minimal to facilitate deposition of the analyte species. A hanging drop electrode is preferred as the microelectrode.

A capillary supplies mercury for drop transfer. Then a hanging drop electrode is formed by rotating the transfer scoop and contacting Hg with platinum wire in a glass tube at the centre. The solution is stirred and the drop can be dislodged by tapping the electrode at the end of analysis. After formation of a drop, a potential, equal to less than the half wave potential ($E_{1/2}$) of the analytical metal ion is applied. Deposition is allowed for 5 minutes but may go up to a maximum of 15 minutes.

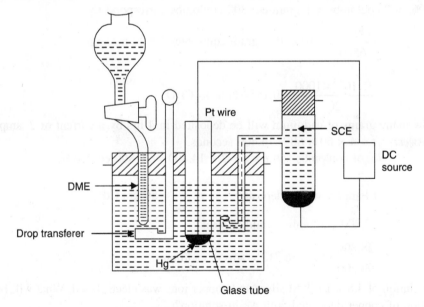

Fig. 48.9 Stripping analysis assembly

(*b*) Voltammetric completion of the analysis: The analyte which is collected in the hanging drop electrode is determined quantitatively by voltammetric analysis. The voltage is decreased at a fixed rate from the cathodic value to the anodic value and the resulting anodic current is recorded as a function of the applied voltage. A plot of current (μa) vs potential with SCE is plotted, e.g. in the analysis of cadmium, the metal was deposited from a 10^{-8} M solution by applying a potential of −0.9 V (vs SCE). It is 0.3 V more negative than $E_{1/2}$ of Cd. After 15 minutes the electrolysis potential was diminished at a rate of 21 mV/ second. A rapid increase in anodic current occurred at about -0.65 V due to the reaction: $Cd\,(Hg)\longrightarrow Cd^{2+}+Hg+2e^{-}$. The current depleted after reaching a maximum. The peak current was proportional to the concentration of cadmium and inversely proportional to the deposition time. A standard curve of the metal is of course necessary. It is thus possible to resolve mixtures by stripping analysis. These techniques find extensive applications in environmental pollution analysis.

PROBLEMS

48.1 In a coulometric determination of zinc, 35.4 ml of a mixture of $(H_2 + O_2)$ was evolved after reduction of the metal. What is the concentration of zinc in the cell?
Answer: In the electrolysis of H_2O, 22.4 litre of H_2 and 11.2 litre of O_2 are liberated. Eq. wt. of zinc = 65.4/2 = 32.7. Hence in total volume (22.4 + 11.2 = 33.6 litres) we have (33.6/32.7) = 0.0354 litres of mixture/x

$$\therefore \quad x = \frac{0.0354 \times 32.7}{33.6} = 0.0342 \text{ gram of zinc metal.}$$

48.2 In a coulometric titration of 20 ml of $K_2Cr_2O_7$ with iron (III), which is generated in solution, took 25 minutes to reduce when 200 mA of current was used. What is the normality of $K_2Cr_2O_7$?
Answer: $Q = i \times t = 25 \times 60 \times 0.200 = 300$ coulombs if $i = 0.200$, A and $t = 25 \times 60 = 1500$ seconds.
If 96, 500 coulombs = 1 gram eq. 300 coulombs correspond to

$$\frac{300 \times 1}{96500} = 3.01 \times 10^{-3} \text{ gram equivalent for 20 ml soln.}$$

$$\frac{3 \times 10^{-3} \times 1000}{20} = 0.1554 \text{ N of } K_2Cr_2O_7$$

48.3 How many grams of cadmium will be deposited in 4 hrs by a current of 2 amperes?
Answer: $t = 4 \times 60 \times 60 = 14,400$ seconds.
No. of coulombs (Q) $i \times t = 2 \times 14,400 = 28,800$ coulombs

I Faraday would deposit $\dfrac{Cd^{2+}}{2} = \dfrac{112}{2} = 56$ gm Cd

So 28,800 coulombs would deposit

$$\frac{28,800}{96,500} \times 56 = 16.712 \text{ grams of cadmium}$$

48.4 A solution of 4.8×10^{-10} M silver and copper ions was electrolysed. What will be the concentration of copper deposited with Ag in solution?
Given $E^0_{Zn} = +0.337$, $E^0_{Zn} = +0.799$, over potential nil)
Answer: We use the Nernst equation

$$E^0_{Zn} + \frac{RT}{nF}\log(Cu^{++}) = E^0_{Zn} + \frac{RT}{nF}\log\left[Ag^+\right]$$

$$0.337 + \frac{0.0591}{nf} \log[Cu^{2+}] = +0.799 + \frac{0.0591}{1} \log(4.8 \times 10^{-10})$$

On solving both sides we have,

$$0.337 + 0.0295 \log[Cu^{++}] = 0.799 - 0.551$$

$$\log[Cu^{++}] = \frac{0.799 - 0.551 - 0.337}{(0.0295)} = \frac{0.089}{0.0295}$$

PROBLEMS FOR PRACTICE

48.5 What over potential of hydrogen would allow Ni to be electrolysed out of a solution, that is 0.010 M in Ni^{+2} and 0.10 M in H^+?

48.6 How many seconds will it take for a current of 0.5 amps to deposit 0.59 gm of silver on the basis of 80% current efficiency?

48.7 A solution of 1×10^{-3} M on Ag^+ contains gold ions. What concentration of gold (III) ions would the silver and gold deposit simultaneously on electrolysis?

48.8 In a coulometric titration of $K_2Cr_2O_7$ how many grams of $K_2Cr_2O_7$ correspond to 1 μa sec?

LITERATURE

1. W.W. Latimer, *The Oxidation States of the Elements and Their Potentials in Aqueous Solutions.* 2nd Ed., Prentice Hall (1952).
2. J.J. Lingane, *Electro Analytical Chemistry,* 2nd Ed. Interscience, NY (1958).
3. A.J. Bard and L.R. Faulkner, *Electroanalytical Chemical,* Vol. 8, Marcel Decker (1975).
4. G.W.C. Milner, G.Philips, *Coulometry in Analytical Chemistry,* Pergamon Press (1967).
5. A.J. Bard, L.R. Faulkner, *Electrochemical Methods,* John Wiley (1980).
6. K. Brainina, E, Neyman, *Electroanalytical Stripping Methods,* John Wiley (1993).

Chapter 49

Chemical Sensors and Biosensors

In last few decades there has been spectacular development in the field of spectral methods as well as in the electrochemical techniques of analysis. These methods provide a very rapid tool for the analysis of analyte (sample) at small concentration. However, in industrial process specially those involving continuous stream of flow there was a need to have rapid methods of detection and determination of a material. With this demand a number of attempts were made to develop devices which could examine qualitatively as well as quantitatively the composition of the sample in motion. Thus for continuous pH measurement glass electrode was used while for the analysis of cations, ion selective electrodes containing liquid ion exchangers were used. So also anions were also analysed similarly.

A further investigation into potentiometric and voltammetric methods of analysis led to development of various gas sensors. Thus for environmental monitoring of dissolved oxygen, carbon dioxide or the oxides of nitrogen electrochemical sensors were most suitable. For the analysis of pharmaceutical preparations or organic compounds, optical sensors were discovered. For rapid change in temperature or heat contents of sample, the thermal sensors were effective. The mass sensitive sensors made use of the piezoelectric devices for measuring small changes in weight.

However, the greatest impact of sensors was witnessed with the discovery of biosensors. These were the sensors making use of enzyme on immobilised substrate or proteins or similar products for analysis. These biosensors although expensive were most effective in clinical chemistry for the analysis of blood or body fluids of human beings. Biocatalytic sensors provided a quick and reliable method for the analysis of constituents at the level of microgram concentration. It varied from 10^{-6} to 10^{-9} gram concentration. The demand for sensitivity measurement *i.e.*, lower detection limit for analysis with utmost precision is growing. For this measurement random background is essential. In instrumental methods for chemical analysis 'noise' is one factor which is influencing the sensitivity of measurements. The elimination of noise in detector or the amplifier of an instrument enhances the sensitivity of measurement. Noise is closely related to the signal e.g. change in intensity of light in the spectrophotometry or variation in diffusion current in the polarography is categorised as the signal. The signal provides analytical information from a system. By improving the signal, sensitivity of measurement is also enhanced. The sensors are the devices that respond to the presence of one or more analytes enabling both qualitative as well as quantitative analysis for which information is obtained continuously and rapidly.

49.1 CLASSIFICATION OF SENSORS

There are variety of sensors available in analytical chemistry. The first sensor which was commonly used was electrochemical sensors. By and large it includes potentiometric, amperometric or conductometric measurements from the signals generated in electrochemical reactions. These are mostly related to the chemical species present in the analyte sample. Next class is optical sensors which originate from spectrophotometric (UV/Visible/IR) or fluorometric measurements of electromagnetic radiation during analysis. These are most frequently used in chemical analysis. The third category is that of thermal sensors which could respond to variation in signal generated by oxidation of an analytical sample or a reaction. The weight measurement is ancient instrumental technique. The mass sensitive sensors include mechano-acoustic devices using piezoelectric devices with mass variation caused by adsorption of analytes on surface of crystal. The group of sensors permit simultaneous monitoring by several devices to improve the selectively and sensitivity of the process during analysis. Finally the last class is biosensors which employ thin layer of substance incorporating an immobilised reagent that contains the biorecognisation of sites. The protein, enzymes, macro molecules serve as the biomolecules. They should display specific interactions with the analyte species e.g. biocatalytical membrane electrodes serve as the biosensors. The selectivity of enzyme-catalysed reactions and electrochemical transducer gives rise to so called 'biosensors' facilitating analysis of biological samples.

Essentially the sensors must be robust and must possess rapid and reproducible response to the analyte. The sensor must be selective and sensitive to any chemical reaction and must possess broad working range in chemical analysis. The drift should be least and it should not be influenced by change of pressure or temperature. These sensors are employed in industrial process stream analysis. The example is the continuous monitoring of air pollution by optical sensors or monitoring of body fluids in patient in a hospital by biosensors.

49.2 SENSITIVITY AND LIMIT OF DETECTION

The value of an instrumental method is based upon what kind of sensitivity it offers to analyst. The sensitivity characterises the response of a module to an input or to a sample. It is a measure of degree of response and denotes function that is defined over particular range of concentration. The limiting value of sensitivity furnishes detection limit. The limit of detection represents the least concentration of substance present in sample. It is the minimum measurable concentration. It is the smallest increment in reading that can be recognised. It can be also compared with the Sandell's sensitivity in absorption spectrophotometry from the calibration plot. It is the detection limit at which the minimum signal or concentration which can be determined at an acceptable level of precision.

49.3 SIGNAL AND NOISE

In analysis, signal carries an information about the analyte while noise is composed of unwanted information leading to degradation of the accuracy and precision of analysis. Further it influences the detection limit. Therefore in order to ensure reliability of measurement the (S/N) signal to noise ratio is ascertained. It gives an idea on quality of analytical method or performance of instrument. So (S/N) = mean/standard deviation. It is reciprocal of relative standard deviation (S/N) = 1/RSD.

(a) *Source:* It is worthwhile considering source of noise. The chemical noise originates due to variables affecting the chemistry of system being analysed while instrumental noise is due to components of an instrument like transducer. Thermal noise is due to thermal agitation of electrons due to numerous reasons. The short noise is possible if electrons or charged particles cross the junction. Flickering noise is typified by having magnitude that is inversely proportional to the frequency of the signal (*i.e.*, 1/*f* noise). It is due to variation in frequency of light or sound. The environmental noise arises from influence of surroundings on the analytical sample.

Signal to noise ratio: This (S/N) must be kept at acceptable level. The incorporation of filters, choppers, modulators, synchronous detectors enhances (S/N) ratio. There are several methods to reduce noise which consequently will lead to increase in magnitude of (S/N) ratio. The methods include grounding and shielding. This is true for electromagnetic radiation related to noise. The use of difference instrumentation as amplifier reduces noise e.g. electrocardiograph utilises advantage of difference amplifiers. Analogue filtering is another way to reduce noise. It enhances (S/N) ratio. The modulation involves low frequency dc signal from transducer which are converted to high frequency (if 1/f noise is less). It gives narrow band after modulation. Finally signal chopping amplifier or lock-in amplifier gives best value of S/N ratio. This is used in recording spectrophotometers.

The chopper amplifier provides a means for controlling signal. It should be done near source e.g. chopper is used in AAS for signal modulation. While lock in amplifier permits recovery of signal even if (S/N) is small. This needs reference signal for purpose of comparison with analyte signal. Only those signals which are locked into reference signal are amplified for which there are various software methods. The (S/N) ratio is controlled to availability of microprocessors and computers. This is done by what is called as ensemble averaging, box car averaging, digital filtering and the correlation methods. The (S/N) ratio is dramatically improved. The correlation method however provides powerful tool for extracting signal. These are carried out by computers.

49.4 DEFINITION OF SENSORS

In order to understand the scientific meaning of the word 'sensors' first we must see the difference between detector, transducer and sensors. They are loosely used synonymously, though there is subtle difference between them. The detector refers to mechanical, electromagnetic or nuclear radiation of particulates or the molecules. The detector system refer to entire assemblies which indicates or record the physical or chemical quantities. While transducer refers to specifically such of the devices which convert information to non-electrical domain of information in electrical field. Phototubes, photomultiple tubes are transducers. The interrelationship between electric output and input radiant power, temperature and force is called as the transducer function. Finally, the term sensor has broad perspective. The class of analytical modes which are capable of monitoring specific chemical species on continuous basis. Glass electrode or ISE is a kind of sensor measuring potential. A sensor consists of transducer coupled with chemically selective recognisation phase. A quartz crystal micro balance (QCM) is an interesting example of mass sensitive sensor like piezoelectric balance. It is based on use of piezoelectric device. On bending quartz crystal its potential changes, so also when voltage is imposed on its surface the potential alters. The crystal in electric circuit oscillates at frequency typical of its mass and shape. The mass is constant. This is called as piezoelectric phenomena. The constant frequency of quartz is the basis of clocks, counters, timers. On coating quartz with polymers it absorbs certain molecules leading to mass increase while its frequency decreases. On desorption of molecule from the surface, crystal restores and reverts to original frequency. If ΔF is change in the frequency and Δ is

change of mass, then $\Delta F = \left(\dfrac{CF^2 \Delta M}{A} \right)$, if M = mass, A = surface area, F = frequency of oscillation and

C = constant. This shows small change in mass can be noted if frequency of crystal is measured. The limit of such sensor is 10^{-12} or ρg. Piezoelectric mass sensor is a good example of transducer. A mass change in electrical quantity or resonance frequency is measured by sensors. A combination of the transducer and selective phase leads to formation of the sensor.

Earlier we had already considered classification of sensors. Based upon this classification we shall now consider some typical examples of the electrochemical sensors, optical sensors, thermal and biosensors.

49.5 ELECTROCHEMICAL SENSORS

Ion selective electrode is a typical example of potentiometric sensors. The solid state redox electrodes as well as field effect transistors (FET) are also electrochemical sensors. Glass electrode used for pH measurements is also used in gas sensors for measuring level of CO, CO_2, NH_3, NO-NO_x, SO_2 and other gases in analytes. Therefore we would first consider membrane indicator electrodes or ISE about which some information has already been considered in earlier chapters. These electrodes are also called p-ion electrode used for measurement of pH, pCa or pNO_3 ions. They are classified in groups as crystalline and noncrystalline electrodes. The first form has two sub classes as single crystalline or polycrystalline electrodes. While the non-crystalline membrane electrodes comprises of glass electrode, liquid or immobilised liquid electrodes. The ion selective membrane electrodes should have minimum solubility and must have excellent electrical conductivity. It should be selective with the analyte, with firm binding by ion exchange complexation and crystallisation. We had considered in greater detail about glass electrodes in earlier chapters. A thin glass membrane at the tip of the electrode responds to changes in pH. A crystalline membrane is fabricated by sealing membrane to one end of tube made of PVC or teflon. Fluoride electrode is made from CaF_3. It is a crystalline electrode. Electrodes based on silver salts are also available. The liquid membranes are formed from immiscible liquids which selectively bound selected ions. Liquid ion exchange MS were used for such liquid membranes electrodes where the porous plastic membrane holds liquid ion exchangers while glass tubing contains (MX) standard solution with silver electrode (Fig. 49.1). M^+ are ions whose activity was to be measured. The liquid cation exchangers or liquid anion exchangers and macrocyclic compounds like crown ether were used, as liquid membranes. For analysis of Ca^{2+}, K^+, NO_3^-, F^-, ClO_4^-, membrane

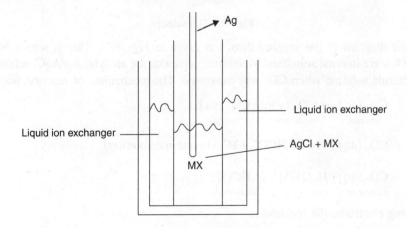

Fig. 49.1 Cell assembly

electrodes were available. Similarly solid state electrodes for the analysis of Ba^{2+}, Cu^{2+}, Pb^{2+}, Ag^+, Cd^{2+}, Cl^-, F^-, I^-, CN^-, SCN^- were also available with inferences. The ion selective field effect transistors (ISFET) were used in computers to control current flow in circuit.

They are sensitive to impurities on the surfaces. In 1970 metal oxide field effect transistors (MOSFET) was fabricated. The ion sensitive surface of ISFETs is selective to pH changes. They are rigid, small, inert to environment, have quick response and low electrical impedance. They were used in dry conditions during analysis.

49.6 GAS SENSORS

It is a molecular selective electrode. The dissolved gases like NH_3, CO_2 were measured by it. They were electrochemical cells composed of ion selective and reference electrodes in solution kept on gas-permeable membrane (see Fig. 49.2).

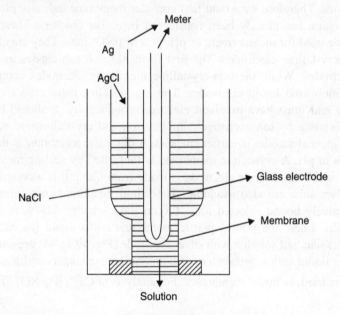

Fig. 49.2 Gas sensor

A schematic diagram of gas sensing device is given in Fig. 49.2. This is sensor for CO_2 where $NaCl + NaHCO_3$ were internal solutions. Membrane separates the analyte. Ag/AgCl reference electrode was used as internal solution when CO_2 was measured. The mechanism of reaction was

$$\underset{\text{external}}{CO_2\left(aq\right)} \rightleftharpoons \underset{\text{membrane}}{CO_2\left(g\right)} \rightleftharpoons \underset{\text{internal}}{CO_2\left(aq\right)} \qquad \qquad ...(49.1)$$

$$CO_2\left(aq\right) + H_2O = HCO_3^- + H^+ \left(\text{in internal solution}\right) \qquad \qquad ...(49.2)$$

Overall:
$$\underset{\text{external soln.}}{CO_2\left(aq\right)} + H_2O = H^+ + \underset{\text{internal soln.}}{HCO_3^-} \qquad \qquad ...(49.3)$$

In nitrate sensing electrode, the reaction is

$$\underset{\text{extn}}{2NO_2\left(aq\right)} + H_2O = NO_2^- + \underset{\text{internal soln.}}{NO_3^-} + 2H^+ \qquad \qquad ...(49.4)$$

49.7 VOLTAMMETRIC SENSORS

It consists of oxygen sensor and in the real sense they were not truly electrodes, but were technically electrometric cells. In oxygen sensor, the DO (*i.e.*, dissolved oxygen) was measured by Clark oxygen sensor, discovered in 1956 (see Fig. 49.3). The cell contained a platinum cathode as working electrode immersed in insulator fitted with silver anode. Both electrodes were kept in cylinder containing buffer solution of potassium chloride. The oxygen permeable membrane was placed at bottom of tube. Electrolyte thickness was as small as 10 μm. On dipping sensor in analyte, oxygen diffused through the membrane and migrated to cathode, to diffuse on electrode. The time needed was 10-20 seconds. The rate of transfer of oxygen across the membrane decides the rate to attain steady equilibrium state. The reactions were

$$\underset{\text{cathode reaction}}{O_2 + 4H^+ + 4e^- = 2H_2O}; \quad \underset{\text{anode reaction}}{Ag^+ + Cl^- = AgCl(s) + e^-} \qquad \qquad ...(49.5)$$

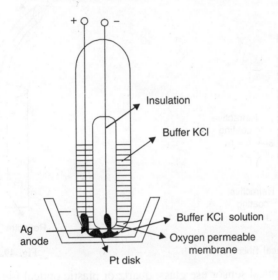

Fig. 49.3 Clark's oxygen sensor

49.8 SOLID STATE ELECTRODE SENSORS

They consist of ZrO_2 crystalline form coated with CaO or Y_2O_3 to render it ionically conducting. It is used to monitor oxygen in combustion gases or molten metal samples. The redox reaction was

$$O_2 + 4e^- \longrightarrow 2O^{2-} \qquad \qquad ...(49.6)$$

It created electrode potential with a Nerstian response to concentration through movement of O^{2-} through the crystalline lattice. The field effect transistors were made sensitive to facilitate gas analysis. Amperometric and conductometric gas sensors have been devised for O_2, H_2, NH_3 and SO_2 analysis.

49.9 OPTICAL SENSORS

Fiber optics were used for transmitting radiation and images from one component to another. They were fine films of glass or plastic to transmit. A 0.05 μm–0.6 cms was the dimension of optical fiber. The bundles of fibers were used for transmission. By total internal reflection the image was transferred. Fiber was coated with such material which had a refractive index lesser than the refractive index of material of construction for the fiber. Glass fiber coating has $\mu = 1.5–1.6$ (Fig. 49.4). A fiber will transmit radiation contained in cone with angle $\hat{\theta}$. The numerical aperture of the fiber provides a measure of the size of the acceptance cone. These fibres can transmit UV, visible, IR radiation. A fiber optic sensor was called as "Optrodes". They consist of a reagent phase immobilised on the end of fibre optic support. Interaction of analyte with ligand lead to change in absorbance which in turn was transmitted to a detector via optical fiber.

Such sensors were cheap, simple and could be easily fabricated. In it the numerical aperture = $n_3 \sin\theta = \sqrt{n_1^2 + n_2^2}$, provided if $n_1 > n_2 > n_3$ (Fig. 49.3).

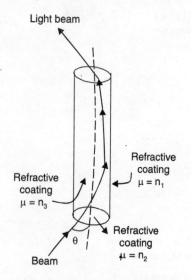

Fig. 49.4 Optical fiber

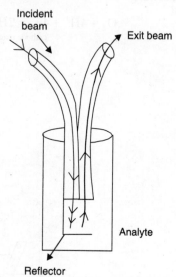

Fig. 49.5 Optical sensor

Alternatively, fiber optical sensor use glass, quartz or plastic optical fibres to transmit incident and attenuated or fluorescent radiation in spectrophotometer or light emitting diode (LED) and detectors and analyte. The end of twin fibres were positioned together in a Y-shaped configuration a short distance above a reflector plot to form a sample cell (Fig. 49.5). The radiation coming from one cable passes through the analyte into which the cell is immersed and was reflected back through the analyte into the end of other cable. The light path in optical cell was double the distance between ends of the cable and plate. The reflected radiation (including one due to the absorbance or fluorescence) from surface was detected. When chromogenic reagents were immobilised on the ends of the fibres in thin layers of the supporting media as cellulose, polyacrylamide, it gave opterodes. The home kits for detecting sugar in blood for diabetic patient or ascertain pregnancy depended on visual observation of resulting colour formed by interaction of analyte with chromogenic reagent via optical fiber. Test slips were used in reflectance technique. Optical sensors were used to monitor pH, metal ion concentration, dissolved gases and organic compounds (preferably latter with biosensors). They have definite merits over electrochemical sensors as not needing electrodes and electrolyte. They provide precious knowledge on spectra in UV, visible or IR region of the wavelength. However, depletion of the immobilised reagent layer, or interference due to light or slow kinetics of the colour formation reaction between analyte and ligand were distinct demerits of the sensors.

49.10 THERMAL SENSORS

The oxidisable gases such as CO_2, CH_4 depended for sensors on measuring a change in resistance of a heated coil due to heat of reaction resulting from oxidation of the gas by absorbed oxygen. Thermal biosensors inclusive of thermistor which have been marketed to measure heats of reactions of enzymes in detection of biological sample substances. In thermal sensors (transducers) wherein the radiation falls upon body was absorbed and the resultant temperature rise was measured e.g. IR radiation. It is satisfactory with power of 10^{-7}–10^{-9} W. The heat capacity of absorbing element was small if the noticeable temperature variation was to be generated. Size and thickness of absorbing media was reduced to concentrate IR-radiation. Temperature change was thousandth of kelvin. Pyroelectric transducers (sensors) were fabricated from single crystalline wafers of pyroelectric materials which were insulators with special thermal and electrical properties. Triglycine sulphate was most popular pyroelectric substance used in transducers. On application of an electric field across dielectric material, polarisation occurs whose magnitude is function of the dielectric constant of the material. It decayed if electric field was removed. However, pyroelectric material retains strong temperature dependent polarisation even after removal of electric field. If pyroelectric material is sandwiched between two electrodes a temperature dependent capacitor was formalised. Changing its temperature changed the charge distribution across the crystal which in turn generated measurable current in circuit which had two capacitors. The magnitude of the current depends upon several factors as were surface area of crystal, rate of change of the polarisation with temperature. Such crystal lose their residual polarisation when heated to curic point.

49.11 BIOSENSORS

Of the several sensors so far we have considered, biosensors have enormous applications in biotechnology. As defined earlier biosensors use thin layer of substance including an immobilised reagent which has biorecognisation sites. The biomolecules were either protein, enzymes or similar macrocycles which were capable of displaying specific biochemical action with sample solution. The reagents were immobilised by the inclusion or binding methods commercially used. Such methods of the entrapment include physical adsorption on surfaces or chemical binding to the surface of transducer or on crosslinked polymeric gel (Fig. 49.6).

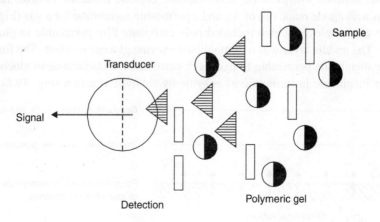

Fig. 49.6 Reactions biosensor

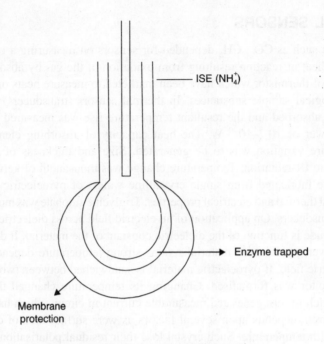

Fig. 49.7 Enzyme based biosensor

Such sensors were easily fabricated in laboratory by covering glass electrode with immobilised enzyme promoting catalytic reaction (Fig. 49.7). Urea coating on NH_3-ISE is used to analyse urea. It was hydrolysed by enzyme. The mechanism of the reaction at pH ~ 7.0–8.0 was as follows:

$$\underset{\text{Urea}}{CO\left(NH_2^+\right)_2} + 2H_2O + H^+ \xrightarrow{\text{ISE}} 2NH_4^+ + HCO_3^- \qquad \qquad ...(49.7)$$

$$NH_4^+ + OH \longrightarrow NH_3 + H_2O \qquad \qquad ...(49.8)$$

A variety of enzyme-based voltameter sensors have to been fabricated e.g., Glucose sensor consists of insulating rod, buffered hydrochloric acid solution, cathode made of Pt-disk, layer of potassium chloride solution with anode made out of Ag and a permeable membrane for a gas (Fig. 49.8). The three layer membrane was employed which included poly carbonate film permeable to glucose, but impervious to protein. The middle layer was a immobilised enzyme glucose oxidase. The final third layer was cellulose acetate membrane permeable to hydrogen peroxide. On immersing in glucose solution, latter diffuses to outer membrane in immobilised enzyme by catalytic reaction (Fig. 49.8).

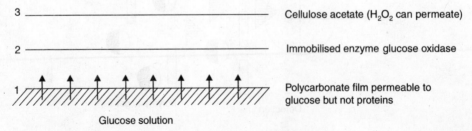

Fig. 49.8 Layers in biosensors

The catalytic reaction between layers 1 and 2 was

$$C_6H_{10}O_6 + O_2 \longrightarrow H_2O_2 + C_6H_8O_4 \text{ (acid)} \qquad \qquad ...(49.9)$$

The hydrogen peroxide so generated travelled to electrode surface (outer). It was then oxidised as

$$H_2O_2 + OH^- \rightleftharpoons O_2(g) + 2H_2O + 2e^- \qquad \qquad ...(49.10)$$

This resulted in a current which varied with glucose concentration in analyte solution. Similar biosensors for fructose, sucrose, lactose, ethanol were fabricated. Thus enzyme-based biosensors as those for cholesterol, glucose from blood, depended on the oxidation of cholesterol or glucose oxidase respectively to produce hydrogen peroxide which was detected by amperiometry technique.

49.12 BIOCATALYTIC BIOSENSORS

Since 1970 attempts were made to prepare such sensors, based upon enzyme catalysed reaction and electrochemical transducer for analysis of biological products. In such sensors, analyte is contacted with immobilised enzyme, permitting catalytic reaction of the analyte. This leads to generation of NH_3, CO_2, hydrogen peroxide or hydrocarbons. Transducer then measures the concentration of the analyte. Membrane electrode, gas-sensors, or voltammetric devices are used. Biosensors using membrane electrodes were most popular with clinical workers, as complex organics could be easily analysed, rapidly. One needed mild conditions of the temperature and pH of analyte. The most impressive aspect was the absence of any interferences. However they were expensive. In immobilised media less enzyme was required which in turn reduced the cost. Two processes of testing were existing. In one analyte is passed on immobilised enzyme bed and then to transducer while in the second porous layer of immobilised enzyme was attached to surface of ion selective electrode. This gives enzyme electrode. The process of diffusion existed for transfer of reactants to membrane. The enzymes are easily immobilised by entrapment or adsorption on porous alumina or copolymerisation of enzyme on appropriate monomer. The kind of electrode made is shown in Figure 49.6 while the mechanism of permeation and catalytic reaction is depicted in Fig. 49.7. The kind of the reaction for detection of urea were shown in equations (49.7) and (49.8) while the reactions of glucose oxidate were indicated in equations (49.9) and (49.10). The enzyme electrode is active at pH > 7.0 because of pH incompatibility of enzyme sensor. The maximum response exist at pH 8.0-10.0. This in turn limits the sensitivity of the electrode. Fixed bed enzyme systems eliminate foresaid problems. Enzyme electrodes based on potentiometric measurements were rare; but those based on voltammetric devices are most common in use in chemical analysis.

49.13 MASS SENSITIVE SENSORS

These were mechano-acoustic devices based on piezoelectric effects caused by the adsorption of sample on a crystal surface. These were piezo electric quartz crystal resonators covered with gas-absorbing organic film. The adsorption of the sample gas led to a change in resonance frequency that could be detected by an oscillator circuit and which was quite sensitive down to 10^{-6} μg concentration of the sample. In earlier sections of this chapter details of weight sensitive sensor based on the applications of the piezoelectric device was described in detail (see 49.5).

49.14 EFFICIENCY OF SENSORS

In clinical analysis one takes the resort of test strips, which gives qualitative information. In order to get the quantitative data one has to use reflectance measurement. The most preferred methods were enzyme based redox measurement made by electrochemical biosensors. The generated gas or hydrogen peroxide was allowed to read with suitable chromogenic reagent to form the optrode. Many times individual sensor provided reliable information. The selectivity was enhanced by using groups of sensors. This was called as 'arrays of sensors'. They could monitor one or more samples of solution using measurement of various parameters by different sensors. Sensors (two electrochemical) were operated at different electrical potentials or in spectrophotometry the frequencies or wavelengths for optodes. Simultaneous measurement of pH or sodium potassium level in blood was monitored by use of three ion-selective field effects transistors (FET) each with different sensitivity tune for three analysis. Using p-Si substrate these three components (viz. pH, Na, K) were monitored with FET sensor utilising n-Si combination transistors.

In conclusion we can say that sensors specially the biosensors have numerous applications in clinical chemistry for the analysis of blood or body fluids. The chemical sensors have enormous industrial uses. The electrochemical and optical sensors have large applications. In comparison thermal and mass sensors have limited utility. The sensitivity and limit of the detection to be sought depends upon kind of analyte used. The consideration of signal to noise (S/N) ratio is of foremost importance. Gas sensors were based on molecular selective electrodes but voltammetric sensors were ideal in biochemical analysis. Solid state sensors had their own limitations, while optical sensor had unlimited applications in analytical chemistry. Biosensors based on biocatalyte reactions were in great demand. We hope to see tremendous development in this field in future years.

LITERATURE

1. D. Kealey, P.J. Haines, *"Instant Notes—Analytical Chemistry"*, Viva books (P) Ltd., New Delhi (2002), pg. 323.
2. D.A. Skoog, J.F. Holler, T.A. Nieman, *"Principles of Instrumental Analysis"*, 5th Ed., Harcourt Brace College Publishers (1998), p. 609, 651.
3. E.P. Serjeant, *"Potentiometry and Potentiometric Titrations"*, Wiley (1984).
4. D.J.G. Iyes, G.J. Janr, *"Reference Electrodes"*, Academic Press, New York (1961).
5. A. Evans, *"Potentiometry and Ion Selective Electrodes"*, Wiley (1987).
6. J. Koryta, K. Stulik, *"Ion Selective Electrodes"*, 2nd Ed., Cambridge University Press (1983).
7. Y. Umezawa, *"CRC Handbook of Ion Selective Electrodes—Selectivity Coefficients"*, Boca Raton, Florida, CRC Press (1990).
8. A.P.E. Turner, I. Karube, G.S. Wilson, *"Ed. Biosensors – Fundamental and Applications"*, Oxford University Press (1987).
9. M.L. Hitchman, *"Measurement of Dissolved Oxygen"*, Wiley (1978).
10. Orion Research, *"Handbook of Electrode Technology"*, Orion Research, Cambridge (1982), pp. 10-13.
11. R.W. Cattrall, *"Chemical Sensors"*, Oxford University Press (1997).
12. F.W. Fifield, D. Kealey, *"Principles and Practice of Analytical Chemistry"*, 5th Ed. Blackwell Science, Oxford U.K. (2000).

Chapter 50

On-line Analysers–Automated Instrumentation Methods

The methods for continuous analysis, control of on-stream process systems, and automatic analysers for routine chemical analysis are becoming popular in India for two reasons. Firstly, they save considerable manpower which in the long run is economical; and as such on-stream analysers help to control the quality of products which are manufactured on a large scale and save time. In the past classical chemical methods were utilised for quantitative chemical analysis. These methods included wet chemical analysis like volumetry or gravimetry. With progress in science, newer and more modern methods of analysis have come into existence. Such methods mainly included characterisation or quantitative chemical analysis of components involving instrumental methods.

Such modern instrumental methods involve the measurement of optical, electrical, and magnetic properties of substances. By and large, optical methods of analysis are most popular, although electroanalytical techniques are gaining considerable importance in chemical analysis. In comparison the methods based upon magnetic measurements are few in number and less commonly used.

50.1 SPECIFICATIONS OF METHODS OF ANALYSIS

Various methods are used in automated methods of analysis. They include ultraviolet, visible and infrared spectroscopy from absorption spectroscopy; similarly atomic absorption spectroscopy, emission spectroscopy and flame photometry or flame emission spectroscopy, fluorescence and phosphorescence methods of analysis and plasma spectroscopy. From emission techniques methods based upon scattering of light such as nephelometry or turbidimetry techniques and Raman spectroscopy are also used. With the advent of knowledge in radioactivity from nuclear energy programmes, neutron activation analysis and isotope dilution methods have come to the forefront in online analysers.

In electroanalytical methods, potentiometry, including pH measurements, conductance, coulometry and electrodeposition and finally polarographic methods have been extensively used. This discussion will not be complete without the mention of polarimetry, thermal and differential thermal analysis and thermometric methods.

50.2 CLASSIFICATION OF AUTOMATED METHODS

There is a great deal of difference in various automation methods. Automatic devices perform specific operations at a given point in analysis. They improve analytical efficiency by performing some of the operations done manually while automated devices control and regulate a process without human intervention. Such instruments control a system based on analysis of results, and are used in process control. Continuous analysers containing sensors, a controller and operater form a part of automated instruments, while automatic instruments rely on the analyst for several steps. They improve precision. Semi automatic instruments conduct one or more specific steps in analysis. The centrifugal analyser is a discrete analyser that analyses several samples in parallel. While flow injection analysis involves injection of samples with a reactant in a flowing stream and measurement of a transient signal. They need a few microliters of sample. Microprocessor controlled instruments collect and process data. The interfacing to a personal computer enhances data processing capabilities. Though instruments give analog signals, computers process digital signals. Networking permits several computers to communicate. A computer sends data to a workstation for processing, this increases the efficiency of a main frame computer workstation.

50.3 AUTOMATED ANALYSIS

The concept of automated analysis entered the field of chemical analysis to fulfil the need for rapid analysis with a large number of samples. In industrial analysis, it is most important to analyse the maximum number of samples in the minimum amount of time. Further, the need was emphasised due to a requirement of a smaller number of trained operators to carry out day to day analysis with a minimum of variation.

The first step in this direction of automated analysis is chromatography. It was the invention of the automatic fraction collector which obviated the need for round the clock personal attention to the chromatographic column. With the development of various auto-analysers it is possible to get 200-300 samples of chromatographic fractions in a few hours. The next invention of a recording spectrophotometer with a strip chart recorder was an important major step in chemical analysis, and with the development of a process infrared analyser the task of the analyst has became quite simple. Analytical absorption spectroscopy has become a more popular tool with chemists with the easy availability of recording devices. The advantages were not so visible in the field of ultraviolet or visible spectrophotometers but such merits were most significant with infrared spectral methods. With the availability of IR spectra from the region of 2500 cm^{-1} to 500 cm^{-1}, it became simpler to characterise an organic compound by identifying the peaks of the functional group.

Another example of automation is that of the automatic titrator. With conventional titration methods involving either acid base, redox, precipitation or complexometric titrations, the uncertainty of end point was eliminated. But with linear titration techniques involving photometry, thermometry, potentiometer conductometric or amperometric or coulometry titration, the detection of end point was simplified with the development of the automatic titrator. The simplest example of this class of titrator is the Karl Fisher titrator for moisture determination involving potentiometry, or use of conductimetric titration with an end point detector with head phones or a magic eye in environmental analysis for sample of water.

With the increased use of automated instruments a beginning was made towards the development of continuous analysers. There are, no doubt, numerous advantages of continuous analysers. However, one must take adequate steps to avoid pitfalls. A typical continuous analyser is depicted in Fig. 50.1.

A fraction of the sample is taken from the main stream into the sampling system by means of an efficient pump, like a proportionating pump which is now a used in high performance liquid chromatography. The next stage involves sample preparation, which may involve addition of chromogenic ligands for colour development or addition of reagents to cause a shift in a physical property such as potential, or conductance. After this there is a measuring system involving transducer and finally an indicator for recording the output or signal.

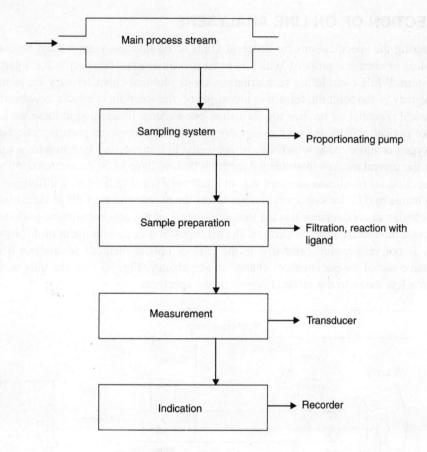

Fig. 50.1 Features of continuous analysis

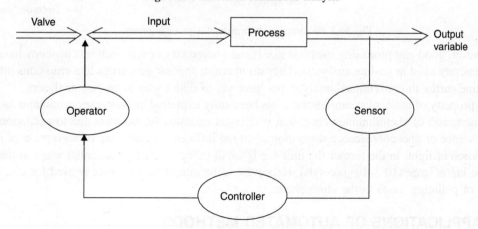

Fig. 50.2 Feedback control

Another example is that of a feedback control 100P shown in Fig. 50.2. The inputs can be controlled and fed to a controller which compares them with a reference. The difference is fed to an operator which controls the value so that it is adjusted to the set point.

50.4 SELECTION OF ON-LINE ANALYSERS

While considering the specifications for different kinds of on-line analysers it must be borne in mind that the physical or chemical property which is to be measured must be specific for a particular component or system. While considering such criteria various physical characteristics are examined. They include the density of the solution, refractive index, pH of the solution, conductance, absorbance of the sample, electrical potential, or nuclear fog formation phenomena. In addition to these we have specific properties like optical activity for optical isomers like carbohydrates, or paramagnetic behaviour for gases like oxygen or nitric oxide which can be measured in a measuring system with a suitable transducer. Out of the several modern instrumental methods that we have so far encountered, all methods can be used for continuous on-stream analysis, e.g. in electroanalytical techniques a difference in potential leading to a change in (H^+) ion concentration has led to the measurement of pH in industrial pH on-line analysers or change in conductance has led to measurement of the concentration of pollutants like SO_2 in environmental monitoring. The property of electro deposition or measurement of diffusion current in polarography is not very useful. Similarly, in the case of optical methods of analysis it has become possible to make use of measurement of change in absorbance (Fig. 50.3) in the ultraviolet or visible region and in a few cases in the infrared region of the spectrum.

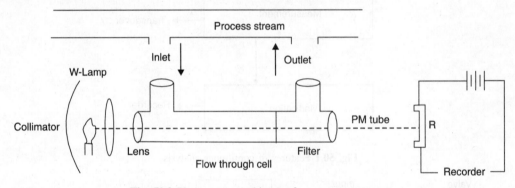

Fig. 50.3 Photometer—turbimeter for process streams

However, good and promising methods like flame photometry or emission spectroscopy have been less extensively used in on-line analysers. They do measure specific properties like emissions intensity for alkaline earths in conventional analysis but have yet to find a way in on-line analysers.

The property of molecular luminescence has been fully exploited in on-stream continuous analysis. The phenomenon of chemiluminescence, that is delayed emission of radiation due to excitation leads to fluorescence or phosphorescence depending upon the difference in time lag of absorption of energy and emission of light. In the former the time lag is small (about 10^{-3}–10^{-6} seconds) while in the latter mode the lag is large (10^{-2}–10^2 seconds). Hence the technique of fluorescence is used for continuous analysis of polluting gases in the atmosphere.

50.5 APPLICATIONS OF AUTOMATED METHODS

As regards applications of infrared measurement, they have the most specific property— perhaps more dependable than any other property for a gas, has been extensively utilised for gas analysis. One cannot find a better method than non dispersive infrared analysis for the measurement of carbon monoxide from the atmosphere. In contrast, in the growing field of atomic absorption spectroscopy in spite of the availability of microprocessors, print out devices, memory profile, this technique has not been used extensively in on-stream analysis. As a mater of fact this is one of the techniques which permits multielement analysis with multiple samples in a record time. With the discovery of newer sources of radiation and monochromatisation, this technique should find its way in on-line analysis. Fortunately it is used as one of the hyphenated techniques.

The sugar industry is an extremely large one but unfortunately the units adopt primitive methods for quality control of their products. On-line analysers fitted with optical activity measurement devices or units involving the measurement of optical rotatory dispersion *i.e.* the change of optical activity with respect to the wavelength of measurement or circular dichromism should find applications in the characterisation of carbohydrates and sugar.

One should be careful while selecting a specific property for measurement. A specific property can be defined as one which is characteristic of or peculiar to one component only.

The next criterion in on-line analysers is the selectivity of measurement. In industrial operations, the cost of analysis is often more important than accuracy. In such situations it is absolutely obligatory to select methods for on-line analysis which compensate for accuracy and precision of measurement. All selective methods can be utilised in either differential analysers or in specific process analysers. A typical module for a selective analyser is depicted in Fig. 50.4.

These kinds of auto analyser are used in the medical field for clinical analysis for the determination of potassium, chloride, carbon dioxide, glucose, urea, nitrogen, albumin and total protein in one single cycle. The main part of such operating systems is the proportionating pump. One can handle a large number of individual samples on the sample turn table. After the mixing coil operations, they are allowed to react with specific chromogenic reagents. An adequate time lag is allowed to mix and develop a colour and then the solution is passed on to the colorimeter and subsequently the reading gets recorded on a strip chart recorder. The same principle is more or less extended in automatic combustion methods for the analysis of carbon, hydrogen, and nitrogen in CHN analysers. Such commercial analysers are available.

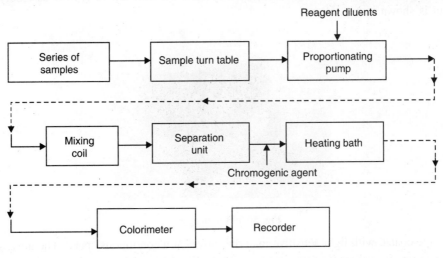

Fig. 50.4 Typical selective modules in an on-line analyser

50.6 ENVIRONMENTAL MONITORING

It is essential to consider some important applications of continuous analysers in environmental monitoring. Environmental pollution analysis is based upon the analysis of a large number of samples with varying complexity. Further, due to periodic variation in the composition of water or air, this task of generalisation of results becomes most difficult.

In water pollution analysis, the exact determination of dissolved oxygen (DO) is most important in finding the biological oxygen demand (BOD) of water. BOD is an indicator of organic and microbiological contamination of water. In such analysis one cannot always resort to Winkler's method for analysis of the dissolved oxygen. Under such circumstances resort to DO analyses (dissolved oxygen meter) is absolutely essential to get authentic information. Such DO analysers coupled with units for the measurement of pH as well as conductivity are indigenously and cheaply available. The fluctuations due to periodic variation in the composition of water can thus be mitigated by use of a continuous on-stream analyser for dissolved oxygen in a water filtration plant or industrial operations involving recycling of waste water. Figure 50.5 indicates the use of such a device.

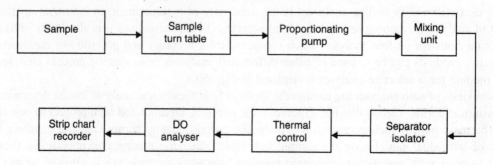

Fig. 50.5 Typical DO on-line analysers for water

As is obvious from the schematic diagram the system permits twenty-four hour monitoring of water samples in an industrial environment. Another example of on-line analysis for gaseous pollutants is the continuous analysis of H_2S from the atmosphere by flow colorimetry. A typical assembly for such measurement is shown in Fig. 50.6.

Fig. 50.6 H_2S analyser

A filter tape coated with light sensitive material moves at a continuous speed. The air is sucked by a pump through an air purifier (to remove SO_2) and is allowed to interact with coated tape to develop colour which is in turn measured by a phototube. This is directly fed to a recorder to give the concentration of H_2S in air (Fig. 50.7).

Fig. 50.7 H_2S analyser working

On-line analysis of SO_2 (sulphur dioxide) gas from polluted air is based upon continuous measurement of changes in conductivity. It is even possible to use differential analysis for such measurements with a variation in the magnitude of conductance. A device to monitor SO_2 has been developed taking care to filter out H_2S, CO_2 and NO–NOx, from the atmosphere before making such measurements. It is possible to measure a large number of samples of air for continuous measurement so long as the absorbing solution in the tank, which normally indicates variation in the magnitude of conductance, is not saturated with SO_2. In the event of supersaturation it will be absolutely necessary to change such a solution before more on-line measurements are carried out (Fig. 50.8).

Fig. 50.8 SO_2 analyser

It is necessary to indicate the future of on-stream analysers in environmental monitoring. In advanced countries, the principle of chemiluminescence specially that of fluorescence quenching has been extensively utilised, for, quantitative chemical analysis of SO_2 and NO–NOx. It has even been extended to the analysis of H_2S by converting it into SO_2 which in turn is excited to SO_2 to measure activity of excited atoms. Unfortunately such instruments are prohibitive in their cost, operation is expensive and spare parts are not easily available.

However, in the years to come with advancement in electronics and especially in instrumentation, there is a bright future for on-line analysers in environmental monitoring. However, it should not be forgotten that better methods of control and abatement of environmental pollution are principally based upon the availability of dependable specific and selective methods of on-line monitoring of environmental pollution.

50.7 FLOW INJECTION ANALYSIS (FIA)

This technique is based on the injection of a liquid sample into a stream of mobile continuous flow of an appropriate liquid. Zone formation is noted due to interaction of the injected sample which in turn is driven towards the detector. The diffusion controlled process promotes mixing of the two liquids. A suitable reaction takes place on mixing. The detector continuously records a suitable physical property such as absorbance, electrode potential, refractive index or electrical conductivity. Such variation in any physical property occurs due to the passage of the sample through the flow cell. We cite two typical examples to demonstrate FIA (Fig. 50.9).

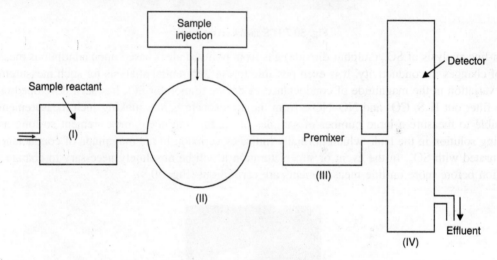

Fig. 50.9 Flow Injection Analyser (FIA) for fluoride

A sample containing zirconium alizarine sulphonate is kept in compartment (I) (Fig. 50.9). A solution containing fluoride is injected from compartment (II) which reacts in the premixer compartment (III) to form $[ZrF_6]^{2-}$ thereby decreasing the absorbance which is in turn monitored by the detector (IV). The excess liquid is drained off as an effluent.

Mercury thiocyanate Hg $(SCN)_2$ is kept in compartment (I) and is subsequently reacted with chloride ion in compartment (II) where there is formation of $HgCl_2$ and free SCN which in turn is allowed to react in chamber (III) with iron (III) to form a coloured species, $[Fe(SCN)_3]^{3-}$. The absorbance of this solution can be measured in compartment (IV) at 480 nm. If the dispersion is uniform then the results are reproducible. The samples are treated sequentially as they flow down the assembly line. A peristaltic pump is used. Less than 50 μl of the sample is injected and is allowed to flow at the rate of 0.8 ml/min. The sample is allowed to stay for a reasonable time in the mixer, so that the path length exceeds 50 cms with 50% of 0.50 mm of tubing inside the mixing chamber. For process analysis a better pump and tubing are needed, which can be controlled by a computer for better results. FIA is useful for various purposes. A sufficient time should be allowed for production of the chromophore in spectrophotometry or significant variation will occur in electrical conductivity in the mixing chamber before the resulting solution is measured. Two-line systems can also be used. Even systems with three tracks have been devised.

50.8 THEORETICAL CONSIDERATIONS IN FIA

The extent of dispersion of dilution is measured in terms of the diffusion coefficient (D). If a sample having an original concentration (C_0) is injected it moves along with the stream forming a dispersive zone. Its shape is dependent upon the geometry of the channel and flow velocity. A typical Gaussian curve as shown in Fig. 50.10 is obtained. When a FIA cell is to be fabricated one must know to what extent the original solution is diluted while flowing through to the detector; one should also be aware of the time between the sample injection point and the read out point. Hence we used dispersion coefficient (D) to decide these dimensions. D is the ratio of the concentration of the sample before and after the dispersion process takes place. For instance:

$$D = C^0/C \qquad \qquad ...(50.1)$$

if C is concentration at any point on the time scale.

Hence when maximum peak height is attained (*i.e.* C^{max}) we have

$$D_s^{max} = \frac{C_s^0}{C_s^{max}} \qquad \qquad ...(50.2)$$

The sample dispersion is rated as limited if $D = 1$–3, medium if $D = 3$–10 and large if $D > 10$. Out of the three, limited dispersion is preferred. A FIA peak occurs due to two processes, one involving the simultaneous physical process of zone dispersion and the second involving the chemical process resulting from reactions between sample and reagent species. A difference in the concentration gradient is thus generated.

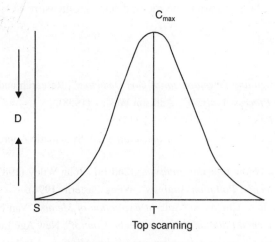

Fig. 50.10 Dispersion curve

50.9 FACTORS AFFECTING PEAK HEIGHT

As a rule the sample volume, channel length, flow rate and channel geometry affect the peak height. We take each of them into consideration very briefly.

We may have a one line or dual line FIA system. If $L = 20$ cm, flow is 1.5 ml/min id = 0.3 mm. We carry out experiments with dye solution and record the spectra. We have a steady state at C_{max}. At this point absorbance so recorded will be for C^0 of dye and $D = 1$. Then we have $C_{max}/C_0 = 1/D_{max}$. This shows that the peak height is directly proportional to the injected volume.

The length of the coil and its internal diameter greatly influence the process of dispersion. If a limited dispersion is required a sample volume of minimum $S_{1/2}$ *i.e.* volume of sample solution required to reach 50% of steady state value equivalent to $D = 2$, should be introduced. Narrow channels can be used for medium dispersion. Hence L must be as short as possible for narrow peaks. $L = 10–100$ cms in usual practice. The dispersion of the sample zone increases with the square root of the distance travelled through a tabular conduit and the square of the radius of tube.

Finally, instead of a coiled tube, a straight tube is usually preferred in FIA experiments, to facilitate mixing radially and secondary flow due to imposed centrifugal force. By the latter process more reagent is exposed to the interacting system and axial dispersion is thereby minimised. FIA has a high precision, better than 1%, due to controlled dispersion of flowing liquid.

50.10 APPLICATIONS OF FLOW INJECTION ANALYSIS

One can use different kinds of systems to demonstrate the use of FIA in industry. One can use 0.1M borax ($Na_2B_4O_7$) solution along with 0.4% bromothymol blue solution in a 1 cm cell to measure absorbance at 580 nm. We can measure various parameters like length, flow, peak height etc. The course of the reaction between $(NH_4)_2 MoO_4$, H_3PO_4, HNO_3 and ascorbic acid can be measured with FIA at 660 nm as the molybdenum blue complex. FIA systems should find numerous applications in environmental science for continuous monitoring e.g. one can detect NO_2 in the atmosphere by allowing NEDA *i.e.* 1-napthyl ethylene diamine hydrochloride, with sodium nitrite solution in the presence of MBT to give a coloured complex which can easily be measured continuously. On a fixed time basis NO_x can be allowed to react with NaOH in a premixing chamber and can be analysed subsequently. Similarly by using PRA i.e the pararosaniline method one can measure SO_2 from the atmosphere on a continuous basis.

LITERATURE

1. S.M. Khopkar, *"The Design and Development of Gas Detectors"*, Research and Industry 29 86 (1984).
2. S. Siggia, *"Continuous Process Analyser"*, Addison Wesley (1958).
3. G.D. Nichols, *"On-line Process Analysers"*, Wiley (1988).
4. H.A. Strobel, W.R. Heineman, *"Chemical Instrumentation"*, A Systematic Approach, 3rd Edition, John Wiley (1989).
5. J. Ruzicka, E.H. Hansen, *"Flow Injection Analysis"*, 2nd Ed., John Wiley (1987).
6. S.M. Khopkar, *"Environmental Pollution Analysis"*, Wiley Eastern (1994).
7. A.K. De, S.M. Khopkar, R.A. Chalmers, *"Solvent Extraction of Metals"*, Van Nostrand, Reinhold (1970).
8. S. M. Khopkar, *"Environmental Pollution Monitoring and Control"*. New Age International Publishers (2004).
9. S.M. Khopkar, *"Analytical Chemistry - Problems and Solution"*, New Age International Publishers (2003).
10. S.M. Khopkar, *"Analytical Chemistry of Macrocyclic and Supramolecular Compounds"*, 2nd Ed., Springer Verlag (2006).
11. S.M. Khopkar, *"Solvent Extraction Separation of Elements with Liquid Ion Exchangers"*, New Age International Publishers (P) Ltd. (2007).
12. S.M. Khopkar, *"Analytical Chemistry – Problems & Solutions"*, 2nd Ed., New Age International Publ. (P) Ltd. (In Press).

Answers to the Problems

CHAPTER 2

2.1 15, 42 accepted; 15, 68 rejected with 15.51 lowest and 15.56 highest

2.2 $A = 107.876$ mean; $S = 0.12$ ppt ± 0.021 and 0.038 amu confidence limit, $B = 107.728$, $S = 0.014$ ppt, 0.002 and 0.041 amu confidence limit

2.3 0.2555 0.00023 0.39 ppt

2.6 No, 0.212 and 0.056, 22.813; 0.022, 0.028, 22.83 14 ppt 8.5 ppt

2.7 $\epsilon = 1.5$ μg. Error negative at low con. + ve error at high con.

2.8 $3.803 \pm 0.003\%$

2.9 $S = 0.031$ 95% confidence limit = 3.22 ± 0.15

2.10 Mean 65.10; median 65.10, $Q = 0.43$, 0.40 No rejection

CHAPTER 3

3.1 $S = 14$, Expected number = 200

3.2 386 g for particle of 0.1 cm in diameter

3.3 $C = 42$, $\sigma_0 = 0.076$ if $n_1 = 3$, $n_2 = 2$, $n_3 = 3$

3.4 1.6×10^{-10}

3.5 0.0020% M

3.6 2.8×10^{-5} ml

CHAPTER 4

4.1 3.07% Na_2O

4.2 10.5% K_2O; 3.67% Na_2O

4.3 68.54% CaO, 6.74% Fe_2O_3

4.4 25.6118 g

4.5 0.5249 g Fe_3O_4

4.6 79.46%

4.8 28.9% NaCl, 71.1% KCl

CHAPTER 5

5.1 250 ml

5.2 49.4% $CaCO_3$, 50.6% $BaCO_3$

5.3 66.25% Na_2CO_3, 24.5% $NaHCO_3$

5.4 0.4334 N

5.5 (SO_3) x = 63.6%, y = 34.9% (H_2SO_4)

5.6 71.26%, 5.40%

5.7 667 ml of 6 N and 333 ml of 3N

CHAPTER 6

6.1 278,63.3, 248.2, 17.01

6.2 31.6, 49.03, 126.9, 17.01

6.3 0.2393

6.4 0.003161 g

6.5 24.00, 0.1699 N

6.6 99.44% 70.48 ml

6.7 0.1053 N

6.8 14.4, 21.6 ml

CHAPTER 7

7.1 6.4%

7.2 53.3% $SrCl_2$

7.3 0.08343 N

7.4 0.03008 N

7.5 354.5 mg, 584.4 mg

7.6 I = 0.043 0.08 ml or 2 ppt

7.7 27.8% NaCl error or 12 ppm

CHAPTER 8

8.1 below pH 2.0

8.3 100 mg

8.4 5.36%

8.6 407.6 ppm

8.7 271 ppm

CHAPTER 9

9.1 $D = 499$

9.2 99.76%

9.3 97.2%

9.4 1247.5

9.5 99.91%

9.6 173, 173

9.7 44.5

9.8 $K = 3.0$

9.9 96.8%, 99.90%, 99.997%

9.10 $E = 63.5\%$, 0.18 mole/litre decrease ~ 9.0 extraction

CHAPTER 11

11.1 $Rs = 1.06$, $N = 3445$, $H = 8.708 \times 10^{-3}$ cm, $t = 35.3$ mts

11.2 $H = 5.36$ cm

CHAPTER 12

12.1 $E = 2.089$, $Dv = 0.478$, $Dw = 0.597$

12.2 $V_{max} = 61.6$ ml

12.3 0.80 g/ml

12.4 0.292 g in 25 ml

12.5 1.7×10^{-2} ppm

12.6 1.7 mg eq/g

12.7 1.59 mg eq/g

12.8 0.62 mg eq/g

12.9 120, 220 ml to each maximum

12.10. 50

CHAPTER 14

14.1 $\phi = 0.69$

14.2 $K_D = 8.35$

14.3   73 ml, 16.2 ml, 21.4 ml

14.4 $V_a = -0.049$, $\phi = 90$ (0.5% gel); $V_a = 0.044$, $\phi = 0.84$ (1% gel)

   $V_a = 0.034$, $\phi = 0.75$ (2%), $V_a = 0.024$

   $\phi = 0.69$ for 3%, $V_a = 0.014$ $\phi = 63$ for 4% gel

CHAPTER 15

15.1   Ag = 0.88, Pb = 0.66, Hg = 0.33

15.2 $A = 0.70$, $B = 0.82$, $C = 0.88$, compd., B

15.3   Compound B, $R_f = 0.79$

15.4   24.92 ml

15.5   3.7 ± 0.4%

CHAPTER 16

16.1   Benzene 31.4%, Toluene = 14.6%, Xylene = 53.9%

16.3   19.7%, 17.4%, 57.4%, 5.5%

16.4   82.4% ethanol, 76.7%

16.5   22.8% hexane; 12.3%, 1-heptanol, 9.9% MBK, 54.9% toluene

16.6   2.24 cm/sec, 2.14 mm

16.7 α for acetone = 1.06

CHAPTER 20

20.1   900 ml

20.2   9.38

20.3   Glycine 0.95×10^{-4} cm^2/V/sec; Serine 14.3×10^{-4} cm/V/sec

20.4   Almost one litre

20.5   All exist as cations; the mobility decreases in order shown

CHAPTER 23

23.1   0.800

23.2 5.76×10^7

23.3   1/2, 4/7

23.4   0.806

23.5   51.3%

23.6   6.67%, 0.477

23.7   33.0, 0.602

23.8 6.5×10^{-5}

23.9   7.0 cm.

23.10 4.24×10^{-5}

CHAPTER 24

24.1 $C_1 = 1.9 \times 10^{-3}$ M; $C_2 = 1.46 \times 10^{-4}$ M

24.2 1.52×10^{-4} M

24.3   9.50

24.4   4.11, 4.11

24.5   Co = 8.21×10^{-3}; Ni = 1.75×10^{-4}

24.6 4.02×10^{-5} ml

24.7   0.186 mg/ml

24.8   0.125 µg

24.9 $7.15 \times 10^{-6}\%$

24.10   1:1

CHAPTER 25

25.1   1:2 Zn (CN)$_2$

25.2   1.70 ppm

25.3   1.72%

25.4   0.248 mg/ml

25.5   Hexachlorobenzene

25.6   Benzylideneaniline

25.8   0.302 M 2,3 dimethylhexane, 0.528 for 2,3 dimethythexane, 0.206 M 2,3 dimethylhexane

CHAPTER 27

27.1   6.4 ppm Hg

27.2   4.3 ppm Ca

27.3   1.250 ppm 0.125%

27.4   0.460 ppm

27.5   53 µg/ml

CHAPTER 28

28.1 12.02 ppm Li

28.2 12 ppm K

28.3 $\lambda = 587.2$ nm

28.4 2.7 ppm

28.5 9.1 ppm

28.6 0.121% Mn

28.7 0.16% Cd

28.8 304.357 nm

CHAPTER 31

31.1 0.8493 µg/ml

31.2 16 mg/100 ml, 32/100 ml

31.3 6.65×10^{-5} M

31.4 11.0 µg/ml

31.5 1.3×10^{-2}%

31.6 8.7×10^{-4} g/ml

CHAPTER 32

32.1 11.19 mg/ml Pb

32.2 2.1 mg/ml Cl$^-$

32.3 53.1%

32.4 35.3 ± 0.7%

32.5 11.0 µg/ml

CHAPTER 34

34.1 0.020 moles/litre

34.2 104° (specific rotation)

34.3 29.29/100 ml, g/100 ml

34.4 26.8 g/100 ml

34.5 −85° 27g/100 ml

34.6 131.2°

34.7 α at 589 nm = 52°, at 486 nm = 80° α, at 434 nm = 104°

34.8   0.60 g/100 ml

34.9   0.67 g/100 ml

CHAPTER 35

35.1   Alanine = 38.46%, β-alanine = 61.53%

35.2   1-butanol 6.91%, 3-butanol 5.86%, 2-butanol 2.1%

35.3   24.5% λ

35.5 $CH_2 = C - CH_2$
$$\quad\quad\quad | \quad\quad |$$
$$\quad\quad\quad O - C = O$$

35.6 C_2H_6O

35.7   6.14% H

CHAPTER 42

42.1   0.366%

42.2   9% mole A, 91% mole B

42.3   39.947

42.4 $A = 15.7$, $B = 24.2$, $C = 6.5$, $D = 53.6$ mole%

42.5 $8-CH_4/2$, $13-CH$, $14-CH_2$, $15-CH_3 \cdot 16-CH_4$, $28-CO$, $44-CO_2$

42.6 $14-CH_2$, $28-CO$, $42-CH_3COOC_5H_7$, $44-CO_2 \cdot 57.5\ C_5H_{11}COO$, $59-CH_3\ COO$, $70-C_5H_{10}$, $71-C_5H_{11}$

CHAPTER 43

43.1   7.2% water: 62.4% $Ca(OH)_2$

43.2 $[CaC_2O_4] = 36\%$, $[MgC_2O_4] = 41\%$, $[H_2O] = 20\%$, $[SiO_2] = 3\%$

43.3 $[CaSO_4 \cdot 2H_2O] = 24.85\%$, $[CaCO_3MgCO_3] = 11.5\%$, $[CaCO_3] = 37.15\%$

43.4 $[Ca(ClO_4)]_2 = 17.4\%$, $CaOCl_2 = 12.9\%$, $Ca(OH)_2 = 20\%$, $CaCO_3 = 7.9\%$
$SiO_2 = 29.8\%$, $H_2O = 12\%$

43.5 $H_2O = 5\%$, water of crystallisation = 49.95%, chemically bound = 3.41%, formula is $MgNH_4PO_4$, $6H_2O$

43.6   91.56%

CHAPTER 44

44.1 2.28×10^3 dps

44.2   7.0% Hg
44.3   1.57 µg/g Se
44.4   0.561%
44.5 6.65×10^6 C/hr/g; 3.04×10^6 C/hr/g

CHAPTER 45

45.1   1.22 V, 0.0683 V
45.2   +1.8498 V
45.3   0.5037 V
45.4   (a) 0.15 V right to left, (b) 0.740 V left to right
45.5   11.56, 0.429 V
45.6   0.188 V
45.9   −0.98 V
45.10 $K_{sp} = 1.79 \times 10^{-10}$

CHAPTER 46

46.1 6.5×10^{-5} cm/sec
46.2 4.02×10^{-5} M
46.3   25.4 µa
46.4   1.53
46.5 1.59×10^{-3} M
46.7 3.75×10^{-4} M
46.8   0.810 M
46.9   12.5 µA
46.10   0.0015 M

CHAPTER 47

47.1 1.26×10^{-5}
47.2   75.8 ohm eq cm^2
47.3   0.127 g NaOH
47.4   0.820 g
47.5   141.7 ohm^{-1}
47.6   396 ohm^{-1}
47.7   0.214 g
47.8 1.88×10^{-5}

CHAPTER 48

48.1 0.0342 g Zn

48.2 0.1554 N

48.3 16.712 gm Cd

48.4 Cu = 1.0×10^{-3} M

48.5 0.255 V

48.6 18.6 mts

48.7 2×10^{-45} M

48.8 5.08×10^{-10} g

APPENDIX-1

International Atomic Weights

Element		Atomic weight	Element		Atomic weight
Aluminium	Al	26.982	Nickel	Ni	58.71
Antimony	Sb	121.75	Niobium	Nb	92.906
Arsenic	As	74.922	Nitrogen	N	14.007
Barium	Ba	137.33	Oxygen	O	15.999
Beryllium	Be	9.0122	Palladium	Pd	106.42
Bismuth	Bi	208.98	Phosphorus	P	30.974
Bromine	Br	79.904	Platinum	Pt	195.08
Cadmium	Cd	112.41	Potassium	K	39.098
Calcium	Ca	40.08	Rubinium	Rb	85.468
Carbon	C	12.011	Selenium	Se	78.96
Cerium	Ce	140.12	Silicon	Si	28.086
Cesium	Cs	132.91	Silver	Ag	107.87
Chlorine	Cl	35.453	Sodium	Na	22.990
Chromium	Cr	51.996	Strontium	Sr	87.62
Cobalt	Co	58.933	Sulfur	S	32.066
Copper	Cu	63.546	Tantalum	Ta	180.95
Gold	Au	196.97	Tellurium	Te	127.60
Hydrogen	H	1.0079	Thallium	Tl	204.38
Iodine	I	126.90	Thorium	Th	232.04
Iron	Fe	55.847	Tin	Sn	118.71
Lead	Pb	207.12	Titanium	Ti	47.88
Lithium	Li	6.941	Tungsten	W	183.85
Magnesium	Mg	24.305	Uranium	U	238.03
Manganese	Mn	54.938	Vanadium	V	50.942
Mercury	Hg	200.59	Zinc	Zn	65.39
Molybdenum	Mo	95.94	Zirconium	Zr	91.224

APPENDIX-2

Selected Standard Electrode Potentials

Half reactions	E^0 Volts
$F_2 + 2H^+ + 2e^- = 2HF$	+3.06
$CO^{3+} + e^- = CO^{2+}$	+1.842
$Ce^4 + e^- = Ce^{3+}$	+1.70
$MnO^{4-} + 4H^+ + 3e^- = MnO_2 + 2H_2O$	+1.695
$Mn^{3+} + e^- = Mn^{2+}$	+1.51
$MnO_4^- + 8H^+ + 6e^- = Mn^{2+} + 4H_2O$	+1.15
$Cl_2 + 2e^- = 2Cl^-$	+1359
$Cr_2O_7{}^{2-} + 14H^+ + 6e^- = 2Cr^{2+} + 7H_2O$	+1.33
$Tl^{3+} + 2e^- = Tl^+$	+1.25
$O_2 + 4H^+ + 4e^- = 2H_2O$	+1.229
$HNO_2 + H^+ + e^- = NO + H_2O$	+1.0
$VO_2^+ + 2H^+ + 2e^- = VO^{2+} + H_2O$	+1.0
$Pd^{2+} + 2e^- = Pd$	+0.987
$2Hg^{2+} + 2e^- = Hg_2^{2+}$	+0.860
$Hg^{2+} + 2e^- = Hg$	+0.854
$Cu^+ + e^- = Cu$	+0.850
$Ag^+ + e^- = Ag$	+0.799
$Fe^{3+} + e^- = Fe^{2+}$	+0.771
$O_2 + 2H^+ + 2e^- = H_2O_2$	+0.682
$I_2 + 2e^- + 2I^-$	+0.619
$MnO^{4-} + e^- = MnO_4^{2-}$	+0.564
$Cu^{2+} + 2e^- = Cu$	+0.337
$Hg_2Cl_2 + 2e^- = 2Hg + 2Cl^-$	+0.268
$Cu^{2+} + 2e^- = Cu$	+0.153

Half reactions	E^0 Volts
$Sn^4 + 2e^- = Sn^{2+}$	+0.150
$S + 2H^+ + 2e^- = H_2O$	+0.141
$2H^+ 2e^- = H_2$	+0.000
$CrO_4^- + 4H_2O + 3e^- = Cr(OH)_3 + 5OH^-$	−0.130
$Sn^{2+} + 2e^- = Sn$	+0.136
$Pb^{2+} + 2e^- = Pb$	−0.126
$Co^{2+} + 2e^- = CO$	−0.277
$Ni^{2+} + 2e^- = Ni$	−0.246
$V^{3+} + e^- = V^{2+}$	−0.255
$HO_2^{2+} + 4H^+ + 2e^- = H^{4+} 2H_2O$	−0.334
$Tl^3 + e^- = Tl$	−0.336
$Cd^{2+} + 2e^- = Cd$	−0.403
$2CO_2 + 2H^+ + 2e^- = H_2C_2O_4$	−0.490
$Cr^{3+} + 3e^- = Cr^{2+}$	−0.410
$Cr^3 + 3e^- = Cr$	−0.440
$Zn^{2+} + 2e^- = Zn$	−0.74
$2H_2O + 2e^- = H^2 + 2OH^-$	−0.828
$Mn^{2+} + 2e^- = Mn$	−1.18
$Al^{3+} + 3e^- = Al$	−1.60
$Mg^{2+} + 2e^- = Mg$	−2.37
$Na + e^- = Na$	−2.714
$Ca^2 + 2e^- = Ca$	−2.87
$K^+ + e^- = K$	−2.925
$Ba^2 + 2e^- = Ba$	−2.925
$Li^+ + e^- = Li$	−3.045

APPENDIX-3

Statistical Tables

Important Physical Constants

Speed of light (c) 2.99792×10^3 ms^{-1}

Planck's constant (h) 6.62608×10^{-34} js

Avagadro's number (N) 6.022137×10^{23} particles mol$^-$

Faraday's Constant (f) 96,485.31 V mol$^-$

Gas Constant (R) $8.31\ 451 \times$ JK^{-1} mol$^-$

Boltzmann's constant (K) 1.38066×10^{-23} JK^{-1}

Mass of an electron (e) 9.109080×10^{-31} kg

Electric charge $- 1.6021 \times 10^{-19}$ C

$$\left(\frac{RT}{F} \right) = 0.0591$$

Values of 't' for various levels of probability

Degree of Freedom	80%	90%	95%	99%	99.9%
1.	3.08	6.31	12.7	63.7	63.7
2.	1.89	2.92	9.92	31.6	31.6
3.	1.64	2.35	5.84	12.9	12.9
4.	1.53	2.13	4.60	8.60	8.60
5.	1.48	2.02	4.03	6.86	6.86
6.	1.44	1.94	3.71	5.96	5.96
7.	1.42	1.90	3.50	5.40	5.40
8.	1.40	1.86	3.36	85.04	5.04
9.	1.38	1.83	3.25	4.78	4.78
10.	1.37	1.81	3.17	4.59	4.59
11.	1.36	1.80	3.11	4.44	4.44
12.	1.36	1.78	3.06	4.32	4.32
13.	1.35	1.77	3.01	4.22	4.22
14.	1.34	1.76	2.98	4.14	1.14
∞	1.29	1.96	2.58	3.29	3.29

Ref.: D.A. Skoog, D.M. West, Analytical Chemistry, Holt Reinhart & Winston p. 71 (1979).

Values of 'Q' for rejection of data at various confidence limits

Number of Observations	90%	96%	99%
3	0.94	−0.98	0.99
4	0.76	0.85	0.93
5	0.64	0.73	0.82
6	0.56	0.64	0.74
7	0.51	0.59	0.68
8	0.47	0.54	0.63
9	0.44	0.51	0.60
10	0.41	0.48	0.57

Ref.: J.S. Fritz, G.H. Schenk–Quantitative Analytical Chemistry 4th ed. Allen & Bacan Inc., p. 38 (1979).

Values of F at 95% confidence level

V_2	V_1					
	2	3	4	5	6	∞
2	19.0	19.16	19.25	19.30	19.33	19.50
3	9.55	9.28	9.12	9.01	8.94	8.53
4	6.94	6.59	6.39	6.26	6.16	5.63
5	5.79	5.41	5.19	5.05	4.95	4.36
∞	3.0	2.60	2.37	2.10	2.10	1-0

Ref.: H.A. Lattinen, W.E. Harris, Chemical Analysis, 2nd ed., p. 545, McGraw Hill & Co. (1975).

Range

The deviation for the determination of
standard deviation from range ($s = K \times$ range)

N	K deviation factor
2	0.89
3	0.59
4	0.49
5	0.43
6	0.40
7	0.37
8	0.35
9	0.34
10	0.33

Ref.: J.S. Fritz, G.H. Schenk–Quantitative Analytical Chemistry, 4th ed., p. 35, Allyn & Bacon in (1979).

Chi square distribution

Degree of Freedom	Probability (P)						
	0.99	0.95	0.90	0.50	0.10	0.01	0.005
1	0.00016	0.0039	0.0158	0.055	2.71	6.63	7.78
2	0.01	0.102	0.211	1.39	4.61	9.21	10.6
3	0.0717	0.352	0.584	2.37	6.25	11.3	12.8
4	0.207	0.711	1.064	3.36	7.78	13.3	14.9
5	0.412	1.15	1.61	4.35	9.24	15.1	16.7
6	0.676	1.64	2.20	5.35	10.6	16.8	18.5
7	0.99	2.17	2.83	6.35	12.0	18.5	20.3
8	1.34	2.73	3.49	7.34	13.4	20.1	22.6
9	1.73	3.33	4.17	8.34	14.7	21.7	23.6
10	2.16	3.94	4.87	9.34	16.0	23.2	23.2
20	7.43	7.26	8.55	14.3	22.3	30.6	32.8
30	13.79	18.49	20.60	29.3	40.3	50.9	53.7

Ref.: H.A. Lattinen, W.E. Harris–Chemical Analysis, 2nd Ed., p. 546, McGraw Hill & Co. (1976).

Confidence levels for different values of Z

Confidence level (%)	Z
50	±0.67
68	±1.0
80	±1.29
90	±1.64
95	±1.96
96	±2.0
99	±2.55
99.9	±3.29

Confidence limit for mean $= x \pm z\sigma$ if σ = standard deviation

Confidence limit for mean $(\mu) = x \pm z\sigma/\sqrt{N}$ (N = nos. of measurements)

Ref.: D.A. Skoog, D.M. West–Analytical Chemistry, 3rd ed., p. 68, Holt Reinhart and Winston (1979).

Values of $C_V/C_s V_s$ against varying R in ion-selective electrode

R	$C_V/C_s V_s$	R	$C_V/C_s V_s$
1.280	0.113	1.560	0.870
1.300	0.140	1.580	0.973
1.320	0.170	1.600	1.086
1.340	0.203	1.620	1.213
1.360	0.240	1.640	1.353
1.380	0.280	1.660	1.510
1.400	0.325	1.680	1.691
1.420	0.373	1.700	1.894
1.440	0.425	1.720	2.126
1.460	0.485	1.740	2.397
1.480	0.548	1.760	2.711
1.500	0.618	1.780	3.088
1.520	0.694	1.800	3.536
1.540	0.778	1.820	3.8015

Ref.: Solving problems in Analytical Chemistry–H. Brewster, John Wiley & Co. (1980).

Index

1, 10-phenanthroline 50
1-10 phenanthroline 26, 81
1-nitro-2-napthol 26
1.2 (pyridylazo) napthol 69
1,1– diphenyl–2 picrylhydrazyl (i.e. DPPH) 402
18C6 97
2, 9 dimethyl 1-10 phenanthroline 26
2-thenoyltrifluoroacetone 95
^{31}P NMR 394
4 (2 pyridyl azo) resorcinol 161
4, 7 diphenyl-1-10 phenanthroline 26
8-hydrocxyquinoline 94
8-hydroquinoline 28
8-hydroxquinoline 81

A

α-naphthylamine 40
α-napthaflavone 47
Abbe's refractometer 372
Absolute and relative difference 8
Absolute method 7
Absolutely stability constant 62
Absorbancy methods 272
Absorption 246
Absorption edge 409
Absorption of chromophores 256
Absorption spectra 246
Absorptivity 380
Abundances of isotopes 457
AC arc circuit 327
AC polarography 515
Acetylacetone 80
Acid chrome black 69
Acid dissociation constant 32

Acid–base indicators 37
Acid–base titration 32
Acid-base titrations 32
Acidity of the metal 78
Acoustic detector 447
Activity coefficient (γ) 137
Activity coefficients 55
Activity corrections 539
Addict with neutral ligands 92
Adjusted retention volume 174, 184
Adsorbents 168
Adsorbtive stripping voltammetry 546
Adsorption 115, 116
Adsorption affinities 219
Adsorption chromatography 116, 167
Adsorption indicators 57
Adsorption isotherms 167
Adsorption linear capacity 168
Adsorption scale 169
Adsorptive stripping voltammetry 546
Advantages of complexometric titrations 73
Affinity chromatography 169, 170
Affinity ligands 169
Air pollutants by IR 295
Aliasing 17
Aliquat 336S, Arquad-6C 91
Aliquot 336 92
Alizarine sulphonate 57
Amberlite LA-1 91
Amberlite LA-I 92
Amperometric titration 519
Amperostatic 543
Analogue filtering 552
Analysis of H$_2$S 566
Analytical chemistry 1

Andrew's titration   47
Angle of incidence   411
Anion exchange   129
Anion extraction   91
Anionic chelates   79
Anionic complexes   91, 103, 149
Anionic indicator   57
Anionic sulphato complexes   150
Anode   484
Anodic currents   508
Anodic oxidation   539
Anodic stripping voltammetry   546
Anthralinic acid   28
Antigen   229
Antistokes   365
Antistokes fluorescence   348
Antistokes lines   302
Appearance of fluorescence   40
Application   109
Applications of fluorimetry   357
Applicator   218
Aquatic environment   104
Arc or spark atomic emission spectroscopy   243
Argentometric methods   53
Artificial transmutations   477
Ascending development   178
Asymmetric potential   494
Atmosphere on Mars   305
Ato   3
Atomic absorption spectra   308
Atomic absorption spectroscopic   243
Atomic absorption spectroscopy   247
Atomic emission detector   192
Atomic emission spectroscopy   243, 325
Atomic fluorescence   347
Atomic force microscopy (AFM)   440, 443
Atomic line broadening   319
Atomisation   311, 337
Atomiser   178
Atomizer   329
Attenuated total reflectance (ATR)   296
Attenuated total reflectance spectrometry   295
Attenuated total reflection   294
Auger electron spectroscopy   428
Auger effect   429
Auger electron   429
Auger emission spectroscopy   429
Auger spectroscopy   438
Autocatalytic effect   46
Automated analysis   562

Automatic titrator   562
Autooxidation   42
Auxiliary complexing agent   66
Auxochrome   257
Average current   508
Average deviation of a single measurement   8
Average deviation of mean (D)   8
Axle   105
Aza crown ethers   99
Azo indicators   38

B

β β′ dichlorodiethyl ether (II)   84
Background emission   331
Balancing method   258
Bandwidth of   251
Barrier layer   260
Base line method   293
Base peak   455
Basic strength   80
Basicity of ligand   82
Basicity of the ligand   78
Basicity of the oxygen   84
Batch extraction   89
Batch method   146
Bathocuporoine   83
Bathophenanthroline   83
Beer's Bougar's law   250
Beer's law   249
Bending   285, 286
Best bonded supports   199
Biamperiometric titration   521
Biamperometric titration   521
Bidentate   79
Binder   178
Bio Gel 0-10   213
Biocatalytic biosensors   559
Biological oxygen demand   566
Bioluminescence   347
Biosensors   550, 551, 557
Biospecific adsorption   169
Birefrigence   380
Bis (2-ethylhexyl) phosphoric acid   91
Bisphenol   99
Blank error   280
Boltzmann distribution equation   309
Bonds   78
Bonner affinity series   138
Born equation   78

Bragg's equation   411, 412, 417
Breakthrough capacity   134
Bromocresol green   39
Bromophenol blue   38
Bromophenol red   38
Bromophyrogallol   72
Bronze   74

C

Calcein   71
Calcium and magnesium in hard water   73
Calix (n) arenes and rotaxanes   97
Calixarenes or calixresorcinol   97
Calmagite   70
Calomel half cell (SCE)   492
Capacitance   530
Capillary electrochromatography   241
Capillary electrophoresis   112, 460
Capillary gel electrophoresis   240
Capillary isoelectric focussing   240
Capillary Isotachophoresis   240
Capillary zone electrophoresis   240
Capture cross section (s)   479
Capture rate   479
Carbon-13 NMR   394
Carrier gas   187, 188
Cartridge   93
Catechol   99
Cathode   484
Cathodic deposition   538
Cathodic stripping voltammetry   546
Cation exchange resins   129
Cationic indicator   57
Cavity of crown ether   101
Cavity size of crown ethers   102
CDTA   62
CE–MS   462
Cell constant   528
Cell potential   42
Cellulose powder   175
Centrifugal (F_e) force   452
Centrifugation   214
Centripetal force (F_m)   452
Ceric sulphate   45
Characterisations of Rotaxanes   108
Characteristic IR absorption peaks   291
Characteristics of solvents used in HPLC   203
Charge to size ratio   224
Charge transfer   185

Charge transfer electrons   255
Charge transfer spectral absorption   257
Chelate   79
Chelate extraction   77
Chelate stability   86
Chelating agents   25
Chelating ligands   79
Chelating resins   129, 153
Chemical analysis   3
Chemical constitution   168
Chemical coulometer   543
Chemical force microscopy   445
Chemical interference in AAS   317
Chemical methods   4
Chemical resistance   135
Chemical selectivity   124
Chemical shift   391
Chemical shift with respect to the oxidation state   434
Chemical shifts in ESCA   432
Chemically bonded supports   199
Chemiluminescence   343, 567
Chi square distribution   585
Chiral chromatography   170
Chiral compounds   170
Chlorophenol red   38
Choice of an organic reagent   26
Choice of fluid   209
Chopper   289, 447
Chopping amplifier   552
Chromatobars   168
Chromatographic column   188
Chromatographic fronts   173
Chromatography   108, 113, 128
Chromogenic   231
Chronoamperometry   490, 506
Chronopotentiometry   490, 506, 523
Chrysodine   50
Chrysodine derivatives   58
Chrysodine indicators   50
Circular birefringence   381
Circular dichromism   381
Circular dichromism (CD)   244, 380
Citric and tartaric acids   151
Classification of chromatographic methods   114
Classification of electrophoretic techniques   226
Classification of extractions   77
Classification of ion selective electrodes   499
Classification of photometers   266
Classification of Rotaxanes   105
Classification of thermal methods   466

Clipping process   105
Coefficient of variation   9
Coextraction   210
Coiled tube   570
Color of resin   133
Colorimetry   248
Colours of the visible region   249
Column length   200
Columnar technique   146
Common ion effect   55
Comparative method   7
Comparison of critical fluids   207
Comparison of polarographic methods   518
Complete extraction   89
Complex with rotaxane   109
Complexation on half wave potential   512
Complexes with dicarboxylic acid   165
Complexing reactions   141
Complexometric titrations   32, 61
Compressibility correction   184
Computerised library database   192
Concentration polarizations   489
Concept of MRI   396
Condensation product   131
Conditions for complexometric titrations   66
Conditions for precipitation   23
Conductance   528
Conductimetry   484, 490, 531
Conductivity detectors   160
Conductometric titrations   531
Confidence levels for different values of Z   586
Confidence limit   10
Conjugate acid   33
Conjugate base   33
Constant current electrolysis   542
Constant potential cathode electrolysis   539
Constant pressure pumps   199
Continuous analysis   561
Continuous electrochromatography   217
Continuous electrophoresis   229, 230
Continuous extraction   89
Controlled potential cathode electrolysis   540
Controlled potential electrogravimetry   28
Controlled potential electrolysis   490
Conventional stripping analysis   547
Coolidge tube   411
Coordination complexes   78
Copolymer of styrene and DVB   132
Copolymer phenol and formaldehyde   132
Coprecipitation   22

Cornu and Littrow mountings   267
Cornu type   267
Coronands   98
Correction   274
Cotton effect   381
Coulometric analysis   543
Coulometric titrations   544
Coulometry   484, 490, 538
Counter anion   101
Counter current extraction   89
Coupling constants (J)   393
Covalent   78
Cresol purple   38
Cresolphthalein, thymolphthalein;
      a napthophthalei   38
Critical angle   370, 372
Critical pressure (P_c)   206
Critical region   206
Critical temperature   206
Cross section of ICP torch   338
Crossed electrophoresis   229
Crosslinking   135
Crown ether   97
Crown ethers   77
Cryptand   98
Cryptand 222   98
Cryptands   77, 97, 99
Crystal field theory   257
Crystal lattice theory   129
Crystal violet   38
CTNB   110
Cupferron   26, 28, 92
Current density   525
Current efficiency   543
Curtain electrochromatography   218
Curtain electrophoresis   230
Cyclic component   107
Cyclic compound   79
Cyclisation   101
Cyptand 222   92

D

d and f electrons   255
d block elements   257
Daughter-ion MS/MS method   463
DB18C6   97
DC arc circuit   327
DC conductivity bridge   535
DC-MS   461

DC18C6 92, 97
Degassers 201
Degree of ionisation 335
Degree of ionization 317
Dehydration 174
Deionisation of water 128, 147
Demasks complexes 73
Densitometric analysis 180
Depth profiling analysis 431
Derivatisation chromophoric groups 193
Derivatisation reactions 193
Derivative thermogravimetric analysis 468
Desalting 178, 213
Descending development 178
Desorption 146
Detection limit 200, 323
Detection limits 316
Detector 200, 340, 552
Detectors 160, 188, 199
Detectors in capillary electrophoresis 240
Determinate errors 7
Determinate errors and indeterminate 6
Development of chromatograph 119
Dextro-rotatory (+) 378
Dialysis 112
Diatomaceous earth 175
Dielectric constant 33, 533
Dielectric constants of common solvents 533
Dielectric material 533
Dielectrophoresis 223
Diethy-dicholorosilane 93
Diethyl ether (I) 84
Differential migration 114, 198
Differential pulse polarography 518
Differential scanning calorimeter 470
Differential scanning calorimetry 466, 473
Differential spectrophotometry 272
Differential thermal analysis 466, 468
Diffraction grating 266
Diffusion coefficient 207, 506
Diffusion coefficient (D) 509
Diffusion current 508
Diffusion current (i_d) 510
Diffusion current drop time 509
Diffusion limiting current 505
Diluent 101, 103
Dimethyl-glyoxime 92
Dimethylaminoazobenzene 38
Dimethyldichlorosilane vapour 177
Dimethylglyoxime 27

Dimethylglyoxine 26
Dinonyl napthalene sulphonate 91
Diphenylamine 50, 59
Diphenylamine sulphonic acid 46
Dispersion forces 185
Displacement development 115, 119
Disproportionation 42
Dissociation constants 34
Dissolved oxygen 48, 513, 566
Distillation 111
Distortion polarisation 533
Distribution coefficient 76
Distribution coefficient of the metal complex 86
Distribution ratio 183
Distribution ratio (D) 76, 85
Dithizone 92
DNN 152
Dolemite 30
Donnan membrane theory 130
Doppler broadening 319
Doppler effect 319, 320, 421
Doppler shift 320
Double Auger process 430
Double beam photometers 262
Double bond equivalent 192
Double bond region 290
Double focussing 451
Double focussing separator 452
Double helix 235
Double layer theory 130
Doublet state 344
Dropping mercury electrode 505
Dropping mercury electrode (DME) 507
DSC 473
Dstribution coefficient 76
Duane-Hunt law 408
Dubosque colorimeter 258
Dubosque colorimeters 258
Dynamic mechanical analysis (DMA) 474
dynamic SIMS 434
Dynamic thermogravimetry 466
Dynode 262

E

Earphones 530
Eddy diffusion 121, 198
EDTA 61
EDTA and the complexones 61
Effect of temperature on fluorescence 358

Effective or conditional stability constant   63
Effective stability constant   67
Efficiency   120
Efficiency of extraction   89
Efficiency of sensors   559
Efficiency of separation   174
Effluent   146
EGTA   62
Electric dipoles   532
Electrocapillary maximum   508
Electrocardiograph   552
Electrochemical cells   484
Electrochemical detectors   201
Electrochemical sensors   550, 551
Electrochemiluminescence   343, 346
Electrochromatography   112, 116, 217
Electrodecantation   112
Electrodeposition step   547
Electrodialysis   112, 220
Electrogravimetry   28
Electrolytic cells   484
Electrolytic conductivity detector   192
Electromagnetic spectrum   245
Electromagnets   388
Electromigration   217, 219
Electron capture detectors   188
Electron densities   82
Electron density   84
Electron double resonance   404
Electron magnetic resonance   400
Electron microprobe analysis (EMA)   435
Electron microscopy   440
Electron nuclear double resonance   404
Electron paramagnetic resonance   400
Electron spectroscopy for chemical analysis   431
Electron spectroscopy for chemical analysis (ESCA)   244, 428
Electron spin resonance   400
Electron spin resonance spectroscopy   244
Electron spray ionisation mass spectrometry   460
Electronegativity   80
Electroneutrality   131
Electronic amplifier   496
Electronic oscillator   402
Electroosmosis   230
Electroosmosis velocity   238
Electroosmotic flow   237
Electrophoresis   223
Electrophoretic mobility   223, 225
Electrophoretic retardation   225

Electrophoretic separation   223
Electrophoretic velocity   225
Electrostatic   78
Electrostatic process   223
Elemental analysis   192
Eluants for anion separations   159
Eluants for cation separation   159
Elution   115, 146
Elution analysis   119
Elution constant (E)   143
Elution development   119
Emission of radiation   247
Emulsion formation   93
End point   31
End point error   31
Energy level diagram   336
Ensemble averaging   552
EnthalpImetric titrations   470
Enthalpimetric titrations   471
Enthalpy   103
Enthalpy ($\Delta H°$)   118
Enthalpy changes   139
Entropy   103, 139
Entropy ($\Delta S°$)   118
Environmental analysis   163
Environmental applications   164
Environmental monitoring   566
Eosine (I)   102
Equation   198
Equilibria for solvation   88
Equilibrium constant   37
Equivalence point   31
Equivalent conductance   529
Equivalent conductivity   528
Equivalent point potential   44
Equivalent weight   51
Erichrome Blue black   69
Eriochrome black T   69
Eriochrome red B   72
Error   6
Errors in fluorimetry   356
Errors in UV/visible spectrophotometry   279
Erythrosine (II)   102
ESCA   432
ESR spectra of hydrogen   403
ESR spectra of semiquinone   403
Europium dipivalo methanato complex   394
Exchange equilibrium   111
Exchange potential   166
Exclusion chromatography   116

External indicators 50
Extraction 77
Extraction by ion pair formation 90
Extraction by solvation 77, 82, 90
Extraction chromatography 93, 175
Extraction equilibrium 85

F

F at 95% confidence level 584
FAB-LSI-MS 458
FAB-LSIMS 459
FAB-MS 459
Factor 185
Factors affecting solubility of precipitate 55
Far infrared region 285
Faradaic current 530
Faraday's law of electrolysis 542
Faraday's laws 490
Fast atom bombardment and liquid secondary
    ion mass 459
Fast neutrons 478
Feldspar 28, 60
Fema 3
Ferroin 46, 50
Fiber optical sensor 556
Fick's diffusion coefficient 506
Fick's law 506
Finger print region 290
First resonance line 336
Fixation pair 328
Flame ionisation detector 188, 208
Flame photometry 243, 308, 328
Flow Injection Analyser (FIA) for fluoride 568
Flow injection analysis 568
Flow programming chromatography 190
Flow rate 198
Fluorescence 244, 344
Fluorescence and phosphorescence 247
Fluorescence detectors 200
Fluorescence quenching 360
Fluorescent indicators 40
Fluorimetric and amperiometric detectors 160
Flushing 513
Flux 478
Foam separation 112
Forensic science 480
Formal potential 46, 487
Formation constant of the metal complex 85
Formation constants 63

Formation of uncharged complex 77
Fourier transform NMR 395
Fraction capacity 124
Fraction collector 147, 562
Fragmentation 451
Fragmentation pattern 455
Fragmentation patterns 455
Free boundary electrochromatography 217
Free boundary electrophoresis 229
Free induction decay 395
Freon 12 513
Frontal analysis 119, 226
Frontal development 115
FT-IR spectroscopy 297
FT-MS 459
Fuel gas 310
Fullerene-stoppers 106
Functions of salting 84

G

Galvanic cells 484
Gas chromatography 112, 183
Gas electrode 485
Gas hold up time 184
Gas ionisation detector 413
Gas sensors 554
Gas solid chromatography 183, 190
Gas-solid chromatography 115
Gaussian error function curve 184
GC–IR 462
GC-MS 462
Gel permeation 212
Geometry of Rotaxane 104
Gibbs free energy of hydration 103
Gibb's phase 76
Glass electrode 493
GLC, HPLC, SFC and SFE 209
Gonimeter 412
Gradient elution 120, 146, 201
Graphite furnace 312
Gravimetric analysis 21
Gravimetry 4
Gross sample 16
Gyromagnetic ratio 384, 387

H

Half cells 484, 486
Half extraction volume 89

Half reactions   42
Half wave potential   511
HDEHP   92, 152
He–Ne laser   300
Head space gas chromatograph   195
Headspace analysis   194
Hehner's cylinder   258
Height equivalent of theoretical plate   186
Height of the theoretical plate   145
Heisenberg principle   420
Heisenberg uncertainty principle   421
Helmholtz double layer   130
Hematite   52
Heterocyclic compounds   80
HETP   145
Hexadentate   79
Hieftje correction   321
High absorbancy   272
High absorbancy method   272
High frequency titrations   532
High molecular weight amine   77
High molecular weight amines (HMWA)   152
High performance liquid chromatography   112
High performance liquid chromatography (HPLC)
   197
High resolution NMR   388, 390
Hilderbrand theory of regular solutions   83
Hollow cathode glow discharge lamp   312
Hollow cathode lamp   308
Holographic gratings   281, 289
Homologous pairs   328
Hooke's law   287
Huge error   10
Hydride generator   339
Hydrodynamic voltammetry   522
Hydrogen bond, dipole-dipole interaction   105
Hydrogen bonding   169
Hydrogen electrode   491
Hydrogen stretching region   290
Hydrolysis   55
Hydrolytic titration   59
Hydrophilic phase as the mobile phase   175
Hydrophilic solvent   175
Hydrophobic   93
Hydrophobic phase as the stationary phase   175
Hydrophobic solvent   175
Hydrous oxides   153
Hyperfine splitting   402
Hyphenated GC-MS   191
Hyphenated methods   165, 458

Hyphenated technique   323, 457
Hyphenated techniques   190, 191, 461

I

I-nitroso-2-naphthol   28
IC-ICP   165
IC-MS   165, 462
ICP-MS   461
Ignition temperature   24
Ilkovic equation   508
Immunoelectrophoresis   229, 233
Immunophoresis   226
Impedance   536
Inclusion and exclusion of solutes   212
Inclusion compounds   105, 111
Indeterminate errors   7
Inductive forces   185
Inductively coupled plasma   337
Inert electrode   485
Infrared absorbance   200
Inhibitors   169
Injection port   187
Inner filter effect   356, 359
Inner transition elements   257
Instability constant   140
Instrumental errors   7
Inter system conversion   355
Interference filters   260
Interfering elements   2
Interferogram   395
Interferometric of refractometer   372
Interferometry   369
Intermolecular hydrogen bonding   175
Internal conversion   344, 355
Internal indicator   46
Internal indicators   50
Internal standard method   331
International atomic weights   580
Interstitial volume   213
Intra-molecular forces   167
Inversion   379
Iodine as Redox Reagent   47
Iodine thiosulphate   47
Iodine-starch complex   47
Ion chromatography   112, 157
Ion collector   450, 454
Ion deformation   78
Ion exchange   128
Ion exchange chromatography   113, 116

Ion exchange membrane 544
Ion exchange methods 112
Ion exchange supports 199
Ion exclusion 157, 165, 214
Ion exclusion and ion retardation 212
Ion pair chromatography 157, 180
Ion pair formation 77, 152
Ion pairing 165
Ion pairs 180
Ion retardation 215
Ion selective electrode 162
Ion selective electrodes 499
Ion separator 450
Ion source 451
Ionic potential 79
Ionic product of water 33
Ionisation chamber 450
Iris diaphragm 263
Isocratic elution 199
Isoelectric focussing 229
Isoelectric point (pI) 224
Isoelectric points 217
Isomeric shift 419, 420, 422
Isotachophoresis 226, 227, 228
Isothermal measurements 473
Isotope dilution 481
Isotope dilution methods 480
Isotopic dilution method 477

J

Jackson candle turbidimeter 367
Job's method of continuous variation 276

K

Karaglonoff equation 524
Karl Fischer method 48
Karl Fischer titration 521
KBr pellet technique 290
Keto enol isomers 27
Kieselguhr 175
Kinetic energy 111
Kinetics of complexometric titration 73
KLL transition 429
Kortum and Seiler 251

L

LA-2 Primene JMT 91

Laevo-rotatory (-) 378
Lake 57
Larmov frequency 384
Laser desorption ionisation 458
Laser mass spectrometry 458
Laser source 300
Law of electroneutrality 131
Law of mass action 137, 141
LC-ESI-MS 458
Leading 187
LIDAR 305
Light filters 260
Limitations of flame emission spectroscopy 334
Limitations of the IR 294
Limiting current 508
LIMS 459
Linear component 107
Linear dynamic range 188, 339
Linear sweep oscillator 388
Liphophobic 93
Liquid anion exchangers 91, 129, 500
Liquid cation exchanger 91
Liquid cation exchangers 129
Liquid chromatography 197
Liquid ion exchange 499, 553
Liquid ion exchangers 90, 129, 151
Liquid membrane electrode 500
Liquid membrane electrodes 495
Liquid partition materials 188
Liquid-liquid extraction 76
Liquid-liquid partition chromatography 172
Littrow type 267
Load capacity 124
Lock-in amplifier 447, 552
Log and step sector methods 328
London 185
London's forces 167
Longer columns 189
Longitudinal diffusion 198, 237
Longitudinal relaxation 387
Lorentz-Lorenze molar refraction (M_r) 373
Low absorbancy 273
Low resolution NMR 390
Lower confidence limit 10
LSD 357
Luminescence 343
Luminol 345, 346
Lunar samples 2
Lunar soil 419

M

Macro 248
Macro sample 3
Macrobicyclic 98
Macrobicyclic compounds 97
Macrobicyclic polyethers 98
Macrocyclic 98
Macrocyclic and supramolecular compounds 97
Magnetic moment 387
Magnetic resonance imaging 395
Magnetic splitting 419, 420, 422
Magnetite 52
Major constituents 3
Malachite green 38
MALDI 458
Mandelic acid 28
Masking agent 62
Mass absorption coefficient 409
Mass absorption coefficient (μ_M) 410
Mass sensitive sensors 559
Mass spectrometer 450
Mass spectrometric detectors 201
Mass spectrometry 244
Mass to charge (m/e) ratio 450
Matrix assisted laser desorption/ionisation 458
Maximum precision method 273
MDOA 92
Mean value and deviation 8
Mechanism for supression 161
Mechanism of extraction 77
Median value 9
Mesityl oxide 96
Metal oxide field effect transistors (MOSFET) 554
Metal separations 148, 175
Metal separations by extraction chromatography 176
Metal separations by ion exchange 151
Metallic electrode 493
Metallochromic indicators 68
Metaloxinates 26
Methyl ferroin 50
Methyl orange 38, 40
Methyl red 38, 39
Methyl violet 38
Methyl yellow 38
Micro 248
Micronutrient in soils 164
Micronutrients 2
Microporous polymers 199
Microprocessor technology 162

Microsample 3
Middle IR region 285
Migration current 509
Migration velocity 223
Minor constituents 3
Miscellor electrokinetic capillary chromatography 237, 241
Mixed indicators 39
MMV transitions 429
Mobile phase 114
Mobile phase ion chromatography 165
Mobile phase is CO_2 208
Mode of lasers and ionisation process 352
Modes of molecular vibrations 286
Modifier 209
Modulation 552
Modulation of energy 420
Mohr's method 53, 56
Mohr's salt 52
Molar absorptivity 270
Molar volume 103
Mole fractions 277
Molecular absorption spectroscopy 243
Molecular emission methods 244
Molecular geometry 111
Molecular polarization 532
Molecular sieve techniques 216
Molecular straws 109
Molecular thread compounds 105
Monochromatic radiation 245
Mössbauer austenitometer 426
Mössbauer isotopes 423
Mössbauer spectroscopy 419
Movable jaw 264
Moving boundary electrophoresis 226
Mull technique 290
Multi element separation 157
Multielement analysis 243
Multielement lamps 313
Murexide 70

N

n − π* 255, 256
N-phenyl N-benzoylhydroxylamine 26
Nature of functional group 134
Near IR region 285
Nebulizers 339
Neel temperature 426
Neocupferron 26

Neodymium YAG source 303
Neodymium/YAG-laser 458
Neoocupoine 83
Nephelometry 244
Nepholometry 364
Nernst distribution law 76
Nernst equation 43, 486
Nernst or Glower's lamp 288
Nessler's reagent 367
Nessler's tubes 258
Neutron activation analysis 477
Neutron capture 478
Neutron flux 478
Nicol analyser 378
Nicol prism 378
Nitriliotriacetic 62
Nitroferroin 50
Nitrogen rule 456
Nitrogen UV laser 458
Nitrogen-Phosphorous detector 192
Nitrophenol 39
NMR of other elements 394
NMR shift reagents 394
Nobel prize 99, 113
Noise 550
Non dispersive infrared analysis 565
Non-chromatographic methods 111
Non-dispersive technique 285
Non-magnetic means 452
Nonaqueous solvents 150
Nonequilibrium mass transfer 198
Normal hydrogen electrode (NHE) 487
Normality 31
Nuclear γ-ray resonance fluorescence 420
Nuclear magnetic resonance 384
Nuclear magnetic resonance spectroscopy 244
Nuclear reactors 477
Nuclear relaxation 387
Nuclear spin quantum numbers 385
Nucleoside 235
Nucleotide 235
Nujol oil 290
Number of fringes N 375
Number of plates 121
Number of theoretical plates (N) 185
Numerical aperture (NA) 278

O

Occluded volume 215

Ohm's law 484
On-line analysers 564
On-line analysers for water 566
On-line analysis of SO_2 567
Optical activity 379
Optical fiber 556
Optical fibres 278
Optical rotary dispersion 244
Optical rotatory dispersion 380
Optical sensor 556
Optical sensors 551, 555, 556
Optically active 378
Optically based sensors 278
Optically transparent electrode 281
Optoacoustic spectroscopy 446
Optrodes 556
ORD and UV spectral relation 381
Organic acids like malonic 151
Organic precipitant 25
Organic precipitants in gravimetric analysis 25
Organophilicity 103
Oriental polarisation 533
Oscillometer 534
Oscillometry 490
Oscilloscope 458
Over voltage 489, 505
Overall feasibility 125
Oxidant 310
Oxidation 42
Oxidation number 42
Oxidising agent 42
Oxygenated solvents 90

P

$\pi \to \pi^*$ 255
π, σ, n-electrons 255
π-electron deficient tetra 105
π-electron rich 105
Paper chromatography 172, 177
Parameters influencing refraction 372
Parent ion 455
Particle induced X-ray 416
Particle size 134
Partition chromatography 116, 172
Partition coefficient 76, 172
Partition ratio 111, 138
Patton and Reeders' Dye 71
Patton-Reeder indicator 73
Peak effluent volume 174

Peak elution volume 144
Peak height 515
Pellicular particles 202
Peptisation 23
Percentage extraction 77
Perchlorate as a counter ion 102
Perfluorokerosine 456
Permanent magnets 388
Permutite 128
Personal error 260
Phase ratio 183
Phenolphthalein 38
Phosphorescence 244, 344
Phosphorescence quantum efficiency 355
Phosphorescence spectroscopy 361
Photoacoustic spectroscopy 244, 446
Photodensitometry 179
Photoelectric filter photometers 248
Photoelectric photometers 248
Photoelectric spectroscopy 436
Photoemissive 260
Photoluminescence 354
Photometric detectors 160
Photometric titration 276
Photometric titrations 273
Photomultiplier tubes 260
Photosensitive materials 262
Phototubes 260
Phthaleins and sulphophthaleins indicators 38
Physical characteristics 564
Physical methods 4
pI 233
pI values of proteins 235
Picric add (IV) 102
Piezoelectric devices 550
Piezoelectric mass sensor 553
Piezoelectric phenomena 552
Piezoelectric transducer 447
Piezoelectric tube 443
Pitfalls in sampling 16
PIXE 406, 416
pK value 277
pKa 275
pKa value of the chelating ligand 80
Planar counter anions 103
Planck's equation 245
Plane polarised light 378
Plasma spectroscopy 335
Plate height 122, 198
Plate theory 142

Podands 98
Polarisation 78
Polarizable electrode 505
Polarized 489
Polarograph 509
Polarographic Maxima 512
Polarography 505
Polarography or voltammetry 484
Polymerisation 85
Polyrotaxanes 105
Polytetrafluoroethylene 93
Porocity 135
Porous layer beads (PLB) 203
Post column detection 194
Post precipitation 23
Potassium permanganate 46
Potassium bromate 46
Potassium dichromate 46
Potential gradient 217
Potential of SCE vs. NHE 492
Potentiometer 496
Potentiometric electrolysis 542
Potentiometric recorders 202
Potentiometric sensors 553
Potentiometric titration 134
Potentiometric titrations 495
Potentiometry 484, 489
Potentiostat 521
Potentiostatic 543
Potentiostats 539
Powder camera (Debye-Scherrer) 417
Precipitation from homogeneous solution 23
Precipitation methods 21
Precipitation titrations 32, 53
Precision and accuracy 6
Precolumn 194
Precolumns 201
Pressure regulator 187
Primary and secondary amines 91
Primary standards 32
Prism 266
Process of extraction 77
Programmed temperature chromatography 190
Projectiles 477
Properties of charged molecule 224
Protective chelating agents 331
Protective colloids 59
Proton shift 392
Pseudo-rotaxane 105
Pulfrich refractometer 372

Pulse dampers   199
Pulse polarography   517
Pump   198
Pyrocatechol violet   69
Pyrocatehol Red   71
Pyroelectric material   557
Pyrolysis   189

Q

'Q' for rejection of data   584
Q-test   10
Q-test for rejection   11
Quadridentate   62
Quadruple   386
Quadrupole   419
Quadrupole analyser   451, 452
Quadrupole splitting   420
Qualitative analysis   2
Quality of cuvettes   280
Quantitative analysis   2
Quantum efficiency   358
Quenching   40, 344, 359
Quenching and inner filter effect   359
Quinaldic acid   26, 28
Quinhydrone electrode   492
Quinoline yellow   47

R

Radial development   178
Radiative scattering   364
Radio frequency generator   340
Radio frequency oscillator   389
Raman radiation   303
Raman scattering   302, 364
Raman shift   301
Raman spectroscopy   244, 300
Raman Stoke's lines   301
Randle and Servek equation   516
Random sampling   18
Range   9
Rapid scan polarography   515
Raster pattern   440
Rate of exchange   136
Rayleigh scattering   248, 301, 364
Reaction gas chromatography   189
Reciprocating piston pumps   199
Reciprocating pumps   201
Recoil free γ-ray resonance absorption   419

Recorder sensitivity   189
Redox indicators   49
Redox reactions   42
Redox titrations   32
Reducing agent   42
Reduction   42
Reduction potentials   42
Reference electrode   492
Reference solution   265
Reflectance spectra   294
Reflectance Spectrophotometry   273
Refractions of atomic groups (n_r)   371
Refractive index   280
Refractive index detectors   200
Refractive index type detectors   202
Refractometry   369
Regeneration   128, 146
Relative concentration error   252
Relative efficiency of detectors in HPLC   201
Relative error in concentration   251
Relative retention   185
Relative stability constants   88
Relative standard deviation   9
Relaxation   255
Remote raman sensing   304
Residual current   508
Resolution, (R)   123
Resolution of compounds   454
Resolution of different analysers   454
Resolution of mixtures   123
Resolution R_s 186
Resonance   37
Resonance effects   81
Resonance fluorescence   347
Resonance lines   309
Resonance maxima   425
Resonance Raman scattering   304
Retardation factor   117, 172, 183
Retention behaviour   187
Retention volume   117, 185
Reverse osmosis   112, 220
Reversed phase extraction
    chromatography   90, 113, 175
Reversible cell over voltage   489
RI detector   369
Ringbom plot   252
Rocking   285, 286
Rotating platinum electrode   521
Rotaxane   97
Rotaxanes   105

RPPC 176
RU lines 328
RU powders 328

S

$\sigma - \sigma^*$ n $- \sigma^*$, n $- \pi^*$ and $\pi - \pi^*$ transitions 255
$\sigma \rightarrow \sigma^*$ transitions 255
Salicyldioxime 28
Salt bridge 490
Salt splitting capacity 134
Salting out agent 82
Salting out agents 84
Salts of heteropoly acids 153
Salts of multivalent metals 153
Sample 16
Sample injection port 198
Sampling gases 17
Sampling of pure liquids 17
Sampling of solids 17
Sampling processes 16
Sampling scheme 17
Sampling tube 17
Sampling unit 16
Sampling units 17
Sand equation 523
Sandell's sensitivity 277, 551
Satellite peak 455
Saturated calomel electrode 490
Saturation zone 449
Scale of selectivity of various ligands 79
Scanning electron microscope 442
Scanning electron microscopy (SEM) 440, 441
Scanning probe microscopes (SPM) 440
scanning probe microscopy 442
Scanning tunneling microscope and AFM 442
Scanning tunneling microscopy (STM) 443
Scattering 248
Scattering of radiation 364
SCE 490
SCE – saturated calomel electrode 43
Scissoring 285, 286
Screw driven pumps 201
Secondary ion mass spectrometry 428, 434
secondary ion mass spectrometry (SIMS) 438
Seeding 21
Selection of such methods 3
Selective means 26
Selective stripping 91
Selectivity 124, 134, 565

Selectivity coefficient 136
Selectivity in complexometric titrations 73
Selectivity of extraction 85
Selectivity sequence for organic cations 143
Self absorption 326
Semimicro 3
Semipermeable membrane 220
Semirotaxane 105
Sensitive to high temperature 136
Sensitivity 316, 551
Sensitivity (S) 278
Sensitivity in GLC 192
Separation but also for quantitative analysis 90
Separation column 198
Separation factor 86, 138, 140
Separation of alkali metals 149
Separation of alkaline earths 149
Separation of anions 149
Separation of lanthanides 148
Separation Science 111
Separations characteristics 124
Separatory funnel 76
Sephadex G 212
Sepharose 212
Sequestering agent 25
Sequestering agents 86
SFE model (online) 210
Shielding parameter 391
Sign of the electrode potential 489
Signal to noise ratio 280, 551, 552
Silicagel 175
Siloxanes 208
Silylation reagents 193
SIMS-MS 461
Simultaneous spectrophotometric 272
Simultaneous spectrophotometry 270, 449
Single beam photometers 262
Single focussing 451
Single focussing centripetal 451
Singlet state 344
Size exclusion 204
Size of cations 101
Size of the chelated complex 81
Size selective exclusion chromatography 212
Slab gel electrophoresis 237
Slope ratio method 276
Slow neutron flux 479
Slow neutrons 477, 478
Snell's law 296
Soft material examination 444

Solid multispace extraction 194
Solid phase microextraction 94
Solid-phase extraction 93
Solubility 111
Solubility of chelate 86
Solubility parameters 83
Solubility product 31
Solvating solvents 175
Solvation phenomena 78
Solvent extraction 76
Solvent matrices in FAB-MS 460
Sorption 146
Sources for excitation 326
SP-MS 461
SPADNS 70
Spark source mass spectrometry 457
SPE disk 94
Speciation 164
Specific conductance 528, 529
Specific eluant 146
Specific property 565
Specific refraction 370
Specific retention volume 174, 185
Specific rotation 370, 380
Specific rotations 378
Specification for gas chromatography columns 186
Spectral bandwidth 280
Spectral interference 330
Spectral interferences 339
Spectral peak 434
Spectrochemical analysis 245
Spectrochemical series 257
Spectroelectrochemistry 281
Spectrophotometer 266
Speed 125
Spin forbidden process 355
Spin label 404
Spin label reagents 402
Spin labelling 404
Spin quantum number 402
Spin spin splitting 391
Spin–lattice 387
Spin–spin 387
Spin–spin splitting 404
Spin-spin relaxation 388
Spin-spin splitting 392
Splitting quadrapole 422
Sprayer 339
Square wave polarography 518
Stabilization factor 92

Stacks of π-electrons 105
Standard addition 8
Standard addition method 349
Standard deviation 9, 18
Standard deviation from range 585
Standard deviation of mean 9
Standard deviation of the peak 186
Standard electrode potential 42
Standard electrode potentials 487, 581
Standard electrode potentials of metals 488
Standard solutions 31
Standard 't' test 11
Starch 47
Stationary phase 90, 114
Stationary phases used in GC 188
Stationary support 117
Statistical methods 8
Statistical tables 583
Stereochemistry 103
Steric 86
Steric hindrance 26
Sterically sensitive position 81
Sterochemical effects 81
Stochiometric equivalence point 42
Stokes 365
Stoke's and antistoke's lines 300
Stoke's fluorescence 343
Stoppers 107
Straight tube 570
Stratified 18
Stratified sampling 17
Stray light 266
Stray light on absorbance 280
Stretching 285, 286
Stripping analysis assembly 547
Stripping voltammetry methods 546
Stroke fluorescence 348
Strongly acidic cation exchanger 133
Strongly basic anion exchange 133
Structure of cation exchange resin 132
Structure of important metallochromic indicators 69
Structure of typical anion exchange resin 132
Styrogel 212
Substoichiometric analysis 481
Substoichiometric extractions 477
Subtractive process 189
Sulphato complex 150
Sulphon black 70
Summit potential 516
Super hyperfine splitting 405

Super saturated solution 21
Superconducting solenoids 388
Supercritical fluid 206
Supercritical fluid extraction 206
Supercritical fluid extraction (SFE) 209
Supersaturation 55
Suppressor 513
Suppressor column 160
Supramolecular compounds 97, 98
Surface activity 111
Surface analysis 244, 428
Surface induced deposition 458
Surfaces of nonconducting or the insulating
      materials 443
Swelling of anion exchange resin 135
Swelling of cation exchange resin 135
Symmetrical stretching H₂O 286
Synergic extraction 77
Synergic solvent extraction 91
Synthesis of calix (4) resorcinarene 101
Synthesis of cryptand 100
Synthesis of cryptand 2, 2, 2 99
Synthesis of DB 18 crown 6 99
Synthesis of rotaxanes 105
Synthesis of supramolecular 99
Syringe 94
Syringe type pumps 199

T

't' for various levels of probability 583
Tables of chi square 10
Tailing 168, 187
Tandem mass spectrometer 462
TBP 92
Temperature for drying 472
Temperature influences 118
Temperature of flames 310
Temperature zones of ICP 339
Template effect 105
Template reaction 106
Templates effect 109
Tertiary amines 91
Tetramethylsilane 389
Thenoyl-trifluroacetone 80
Thenoyltrifluoro acetone (TTA) 92
Thermal conductivity detector 188
Thermal conductivity of carrier gases 189
Thermal diffusion 112
Thermal sensors 557

Thermistors 188
Thermo mechanical analysis 474
Thermoanalytical methods 466
Thermocouple or bolometer 288
Thermodynamic stability constant 138
Thermogram 466
Thermogravimetric analysis 466
Thermomechanical analysis 466
Thermometric titration 466
Thermometric titrations 471
Thia crown 99
Thief 17
Thin layer chromatography (TLC) 178
Thorin 70
Threading mechanism 105
Three dimensional NMR 395
Three dimensional structures 104
Thymol blue 38
Thymolphthalexone 72
Time of flight analysers 451, 452
Time of flight analysis 435
Titration curve 37, 54
Titration curves 34
Titration curves in redox reaction 42
Titrations by turbidity 56
Titrimetry 31
TOA 92
TOPO 92
Torque 400
Total exchange capacity 130, 134
Total ion concentration 137
Total salt concentration 147
Trace 248
Trace concentrations 3
Transducer 552
Transition elements 80
Transition elements and inner transition
      elements 90
Transitions of electrons in ESCA 432
Transmission electron microscopy 435
Transmittance 255
Transportation 19
Transverse relaxation 387
Tridentate 79
Trien 62
Triocylamine (TOA) 91
Triodide 47
Triphenylemethane 50
Triphenylmethane indicators 38
Triple bond region 290

Triplet state   344
Trisooctylamine (TIOA)   91
TropeolinOO   38
TTA + TBP   92
Tubular columns   115
Tunable dye laser   350
Tuneable dye lasers   352
Tungsten lamp   267
Turbidimetry   244, 364, 366
Twisting   285
Twisting out   286
Two dimensional FT–NMR   395
Tyndall meters   366
Tyndall scattering   364
Types of electron spectroscopy   433
Typical eluants   159
Typical properties of supercritical fluids   207
Typical solvent extraction systems   92

U

Ultraviolet photoelectron spectroscopy   436
Uncharged complex   77
Upper atmosphere   345
Upper confidence limit   10
UV detector   200

V

Vacuum fusion   111
Valence electrons   255
Value of pH 1/2 (pH half)   87
Values of α_4 for EDTA   63
Values of $C_v/C_s V_s$ against varying R   586
Van Deemeter   198
Van Deemeter equation   198
Van Deemters equation   122
van der Waal's   185
van der Waal's adsorption   137
van der Waal's forces   93
van der Waal's interaction, or dipole induced dipole
     167
Varamine blue   70
Variable resistance   263
Variance   18
Various amperometric titrations   520
various detectors in GLC   193
Various refractometers   373
Various spectrophotometers   269
Vibrational absorption   247

Vibrational frequencies   287
Vibrational modes   288
Vibrational relaxation   344, 355
Vibrations   285
Viscosity   174
Viscous drag   224
Visual colorimetry   248
Void volume   147
Volatility   111, 188
Volhard's method   56
Voltage ramp   508
Voltaic cell   260
Voltammetric completion of the analysis   547
Voltammetric sensors   555
Voltammetry   108
Voltammetry and polarography   489
Volume distribution coefficient (DV)   144
Volumetric methods   31

W

Wagging   285
Wagging out   286
Washing   24
Washing the precipitate   24
Water analysis   48, 57
Wave number   245, 287
Wavelength   245
Weakly acidic cation exchange   133
Weight distribution coefficient (DW)   144
Well water   74
Weston cell   497
What column capacity   118
Wheatstone bridge   530
Wheel   105
Winkler's method   48

X

X-fluorescence   406
X-ray absorption   406, 414
X-ray diffraction   406
X-ray emission   406
X-ray emissions   414
X-ray fluorescence   414, 429
X-ray photoelectron spectroscopy   244, 428, 438
X-ray spectroscopy   406
X-rays absorption   409
Xerogels   212
Xylenecyanol FF (blue)   40

Xylenol orange   69

Y

Yoe's mole ratio method   276

Z

Zeeman background correction   321
Zeeman effect   319
Zeeman's effect   320

Zeolite   128
Zeolites   215
Zeta potential   225
Zincon   71
Zone electrochromatography   217
Zone electrophoresis   226, 229
Zone levelling   111
Zone refining   111
Zone spreading   119, 121
Zorbax   203